Table of Atomic Masses*

Element	Symbol	Atomic Number	Atomic Mass
...um	Ac	89	(227)†
...inum	Al	13	26.98
...icium	Am	95	(243)
...ony	Sb	51	121.8
...	Ar	18	39.95
...ic	As	33	74.92
...	At	85	(210)
...m	Ba	56	137.3
...	Bk	97	(247)
...lium	Be	4	9.012
...uth	Bi	83	209.0
Boron	B	5	10.81
Bromine	Br	35	79.90
Cadmium	Cd	48	112.4
Calcium	Ca	20	40.08
Californium	Cf	98	(251)
Carbon	C	6	12.01
Cerium	Ce	58	140.1
Cesium	Cs	55	132.9
Chlorine	Cl	17	35.45
Chromium	Cr	24	52.00
Cobalt	Co	27	58.93
Copper	Cu	29	63.55
Curium	Cm	96	(247)
Dysprosium	Dy	66	162.5
Einsteinium	Es	99	(252)
Erbium	Er	68	167.3
Europium	Eu	63	152.0
Fermium	Fm	100	(257)
Fluorine	F	9	19.00
Francium	Fr	87	(223)
Gadolinium	Gd	64	157.3
Gallium	Ga	31	69.72
Germanium	Ge	32	72.59
Gold	Au	79	197.0

Element	Symbol	Atomic Number	Atomic Mass
Hafnium	Hf	72	178.5
Helium	He	2	4.003
Holmium	Ho	67	164.9
Hydrogen	H	1	1.008
Indium	In	49	114.8
Iodine	I	53	126.9
Iridium	Ir	77	192.2
Iron	Fe	26	55.85
Krypton	Kr	36	83.80
Lanthanum	La	57	138.9
Lawrencium	Lr	103	(260)
Lead	Pb	82	207.2
Lithium	Li	3	6.941
Lutetium	Lu	71	175.0
Magnesium	Mg	12	24.31
Manganese	Mn	25	54.94
Mendelevium	Md	101	(258)
Mercury	Hg	80	200.6
Molybdenum	Mo	42	95.94
Neodymium	Nd	60	144.2
Neon	Ne	10	20.18
Neptunium	Np	93	(237)
Nickel	Ni	28	58.69
Niobium	Nb	41	92.91
Nitrogen	N	7	14.01
Nobelium	No	102	(259)
Osmium	Os	76	190.2
Oxygen	O	8	16.00
Palladium	Pd	46	106.4
Phosphorus	P	15	30.97
Platinum	Pt	78	195.1
Plutonium	Pu	94	(244)
Polonium	Po	84	(209)
Potassium	K	19	39.10
Praseodymium	Pr	59	140.9

Element	Symbol	Atomic Number	Atomic Mass
Promethium	Pm	61	(145)
Protactinium	Pa	91	(231)
Radium	Ra	88	226
Radon	Rn	86	(222)
Rhenium	Re	75	186.2
Rhodium	Rh	45	102.9
Rubidium	Rb	37	85.47
Ruthenium	Ru	44	101.1
Samarium	Sm	62	150.4
Scandium	Sc	21	44.96
Selenium	Se	34	78.96
Silicon	Si	14	28.09
Silver	Ag	47	107.9
Sodium	Na	11	22.99
Strontium	Sr	38	87.62
Sulfur	S	16	32.07
Tantalum	Ta	73	180.9
Technetium	Tc	43	(98)
Tellurium	Te	52	127.6
Terbium	Tb	65	158.9
Thallium	Tl	81	204.4
Thorium	Th	90	232.0
Thulium	Tm	69	168.9
Tin	Sn	50	118.7
Titanium	Ti	22	47.88
Tungsten	W	74	183.9
Uranium	U	92	238.0
Vanadium	V	23	50.94
Xenon	Xe	54	131.3
Ytterbium	Yb	70	173.0
Yttrium	Y	39	88.91
Zinc	Zn	30	65.38
Zirconium	Zr	40	91.22

*The values given here are to four significant figures.
†A value given in parentheses denotes the mass of the longest-lived isotope.

Applied Chemistry

Applied Chemistry

Third Edition

William R. Stine
Wilkes University

Contributors:

Terese M. Wignot
(Chapter Twenty-Three)

Edward B. Stockham
(Chapter Twenty-Six)

D. C. Heath and Company
Lexington, Massachusetts   Toronto

Address editorial correspondence to:
D. C. Heath and Company
125 Spring Street
Lexington, MA 02173

To: Echo Sierra, Charlie Romeo, and Bravo

Acquisitions Editor: Kent Porter Hamann
Developmental Editor: Betty B. Hoskins
Production Editor: Elizabeth Gale
Designer: Henry Rachlin
Photo Researcher: Susan Doheny
Production Coordinator: Richard Tonachel
Permissions Editor: Margaret Roll

Preface

The third edition of *Applied Chemistry* is designed to meet the needs of nonscience students taking an introductory chemistry course. These students do not require an extensively detailed and quantitative presentation of chemistry, but can benefit greatly from a clear understanding of the chemistry surrounding them in everyday life. Finding solutions to energy and environmental problems and maintaining good health are important to everyone, and subjects like these are the focus of this book. The third edition continues to emphasize timely and interesting applications of chemistry and to present chemical concepts on a need-to-know basis.

Organization and Expanded, Updated Coverage

In response to the reviewers of the third edition manuscript who include more coverage of basic principles in their courses, I have added four new chapters to Part I (Atoms, Chemical Reactions, Mole and Mass Relationships, and Polymers) and have upgraded three other chapters (The Periodic Table, Organic Chemistry, and Solution Chemistry). Also, the two chapters on nuclear chemistry were repositioned as a transition between Parts I and II. The most significant additions to these chapters on nuclear chemistry are discussions of radon and updates on the accidents at Three Mile Island and Chernobyl.

Whenever possible, applications are integrated within the chapters on chemical principles. While I have responded to the desire for greater coverage of principles, I did not want to sacrifice the coverage of applications and have not done so.

One unusual aspect of the Part I organization is the coverage of organic chemistry (Chapter 6) prior to the chapter on solution chemistry and acids and bases (Chapter 7). With this arrangement, organic acids can be included in Chapter 7, since the carboxylic acids such as carbonic, acetic, lactic, citric, and malic acids are often encountered by the consumer.

With Part I as background, the four chapters of Part II cover energy and the environment. As in earlier editions, environmental issues are also integrated into many other parts of the text. Some specific aspects of air and water pollution are addressed in Chapter 14, which has been considerably updated. The extensive section on chlorofluorocarbons and stratospheric ozone depletion in this chapter is completely new.

Part III on agricultural chemistry and Part IV on food chemistry are unique features of the text. With much helpful feedback from previous users and reviewers, significant updating has been done and includes, for example, reference to newer sweeteners (such as aspartame) and fat substitutes (such as Simplesse) and discussions of the roles of acidity and carbonation in winemaking.

Part V addresses the importance of consumer chemistry: products that are used on a daily basis at home, such as soaps, synthetic detergents, and toothpastes; drugs that are used for curing disease and for treating symptoms of disease; and steroids as body substances and as prescription drugs. Topics have been revised to reflect current concerns. As a result, I discuss the degree to which phosphates in detergents contribute to environmental pollution and I provide new information on the chemistry of tooth decay, materials for filling teeth, and the controversy over fluoridation of community water supplies.

While the chapter on immunochemistry (Chapter 23) covers the recurring problem of influenza, it also concentrates on the acquired immune deficiency syndrome (AIDS). Chemotherapy and drugs are presented in two chapters (Chapters 24 and 25), the first of which discusses the use of drugs as therapeutic agents in the treatment of disease. The second chapter discusses two other uses of drugs: as treatments for symptoms (for example, aspirin and antacids) and as chemicals that are abused and are potentially addictive. New information on the role of neurotransmitters clarifies the mechanisms of drug abuse and addiction. Finally, the chapter on steroids and contraception (Chapter 26) covers three key issues in steroid chemistry: the mode of operation of various types of oral contraceptives, the use and abuse of anabolic steroids, and the importance and dangers of cholesterol in the human body.

Flexibility of Course Outlines

Feedback from users and reviewers indicates that most of the courses using this text are one-semester courses, in which the instructor has to carefully select a manageable number of topics to cover in a limited amount of time. Therefore, throughout the development and revision of this book, I have devoted considerable attention to maximizing the flexibility of the coverage so that instructors can adapt it to their own styles and emphases in teaching. *Applied Chemistry* offers applications in sufficient breadth and depth of coverage to allow even the uninitiated instructor to handle each subject comfortably.

The Organizational Chart on p. ix offers a guide to the essential background chapters for each topic. While this chart does reflect author preferences, I hope it will assist each instructor in deciding how to select topics within the constraints of time to effectively fulfill the particular course objectives.

Because of the desire to maximize flexibility, certain information is repeated in more than one chapter of the text. Such repetition helps to reinforce important concepts, and makes it easier to move from chapter to chapter without undue concern about missing needed background. The more thorough coverage of applications does not make *Applied Chemistry* more difficult to read, but gives both students and instructors a better opportunity to understand each topic.

Learning Aids

Each section of each chapter in the third edition has learning goals to focus student learning. At the end of each chapter, there is a summary to reinforce material covered in the chapter, an expanded selection of problems to test recall and comprehension, and numbered references. The many citations of chemical literature allow instructors and interested students to explore topics in more detail. Users of previous editions told us they found this feature useful.

At the end of the textbook is a glossary along with answers to selected even-numbered problems.

I have retained the rather informal conversational style of writing used in earlier editions. Within the chapters, marginal notes signal cross-references to concepts and applications, emphasize various rules and terminology, and clarify certain facts without distracting from the text.

Acknowledgments

Besides recognizing the patience, encouragement, and assistance from those individuals to whom the book is dedicated, I wish to express my special feelings of appreciation to those who have taken the time to comment on the text at several stages of its development.

Parts of the manuscript for the new edition were carefully reviewed by professors Pamela M. Aker, University of Pittsburgh; Phrosene Chimiklis, Victor Valley Community College; Henry O. Daley Jr., Bridgewater State College; Salim M. Diab, College of St. Francis; David Frank, California State University-Fresno; Cleta Kay Hanebuth, University of South Alabama; Jeffrey R. Pribyl, Mankato State University; Merlyn D. Schuh, Davidson College; Julie Smist, Springfield College; Robert L. Soulen, Southwestern University; Dennis L. Stevens, University of Nevada-Las Vegas; Judith A. Strong, Moorhead State University; Harold I. Weingarten, Louisiana State University. Their thoughtful comments were extremely valuable in molding the finished product. The very thorough review by Steven H. Albrecht, Oregon State University was particularly helpful. Many thanks to Charlotte Hoffman and Mark Bauman for their many contributions to the project.

Users of earlier editions, including my own students, provided many helpful comments. Above all, it is students like Tony DiMichele who make the effort all worthwhile.

The excellent developmental editing of Betty Hoskins has greatly improved the organization and readability of the text. Only I will ever know how much the project was improved by her efforts.

A special thanks to Kent Porter Hamann and Joanne Williams for their support and encouragement, and especially to Liz Gale. It has been a joy to work with you all.

W. R. S.

Contents

Part II   Energy and Environment

Basic Principles
of Chemistry

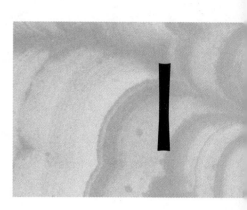

1

Chemistry, a Science for Today and Tomorrow

1

T he time is 7:30 A.M. Your clock radio has just started to talk, waking you out of a sound sleep. The announcers have just finished the news, and summarize the weather by saying that it is going to be a great day.

But what do they know? They don't have a chemistry class to go to at 9 A.M. So you push the snooze alarm for another 5 minutes of sleep. You put your head down on your pillow stuffed with polyurethane foam and snuggle under the polyester sheets. After three more bouts with the alarm, you find yourself staring into a mirror, brushing your teeth with a fluoride toothpaste. Since you stayed up late the night before studying chemistry, you have a slight headache and you decide to take some aspirin or acetaminophen (the active ingredient of Tylenol). You jump into the shower and scrub with your favorite coconut oil soap, and head off for breakfast.

Your breakfast consists of fruit juice (fructose, citric acid, orange flavoring), eggs (protein, fat, cholesterol), toast (starch, protein), cereal (carbohydrate, protein), bacon (fat, protein), and coffee (flavoring, caffeine) with milk (lactose, casein, butterfat) and sugar (sucrose). You sweeten your cereal with sugar, Sweet N' Low (saccharin), or Equal (NutraSweet). You put butter or hydrogenated margarine on your toast. The butter comes from milk that came from a cow that was fed with agricultural products grown with fertilizers. The margarine is derived from soybean oil hardened with hydrogen gas.

The eggs are cooked in a metal pan coated with nonstick Teflon. The toaster uses electricity from the local utility company that burns coal or uses nuclear fission as a source of energy to generate electricity. You sit on a plastic, wood, or metal chair while you eat off melamine-formaldehyde dishes at a table with a Formica surface. After you sit down, you realize that you need eating utensils, and get a silverplated knife and fork (nickel coated with silver).

After breakfast, you dress in clothing made of cotton, acrylics, and polyesters. You put on your glasses consisting of plastic frames and plastic lenses, and slip on your 10-karat gold high school class ring (42% gold, 35% copper, 13% silver, 7% zinc, and 3% copper). As you prepare to leave for class, you recall a matter that will require your attention later in the day, and you make yourself a note on chemically processed paper using a pencil made of wood and graphite (carbon) or a pen containing ink. You dispose of some old notes in a polyethylene waste basket and head out to the car.

You look with pride at your shiny car, washed the previous day with a detergent and protected by wax. It is an older car that has been protected from corrosion with rustproofing materials. You turn the key and electrons flow from the electrochemical cell (the battery) to the starter and sparks begin to ignite the hydrocarbons that are flowing to the engine from the gas tank. Octane boosters added to the gasoline make the car run smoothly. Since your car sat outside in the cold overnight, you use the windshield washer containing soap, water, and methanol to remove a coating of frost on the windows. Lubricating oils allow you to start the car easily regardless of temperature. Antifreeze in the coolant system has protected the engine from the cold and now protects it from the heat that is generated by burning the fuel.

Since you got off to a late start, you are rather impatient as you find yourself delayed at a traffic light, even though you are sitting on a comfortable seat cushioned with polyurethane foam and covered with a polyester fabric. You notice the envelope you placed on the dashboard the day before. You admire the colorful stamp held in place on the letter by an adhesive made of carbohydrate and you grumble at the sticky adhesive that makes the envelope hard to open. Just then you notice that the traffic light is now shining through the green Plexiglas covering; it is your turn to go. The car in front of you is making a turn, so you go around it by turning your plastic steering wheel and pointing your vulcanized rubber tires in the proper direction.

You find a place to park, and gather up your book bag made of polyvinyl chloride with plasticizer added. You then walk across the asphalt parking lot, enter the door, and go to the classroom where you are about to listen to a chemistry lecture. And you think to yourself, "Why do I have to learn chemistry? What does it have to do with me anyway?"

1.1 Introduction

GOAL: To illustrate the wide range of applications of chemistry in everyday life.

The introductory story shows you, the student, that the science of chemistry is concerned with all matter, including all the objects and all the substances you encounter in daily life. The forms in which matter exists, the changes it undergoes, and the energy involved in those changes all are the subjects of chemistry.

Not only the objects you encounter, but also the human body is the subject of chemical study. It is a complex machine that continually consumes chemicals from the air and from food and processes them into new forms. As it converts and releases energy, the body grows and maintains itself. The body also encounters additives in foods at every meal; prescription and over-the-counter drugs are consumed for various ailments and symptoms. Evaluating the claims of advertisers, and avoiding potentially harmful so-called recreational and body-building chemicals are part of your study too. Chemistry is so much a part of life, no one can responsibly neglect having some knowledge of the subject.

Since chemistry is about matter, it is a broad field. Chemists work in academic research as well as in industrial research and development. Some develop chemical reactions to be used in manufacturing processes. Some are involved in purification of water, food production and preservation, development of prescription and over-the-counter drugs, refinement of petroleum, production of plastics, gasification of coal, and other applications. Lawn services provide chemical fertilizers and weed killers, and home gardeners use composting methods to develop new rich soil. This textbook addresses many industrial processes that prepare products for you, the consumer. The author hopes you will understand enough fundamental facts and enough applications of chemistry to be better able to deal with many consumer issues.

Science and technology are the basis of our complex civilization, yet they can also cause problems. Desired products may result in undesired by-products, and the manufacturing processes may produce pollutants. Rather than applying a series of technical fixes, understanding the science often leads to solutions that go to the source of the difficulties. Throughout this textbook you will find examples of products that were synthesized, marketed, and then replaced as they became obsolete or were banned. Such trends in the use of chemicals are caused by social as well as scientific factors. Wherever possible, this textbook considers both the social ''why'' and the scientific ''what.''

In addition to scientific data, two types of calculations aid decision making about chemical products. One is the risk/benefit ratio, the other is the cost/benefit ratio. No activity on earth is entirely risk-free, but humans must make choices about the amount of risk they are willing to build into an industrial or public-funded process. Sometimes these levels of risk are set by government agencies, sometimes by companies and professional societies. The Environmental Protection Agency (EPA), for example, regulates the amount of sulfur dioxide to be released by a coal-burning plant that generates electricity. Levels may be set differently for different parts of the country, and for plants of different ages (see Chapter 12).

Low-sulfur coal costs more, but releases less sulfur dioxide. Scrubbers that capture sulfur dioxide from power plants add greatly to the costs of operation. These costs are passed on to the consumer. This illustrates the cost/benefit ratio. Are the benefits worth the added cost, and who receives the benefits and who the cost?

1.2 Properties of Matter

GOALS: To compare physical and chemical properties.
To define energy and illustrate some of its forms.
To know where to locate information about the elements.
To see the distinction between organic and inorganic chemistry.
To become acquainted with the range of topics covered in this text.

In chemistry, matter is recognized and classified according to its characteristic physical or chemical properties. **Physical properties** include the state (gas, liquid, or solid), odor, color, consistency (viscous, crystalline, powdery, amorphous, etc.), and solubility. All of these are readily observable. The statement that table salt is a white, crystalline solid specifies three physical properties of this substance—white, crystalline, and solid.

Some physical properties, such as the temperature at which melting occurs (melting point), the temperature at which a liquid becomes a gas (boiling point), and the amount of matter per unit of volume (density), generally require more sophisticated measurements for accurate determination. However, even simple observations can provide crude information about these properties. For example, a helium-filled balloon rises because helium is less dense than air. Butter and margarine retain their shapes on the table but melt in the mouth because their melting points are above room temperature but below body temperature.

Chemical properties describe the ability of a substance to react with other matter or to change into a new substance. The tendencies for iron to rust and for paper to burn are familiar chemical properties that describe reactions with oxygen.

Matter undergoes changes that are classified as physical or chemical. During a **physical change,** the state or form of a substance changes while the composition remains the same: water freezes, coal is crushed into powder, and metal expands. During a **chemical change,** composition changes as substances react together or decompose: iron rusts (Chapter 4), coal burns (Chapter 12), the sugar glucose becomes alcohol by fermentation (Chapter 17).

Atoms and Elements, Molecules and Compounds

The basic units of all matter are called elements. At this time 109 elements are known. A complete, alphabetical table of the elements and their universal abbreviations, called symbols, appears on the page facing the inside front cover of this book. The table on the inside front cover is called the periodic table; it will be discussed in Chapter 3.

Every element is composed of atoms. It is the structure of its atoms that makes each element unique and determines its chemical and physical properties. In Chapter 2 we will study the structure of the atom.

When two atoms (of the same or different kinds) react, a molecule is formed. That is, elements react to form compounds. These compounds, too, have specific physical and chemical properties. Because matter is composed of atoms, precise relationships exist between substances that react to form new products. In Chapter 5 we will see that the number of atoms that will react in a particular way is precise and predictable. Many chemical reactions are carried out in solution, where solubility, pH, and concentration are important variables (Chapter 7).

The nuclei of some atoms undergo changes that provide useful techniques in medicine and research (Chapter 10), as well as controversial sources of electrical power and the potential of nuclear bombs (Chapter 11).

Energy

Energy is a property of matter, and thus a part of the study of chemistry. It is an agent for initiating physical and chemical changes. It also exists as heat that may be required to start a reaction, or it may be one of the products of reaction.

Simply defined as the ability to do work, energy exists in many forms. These forms can interconvert, a most important fact for the application of chemistry. When paper or fuel such as wood or oil is burned, stored chemical energy is converted into heat energy. Heat energy can be used to boil water, converting it to the gaseous state of steam, which can drive a turbine. This mechanical energy can be used to generate electrical energy. When a lamp is lit, electrical energy is converted to radiant energy and heat. When an appliance such as a vacuum cleaner is turned on, electrical energy is converted to mechanical energy. Solar energy is produced by nuclear fusion reactions that take place in the sun. Geothermal energy is heat energy produced from nuclear reactions deep in the earth acting on core elements and water.

At present, our society has serious questions about the best sources of energy that also would minimize pollution. Alternatives such as coal, oil, natural gas, solar energy, and nuclear energy can be evaluated only by knowing their energy content and their properties. For example, liquid fuels for transportation are a particular concern (Chapter 12). The vehicle in the introductory story is powered by gasoline. Once the pollution problems of leaded gasoline were recognized, blends of unleaded gasoline and other substances were developed to give enhanced octane ratings while stretching petroleum supplies. These and other alternative energy sources for the future are presented in Chapter 13. Electric cars are possible as an antipollution device in urban areas, and new batteries (electrochemical cells) are being designed for this purpose. Still, the chemical nature of acid rain, the greenhouse effect, smog covers, and stratospheric ozone holes remain serious problems (Chapter 14). Understanding the basic chemistry leads to testable ideas about preventing and reversing damage.

Carbon Compounds

Carbon compounds are discussed throughout the text. **Organic chemistry** is the study of compounds containing the element carbon. **Inorganic chemistry** deals with all elements and compounds other than those classified as organic. The number of organic compounds far exceeds the number of inorganic compounds.

The fossil fuels (coal, oil, and gas) are the primary sources of carbon compounds. Organic compounds are first introduced in Chapter 6, with a discussion of oil and gas. Coal and coal by-products are described in detail in Chapter 12.

A major application of carbon compounds is in making natural and synthetic polymers, including proteins and plastics, discussed in Chapter 8. Chapter 9 describes the chemistry of the carbohydrate polymers, such as starch, from their building blocks, the sugars. These chapters lead to the application of carbon chemistry to foods in Part IV, from fermented juices to bakery products, and from dairy products to pickled foods, processed fats, and oils. The chemistry of fermentation is used to manufacture alcoholic beverages including beer, wine, and distilled drinks (Chapter 17). Similar chemical processes are used to produce leavened bakery goods (Chapter 18) and to make dairy products such as cheeses and yogurt (Chapter 19). Preserved foods, such as corned beef, pickles, and sauerkraut, are discussed together with methods of canning. Food additives including sweeteners, colorings, flavor enhancers, and preservatives (Chapter 20) are commonly used. Finally fat sources, both plant and animal, are used to make liquid cooking fats, margarines, and shortenings (Chapter 21).

The agricultural production of food is based on chemistry, as is shown in Part III. Soil conditions and fertilizers, both organic and synthetic, contribute to the productivity of both the commercial farmer and the home gardener (Chapter 15). Insect pests that do massive damage to crops can be killed off with insecticides, but not without environmental consequences. Understanding insects themselves and the ways in which they use chemicals for communication leads to new forms of pest control, without destroying helpful insects and other creatures (Chapter 16).

Part V looks at health concerns. We acknowledge the many consumer products in our homes, including soaps and detergents, fabric softeners, drain and oven cleaners, and dental products (Chapter 22) that are marketed in response to consumer demands. Environmental concerns have shaped many of these products and have led to continued research and development.

The chapter on immunochemistry (Chapter 23) explains how we resist infectious diseases, even though microbes coexist with us at all times. But we also should understand why repeated bouts of influenza are possible and why worldwide epidemics occur. The current threat to the immune system is the AIDS virus. Its chemical structure is better understood than are ways to block its effects; current research will be described.

Chemists have important roles in producing vaccines (Chapter 23) and drugs, both for curing diseases (Chapter 24) and for treating symptoms (Chapter 25). Over-the-counter painkillers (analgesics), combination remedies, and antacids are widely used. It is important to understand the chemistry of how they work as well as potential disadvantages. Antibiotics that provide ''miracle cures'' for many bacterial infections also can result in development of resistant microbes, so more effective chemical forms are constantly being sought in the ''war between the bugs and the drugs.'' Particularly, the serious hazards of illegal drugs, with adverse effects on the central nervous system, are important to understand (Chapter 25). Finally, the complex steroid molecules act on the brain and reproductive organs, providing ways to understand human cycles and contraception (Chapter 26). The best-known steroid is cholesterol, and its roles in hormone synthesis as well as in heart disease are important examples of the relevance of chemistry to daily life.

1.3 The Mathematics of Chemistry

GOALS: To study the factor-label method for interconverting units.
To know the common units of measurement used in science.
To be able to interconvert among metric units and between metric and English units.
To study the way to write very large and very small numbers.

Throughout later chapters, margin notes will refer you back to Sections 1.3 through 1.6 for necessary background information.

Certain mathematical expressions, units, and manipulations are often used in chemistry. Because matter is composed of atoms, precise relationships exist between starting materials and products in chemical conversions. On both individual consumer and industrial scales, substances can be measured in exact amounts and allowed to react, predicting the amount of products to be obtained. When you understand the logic of chemical methods of measurement, you can solve common problems.

Factor-Label Analysis

Chemical calculations frequently require the conversion of one unit into another. Such conversions can be accomplished efficiently by using factor-label analysis, also called unit analysis. This method stresses the unit component of the quantity to be converted. It is based on the fact that a value can be multiplied by a factor whose numerator and denominator are equal, but whose units are different. After canceling like units, the result of multiplication has the same value, but its units are changed. Consider the following conversion example.

EXAMPLE 1.1:

Eggs cost 85 cents per dozen. What is the cost of 6 dozen eggs?

Solution:

The problem asks for 6 dozen eggs to be converted into cost (cents). Although the answer to this problem may be obvious, the calculation illustrates the factor-label method of conversion. The key to the use of the method for this particular problem is the equality:

$$1 \text{ dozen eggs} = 85 \text{ cents}$$

This allows us to formulate a conversion factor in which the numerator and denominator are equal, for use in the following way:

First, the quantity to   is multiplied by a   to produce the
be converted   conversion factor   desired quantity

$$6 \text{ dozen eggs} \times \text{conversion factor} = ? \text{ cents}$$

Second, the conversion factor is arranged so that the unit on the quantity to be converted is canceled.

$$6 \text{ dozen eggs} \times \frac{\text{cents}}{\text{dozen eggs}} = ? \text{ cents}$$

Third, the numbers are added to the conversion factor so that it represents a true statement, with the numerator equal to the denominator.

$$6 \text{ dozen eggs} \times \frac{85 \text{ cents}}{1 \text{ dozen eggs}} = 510 \text{ cents}$$

Similarly, simple English unit interconversions can be done using the factor-label method.

EXAMPLE 1.2:

Convert 42 inches to yards.

Solution:

Determine the proper form of the conversion factor that can be derived from the following equality:

$$1 \text{ yd} = 36 \text{ in.}$$

Therefore,

$$42 \text{ in.} \times \frac{1 \text{ yd}}{36 \text{ in.}} = 1.2 \text{ yd}$$

Not all conversions are accomplished this easily. Several steps may be required to arrive at the desired unit.

EXAMPLE 1.3:

Calculate the volume, in milliliters, of 1.4 quarts of water.

Solution:

We seldom have a simple equality available to show the relationship between quarts and milliliters. Therefore, we must use the following equalities to formulate a sequence of two conversion factors:

$$0.946 \text{ L} = 1 \text{ qt}$$
$$1000 \text{ mL} = 1 \text{ L}$$

Considering only the original value (1.4 qt) and the desired change of units, we will want to set up the conversion as follows:

$$1.4 \text{ qt} \times \frac{\text{L}}{\text{qt}} \times \frac{\text{mL}}{\text{L}} = ? \text{ mL}$$

Now adding the numbers from the two equalities given previously, the calculation is completed as follows:

$$1.4 \text{ qt} \times \frac{0.946 \text{ L}}{1 \text{ qt}} \times \frac{1000 \text{ mL}}{1 \text{ L}} = 1.3 \times 10^3 \text{ mL}$$

See later discussion for details on exponential notation.

Any number of conversion steps may be required to produce the final quantity. If each conversion factor is correct, with units canceling and with the numerator equal in value to the denominator, the answer will be correct.

Consider the following equalities between ''nonsense'' units:

$$1 \text{ zetter} = 2.5 \text{ luks}$$
$$3 \text{ parms} = 1 \text{ talp}$$
$$100 \text{ luks} = 8 \text{ parms}$$

EXAMPLE 1.4:

How many talps are equivalent to 5 zetters?

Solution:

Restating the problem:

$$5 \text{ zetters} \times \text{conversion factor(s)} = ? \text{ talps}$$

To cancel zetters:

$$5 \text{ zetters} \times \frac{\text{luks}}{\text{zetters}}$$

To cancel luks:

$$5 \text{ zetters} \times \frac{\text{luks}}{\text{zetters}} \times \frac{\text{parms}}{\text{luks}}$$

To cancel parms and produce the desired talps:

$$5 \text{ zetters} \times \frac{\text{luks}}{\text{zetters}} \times \frac{\text{parms}}{\text{luks}} \times \frac{\text{talps}}{\text{parms}} = ? \text{ talps}$$

We then insert the correct numbers from the equalities given previously to arrive at:

$$5 \text{ zetters} \times \frac{2.5 \text{ luks}}{1 \text{ zetter}} \times \frac{8 \text{ parms}}{100 \text{ luks}} \times \frac{1 \text{ talp}}{3 \text{ parms}} = 0.33 \text{ talps}$$

The Metric System

The metric system is the measurement system of science and technology. We in the United States are more familiar with the so-called English system that uses the foot as the unit of length, the pound as the unit of weight, and the gallon as the unit of volume. Elsewhere in the world, including all of Great Britain, the metric system is used exclusively (Figure 1.1).

More and more products in the United States show metric measurements. How about that two-liter soda you bought recently? How does it compare in size with a half-gallon of milk? (Since Table 1.1 shows that 1 gal is 3.78 L, a half-gallon is slightly less than 2 L.)

The popularity of the metric system is due to its decimal nature. All its units may be interconverted by multiplying or dividing by 10^{exponent} as indicated in Table 1.2. (Exponential notation is discussed later.) In other words, you can convert from meters to centimeters (1/100) by moving the decimal point two places. Compare the memorization required to convert inches to feet (divide by 12), or feet to yards (divide by 3), or feet to miles (divide by 5,280).

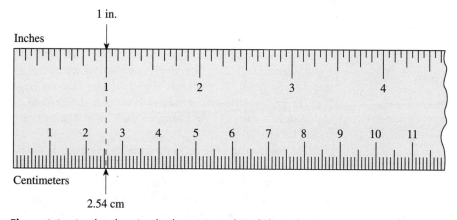

Figure 1.1 A ruler showing both metric and English graduations.

Table 1.1	Some English-metric conversions

Length

1 mile (mi)	= 1.61 kilometers (km)
1 yard (yd)	= 0.914 meter (m)
1 inch (in.)	= 2.54 centimeters (cm)

Mass

1 pound (lb)	= 454 grams (g)
1 pound (lb)	= 0.454 kilogram (kg)
1 ounce (oz)	= 28.4 grams (g)

Volume

1 gallon (gal)	= 3.78 liters (L)
1 quart (qt)	= 0.946 liter (L)
1 pint (pt)	= 0.473 liter (L)
1 fluid ounce (fl oz)	= 29.6 milliliters (mL)

Table 1.2	Prefixes used in the metric system

Multiple	Prefix	Abbreviation
10^9	Giga-	G
10^6	Mega-	M
10^3	Kilo-	k
10^2	Hecto-	h
10^1	Deka-	da
10^{-1}	Deci-	d
10^{-2}	Centi-	c
10^{-3}	Milli-	m
10^{-6}	Micro-	μ
10^{-9}	Nano-	n

One meter is officially defined as 1,650,764 times the wavelength of the orange-red light emitted by krypton-86.

The metric unit of length is the meter (m). One-tenth of a meter is a decimeter (dm), a unit seldom used. One-hundredth of a meter is a centimeter (cm), and one-thousandth is a millimeter (mm). The meter is slightly longer than the English yard (Table 1.1).

The basic metric unit of volume is the liter (L), defined as the volume of a cube which is 1/10th of a meter (1 dm) long on each side. The metric liter is slightly larger than a quart (Table 1.1).

The metric unit of mass is the gram (g), defined as the mass of 1/1000th of a liter, or 0.001 L, or 1 milliliter (mL), of pure water at a temperature of 4 °C. The gram, a small quantity, is commonly used in the scientific laboratory. On a larger scale, the kilogram (kg) is often used. It is defined as the mass of 1 L of water at 4 °C; it equals 1000 g, about 2.2 lb (Table 1.1). Therefore, 1 g is equal to 0.001 kg.

The metric unit of energy most often used in chemistry is the calorie. It is defined as the amount of heat required to raise the temperature of 1 g of water 1 °C (from 15-16 °C).

You are probably more familiar with the Calorie (with a capital C) used by nutritionists to express the energy content of foods. One Calorie—sometimes called the large calorie—equals 1000 calories, or 1 kilocalorie (kcal). Energy units are described in detail in Chapter 12.

EXAMPLE 1.5:

What is the metric equivalent, in centimeters, of 100 in.? See Figure 1.1.

Solution:

Visually, you can see the answer in Figure 1.1. From Table 1.1, notice that 1 in. equals 2.54 cm. Therefore,

$$100 \text{ in.} \times \frac{2.54 \text{ cm}}{1 \text{ in.}} = 254 \text{ cm}$$

EXAMPLE 1.6:

What is the weight, in kilograms, of a 110-lb person?

Solution:

From Table 1.2, 1 lb is equivalent to 0.454 kg. Therefore,

$$110 \text{ lb} \times \frac{0.454 \text{ kg}}{1 \text{ lb}} = 49.9 \text{ kg}$$

EXAMPLE 1.7:

What is the volume, in gallons, of a 2.0-L container?

Solution:

From Table 1.2, 1 gal = 3.78 L. Therefore,

$$2.0 \text{ L} \times \frac{1 \text{ gal}}{3.78 \text{ L}} = 0.53 \text{ gal}$$

EXAMPLE 1.8:

What is the length, in inches, of a 5.00-m stick?

Solution:

Recall that 1 yd = 36 in. and that 1 yd = 0.914 m (Table 1.1). Therefore,

$$5 \text{ m} \times \frac{1 \text{ yd}}{0.914 \text{ m}} \times \frac{36 \text{ in.}}{1 \text{ yd}} = 197 \text{ in.}$$

In the metric system, as illustrated, any unit can be converted into a larger or smaller unit by addition of a prefix. Each prefix changes the basic measurement by a multiple of the number 10 raised to some power as indicated in Table 1.2. For example, 1 km = 10^3 m; 1 kg = 10^3 g. Similarly, 1 mm = 10^{-3} m; 1 mg = 10^{-3} g.

EXAMPLE 1.9:

What is the milligram equivalent of 50 g?

Solution:

From Table 1.2, 1 mg = 10^{-3} g. Therefore,

$$50 \text{ g} \times \frac{1 \text{ mg}}{10^{-3} \text{ g}} = 50 \times 10^{+3} \text{ mg} = 50,000 \text{ mg}$$

EXAMPLE 1.10:

Convert 8000 μL into liters.

Solution:

From Table 1.2, 1 μL = 10^{-6} L. Therefore,

$$8000 \text{ μL} \times \frac{10^{-6} \text{ L}}{1 \text{ μL}} = 8000 \times 10^{-6} \text{ L} = 0.008 \text{ L}$$

EXAMPLE 1.11:

The electric car, or electric vehicle (EV), is under development for use in urban areas. The Impact from General Motors is expected to have a range of 120 mi at 55 miles per hour (mph). Convert this range to kilometers. (The EV is described in Chapter 13.)

Solution:

From Table 1.1, 1 mi = 1.61 km. Therefore,

$$120 \text{ mi} \times \frac{1.61 \text{ km}}{1 \text{ mi}} = 193.2 \text{ km}$$

EXAMPLE 1.12:

The hydrodynamic energy of falling water can be used to generate electrical energy. As we will see in Chapter 13, the difference in height of low and high tides can be used in a few places in the world. The first tidal power plant is across the Rance River in France. Tides often exceed 12.2 m. How many yards is that?

Solution:

From Table 1.1, 1 yd = 0.914 m. Therefore,

$$12.2 \text{ m} = \frac{1 \text{ yd}}{0.914 \text{ m}} = 13.3 \text{ yd}$$

EXAMPLE 1.13:

The Chrysler Concorde is rated at 214 horsepower (hp). Express this power rating in kilowatts (kW).

Solution:

Given that 1 hp = 746 W,

$$214 \text{ hp} \times \frac{746 \text{ W}}{1 \text{ hp}} \times \frac{1 \text{ kW}}{1000 \text{ W}} = 160 \text{ kW}$$

EXAMPLE 1.14:

A jar of mayonnaise contains 8 fl oz. It is said to contain 16 servings (1 tablespoon each), with each serving containing 70 mg of sodium (Na). How many grams of sodium are present in the jar?

Solution:

Given that 1 serving = 70 mg,

$$16 \text{ servings} \times \frac{70 \text{ mg}}{1 \text{ serving}} \times \frac{1 \text{ g}}{1000 \text{ mg}} = 1.12 \text{ g}$$

The International System

A modified form of the metric system is used increasingly. The International System of Units, or Système International (SI), has some units used only in more advanced studies. The metric and SI units of mass and length are the same.

The units of volume are different in the SI, but in a very subtle way. Whereas the basic unit of volume in the metric system is the liter, in the SI, it is the cubic meter (m^3). Since the cubic meter is a very large volume, chemists continue to express volumes in liters or milliliters. One milliliter is equivalent to 1 cubic centimeter (cm^3 or cc).

Exponential Notation

Scientific measurements and calculations often involve very large and very small numbers. A shorthand expression of these numbers is often useful. This expression, exponential notation, takes the following form:

$$N \times 10^{exponent}$$

where N is usually a number between 1 and 10, and the exponent is a whole number. The exponent may be either positive or negative.

The notation is summarized in Table 1.3. It is easily understood once you recognize that the exponent does nothing more than describe the location of the decimal point. For example, the number 1 million can be written in exponential notation by moving the decimal point six places *to the left* and using the exponent +6.

$$1 \text{ million} = 1{,}000{,}000 = 1.0 \times 10^{6}$$

move decimal
6 places to left

One millionth can also be expressed in exponential notation by moving the decimal point *to the right* six places and using the exponent −6.

$$1 \text{ millionth} = 0.000001 = 1.0 \times 10^{-6}$$

move decimal
6 places to right

In other words, when a number has been converted to exponential notation, a positive exponent signifies that the decimal *has been moved* to the left, whereas a negative exponent indicates that the decimal *has been moved* to the right.

Table 1.3   Exponential notation

Number	Exponential Notation
1,000,000,000	1.0×10^{9}
1,000,000	1.0×10^{6}
10,000	1.0×10^{4}
1,000	1.0×10^{3}
100	1.0×10^{2}
10	1.0×10^{1}
1	1.0×10^{0}
0.1	1.0×10^{-1}
0.01	1.0×10^{-2}
0.001	1.0×10^{-3}
0.000 1	1.0×10^{-4}
0.000 001	1.0×10^{-6}
0.000 000 001	1.0×10^{-9}

You may commit these rules to memory, or simply recall that positive exponents are used for large numbers and negative exponents are used for small numbers. This fact is a guide when you are converting in either direction. Examples follow.

EXAMPLE 1.15:

Express 62,000 in exponential notation.

Solution:

Since the number N in the exponential is normally given between 1 and 10, it is necessary to move the decimal four places to the left, and use the exponential $+4$. Therefore, the correct answer is 6.2×10^4. Notations such as 62×10^3 or 0.62×10^5 are also correct, but are not usually used, since N of each expression does not follow convention.

EXAMPLE 1.16:

Express 0.000 029 in exponential notation.

Solution:

To give N between 1 and 10, move the decimal five places to the right, which requires the exponent -5. Therefore, the correct answer is 2.9×10^{-5}.

EXAMPLE 1.17:

Express 2.36×10^2 as a common number.

Solution:

236

EXAMPLE 1.18:

Express 8.6×10^{-6} as a common number.

Solution:

0.000 008 6

EXAMPLE 1.19:

Express 6.02×10^{23} (Avogadro's number) as a common number.

Solution:

602,000,000,000,000,000,000,000

1.4 Mass and Weight

GOAL: To appreciate the distinction between mass and weight.

As we try to understand various topics, knowing the mass of an object or its component parts is often useful. Examples range from the minute, such as the masses of atoms described in atomic mass units, to huge quantities such as the mass of coal required to supply a power plant. The terms mass and weight are often used interchangeably, but strictly speaking, they do not have the same meaning. The mass of an object is a constant property that does not depend on the location of the object. The weight of an object is the amount of force required to start or stop the motion of the object. Thus it takes very little force to pick up a pencil, whereas a much greater force is required to move a boulder. On the earth, we use the force of gravitational attraction as the weight of an object. We describe astronauts in space as experiencing weightlessness without gravity; but their masses have not changed.

The term *weight* is often used when *mass* is correct. The mass of an object is determined by comparison with a standard of known mass. In the metric system, the gram or some convenient multiple, for example, the kilogram, is used to describe mass.

1.5 Density

GOAL: To define density and compare it with specific gravity.
To learn a practical application of specific gravity measurement.

Density is defined as mass per unit volume. It is an important physical property of matter.

$$\text{density} = \frac{\text{mass}}{\text{volume}}$$

The density of a solid or liquid substance is normally expressed in the units grams per milliliter (g/mL), or the equivalent grams per cubic centimeter (g/cc); it is the mass of 1 mL of that substance. The density of a gas is more commonly expressed in the units grams per liter (g/L), stating the mass of 1 L of the substance.

Most frequently the density of a substance is useful for determining the mass of a sample from its volume.

$$\text{density} = \frac{\text{mass}}{\text{volume}} \; ; \; \text{mass} = \text{volume} \times \text{density}$$

For example, the element mercury is a liquid with a density of 13.6 g/mL: the mass of a 4.2-mL sample can be easily calculated by cross-multiplying the previous equation.

$$\text{mass} = \text{volume} \times \text{density} = 4.2 \text{ mL} \times \frac{13.6 \text{ g}}{\text{mL}} = 57.1 \text{ g}$$

This application is especially useful for volatile or hazardous liquids that can be measured more readily by volumetric (volume-measuring) methods than by weighing.

EXAMPLE 1.20:

A 50-mL sample of a liquid weighs 45 g. What is its density?

Solution:

$$d = \frac{m}{v} = \frac{45 \text{ g}}{50 \text{ mL}} = 0.90 \text{ g/mL}$$

EXAMPLE 1.21:

What is the mass of a 200-mL sample of the liquid in Example 1.20?

Solution:

$$d = \frac{m}{v} \; ; m = d \times v$$

$$m = \frac{0.90 \text{ g}}{\text{mL}} \times 200 \text{ mL} = 180 \text{ g}$$

EXAMPLE 1.22:

The density of mercury is 13.6 g/mL. What is the volume of 100 g of this liquid?

Solution:

$$d = \frac{m}{v} \; ; v = \frac{m}{d} = \frac{100 \text{ g}}{13.6 \text{ g/mL}} = 7.35 \text{ mL}$$

Specific Gravity

Specific gravity is a comparison of the density of a substance with the density of a standard, usually water at 4 °C.

$$\text{specific gravity} = \frac{\text{density of substance}}{\text{density of water at 4 °C}}$$

The progress of fermentation during wine making can be monitored by measuring specific gravity (Figure 17.6).

Since both densities are expressed in the same units, specific gravity is a number without units. Furthermore, since the density of water in the metric system is 1.00 g/mL, specific gravities are numerically equal to densities.

For example, since the density of aluminum is 2.7 g/mL, the specific gravity of aluminum is 2.7.

$$\text{specific gravity of Al} = \frac{\text{density of Al}}{\text{density of water at 4 °C}} = \frac{2.7 \text{ g/mL}}{1.0 \text{ g/mL}} = 2.7$$

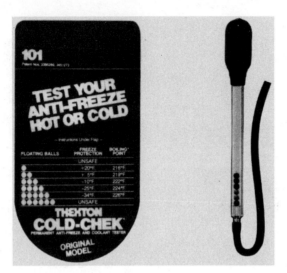

Figure 1.2
A hydrometer for checking the coolant in an automobile radiator. A sample of coolant is removed from the radiator by squeezing the bulb and placing the rubber tubing under the surface of the coolant. The degree of protection can be determined by seeing how many of the balls float on the mixture. The specific gravity of the balls increases from top to bottom. The higher-specific-gravity balls sink, and lower-specific-gravity balls float. More floating balls indicates a water-antifreeze mixture containing more antifreeze.

The specific gravity of the coolant in an automobile radiator can be checked by using a hydrometer. Ethylene glycol (specific gravity 1.11) and water (specific gravity 1) are mixed together in nearly equal amounts for use as a coolant. The mixture can be tested using a device as shown in Figure 1.2 to ensure proper protection against freezing.

1.6 Temperature Scales

GOAL: To understand the Celsius, Fahrenheit, and Kelvin scales.

The Fahrenheit temperature scale is commonly used in the United States, but two other scales, the Celsius and Kelvin (or absolute) scales, are used for scientific measurement. All three are based on the same two fixed points, the boiling point and freezing point of water, as shown in Figure 1.3. Notice that the degree sign (°) is not used in expressing temperature on the Kelvin scale.

Figure 1.3
The three common temperature scales. All temperatures shown are for pure water. Note that the degree sign is omitted for temperatures given on the Kelvin scale.

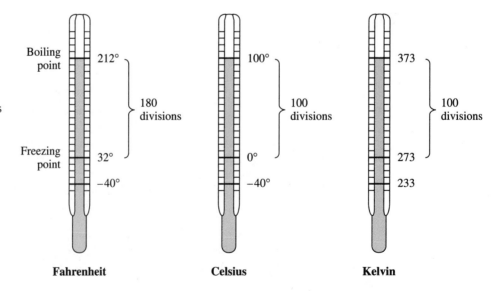

Table 1.4   Temperature conversion factors

To Convert	Into	Conversion Formula
°F	°C	°C = ⅝ (°F − 32)
°C	°F	°F = ⅑ °C + 32
°C	K	K = °C + 273
K	°C	°C = K − 273

The value of the freezing point of water on the Kelvin scale is due to the point assigned on this scale as zero. Zero Kelvin is the temperature at which all molecular motion would cease.

Figure 1.3 shows that the Fahrenheit scale divides the distance between the boiling and freezing points for water into 180 (whole-number) divisions, whereas the Celsius and Kelvin scales use 100 divisions. Therefore, a temperature change of 1 ° Fahrenheit represents a smaller change in temperature than a change of 1 ° Celsius or Kelvin. It should be noted that:

1. A Kelvin temperature reading is always higher than the corresponding Celsius reading by a factor of 273.

2. At −40 °C, the temperature is also −40 °F. At all temperatures above that level, the Celsius temperature reading is lower than the corresponding Fahrenheit reading.

3. Since 180 Fahrenheit divisions is equivalent to 100 Celsius divisions, a change of 1 Fahrenheit degree is equivalent to a change of 5/9 (100/180) of a Celsius degree.

All these factors lead to the conversion formulas shown in Table 1.4. Recall that the degree sign is omitted when expressing temperatures on the Kelvin scale. For example, to find the Celsius temperature equivalent to 70 °F:

$$°C = \frac{5}{9}(70 - 32) = \frac{5}{9}(38) = 21 \ °C$$

To express this same temperature on the Kelvin temperature scale:

$$K = 21° + 273 = 294 \ K$$

Therefore,

$$70 \ °F = 21 \ °C = 294 \ K$$

The interconversion between Celsius and Kelvin is simple since it involves only addition or subtraction of 273. Memorizing the two formulas for interconversion between Fahrenheit and Celsius may seem necessary, but can be reasoned easily. It is true that the fractions 9/5 and 5/9 must be learned along with the number 32, but which fraction is used, and whether 32 must be added or subtracted and when, depend on which conversion you are making. Since 0 °C and 32 °F both represent the same actual temperature, 32 must be added to a Celsius temperature when converting to Fahrenheit. Similarly, it is necessary to multiply by 9/5, rather than 5/9, to get a number larger than the value on the Celsius scale. It is then only necessary to decide whether to multiply by 9/5 before or after adding 32. These two possibilities are as follows:

$$°F = \frac{9}{5}°C + 32 \tag{1}$$

$$°F = \frac{9}{5}(°C + 32) \tag{2}$$

In equation (1), the multiplication is done before adding 32, and in (2), it is done after. Which one is correct?

You can answer the question by trying the conversion from 0 °C into °F. You already know the correct answer. Which equation gives this answer? Equation (1). Equation (2) is incorrect since it would give an answer much larger than 32 °F. Similar reasoning can be used to help recall the equation for conversion from °F to °C.

EXAMPLE 1.23:

Convert 75 °F into °C and K.

Solution:

$$°C = \frac{5}{9}(75° - 32°) = \frac{5}{9}(43°) = 24 °C$$
$$K = 24° + 273 = 297 K$$

Therefore,

$$75 °F = 24 °C = 297 K$$

EXAMPLE 1.24:

Convert −10 °C into °F and K.

Solution:

$$°F = \frac{9}{5}(°C) + 32° = \frac{9}{5}(-10°) + 32° = 14°F$$
$$K = °C + 273 = -10° + 273 = 263 K$$

EXAMPLE 1.25:

A recipe calls for cooking a turkey at 325 °F. Convert the oven temperature into °C and K.

Solution:

$$°C = \frac{5}{9}(°F - 32°) = \frac{5}{9}(325° - 32°) = \frac{5}{9}(293°) = 163°C$$

$$K = °C + 273 = 163° + 273 = 436 \text{ K}$$

EXAMPLE 1.26:

When a cake is baked, the batter rises partially because of the expansion of trapped air due to the increased temperature, a phenomenon known as oven spring (see Chapter 18). Assuming room temperature is 68 °F and the oven is set at 375 °F, describe the temperature change in °C and K.

Solution:

Before the cake goes in the oven:

$$°C = \frac{5}{9}(°F - 32°) = \frac{5}{9}(68° - 32°) = \frac{5}{9}(36°) = 20 \text{ °C}$$

$$K = °C + 273 = 20° + 273 = 293 \text{ K}$$

After the cake goes in the oven:

$$°C = \frac{5}{9}(°F - 32°) = \frac{5}{9}(375° - 32°) = \frac{5}{9}(343°) = 190°C$$

$$K = °C + 273 = 190° + 273 = 463 \text{ K}$$

Therefore, the temperature changes are 170 °C and 170 K. This equivalence also emphasizes the identical graduations on the Celsius and Kelvin scales (Figure 1.3). Similarly, the figure shows that a 100° change on the Celsius (or Kelvin) scale corresponds to a 180° change on the Fahrenheit scale. Therefore, we can apply factor-label analysis to determine the temperature changes directly.

$$\text{Temperature change} = 375 \text{ °F} - 68 \text{ °F} = 307 \text{ °F}$$

$$= 307 \text{ °F} \times \frac{100 \text{ °C change}}{180 \text{ °F change}} = 170 \text{ °C} = 170 \text{ K}$$

SUMMARY

The study of chemistry includes the structure and changes of matter, and the energy changes that accompany them. Chemistry is a part of everyday life, the air we breathe, the food and drink we consume, the clothes we wear, and all consumer products that we use. Chemical changes are many and varied, but all involve a change of composition of matter, not merely a change of state or form, that is known as a physical change.

Factor-label analysis is a useful means for changing units on numbers, and for solving other problems. The metric system is used worldwide, particularly in science. With this and other units of measurement, exponential notation often is helpful for writing very small and very large numbers.

The mass of an object is a constant property that does not depend on the location of the object. The weight of an object is the amount of force required to start or stop the motion of the object. The term *weight* is often used when the term *mass* is correct. Density is defined as the mass per unit volume of a substance. The specific gravity of a substance is numerically equal to its density. The latter is a ratio of densities, and so has no units.

The Fahrenheit, Celsius, and Kelvin temperature scales are all based on the same two fixed points, the boiling point and freezing point of water. The value of the freezing point of water on the Kelvin scale is due to the point assigned on this scale as zero. Zero Kelvin is the temperature at which all molecular motion would cease.

PROBLEMS

1. Use factor-label analysis to solve each of the following.
   a. Convert 15,000 in. into mi.
   b. How many talps are contained in 75 parms?
   c. How many zetters are equivalent to 55.0 talps?

2. a. Convert 8 oz into grams.
   b. Convert 2.5 L into quarts.
   c. Convert 100 mi into kilometers.
   d. Convert 500 cm into feet.
   e. What is the English equivalent, in quarts, of 1 L?
   f. How many pounds are equivalent to 1 kg?
   g. What is the length, in inches, of a meter stick?

3. a. Convert 8.4 m into centimeters.
   b. Convert 5180 km into meters.
   c. Convert 0.27 L into milliliters.
   d. Convert 150 Calories into kilocalories.
   e. Convert 49 mg into grams.
   f. Convert 2.6×10^4 µg into grams.

   g. Convert 21 Msec into seconds.
   h. Convert 0.054 hm into meters.

4. a. Express 0.02 in exponential notation.
   b. Express 0.000 83 in exponential notation.
   c. Express 907 in exponential notation.
   d. Express 9.6×10^{-3} as a common number.
   e. Express 1.85×10^4 as a common number.

5. a. A solid having a volume of 4.5 mL has a mass of 126.0 g. What is its density?
   b. The density of an acid solution is 1.15 g/mL. What volume of the solution must be used in an experiment requiring 200 g?
   c. An alcohol has a density of 0.88 g/mL. What is the mass of a 25-mL sample?

6. a. Express 59 °F as °C.
   b. Express −10 °C as °F.
   c. Express 25 °C, 50 °C, and −15 °C as K.

Atoms

2

Every element is composed of atoms. The structure of its atoms makes each element unique and determines the chemical and physical properties of that element. In this chapter we will study first how we came to believe in atoms. Then we will look at the component parts and properties of atoms to see how modern atomic theory developed.

In later chapters we will study the relationships between the structure of atoms and the properties of elements, both when the elements occur in pure form and when they are combined with one another.

2.1 History of Atomic Theory

GOAL: To describe highlights of the historical development of current atomic theory.

In the period 450–400 B.C., the Greek philosophers Leucippus and Democritus proposed the idea of atoms as the building blocks of all matter. Democritus coined the term *atomos*, meaning ''indivisible.'' Many sophisticated experiments conducted over the last 100 years led us to conclude that the original idea of the atom is correct. But for almost 2000 years the ideas of the Greek philosophers were not widely accepted.

The French chemist Antoine Lavoisier is often described as ''the father of chemistry'' (Figure 2.1). In the late 1700s he performed studies leading to his statement of the *law of conservation of mass*. This law states that the total mass of the products of a chemical reaction is equal to the total mass of the starting materials (known as reactants). In other words, it suggests that chemical reactions do not cause the creation or destruction of matter. Instead, reactions cause a change in matter from one form to another without any gain or loss of mass. For example, if a piece of wood is burned, its components are converted to measurable products, including water (H_2O), carbon dioxide (CO_2), and ash. Although the piece of wood disappears, a careful weighing of the products, including water vapor and carbon dioxide gas, accounts for all of the mass of the original wood.

Scientists who followed Lavoisier incorporated his findings and conclusions into a picture consistent with other knowledge of the behavior of atoms. At the end of the eighteenth century, another French chemist, Joseph Proust (1754–1826), studied various chemical reactions. From them he formulated the *law of definite proportions* (also known as the *law of constant composition*), which states that the percentage by mass of the component elements of a compound is always the same. The water molecule illustrates this law: water in ice, rain, seawater, and water vapor always has the formula H_2O.

John Dalton (1766–1844), an English chemist (Figure 2.2), performed studies in the early 1800s that led him to formulate the *law of multiple proportions*. It states that if two elements combine to form more than one compound, the masses of the elements that combine are found in the ratio of small whole numbers. For example, nitrogen and oxygen combine to form the so-called *oxides of nitrogen* with formulas such as NO, N_2O, and NO_2 (Figure 2.3).

Dalton also proposed a theory in which he characterized the properties of matter by combining the ideas of conservation of mass, definite composition, and multiple proportions with the Greeks' concept of the atom. Known as Dalton's *atomic theory*, it states that:

- Every element is composed of small particles called atoms.
- All atoms of any element are the same. The properties of the atoms of one element are unlike those of other elements.
- Atoms cannot be created, divided, or converted to atoms of other elements.
- Atoms combine in fixed, whole-number ratios to form compounds.
- Chemical changes include changes in the grouping of atoms that make up elements and compounds without the creation or destruction of atoms.

Figure 2.1 Antoine Lavoisier (1743–1794) is often called the father of chemistry.

Figure 2.2 John Dalton (1766–1844) proposed his atomic theory, a major step leading to modern theory of atomic structure.

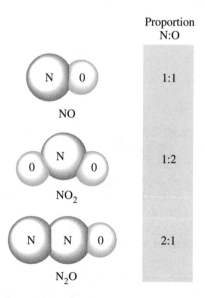

Figure 2.3 The three oxides of nitrogen illustrate Dalton's *law of multiple proportions*.

Concern has emerged over the presence of radon gas in homes. In Chapter 10 we will see how radon atoms form naturally by radioactive decay of uranium atoms in the soil.

The five components of Dalton's theory have withstood the test of almost two centuries of experimental work. Only the third statement requires amplification since nuclear reactions, discussed in Chapters 10 and 11, were unknown in Dalton's time. Nuclear reactions, such as radioactive decay, are now known to occur and most result in a conversion of one element into another. This may seem to represent an exception to Dalton's theory, but most scientists regard nuclear reactions and chemical reactions as distinctly different phenomena.

Once the existence of atoms was accepted, the next mystery to be solved was their internal structure. This is the subject of the following sections.

2.2 The Structure of the Atom

GOALS: To understand the basic structure of an atom.
　　　　To know the characteristics of the subatomic particles.
　　　　To appreciate the use of atomic mass units.

Although the existence of atoms has been suggested for many centuries, the picture of an atom has changed markedly. The development of new techniques led to new data and models for studying its structure and properties. Table 2.1 is a summary of key developments leading to today's description of the atom.

Atoms consist of three basic units: **protons, electrons,** and **neutrons.** The locations and important characteristics of these three subatomic particles are listed in Table 2.2.

Table 2.1　A brief history of atomic theory

400–500 B.C.	The idea existed among Greek thinkers that matter consisted of basic, indivisible particles. In the fifth century B.C. the Greek scholar Democritus named these particles ''atoms'' from the Greek word for indivisible. All substances were thought to be some combination of just four elements: earth, air, water, and fire. These ideas remained essentially unchanged until early in the nineteenth century.
1803–1808	The laws of conservation of mass (Lavoisier), of constant composition or definite proportions (Proust), and of multiple proportions led the English chemist John Dalton (1766–1844) to propose a detailed theory describing the composition of elements and the ability of atoms to combine to form compounds.
1834	Michael Faraday (1791–1867), an English chemist, working with electric currents in solutions, proposed the existence of a particle common to electricity and matter.
1859–1900	The work of the English physicist Sir William Crookes (1832–1919) resulted in the discovery of negatively charged particles given off by atoms in cathode ray tubes. Since most matter is electrically neutral, this observation led to the assumption that a positively charged particle also must be present.
1897	J. J. Thomson (1856–1940), an English physicist, was credited with the actual discovery of the electron when he determined the charge-to-mass ratio of the particles in cathode rays. These particles were named electrons in 1900. Thomson received the Nobel Prize in 1906 for the discovery.
1909	American physicist Robert Millikan's (1868–1953) studies of oil drops led to the determination of the actual charge on an electron. The work of Thomson and Millikan allowed calculation of the electron mass.

1910–1911	Sir Ernest Rutherford (1871–1937), an English physicist who received a Nobel Prize in 1908 for his work on atomic structure, studied the deflection of alpha particles from their paths when they were accelerated toward thin sheets of gold foil. His observations resulted in the idea of a nuclear charge and mass concentrated in a small, central region with most of the volume of the atom made up of empty space in which the electrons move.
circa 1911	The existence of positively charged particles (protons) in atoms was confirmed.
1913	The Danish physicist Niels Bohr (1885–1962) proposed a model of the atom to explain the radiation emitted by hydrogen atoms. This was a Rutherford-like nuclear atom in which the electrons travel outside the nucleus in orbits of given radii representing specific energies. When electrons move from higher to lower energy levels (or orbits), the observed radiation is emitted.
1924	The French physicist Louis de Broglie (1892–1977) proposed that electrons could have wave-like properties as well as particle properties.
1926	The Austrian physicist Erwin Schrödinger (1887–1961) expressed the positions and energies of electrons in an atom as a mathematical equation. The solutions to this equation, called quantum numbers, represent the different orbitals that electrons can occupy in an atom. Schrödinger received the Nobel Prize in 1933 for his work.
1932	The English physicist James C. Chadwick (1891–1974) discovered the neutron.

Table 2.2 Characteristics of subatomic particles: protons, neutrons, and electrons

Particle	Location	Charge	Exact Mass (amu)*	Approximate Mass (amu)
Proton	Nucleus	+1	1.0078252	1
Neutron	Nucleus	0	1.0086654	1
Electron	Outside the nucleus	−1	0.0005468	0

*amu = 1 atomic mass unit = 1.66×10^{-24} g (see Section 2.2 for discussion).

Protons and neutrons are located in a small, dense region at the center of the atom known as the **nucleus.** The nucleus accounts for only a small part of the total volume of the atom, having a diameter approximately 1/10,000 that of the entire atom. (Atomic diameters are in the range 10^{-7} to 10^{-8} cm. Nuclear diameters approximate 10^{-12} cm.) As shown in Table 2.2, protons and neutrons account for most of the mass of the atom; the mass of an electron is negligible compared to the masses of protons and neutrons.

Atoms, in their simplest uncombined form, are electrically neutral. Therefore, the number of positively charged protons in an atom must be equal to the number of negatively charged electrons. The electrons are located outside the nucleus in regions known as **orbitals.** The space in which the electrons are found makes up most of the volume of the atom and is large compared to the size of the electron. Therefore, most of the volume of the atom is empty space.

The masses of atoms and subatomic particles are usually given in **atomic mass units** or **amu.** Using normal metric units, the mass of a single hydrogen atom is 1.67×10^{-24} g. Such a small number is difficult to use in common mathematical problems, so the atomic mass unit system was devised. On this relative scale, 1 amu is defined as $\frac{1}{12}$ the mass of a carbon atom with a nucleus that contains six protons and six neutrons, identified as carbon-12. Therefore, the atomic mass of carbon-12 is 12 amu. This number is exactly 1.66×10^{-24} g. The mass of a hydrogen atom with no neutrons in the nucleus is approximately 1 amu, due mostly to the one proton. The more exact value of the mass of the hydrogen atom with no neutrons is 1.00768 amu.

2.3   Atomic Number and Mass Number

GOALS: To know the significance of the atomic number and the mass number of an atom.
To define and give examples of isotopes.
To understand the meaning of average atomic mass.

Each atom has two properties given by its **atomic number, Z,** and a **mass number, A.** The atomic number of an atom is the number of protons in the nucleus. This property is the same for all atoms of a given element. For example, all atoms of the element carbon have six protons in the nucleus: for carbon, $Z = 6$. An atom with eight protons in its nucleus, $Z = 8$, is always an atom of the element oxygen. Since atoms are electrically neutral, the atomic number also gives the number of electrons. All carbon atoms have six electrons; oxygen atoms have eight electrons. Consult the listing of the elements facing the inside front cover of the text.

The mass number, A, for any element is the total (whole-number) units of mass present in the nuclei of its atoms. Protons and neutrons have essentially the same mass. The mass number is therefore the sum of the number of protons and neutrons in the nucleus.

Isotopes

The mass numbers of atoms of the same element can be different. All atoms of the same element must have the same number of protons, but the number of neutrons can differ. Atoms having the same atomic number but different mass numbers are called **isotopes.** The element carbon, for example, has isotopes with 6, 7, and 8 neutrons in the nuclei. All these isotopes have six protons, as illustrated in Table 2.3.

Table 2.3   The isotopes of carbon

Atomic Number (Z)	Mass Number (A)	Common Name	Composition
6	12	Carbon-12	6 protons 6 electrons 6 neutrons
6	13	Carbon-13	6 protons 6 electrons 7 neutrons
6	14	Carbon-14	6 protons 6 electrons 8 neutrons

Table 2.4 The isotopes of hydrogen

Hydrogen Isotope	Z	A	Number of Protons	Number of Neutrons	Number of Electrons	Common Name
^{1}H	1	1	1	0	1	Protium (ordinary hydrogen)
^{2}H	1	2	1	1	1	Deuterium (heavy hydrogen)
^{3}H	1	3	1	2	1	Tritium (radioactive hydrogen)

Isotopes are commonly represented by writing the atomic symbol for the element with a superscript for the mass number and a subscript indicating the atomic number:

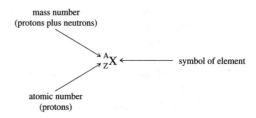

Since the atomic number is fixed for each element, the subscript may be omitted. The three isotopes of carbon can be symbolized as:

$$^{12}_{6}C \qquad ^{13}_{6}C \qquad ^{14}_{6}C \qquad or \qquad ^{12}C \qquad ^{13}C \qquad ^{14}C$$

These isotopes are referred to as carbon-12, carbon-13, and carbon-14.

When the atomic number is shown, it is easier to determine the number of neutrons present. The difference between the mass number and the atomic number, A minus Z, is equal to the number of neutrons.

number of neutrons = A − Z

Hydrogen with an atomic number of 1 also has three isotopes. The composition of these isotopes is shown in Table 2.4.

Average Atomic Mass: Atomic Weight

When you consult a table of the elements, you will find carbon listed with an **average atomic mass** of 12.011 rather than 12.000 amu. This seemingly small difference is actually quite significant.

6
C

12.01

The element carbon found in nature is a mixture of all three of the isotopes of carbon. The predominant form is ^{12}C, accounting for about 98.9% of all carbon; ^{13}C accounts for about 1.1% and ^{14}C is present in only trace amounts. However, the presence of ^{13}C (and the tiny amount of ^{14}C) causes the average atomic mass of carbon to be greater than 12.000 amu.

The term **atomic weight** has the same meaning as average atomic mass, although the latter is more precise and will be used in all later discussions. Atomic mass describes the

The low abundance of tritium is a result of its instability, which causes it to undergo radioactive decay. See Chapter 10 for details.

mass of specific isotopes. Average atomic mass reflects the natural abundance of each of its isotopes.

As another example, hydrogen has three isotopes. The most abundant is ^1H (99.9%); ^2H (0.01%) is known as **deuterium.** The third isotope, ^3H, known as **tritium,** is normally present only in trace amounts.

2.4 Electron Configuration

GOALS: To understand the difference between the Bohr atom and the wave mechanical model.
To draw models of Bohr atoms for simple elements.
To distinguish between Bohr orbits and orbitals.

The particles of the atomic nucleus are important, but the properties of any element are much more dependent on the arrangement of the electrons outside the nucleus. Electrons have energy that allows them to move continuously in the extranuclear region of an atom. The attractive pull of the nucleus helps to hold the electrons in the atom. Two theoretical models that have been used extensively to describe electrons are the Bohr atom and the wave mechanical atom.

The Bohr Atom

Figure 2.4 Niels Bohr proposed a model of the atom, known as the Bohr atom, in 1913. He is pictured here at age 37 when he was awarded the Nobel Prize in 1922 for his work.

The first detailed description of the arrangement of electrons in atoms was formulated in 1913 by the Danish physicist Niels Bohr (Figure 2.4). The **Bohr atom** is generally regarded as outdated and oversimplified, but this model is still useful for indicating the general arrangements of electrons in small atoms.

In the Bohr atom, electrons move about the nucleus in circular orbits at fixed distances directly related to the energy of the electrons. The greater the distance from the nucleus, the more energy the electrons contain. Furthermore, he proposed that electron energies could occur only as discrete values, called **energy levels.** The levels are like steps on a flight of stairs (Figure 2.5). Electrons may be found on any of the steps but not in between.

In other words, the energy of an electron is **quantized.** When an electron absorbs a specific amount of energy, called a **quantum,** it may move from a low energy level to a higher energy level. The electron may then return to the lower energy level. When this happens, energy is released, some as visible radiation.

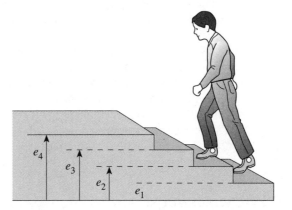

Figure 2.5 The flight of steps is analogous to quantized energy levels (e_1, e_2, etc.). The figure is able to stand on any of the steps but not in the space in between. Similarly, electrons may have only discrete energy values.

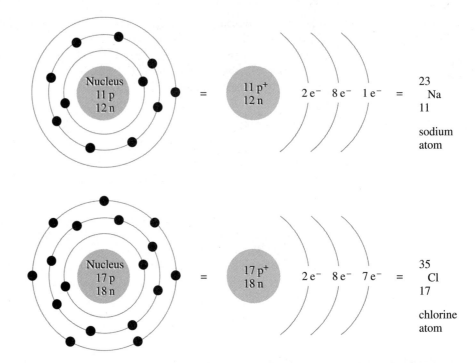

Figure 2.6   Bohr models of sodium and chlorine atoms. Each circular orbit represents an energy level in which electrons can be accommodated. Since it is now known that the electron energy levels are not really circular, the abbreviations in the center column are preferred.

Before an electron absorbs energy, it is said to be in its **ground state.** After absorption, the electron exists in an **excited state.** It can return to its ground state by emitting a quantum of energy. The effects of heating an element in a flame are explained by this model. The heat causes an electron to be excited. When the electron returns to its ground state, radiation energy is released. Thus a compound of sodium emits a bright yellow light. A violet flame is emitted by potassium salts. Lithium and strontium salts give brilliant red flames. Barium compounds give a green flame. A blue color is observed in the flame of copper salts. Fireworks displays are based on this type of excitation-emission process to give the wide variety of colors.

In the Bohr atom, each of the energy levels is pictured as a circular orbit about the nucleus (Figure 2.6). According to the model, each energy level can contain only a limited number of electrons. The higher the energy level (that is, the farther it is from the nucleus), the larger it is and, therefore, the more electrons it can hold. The capacity of Bohr energy levels, also called **principal energy levels,** to contain electrons is given in Table 2.5. The energy levels are designated by n, and the capacity of each level for electrons is given by $2n^2$. The actual number of electrons levels off at 32 and eventually decreases because no known elements have enough electrons to fill the outermost levels.

The Wave Mechanical Atom

In 1926 Erwin Schrödinger at the University of Zurich refined the picture of the Bohr atom. Bohr inaccurately described electrons as moving in circular orbits at discrete distances from the nucleus. Schrödinger proposed that electrons move with different energies

Table 2.5   The capacity of the Bohr energy levels

Energy Level (n)	Predicted Maximum Number of Electrons ($2n^2$)	Actual Number of Electrons
1	2	2
2	8	8
3	18	18
4	32	32
5	50	32
6	72	18
7	98	8

in three-dimensional regions in space. He called the regions **orbitals.** Such regions are electron clouds defined by the constant movement of the negatively charged electrons throughout the space available to them.

The possible energies of electrons are found by solving a complex mathematical expression called the Schrödinger wave equation, which relates the energy of an electron to the probability of finding that electron in a given position. The solutions to the Schrödinger equation define the regions in space that we will call orbitals.

The orbitals that will concern us most are designated as *s* and *p* **orbitals.** The shapes of these orbitals are pictured in Figure 2.7. The *s* orbital has a spherical shape oriented around the nucleus at the center. Recall that the shape is a mathematical solution to the Schrödinger equation, which predicts the probability of finding an electron at any particular point in space. The spherical *s* orbital represents the space where there is a 90% probability of finding a particular electron.

For the hydrogen atom, the single electron is located in a **1***s* orbital, where the number 1 signifies that the electron is in the first (and lowest) energy level. The energy levels are commonly represented by the symbol *n*, as in the Bohr model (see Table 2.5). Orbitals of the *s* type with higher energies are designated as **2***s* ($n = 2$), **3***s* ($n = 3$), and so forth.

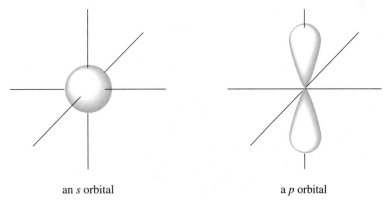

an *s* orbital   a *p* orbital

Figure 2.7   Orbitals. The nucleus is at the center of the three-dimensional axis system in each orbital. The *s* orbital has a spherical shape. The *p* orbital has two lobes, giving it a dumbbell shape. The volumes represented by these shapes are occupied by the electrons that are characterized as *s* and *p* electrons. Each orbital can hold two electrons.

Figure 2.8   The 1s orbital of hydrogen is defined by a sphere in which there is a 90% probability of finding its single electron. The intensity of the color gives a better indication of the relative probability of finding an electron at any particular point.

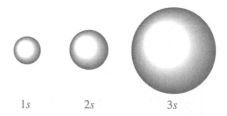

1s 2s 3s

Figure 2.9   The relative sizes of the 1s, 2s, and 3s orbitals of hydrogen are represented.

While it is convenient to draw a sphere to represent an s orbital, a probability map, such as that shown in Figure 2.8, is more accurate. The s orbitals at the higher energy levels have larger volumes (Figure 2.9) within which electrons may be found.

Orbitals of the p type have a dumbbell or hourglass shape to represent the region where there is a high probability of finding an electron. The energy of electrons in p orbitals is greater than the energy of electrons in the corresponding (same energy level) s orbital. As represented in Figure 2.7, the orbital is oriented symmetrically around the nucleus with two separate *lobes*. Notice that the two lobes do *not* represent two orbitals; they are two halves of a single p orbital.

2.5   The Filling of Orbitals: The First 18 Elements

GOALS: To explain the meaning of an electron configuration.
        To arrange the atomic sublevels in order of increasing energy.
        To write an electron configuration for elements 1 through 18.
        To distinguish between sublevels and orbitals.

To illustrate the relationship between orbitals and electrons, consider the location of the electrons in the first 18 elements, Z = 1 to Z = 18. A few rules must be followed for placing the electrons in their proper orbitals. The rules will be introduced as needed, and summarized in the margin notes.

Rule One: Fill the available orbital of lowest energy first. The value of n gives the number of types of orbitals present.

Hydrogen, Z = 1, has only a single electron. According to the first rule for filling orbitals, *the electron goes into the available orbital of lowest energy.* For the first energy level ($n = 1$), only the 1s orbital exists. The value of n also tells us how many different types of orbitals are available in that energy level. We can represent the location of the electron of hydrogen as:

$$\frac{\uparrow}{1s} \text{ equals } 1s^{1}$$

This representation is known as the **electron configuration.** In the second representation, to be read as ''one-*s*-one,'' the superscript indicates one electron is in the $1s$ orbital.

The next element, helium (He, Z = 2), has atoms with two electrons. These electrons also will exist in the orbital of lowest energy. But the second rule of orbital filling says *to exist in the same orbital, two electrons must have opposing spins.* In a representation of the electron configuration, these opposing spins are shown by arrows pointing in opposite directions, and the electrons are said to be *paired.*

Rule Two: When two electrons are present in the same orbital, they must have opposing spins.

$$\frac{\uparrow\downarrow}{1s} \text{ equals } 1s^2$$

With two electrons present, the first energy level is filled. This illustrates the third rule of orbital filling: *the capacity of an energy level is given by* $2n^2$. Since $n = 1$ for the first energy level, $2n^2 = 2$, which tells us that the $1s$ level has a capacity for only two electrons. For a lithium atom, Z = 3, the third electron must exist in the next energy level, for which $n = 2$, because the first energy level is filled.

Rule Three: The capacity of an energy level for electrons is given by $2n^2$, where n is the principal energy level.

Recall the additional statement with Rule One, which says that the value of n gives the number of available types of orbitals, such as s and p orbitals. In the second energy level ($n = 2$), two types of orbitals are available, so the third electron will exist in the orbital of lowest energy. Although it was not stated previously, the s orbital in any energy level has a lower energy than the p orbitals in the same energy level. Therefore, the third electron of lithium enters the $2s$ orbital (rather than the $2p$) and the electron configuration of a lithium atom is:

$$\frac{\uparrow\downarrow}{1s} \quad \frac{\uparrow}{2s} \text{ equals } 1s^2 2s^1$$

For beryllium (Be), Z = 4, the $2s$ orbital is filled and the two electrons are paired giving an electron configuration of $1s^2 2s^2$.

For boron (B), Z = 5, the first four electrons exist as they do for beryllium, but now a $2p$ orbital is used to accommodate the additional electron. This requires a fourth rule of orbital filling that describes the *numbers* of orbitals of each type that are available in any energy level. Rule Four says *that 1 orbital of the* s *type and 3 orbitals of the* p *type are available to accommodate electrons in any energy level.* In the following section, as we go beyond the first 18 elements, this rule will be extended to add 5 orbitals of the d type and 7 orbitals of the f type.

Rule Four: In any energy level, there are 1 orbital of the s type, 3 orbitals of the p type, 5 orbitals of the d type, and 7 orbitals of the f type available to accommodate electrons. The application of this rule is governed by Rule One.

Although three distinct $2p$ orbitals exist, each p orbital in a set has the same shape and the same energy. However, the three orbitals have a different orientation in space, as shown in Figure 2.10. Each orbital is aligned along an axis of a three-dimensional axis system; each is identified in relationship to its location, giving rise to the $2p_x$, the $2p_y$, and the $2p_z$ orbitals. Each p orbital may contain two electrons.

For boron, the configuration could be designated as $1s^2 2s^2 2p^1$, or as $1s^2 2s^2 2p_x^1$. In diagram form, this is represented by:

$$\frac{\uparrow\downarrow}{1s} \quad \frac{\uparrow\downarrow}{2s} \quad \frac{\uparrow}{\underset{2p}{\underline{}\,\underline{}}} \text{ equals } 1s^2 2s^2 2p^1$$

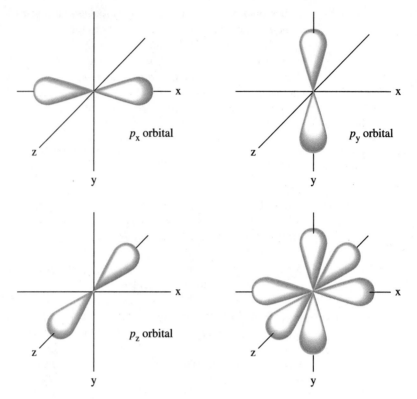

Figure 2.10 The *p* orbitals shown separately and in combination. Each orbital is identified by the axis that runs through its lobes.

Moving on to carbon, $Z = 6$, another rule for orbital filling is needed: as equivalent orbitals are filled, *electrons remain unpaired as long as orbitals of the same energy remain vacant*. As soon as it becomes necessary either to pair up electrons or place electrons in a higher energy level, pairing will occur.

By applying all of the rules, the following electron configurations are correct for carbon, nitrogen, and oxygen.

Carbon ($Z = 6$):

$$\underset{1s}{\underline{\uparrow\downarrow}} \quad \underset{2s}{\underline{\uparrow\downarrow}} \quad \underset{\underset{2p}{\underline{\quad}}}{\underset{\underline{\uparrow}}{\underline{\uparrow}}} \text{ equals } 1s^2 2s^2 2p^2$$

In more detail, this configuration can be specified as $1s^2 2s^2 2p_x^1 2p_y^1$. Both $2p$ levels and the electrons they contain have the same energy.

Nitrogen ($Z = 7$):

$$\underset{1s}{\underline{\uparrow\downarrow}} \quad \underset{2s}{\underline{\uparrow\downarrow}} \quad \underset{\underset{2p}{\underline{\uparrow}}}{\underset{\underline{\uparrow}}{\underline{\uparrow}}} \text{ equals } 1s^2 2s^2 2p^3$$

In more precise terms, this configuration can be specified as $1s^2 2s^2 2p_x^1 2p_y^1 2p_z^1$.

Oxygen (Z = 8):

$$\underset{1s}{\uparrow\downarrow}\ \underset{2s}{\uparrow\downarrow}\ \underset{2p}{\overset{\textstyle\uparrow\downarrow}{\underset{\textstyle\uparrow}{\uparrow}}}\ \text{equals}\ 1s^2\,2s^2\,2p^4$$

In more detail, this configuration can be specified as $1s^2 2s^2 2p_x^2 2p_y^1 2p_z^1$.

Still larger atoms have electrons in the third principal energy level, $n = 3$. Consider silicon (Si) and sulfur (S); their configurations also are readily predicted by Rules One through Five.

The two unpaired electrons of the oxygen atom account for its ability to bond to two hydrogen atoms to form H_2O. See Chapter 3 for details.

Silicon (Z = 14):

$$\underset{1s}{\uparrow\downarrow}\ \underset{2s}{\uparrow\downarrow}\ \underset{\underset{2p}{\uparrow\downarrow}}{\overset{\uparrow\downarrow}{\uparrow\downarrow}}\ \underset{3s}{\uparrow\downarrow}\ \underset{3p}{\uparrow}$$

This configuration can be described as $1s^2 2s^2 2p^6 3s^2 3p^2$, or in more detail as $1s^2 2s^2 2p^6 3s^2 3p_x^1 3p_y^1$. Note the similarity of this configuration to that of carbon, which sits directly above silicon in the periodic table. This type of relationship will be discussed later.

Sulfur (Z = 16):

$$\underset{1s}{\uparrow\downarrow}\ \underset{2s}{\uparrow\downarrow}\ \underset{\underset{2p}{\uparrow\downarrow}}{\overset{\uparrow\downarrow}{\uparrow\downarrow}}\ \underset{3s}{\uparrow\downarrow}\ \underset{3p}{\overset{\uparrow\downarrow}{\uparrow}}$$

This configuration can be described as $1s^2 2s^2 2p^6 3s^2 3p^4$, or in more precise terms as $1s^2 2s^2 2p^6 3s^2 3p_x^2 3p_y^1 3p_z^1$. Note the similarity of this configuration to that of oxygen, which sits directly above sulfur in the periodic table.

Sublevels versus Orbitals

Orbitals such as the 2*s* and the 3*p* represent not only a region in space, but also an energy sublevel of the principal energy level. In other words, we can speak of the 2*s* sublevel or the 3*p* sublevel. But the terms orbital and sublevel really do not mean the same thing. The distinction is that *orbitals are contained within sublevels.* Thus, we can refer to the 2*s* sublevel and the 2*s* orbital and mean the same thing, since there is only one orbital in the 2*s* sublevel. Any reference to the 2*p* sublevel, however, should be understood to describe a set of three *p* orbitals—$2p_x$, $2p_y$, and $2p_z$. Stated another way, a sublevel is a *type* of orbital, not an orbital.

2.6 The Filling of Orbitals: Elements 19 Through 109

GOALS: To write an electron configuration for elements 19 through 109.
       To use a block diagram for predicting energies of sublevels.

Element number 18 is argon, which has the electron configuration $1s^2 2s^2 2p^6 3s^2 3p^6$. After filling the 3*p* sublevel, the next (19th) electron must enter the next higher sublevel. Accord-

Figure 2.11
Diagrams of the *d* orbitals.

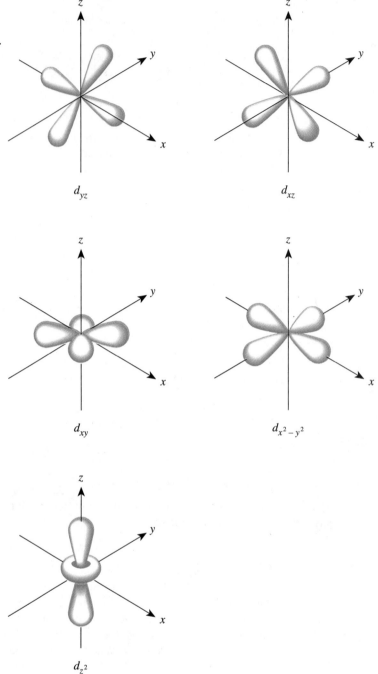

ing to Rule One, when $n = 3$, there is a third type of sublevel to accommodate electrons; it is the 3*d* (Figure 2.11), which consists of 5 orbitals (from Rule Four). Although we would expect the nineteenth electron to enter this 3*d* sublevel, it does not. Instead, it goes into the 4*s* sublevel. This unexpected behavior can be attributed to the energy of the 4*s* sublevel, which is lower than the energy of the 3*d* sublevel.

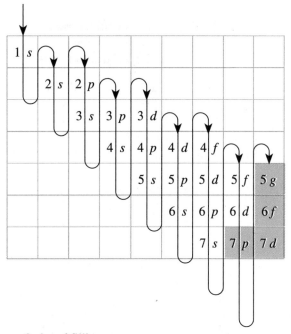

Order of filling:
1*s* 2*s* 2*p* 3*s* 3*p* 4*s* 3*d* 4*p* 5*s* 4*d* 5*p* 6*s* 4*f* 5*d* 6*p* 7*s* 6*d*

Figure 2.12 A block diagram for predicting the order in which orbitals are filled. Note that those orbitals shaded in color are theoretically possible, but they are not necessary for the 109 known elements.

To predict the relative energies of the sublevels, and therefore the order in which orbitals are filled, it is necessary to have some way to apply Rule One with accuracy. This can be accomplished by preparing and using a block diagram like that shown in Figure 2.12. By following the arrows from top to bottom of the diagram, you will obtain the correct order of increasing energy of the sublevels. It is not necessary to memorize this diagram if you take a few minutes to see how it is prepared. This will allow you to make such a diagram, or a portion of it, when needed. Notice that the block diagram correctly predicts that the 4*s* sublevel will be filled before the 3*d* sublevel.

For barium, Z = 56, the configuration is predicted by the block diagram to be:

$$1s^2 2s^2 2p^6 3s^2 3p^6 4s^2 3d^{10} 4p^6 5s^2 4d^{10} 5p^6 6s^2$$

Here again, we find that the 4*s* orbital precedes the 3*d*, but we also find that 5*s* precedes 4*d*. Note that the 6*s* orbital even precedes the 4*f*.

An Exception to the Rules

Table 2.6 gives the electron configurations of the first 40 elements. Careful inspection of the table will show two slight exceptions to the configurations predicted using the two methods described. Experimental evidence indicates that filled and half-filled sublevels add to the stability of an atom and, in several cases, the electrons are rearranged to achieve this extra stability.

Table 2.6 **Electron configurations of the first 40 elements. The external orbitals of elements that are exceptions to the normal rules for filling orbitals are boldfaced. See text for discussion.**

Atomic Number	Element	Electron Configuration
1	H	$1s^1$
2	He	$1s^2$
3	Li	$1s^22s^1$
4	Be	$1s^22s^2$
5	B	$1s^22s^22p^1$
6	C	$1s^22s^22p^2$
7	N	$1s^22s^22p^3$
8	O	$1s^22s^22p^4$
9	F	$1s^22s^22p^5$
10	Ne	$1s^22s^22p^6$
11	Na	$1s^22s^22p^63s^1$
12	Mg	$1s^22s^22p^63s^2$
13	Al	$1s^22s^22p^63s^23p^1$
14	Si	$1s^22s^22p^63s^23p^2$
15	P	$1s^22s^22p^63s^23p^3$
16	S	$1s^22s^22p^63s^23p^4$
17	Cl	$1s^22s^22p^63s^23p^5$
18	Ar	$1s^22s^22p^63s^23p^6$
19	K	$1s^22s^22p^63s^23p^64s^1$
20	Ca	$1s^22s^22p^63s^23p^64s^2$
21	Sc	$1s^22s^22p^63s^23p^64s^23d^1$
22	Ti	$1s^22s^22p^63s^23p^64s^23d^2$
23	V	$1s^22s^22p^63s^23p^64s^23d^3$
24	Cr	$1s^22s^22p^63s^23p^6\mathbf{4s^13d^5}$
25	Mn	$1s^22s^22p^63s^23p^64s^23d^5$
26	Fe	$1s^22s^22p^63s^23p^64s^23d^6$
27	Co	$1s^22s^22p^63s^23p^64s^23d^7$
28	Ni	$1s^22s^22p^63s^23p^64s^23d^8$
29	Cu	$1s^22s^22p^63s^23p^6\mathbf{4s^13d^{10}}$
30	Zn	$1s^22s^22p^63s^23p^64s^23d^{10}$
31	Ga	$1s^22s^22p^63s^23p^64s^23d^{10}4p^1$
32	Ge	$1s^22s^22p^63s^23p^64s^23d^{10}4p^2$
33	As	$1s^22s^22p^63s^23p^64s^23d^{10}4p^3$
34	Se	$1s^22s^22p^63s^23p^64s^23d^{10}4p^4$
35	Br	$1s^22s^22p^63s^23p^64s^23d^{10}4p^5$
36	Kr	$1s^22s^22p^63s^23p^64s^23d^{10}4p^6$
37	Rb	$1s^22s^22p^63s^23p^64s^23d^{10}4p^65s^1$
38	Sr	$1s^22s^22p^63s^23p^64s^23d^{10}4p^65s^2$
39	Y	$1s^22s^22p^63s^23p^64s^23d^{10}4p^65s^24d^1$
40	Zr	$1s^22s^22p^63s^23p^64s^23d^{10}4p^65s^24d^2$

2.7 Atomic Theory and the Periodic Table

GOAL: To see the relationship between electron configuration and the arrangement of elements in the periodic table.

As you proceed to develop greater understanding of elements and compounds, you will have frequent occasion to refer to the periodic table of the elements. You will find the table presented in a variety of ways depending on the information that is being communicated. The complete rationale for the organization of the periodic table will be given in Chapter 3. The table is not just a list of elements, symbols, and atomic mass information, but rather it is an organization of the elements based on regular (periodic) repetition of the properties of the elements. For example, we find that the elements in group VIIIA (Figure 2.13) in the far right column of the table are all gases, commonly called noble gases, that are chemically inert.

When the periodic table was first proposed, the properties of many elements were known and were used as the justification for such a tabular arrangement. But now it turns out that this arrangement of the elements also corresponds to the periodic repetition of the electron configurations of the elements. For example, the elements in group IA (the far left column) all have an electron configuration in which the outer sublevel contains a single s electron.

Lithium (Li)	$1s^2 2s^1$
Sodium (Na)	$1s^2 2s^2 2p^6 3s^1$
Potassium (K)	$1s^2 2s^2 2p^6 3s^2 3p^6 4s^1$
Rubidium (Rb)	$1s^2 2s^2 2p^6 3s^2 3p^6 4s23d^{10}4p^6 5s^1$
Cesium (Cs)	$1s^2 2s^2 2p^6 3s^2 3p^6 4s23d^{10}4p^6 5s^2 4d^{10}5p^6 6s^1$

Representative elements — Transition elements — Representative elements

1 IA ns^1																	18 VIIIA ns^2np^8
1 H $1s^3$	2 IIA ns^2											13 IIIA ns^2np^1	14 IVA ns^2np^2	15 VA ns^2np^3	16 VIA ns^2np^4	17 VIIA ns^2np^5	2 He $1s^2$
3 Li $2s^1$	4 Be $2s^2$											5 B $2s^22p^1$	6 C $2s^22p^2$	7 N $2s^22p^3$	8 O $2s^22p^4$	9 F $2s^22p^5$	10 Ne $2s^22p^6$
11 Na $3s^1$	12 Mg $3s^2$	3 IIIB	4 IVB	5 VB	6 VIB	7 VIIB	8	9 VIIIB	10	11 IB	12 IIB	13 Al $3s^23p^1$	14 Si $3s^23p^2$	15 P $3s^23p^3$	16 S $3s^23p^4$	17 Cl $3s^23p^5$	18 Ar $3s^23p^6$
19 K $4s^1$	20 Ca $4s^2$	21 Sc $4s^23d^1$	22 Ti $4s^23d^2$	23 V $4s^23d^3$	24 Cr $4s^13d^5$	25 Mn $4s^23d^5$	26 Fe $4s^23d^6$	27 Co $4s^23d^7$	28 Ni $4s^23d^8$	29 Cu $4s^13d^{10}$	30 Zn $4s^23d^{10}$	31 Ga $4s^24p^1$	32 Ge $4s^24p^2$	33 As $4s^24p^3$	34 Se $4s^24p^4$	35 Br $4s^24p^5$	36 Kr $4s^24p^6$
37 Rb $5s^1$	38 Sr $5s^2$	39 Y $5s^24d^1$	40 Zr $5s^24d^2$	41 Nb $5s^14d^4$	42 Mo $5s^14d^5$	43 Tc $5s^24d^6$	44 Ru $5s^14d^7$	45 Rh $5s^14d^8$	46 Pd $5s^04d^{10}$	47 Ag $5s^14d^{10}$	48 Cd $5s^24d^{10}$	49 In $5s^25p^1$	50 Sn $5s^25p^2$	51 Sb $5s^25p^3$	52 Te $5s^25p^4$	53 I $5s^25p^5$	54 Xe $5s^25p^6$
55 Cs $6s^1$	56 Ba $6s^2$	57 La $6s^25d^1$	72 Hf $6s^25d^2$	73 Ta $6s^25d^3$	74 W $6s^25d^4$	75 Re $6s^25d^5$	76 Os $6s^25d^6$	77 Ir $6s^25d^7$	78 Pt $6s^15d^9$	79 Au $6s^15d^{10}$	80 Hg $6s^25d^{10}$	81 Tl $6s^26p^1$	82 Pb $6s^26p^2$	83 Bi $6s^26p^3$	84 Po $6s^26p^4$	85 At $6s^26p^5$	86 Rn $6s^26p^6$
87 Fr $7s^1$	88 Ra $7s^2$	89 Ac $7s^26d^1$	104 Unq $7s^26d^2$	105 Unp $7s^26d^3$	106 Unh $7s^26d^4$	107 Uns $7s^26d^5$	108 Uno $7s^26d^6$	109 Une $7s^26d^7$									

Figure 2.13 The periodic table, including partial electron configuration for the elements shown.

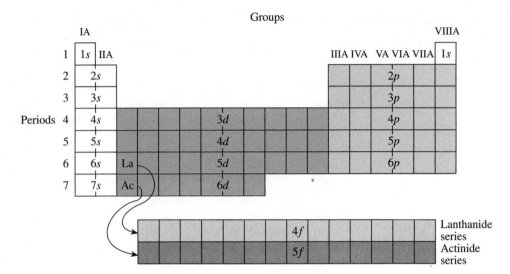

Figure 2.14   The periodic table showing the sublevels to which electrons are being added in each segment. This diagram emphasizes that each *s* sublevel can hold 2 electrons, each *p* sublevel can hold 6 electrons, each *d* sublevel can contain 10 electrons, and each *f* sublevel can contain 14 electrons.

The properties of the IA elements are all quite similar. The similarity is directly related to the common feature of their electron configurations. For all elements, the properties are related to the electron configuration in the outer energy level, known as the **valence level.** Once the levels below the valence level are filled, that part of the atomic structure is so stable that it remains undisturbed and unreactive. But the valence level is where the action is. It is this part of the electron configuration that causes us to study the whole matter in the first place. Figure 2.13 is a version of the periodic table showing the configuration of the outer electrons of many of the elements. A summary of the configuration of the outer electrons is also given in the period table shown in Figure 2.14, which indicates the sublevel that is being filled in each section of the periodic table.

The significance of electrons in the valence levels, called **valence electrons,** will be discussed in detail in Chapter 3.

SUMMARY

Atoms are composed of protons and neutrons, which are located in the nucleus, and electrons, which travel outside the nucleus with different amounts of energy. Atoms are identified by an atomic number, Z, and a mass number, A. The Z indicates the number of protons in the nucleus, and A gives the total units of mass in the nucleus. Since atoms are electrically neutral, the number of protons present in the nucleus must be identical to the number of electrons traveling about that nucleus.

Although atoms of the same element all have the same number of protons in the nucleus, most elements exist in several forms, called isotopes, that differ in the number of neutrons. The *average* atomic mass (also known as the atomic weight) is calculated considering the natural abundance of each isotope.

The Bohr model of the atom includes energy levels, which are retained in the modern theory. The Bohr model also pictures the energy levels as circular orbits, which is not correct. Modern quantum theory has replaced the orbits with orbitals of various shapes and locations.

The arrangement of the electrons around the nucleus of an atom is called the electron configuration. The elec-

trons are present in different energy levels and different types of orbitals (sublevels) depending on the energy they possess. When the atom is most stable, each electron occupies the position that allows it to have the least energy.

PROBLEMS

1. Define each of the following terms and explain how they are related.

   a. Atom
   b. Nucleus
   c. Atomic number
   d. Mass number
   e. Sublevel
   f. Orbital

2. Give a definition or example for each of the following.

   a. Isotope
   b. Deuterium
   c. Tritium
   d. Average atomic mass
   e. Atomic weight
   f. Oxides of nitrogen
   g. Ground state
   h. Excited state
   i. Valence electrons

3. Describe protons, neutrons, and electrons as to location in an atom, electrical charge, and mass.

4. Consider the components of Dalton's atomic theory and tell how each of the following relates to part(s) of the theory.

   a. Law of conservation of mass from Lavoisier
   b. Law of definite proportions (constant composition) from Proust
   c. Law of multiple proportions from Dalton

5. Regarding atomic mass units (amu):

   a. Give a definition.
   b. Explain why this system of units is useful.

6. Using the periodic table on the inside front cover of the book, list and identify each number in the block for each of the following elements.

   a. C
   b. Al
   c. Ba
   d. U

7. What characteristics do all atoms of the same element have in common? How can atoms of the same element differ?

8. How are average atomic masses influenced by isotopes?

9. State the number of protons, neutrons, and electrons in an atom of the most common isotope of each of the following elements. Assume the average atomic mass in the periodic table is closest to the mass number of the most abundant isotope. (Consult the table of the elements on the inside back cover for any symbols that you do not know.)

   a. Oxygen
   b. Copper
   c. Iron
   d. Thallium
   e. Tin
   f. Potassium

10. Compare and contrast the Bohr atom and the wave mechanical atomic model accepted today.

11. Explain the statement: Most of the volume of an atom can be pictured as empty space.

12. a. What is meant by the term orbital?
    b. What types of orbitals occur in atoms known at the present time?
    c. When do the terms orbital and sublevel refer to the same thing?
    d. When do the terms orbital and sublevel *not* refer to the same thing?

13. Write symbols, including both atomic numbers and mass numbers, for each of the following.

    a. Phosphorus-31
    b. Potassium-40
    c. Magnesium-25
    d. Manganese-55
    e. Iodine-131
    f. Lead-207
    g. Uranium-235
    h. Neon-20
    i. Iron-56
    j. Technetium-99

14. How many neutrons are present in each of the following isotopes? What is the name of each isotope? (Consult a table of the elements as necessary.)

    a. $^{14}_{7}N$
    b. $^{23}_{11}Na$
    c. ^{19}F

d. ^{39}K

e. ^{35}Cl

15. How many neutrons are present in each of the following isotopes? (Consult a table of the elements as necessary.)

a. Zinc-66

b. Copper-64

c. Gold-197

d. Silver-108

e. Radon-222

f. Radium-226

16. Using Table 2.5 as a reference, sketch a Bohr diagram for an atom of each of the following elements.

a. C

b. F

c. Na

d. Al

e. Ca

17. Which of the following represent correct electron configurations? How would you correct any incorrect configuration(s)?

a. ↑↓ ↑↓ ↑↓
    1s   2s   ↑

        ___
         2p

b. ↑↓ ↑↓ ↑↓
    1s   2s   ↑↓
              ↑
             2p

c. ↑↓ ↑↓ ↑↓ ↑
    1s   2s   ↑↓ 3s
              ↑↓
              2p

d. ↑↑ ↑↑ ↑
    1s   2s   ↑
              ↑
             2p

18. Consult Figure 2.14 to determine the electron configuration (for example, $5p^4$) of the last electron to enter each of the following elements.

a. As

b. Ca

c. Mn

d. I

e. U

f. Kr

19. Identify the elements whose neutral atoms have the following electron configurations.

a. $1s^2 2s^2 2p^6$

b. $1s^2 2s^2 2p^1$

c. $1s^2 2s^2 2p^6 3s^2 3p^6 4s^2 3d^5$

d. $1s^2 2s^2 2p^6 3s^2 3p_x^1 3p_y^1$

20. Write electron configurations for each of the following elements.

a. Beryllium

b. Bromine

c. Magnesium

d. Manganese

e. Mercury

f. Phosphorus

g. Potassium

h. Lead

Chemical Bonding: The Periodic Table

3

Just as atoms are the basic units of elements, molecules are the basic units of chemical compounds. In this chapter we will consider how atoms bond together to form chemical compounds such as salts. We will find that the periodic table can be used to predict bonding. Also, we will consider how compounds are named.

3.1 Salts

GOAL: To write the formula for a salt, given the charges on the cation and the anion.

In a neutral atom, the number of protons equals the number of electrons. If the numbers of protons and electrons are unequal, the atom exists as a charged particle, called an **ion.** If the protons outnumber the electrons, a positively charged ion, or **cation,** results. A negatively charged particle, called an **anion,** exists when the electrons outnumber the protons.

Ions of opposite charge may combine to form neutral compounds called salts. For example, common table salt is sodium chloride, NaCl, which consists of the ions Na^+ and Cl^-. As you progress through later sections of this chapter you will come to appreciate why these ions exist. Since sodium ion exists as Na^+, rather than Na^{2+} or Na^{3+}, the $+1$ charge on the ion reveals how sodium ion can combine with other ions. For example, if sodium ion and chloride ion, Cl^-, are to combine to form a salt with no net charge, there must be one sodium ion and one chloride ion. Similarly, aluminum chloride would form by the combination of Al^{3+} cation and Cl^- anions to give $AlCl_3$, which also has a net charge of zero. Subscripts signify the presence of two or more atoms of an element in a compound.

In the examples below, the anions and their names appear in boldface type. Some of the examples are salts of *polyatomic ions;* these are ions in which two or more elements are combined. Some polyatomic anions are hydroxide, nitrate, sulfate, carbonate, and bicarbonate ions. The ammonium ion is an example of a polyatomic cation. If such an ion occurs more than once in a compound, as in aluminum hydroxide, the polyatomic unit is enclosed in parentheses and followed by the appropriate subscript.

$Na^+ + Cl^- = NaCl$	Sodium **chloride**
$2\,K^+ + S^{2-} = K_2S$	Potassium **sulfide**
$Mg^{2+} + 2\,Br^- = MgBr_2$	Magnesium **bromide**
$Ca^{2+} + O^{2-} = CaO$	Calcium **oxide**
$Al^{3+} + 3\,OH^- = Al(OH)_3$	Aluminum **hydroxide**
$2\,Na^+ + CO_3^{2-} = Na_2CO_3$	Sodium **carbonate**
$Ba^{2+} + SO_4^{2-} = BaSO_4$	Barium **sulfate**
$Zn^{2+} + HPO_4^{2-} = ZnHPO_4$	Zinc **monohydrogen phosphate**
$Ca^{2+} + 2\,NO_3^- = Ca(NO_3)_2$	Calcium **nitrate**
$Cd^{2+} + 2\,HCO_3^- = Cd(HCO_3)_2$	Cadmium **bicarbonate**
$2\,Al^{3+} + 3\,SO_4^{2-} = Al_2(SO_4)_3$	Aluminum **sulfate**
$NH_4^+ + I^- = NH_4I$	Ammonium **iodide**
$3\,NH_4^+ + PO_4^{3-} = (NH_4)_3PO_4$	Ammonium **phosphate**
$2\,NH_4^+ + HPO_4^{2-} = (NH_4)_2HPO_4$	Ammonium **monohydrogen phosphate**
$NH_4^+ + H_2PO_4^- = NH_4H_2PO_4$	Ammonium **dihydrogen phosphate**
$Sn^{4+} + 4\,NO_3^- = Sn(NO_3)_4$	Tin(IV) **nitrate** or stannic **nitrate**

Practice writing formulas for compounds by combining the cations and anions that are listed. As you do so, give some thought to the formulas and names. You will then be better able to appreciate some important information discussed in later sections, and you will learn these ions without conscious memorization. After repeated exposure to many common ions, one automatically commits some facts to memory, and this is very helpful.

The periodic table is a source of information for explaining or predicting the structures of many ions and compounds. We will examine it with one major goal in mind: to understand how elements combine to form chemical compounds.

3.2 **The Periodic Table**

GOALS: To describe the organization of the periodic table.
To appreciate the reasons for placing the elements within each group in the periodic table.
To understand the basis of the octet rule and use it to predict the valence of many ions or atoms that participate in covalent bonding.

The periodic table dates back to 1869, when the Russian chemist Dmitri Mendeleev and the German chemist Lothar Meyer recognized that certain elements showed similar physical and chemical properties. When the elements were arranged in order of increasing atomic weight, those with very similar properties showed up periodically at regular intervals. This regular behavior, or *periodicity,* is the basis for the periodic table of the elements (Figure 3.1).

Periodicity seems obvious to chemists now, but in 1869, only about 60 elements were known and the concept of atomic number was not yet appreciated. Mendeleev boldly predicted the existence of elements not yet discovered. Later experiments verified his hypotheses.

Representative elements		Transition elements										Representative elements					

																	18 VIIIA
1 1 IA	2 IIA	atomic number — 1	symbol of the element — H	average atomic mass — 1.008								13 IIIA	14 IVA	15 VA	16 VIA	17 VIIA	2 **He** 4.003
2 3 **Li** 6.941	4 **Be** 9.012											5 **B** 10.81	6 **C** 12.01	7 **N** 14.01	8 **O** 16.00	9 **F** 19.00	10 **Ne** 20.18
3 11 **Na** 22.99	12 **Mg** 24.31	3 IIIB	4 IVB	5 VB	6 VIB	7 VIIB	8	9 VIIIB	10	11 IB	12 IIB	13 **Al** 26.98	14 **Si** 28.09	15 **P** 30.97	16 **S** 32.07	17 **Cl** 35.45	18 **Ar** 39.95
4 19 **K** 39.10	20 **Ca** 40.08	21 **Sc** 44.96	22 **Ti** 47.88	23 **V** 50.94	24 **Cr** 52.00	25 **Mn** 54.94	26 **Fe** 55.85	27 **Co** 58.93	28 **Ni** 58.69	29 **Cu** 63.55	30 **Zn** 65.38	31 **Ga** 69.72	32 **Ge** 72.59	33 **As** 74.92	34 **Se** 78.96	35 **Br** 79.90	36 **Kr** 83.80
5 37 **Rb** 85.47	38 **Sr** 87.62	39 **Y** 88.91	40 **Zr** 91.22	41 **Nb** 92.91	42 **Mo** 95.94	43 **Tc*** (98)	44 **Ru** 101.1	45 **Rh** 102.9	46 **Pd** 106.4	47 **Ag** 107.9	48 **Cd** 112.4	49 **In** 114.8	50 **Sn** 118.7	51 **Sb** 121.8	52 **Te** 127.6	53 **I** 126.9	54 **Xe** 131.3
6 55 **Cs** 132.9	56 **Ba** 137.3	57 **La** 138.9	72 **Hf** 178.5	73 **Ta** 180.9	74 **W** 183.9	75 **Re** 186.2	76 **Os** 190.2	77 **Ir** 192.2	78 **Pt** 195.1	79 **Au** 197.0	80 **Hg** 200.6	81 **Tl** 204.4	82 **Pb** 207.2	83 **Bi** 209.0	84 **Po*** (209)	85 **At*** (210)	86 **Rn*** (222)
7 87 **Fr*** [223]	88 **Ra*** 226.0	89 **Ac*** 227.0	104 **Unq*** [261]	105 **Unp*** [262]	106 **Unh*** [263]	107 **Uns*** [262]	108 **Uno*** 265	109 **Une*** [266]									

Periods

Inner transition elements

Lanthanides

58 **Ce** 140.1	59 **Pr** 140.9	60 **Nd** 144.2	61 **Pm*** (145)	62 **Sm** (145)	63 **Eu** 152.0	64 **Gd** 157.3	65 **Tb** 158.9	66 **Dy** 162.5	67 **Ho** 164.9	68 **Er** 167.3	69 **Tm** 168.9	70 **Yb** 173	71 **Lu** 175.0

Actinides

90 **Th*** 232.0	91 **Pa*** [231]	92 **U*** 238.0	93 **Np*** [237]	94 **Pu*** [244]	95 **Am*** [243]	96 **Cm*** [247]	97 **Bk*** [247]	98 **Cf*** [251]	99 **Es*** [252]	100 **Fm*** [257]	101 **Md*** [258]	102 **No*** [259]	103 **Lr*** [260]

*All isotopes are radioactive.

[ ] Indicates mass number of the isotope of longest known half-life.

Group numbers 1–18 (in blue) represent the system recommended by the International Union of Pure and Applied Chemistry.

☐ Metals
▨ Metalloids
▨ Nonmetals

Figure 3.1 The periodic table.

Within the periodic table, the vertical columns are called **groups** and the horizontal rows are known as **periods.** The groups can be numbered in two different ways. In the United States, the traditional method uses IA, IIA, IIIB, and so on. A newer approach numbers the groups consecutively from 1 through 18, but the more traditional system is used throughout the following discussion.

The elements within each group exhibit similar characteristics. Consider two examples of periodicity. The elements in group IA on the far left are known as the *alkali metals.* They are all very reactive, metallic substances whose ions have a plus one charge (Na^+, K^+). At the extreme right of the periodic table, the group VIIIA elements, known as the *noble gases,* are normally inert substances that are rarely combined with other elements.

The periods are numbered from 1 through 7; they differ in length as required by the periodicity of properties (Table 3.1). Beginning with period 4, the length increases dramatically to accommodate the *transition elements.* Elements 57 through 71 are regarded as part of the sixth period and elements 89 through 103 as part of the seventh period. These two sets of elements are tabulated separately for convenience and because of structural similarities within each series. In our consideration of the elements, we will concen-

Table 3.1 Periodicity of elements in the periodic table

Period	Number of Elements
1	2
2	8
3	8
4	18
5	18
6	32
7	32

trate on those in groups IA through VIIIA. The transition elements, the lanthanide series (atomic numbers 57–71), and the actinide series (atomic numbers 89–103) show more complex behavior and will not be considered here.

Hydrogen has certain features of both group IA and group VIIA elements, yet its properties are quite different from these elements. In many textbooks, hydrogen is shown in either group IA or VIIA, but it is shown separately here.

Valence Electrons and the Periodic Table

As described in Section 2.7, Figure 2.13, the electron configurations of the noble gases are all np^6. For example, neon is $2s^22p^6$.

To appreciate the periodic behavior of the elements, we must now consider the role of electrons in an atom. It is primarily the arrangement of electrons that accounts for the properties of each element.

The properties of the elements in group VIIIA help us to understand the behavior of the other elements. The group VIIIA elements have little tendency to combine with any other elements. For this reason, they are known as the *noble gases,* or *inert gases.* This behavior of the noble gases is the result of an unusually stable arrangement of electrons. Therefore, they have little tendency to give up electrons (to form cations) or accept electrons (to form anions).

The Bohr models of two of the noble gases, neon (atomic number 10) and argon (18), are compared with IA elements sodium (11) and potassium (19) in Figure 3.2. Each partial circle signifies an energy level where electrons are located. These energy levels are arranged in order of increasing energy. The first two electrons are found in the lowest energy level. The remaining electrons are located in outer (higher) energy levels. The outermost level is called the **valence level,** and the electrons in this level are called **valence electrons.**

Each energy level has a certain capacity for electrons. Each of the elements in groups IA through VIIIA has a simple electron configuration in which all of the lower levels are filled and the *outer shell contains the number of electrons given by the group number.* Sodium appears in group IA and has one electron in its outer level. Chlorine is in group VIIA and has seven electrons in its outer level. The atoms of the noble gases (group VIIIA) all have eight valence electrons, except for helium, which has only two.

In view of the inert (very stable) nature of the noble gases, it is clear that the optimum number of valence electrons is eight—an *octet* of electrons. This fact is the basis of the **octet rule,** which states that any element is most stable when it has a set of eight valence electrons. (Recall again that we are restricting our discussion to the elements in groups IA–VIIIA.)

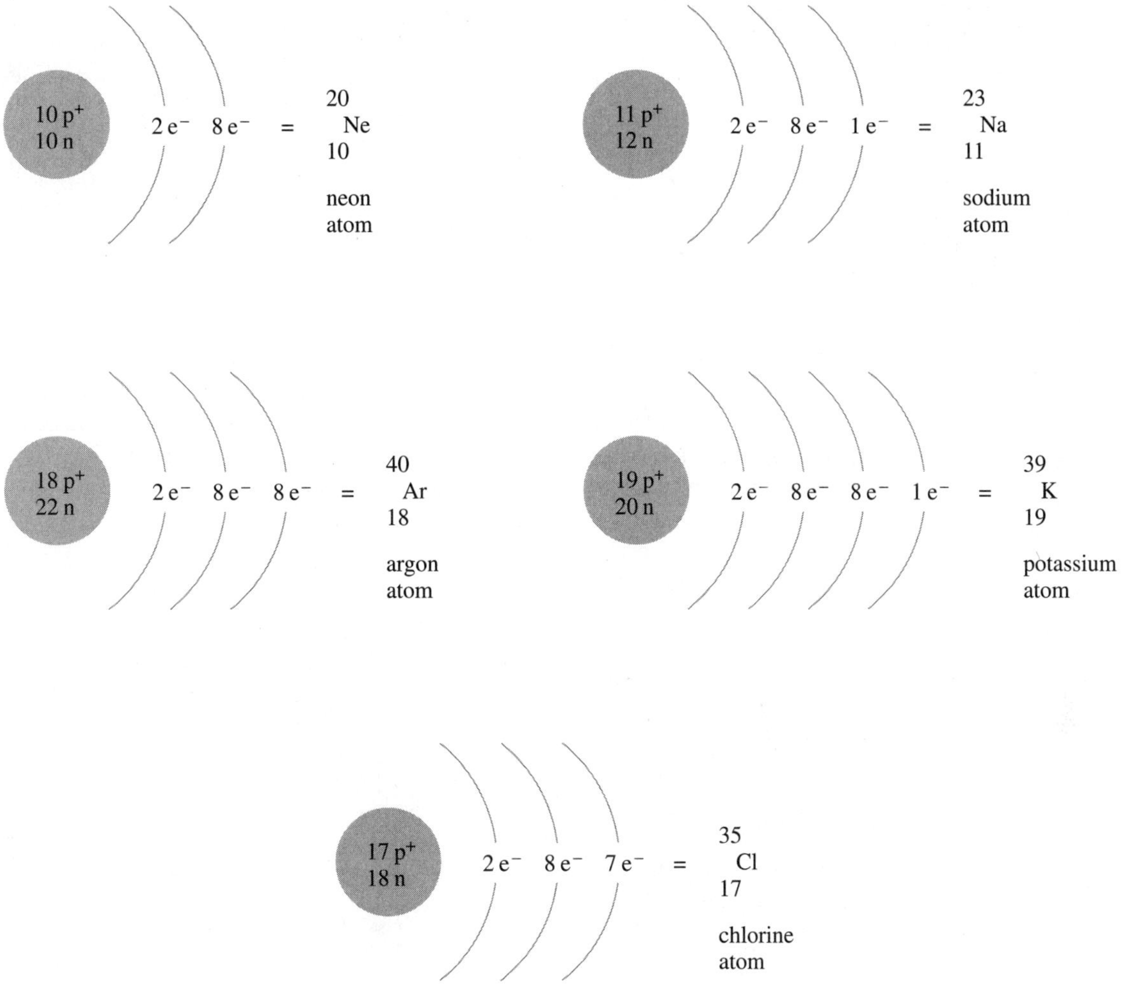

Figure 3.2   Bohr models of several common atoms illustrate the capacity of the various energy levels. For the noble gas neon, the valence level is filled with an octet of electrons. When an additional electron is present, as in sodium, the electron is found in a higher energy level, which then becomes the valence level. The same difference is seen between the noble gas argon and the next higher element, potassium. Both sodium and potassium are group IA elements. The element chlorine is in group VIIA, which is consistent with having seven valence electrons.

Ionization and Ionic Compound Formation

Let us take a closer look at the sodium and chlorine atoms as they combine to form sodium chloride, which consists of Na^+ and Cl^-. As sodium (group IA) becomes a cation, it loses its one valence electron and attains an electron arrangement equivalent to that in the noble gas neon.

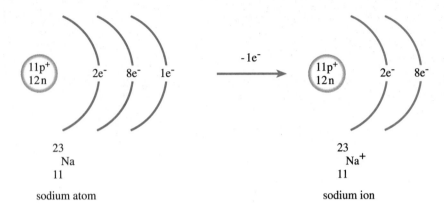

sodium atom sodium ion

Similarly, the chlorine (group VIIA) atom achieves the electron structure of the noble gas argon by gaining one electron and forming an anion—the chloride ion.

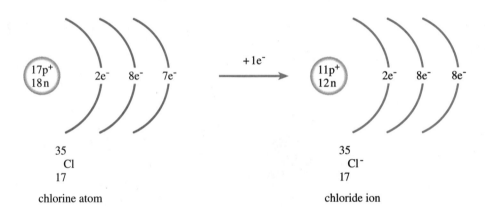

chlorine atom chloride ion

The formation of each of these ions involves a chemical change in the elements, showing how important the valence electrons are in controlling the chemical properties, or chemical reactions, of the various elements. The noble gases are also said to have zero valence electrons since they neither take on nor give off valence electrons. Both expressions, zero valence electrons and eight valence electrons, represent the same stable structure of the atom.

The noble gas helium, He, is an exception to the octet rule, since the first shell of electrons has a maximum capacity of only two electrons. The helium structure is very stable, as indicated by the fact that Li^+, the stable ion of lithium, has the same electron structure as He. In other words, the Li (atomic number 3, group IA) *atom* has a total of three electrons. When the valence electron is lost from the outer shell, the arrangement of electrons becomes equivalent to He. Whereas a lithium *ion* (cation) is often found combined with an anion, the cation is resistant to any chemical change, just like the helium atom.

Ions such as Li^+, Na^+, and K^+ are called **monovalent cations,** signifying the loss of one valence electron. The elements in group IA are known as the **alkali metals;** all readily form monovalent cations. Ions such as Li^{2+}, Na^{2+}, and K^{2+} do not exist, since none conforms to a noble gas structure.

The elements in group IIA are called the **alkaline earth metals.** When ionized, they exist as divalent cations due to the loss of their two valence electrons. Examples are Mg^{2+}, Ca^{2+}, and Ba^{2+}.

Group VIIA elements are called **halogens.** Since each halogen atom has seven valence electrons, they tend to gain one additional valence electron and are commonly found as **monovalent anions** when combined with other elements. Examples are sodium iodide, NaI, calcium bromide, $CaBr_2$, and aluminum chloride, $AlCl_3$. The rules for naming compounds are discussed in Sections 3.6 through 3.8.

3.3   Electron Dot Symbolism

GOAL: To use the electron dot formula to represent the number of valence electrons.

Another way of predicting what sort of ion is most likely to form in a reaction is to draw an *electron dot formula,* illustrated here for each element in the second period:

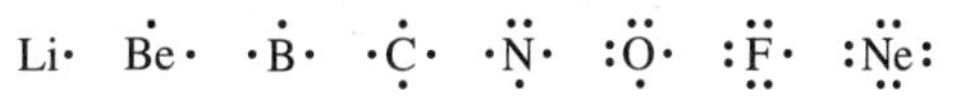

$$\text{Li}\cdot \quad \dot{\text{Be}}\cdot \quad \cdot\dot{\text{B}}\cdot \quad \cdot\dot{\overset{\cdot}{\text{C}}}\cdot \quad \cdot\dot{\overset{\cdot\cdot}{\text{N}}}\cdot \quad :\dot{\overset{\cdot\cdot}{\text{O}}}\cdot \quad :\dot{\overset{\cdot\cdot}{\text{F}}}\cdot \quad :\overset{\cdot\cdot}{\underset{\cdot\cdot}{\text{Ne}}}:$$

In this representation, only the valence electrons are included. We see once again that *the number of valence electrons corresponds to the group number* of each element. Within the valence level, four distinct sublevels are available to hold the valence electrons. The electrons occupy separate sublevels when possible. Thus carbon (group IVA) is shown with its four valence electrons unpaired. However, some of the valence electrons for the elements in groups VA through VIIIA must be paired, until finally all of the electrons are paired in noble gases, such as neon.

Knowing the characteristic electron arrangements, we can readily predict how many electrons must be lost or gained by each element to form a structure like that of a noble gas.

3.4   Covalent Bonding

GOALS: To appreciate the nature of covalent bonds.
           To consider examples of combined ionic-covalent compounds.

Thus far, we have considered only those elements near the left and right sides of the periodic table. As we continue to discuss groups IA through VIIIA, we find that certain elements do not form simple ions. Carbon and nitrogen illustrate another type of behavior when they combine with other atoms. The electron dot structures for these two elements are as follows:

$$\cdot\dot{\underset{\cdot}{\text{C}}}\cdot \qquad \cdot\dot{\overset{\cdot\cdot}{\text{N}}}\cdot$$

group IVA   group VA

From the dot structure for the carbon atom, you would be likely to predict that it must either gain or lose four electrons to achieve a noble gas structure. Both possibilities are very unlikely. As an atom loses electrons to form a cation, removing each successive electron becomes more difficult because the increasing positive charge causes the atom to have a steadily greater attraction for its electrons. The same sort of problem arises as an atom takes on electrons to become an anion; the increasing negative charge tends to repel additional electrons. The gain or loss of one or two electrons occurs readily to form monovalent and divalent ions, but ions with a higher valence are less common.

Some atoms, instead of forming ions, achieve a structure like that of a noble gas in an alternative way—by forming **covalent bonds,** as illustrated for methane, CH_4.

$$\cdot \overset{\cdot}{\underset{\cdot}{C}} \cdot \; + \; 4\,H\cdot \; \longrightarrow \; H\!:\!\overset{..}{\underset{..}{C}}\!:\!H \; = \; H-\overset{\overset{\textstyle H}{|}}{\underset{\underset{\textstyle H}{|}}{C}}-H$$

methane

As the methane molecule forms, the carbon atom forms an electron structure like that of neon because it effectively gains four electrons. Each of the lines between the atoms in the second drawing of the methane molecule represents two electrons being shared by the carbon atom and a hydrogen atom. This *sharing of electrons* constitutes a bond, known as a covalent bond, between the two atoms.

The carbon atom in methane is *tetravalent* (having a valence of four) due to the formation of four bonds. The hydrogen atoms in methane are *monovalent,* since each acquires one electron by covalent bond formation and forms an electron structure like that of the noble gas helium.

Three additional electrons are required for a nitrogen atom to achieve a noble gas structure. Consequently, three hydrogen atoms can join nitrogen to form the ammonia molecule; the nitrogen atom in this arrangement is trivalent. Here again, nitrogen and hydrogen atoms *share* valence electrons so that each atom forms a noble gas structure.

$$\cdot \overset{..}{\underset{\cdot}{N}} \cdot \; + \; 3\,H\cdot \; \longrightarrow \; H\!:\!\overset{..}{\underset{..}{N}}\!:\!H \; = \; H-\overset{\overset{\textstyle ..}{N}}{\underset{\underset{\textstyle H}{|}}{}}-H$$

ammonia

The same kind of covalent bonding is present in many other compounds, such as water, H_2O, and hydrogen fluoride, HF. These are not ionic compounds. They contain covalent O−H and F−H bonds as shown:

$$:\!\overset{..}{\underset{|}{O}}\!-\!H \qquad\qquad :\!\overset{..}{\underset{..}{F}}\!-\!H$$
$$H$$

water   hydrogen fluoride

Many ionic compounds also contain covalent bonds. Sodium hydroxide is a simple example; it consists of the sodium ion and the hydroxide ion, but within the hydroxide ion, there is a covalent oxygen-hydrogen bond:

$$Na^+ \quad {}^-\!:\!\overset{..}{\underset{..}{O}}\!-\!H$$

sodium hydroxide

Another example of an ion containing covalent bonds is the ammonium ion:

$$H-\overset{\overset{\textstyle H}{|}}{\underset{\underset{\textstyle H}{|}}{N^+}}-H$$

ammonium ion

3.5 Valence

GOAL: To understand the different meanings of valence.

The term **valence** has been used several times and explained by example. There are two ways to look at valence. For compounds that form ions, valence describes the charge on the cation or anion resulting from the loss or gain of electrons to form a filled valence level. Thus, for calcium chloride, $CaCl_2$, a divalent cation (Ca^{2+}) combines with a monovalent anion (Cl^-) in a 1:2 ratio. When considering covalent compounds, valence refers to the number of atoms or groups of atoms attached to a particular atom. The oxygen atom of water, with two hydrogen atoms attached, is an example. In both instances, *valence is the combining capacity of an atom or ion;* this more general definition applies to both ionic and covalent compounds.

3.6 Nomenclature of Ionic Compounds

GOALS: To be aware of the distinction between metals and nonmetals.
To know the location of metalloids in the periodic table.
To consider many simple and polyatomic ions and learn how their compounds are named.

We have seen several examples of ionic compounds, including some with covalent bonds. Let us examine the rules that are followed in naming these substances. The periodic table in Figure 3.1 classifies each of the elements as a metal, metalloid (also called a semimetal), or nonmetal. **Metals** have the following properties: luster, a high ability to conduct heat and electricity, and a tendency to lose electrons to form cations. The **nonmetals** have the opposite properties. Except for the noble gases, nonmetals tend to gain electrons to form anions. The **metalloids** appear in the periodic table in a kind of "staircase arrangement" between the metals and nonmetals, and they exhibit properties intermediate between metals and nonmetals.

The alkali metals (group IA) and the alkaline earth metals (group IIA) appear only as cations in ionic compounds. These cations are named as if the element itself is present. Examples follow.

NaI	sodium iodide
$BaSO_4$	barium sulfate
$MgCO_3$	magnesium carbonate
$KHCO_3$	potassium bicarbonate
$Ca_3(PO_4)_2$	calcium phosphate

For the simple anions consisting of only one element, the names are formulated in a different way. As we have already seen, the element chlorine often exists as the ion known as chloride, Cl^-. The ending *-ine* is replaced by *-ide,* as in the following:

NaF	sodium fluoride
$CaCl_2$	calcium chloride
$AlBr_3$	aluminum bromide

Oxygen forms the divalent anion called oxide, as in sodium oxide (Na_2O). Sulfur forms sulfide as in potassium sulfide (K_2S). For each of these anions, the ending on the name of the element is replaced by *-ide.* Table 3.2 lists many of the common ions.

Table 3.2 Some common simple cations and anions

Group	Element	Name of Ion	Symbol of Ion
	Hydrogen	Hydrogen	H^+
	Hydrogen	Hydride	H^-
IA	Lithium	Lithium	Li^+
IA	Potassium	Potassium	K^+
IIA	Magnesium	Magnesium	Mg^{2+}
IIA	Calcium	Calcium	Ca^{2+}
IIA	Barium	Barium	Ba^{2+}
IIIA	Aluminum	Aluminum	Al^{3+}
IVA	Carbon	Carbide	C^{4-}
VA	Nitrogen	Nitride	N^{3-}
VIA	Oxygen	Oxide	O^{2-}
VIA	Sulfur	Sulfide	S^{2-}
VIIA	Fluorine	Fluoride	F^-
VIIA	Chlorine	Chloride	Cl^-
VIIA	Bromine	Bromide	Br^-
VIIA	Iodine	Iodide	I^-
IB	Silver	Silver	Ag^+
IB	Copper	Copper(I) (cuprous ion)	Cu^+
		Copper(II) (cupric ion)	Cu^{2+}
IIB	Zinc	Zinc	Zn^{2+}
VIIIB	Iron	Iron(II) (ferrous ion)	Fe^{2+}
		Iron(III) (ferric ion)	Fe^{3+}
IVA	Tin	Tin(II) (stannous ion)	Sn^{2+}
		Tin(IV) (stannic ion)	Sn^{4+}

Notice from Table 3.2 that some elements can form ions with different valences. Examples are copper, iron, and nickel. The tendency to form different ions is common among the *transition elements* and is attributable to their complex electron structures. Tin and lead are also common elements with the potential to form different ions, even though they are in group IVA.

Examples of elements that can exist in different valence states are Fe^{2+} called iron(II) and Fe^{3+} called iron(III). The Roman numerals tell the valence state. Using an older system of naming, iron(II) is also known as ferrous ion and iron(III) as ferric ion from the Latin *ferrum,* "iron." In this system, an ion with an *-ous* ending has a lower valence state than one ending with *-ic.*

Tin (Sn) from Latin, *stannum;* lead (Pb) from Latin, *plumbum;* iron (Fe) from Latin, *ferrum.*

Polyatomic Ions

Two common polyatomic ions are ammonium and hydroxide, described earlier as containing covalent bonds between the atoms within the ions. Most of the common polyatomic ions contain oxygen as one of the atoms. Very often more than one oxygen atom is present.

Table 3.3 contains a list of many common polyatomic ions and their names. Since oxygen and hydrogen atoms are so prevalent, the ions are tabulated according to the other element that is present. There is no simple logic to the names that are used, except that the *-ate* or *-ite* ending signifies the presence of oxygen. It is recommended that you learn some of the most common of these ions, particularly those in boldface type.

Table 3.3 Some common polyatomic ions

Element Present Other than O or H	Formula	Name of Ion
Sulfur	SO_4^{2-}	**Sulfate**
	HSO_4^-	Hydrogen sulfate or bisulfate
	SO_3^{2-}	Sulfite
	HSO_3^-	Hydrogen sulfite or bisulfite
Nitrogen	NO_3^-	**Nitrate**
	NO_2^-	**Nitrite**
	NH_4^+	**Ammonium**
Oxygen	OH^-	**Hydroxide**
	H_3O^+	Hydronium
Carbon	CO_3^{2-}	**Carbonate**
	HCO_3^-	**Hydrogen carbonate or bicarbonate**
	$CH_3CO_2^-$ or $C_2H_3O_2^-$	Acetate
	CN^-	Cyanide
Phosphorus	PO_4^{3-}	Phosphate
	HPO_4^{2-}	Monohydrogen phosphate
	$H_2PO_4^-$	Dihydrogen phosphate
Bromine	BrO_3^-	Bromate

3.7 Formulas and Names of Covalent Compounds

GOALS: To consider the nature of some important diatomic molecules.
To study the naming of covalent compounds, including the use of electronegativities in formulating names.

Some elements exist in nature as *diatomic molecules* rather than as single atoms. Examples are the halogens (group VIIA) and the gases hydrogen (IA), oxygen (VIA), and nitrogen (VA):

$$F_2 \quad Cl_2 \quad Br_2 \quad I_2$$
$$H_2 \quad O_2 \quad N_2$$

The electron dot symbolism shows how these covalent molecules form as the atoms combine to achieve noble gas structures:

$$H\cdot \; + \; H\cdot \; = \; H{:}H \quad or \quad H{-}H$$
$$:\!\ddot{C}l\cdot \; + \; :\!\ddot{C}l\cdot \; = \; :\!\ddot{C}l{:}\ddot{C}l{:} \quad or \quad :\!\ddot{C}l{-}\ddot{C}l{:}$$

If we try to extend the process of pairing electrons to form covalent bonds, we do not always succeed in achieving a noble gas structure (an octet) for each atom. Thus, if we combine two nitrogen atoms, we might write

$$\cdot\ddot{N}\cdot \; + \; \cdot\ddot{N}\cdot \; = \; \cdot\ddot{N}{:}\ddot{N}\cdot \quad \text{incorrect}$$

But this structure does not have an octet for either atom. However, an octet may be achieved by sharing of two more pairs of electrons to give

$$:N:::N: \quad \text{or} \quad :N\equiv N: \quad \text{correct}$$

The triple bond resulting from the sharing of three pairs of electrons gives each atom an electron structure corresponding to the noble gas neon.

We might expect the oxygen molecule to exist as shown in the following structure. But, oxygen does not obey the octet rule, existing in a more complex electron structure that is beyond the scope of our coverage.

$$:\overset{\cdot}{\underset{\cdot\cdot}{O}}\cdot \; + \; :\overset{\cdot}{\underset{\cdot\cdot}{O}}\cdot \; = \; :\overset{}{\underset{\cdot\cdot}{O}}::\overset{}{\underset{\cdot\cdot}{O}}: \quad \text{or} \quad :\overset{}{\underset{\cdot\cdot}{O}}=\overset{}{\underset{\cdot\cdot}{O}}: \quad \text{incorrect}$$

Two different nonmetallic elements can also share electrons and form compounds. These covalent compounds are often observed with such elements as hydrogen, boron, carbon, nitrogen, phosphorus, sulfur, oxygen, and the halogens. Examples follow:

HCl	hydrogen chloride
CO_2	carbon dioxide
P_2O_5	diphosphorus pentoxide
BF_3	boron trifluoride
CCl_4	carbon tetrachloride
N_2O	dinitrogen monoxide
SO_3	sulfur trioxide

Two things you should be aware of when considering these formulas and the names used to describe them are the order of appearance of the elements in each formula and the number of atoms of each element. To understand the order in which elements are written in formulas of covalent compounds, we have to consider a property of the elements known as electronegativity. **Electronegativity** is the relative attraction of an atom for electrons in a molecule. An atom with a relatively strong attraction for electrons has a high electronegativity. Electronegativity values are not assigned to the noble gases since they are not normally found in combined form.

Electronegativity values are included for each element in Figure 3.3. Notice that they increase from left to right and from bottom to top of the periodic table, as emphasized by the intensity of the color. Fluorine has the highest possible electronegativity with an assigned value of 4.0.*

Electronegativity is the property used in writing formulas and assigning names to many covalent molecules. In each of the previous examples, such as carbon dioxide, CO_2, the first element in the formula has the lower electronegativity. Now go back to the examples and see that this trend is used in all cases.

---

*The variation in electronegativity among the elements is related to the charge of the nucleus (the number of protons) and the distance between the nucleus and the outer electrons. Since the positive nuclear charge in an atom steadily increases in going from left to right across a period, the attraction for electrons increases as a result. Upon moving down the periodic table from F to Cl, the nuclear charge also increases, but this change is more than offset by the increased distance between the nucleus and the valence electrons. Similarly, electronegativity decreases moving down the entire series F, Cl, Br, and I. The significance of electronegativity will be illustrated in Chapter 7.

EN increases →

EN decreases ↓

1	2	3	4	5	6	7	8	9	10	11	12	13	14	15	16	17
1 H 2.1																
3 Li 1.0	4 Be 1.5											5 B 2.0	6 C 2.5	7 N 3.0	8 O 3.5	9 F 4.0
11 Na 0.9	12 Mg 1.2											13 Al 1.5	14 Si 1.8	15 P 2.1	16 S 2.5	17 Cl 3.0
19 K 0.8	20 Ca 1.0	21 Sc 1.3	22 Ti 1.5	23 V 1.6	24 Cr 1.6	25 Mn 1.5	26 Fe 1.7	27 Co 1.8	28 Ni 1.8	29 Cu 1.9	30 Zn 1.6	31 Ga 1.6	32 Ge 1.9	33 As 2.0	34 Se 2.4	35 Br 2.8
37 Rb 0.8	38 Sr 1.0	39 Y 1.2	40 Zr 1.4	41 Nb 1.6	42 Mo 1.8	43 Tc 1.9	44 Ru 2.2	45 Rh 2.2	46 Pd 2.2	47 Ag 1.9	48 Cd 1.7	49 In 1.7	50 Sn 1.8	51 Sb 1.9	52 Te 2.1	53 I 2.5
55 Cs 0.7	56 Ba 0.9	57–71 La-Lu 1.1–1.2	72 Hf 1.3	73 Ta 1.5	74 W 1.7	75 Re 1.9	76 Os 2.2	77 Ir 2.2	78 Pt 2.2	79 Au 2.4	80 Hg 1.9	81 Tl 1.8	82 Pb 1.8	83 Bi 1.9	84 Po 2.0	85 At 2.2
87 Fr 0.7	88 Ra 0.9	89 Ac- 1.1–1.7														

Figure 3.3   Electronegativity values of the elements. The general trend is toward decreasing electronegativity in the direction from top to bottom of the periodic table, and toward increasing electronegativity from left to right as indicated by the intensity of the color.

The number of atoms of each element present in a covalent compound also is important to its name. This is communicated by using the Greek prefixes *mono-*, *di-*, *tri-*, and so forth. The prefix mono- is usually omitted for the *first* element in the molecule unless necessary to avoid confusion. Thus, N_2O is known as dinitrogen monoxide and NO_2 is nitrogen dioxide. Note also the suffix *-ide*. For CO and CO_2, the name carbon oxide would be ambiguous. Therefore, CO is identified as carbon monoxide and CO_2 is carbon dioxide.

Compounds containing hydrogen as one of the elements, such as HCl, H_2S, NH_3, and H_2O, are usually identified in ways that are not consistent with the rules described above. The name hydrogen chloride is logical and unambiguous. But the usual name hydrogen sulfide for H_2S fails to specify the presence of two hydrogen atoms, and fails to mention the number of sulfur atoms. In this case, as for ammonia (NH_3) and water (H_2O), the common names were given long before any system was devised for naming covalent molecules. These names are now considered as officially acceptable. If you want to bring a smile to your instructor's face, try referring to water as hydrogen oxide.

3.8   Rules for Naming Chemical Compounds

GOAL: To study the logic of writing names from formulas and formulas from names.

Now that we have considered ionic, covalent, and mixed ionic-covalent compounds, you can demonstrate the ability to (a) name a compound from its chemical formula and (b) write a correct formula for a given chemical name. A logical approach is suggested to get you started. Then you may wish to vary the order in which you approach the problem.

Writing a Name from a Formula

If you are trying to create a name from a formula, use any of the following steps as appropriate.

Step 1:   Recognize whether any of the elements present are metals by consulting Figure 3.1. Omit step 2 for metal-containing compounds.

Step 2:   If only nonmetals (or metalloids) are present, determine if a polyatomic cation (usually ammonium ion, NH_4^+) is present. If such a cation is present, name the cation and proceed to step 4. If there is no cation, formulate a name as illustrated for compounds such as carbon dioxide, dinitrogen monoxide, etc. Omit steps 3 and 4 for covalent compounds.

Step 3:   If a metal is present, determine if it is one that may exhibit different valences, such as iron or tin. If so, name the metal cation using a Roman numeral in parentheses, as in iron(II). If not, simply identify the metal cation.

Step 4:   Identify the anion. Look for the possibility of a polyatomic anion.

Try these rules on the following examples, which illustrate the various combinations. Solutions are given for each problem, but try to solve them on your own before consulting the answers.

EXAMPLE 3.1:

Give names for each of the following compounds. Consult a periodic table whenever necessary.

a. $Ca(OH)_2$ b. $FeCl_3$ c. NH_4NO_3 d. PCl_3

Solution:

For $Ca(OH)_2$:
From Table 3.1, we recognize that calcium is a metallic element. Step 2 does not apply to such a compound. Since calcium is in group IIA, it can only exist in ionic form as Ca^{2+}. Therefore, we do not have to specify the valence of calcium in the name. The anion is the polyatomic anion, hydroxide ion. Therefore, the correct name is calcium hydroxide. It is not necessary to specify the presence of two hydroxide ions; that is required by the valence of the calcium.

For $FeCl_3$:
The metallic element iron is present. It is a transition element that can exist in two common ionic forms, Fe^{2+} and Fe^{3+}. Since chloride ion is always monovalent, Cl^-, the iron must be Fe^{3+}, leading us to the name iron(III) chloride. The name ferric chloride is also acceptable.

For NH_4NO_3:
No metallic elements are present in this compound, but both the cation and the anion are monovalent polyatomic ions. The compound is ammonium nitrate.

For PCl_3:
No metallic elements are present, and neither are there any polyatomic ions. This is a purely covalent compound named as phosphorus trichloride. The prefix *mono-* is not required.

Writing a Formula from a Name

The following steps should allow you to write a name from a formula:

> **Step 1:** Scan the name for a metallic element and for any polyatomic ion(s) such as nitrate, sulfate, or ammonium. Are you dealing with a purely covalent compound, or are ions present? If the compound is ionic, omit step 2.
>
> **Step 2:** If the compound is covalent, determine the correct symbols for the elements involved. Write the formula using subscripts to show the number of atoms present for each element. Recall that the prefix *mono-* is normally omitted for the first element of a covalent compound, but should be used for the second element.
>
> **Step 3:** If the compound contains ions, identify the correct symbols and charges for both the cation and the anion. If the cation can have different valences, such as copper, the name should tell you the charge of this ion. Write the formula indicating the proper number of each ion required to give a net charge of zero. If a polyatomic ion is repeated, enclose the entire formula of the ion in parentheses followed by a subscript. Examples are $(NH_4)_2S$ and $Ba(NO_3)_2$.

Try these rules on the following examples that illustrate the various combinations. Solutions are given for each problem, but try to solve them on your own before consulting the answers.

EXAMPLE 3.2:

Give names for each of the following compounds. Consult a periodic table whenever necessary.

a. Diphosphorus pentoxide   b. Tin(IV) iodide

c. Potassium phosphate   d. Ammonium sulfate

Solution:

For diphosphorus pentoxide:
Only nonmetals are present; there are no polyatomic ions. As a purely covalent compound of phosphorus (P) and oxygen, the prefixes *di-* and *penta-* give us the information to write P_2O_5.

For tin(IV) iodide:
This is a metal salt composed of Sn^{4+} and I^- (Table 3.2). Therefore, the correct formula is SnI_4.

For potassium phosphate:
This is a metal salt of K^+ and the polyatomic PO_4^{3-}. The formula is K_3PO_4.

For ammonium sulfate:
There are no metallic elements, but both the cation NH_4^+ and the anion SO_4^{2-} are polyatomic. The formula is $(NH_4)_2SO_4$. Note that the ammonium ion is enclosed in parentheses followed by the subscript 2 in order to create a balance of positive and negative charges.

SUMMARY

Formulas can be written for ionic compounds by knowing the charges on the individual cations and anions. The periodic table can be used to determine the charges on ions. The periodic recurrence of properties of elements, summarized in the periodic table, relates to electron structure. The unique stability of the noble gases serves as the point of reference in the arrangement of the electrons. The partial circle designation, the electron dot symbolism, and the octet rule all emphasize the importance of the valence electrons in determining the ability of individual atoms to form ions or to bond covalently to other atoms.

To name chemical compounds when formulas are given or to write formulas when names are given, recognize the presence or absence of metallic elements, polyatomic ions, and elements capable of different valence states. For ionic compounds, the charges must be balanced when writing formulas. Prefixes, such as *mono-, di-,* and so forth, are used when naming purely covalent compounds, but not when naming ionic compounds.

PROBLEMS

1. Write equations for formation of each of the following compounds from the component ions, as illustrated in Section 3.1. Consult the tables of simple ions (Table 3.2) and polyatomic ions (Table 3.3) or the periodic table (Figure 3.1) as necessary.

   a. Lithium chloride
   b. Sodium hydroxide
   c. Potassium sulfide
   d. Lithium oxide
   e. Sodium carbonate
   f. Potassium sulfate
   g. Barium sulfate
   h. Ammonium nitrate
   i. Zinc monohydrogen phosphate
   j. Strontium nitrate
   k. Calcium nitrate
   l. Magnesium iodide
   m. Cadmium bicarbonate
   n. Sodium bisulfite
   o. Aluminum sulfate
   p. Potassium hydroxide
   q. Ammonium phosphate
   r. Ammonium monohydrogen phosphate
   s. Ammonium dihydrogen phosphate
   t. Cupric sulfate
   u. Tin(IV) nitrate
   v. Potassium bromate
   w. Sodium nitrite
   x. Lead(II) chloride
   y. Stannous fluoride
   z. Silver chloride

2. Give definitions or examples of each of the following.

   a. Valence
   b. Valence electron
   c. Period
   d. Group
   e. Octet rule
   f. Noble gas
   g. Alkali metal
   h. Alkaline earth metal
   i. Halogen
   j. Transition element
   k. Diatomic molecule
   l. Polyatomic ion
   m. Metal
   n. Nonmetal
   o. Metalloid
   p. Electronegativity
   q. Triple bond

3. Give an example of each of the following.

   a. Monovalent cation
   b. Divalent cation
   c. Trivalent cation
   d. Monovalent anion
   e. Divalent anion
   f. Trivalent anion

4. The inertness of the group VIIIA elements is an important key to understanding the behavior of other elements. Explain.

5. What is the charge on the ions formed by the group IIA elements?

6. Why do the group VIIA elements all have similar properties?

7. What noble gas structure is found in each of the ions in the following compounds?

   a. LiI
   b. $CaBr_2$
   c. Na_2Te
   d. BaO

8. Explain the relationship between the group number and the valence of elements in groups IA through VIIIA (A groups only).

9. Answer the following questions as they relate to the second and fourth periods in the periodic table.

   a. How many elements appear in the period?
   b. How many electrons are found in the neutral atom of the group VIA element in the period?
   c. How many valence electrons are found in the neutral atom of the group VIA element?
   d. What is the group VIA element?
   e. Give the formula for the sodium salt of the group VIA element.
   f. Draw the electron dot symbol for the group VIA element.
   g. Is the octet rule adhered to in the group VIA element or ion?
   h. Give the formula for the covalent carbon compound of the group VIIA element.
   i. Draw the electron dot symbol for the covalent carbon compound of the group VIIA element.

10. By what two processes can atoms achieve a noble gas structure? Give an example of each.

11. Give names for the following compounds. Consult a periodic table whenever necessary.

    a. LiOH
    b. P_2O_5
    c. $SnCl_4$
    d. SO_2
    e. $Ca(NO_3)_2$
    f. Ag_3PO_4
    g. CuCl
    h. $Mg(OH)_2$
    i. SiO_2
    j. $(NH_4)_2S$
    k. Na_2CO_3
    l. PCl_5
    m. $Ba(OH)_2$
    n. $Al(CN)_3$
    o. $NaHCO_3$
    p. $CaCO_3$
    q. KNO_2
    r. Na_2SO_3
    s. NH_3
    t. $(NH_4)_2SO_4$
    u. Na_2HPO_4
    v. IBr
    w. $KBrO_3$
    x. K_2S
    y. CO_2
    z. $Ca(HCO_3)_2$

12. Give formulas for each of the following compounds. Consult a periodic table whenever necessary.

    a. Iodine monochloride
    b. Barium oxide
    c. Ammonium bromide
    d. Hydrogen sulfide
    e. Copper(I) nitrite
    f. Silver nitrate
    g. Potassium bicarbonate
    h. Aluminum sulfate
    i. Tin(IV) oxide
    j. Potassium dihydrogen phosphate
    k. Magnesium cyanide
    l. Calcium chloride
    m. Strontium carbonate
    n. Sodium nitrite
    o. Sodium hydrogen sulfate
    p. Sulfur trioxide
    q. Silicon tetrachloride
    r. Aluminum hydroxide
    s. Iron(III) sulfate
    t. Ammonium sulfite

Chemical Reactions

4

When John Dalton formulated his atomic theory in the early 1800s (Section 2.1), he stated that chemical changes alter the grouping of atoms in elements and compounds without the creation or destruction of atoms. Except for nuclear processes, discussed in Chapters 10 and 11, his words are applicable today.

In this chapter we will look at chemical changes, or chemical reactions, to see some of the ways in which elements and compounds may be altered and how chemists describe these changes. Oxidation-reduction (redox) reactions are given special attention. In Chapter 5, the concept of the mole and the subject of stoichiometry are introduced so that you can understand the thought process that chemists use to predict the quantitative results of chemical reactions.

4.1 Chemical Equations

GOAL: To write balanced chemical equations, including net equations.

Chemical reactions are represented by chemical equations. A chemical equation is a shorthand expression that describes a reaction. Using symbols and formulas, we can represent the qualitative changes in atoms and molecules, and at the same time describe the quantitative relationships among the *reactants* (starting materials) and the products.

One important type of reaction is *combustion*. Molecules of a fuel undergo combustion when they react with oxygen molecules at high temperature. Accordingly, the combustion of methane (the major component of natural gas) can be represented by the following equation:

$$\underbrace{\underset{\text{methane}}{CH_4} + \underset{\text{oxygen}}{O_2}}_{\text{reactants}} \longrightarrow \underbrace{\underset{\substack{\text{carbon} \\ \text{dioxide}}}{CO_2} + \underset{\text{water}}{H_2O}}_{\text{products}} \quad \text{(unbalanced)}$$

This equation tells us that methane reacts with oxygen to form carbon dioxide and water. Although this form of the equation corresponds exactly to these words, it is not a correct

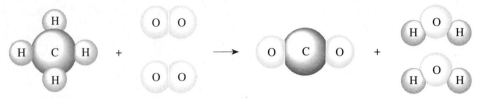

Figure 4.1   The combustion of methane represented by molecular models.

chemical equation because it is not balanced. That is, on the left side of the equation there are two oxygen atoms, and on the right side of the equation there are three oxygen atoms. The numbers of hydrogens are also unequal.

Since chemical reactions do not create or destroy atoms, the unbalanced equation violates the law of conservation of mass. Therefore, the equation is modified by addition of *coefficients* to the left of some of the formulas as shown.

$$CH_4 + 2\,O_2 \rightarrow CO_2 + 2\,H_2O \text{ (balanced)}$$

To determine if the equation is correctly balanced, check to see that each type of atom appears in equal numbers on both sides of the equation. According to the balanced equation, the combustion of one methane molecule requires two oxygen molecules and causes the formation of one carbon dioxide molecule and two water molecules. Notice that only coefficients are changed; subscripts may not be changed since this would alter the formulas of the compounds involved in the reaction.

The same reaction is illustrated in Figure 4.1 in model form. In all equations, an arrow is used to indicate the flow from reactants to products.

Another important reaction is involved in the *refining* of iron ore to produce iron and steel. Much of the iron in the ore exists as iron oxides, such as Fe_2O_3. In a blast furnace (Figure 4.2), carbon monoxide reacts with iron(III) oxide through a complex process that can be represented by the following equation.

$$\underset{\substack{\text{iron(III)}\\\text{oxide}}}{Fe_2O_3} + \underset{\substack{\text{carbon}\\\text{monoxide}}}{CO} \rightarrow \underset{\text{iron}}{Fe} + \underset{\substack{\text{carbon}\\\text{dioxide}}}{CO_2} \quad \text{(unbalanced)}$$

Here again, we have an equation that is not balanced. This time let us consider the thought process required to balance this and most other chemical equations. There is no single approach that must be followed when balancing equations. Very often, a quick inspection of the equation will allow you to balance it in one easy step; sometimes several steps are needed. We might begin by balancing the number of iron (Fe) atoms on the two sides of the equation by adding the coefficient 2 to the product Fe:

$$Fe_2O_3 + CO \rightarrow 2\,Fe + CO_2 \text{ (unbalanced)}$$

This equation is still not balanced since the number of oxygen atoms is not the same on both sides. To accomplish this, we add the coefficient 2 to the product CO_2 to give four oxygen atoms on both sides:

$$Fe_2O_3 + CO \rightarrow 2\,Fe + 2\,CO_2 \text{ (unbalanced)}$$

Now the iron and oxygen atoms are balanced, but the carbon atoms are not. We can then balance the number of carbon atoms by using the coefficient 2 for CO:

$$Fe_2O_3 + 2\,CO \rightarrow 2\,Fe + 2\,CO_2 \text{ (unbalanced)}$$

Figure 4.2 A blast furnace is used to produce molten iron from iron ore by reaction of iron oxides with carbon monoxide. The molten iron can be poured into a suitable mold of desired shape. Carbon monoxide is formed in a blast furnace by the following reactions, where carbon is introduced in the form of coke, obtained from coal (see Section 12.7).

$$C + O_2 \rightarrow CO_2$$
$$CO_2 + C \rightarrow 2\,CO$$

But now the oxygen atoms are no longer balanced. To achieve a balanced equation finally, we must change the coefficients of both CO and CO_2 to give:

$$Fe_2O_3 + 3\,CO \rightarrow 2\,Fe + 3\,CO_2 \text{ (balanced)}$$

This was a relatively difficult equation to balance. Up until the last step, the addition of coefficients should have seemed reasonable and logical, but the last step was not obvious. The key to it came from the fact that there had to be a ratio of $Fe:Fe_2O_3$ of 2:1 and a ratio of $CO_2:CO$ of 1:1 to balance the Fe and C atoms. This method allows one to apply trial and error until success is achieved. It is also possible to use the approach that will be illustrated in Section 4.2.

When writing chemical equations, a number of additional symbols may be used, although they are not always required. The common symbols are given in Table 4.1.

The decomposition of potassium chlorate provides an example of how such symbols are used.

$$KClO_3(s) \xrightarrow{\Delta} KCl(s) + O_2(g) \text{ (unbalanced)}$$

potassium   potassium
chlorate   chloride

Table 4.1   Common symbols used in writing chemical equations

Symbol	Meaning
Δ	Heat is added.
(aq)	Substance is dissolved in an *aqueous* (water) solution.
(s)	Substance exists as a solid.
(l)	Substance exists as a liquid.
(g)	Substance exists as a gas.

According to this unbalanced equation, solid potassium chlorate can be heated to form solid potassium chloride and oxygen gas. This reaction can be used in the laboratory as a convenient way to generate oxygen. To balance this equation, we can use a common trick. Since only the oxygen atoms are out of balance, we note that the common multiple of subscripts of the oxygen atoms (3 in $KClO_3$ and 2 in O_2) is six. Therefore, we can balance the oxygen atoms by adding coefficients to obtain six oxygen atoms on each side. This affects the KCl at the same time to yield:

$$2\ KClO_3(s) \overset{\Delta}{\rightarrow} 2\ KCl(s) + 3\ O_2(g) \quad \text{(balanced)}$$

Other symbols are illustrated by the following balanced equations.

$$NaOH(aq) + HCl(aq) \rightarrow NaCl(aq) + H_2O(l)$$
$$KCl(aq) + AgNO_3(aq) \rightarrow AgCl(s) + KNO_3(aq)$$

The first equation says that aqueous solutions of NaOH and HCl can be combined to form a solution of NaCl and water. The second equation shows that aqueous solutions of KCl and $AgNO_3$ may be combined to form AgCl solid and KNO_3, which dissolves in the solution. When this reaction is carried out, a white precipitate of solid silver chloride, AgCl(s), is observed (Figure 4.3).

Net Equations

The two reactions just discussed may also be written in abbreviated form. In the first equation, HCl, NaOH, and NaCl in water are all ionic compounds, and their ions separate (dissociate) from one another when these compounds are present in a water solution. If we regard these ionic compounds as:

$$HCl = H^+ + Cl^-$$
$$NaOH = Na^+ + OH^-$$
$$NaCl = Na^+ + Cl^-$$

then we may write the reaction of HCl with NaOH as:

$$Na^+(aq) + OH^-(aq) + H^+(aq) + Cl^-(aq) \rightarrow Na^+(aq) + Cl^-(aq) + H_2O(l)$$

or simply:

$$Na^+ + OH^- + H^+ + Cl^- \rightarrow Na^+ + Cl^- + H_2O$$

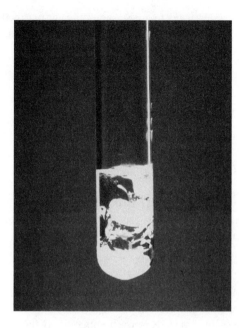

Figure 4.3
As a water solution of KCl is added to a water solution of $AgNO_3$, a precipitate of AgCl is formed. The $NaNO_3$ formed by the reaction remains in solution and is symbolized by $NaNO_3$ (aq).

Notice, however, that both Na^+ and Cl^- appear on both sides of the equation. Since neither of these ions is changed by the overall reaction, they are known as *spectator ions.* Therefore, it is possible to omit the spectator ions from the equation and write a *net equation:*

$$OH^- + H^+ \rightarrow H_2O$$

The water molecule is covalent with only a slight tendency to ionize.

As for the reaction of KCl and $AgNO_3$, these two reactants and the KNO_3 product are all ionic compounds that are readily dissolved in water. In addition, AgCl is an ionic compound, but since it is insoluble, it does not dissociate into Ag^+ and Cl^-. We may expand the equation to show each of the dissociated ions:

$$K^+(aq) + Cl^-(aq) + Ag^+(aq) + NO_3^-(aq) \rightarrow AgCl(s) + K^+(aq) + NO_3^-(aq)$$

and then simply by recognizing K^+ and NO_3^- as spectator ions to give the net equation:

$$Cl^-(aq) + Ag^+(aq) \rightarrow AgCl(s)$$

This equation may be simplified even further as:

$$Cl^- + Ag^+ \rightarrow AgCl$$

Reversible Reactions

Up to now we have considered reactions as though they all go to completion and only flow in one direction from reactants to products. For a great many reactions, however, these assumptions are not valid. The reactions do not go to completion because they are *reversible;* that is, as products form, they can react to produce the starting materials. In writing an equation, double arrows show that the reaction occurs in both the forward direction (left to right) and the reverse direction (right to left).

The Haber process is used to make ammonia, primarily for use in fertilizers (see Section 15.1). The reaction of nitrogen gas (N_2) with hydrogen gas (H_2) is reversible, as indicated by the two arrows.

$$N_2(g) \ + \ 3\,H_2\,(g) \rightleftarrows 2\,NH_3\,(g)$$
$$\text{nitrogen} \qquad \text{hydrogen} \qquad\qquad \text{ammonia}$$

Fritz Haber received the Nobel Prize in 1918 for his work on the synthesis of ammonia.

This reaction could just as well be written in the opposite direction to indicate the same information:

$$2\,NH_3(g) \rightleftarrows N_2(g) \ + \ 3\,H_2\,(g)$$

Usually, a reaction is favored to go more in one direction than the other. For example, the formation of ammonia from nitrogen and hydrogen is favored by high pressure. This could be represented by varying the lengths of the arrows.

$$N_2(g) \ + \ 3\,H_2(g) \xrightleftharpoons{\text{pressure}} 2\,NH_3(g)$$

If one combines nitrogen gas and hydrogen gas and then applies the necessary heat and pressure, the reaction proceeds to form ammonia. If the reaction were not reversible, you might logically expect it to continue until either the N_2 or the H_2, or both, were used up. Then the reaction would stop. But since this reaction is reversible, the formation of more NH_3 will stop at some point, because a condition will be reached when the formation and breakdown of NH_3 balance one another. Both the forward and reverse reactions will continue, but the amount of each of the three gases will remain constant. Such a condition is called an **equilibrium.** Since the forward and reverse reactions continue to occur, the condition may even be called a *dynamic equilibrium* to emphasize that the reaction continues in each direction, without any further net change. Nevertheless, once the mixture is cooled, the reaction ceases and a stable sample of ammonia can be obtained.

Summary of Rules for Writing Equations

Before continuing, let us review what has been considered so far. Equations are written to signify the events in chemical reactions. All reactants and products must be known. The reactants are written on the left side of an equation and are separated by an arrow from products on the right side of the equation. For reversible reactions, two arrows are used; their length indicates the preferred direction of the reaction (if the information is known).

When ionic compounds are involved and spectator ions can be recognized, a net equation can be written to emphasize only those components of the reactants that are directly involved in the formation of products. The symbols (g), (l), (s), (aq), and Δ are used to signify a gas, liquid, solid, aqueous solution, and added heat, respectively.

All equations must be balanced. The coefficients in a balanced chemical equation state the simplest whole-number ratio in which substances react and form in a chemical change. Before an equation is balanced, all reactants and products must be included with their correct formulas. Balancing is done by adding coefficients to the left of formulas for the reactants and products, one element at a time, until all are balanced. Only coefficients may be changed; subscripts may not be changed since this would alter the formulas of the compounds involved. As each element is balanced, any elements that have already been balanced should be checked to be sure they have not become unbalanced.

4.2   Oxidation-Reduction Reactions: Redox

GOALS: To recognize redox, and oxidizing and reducing agents.
To understand the corrosion of iron and how it can be prevented.
To appreciate why aluminum and chromium resist corrosion.
To understand the components of a galvanic cell.
To differentiate between galvanic cells and electrolytic cells.
To know how electrical and chemical energy can be interconverted.
To predict the direction of a spontaneous redox reaction.
To appreciate the electrolysis of water to generate oxygen and hydrogen.
To understand the purpose and process of electroplating.

Of all the chemical reactions that we might study, one of the most common types is the *oxidation-reduction,* or *redox* reaction. These two classes of reactions must be studied together since they cannot take place separately.

The term *oxidation* was first used to describe the reaction of any substance with oxygen, such as the combustion of methane mentioned earlier

$$CH_4 + 2\,O_2 \rightarrow CO_2 + 2\,H_2O$$

or the rusting of iron

$$4\,Fe + 3\,O_2 + 3\,H_2O \rightarrow 2\,Fe_2O_3 \cdot 3\,H_2O$$

The product $Fe_2O_3 \cdot 3\,H_2O$ is equivalent to $2Fe(OH)_3$. The three water molecules in the first formula are known as *waters of hydration.*

In these reactions, atoms have chemically combined with oxygen atoms and may be said to have been *oxidized.* Conversely, if an element that is combined with oxygen reacts with a loss of that oxygen, it is said to be *reduced.*

$$Fe_2O_3 + 3\,CO \rightarrow 2\,Fe + 3\,CO_2$$

In this reaction, the iron has been reduced, but the carbon that enters as carbon monoxide is oxidized. This illustrates that oxidation and reduction always occur together. Each process depends on the other, and the word redox has evolved as a means of emphasizing the mutual dependence.

Redox reactions do not always include a gain or loss of oxygen atoms. In some instances, they occur with no involvement of oxygen atoms. Therefore, we must have some other means of recognizing redox.

Oxidation is the process in which an element loses electrons. The name is derived from the fact that any element, when it combines with oxygen, loses electrons. The lost electrons are transferred to the oxygen atom.

Consider the oxidation of the element magnesium to form magnesium oxide.

$$2\,Mg + O_2 \rightarrow 2\,MgO$$

In this reaction, the neutral magnesium atom is converted to Mg^{2+} ion when it becomes part of the compound MgO. When this change takes place, the neutral, uncharged magnesium atom (Mg^0) loses two electrons and is therefore oxidized.

$$Mg^0 \xrightarrow{\text{loss of 2 electrons}} Mg^{2+}$$

The oxidation of silver atoms occurs in the following reaction despite the fact that oxygen is not involved.

$$2\,Ag + S \rightarrow Ag_2S$$

The neutral silver atoms are converted to Ag^+ ions, a loss of one electron per atom. Sulfur atoms are converted to sulfide ions, S^{2-}. Since electrons are gained by the sulfur, it is reduced. Overall, we conclude that the electrons lost by the silver atoms are gained by the sulfur atoms, again illustrating the interdependence of oxidation and reduction.

Any substance that is oxidized provides the electrons required for reduction to take place. For this reason, the substance that is oxidized is called the *reducing agent*. Similarly, by accepting electrons, a substance that can be reduced causes oxidation to take place and is called the *oxidizing agent*.

$$2\,Mg + O_2 \rightarrow 2\,MgO$$

Mg: substance oxidized; reducing agent
O_2: substance reduced; oxidizing agent

Furthermore, when a redox reaction takes place, *the number of electrons lost by the oxidized substance is equal to the number of electrons gained by the substance that is reduced.* When the equation for a redox reaction is correctly balanced, not only must the atoms balance, but the electron transfers represented by the oxidation and the reduction must be equivalent.

When one magnesium atom is oxidized, it loses two electrons. When the two oxygen atoms of O_2 are reduced to oxide ions, O^{2-}, each atom gains two electrons for a total of four electrons gained. Therefore, the balanced equation must indicate that two magnesium atoms, each losing two electrons for a total loss of four electrons, are required to react with one molecule of oxygen gas, which will gain the four electrons.

$$2\,Mg + O_2 \rightarrow 2\,MgO$$

Two units of magnesium oxide are produced by the reaction of the two magnesium atoms and one oxygen molecule. The equation is correctly balanced not only because the atoms balance on both sides, but also because equivalent electron transfers are represented.

Consider the redox reaction of copper with silver nitrate.

$$Cu + AgNO_3 \rightarrow Cu(NO_3)_2 + Ag \qquad \text{(unbalanced)}$$

Here, the copper atom is oxidized to copper(II) ion by losing two electrons; the silver ion is reduced to free silver by gaining one electron. For the electron transfers to be equivalent, two silver ions will be required to react with one copper atom.

$$Cu + 2\,AgNO_3 \rightarrow Cu(NO_3)_2 + 2\,Ag$$

The copper serves as the reducing agent since it is oxidized. Since silver ion is reduced, it is the oxidizing agent. Since the two nitrate salts in this equation are water-soluble and the nitrate ions are spectator ions, with no involvement in the reaction, a net equation can be written to emphasize the participants in the redox:

$$Cu + 2\,Ag^+ \rightarrow Cu^{2+} + 2\,Ag$$

Or, to emphasize even more strongly that redox is involved, the net equation can be rewritten as follows, where the superscript zero is used to signify the neutral, uncharged atoms of an element.

$$Cu^0 + 2\,Ag^+ \rightarrow Cu^{2+} + 2\,Ag^0$$

Corrosion

When corrosion occurs, metals are converted into compounds, usually oxides. The rusting of iron is a form of corrosion in which iron metal is converted into a hydrated oxide of iron. This process requires both moisture and oxygen. It was described by a single reaction at the start of Section 4.2:

$$4\,Fe + 3\,O_2 + 3\,H_2O \rightarrow 2\,Fe_2O_3{\cdot}3\,H_2O$$

This single equation is a summary of a set of two reactions, with oxidation of iron (the reducing agent) by oxygen (the oxidizing agent), which is reduced, taking place in each reaction:

$$2\,Fe + O_2 + 2\,H_2O \rightarrow 2\,Fe(OH)_2$$
$$4\,Fe(OH)_2 + O_2 + 2\,H_2O \rightarrow 4\,Fe(OH)_3 \text{ (equals } Fe_2O_3{\cdot}3\,H_2O)$$

Rust does not adhere to the surface of the metal the way some oxide coatings do. Since the rust tends to flake off, the surface of the iron is exposed to further oxidation. Rusting can be prevented by painting to cover the metal surface or by using rustproofing materials that prevent water from reaching the surface of the metal (Figure 4.4).

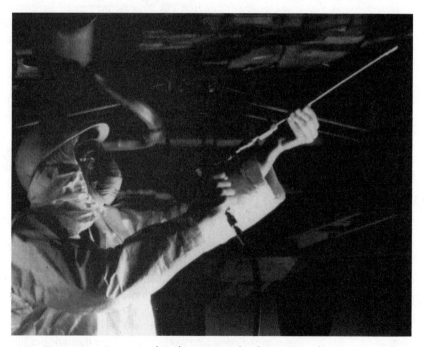

Figure 4.4   Since water is required in the reaction for the rusting of iron, rustproofing materials are sprayed onto metal surfaces and into spaces within car doors, fenders, and elsewhere to prevent the entry and accumulation of moisture.

Figure 4.5 Many aircraft have been in service for decades. This is evidence of the resistance of aluminum to corrosion. Add a new engine and a new paint job, and they can run indefinitely.

The rusting of iron can also be prevented by **galvanization.** This process involves coating the surface of the iron with another metal, usually zinc. Since zinc is oxidized more readily than iron, the iron metal is protected as long as the zinc coating remains intact.''

The presence of chromium as an ingredient of steel also inhibits rusting and gives a product known as *stainless steel.* Stainless steel also contains some nickel, but it is the formation of a coating of chromium oxide that prevents the iron from becoming oxidized.

In a similar fashion, aluminum metal resists corrosion by undergoing oxidation on the surface of the metal. The aluminum oxide, Al_2O_3, coating that forms on the surface adheres tightly and protects the aluminum below from undergoing further oxidation. Airplanes are constructed primarily of aluminum both because of its light weight and its resistance to corrosion (Figure 4.5).

Applications of Redox Reactions: Electrochemistry

Since redox reactions involve the transfer of electrons, and since a flow of electrons constitutes an electric current, redox reactions can be used to produce an electric current. Conversely, an electric current can be used to drive certain chemical reactions.

When an oxidizing agent reacts with a reducing agent, energy is released in the form of heat. If these same reactants are placed into separate solutions containing ions with a conducting wire joining them externally, the transfer of electrons (e^-) from the reducing agent to the oxidizing agent takes place through the wire; that is, current flows. Such an arrangement, like that in Figure 4.6, is called an **electrochemical cell** or **galvanic cell,** where the redox reaction between zinc metal and copper ions can take place.

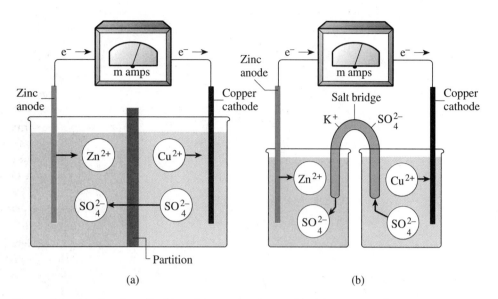

Figure 4.6   A galvanic cell with a zinc anode where oxidation occurs and a copper cathode where reduction occurs. In both (a) and (b), the two electrodes are joined externally so that current (electrons) can flow from the zinc to the copper and be detected by the ammeter. The two compartments must be connected by (a) a porous partition or (b) a salt bridge to allow for the movement of sulfate ions as shown. See text for discussion.

In this cell, the two compartments are joined in two ways to make a complete circuit. On the left, a strip of zinc metal is in contact with a solution of zinc ions (introduced as $ZnSO_4$). On the right, a strip of copper metal is immersed in a solution of $CuSO_4$. The two strips of metal are joined by a wire running through an ammeter (to measure the flow of electric current). On the left, zinc metal is converted to Zn^{2+} while on the right, Cu^{2+} is converted to copper metal. The solutions also are joined by a porous partition or salt bridge.

Taken together, the two reactions in the galvanic cell are represented by the following equation:

$$Zn + Cu^{2+} \rightarrow Zn^{2+} + Cu$$

Since Zn and Cu^{2+} are not in contact with one another, electrons released by Zn can reach Cu^{2+} only by traveling through the wire. There they are detected with an ammeter as a flow of electric current. The Zn and Cu^{2+} cannot be in the same solution; if they were, the electrons would not have to flow through the wire in order to transfer from the zinc to the copper ions. But the partition that separates the two solutions must allow for the movement of SO_4^{2-} ions since Zn^{2+} ions are formed from Zn on the left and Cu^{2+} are reduced to Cu on the right. This requires that anions move from right to left in the apparatus so that the number of positive and negative charges remains balanced *within* each compartment.

The two strips of metal where oxidation and reduction take place are called **electrodes.** The zinc electrode or **anode** is *the site of oxidation.* Anions move toward this electrode since cations are formed here. The copper electrode, *the site of reduction,* is called the **cathode.** Since electrons flow to this electrode from outside the solution, it attracts cations,

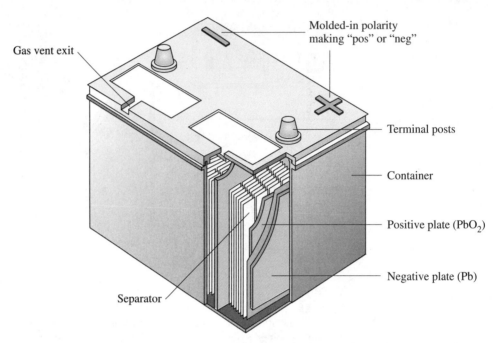

Gas vent exit

Molded-in polarity
making "pos" or "neg"

Terminal posts

Container

Positive plate (PbO_2)

Negative plate (Pb)

Separator

Figure 4.7   The lead storage battery, also known as the lead-acid battery, is commonly
used in automobiles. It consists of a set of lead anodes and lead (IV) oxide
cathodes. See Section 13.12 for details.

such as Cu^{2+}. We can write the two separate reactions, known as **half-reactions,** as
follows:

$$\text{oxidation (at anode)}$$
$$Zn \rightarrow Zn^{2+} + 2\,e^-$$
$$\text{reduction (at cathode)}$$
$$Cu^{2+} + 2\,e^- \rightarrow Cu$$

If the ammeter in this arrangement were replaced by a motor, the electric energy could be
used to run the motor and perform some kind of work. The electric current could also
power a light bulb or a flashlight. In other words, chemical energy can be converted into
electrical energy.

   Batteries are electrochemical cells. In some cases, a battery is a series of electrochemical cells arranged so that the total voltage available is the sum of that available from each
individual cell. The lead storage battery (Figure 4.7) used in automobiles is the most
common example. This device and other common batteries are discussed in Section 13.12.

   The rusting of steel in automobiles is an example of redox, and it is also an electrochemical process that is encouraged by the presence of salts. Salts such as NaCl and $CaCl_2$
are used for melting snow and ice on highways in many locations. The ions released from
these salts assist both oxidation and reduction, which can take place some distance apart
on the surface of the iron. The half-reactions for the initial redox step in rusting are the
following:

$$Fe \rightarrow Fe^{2+} + 2\,e^- \qquad \text{oxidation}$$
$$O_2 + 2\,H_2O + 4\,e^- \rightarrow 4\,OH^- \qquad \text{reduction}$$

The electrons released by the oxidation of Fe can be conducted through the iron metal to another location on the automobile where the reduction reaction takes place. The anions of a salt will encourage the formation of Fe^{2+}, and the cations of the salt will assist in the formation of OH^- ions. What is observed is the loss of iron from the anode, which causes the metal to disappear slowly and be replaced by iron oxide. The importance of water (moisture) must not be overlooked in rusting. If moisture can be excluded, the reduction half-reaction does not take place. The strategy of rustproofing is to prevent corrosion due to moisture getting inside or underneath a car.

The Flow of Electrons

In considering any possible redox reaction for use in an electrochemical cell, we need to know if a reaction will actually occur. The following reaction has been considered:

$$Zn + Cu^{2+} \rightarrow Zn^{2+} + Cu$$

By knowing that this is a real reaction, we can conclude that Cu^{2+} is a stronger oxidizing agent than Zn^{2+} and that Zn is a stronger reducing agent than Cu. If these two statements were not true, the reaction would, in fact, go in the opposite direction. Similarly, from observations of other redox reactions, Table 4.2 can be used in predicting which ones will occur. By locating the potential oxidizing and reducing agents, we conclude that the process will occur if the oxidizing agent appears above the reducing agent in the table. A strong reducing agent easily gives up electrons to become a weaker oxidizing agent. A strong oxidizing agent readily takes on electrons to form a weaker reducing agent.

Table 4.2   Relative strengths of oxidizing and reducing agents

	Oxidizing Agents	Reducing Agents	
Strong	F_2	F^-	Weak
	Cl_2	Cl^-	
	Br_2	Br^-	
	Ag^+	Ag	
	Fe^{3+}	Fe^{2+}	
	I_2	I^-	
	Cu^{2+}	Cu	
	Pb^{2+}	Pb	
	Sn^{2+}	Sn	
	Ni^{2+}	Ni	
	Fe^{2+}	Fe	
	Cr^{3+}	Cr	
	Zn^{2+}	Zn	
	Al^{3+}	Al	
	Mg^{2+}	Mg	
	Na^+	Na	
	Ca^{2+}	Ca	
	K^+	K	
Weak	Li^+	Li	Strong

If we emphasize the oxidation half-reaction, we could say that Cu^{2+} can oxidize Zn metal, but Zn^{2+} cannot oxidize Cu metal. Or, if we emphasize reduction, we could say that Zn metal can reduce Cu^{2+}, but Cu metal cannot reduce Zn^{2+}. In other words, the reducing agent must appear in the table below the element or ion that is going to be reduced.

Similarly, if silver metal (Ag) is placed in a solution of Pb^{2+}, we can predict from Table 4.2 that no reaction would take place. But if Ag^+ and Pb are combined, the following reaction will be observed.

$$2\,Ag^+ + Pb \rightarrow 2\,Ag + Pb^{2+}$$

Electrolysis

When electric energy from outside the system is used to drive chemical reactions, the process is called **electrolysis.** If current is supplied to an electrochemical cell from an outside source, the electrochemical process can be forced to move in either direction, depending on the direction of flow of the current. Instead of a *spontaneous* set of oxidation and reduction reactions that produces electrons in a galvanic cell, electrons can be forced to flow from the copper toward the zinc. By forcing electrons toward the zinc electrode, zinc ions in the surrounding solution would be reduced, thereby causing the zinc to become a cathode. The copper electrode then becomes the anode, where copper metal is oxidized to Cu^{2+}. This *nonspontaneous* redox process is an example of electrolysis.

Another example of electrolysis is the production of oxygen gas and hydrogen gas by supplying electric current through inert electrodes to an aqueous solution. The process is illustrated in Figure 4.8. Platinum (Pt) electrodes are immersed in a solution of Na_2SO_4. Both the sodium ions and the sulfate ions are so resistant to reduction and oxidation that only water molecules undergo redox at each electrode. As always, oxidation takes place at the anode, where oxygen gas and hydrogen ions form and electrons are released to the anode.

$$2\,H_2O \rightarrow O_2(g) + 4\,H^+ + 4\,e^-$$

The reduction of water molecules occurs simultaneously at the cathode as water molecules gain electrons from the electrode to form hydrogen gas and hydroxide ions.

$$4\,H_2O + 4\,e^- \rightarrow 2\,H_2(g) + 4\,OH^-$$

This reduction half-reaction is represented by an equation in which all of the coefficients have been doubled, so that the number of electrons transferred is the same as the number of electrons involved in the oxidation. If these two half-reactions are combined, the balanced equation appears as:

$$6\,H_2O \rightarrow O_2(g) + 2\,H_2(g) + 4\,OH^- + 4\,H^+$$

As they form, H^+ and the OH^- will combine to form water. Therefore, the overall equation for the electrolysis of water should be represented as:

$$6\,H_2O \rightarrow O_2(g) + 2\,H_2(g) + 4\,H_2O$$

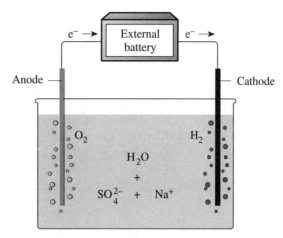

Figure 4.8 The electrolysis of water using an external source of electric current. Water molecules are simultaneously oxidized and reduced in such an apparatus, where the electrodes are made of an inert material such as platinum that will not undergo oxidation or reduction. Inert ions, such as sodium and sulfate ions, must be present to support the formation of ions in redox half-reactions.

or simply as:

$$2\ H_2O \rightarrow O_2(g) + 2\ H_2(g)$$

Even though the sodium and sulfate ions are not changed by the reaction, both must be present to support the electrochemical reaction. Since hydrogen ions are produced at the anode, sulfate ions must be present to maintain a balance of charges at the surface of the electrode. Similarly, sodium ions are required to support the formation of hydroxide ions at the cathode.

Electroplating

One final example of electrolysis is known as **electroplating,** in which current flow causes the ions of a metal to be reduced to the metal. The metal then deposits as a coating on the surface of a cathode; this process is called plating out. Jewelry, ''silverware,'' and ''tin cans'' are commonly produced by electroplating. Jewelry is often plated with gold or silver. Sterling silver eating utensils, which usually consist of 92% silver, with 8% nickel added for hardness, are considered expensive for everyday use. But silverplated utensils, which are made mostly of nickel and have a thin coating of silver, are more reasonably priced. The process for plating with silver is symbolized in Figure 4.9.

Tin cans are actually made of steel, which is then coated with tin to resist corrosion. Coating with tin is commonly known as *galvanization.* Galvanized iron is used in many applications where corrosion can be a problem.

''Chrome'' bumpers and grillwork on automobiles consist of steel coated with chromium deposited by electroplating. In recent times, plastics have replaced most of the metal used in this application to allow for longer life and lighter weight.

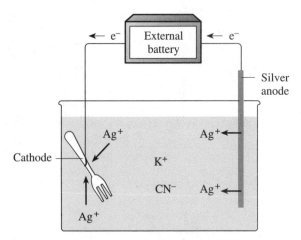

Figure 4.9 Silverplating. A fork made of nickel is given a silver coating by electrolysis of a solution of KCN and AgCN. The fork is the cathode and the anode is silver. Silver ions are produced at the anode to replace Ag^+ ions that are reduced at the cathode. The presence of cyanide ions causes the silver to plate out as a fine-grained deposit with a much smoother consistency than if nitrate or other anion is used.

Electroplating is also routinely used for the purification of metals that are available only in an impure state from ores. An important example is copper, which can be dissolved into solution from its ore as Cu^{2+} and then plated out on copper electrodes in very pure form (Figure 4.10).

Figure 4.10 Cathodes made of pure copper can be produced by electrolysis of solutions of copper salts. Sheets of solid copper produced by electrolysis are shown hanging above the bath from which they have been removed.

Metabolism: The Ultimate Oxidation

The following oxidation is of biochemical importance.

$$C_6H_{12}O_6 + 6\,O_2 \rightarrow 6\,CO_2 + 6\,H_2O$$
glucose

Glucose is the most common sugar found in nature; it is used as a building block of starch and other carbohydrates, and as a source of energy by most organisms. Like the rusting of iron, the oxidation of glucose is not a single redox reaction; instead it involves a series of many reactions, some of which are redox reactions (see Section 17.1). Several oxidizing agents are involved in the overall process. Oxygen is the ultimate oxidizing agent that causes other compounds to become oxidized. Despite the complexity of this process, it illustrates that the fully oxidized form of carbon is carbon dioxide, regardless of its source. The structure of carbon dioxide is discussed in Chapter 6. Similarly, water is the fully oxidized form of hydrogen.

SUMMARY

Chemical equations can be written to represent chemical reactions. These equations must be balanced by the addition of coefficients to reflect the quantitative changes that may occur. Net equations may be written by omitting spectator ions. The symbols (g), (l), (s), (aq), and Δ are used to signify a gas, liquid, solid, aqueous solution, and added heat, respectively. The notation NaCl (aq) is used for sodium chloride dissolved in aqueous solution.

Redox reactions are an important class of chemical reactions in which a reducing agent is oxidized and an oxidizing agent is reduced. Correct redox equations have a balance of both the atoms and the electron transfers that occur between the oxidizing and reducing agents. Corrosion of iron and other metals is an important example of a redox reaction. The rusting of iron can be prevented by excluding moisture from contact with iron by galvanization (coating with zinc) or addition of chromium (as in stainless steel). Aluminum and chromium interfere with corrosion by forming oxide coatings.

Spontaneous redox occurs in a galvanic cell in the direction that is dependent on the strength of the oxidizing and reducing agents. The direction of reaction can be reversed by imposing an external current flow on the system, thus converting it into an electrolytic cell. If inert electrodes and inert salts are used, water molecules can undergo electrolysis to yield oxygen and hydrogen. Jewelry, kitchen utensils, and other items can be electroplated if they are used as the cathode in an electrolytic cell. Galvanization is the same process; it is used to produce tin cans, which are iron cans coated with tin to prevent corrosion. Highly purified copper and other metals can be prepared by electrolysis of salts of the metals.

PROBLEMS

1. Give a definition or an example of each of the following.

   a. Combustion
   b. Reactant
   c. Aqueous
   d. Net equation
   e. Spectator ions
   f. Reversible reactions
   g. Dynamic equilibrium
   h. Oxidation
   i. Reduction
   j. Oxidizing agent
   k. Reducing agent
   l. Corrosion
   m. Galvanization
   n. Stainless steel
   o. Half-reaction
   p. Electrolysis
   q. Electroplating

2. Write complete and balanced equations for each of the following reactions. Write the equations in simple form and again using the appropriate symbols to signify solids (s), liquids (l), gases (g), and aqueous solutions (aq).

   a. Combustion of methane
   b. $H_2 + O_2 \rightarrow H_2O$
   c. $NaOH + HCl$ in aqueous solution
   d. $KClO_3 \overset{\Delta}{\rightarrow} KCl$
   e. Formation of solid AgCl from soluble salts in aqueous solution
   f. $Ca + H_2O \rightarrow H_2 + Ca(OH)_2$
   g. Synthesis of NH_3 from hydrogen gas and nitrogen gas
   h. $SO_2 + O_2 \rightarrow SO_3$ (all are gases)
   i. Combustion of magnesium metal in oxygen gas
   j. $Al(OH)_3 + H_2SO_4 \rightarrow Al_2(SO_4)_3 + H_2O$ in aqueous solution
   k. $Fe_2O_3 + CO \rightarrow Fe + CO_2$
   l. $Cu + AgNO_3 \rightarrow Ag + Cu(NO_3)_2$ in aqueous solution

3. In Problem 2, which are redox reactions? Where redox is involved, what is the oxidizing agent, and what is the reducing agent?

4. It has been said that oxidation and reduction are complementary processes. Explain.

5. How is it possible for oxidation and reduction to take place at separate locations, even though each process drives the other?

6. Give two important reasons for using aluminum in the construction of airplanes.

7. Distinguish between the two types of electrochemical cells: the galvanic cell and the electrolytic cell.

8. Explain the movement of sulfate ions in Figure 4.6.

9. What changes always take place at an anode? At a cathode?

10. Why is a salt required in an electrolytic cell?

11. When a piece of jewelry is being electroplated, is the piece used as the anode or the cathode? Explain.

12. How does galvanization inhibit corrosion?

13. Why is stainless steel resistant to corrosion?

14. Why does aluminum resist corrosion?

15. During the rusting of iron, what is oxidized and what is reduced? Write the reaction for each.

16. How does salt (used to melt ice and snow) assist in rusting an automobile?

Mole and Mass Relationships

5

When the time comes to carry out a chemical reaction, it is necessary to understand certain quantitative relationships among the reactants and products. In this chapter, the concept of the mole and the subject of stoichiometry are introduced so that you can understand the thought process that chemists use to plan the conduct of reactions and to predict the quantitative results.

5.1 Molecular Mass

GOAL: To understand how to determine molecular mass and formula mass.

As noted in Section 2.3, we may state the atomic mass of a particular isotope or the average atomic mass of all natural isotopes for any element. In speaking of individual atoms, the usual unit of mass is the atomic mass unit (amu). For example, the average atomic mass of carbon is shown in the periodic table as 12.011 amu.

Now that we are considering reactions among several chemical *compounds,* we need information about the masses of molecules, not just atoms. The molecules of any compound are made up of a specific number of atoms of each element according to the law of constant composition. Therefore, the mass associated with any molecular formula can be determined by adding the atomic masses of the component atoms.

The water molecule is an example. The mass of a molecule of water, called the **molecular mass,** can be obtained by adding the atomic masses of the two hydrogen atoms and the one oxygen atom of H_2O.

$$1 \text{ molecule of water } (H_2O) = \begin{array}{c} 2 \text{ hydrogen atoms} \\ + \\ 1 \text{ oxygen atom} \end{array}$$

$$= \begin{array}{c} 2 \times \text{atomic mass of hydrogen} \\ + \\ 1 \times \text{atomic mass of oxygen} \end{array}$$

$$= \begin{array}{c} 2 \ (1.0 \text{ amu}) \\ + \\ 1 \ (16.0 \text{ amu}) \end{array} = \begin{array}{c} 2.0 \text{ amu} \\ + \\ 16.0 \text{ amu} \end{array} = 18.0 \text{ amu}$$

This tells us that the molecular mass of water is 18 amu.

This approach can be used to determine the molecular mass of any other substance, even complex compounds such as glucose, $C_6H_{12}O_6$.

$$1 \text{ molecule of glucose } (C_6H_{12}O_6) = \begin{array}{c} 6 \text{ carbon atoms} \\ + \\ 12 \text{ hydrogen atoms} \\ + \\ 6 \text{ oxygen atoms} \end{array}$$

$$= \begin{array}{c} 6 \times \text{atomic mass of carbon} \\ + \\ 12 \times \text{atomic mass of hydrogen} \\ + \\ 6 \times \text{atomic mass of oxygen} \end{array}$$

$$= \begin{array}{c} 6 (12.0 \text{ amu}) \\ + \\ 12 (1.0 \text{ amu}) \\ + \\ 6 (16.0 \text{ amu}) \end{array} = \begin{array}{c} 72.0 \text{ amu} \\ + \\ 12.0 \text{ amu} \\ + \\ 96.0 \text{ amu} \end{array} = 180 \text{ amu}$$

Therefore, the molecular mass of glucose is 180 amu.

For some substances, the molecule is not a meaningful unit. An ionic compound such as NaCl does not exist as discrete molecules, but rather as the separated ions Na^+ and Cl^-, either in solution or in a crystal structure in which each Cl^- is surrounded by six Na^+ and each Na^+ is surrounded by six Cl^-. It is for this reason that the phrase *formula unit* may be used instead of the word molecule. Nevertheless, one chloride ion is always present for every sodium ion in any sample of sodium chloride. Although we should always describe sodium chloride in terms of formula units and *formula mass,* the phrase molecular mass is also acceptable.

The crystal structure of NaCl is described in Chapter 7.

$$1 \text{ formula unit of sodium chloride } (NaCl) = \begin{array}{c} 1 \text{ sodium ion} \\ + \\ 1 \text{ chloride ion} \end{array}$$

$$\begin{array}{c} 1 \times \text{atomic mass of sodium} \\ + \\ 1 \times \text{atomic mass of chlorine} \end{array}$$

$$\begin{array}{c} 1 (23.0 \text{ amu}) \\ + \\ 1 (35.5 \text{ amu}) \end{array} = \begin{array}{c} 23.0 \text{ amu} \\ + \\ 35.5 \text{ amu} \end{array} = 58.5 \text{ amu}$$

Therefore, the formula mass of sodium chloride is 58.5 amu.

5.2 The Mole

GOAL: To understand the concept of the mole and why it is useful.

The atomic mass unit scale converts cumbersome numbers into values that are more easily handled. For example, one atom of ^{12}C with a mass of 19.92×10^{-24} g becomes one atom of ^{12}C with a mass of 12.00 amu after conversion to the atomic mass scale. But the atomic mass unit still does not make the weighing of atoms and molecules practical in the chemistry laboratory. For this purpose, chemists have devised another measurement unit that relates the number of atoms or molecules in a sample to the mass of that sample in grams. This unit is the mole.

A **mole** (abbreviated **mol**) is that quantity of a substance (element, compound, ion, etc.) that contains the *same number* of particles or formula units as there are atoms of ^{12}C in exactly 12 g of a pure sample of ^{12}C.

Recall that *one ^{12}C atom has a mass of 12 amu, or 19.92×10^{-24} g*. This fact allows us to calculate the number of ^{12}C atoms in a sample with a mass of 12 g—the quantity defined as 1 mol.*

$$12 \text{ g } ^{12}C \times \frac{1 \text{ atom } ^{12}C}{19.92 \times 10^{-24} \text{ g } ^{12}C} = 6.02 \times 10^{23} \text{ atoms } ^{12}C$$

This calculation says that there are 6.02×10^{23} atoms in 12 g of ^{12}C. Similarly, 1 mol of any other substance also contains 6.02×10^{23} units.

The number 6.02×10^{23} is known as *Avogadro's number,* in recognition of Amadeo Avogadro's work in developing the mole concept. The abbreviation N_A is commonly used.

$$\text{Avogadro's number} = N_A = 6.02 \times 10^{23} = \text{the number of units}$$
$$\text{of any substance}$$
$$\text{present in 1 mol}$$

In other words, 6.02×10^{23} units of anything is equal to 1 mol. Thus, 6.02×10^{23} molecules of water is 1 mol of water. Similarly, 6.02×10^{23} atoms of silver is 1 mol of silver. It is even true, by definition, that 6.02×10^{23} baseballs is 1 mol of baseballs. The mole could conceivably be applied to any substance, but it is normally used only in the sciences, where the extremely small size and mass of atoms and molecules make it a convenient quantity with which to work. In this context, the following relationships are useful:

$$\begin{matrix} \textbf{1 mol of} \\ \textbf{any element} \end{matrix} = \begin{matrix} \textbf{6.02} \times \textbf{10}^{23} \textbf{ atoms} \\ \textbf{of that element} \end{matrix}$$

$$\begin{matrix} \textbf{1 mol of} \\ \textbf{any compound} \end{matrix} = \begin{matrix} \textbf{6.02} \times \textbf{10}^{23} \textbf{ molecules} \\ \textbf{or formula units} \\ \textbf{of that compound} \end{matrix}$$

---

*If

$$1 \text{ atom } ^{12}C = 19.92 \times 10^{-24} \text{ g } ^{12}C$$

then

$$\frac{1 \text{ atom } ^{12}C}{19.92 \times 10^{-24} \text{ g}} = 1$$

Since this ratio is equal to 1, it can be used as a conversion factor to be multiplied by any other number without changing the size of that number. This procedure of taking an equality and using it as a conversion factor is the basis of the factor-label analysis method. This method was discussed in Section 1.3.

5.3 **Molar Mass**

GOALS: To understand how to determine molar mass.
To interconvert the mass and the number of moles of any substance.

Although a mole of any element or any compound contains the same number of units, the mass of a mole, or **molar mass,** depends on the particular substance. When dealing with an element (rather than compound) the molar mass is numerically equal to the atomic mass. The only difference is that the units are changed from atomic mass units to grams. For example, carbon has an atomic mass of 12.011 amu and a molar mass of 12.011 g.

An exception to this rule occurs when the elements exist as diatomic molecules, such as H_2, N_2, O_2, F_2, Br_2, Cl_2, and I_2. For these substances, 1 mol contains 2 mol of atoms and the molar mass is twice the atomic mass. For example, the atomic mass of oxygen is 16 amu, whereas the molar mass is 32 g.

Molecular masses are expressed in atomic mass units. Molar masses are given in grams.

For all compounds, including the diatomic elements, the *molecular mass* is determined by totalling the atomic masses of the component atoms expressed in atomic mass units. The molar mass is determined in the same manner, except that the units are grams. Refer to Section 5.1 for a review of these calculations.

Numerically, the molecular mass and the molar mass are the same for any substance. Therefore, water has a molecular mass of 18 amu and a molar mass of 18 g. Carbon monoxide has a molecular mass of 28 amu and a molar mass of 28 g.

For sodium chloride, where the formula unit is more meaningful than the molecule, the formula mass and the molar mass are numerically equal—NaCl has a formula mass of 58.5 amu (calculated earlier) and a molar mass of 58.5 g.

Figure 5.1   A mole of each of seven common substances. The graduated cylinder in the background contains water (18 g, 18 mL). Starting from the front left and moving clockwise is rock salt (NaCl, 58 g), glucose ($C_6H_{12}O_6$, 148 g), zinc metal (65 g), aspirin ($C_9H_8O_4$, 554 tablets each containing 325 mg), powdered charcoal (12 g of carbon), and mothballs (128 g of naphthalene, $C_{10}H_8$).

Although the mole concept may seem abstract, you may perhaps appreciate its value from the items shown in Figure 5.1. One mole of each substance is present and the mass of each could be readily measured.

How Many Moles?

If we are going to use the mole as a convenient unit for measuring the amount of material, then we have to have some way of determining how many moles of any substance are present in a sample. We cannot directly measure the number of moles, but we can always determine the mass of a sample and calculate the number of moles from the mass. The calculation is carried out by the factor-label method using molar mass. Consider 1 mol of water:

$$1 \text{ mol } H_2O = 18 \text{ g } H_2O$$

This equality gives us access to two conversion factors that may be used for converting between moles and grams of water.

$$\frac{1 \text{ mol } H_2O}{18 \text{ g } H_2O} = 1 \quad \text{and} \quad \frac{18 \text{ g } H_2O}{1 \text{ mol } H_2O} = 1$$

For example, if you have a sample of 10 g of water, the first conversion factor can be used to find the number of moles present:

$$10 \text{ g } H_2O \times \frac{1 \text{ mol } H_2O}{18 \text{ g } H_2O} = 0.56 \text{ mol}$$

If we have a sample of 100 g of NaCl, we could use the molar mass of this salt as a conversion factor to calculate the number of moles:

$$100 \text{ g NaCl} \times \frac{1 \text{ mol NaCl}}{58.5 \text{ g NaCl}} = 1.71 \text{ mol}$$

Suppose now that we wanted to know the mass of 0.75 mol of ethyl alcohol (C_2H_5OH). We must first add the atomic masses for this compound to arrive at a molecular mass of 46 amu or a molar mass of 46 g. The factor-label method allows for the calculation of the mass of 0.75 mol of C_2H_5OH:

$$0.75 \text{ mol } C_2H_5OH \times \frac{46 \text{ g } C_2H_5OH}{1 \text{ mol } C_2H_5OH} = 34.5 \text{ g}$$

Notice that this form of the conversion factor is the reciprocal of that used when converting from mass to moles. This form causes the units of ''moles of C_2H_5OH'' to cancel, giving an answer with the proper units of mass.

Both the conversion from moles to grams and the conversion from grams to moles are required to describe the quantitative relationships involved in chemical reactions. This is illustrated in the following section.

5.4   Mole and Mass Relationships in Chemical Reactions

GOALS: To understand the mole ratio in which substances react and form as expressed by
a balanced chemical equation.
To convert the mass of one substance into the corresponding mass of another in
the same reaction using the mole-ratio method.
To understand the significance of excess reagents in chemical reactions.
To determine the limiting reagent in a chemical reaction.
To determine the excess reagent in a chemical reaction from the quantities used.
To determine the amount of each substance present when a chemical reaction
stops.

Up to this point, our discussions of chemical reactions have focused on writing and balancing equations and on characterizing certain types of reactions. Now let us consider the mass and mole relationships among substances that enter into chemical reactions. In this section, we will consider the topic of **stoichiometry**, which is the application of the mole concept to determining the quantitative results of chemical reactions. The term stoichiometry originates from the Greek *stoicheion,* ''element,'' and *metron,* ''measure.''

In studying stoichiometry, we explore questions such as, How much of a particular product can be formed by a reaction? or, If we have only a limited amount of reactant to carry out a reaction, how will this determine the amount of product that will form? or, If we have to synthesize a chemical by carrying out a chemical reaction, how much of each of the starting materials will be required to achieve the desired result? All of these considerations fall under the heading of stoichiometry.

Equation Relationships

The coefficients in a balanced chemical equation tell the simplest mole ratio in which substances react and form in a chemical change. For example, the reaction of zinc metal with hydrochloric acid and the quantities represented by the balanced equation are shown below, using (a) a standard chemical equation, (b) the mole relationships, and (c) the mass relationships calculated from the molar masses.

a.	Zn	$+$	$2\ HCl$	$\rightarrow$	$ZnCl_2$	$+$	H_2
b.	1 mol	reacts with	2 mol	to produce	1 mol	and	1 mol
c.	65 g	reacts with	73 g	to produce	136 g	and	2 g

Note the adherence to the
law of conservation of
mass (Section 2.1) in c.

But rarely are chemical reactions carried out in the laboratory using exactly the number of moles of each substance stated by an equation. It is not necessary to use exactly 65 g (1 mol) of zinc every time this reaction is performed, and it is unrealistic to assume that 136 g of zinc chloride will always be the amount of product that is desired.

Many other workable mole and mass relationships can be deduced from the balanced equation. For example, 10 mol of HCl must react for 5 mol of hydrogen gas to be produced. Or, 6.5 g of Zn will react with 7.3 g of HCl to produce 13.6 g of $ZnCl_2$ and 0.2 g of H_2.

Zn	+	2 HCl	→	ZnCl$_2$	+	H$_2$
1 mol	reacts with	2 mol	to produce	1 mol	and	1 mol

or,

		10 mol				5 mol

or,

65g	+	73 g	→	136 g	+	2g

or,

6.5g		7.3g				

Suppose we have to produce 200 g of zinc chloride by this reaction. In this case, the quantity of zinc (or HCl) required is not immediately obvious. But if we understand the equation and the method of factor-label analysis, the required mass of each reactant can be readily calculated. A *mole-ratio method* is commonly used for this type of *mass-to-mass* calculation. Thus, a known mass of one substance in a chemical reaction (for example, 200 g of ZnCl$_2$) is used to determine the mass of another substance involved in the same reaction.

The mole-ratio method uses the mole relationship expressed by a balanced equation as its conversion factor. The method is illustrated by the following example.

EXAMPLE 5.1:

Determine the mass of zinc metal that must react with hydrochloric acid to form 200 g of zinc chloride.

Solution:

1. The 200 g of zinc chloride specified in the problem is converted into moles using the molar mass of ZnCl$_2$.

$$200 \text{ g ZnCl}_2 \times \frac{1 \text{ mol ZnCl}_2}{136 \text{ g ZnCl}_2} = 1.47 \text{ mol ZnCl}_2$$

2. The quantity of zinc chloride desired, now expressed in moles, is converted into the number of moles of zinc required for its production using the mole relationship of the balanced equation.

$$1.47 \text{ mol ZnCl}_2 \times \frac{1 \text{ mol Zn}}{1 \text{ mol ZnCl}_2} = 1.47 \text{ mol Zn required}$$

3. The number of moles of zinc that are required is then converted into the corresponding mass using the molar mass of Zn.

$$1.47 \text{ mol Zn} \times \frac{65 \text{ g Zn}}{1 \text{ mol Zn}} = 96 \text{ g Zn required}$$

These steps of the mole-ratio method can be written more efficiently as one calculation with three conversion factors as follows.

$$200 \text{ g ZnCl}_2 \times \frac{1 \text{ mol ZnCl}_2}{136 \text{ g ZnCl}_2} \times \frac{1 \text{ mol Zn}}{1 \text{ mol ZnCl}_2} \times \frac{65 \text{ g Zn}}{1 \text{ mol Zn}} = 96 \text{ g Zn}$$

(desired)   (1)   (2)   (3)   (required)

In this calculation, conversion factor (1) comes from the molar mass of $ZnCl_2$ and converts the mass of $ZnCl_2$ into moles of $ZnCl_2$. Conversion factor (2) comes from the balanced equation, which shows that 1 mol of Zn causes formation of 1 mol of $ZnCl_2$ and converts moles of $ZnCl_2$ into moles of Zn. Conversion factor (3) comes from the molar mass of zinc and converts moles of Zn into grams of Zn.

The solution reveals that 96 g of zinc metal must be reacted if 200 g of zinc chloride is to be produced. This method is summarized in Table 5.1.

The mole-ratio method is further illustrated by the following examples.

EXAMPLE 5.2:

What mass of sulfuric acid can be neutralized by 150 g of sodium hydroxide, according to the following *unbalanced* equation? Neutralization, the reaction of an acid with a base to form a salt, is discussed in Chapter 7.

$$\underset{\text{base}}{NaOH} + \underset{\text{acid}}{H_2SO_4} \rightarrow \underset{\text{salt}}{Na_2SO_4} + \underset{\text{water}}{H_2O}$$

Solution:

1. Write a balanced equation for the reaction.

$$2 NaOH + H_2SO_4 \rightarrow Na_2SO_4 + 2H_2O$$

2. Convert the mass given into moles.

$$150 \text{ g NaOH} \times \frac{1 \text{ mol NaOH}}{40 \text{ g NaOH}} = 3.75 \text{ mol NaOH}$$

3. Convert moles of NaOH into moles of H_2SO_4 using the mole ratio from the chemical equation.

$$3.75 \text{ mol NaOH} \times \frac{1 \text{ mol } H_2SO_4}{2 \text{ mol NaOH}} = 1.88 \text{ mol } H_2SO_4 \text{ (can be reacted)}$$

4. Moles of the substance H_2SO_4 are converted to the corresponding mass.

$$1.88 \text{ mol } H_2SO_4 \times \frac{98 \text{ g } H_2SO_4}{1 \text{ mol } H_2SO_4} = 184 \text{ g } H_2SO_4$$

Or, the calculation may be written in the combined form, as follows.

$$150 \text{ g NaOH} \times \frac{1 \text{ mol NaOH}}{40 \text{ g NaOH}} \times \frac{1 \text{ mol H}_2\text{SO}_4}{2 \text{ mol NaOH}} \times \frac{98 \text{ g H}_2\text{SO}_4}{1 \text{ mol H}_2\text{SO}_4} = 184 \text{ g H}_2\text{SO}_4$$

Again, the solution says that 184 g of sulfuric acid can react with (be neutralized by) 150 g of sodium hydroxide.

EXAMPLE 5.3:

Determine the mass of potassium chlorate that must be decomposed for 25 g of oxygen gas to be liberated.

Solution:

1.
$$2KClO_3 \xrightarrow{\Delta} 2KCl + 3O_2$$

2.
$$25 \text{ g O}_2 \times \frac{1 \text{ mol O}_2}{32 \text{ g O}_2} = 0.78 \text{ mol O}_2$$

3.
$$0.78 \text{ mol O}_2 \times \frac{2 \text{ mol KClO}_3}{3 \text{ mol O}_2} = 0.52 \text{ mol KClO}_3$$

4.
$$0.52 \text{ mol KClO}_3 \times \frac{122.5 \text{ g KClO}_3}{1 \text{ mol KClO}_3} = 64 \text{ g KClO}_3$$

Or,

$$25 \text{ g O}_2 \times \frac{1 \text{ mol O}_2}{32 \text{ g O}_2} \times \frac{2 \text{ mol KClO}_3}{3 \text{ mol O}_2} \times \frac{122.5 \text{ g KClO}_3}{1 \text{ mol KClO}_3} = 64 \text{ g KClO}_3$$

Table 5.1   Mass-to-mass relationship calculations by the mole-ratio method

1. Write the balanced equation for the reaction.
2. Convert mass given in problem to moles.
3. Convert moles of substance just calculated into moles of substance sought by factor-label analysis using the mole ratio from the balanced equation.
4. Convert moles of substance sought into its corresponding mass.

Excess and Limiting Reagents

In most cases, the reagents used to carry out a chemical reaction in the laboratory are not combined in the exact ratio in which they react together. Frequently an excess of one starting material is used to ensure the complete conversion of the other reactant(s) into the desired product(s). For example, when silver metal is to be collected by reacting copper metal with silver nitrate, an excess of copper metal may be added to the silver nitrate to ensure maximum recovery of the silver.

The equation for this displacement indicates that if 64 g of copper is reacted with 340 g of silver nitrate, 216 g of silver metal can be produced.

Cu	+	2 AgNO$_3$	→	2 Ag	+	Cu(NO$_3$)$_2$
1 mol	reacts with	2 mol	to produce	2 mol	and	1 mol
64 g	reacts with	340 g	to produce	216 g	and	188 g

If 340 g of silver nitrate were to be reacted, a quantity of copper greater than 64 g would normally be used. Use of excess copper ensures complete reaction of the expensive silver nitrate and complete recovery of the silver metal. All of the silver nitrate would react. Excess copper would be present, along with the products, at the end of the reaction, and it can be recovered and used again. In this situation, the copper is said to be *in excess.* Since silver nitrate is completely consumed while some copper remains, the amount of product that forms is determined by the amount of silver nitrate present. For this reason, silver nitrate is called the *limiting reagent.* The reaction stops when the limiting reagent has been consumed and only excess reagent and products remain. The following examples further illustrate the processes involved with excess and limiting reagents.

EXAMPLE 5.4:

Consider the addition of 1 mol of lead nitrate to 3 mol of sodium chloride.

a. Determine the reagent that is in excess and the amount of that excess.
b. Determine the limiting reagent.
c. Determine the quantity of each reactant that will be consumed in the reaction.
d. Determine the quantity of each product that will be formed.

Solution:

The balanced equation:

Pb(NO$_3$)$_2$	+	2 NaCl	→	PbCl$_2$	+	2 NaNO$_3$
1 mol	reacts with	2 mol	to produce	1 mol	and	2 mol

a. One mole of Pb(NO$_3$)$_2$ can react with only 2 mol of NaCl, not with the entire 3 mol that are added. Therefore, 1 mol of NaCl is in excess.

b. All of the $Pb(NO_3)_2$ can react. It is therefore the limiting reagent.

c. From the equation and the quantities being used, 1 mol of $Pb(NO_3)_2$ will react with 2 mol of NaCl.

d. From the mole relationship of the reaction and the quantities being used, 1 mol of $PbCl_2$ and 2 mol of $NaNO_3$ will be produced.

EXAMPLE 5.5:

Determine the excess reagent and the limiting reagent when 50 g of Al_2O_3 is added to 150 g of H_2SO_4.

Solution:

$$Al_2O_3 \quad + \quad 3H_2SO_4 \quad \rightarrow \quad Al_2(SO_4)_3 \quad + \quad 3H_2O$$

1 mol   reacts   3 mol   to   1 mol   and   3 mol
       with   produce

Amounts used: $50 \text{ g } Al_2O_3 \times \dfrac{1 \text{ mol}}{102 \text{ g}} = 0.49 \text{ mol } Al_2O_3$

$$150 \text{ g } H_2SO_4 \times \dfrac{1 \text{ mol}}{98 \text{ g}} = 1.53 \text{ mol } H_2SO_4$$

According to the balanced equation, 1 mol of Al_2O_3 can react with 3 mol of H_2SO_4. Therefore, 0.49 mol of Al_2O_3 can react with 1.47 mol of H_2SO_4, determined as follows.

$$0.49 \text{ mol } Al_2O_3 \times \dfrac{3 \text{ mol } H_2SO_4}{1 \text{ mol } Al_2O_3} = 1.47 \text{ mol } H_2SO_4$$
$$\text{(can react)}$$

The H_2SO_4 is in excess of the amount that can react. The Al_2O_3 is the limiting reagent. The reaction will involve the following quantities.

$$0.49 \text{ mol } Al_2O_3 + 1.47 \text{ mol } H_2SO_4 \rightarrow 0.49 \text{ mol } Al_2(SO_4)_3 + 1.47 \text{ mol } H_2O$$

This would leave 0.06 (1.53 − 1.47) mol of H_2SO_4 unreacted.

Using molar masses, these quantities can also be expressed in grams.

$$50 \text{ g } Al_2O_3 + 144 \text{ g } H_2SO_4 \rightarrow 168 \text{ g } Al_2(SO_4)_3 + 26 \text{ g } H_2O$$

Unreacted $H_2SO_4 = 6$ g, determined as follows:

$$0.06 \text{ mol } H_2SO_4 \times \dfrac{98 \text{ g}}{1 \text{ mol}} = 6 \text{ g}$$

SUMMARY

The mass of a molecule, known as the molecular mass, can be determined by adding the atomic masses of the component atoms. For substances that do not exist as molecular units, such as NaCl, the sum of the atomic masses is known as the formula mass.

A mole of any substance contains Avogadro's number, 6.02×10^{23}, of units of that substance. One mole of any element contains 6.02×10^{23} atoms of that element, and 1 mol of any compound contains 6.02×10^{23} molecules or formula units of that compound. Except for diatomic elements, the molar mass is numerically equal to the atomic mass of an element or the molecular mass of a compound. The number of moles and the mass of a substance are related by the molar mass, so that either one can be calculated from the other using the factor-label method.

The coefficients in a balanced chemical equation state the simplest whole-number ratio in which substances react and form in a chemical change. This mole ratio and the mass ratio that it represents constitute the stoichiometry of the reaction.

The mole-ratio method is commonly used to calculate mass relationships in chemical reactions. It uses the mole relationship expressed by a balanced equation as its conversion factor.

When the reactants in a chemical change are not combined in the exact ratio required by the balanced equation, the substance that will be entirely converted into products is called the limiting reagent. The limiting reagent determines the amount of product(s) formed. A reaction stops when the limiting reagent has been consumed. Only products and reagent(s) used in excess remain.

PROBLEMS

1. Give a definition or an example of each of the following.

   a. Molecular mass
   b. Molar mass
   c. Mole
   d. Limiting reagent

2. Determine the molecular mass (or formula mass) and the molar mass of each of the following.

   a. H_2S
   b. $CaCl_2$
   c. Ethyl alcohol, C_2H_5OH
   d. O_2
   e. Table sugar (sucrose), $C_{12}H_{22}O_{11}$
   f. $Cu(NO_3)_2$

3. Determine the number of moles in each of the following.

   a. 98 g H_2SO_4
   b. 17 g BF_3
   c. 100 g octane, C_8H_{18}
   d. 1.7 g $NaNO_3$
   e. 25 g Br_2
   f. 26 g styrene, C_8H_8

4. Determine the mass of each of the following.

   a. 0.2 mol H_2O
   b. 2.2 mol MnO_2
   c. 6 mol $MgBr_2$
   d. 0.06 mol H_2SO_4

5. Consider the reaction of aluminum with chlorine, Cl_2, to form aluminum chloride.

   a. Write a balanced equation for the reaction.
   b. Calculate the number of moles of chlorine required to react with 2.0 mol of aluminum.
   c. Calculate the number of moles of chlorine required to react with 5.0 mol of aluminum.
   d. Calculate the number of moles of aluminum chloride produced by the reaction of 3.0 mol of chlorine.
   e. Calculate the number of moles of aluminum chloride produced by the reaction of 7.5 mol of chlorine.
   f. Calculate the mass of aluminum that is required to produce 12.0 mol of aluminum chloride.
   g. Calculate the mass of chlorine required to react with 10.8 g of aluminum.

6. a. Write a complete and balanced equation for the reaction of silver nitrate, $AgNO_3$, with magnesium chloride.
   b. How many moles of silver nitrate are required to produce 5.0 mol of magnesium nitrate?
   c. How many moles of magnesium nitrate are produced by the reaction of 7.0 mol of silver nitrate?
   d. What mass of silver nitrate is required for 37.8 g of magnesium nitrate to be produced?
   e. What mass of silver chloride is produced by the reaction of 47.5 g of magnesium chloride?

7. If 3.6 g of magnesium is burned in air, what mass of magnesium oxide is formed?

8. Calcium carbide, CaC_2, reacts with water to produce acetylene gas, C_2H_2, according to the following unbalanced

equation.

$$CaC_2 + H_2O \rightarrow Ca(OH)_2 + C_2H_2$$

   a. How many moles of acetylene are produced by the reaction of 5.0 mol of calcium carbide?

   b. What mass of water must react with calcium carbide for 3.9 g of acetylene gas to form?

9. Solid barium carbonate can be prepared in the laboratory by reacting barium chloride with sodium carbonate. What mass of barium chloride must be used if 9.85 g of barium carbonate is desired?

10. The reaction of the air pollutant, sulfur dioxide, with oxygen in the air forms an even more noxious gas, sulfur trioxide. A student combines 64 g of SO_2 and 64 g of O_2.

   a. Write a complete and balanced equation for the oxidation of SO_2.

   b. How many moles of sulfur dioxide are being used?

   c. How many moles of oxygen are being used?

   d. Which is the limiting reagent?

   e. Which reagent is in excess? By how many moles? By how many grams?

   f. How many moles of each substance will react?

   g. What mass of each substance will react?

   h. How many grams of the limiting reagent would be required for both reactants to be consumed totally?

11. 153.3 g of HCl is combined with 101.4 g of $Al(OH)_3$.

$$HCl + Al(OH)_3 \rightarrow AlCl_3 + H_2O \text{ (unbalanced)}$$

   a. How many moles of HCl are being used?

   b. How many moles of $Al(OH)_3$ are being used?

   c. How many moles of $Al(OH)_3$ are required to react with the 153.3 g of HCl?

   d. Which is the limiting reagent?

   e. Which reagent is in excess? By how many moles? By how many grams?

   f. What mass of each reagent will react?

   g. What mass of aluminum chloride will be formed by the reaction?

12. If 23.40 g of NaCl is added to 26.48 g of $Pb(NO_3)_2$,

   a. Which reagent is in excess?

   b. How many moles of each reagent will react?

   c. What mass of each reagent will react?

   d. What mass of $PbCl_2$ will be produced?

13. What mass of oxygen is required to react with 10.8 g of aluminum to form Al_2O_3? What mass of aluminum oxide will be formed by this reaction?

14. A strip of copper with a mass of 0.64 g is reacted with an excess of silver nitrate.

   a. How many moles of silver can be produced?

   b. What mass of silver can be recovered?

15. During photosynthesis, carbon dioxide and water are converted into glucose and oxygen.

$$CO_2 + H_2O \rightarrow C_6H_{12}O_6 + O_2 \text{ (unbalanced)}$$

   a. What mass of carbon dioxide is required to produce 0.90 kg of glucose?

   b. What mass of oxygen gas is released by the reaction specified in part (a)?

16. When sodium metal reacts with water, sodium hydroxide and hydrogen gas are produced.

   a. What mass of sodium is required for 12.0 g of sodium hydroxide to be produced?

   b. What mass of water will be consumed by the reaction of 6.9 g of sodium?

17. Hydrogen gas was prepared in the laboratory by reacting zinc metal with sulfuric acid, H_2SO_4. What mass of the acid must be used to react completely with 19.5 g of zinc?

18. What mass of silver chloride will be prepared by the reaction of 85.0 g of silver nitrate with an excess of barium chloride?

19. Heating a sample of potassium chlorate, $KClO_3$, released 5.24 g of oxygen gas.

   a. How many molecules of oxygen were released?

   b. What mass of potassium chlorate was present in the original mixture?

20. A mixture of 8.0 g of hydrogen gas and 8.0 g of oxygen gas is exploded to form water vapor.

   a. Which reagent is in excess?

   b. What mass of the excess reagent will remain at the end of the reaction?

   c. What mass of water is formed by the reaction?

21. What mass of calcium chloride will be produced when 5.0 mol of calcium hydroxide is added to 5.0 mol of hydrochloric acid?

22. A student adds 100 g of sodium hydroxide to 1 mol of sulfuric acid. What mass of sodium sulfate will be produced?

23. What mass of copper must be added to a solution containing 17.0 g of silver nitrate if 0.5 g of copper is to remain as excess?

24. 10.0 g of Zn is added to 100 mL of a solution of HCl in water that contains 0.40 mol of HCl.

   a. How many moles of Zn are being used?

   b. How many moles of HCl are required to react with the 10.0 g of Zn?

   c. How many moles of Zn are required to react with the available HCl solution?

   d. Which reagent is in excess? How many moles of this reagent will remain at the end of the reaction?

   e. How many moles of each substance will react?

   f. How many moles of H_2 will be produced by the reaction?

   g. What mass of $ZnCl_2$ will be produced by the reaction?

Organic Chemistry

6

In earlier chapters we considered the formulas, names, and reactions of many chemical compounds, both ionic and covalent. In this chapter we will concentrate on compounds of carbon, most of which are classified as organic. This suggests an association between organic compounds and living organisms. Fats (see Chapter 21), carbohydrates (see Chapter 9), and proteins (see Chapter 8) are examples of organic compounds present in the cells of all organisms.

Petroleum and coal are fossil fuels (see Chapters 12 and 13) formed long ago from decayed organic matter, both plant and animal. Many organic chemicals can be obtained directly from petroleum; many others are synthesized from starting materials that are derived from petroleum. By comparison, inorganic chemicals are derived from minerals.

Entire books and courses are devoted to the study of organic chemistry even at an elementary level, but we can do little more than scratch the surface. But an educated consumer can reasonably expect to be able to understand organic formulas and names, and to differentiate among the many types of organic compounds.

Although organic and inorganic compounds may be studied as two separate classes, there is much interplay between them. For one thing, many compounds have both an organic and an inorganic component. In addition, metals (inorganic) form many compounds by combining with organic compounds, and reactions of organic compounds are usually reactions with inorganic compounds.

Nature illustrates the strong connection between organic and inorganic molecules. Plants consume inorganic nutrients from the soil and the air (oxygen gas) and synthesize organic substances such as carbohydrates and proteins. Animals consume these substances and generate waste. Sometimes, this waste can be returned to the soil as fertilizer in the form of manure or compost, where soil microorganisms convert it back from organic matter into inorganic ions such as nitrate and phosphate (see Chapter 15 for details). Therefore, the distinction between organic and inorganic compounds is clear, but neither can be studied without an appreciation of both.

Covalent bonds are the principal feature of compounds containing carbon. There is no simple ion of carbon, but carbon does exist in some important combination ions, such as bicarbonate and carbonate, both derived from carbonic acid, H_2CO_3.

$$\underset{\substack{\text{carbonic}\\\text{acid}}}{HO-\overset{\overset{\displaystyle O}{\|}}{C}-OH} \qquad \underset{\substack{\text{bicarbonate}\\\text{ion}}}{HO-\overset{\overset{\displaystyle O}{\|}}{C}-O^-} \qquad \underset{\substack{\text{carbonate}\\\text{ion}}}{{}^-O-\overset{\overset{\displaystyle O}{\|}}{C}-O^-}$$

In each of these, the carbon-oxygen bonds are covalent. All are regarded as inorganic compounds, like all of the salts mentioned in earlier chapters. More often, carbon appears in organic compounds, which are characterized by covalent carbon-hydrogen bonds. Even more important, almost all organic compounds contain carbon-carbon bonds, which may be single, double, or even triple bonds. The following compounds illustrate the bonding.

$$\underset{\text{ethane}}{H-\overset{\overset{\displaystyle H}{|}}{\underset{\underset{\displaystyle H}{|}}{C}}-\overset{\overset{\displaystyle H}{|}}{\underset{\underset{\displaystyle H}{|}}{C}}-H} \qquad \underset{\text{ethylene}}{\overset{H}{\underset{H}{\diagdown}}C=C\overset{H}{\underset{H}{\diagup}}} \qquad \underset{\text{acetylene}}{H-C\equiv C-H}$$

In all stable compounds of carbon, the magic number is four. That is, carbon almost always appears with four covalent bonds; it is therefore said to be **tetravalent.** In this arrangement, every carbon atom obeys the octet rule. Even carbon dioxide has four covalent bonds.

$$:\!\overset{..}{O}\!=\!C\!=\!\overset{..}{O}\!: \qquad \text{carbon dioxide}$$

Organic chemicals are abundant. They include (1) synthetic chemicals, many of which are synthesized from (2) petroleum and coal, which were synthesized long ago from formerly living organisms; and (3) compounds synthesized by living organisms, both plants and animals. Among the synthetic organic chemicals are fibers, paints, soaps, detergents, drugs, plastics, and many other substances that are important to the consumer. Organic chemicals are the main components of foods, including fats, carbohydrates, and proteins. All living organisms consist of these same substances.

The literally millions of organic compounds make organic chemistry a major subdiscipline of chemistry. Fortunately, they have common features, so you do not have to learn many of them to understand the meaning of organic formulas. Furthermore, each organic compound may undergo many chemical reactions. However, most reactions are not specific to just a single compound, so that it is not necessary to learn hundreds of different chemical reactions. For example, although thousands of compounds might be classified as organic acids, all undergo a similar change when they react with bases.

6.1 · Methane and Its Derivatives

GOALS: To appreciate the tetrahedral nature of saturated carbon atoms.
To understand and write condensed formulas and bond-line formulas.
To write IUPAC names for constitutional isomers of alkanes.
To appreciate the effect of increasing molecular weight and branching on the boiling points of hydrocarbons.
To understand the differences among the various petroleum fractions, and relate their physical properties and uses.
To describe the importance of cracking and catalytic reforming in the production of gasoline.
To understand the octane rating scale, and appreciate the relationship between structure and octane of hydrocarbons.

The simplest organic compounds contain only carbon and hydrogen, and are known as **hydrocarbons.** The simplest hydrocarbon, **methane,** contains only a single carbon atom. Since carbon must be tetravalent, the formula must be CH_4. This formula also acknowledges that hydrogen is **monovalent,** since it forms only one covalent bond at any time.

$$H\!:\!\overset{\overset{H}{\cdot\cdot}}{\underset{\cdot\cdot}{C}}\!:\!H \;=\; H-\overset{\overset{\textstyle H}{|}}{\underset{\underset{\textstyle H}{|}}{C}}-H \;=\; CH_4$$

methane

Some other atoms often present in organic compounds, and their normal valences are listed in Table 6.1. Examples of such atoms in compounds derived from methane are methyl chloride, methyl alcohol (methanol), methyl amine, and dimethyl ether.

$$H-\overset{\overset{\textstyle H}{|}}{\underset{\underset{\textstyle H}{|}}{C}}-H = CH_4 \qquad H-\overset{\overset{\textstyle H}{|}}{\underset{\underset{\textstyle H}{|}}{C}}-Cl = CH_3Cl \qquad H-\overset{\overset{\textstyle H}{|}}{\underset{\underset{\textstyle H}{|}}{C}}-O-H = CH_3OH$$

methane   methyl chloride   methyl alcohol
                                                                                                    (methanol)

C is tetravalent   C is tetravalent   C is tetravalent
H is monovalent   H is monovalent   H is monovalent
                                   Cl is monovalent   O is divalent

$$H-\overset{\overset{\textstyle H}{|}}{\underset{\underset{\textstyle H}{|}}{C}}-\overset{\overset{\textstyle H}{\diagup}}{\underset{\underset{\textstyle H}{\diagdown}}{N}} = CH_3NH_2 \qquad H-\overset{\overset{\textstyle H}{|}}{\underset{\underset{\textstyle H}{|}}{C}}-O-\overset{\overset{\textstyle H}{|}}{\underset{\underset{\textstyle H}{|}}{C}}-H = CH_3OCH_3$$

methyl amine   dimethyl ether

C is tetravalent   C is tetravalent
H is monovalent   H is monovalent
N is trivalent   O is divalent

Table 6.1   Normal valence of atoms common to organic compounds

Element	Normal Valence
H	1
C	4
O	2
N	3
X (F, Cl, Br, I)	1
S	2
P	3 or 5

Not only is carbon tetravalent, but the four hydrogens of methane occupy the four corners of a tetrahedron. This **tetrahedral** arrangement allows the four atoms to be separated by the greatest possible distance (Figure 6.1).

Higher Hydrocarbons

The hydrocarbon series continues with **ethane, propane,** and **butane,** chains of two, three, and four carbons.

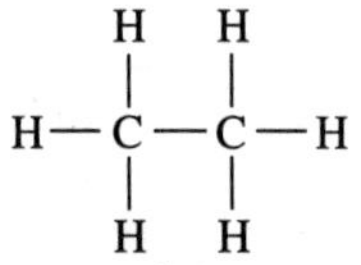

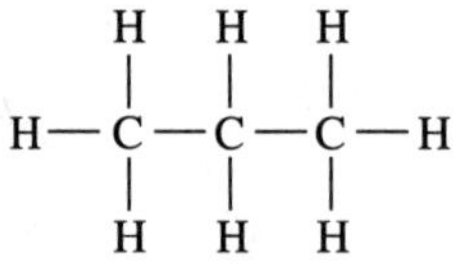

Ball-and-stick models of these four compounds are shown in Figure 6.2.

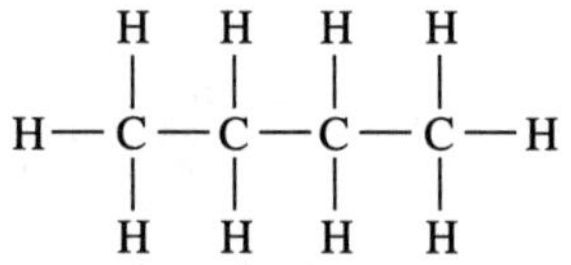

At four carbons, a variation in structure is possible. The carbon skeleton can be a simple chain or could consist of a three-carbon chain with the fourth carbon as a branch. This compound is a **constitutional isomer** (other form) of butane known as **isobutane** or 2-methyl propane; it has the same molecular formula (C_4H_{10}) as the structure with an unbranched chain. Constitutional isomers differ from one another in their connectivity; that is, in the way that the carbon atoms are bonded together.

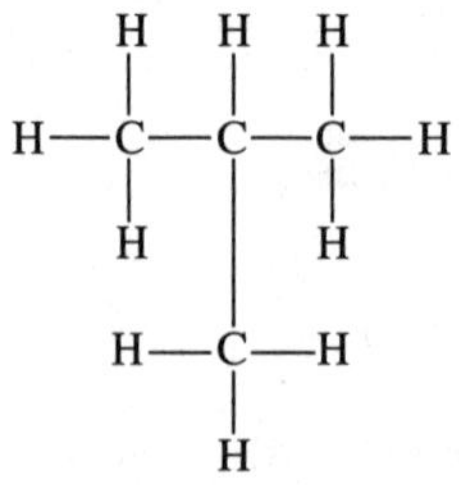

isobutane
(2-methylpropane)

Butane (unbranched) and isobutane can also be compared from their ball-and-stick models (Figure 6.3).

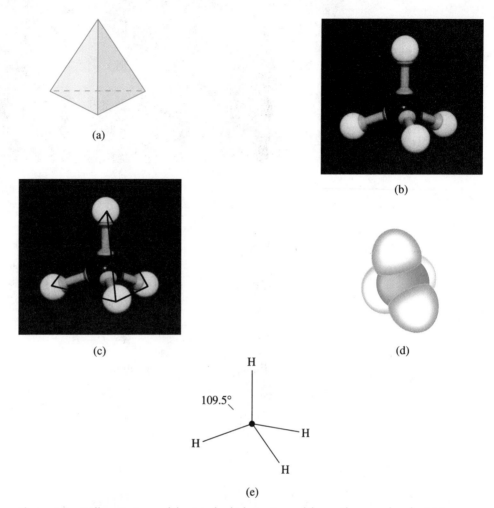

Figure 6.1   Different views of the tetrahedral structure of the methane molecule: (a) is a tetrahedron; (b) is a ball-and-stick model of methane; (c) shows how methane conforms to the tetrahedral shape with the four hydrogen atoms at the corners and the carbon atom in the center; (d) a space-filling model of methane; and (e) emphasizes the tetrahedral angle of 109.5 °, the maximum separation possible for four atoms around a central atom.

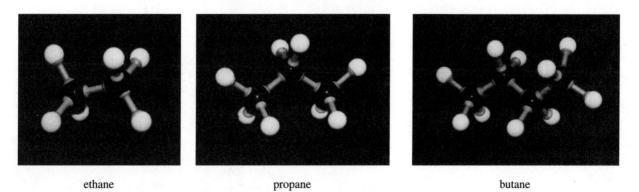

Figure 6.2   Ball-and-stick models of ethane, propane, and butane.

butane   isobutane

Figure 6.3   Ball-and-stick models of butane and isobutane (2-methylpropane).

When there are five carbon atoms in the molecule, the number of possible constitutional isomers becomes three. The five-carbon hydrocarbons are called **pentanes,** C_5H_{12}.

The six-carbon hydrocarbons are called **hexanes;** five isomeric forms (I–V) are possible as shown; all have the molecular formula C_6H_{14}.

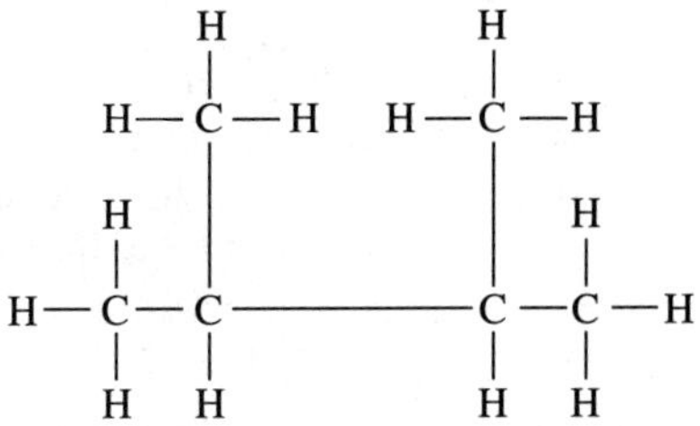

Several shorthand notations for representing these five isomers have been devised, since the complete structural formulas become unwieldy. Consider compound I.

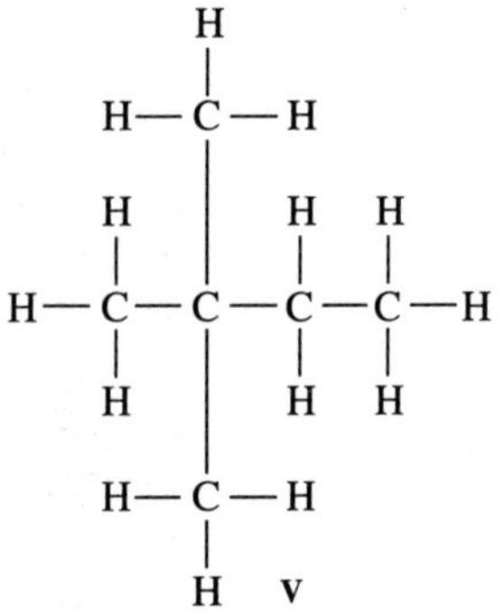

Compound I

a. Carbon skeleton $C-C-C-C-C-C$

b. Bond-line formula

c. Condensed formula $CH_3CH_2CH_2CH_2CH_2CH_3$

d. Abbreviated condensed formula $CH_3(CH_2)_4CH_3$

The a representation shows only the carbon skeleton. It is easy to write, but it is incomplete and should be avoided.

The abbreviation in b, the **bond-line formula,** is frequently used for compounds containing long carbon chains. The end of each line in this abbreviation represents a carbon atom. Recalling that carbon is tetravalent, we see that this abbreviation assumes each of the two terminal carbon atoms carries three hydrogen atoms. Each of the carbon atoms within the chain has two hydrogen atoms attached.

Notation c, a **condensed formula,** seems more cumbersome to write. However, it shows a chain of carbon atoms with hydrogens added to make each carbon tetravalent. Note also that although the bonds normally are not shown when writing a condensed formula, occasionally some bonds are written for emphasis or clarity.

Formula d is a further abbreviation of c in which the repeating CH_2 (methylene) groups are combined and counted; the number of groups is given by the subscript after the parentheses.

Next consider compound II.

Compound II

a.

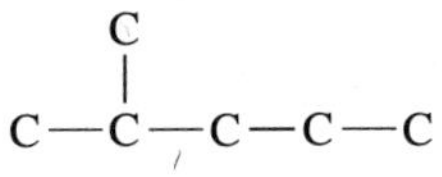

b.

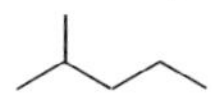

c. 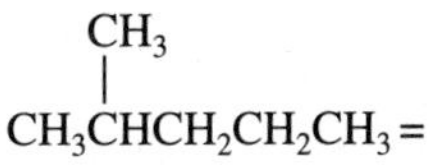
$CH_3CHCH_2CH_2CH_3 = CH_3CH(CH_3)CH_2CH_2CH_3 = (CH_3)_2CHCH_2CH_2CH_3$

d. 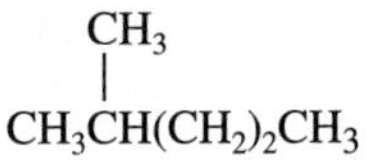
$CH_3CH(CH_2)_2CH_3$

Compounds I and II are isomers, one with an unbranched chain, one with a CH_3 (methyl) group as a branch off the main chain. Both have the molecular formula C_6H_{14}. The bond-line formula, b, includes a line intersecting the second carbon to represent the methyl group. A condensed formula can be written in more than one way, as shown in c. The first representation is easier to write and to interpret. The other two drawings are called *single-line formulas.* Placing the methyl branch in parentheses is acceptable, but potentially confusing. For example, the third formula in c should not be equated to the impossible structure on the right below.

$$(CH_3)_2CHCH_2CH_2CH_3 \neq CH_3CH_3CHCH_2CH_2CH_3$$

pentavalent carbon

trivalent carbon

The confusion is readily avoided if you remember that carbon is always tetravalent. We will use parentheses only to abbreviate chains of repeating methylene groups as in formula d for compound II.

Structural formulas for compounds III through V are also shown. When methyl groups or other branches are present, it does not matter if groups are written above or below the chain. Since the four atoms attached to each carbon describe a tetrahedral shape, all four positions around each carbon are equal. Abbreviated condensed formulas are not possible since there are no repeating CH_2 (methylene) groups in these compounds.

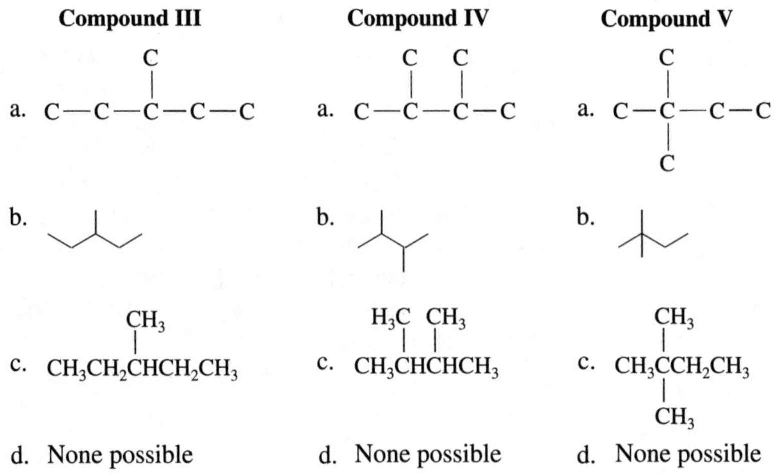

Compound III	Compound IV	Compound V
a. (structure)	a. (structure)	a. (structure)
b. (bond-line)	b. (bond-line)	b. (bond-line)
c. $CH_3CH_2CHCH_2CH_3$ with CH_3 branch	c. $CH_3CHCHCH_3$ with H_3C CH_3 branches	c. $CH_3CCH_2CH_3$ with two CH_3 branches
d. None possible	d. None possible	d. None possible

With increasing numbers of carbon atoms in the molecule, more patterns of branching are possible; the number of isomers increases dramatically. The formula C_6H_{14} has five possible isomers. The number of isomers increases to 18 for C_8H_{18} and to 75 for $C_{10}H_{22}$.

Nomenclature of Hydrocarbons and Their Derivatives

Table 6.2 contains the names of some unbranched hydrocarbon isomers. Other compounds, including the branched hydrocarbons, are named as derivatives of the unbranched hydrocarbons. Some organic compounds were assigned common or trivial names in the past;

Table 6.2   Unbranched hydrocarbons

	Formula	Name	Boiling Point (°C)	Melting Point (°C)
CH_4	CH_4	Methane	−162	−183
C_2H_6	CH_3CH_3	Ethane	−88	−172
C_3H_8	$CH_3CH_2CH_3$	Propane	−42	−187
C_4H_{10}	$CH_3(CH_2)_2CH_3$	Butane	0	−138
C_5H_{12}	$CH_3(CH_2)_3CH_3$	Pentane	36	−130
C_6H_{14}	$CH_3(CH_2)_4CH_3$	Hexane	69	−95
C_7H_{16}	$CH_3(CH_2)_5CH_3$	Heptane	98	−91
C_8H_{18}	$CH_3(CH_2)_6CH_3$	Octane	126	−57
C_9H_{20}	$CH_3(CH_2)_7CH_3$	Nonane	151	−54
$C_{10}H_{22}$	$CH_3(CH_2)_8CH_3$	Decane	174	−30
$C_{12}H_{26}$	$CH_3(CH_2)_{10}CH_3$	Dodecane	216	−10
$C_{14}H_{30}$	$CH_3(CH_2)_{12}CH_3$	Tetradecane	252	6
$C_{16}H_{34}$	$CH_3(CH_2)_{14}CH_3$	Hexadecane	280	18
$C_{18}H_{38}$	$CH_3(CH_2)_{16}CH_3$	Octadecane	308	28
$C_{20}H_{42}$	$CH_3(CH_2)_{18}CH_3$	Eicosane	343	37
$C_{22}H_{46}$	$CH_3(CH_2)_{20}CH_3$	Docosane	369	44
$C_{24}H_{50}$	$CH_3(CH_2)_{22}CH_3$	Tetracosane	391	54
$C_{30}H_{62}$	$CH_3(CH_2)_{28}CH_3$	Triacontane	450	66

many of these names are still in use, such as isobutane. Most of the time, however, we use systematic names known as IUPAC names. The International Union of Pure and Applied Chemistry (IUPAC) has devised rules for naming all organic compounds. When the rules are followed, each compound has its own name, and there is no confusion about the meaning of the names.

Each hydrocarbon name in the table has the ending *-ane*. Thus, the number of carbons is given by prefixes meth- for one carbon, eth- for two carbons, prop- for two carbons, oct- for eight carbons, and so forth.

Branched hydrocarbons are named using the following rules.

Rule 1. Locate the longest continuous carbon chain and use it as the base name of the compound.

Using this rule, the following compound will be named as a derivative of hexane since the longest chain is six carbons, no matter how the structure is written.

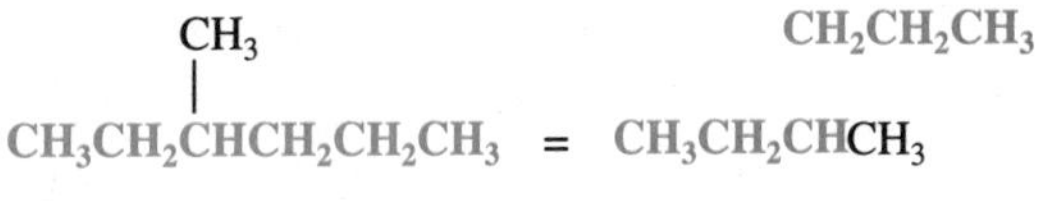

$$CH_3CH_2CHCH_2CH_2CH_3 \ = \ CH_3CH_2CHCH_3$$

with CH_3 branch on the third carbon, equal to the form with $CH_2CH_2CH_3$ branch.

Rule 2. Number the chain beginning from the end nearest to the branch. Identify the branch group and its location.

Applying this rule to the following compound, the correct name is 3-methylhexane.

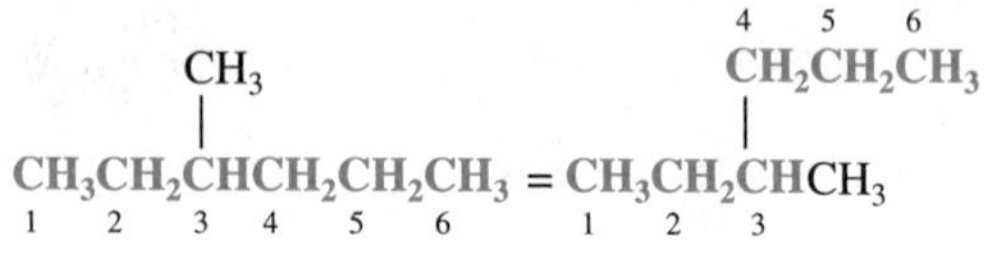

$$\begin{array}{c} \overset{\displaystyle CH_3}{\underset{\displaystyle |}{}} \\ CH_3CH_2CHCH_2CH_2CH_3 \\ {\scriptstyle 1 \quad 2 \quad 3 \quad 4 \quad 5 \quad 6} \end{array} = \begin{array}{c} \overset{\scriptstyle 4 \quad 5 \quad 6}{CH_2CH_2CH_3} \\ | \\ CH_3CH_2CHCH_3 \\ {\scriptstyle 1 \quad 2 \quad 3} \end{array} \qquad \text{3-methylhexane}$$

This name also illustrates the next rule of the IUPAC system.

Rule 3. **Groups derived from hydrocarbons are named by dropping the *-ane* ending and replacing it with *-yl*.**

Therefore, methane (CH_4) becomes methyl (CH_3-) when a hydrogen atom is removed and the group is a branch on a hydrocarbon chain. Ethane (CH_3CH_3) becomes ethyl (CH_3CH_2-). Since a hydrogen atom from carbon 3 is removed and is substituted by the methyl group, we describe this methyl group as a *substituent* in compound III.

Rule 4. **a. When two or more substituents are present, identify and locate each.**

The following compound illustrates this rule and rule 2 (which requires numbering from right to left).

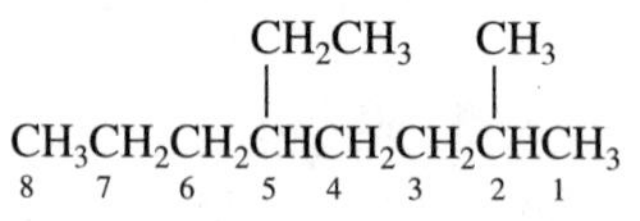

$$\begin{array}{c} \overset{\displaystyle CH_2CH_3}{|} \qquad \overset{\displaystyle CH_3}{|} \\ CH_3CH_2CH_2CHCH_2CH_2CHCH_3 \\ {\scriptstyle 8 \quad 7 \quad 6 \quad 5 \quad 4 \quad 3 \quad 2 \quad 1} \end{array} \qquad \text{5-ethyl-2-methyloctane}$$

b. The substituents should be listed in alphabetical order.

Therefore, *ethyl* (derived from ethane) precedes *methyl* in the name of this compound.

Rule 5. **When two or more substituents are found on the same carbon atom of the chain, use the number twice. If the same substituent is repeated, use a prefix, such as *di-, tri-, tetra-, penta-, hexa-, hepta-, octa-, nona-*, and so forth. These prefixes are not considered when placing substituents in alphabetical order. Commas separate numbers, and hyphens separate numbers from letters.**

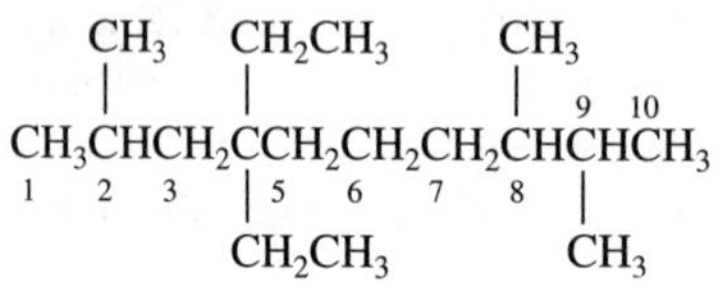

$$\begin{array}{c} \overset{\displaystyle CH_3}{|} \qquad \overset{\displaystyle CH_2CH_3}{|} \qquad \overset{\displaystyle CH_3}{|} \\ CH_3CHCH_2CCH_2CH_2CH_2CHCHCH_3 \\ {\scriptstyle 1 \quad 2 \quad 3 \;\; | \;\; 5 \quad 6 \quad 7 \quad 8 \;\; |} \qquad {\scriptstyle 9 \;\; 10} \\ \quad\;\; CH_2CH_3 \qquad\qquad CH_3 \end{array} \qquad \text{4,4-diethyl-2,8,9-trimethyldecane}$$

These same rules can be applied to derivatives of hydrocarbons as well. For example, if one or more hydrogen atoms are replaced by halogen atoms such as bromine or chlorine, the location and identity are easily specified.

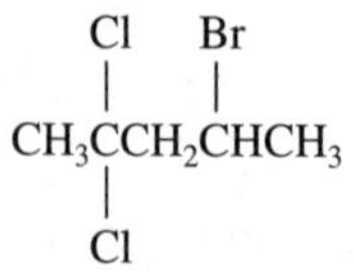

$$\begin{array}{c} \overset{\displaystyle Cl}{|} \qquad \overset{\displaystyle Br}{|} \\ CH_3CCH_2CHCH_3 \\ | \\ Cl \end{array} \qquad \text{2,2-dichloro-4-bromopentane}$$

For some derivatives, the nomenclature is more complicated. Examples will be discussed in later sections.

Properties and Uses of Hydrocarbons

Table 6.2 also shows the regular increase in the boiling point with increasing chain length. There is also a steady increase in melting point, with a few irregularities. The early members of the series are gases far below room temperature, and the last few entries in the table are solids at room temperature (20 °C = 68 °F).

Branching causes a lowering of both the boiling points (bp) and the melting points (mp). For example, branched butane, called isobutane or 2-methylpropane, has a bp of -12 °C and an mp of -159 °C compared to 0 °C and -138 °C for unbranched butane.

Petroleum

Crude oil is a complex mixture of hydrocarbons, including the linear and branched forms already covered. Other variations, such as cyclic and unsaturated hydrocarbons, will be discussed in the following sections. A primary step in refining petroleum is fractional distillation, which separates the components of a mixture according to boiling points. The *volatile* (low bp) components are removed first. As the distillation is carried to completion, not everything distills. The high-boiling residue has its uses also, as in making paraffin, lubricating greases (petroleum jelly), and asphalt. The principal fractions and their uses are shown in Table 6.3.

Let us consider the difference between gasoline and kerosene. There is some overlap in the boiling points of these two petroleum fractions; but gasoline contains some smaller, more volatile hydrocarbons that are not present in kerosene. The lower boiling temperatures of gasoline components allow it to vaporize more readily than kerosene. This makes for easier starting of automobile engines, even in cold weather (see Section 13.3). On the other hand, kerosene is used as diesel fuel (Figure 6.4), jet fuel, and home heating oil. Diesel engines are more difficult to start in cold weather, but run at higher efficiency than ordinary automobile engines.

Kerosene can be burned safely in heaters due to its low volatility. The higher volatility of gasoline makes it very dangerous to use in place of kerosene. Similarly, starter fluid for charcoal fires (Figure 6.5) can be used with relative safety due to the lower volatility of kerosene; gasoline is too volatile to be used safely for this purpose, since its vapor ignites and spreads much more quickly.

Table 6.3 shows an overlap of carbon chain length and of boiling points among the several fractions. Petroleum refiners experience varying demands for their products, giving

Table 6.3 Petroleum fractions

Fraction	*Range of Hydrocarbons*	*Approximate Boiling Range (°C)*	*Uses*
Gas	C_1–C_4	Below 20	Fuel, plastics
Petroleum ether	C_5–C_6	30–60	Solvents
Ligroin	C_6–C_8	60–120	Solvents
Gasoline	C_5–C_{12}	40–205	Fuel
Kerosene	C_{11}–C_{18}	175–325	Diesel fuel, jet fuel, home heating
Lubricating oil	C_{17}–higher	Above 300	Lubricants (greases)
Residue	C_{20}–higher	Above 350	Asphalt

Figure 6.4   An important consumer of diesel fuel.

Figure 6.5   Charcoal starter fluid
is kerosene, which is much less
volatile than gasoline and can,
therefore, be ignited much more
safely. This product is labeled
"odorless," but most charcoal
lighter fluids have a slight odor.

them flexibility in deciding how much of each fraction to separate in response to demand.
Even more flexibility is possible due to **cracking.** In this process, the refiner can heat a
fraction, such as kerosene, with a suitable catalyst and cause the relatively long chains to
break into smaller fragments with lower boiling points (Figure 6.6). In this way, gasoline
can be produced from kerosene. The process also is illustrated by the cracking of butane,

Figure 6.6   A catalytic cracker.

which produces ethylene and hydrogen. Although ethylene is not used as a fuel, it is in demand as a starting material for the production of polyethylene and other plastics (see Chapter 8).

$$CH_3CH_2CH_2CH_3 \xrightarrow[\text{high temperature}]{\text{catalyst}} 2\ CH_2 = CH_2 + H_2$$
$$\underset{\text{butane}}{} \qquad\qquad \underset{\text{ethylene}}{} \quad \underset{\text{hydrogen}}{}$$

Besides simple cracking, *catalytic reforming* can be carried out on petroleum fractions. This process converts some long-chain hydrocarbon molecules to branched isomers, thus improving their efficiency for burning as gasoline. The efficiency is described quantitatively by the **octane rating** scale, which arbitrarily gives unbranched heptane (C_7H_{16}) a rating of zero and isooctane a rating of 100. (The word ''isooctane'' is an unsystematic name used only in this context. The IUPAC name for this compound is 2,2,4-trimethylpentane.) Figure 6.7 compares the structures of the two compounds.

$$
\begin{array}{c}
CH_3 \quad CH_3 \\
| \qquad | \\
CH_3CCH_2CHCH_3 \\
| \\
CH_3
\end{array}
$$

2,2,4-trimethylpentane
(isooctane)
octane rating = 100

$CH_3CH_2CH_2CH_2CH_2CH_2CH_3$ heptane
octane rating = 0

In other words, unbranched heptane and other similar hydrocarbons cause engines to knock badly when used alone as gasoline, whereas highly branched isomers burn smoothly. The importance of the shapes of the molecules is best appreciated from the space-filling models in Figure 6.7. Because the highly branched hydrocarbons make superior fuels, catalytic cracking can improve the octane rating dramatically; some highly branched isomers have octane ratings greater than 100. Unbranched octane has a rating of −19.

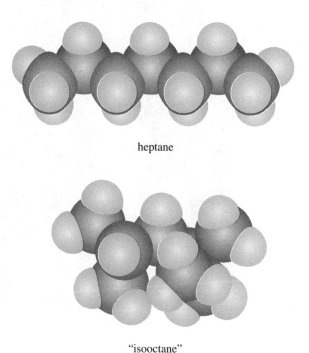

heptane

"isooctane"

Figure 6.7   Space-filling models of heptane (unbranched) and isooctane emphasize the shapes of the molecules that cause them to ignite differently in an automobile engine. The more extended heptane burns more rapidly, even explosively, which causes engine knock or ping. The highly branched isooctane, with less of its surface exposed, burns more slowly and evenly, from the outside inward, leading to increased power, particularly under load.

Tetraethyllead, Pb(CH$_2$CH$_3$)$_4$, was used for many years to improve octane rating. To reduce pollution, automobile engines are now being engineered to use gasolines without lead. Other additives and fuel mixtures are being developed as well (see Chapter 13).

Catalytic reforming also can be used to convert saturated hydrocarbons to aromatic hydrocarbons (see section on Cyclic Hydrocarbons). This is useful since aromatic hydrocarbons have high octane ratings.

Natural Gas

The chief component (about 95%) of natural gas is methane, CH$_4$. Lesser amounts of ethane, propane, and the butanes are also present. Natural gas is collected from subterranean deposits, put under pressure, and piped nationwide for use as heating fuel and in industrial applications. Because propane and butanes have higher boiling points than methane and ethane, a propane-butane fraction can be removed from natural gas. When this fraction is cooled under pressure, it liquefies. The mixture is then compressed into cylinders and sold as *bottled gas.*

6.2   Alkanes, Alkenes, and Alkynes

GOALS: To write condensed formulas and bond-line formulas for unsaturated hydrocarbons and derivatives, including *cis* and *trans* isomers.
To recognize benzene and related compounds.

Thus far we have concentrated on hydrocarbons with only carbon-carbon single bonds. These compounds are called **alkanes;** they are occasionally known by an older name, *paraffins.* Hydrocarbons also exist as **alkenes** and **alkynes,** which contain carbon-carbon double and triple bonds, respectively. Compounds with double or triple bonds are termed **unsaturated,** whereas compounds with only single bonds are termed **saturated.** The heavily advertised polyunsaturated cooking oils contain many carbon-carbon double bonds (see Chapter 21).

The simplest unsaturated compounds are *ethylene* and *acetylene.* These are common names; the corresponding IUPAC names are *ethene* and *ethyne.* As the formula for ethylene indicates, each carbon shares four valence electrons with the other carbon atom, since each atom must be tetravalent. The conversion from ethane (an alkane) to ethylene (an alkene) is accompanied by a loss of two hydrogen atoms from adjacent atoms. The same is true for the conversion from ethylene to acetylene. In acetylene, each carbon is sharing six valence electrons with the other carbon atom.

ethylene (common name)
ethene (IUPAC name)

acetylene (common name)
ethyne (IUPAC name)

Geometric Isomers

The alkene structure is a particularly common one; it also shows a different type of isomerism. The compound dichloroethene can exist as any of three different forms as shown. First there is the distinction between 1,1 and 1,2 constitutional isomers (Section 6.1), in which the two chlorine atoms may be located on the same or adjacent carbon atoms. This type of isomerism is the same as that observed for the alkanes. But when a double bond is present, **stereoisomerism** is possible. Here, isomers differ only in the three-dimensional arrangement of atoms in space. For alkenes, the specific form of stereoisomerism is called **geometric isomerism,** in which the isomers differ only in the relative location of atoms or groups attached to the double bond. This type of isomerism, illustrated by *cis*-1,2-dichloroethene and *trans*-1,2-dichloroethene, is often described as **cis-trans isomerism** (Figure 6.8).

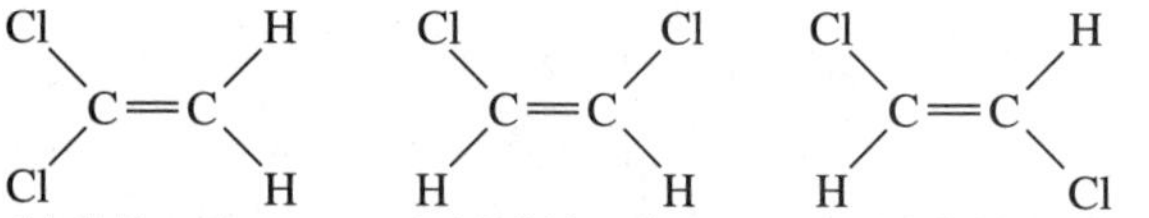

1,1-dichloroethene   *cis*-1,2-dichloroethene   *trans*-1,2-dichloroethene

Two groups on the same side of the double bond are said to be *cis* (Latin, "on the side"). Groups on opposite sides of the double bond are said to be *trans* (Latin, "across"). Such isomers are possible only because of rigidity in the molecule at the double bond. If either the *cis* or the *trans* isomer could rotate around the double bond, it could be interconverted. But such rotation is restricted unless heat or light energy is provided to cause the double bond to break. *Cis* and *trans* isomers are distinctly different compounds with different physical properties such as boiling point and solubility. These properties allow us to separate them.

The ball-and-stick models for the geometric isomers, shown in Figure 6.8, illustrate the difference between alkanes and alkenes. The two arrangements, known as *conformations,* of 1,2-dichloroethane are not regarded as isomers; they can be interconverted by simple rotation about the carbon-carbon single bond, and cannot be separated experimentally. In the alkene model, the second bond between the carbons shows how rotation is restricted and interconversion is prevented. Unfortunately, the model does not accurately depict the nature of the two bonds of the double bond, but this topic is beyond our scope of coverage.

Cis and *trans* isomers also occur in hydrocarbons. For example, the compound 2-butene can exist as *cis*-2-butene or *trans*-2-butene. (The derivation of the name 2-butene will be discussed in Section 6.3.)

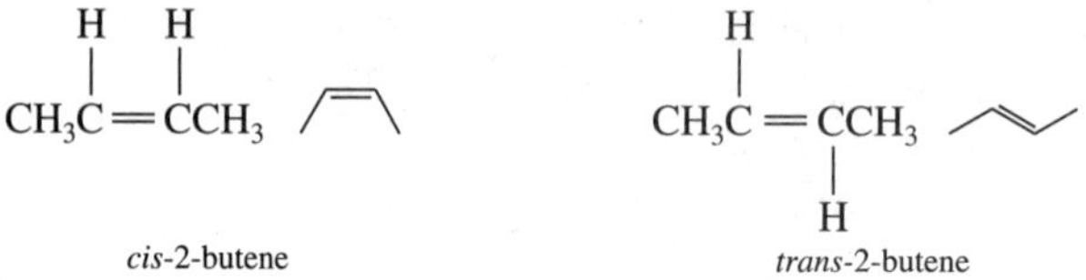

cis-2-butene   *trans*-2-butene

The abbreviated form of each isomer can be rewritten to emphasize the *cis* and *trans* orientation.

cis-2-butene   *trans*-2-butene

The structures on the left for each isomer emphasize the *cis* and *trans* orientation of the carbon atoms that are attached to the carbon-carbon bond. The structures on the right show the *cis* and *trans* orientation of the hydrogen atoms, which are normally omitted from bond-line formulas.

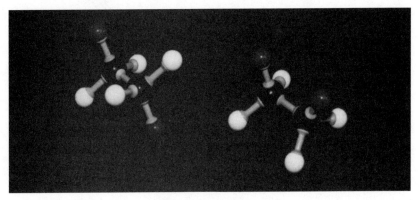

1,2-dichloroethane
(no geometric isomers)

1,2-dichloroethane
(two geometric isomers)

Figure 6.8   Ball-and-stick models of a saturated compound and an alkene show why geometric isomers are possible for alkenes.

Cyclic Hydrocarbons

Compounds such as cyclohexane and benzene are also found in petroleum. Cyclohexane and other cycloalkanes (called naphthenes) are particularly abundant in California petroleum. These compounds differ from the open-chain analogs in that the ends are joined, but most of their properties are similar. The bond-line formula is generally used to represent ring systems, when it is called a **polygon formula.**

cyclohexane   benzene

The benzene structure is common to many organic compounds and has some unique chemical properties. For this reason, benzene and related compounds (Figure 6.9), such as

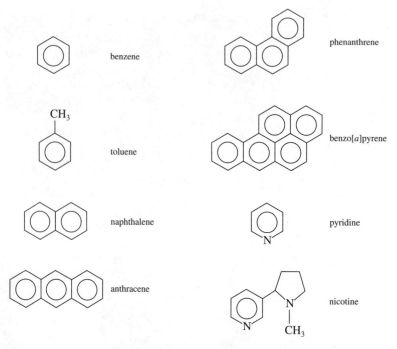

Figure 6.9   Some common aromatic compounds.

toluene, naphthalene, benzo[*a*]pyrene, pyridine, and nicotine, are given the special designation of **aromatic compounds.** The properties of these compounds justify the use of abbreviations for the multiple double bonds in the rings. A circle in the center of a ring signifies that each atom is part of a continuous, unsaturated structure.

Cyclopentene and cyclohexene, each with just one double bond in the ring, are not regarded as aromatic.

6.3   Functional Groups

GOALS: To see a variety of functional groups and how they may influence IUPAC names.
To appreciate the use of the symbol R in writing organic compounds.
To recognize and draw structural formulas for alcohols, acids, and esters, and to appreciate the relationships among these three classes of compounds.

Much of the importance of hydrocarbons arises from their role as building blocks for formation of other organic compounds. Usually, the chemical properties, that is, the chemical reactions, of organic compounds involve only the groups or atoms that are attached to the hydrocarbon structure. For this reason, we refer to the hydrocarbon portion of an organic molecule as the *backbone* or skeleton to which are attached the **functional groups** that account for most of the chemical properties of each compound.

The list of possible types of compounds is large; some of the oxygen- and nitrogen-containing functional groups are shown in Table 6.4. The symbol R is a common shorthand notation used to designate a *radical,* which is formed from any molecule by removing a hydrogen. For example, if a hydrogen atom is removed from methane, CH_4, the result is a methyl radical, CH_3. Similarly, if a hydrogen atom is removed from ethane, the resultant radical is called an *ethyl* radical.

The methyl radical has a trivalent carbon and must be combined with some atom, such as chlorine, to give a stable molecule. Once the radical is attached to a chlorine atom or other functional group, it is referred to as a methyl *group.* Similarly, the ethyl radical becomes an ethyl group when attached to a functional group or a hydrocarbon skeleton.

The methyl radical also can combine with a functional group hydroxyl, OH, to form methyl alcohol, CH_3OH. (Its IUPAC name is methanol.) More generally, if a hydrogen atom is removed from an alkane, the resultant particle is called an *alkyl radical.* The radical is called an *alkyl group* once it is attached to a functional group or another hydrocarbon. Organic chemists use the symbol R in writing compounds ranging from simple ones like methyl alcohol to complex ones such as cholesterol.

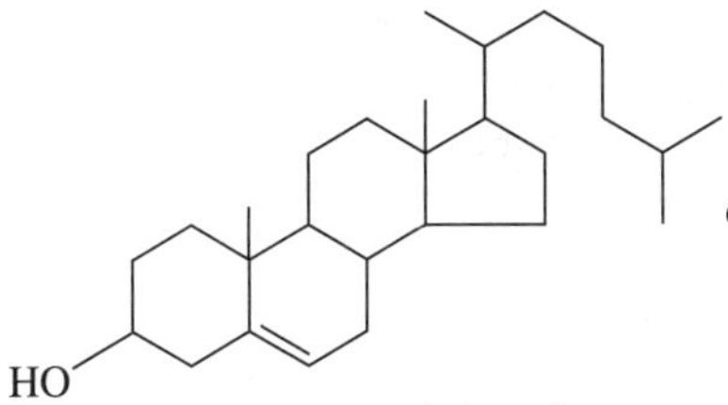

cholesterol

Some of the functional groups listed in Table 6.4 will be discussed in the final section of this chapter.

Table 6.4 Common functional groups containing oxygen and nitrogen

Group	General Formula	Type of Compound
—OH	ROH	Alcohol
—OR'	ROR'	Ether
—C(=O)—H	R—C(=O)—H	Aldehyde
—C(=O)—R'	R—C(=O)—R	Ketone
—C(=O)—OH (—COOH)	R—C(=O)—OH (RCOOH)	Acid
—C(=O)—OR'	R—C(=O)—OR'	Ester

Table 6.4   Common functional groups containing oxygen and nitrogen (*cont.*)

Group	General Formula	Type of Compound
$\overset{\displaystyle O}{\overset{\|}{-C-NH_2}}$	$\overset{\displaystyle O}{\overset{\|}{R-C-NH_2}}$	Amide
$-O-O-H$	$R-O-O-H$	Hydroperoxide
$-NH_2$	$R-NH_2$	Amine
$-NH_2 + -COOH$	$\underset{\underset{\displaystyle NH_2}{\|}}{RCH-COOH}$	Amino acid
$-C\equiv N$	$R-C\equiv N$	Nitrile
$-N=C=O$	$R-N=C=O$	Isocyanate

Nomenclature: The Effect of Functional Groups

The rules for naming hydrocarbons were introduced earlier. Since we will study the role of organic compounds in a variety of applications, let us see how the IUPAC system describes unsaturation and functional groups.

The compound 2-methylhexane is a straightforward illustration of how alkanes are named, as discussed in Section 6.1. The compound 5-methyl-2-hexene illustrates how the rules change when a double bond is present.

$$\underset{CH_3CH_2CH_2CH_2CHCH_3}{\overset{\overset{\displaystyle CH_3}{\|}}{}} \quad \text{2-methylhexane}$$

$$\underset{CH_3CH=CHCH_2CHCH_3}{\overset{\overset{\displaystyle CH_3}{\|}}{}} \quad \text{5-methyl-2-hexene}$$

First of all, the presence of the double bond is indicated by changing the ending of the name from -*ane* to -*ene*. Second, the position of the double bond after the second carbon in the chain is indicated by the 2, counting from left to right as written. Third, notice that the chain is numbered from right to left for 2-methylhexane and from left to right for 5-methyl-2-hexene. When a double bond is present in the molecule, it is given priority and the chain is numbered from the end closest to it. If more than one double bond is present, it is necessary to locate each and tell how many are present.

$$CH_2=CHCH=CH_2 \quad \text{1,3-butadiene}$$

$$CH_2=CHCH_2CH=CH_2 \quad \text{1,4-pentadiene}$$

$$\underset{\underset{\displaystyle CH_3}{\|}}{CH_2=CCH=CH_2} \quad \begin{array}{l}\text{2-methyl-1,3-butadiene}\\ \text{(common name isoprene)}\end{array}$$

Other functional groups, such as an acid group, take priority over the double bond in numbering the chain. Then, the numbering is from right to left, as it is for 2-methylhexane. Notice that the 5 specifies the location of the double bond between carbons 5 and 6.

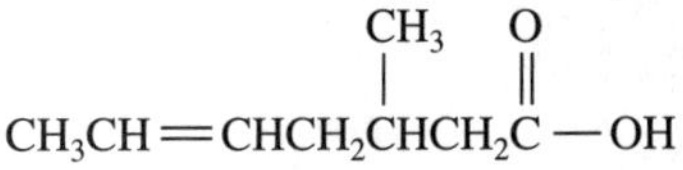

$$CH_3CH = CHCH_2 \overset{\overset{\displaystyle CH_3}{|}}{C}HCH_2 \overset{\overset{\displaystyle O}{\|}}{C} - OH \qquad \text{3-methyl-5-heptenoic acid}$$

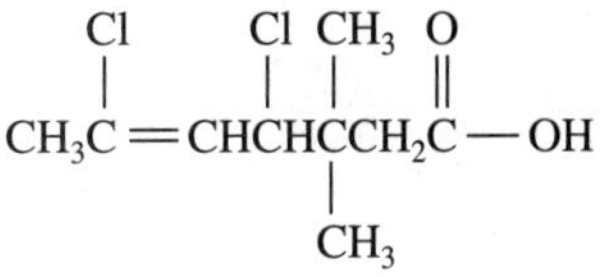

$$\overset{\overset{\displaystyle Cl}{|}}{CH_3}C = \overset{\overset{\displaystyle Cl}{|}}{C}H\overset{\overset{\displaystyle CH_3}{|}}{C}H\underset{\underset{\displaystyle CH_3}{|}}{C}CH_2 \overset{\overset{\displaystyle O}{\|}}{C} - OH \qquad \begin{array}{l}\text{4,6-dichloro-3,3-dimethyl-}\\ \text{5-heptenoic acid}\end{array}$$

The position of the acid group is not specified in the name, since it must end the chain in order for the carbon atom of the group to be tetravalent. The ending *-oic* is used when naming acids.

Alcohols, Acids, and Esters

Alcohols are characterized by the presence of the OH functional group. The same is true of organic acids. Consider the 2-carbon compound in each class.

$$\begin{array}{cc} R - OH & CH_3CH_2 - OH \\ \text{general formula} & \text{ethanol} \\ \text{of an alcohol} & \text{(common name} \\ & \text{ethyl alcohol)} \end{array}$$

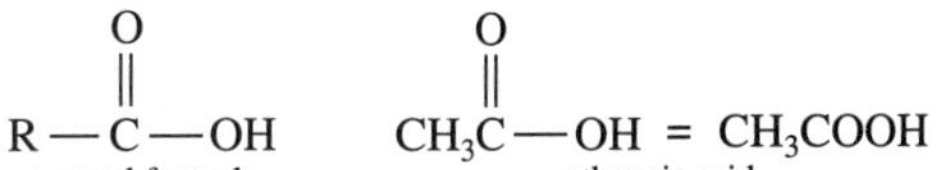

$$\begin{array}{cc} \overset{\overset{\displaystyle O}{\|}}{R - C} - OH & \overset{\overset{\displaystyle O}{\|}}{CH_3C} - OH = CH_3COOH \\ \text{general formula} & \text{ethanoic acid} \\ \text{of an alcohol} & \text{(common name} \\ & \text{acetic acid)} \end{array}$$

These are entirely different compounds with different properties. Most people have an acquaintance with the flavor of ethyl alcohol and acetic acid as important components of alcoholic beverages and vinegar, respectively. Acids are often characterized as having a sour taste, a property not associated with alcohols. The major structural difference is the presence of the *carbonyl group* in acetic acid.

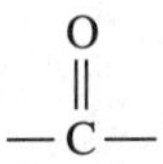

$$\overset{\overset{\displaystyle O}{\|}}{- C -} \qquad \text{the carbonyl group}$$

The carbonyl group coupled with the OH group is called a *carboxyl group,* and it is this combination that is required for the compound to be an acid (Figure 6.10). In fact, this type of acid is often called a *carboxylic acid* to distinguish it from other kinds of acids.

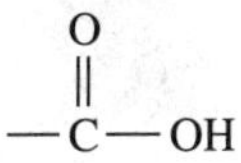

$$\overset{\overset{\displaystyle O}{\|}}{- C - OH} \qquad \text{the carboxyl group}$$

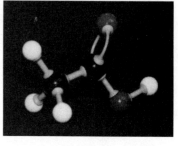

ethanol   acetic acid

Figure 6.10   The carboxyl group in acetic acid gives the compound properties that are quite different from ethyl alcohol with a simple −OH group.

The nature of the carboxyl group causes acetic acid to yield a hydrogen ion, H^+. In doing so, it conforms to the definition of an acid as a hydrogen ion donor as described in Chapter 7.

Acids are important in
many ways, including:
agriculture (Chapter 15)
winemaking (Chapter 17)
baking (Chapter 18)
food preservation
(Chapter 19)
fats (Chapter 21)
antacids (Chapter 25)

$$CH_3\overset{\overset{\displaystyle O}{\|}}{C}-OH \; \rightleftharpoons \; CH_3\overset{\overset{\displaystyle O}{\|}}{C}-O^- + H^+$$

In contrast, ethyl alcohol has little tendency to release a hydrogen ion from its OH group.

Esters are an interesting class of compounds in which the OH hydrogen atom is replaced by an alkyl group.

$$R-\overset{\overset{\displaystyle O}{\|}}{C}-OR' \qquad CH_3\overset{\overset{\displaystyle O}{\|}}{C}-OCH_2CH_3$$

general formula   ethyl acetate
of an ester

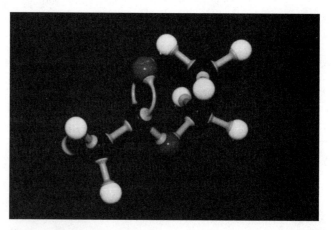

Figure 6.11   Ethyl acetate.

In a general formula for an ester, the R group on the left is not necessarily the same as the one on the right, so they are designated as R and R′. An example is ethyl acetate (Figure 6.11), which is used as a food flavoring and as a solvent in nail polish and polish remover.

Most esters have strong but pleasant aromas. Bananas have a distinctive odor that is largely due to the ester isopentyl acetate.

$$
\begin{array}{cc}
O & CH_3 \\
\| & | \\
CH_3C - OCH_2CH_2CHCH_3 &
\end{array}
\qquad
\begin{array}{l}
\text{isopentyl acetate} \\
\text{(present in banana oil)}
\end{array}
$$

One important aspect of the chemistry of esters is that they are synthesized by combining an acid with an alcohol.

$$
\underset{\text{an acid}}{R - \overset{\displaystyle O}{\overset{\|}{C}} - OH} + \underset{\text{an alcohol}}{R' - OH} \;\rangle\; \underset{\text{an ester}}{R - \overset{\displaystyle O}{\overset{\|}{C}} - O - R'} + \underset{\text{water}}{HOH}
$$

$$
\underset{\text{acetic acid}}{CH_3C - OH} + \underset{\textbf{ethyl alcohol}}{CH_3CH_2OH} \;\rangle\; \underset{\textbf{ethyl} \text{ acetate}}{CH_3C - OCH_2CH_3} + HOH
$$

Fats are an important group of esters, discussed in Chapter 21.

Thus an ester is made up of two parts, one supplied by the carboxylic acid and the other by the alcohol, as emphasized by the color. Even the water that is formed as a byproduct is derived from the two starting materials.

SUMMARY

Organic chemistry is the chemistry of carbon. The carbon atom shows much versatility in bonding to other atoms and to other carbon atoms. The bonding between carbon atoms may include single, double, and triple bonds, but in all cases, the octet rule is obeyed, and carbon is tetravalent.

Among the hydrocarbons, many constitutional isomers result from the branching of the carbon chains. Structural formulas of the isomers can be drawn using several abbreviations, including bond-line formulas and condensed formulas. As carbon chains increase in length, the boiling temperatures are correspondingly higher, with exceptions for branching, which lowers the boiling point.

The boiling point is the basis for classifying and separating the common hydrocarbons obtained from petroleum such as gasoline, kerosene, and asphalt. Catalytic cracking and reforming are processes for increasing the utility of natural hydrocarbons. Reforming tends to increase the degree of branching; this leads to an improved octane rating due to smoother burning.

Unsaturation introduces the potential for geometric isomerism—the existence of *cis-trans* isomers. Ring compounds, including highly unsaturated aromatic hydrocarbons, are also common among organic compounds. The presence of other atoms and functional groups increases the variety of organic compounds, which includes alcohols, carboxylic acids, and esters.

Many organic compounds have common or trivial names, but all have IUPAC names. When the appropriate rules are followed in assigning IUPAC names, each compound has its own name.

PROBLEMS

1. Give a definition or example for each of the following.

   a. Valence
   b. Covalent bond
   c. Methane
   d. Bottled gas
   e. Volatile
   f. Polygon formula
   g. Cracking
   h. *cis*
   i. *trans*
   j. Constitutional isomer
   k. Geometric isomer
   l. Stereoisomer
   m. Single-line formula
   n. Bond-line formula
   o. Condensed formula
   p. Alkane
   q. Alkene
   r. Alkyne
   s. Paraffin
   t. Unsaturated
   u. Saturated
   v. Functional group
   w. Alcohol
   x. Carboxylic acid
   y. Ester

2. Name each of the following compounds. Be sure to identify *cis* and *trans* isomers where they exist.

   a. $CH_3CH_2CH_2CH_2CH_2CH_3$

   b. $CH_3CHCH_2CH_2CH_2CH_3$
   $\quad\quad\ \ |$
   $\quad\quad CH_3$

   c. $CH_3CHCH_2CHCH_2CH_3$
   $\quad\quad\ \ |\quad\quad\ |$
   $\quad\quad CH_3\quad Cl$

   d. $CH_3CHCH_2CHCH_2CH_3$
   $\quad\quad\ \ |\quad\quad\ \ |$
   $\quad\quad CH_3\quad CH_3$

   e. $CH_3CCH_2CHCH_3$
   $\quad\quad\ \ |\quad\quad |$
   $\quad\quad CH_3\ \ Cl$

   f. $CH_3CCH_2CHCH_3$ with Cl on upper C and Cl shown
   $\quad\quad Cl$
   $\quad\quad\ |$
   $\quad CH_3CCH_2CHCH_3$
   $\quad\quad\ |\quad\quad\ |$
   $\quad\quad Cl\quad\ Cl$

   g. $CH_3C{=}CCH_2CH_3$
   with H H above the double bond carbons

   h. $CH_3C{=}CCH_2CH_3$
   with H above and H below

   i. $CH_2{=}CHCH_2CH_3$

3. Draw condensed and bond-line formulas for each of the following.

   a. Pentane
   b. 3-Methylpentane
   c. 2,3-Dimethylpentane

   d. 2,2,4-Trimethylpentane
   e. 3-Ethylhexane
   f. 3-Chlorooctane
   g. 2-Hexene (show both *cis* and *trans* isomers)
   h. 2-Bromo-1-pentene
   i. 2-Bromo-2-pentene (show both *cis* and *trans* isomers)

4. Name each of the following compounds.

5. Why is carbon tetravalent?

6. List all of the unbranched hydrocarbons from Table 6.2 that might be found in the kerosene fraction of crude oil. What other hydrocarbons might be present?

7. List all of the unbranched hydrocarbons from Table 6.2 that might be found in gasoline but not in kerosene. How are these compounds useful in gasoline?

8. Why is the tetrahedral angle preferred for saturated carbon atoms?

9. Write structures for an alcohol and an acid, each containing four carbons, and for the ester that would form by reacting the two compounds.

10. Why does "isooctane" have a higher octane rating than unbranched octane?

11. What is the value of catalytic reforming?

12. Describe the difference between cyclohexane and benzene.

13. What is the structural difference between an alcohol and a carboxylic acid?

14. What is the structural difference between an ester and a carboxylic acid?

15. Give an example of an ether and an ester, each containing six carbon atoms. Refer to Table 6.4 as necessary.

16. What is the difference in composition between natural gas and bottled gas?

17. Draw condensed and bond-line formulas for the 18 isomers with the molecular formula C_8H_{18}. Give IUPAC names for each.

Solution Chemistry: Acids and Bases

7

In this chapter we will study the nature of solutions. We may commonly think of preparing solutions by dissolving solids in water, but they may take many forms, including various combinations of solids, liquids, and gases. As with most other topics in chemistry, solutions have a language of their own. Thus, we must consider the many kinds, why they form, what factors influence their stability, and how they are described both qualitatively and quantitatively. Among the most important chemical compounds are acids and bases; both form solutions with important properties.

7.1 The Language of Solutions

GOALS: To understand the terminology of solutions.
To recognize the many forms that solutions can take.

Most of the substances that we encounter in everyday life are mixtures. Many are homogeneous mixtures from which the components do not separate; these are called solutions. Each substance in a solution is designated as a *solute* or a *solvent*. A **solute** is a substance that is dissolved in a **solvent,** and their combination is the solution. Simple examples are salt water, in which the salt is the solute and the water is the solvent, and household ammonia, in which ammonia gas is the solute and water is the solvent (Table 7.1).

The distinction between solutes and solvents is not always so clear. When alcohol and water are mixed, it is not obvious which liquid is dissolving in which. The same is true for mixtures of gases, such as air. Usually, the solvent is regarded as the more abundant component. By this reasoning, nitrogen might be regarded as the solvent in air. But this means of classification is not always logical. Water, for example, is capable of dissolving more than its own mass of certain solids, particularly at high temperatures. Even at only 25 °C, 100 g of water will dissolve 97 g of ammonium bromide, NH_4Br.

Table 7.1	Common solutions		
Type	*Solution*	*Solute(s)*	*Solvent*
Gas-gas	Air	Oxygen, carbon dioxide, water vapor, etc.	Nitrogen
Gas-liquid	Carbonated soda	Sugar, flavoring, carbon dioxide	Water
Gas-liquid	Household ammonia	NH_3 gas	Water
Liquid-liquid	Vinegar	Acetic acid	Water
Liquid-liquid	Rubbing alcohol	Water	Isopropyl alcohol
Solid-liquid	Tincture of iodine	Iodine	Alcohol
Solid-solid	Steel	Carbon	Iron

The most common examples of solutions are mixtures of gases (air), gases dissolved in liquids (carbonated soda, ammonia water), liquids in liquids (alcohol-water), solids in liquids (salt water), and solids in solids (alloys). Solutions in which water is the solvent are called *aqueous solutions.*

Most of these solutions are familiar. **Alloys** may be unfamiliar; they are solutions formed when two or more solids, usually metals, are melted, mixed, and cooled to obtain a homogeneous solid. Common examples are sterling silver (92% silver, 8% copper), brass (copper, zinc, variable composition), bronze (copper, tin, variable composition), stainless steel (minimum 10% chromium, iron), pewter (mostly tin, small amounts of copper, bismuth, antimony), and solder (tin, lead, variable composition).

Alloys of gold are common; they vary in appearance depending on their composition. Gold does not always have a gold color. Yellow gold is an alloy of gold and copper; white gold is an alloy of gold, zinc, copper, and nickel; green gold is an alloy of gold and silver; and pink gold is an alloy of gold, silver, and copper. See Section 7.4 for further discussion of gold alloys.

Dentists use complex alloys for filling teeth. They are called *amalgams* to signify that mercury is one component. Mercury is an unusual metal because it is a liquid at room temperature. Silver (approximately 70%) and tin (approximately 25%) are the other main components of dental amalgams, together with smaller amounts of copper and zinc. Details are presented in Chapter 22.

7.2 Water

GOALS: To appreciate the nature of the water molecule and the importance of electronegativity in influencing its properties.
To understand hydrogen bonding and solvation of dissolved solutes.

The most common household and laboratory solvent is water. Its ability to dissolve so many substances is due to its polar structure, shown in Figure 7.1.

The different electronegativities (Section 3.7), or attractions of the hydrogen and oxygen atoms for the electrons shared in their covalent bonds, produce distinct positive and

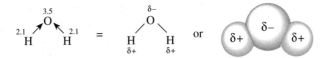

Figure 7.1   The polarity of water. Due to the large difference in electronegativities of oxygen (3.5) and hydrogen (2.1), each H-O bond is quite polar. The partial charges, represented as $\delta+$ and $\delta-$, allow for attractions between water molecules and ions, thus causing many salts to be water soluble.

negative regions (poles) in the water molecule. The bonding electrons are not shared equally by the oxygen and hydrogen atoms, but are pulled in the direction of the oxygen atoms. This results in the hydrogen and oxygen atoms each carrying a small amount of charge, a *partial charge* symbolized by $\delta+$ and $\delta-$, respectively.

The polarity of water enables it to dissolve other polar or ionic substances. When attractive forces between water and the solute are greater than the forces within the solute itself, the substance dissolves. For example, water molecules attract the ions in the sodium chloride crystals (Figure 7.2) and pull them away from their solid, crystalline structure. This process is called **solvation,** and the ions are said to be *solvated* by the solvent molecules (Figure 7.3). When water is the solvent, the solvation process is commonly called **hydration** and the ions are said to be *hydrated* by the water molecules.

A crystal of sodium chloride is typical of most salts. It has a three-dimensional structure, called a *lattice structure.* The sodium and chloride ions alternate so that each is surrounded by the ions of opposite charge. The amount of attraction between the cations and anions, known as the *lattice energy,* varies from one salt to another. The attractions within the crystals of sodium chloride are strong, but they are easily disrupted by interactions with water. Solubility results because the energy that is released when the sodium

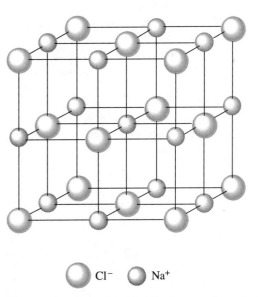

Cl^- Na^+

Figure 7.2   The crystal lattice of sodium chloride. The larger spheres (color) represent Cl^- and the smaller ones Na^+.

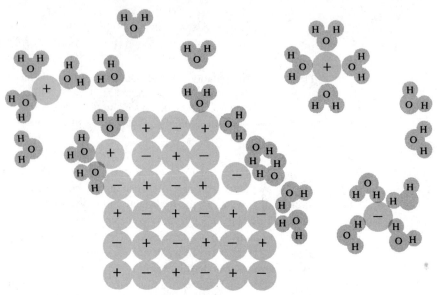

Figure 7.3   The dissolving of sodium chloride in water. Spheres with positive and negative charges represent sodium and chloride ions, respectively. Water molecules hydrate the ions on the surface of a salt crystal and move them out into the solution.

and chloride ions are hydrated is greater than the energy required to overcome the ionic attractions within the lattice.

This relationship does not hold for the salt silver chloride, AgCl. In the silver chloride crystal, the energy required to overcome the ionic attractions is greater than the energy released when the ions are hydrated; therefore, the solid does not dissolve in water (Figure 4.3). Substances such as grease and oil, which lack polar or ionic sites, are not attracted by water molecules and are insoluble in water.

The polarity of the water molecule results in a strong attraction between the hydrogen atoms in one molecule and the oxygen atoms in other molecules. Whereas this kind of attraction, called a *hydrogen bond,* is not as strong as an ionic or covalent bond, it is responsible for several unusual properties of water. For example, the attraction of a hydrogen atom in one water molecule for the oxygen atom in another water molecule (Figure 7.4) gives water its liquid form at room temperature and its unusually high boiling point

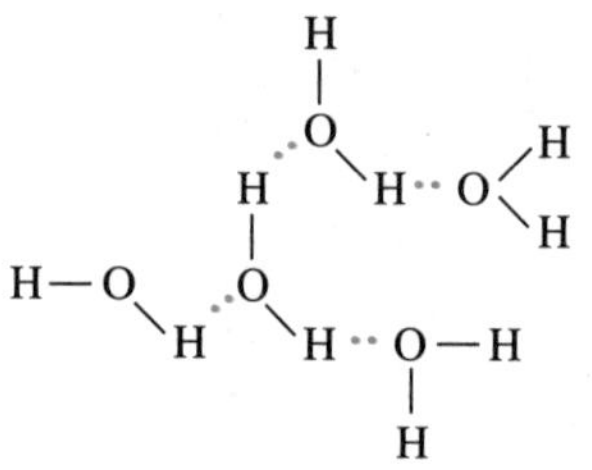

Figure 7.4   Due to the polarity of the H-O bonds, water molecules tend to cluster together due to attractions known as *hydrogen bonds.*

for a compound of such a low molar mass. It is common for water molecules to hydrogen bond together into groups of four to eight attached molecules. This grouping together gives water characteristics similar to those of substances with higher molar masses.

7.3   Solubility

GOAL: To consider the factors that influence solubility.

The facts that oil and water do not mix, salts and sugars dissolve in water, and turpentine can be used to clean a brush used with oil paints are simple examples of solubility phenomena. The **solubility** of a solute in a solvent is the maximum amount of solute that can be dissolved in a specified amount of solvent or solution at a stated temperature. Solubilities, as represented in Table 7.2, are most commonly expressed as the number of grams of solute that can be dissolved in 100 mL or 100 g of the solvent at a given temperature.

An arbitrary value of 3 g of solute per 100 g of solvent is used to quantify solubility. If 3 g or more of solute goes into solution, it is regarded as soluble; if less than 3 g dissolves, the solute is said to be insoluble. As is seen in Table 7.2, most solutes are either much more or much less soluble than the 3 g level, so usually the distinction is clear. The term *miscible* is generally used to describe liquids that are mutually soluble. Liquids that do not dissolve in each other are *immiscible*.

Factors Affecting Solubility

The solubility of a solute in a solvent is influenced by the properties of the substances themselves, the temperature of the solution, and, for gaseous solutes, the pressure on the solution.

Solute/Solvent Structure

A general rule for determining the ability of a solvent to dissolve a specific solute is ''Like dissolves like.'' That is, polar and ionic compounds dissolve in polar solvents; nonpolar substances dissolve in nonpolar solvents. Three common solvents, in order of decreasing polarity, are water, ethyl alcohol, and carbon tetrachloride. Water is an excellent solvent

Table 7.2   Typical solubility data

Solute	Solvent	Solubility
NH_4Br	H_2O	97 g per 100 mL H_2O at 25 °C
$BaCl_2$	H_2O	37.5 g per 100 mL H_2O at 25 °C
$MgSO_4$	H_2O	26 g per 100 mL H_2O at 0 °C 73.8 g per 100 mL H_2O at 100 °C
I_2	H_2O CCl_4	0.03 g per 100 mL H_2O at 25 °C 2.91 g per 100 mL CCl_4 at 25 °C

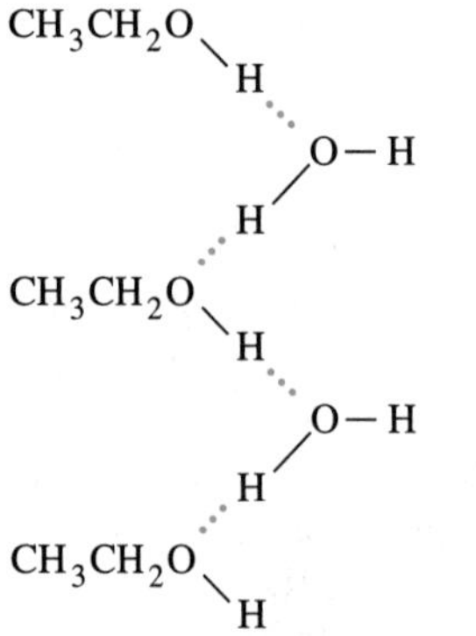

CH$_3$CH$_2$O

H

O—H

H

CH$_3$CH$_2$O

H

O—H

H

CH$_3$CH$_2$O

H

Figure 7.5 Hydrogen bonding between ethyl alcohol and water.

Figure 7.6 Permanent antifreeze, ethylene glycol.

for polar or ionic solids such as sodium chloride and sugar. Carbon tetrachloride readily dissolves nonpolar greases and oils. Ethyl alcohol, which is moderately polar, can dissolve water or carbon tetrachloride, as well as sugar and a small amount of sodium chloride.

The formation of hydrogen bonds between water and a solute also is a factor in increasing solubility. Such attractions can form between the hydrogen atoms in water and the oxygen atoms in organic alcohols, compounds that otherwise might not dissolve because of the nonpolar character of the hydrocarbon chain. The attraction between ethyl alcohol and water molecules is represented in Figure 7.5.

Similarly, ethylene glycol molecules interact very strongly with themselves and with water molecules. The latter leads to high solubility in water. The attraction between molecules of ethylene glycol makes it difficult to separate them.

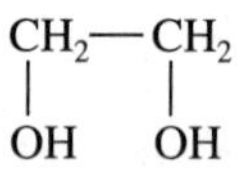

CH$_2$—CH$_2$
| | ethylene glycol
OH   OH

Since they must be separated for ethylene glycol to boil and vaporize, the boiling point is high (198 °C). The high boiling point plus the high water solubility explain why ethylene glycol makes an excellent permanent antifreeze (Figure 7.6).

Temperature

In general, the solubility of a solid solute in a liquid solvent increases as the temperature of the solution is increased. Since the dissolution of most solids requires an input of energy to break the solid structure apart, an increase in temperature facilitates the process. An *endothermic process* is one in which energy is absorbed and a cooling of the solution may be observed. An *exothermic process* is one that releases energy as heat.

Solids that dissolve exothermically show a decrease in solubility with an increase in the temperature of a solution. The relationship between temperature and solubility in water is shown for several solids in Figure 7.7.

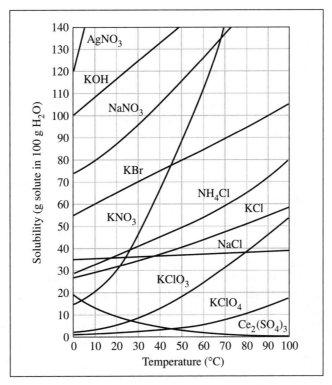

Figure 7.7   The effect of temperature on the solubility of several salts.

The solubility of a gaseous solute in a liquid solvent generally decreases as the temperature is raised. With an increase in temperature, the energy of the gaseous molecules is increased, and attractive forces between the solute and solvent molecules are broken. Gas molecules are then increasingly free to escape from a liquid. When a solution freezes, however, solubility sharply decreases. A carbonated soft drink placed in a freezer may explode as the pressure builds up to hazardous levels.

A considerable amount of gas can be dissolved in most liquids. A simple example is air dissolved in water. Heating a container of water results in vigorous bubbling long before the water is hot enough to boil, since air escapes as heating decreases its solubility.

Pressure

The solubility of liquids and solids is affected little by pressure. The properties of a gas, including its solubility in a liquid, are greatly influenced by the pressure applied to it. An increase in the pressure increases contact between the gas and a solvent, so gas molecules can more easily enter the liquid. Therefore, the amount of gas that can be dissolved increases as the pressure above the solution increases.

Carbon dioxide, a gas only slightly soluble in water under normal conditions, is forced into carbonated beverages under pressures between 5 and 10 atmospheres. When removal of the bottle cap or can tab reduces the pressure to approximately 1 atmosphere, the gas rapidly decreases in solubility. This rapid escape of the dissolved gas, as its solubility is decreased by the decrease in pressure, is seen as effervescence.

7.4 Concentration

GOAL: To perform simple calculations involving the common units of concentration.

The **concentration** of a solution is an expression of the amount of solute dissolved in a given quantity of the solvent, or present in a given quantity of the solution. For example, the statements that "a solution contains 10.0 g of sodium chloride dissolved in 100 mL of water" and "a solution contains 0.50 mol of glucose in 1 L of solution" are expressions of concentration.

Several general terms are commonly used to describe the amount of solute present. A **concentrated** solution is one in which (1) the amount of solute is large compared with the amount of solvent, or (2) the amount of solute is large compared with the total amount that can be dissolved in a given amount of solvent. For example, since it is possible to dissolve up to 97 g of ammonium bromide in 100 mL of water at 25 °C (see Table 7.3), a solution prepared by dissolving 90 g of the salt in 100 mL of water might be labeled as concentrated. This amount of ammonium bromide is comparable to the amount of water used and is close to the total solubility of the salt in the solvent.

A solution in which the amount of solute is small compared with the amount of solvent or compared with the amount of solute that could be dissolved is called **dilute.** Since it is possible to dissolve up to 97 g of ammonium bromide in 100 mL of water, a solution prepared by dissolving only 1 g of the salt in 100 mL of water might be labeled dilute.

Any solution that contains the maximum amount of solute that can be dissolved in a solvent at a given temperature and pressure is called **saturated.** Saturated solutions of ammonium bromide can be prepared by dissolving 97 g of the salt in 100 mL of water, by dissolving 970 g of the salt in 1 L of water, or by dissolving 9.7 g of ammonium bromide in 10 mL of water, at 25 °C.

Saturated solutions can be identified visually by the presence of solid on the bottom of the container.

Any solution that contains less than the maximum amount of solute dissolved in a given amount of solvent is referred to as **unsaturated.** Therefore, any solution of ammonium bromide dissolved in water that contains less than 97 g of the salt for every 100 mL of water used is unsaturated. Both a solution containing 90 g of ammonium bromide per 100 mL of solution (concentrated) and one containing 1 g of the salt per 100 mL of solution (dilute) are unsaturated.

Sometimes solutions may contain more than the normal maximum amount of dissolved solute. They are said to be **supersaturated** (Figure 7.8). These solutions, which

Figure 7.8   The behavior of a supersaturated solution. In the left photo, a seed crystal is being added to a supersaturated solution of sodium acetate in water. Crystallization begins (middle photo) and continues as the solution converts to saturated by formation of solid sodium acetate (right photo).

can result when a saturated solution cools slowly, are extremely unstable. If disturbed in any way, they readily revert to the point of saturation by forming crystals.

Percentage Concentration

The concentrations of household, pharmaceutical, and biological solutions are frequently expressed as percentage values. Solutions labeled 3% hydrogen peroxide, 97% inert ingredients, 14% alcohol, 10% iodine, 0.67% sodium chloride, and 4% boric acid are common. The specific meaning of these percentages, however, varies with the type of solute and solvent.

Mass Percent

Mass percent (% m/m, also called weight percent) is the mass of solute present in 100 g of *solution*. For example, a 2% (m/m) solution of NaCl in water would contain 2 g of NaCl in every 100 g of the solution; the other 98 g would be water. The mass percent for a solution of known composition can be calculated from:

$$\text{Mass percent solute} = \frac{\text{mass of solute}}{\text{total mass of solution}} \times 100$$

The ratio is expressed *per gram of solution,* so the factor of 100 is required to describe the mass of solute *per 100 g of solution.* The mass of solvent is the difference between the mass of solution and the mass of the solute; it is unspecified, but it is needed to prepare a solution of a given mass percent. Consider the following examples.

EXAMPLE 7.1:

What mass of sodium chloride is required to prepare 500 g of a solution of sodium chloride in water that is 2.0% (m/m)?

Solution:

A 2% (m/m) solution requires 2.0 g of the solute sodium chloride for every 100 g of solution prepared. Therefore, 500 g of solution will require 10.0 g of NaCl.

$$500 \text{ g solution} \times \frac{2.0 \text{ g NaCl}}{100 \text{ g solution}} = 10.0 \text{ g NaCl}$$

The ratio in this expression is equal to the mass percent of NaCl, which is 2 g NaCl per 100 g solution. The desired solution would be prepared by combining 10.0 g of NaCl with 490 g of water.

EXAMPLE 7.2:

An experiment requires 12.0 g of sodium hydroxide. What mass of solution 10.0% (m/m) must be used?

Solution:

The solution provided contains 10.0 g of sodium hydroxide in every 100 g. Therefore, 120 g of solution is required.

$$12.0 \text{ g NaOH} \times \frac{100 \text{ g solution}}{10.0 \text{ g NaOH}} = 120 \text{ g solution}$$

Here again, the *ratio* is the mass percent of NaOH (in reciprocal form) used as a conversion factor in the factor-label method (Section 1.3). If it is necessary to prepare the solution, this can be done by combining 12.0 g of NaOH and 108 g (120 − 12) of water.

Mass/Volume Percent

The percentage concentration of a solution of a solid solute in a liquid solvent, such as sodium chloride in water, can also be given as a *mass to volume percent* (% m/v)—the mass of solute dissolved in 100 mL of the *solution.* For example, 0.67% (m/v) aqueous sodium chloride is a solution that contains 0.67 g of sodium chloride in every 100 mL of solution. A 5% (m/v) solution of sucrose in water contains 5 g of sucrose dissolved in every 100 mL of solution. The mass/volume percent of a solution of known composition is calculated from:

$$\text{Mass/volume percent solute} = \frac{\text{mass of solute}}{\text{volume of solution (mL)}} \times 100$$

In this expression, the ratio is expressed *per milliliter of solution.* Therefore, the factor of 100 is required to describe the mass of solute *per 100 mL of solution.* Notice that the volume of solvent is not specified and is not required to describe or prepare the solution.

EXAMPLE 7.3:

What mass of copper sulfate, $CuSO_4$, is required to prepare 500 mL of a solution in water that is 2.0% (m/v)?

Solution:

A 2% (m/v) solution requires 2.0 g of the solute copper sulfate for every 100 mL of solution prepared. Therefore, 500 mL of solution will require 10.0 g of $CuSO_4$.

$$500 \text{ mL solution} \times \frac{2.0 \text{ g CuSO}_4}{100 \text{ mL solution}} = 10.0 \text{ g CuSO}_4$$

The desired solution would be prepared by placing 10.0 g of $CuSO_4$ into a volumetric flask, adding enough water to dissolve the salt, and then adding additional water to bring the total volume of solution to 500 mL. The process is illustrated in Figure 7.9.

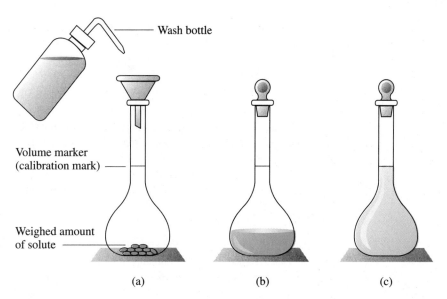

Figure 7.9 Preparation of 500 mL of a 2% (m/v) solution requires 10.0 g of $CuSO_4$. See Example 7.3 for calculation. To prepare this solution, (a) 10.0 g of $CuSO_4$ is placed in a 500-mL volumetric flask. (b) A small amount of water is added to dissolve the solid. (c) More water is then added until the total volume of the solution is exactly 500 mL.

Wash bottle

Volume marker (calibration mark)

Weighed amount of solute

(a)   (b)   (c)

EXAMPLE 7.4:

An experiment requires 12.0 g of sodium hydroxide. What volume of a solution 10.0% (m/v) must be used?

Solution:

The solution provided contains 10.0 g of sodium hydroxide in every 100 mL. Therefore, 120 mL of solution is required.

$$12.0 \text{ g NaOH} \times \frac{100 \text{ mL solution}}{10.0 \text{ g NaOH}} = 120 \text{ mL solution}$$

Comparison of Mass Percent and Mass/Volume Percent

Before considering other units of concentration, let us see how mass percent and mass/volume percent differ. Consider a solution that is 10.0% (m/m) NaCl:

$$10.0\% \text{ (m/m) NaCl} = \frac{10.0 \text{ g NaCl}}{100 \text{ g solution}}$$

By knowing that the density of this solution, 1.07 g/mL at 25 °C, we can use it to convert this relationship to the corresponding mass/volume percent:

$$\% \text{ (m/v)} = \frac{10.0 \text{ g NaCl}}{100 \text{ g solution} \times \dfrac{1 \text{ mL solution}}{1.07 \text{ g solution}}}$$

$$= 10.7\% \text{ (m/v)} = \frac{10.7 \text{ g NaCl}}{100 \text{ mL solution}}$$

As this calculation shows, the mass/volume percent and the mass percent are related according to:

$$\% \ (m/v) = \% \ (m/m) \times density$$

$$10.7\% \ (m/v) = 10.0\% \ (m/m) \times 1.07 \ g/mL$$

This relationship shows that the mass percent and the mass/volume percent are equal when the density is 1 g/mL; this is approximately true for pure water and for any very dilute aqueous solution. The significance of this is illustrated in the following section. As the concentration of a salt increases in water, the density increases and the mass/volume percent becomes correspondingly greater than the mass percent.

Parts per Million and Parts per Billion

For very dilute solutions, as with trace impurities in water and with many environmental samples (see Chapter 14), the concentration is often expressed in *parts per million* (ppm). This unit of concentration is defined according to:

$$ppm \ solute = \frac{mass \ of \ solute}{total \ mass \ of \ solution} \times 10^6$$

Recall that mass percent is defined by a similar equation except that it is the mass per 100 g of solution, whereas the concentration in parts per million (10^6) is the mass per 1 million g of solution. From the respective definitions of mass percent and parts per million, the two units of concentration are related according to:

$$ppm = mass \ percent \times 10^4$$

When the concentration of solute is even smaller, the concentration is often expressed in *parts per billion* (ppb). This unit of concentration is defined according to:

$$ppb \ solute = \frac{mass \ of \ solute}{total \ mass \ of \ solution} \times 10^9$$

In this case, the concentration represents the mass of solute in 1 billion (10^9) g of solution. The mass percent and the concentration in parts per billion are related according to:

$$ppb = mass \ percent \times 10^7$$

The concentrations of aqueous solutions that are normally given in ppm and ppb are very low, with densities close to 1 g/mL, where there is no significant difference between mass/volume percent and mass percent. The following examples illustrate the value of expressing concentrations in ppm or ppb.

EXAMPLE 7.5:

What is the concentration in ppm of a solution that is 0.0027% (m/v)?

Solution:

Since mass percent and mass/volume percent are approximately equal when the concentration is very low, the concentration can be calculated to be 27 ppm, a number that is more easily communicated than the very small number for the corresponding mass percent.

$$\text{ppm} = 0.0027 \times 10^4 = 27 \text{ ppm}$$

EXAMPLE 7.6:

What is the concentration in ppb of a solution that is 0.00000084% (m/m)?

Solution: $\qquad \text{ppb} = 0.00000084 \times 10^7 = 8.4 \text{ ppb}$

Volume Percent and Proof

The percentage concentrations of solutions containing liquid solutes dissolved in liquid solvents, such as solutions of alcohols in water, can be expressed as *volume to volume percent* (% v/v), simply known as the *volume percent*. Such concentrations state the number of milliliters of solute dissolved in 100 mL of the solution. The volume percent may be calculated from:

$$\text{Volume percent solute} = \frac{\text{volume of solute (mL)}}{\text{volume of solution (mL)}} \times 100$$

For example, a solution labeled 14% (v/v) ethyl alcohol contains 14 mL of the alcohol in every 100 mL of the solution. A 70% rubbing alcohol (isopropyl alcohol) solution contains 70 mL of the alcohol in every 100 mL of solution (Figure 7.10).

Figure 7.10 Rubbing alcohol usually contains 70% (v/v) isopropyl alcohol.

The alcohol content of some liquids is given in *proof,* for which the symbol $\sim$ is sometimes used. This unit of concentration is usually restricted to distilled alcoholic beverages, such as vodka, gin, and various types of whiskey. The concentration of alcohol described in proof is double the percent by volume. Therefore, 100% alcohol is the same as 200$\sim$. A typical gin might be 80$\sim$ or 40% alcohol. No doubt, using proof as the unit of concentration fools some consumers into thinking they are getting much more alcohol for their money.

EXAMPLE 7.7:

What volume of alcohol is present in a 1.0-L bottle of rubbing alcohol that is 70% (v/v) isopropyl alcohol?

Solution:

The rubbing alcohol contains 70 mL of isopropyl alcohol in every 100 mL of solution. The bottle has a volume of 1.0 L, or 1000 mL. Therefore, the bottle contains 700 mL of isopropyl alcohol.

$$1000 \text{ mL solution} \times \frac{70 \text{ mL alcohol}}{100 \text{ mL solution}} = 700 \text{ mL alcohol}$$

EXAMPLE 7.8:

What volume of ethyl alcohol (ethanol) is required to prepare 60.0 mL of a solution of the alcohol in water that is 3.8% (v/v)?

Solution:

The desired solution is to contain 3.8 mL of ethanol in 100 mL of solution. To prepare 60.0 mL of solution would require 2.3 mL of ethanol.

$$60.0 \text{ mL solution} \times \frac{3.8 \text{ mL ethanol}}{100 \text{ mL solution}} = 2.3 \text{ mL ethanol}$$

Karat

The *carat* is a unit of mass used for precious stones; it is equal to 200 mg. The *karat* is a unit of concentration, used to describe the amount of gold in an alloy. Pure gold is extremely soft; one ounce can be stretched into wire several miles long or it can be flattened into gold leaf with an area of 30 square meters. Because of the softness, jewelers combine gold into alloys with other metals for greater durability. The gold content of an alloy is expressed in 24ths or parts per 24. Pure gold is 24 karat, whereas 18-karat gold is 18/24 or 75% (m/m) gold.

Yellow gold is commonly used for making coins, rings, and other jewelry; it is usually 22 karat (approximately 90%) gold with copper added for durability. White gold is usually 18-karat gold; the color is largely due to nickel (15–20%), although a small amount of zinc is ordinarily present. The color of pink gold is primarily due to copper (about 25%), with some silver added to improve the ease of working the alloy into various shapes. Green gold derives its color from the presence of silver (usually 25%); a small amount of copper is sometimes present.

Molarity

Molarity, the concentration expression most frequently used in the chemical laboratory, is based on the chemical unit of measurement, the mole (Section 5.2). The **molarity** of a solution is defined as the number of moles of solute dissolved in one liter of solution.

$$\text{Molarity} = \frac{\text{moles of solute}}{\text{liters of solution}}; \; M = \frac{\text{mol}}{L}$$

Solutions that contain 1 mol of solute dissolved in 1 L of solution, 0.5 mol of solute dissolved in 500 mL (0.5 L) of solution, and 6 mol of solute in 6 L of solution are all one molar (1 M).

$$M = \frac{1 \text{ mol}}{1 \text{ L}} = 1 \text{ molar}$$

$$M = \frac{0.5 \text{ mol}}{0.5 \text{ L}} = 1 \text{ molar}$$

$$M = \frac{6 \text{ mol}}{6 \text{ L}} = 1 \text{ molar}$$

Similarly, a solution labeled 0.25 M is known to contain 0.25 mol of the solute in 1 L of solution. Therefore, 200 mL of this solution would contain 0.05 mol of solute and 2.0 L would contain 0.50 mol.

$$0.25 \text{ M} = \frac{0.25 \text{ mol}}{1.0 \text{ L}} = \frac{0.50 \text{ mol}}{2.0 \text{ L}} = \frac{0.05 \text{ mol}}{0.200 \text{ L}}$$

EXAMPLE 7.9:

What is the molarity of a solution prepared by dissolving 1.5 mol of sodium chloride in enough water to prepare 500 mL of solution?

Solution

$$M = \frac{\text{mol NaCl}}{\text{L solution}}$$

$$M = \frac{1.5 \text{ mol NaCl}}{0.500 \text{ L solution}} = 3.0 \text{ M}$$

The solution is said to be 3.0 molar (3 M); each liter of the solution contains 3.0 mol of sodium chloride.

EXAMPLE 7.10:

Calculate the molarity of a solution for which enough water is added to 8.0 g of sodium hydroxide to prepare 100 mL (0.100 L) of solution. The molar mass of NaOH is 40 g/mol.

Solution:

Using the molar mass as a conversion factor (see Section 5.3):

$$M = \frac{8.0 \text{ g NaOH} \times \dfrac{1 \text{ mol}}{40 \text{ g}}}{0.100 \text{ L solution}} = 2.0 \text{ M}$$

EXAMPLE 7.11:

What mass of sodium sulfate is required to prepare 100 mL (0.100 L) of a 0.50 M solution of sodium sulfate in water? The molar mass of Na_2SO_4 is 142 g/mol.

Solution

$$M = \frac{\text{mol } Na_2SO_4}{\text{L solution}}$$

Or,

$$\text{mol } Na_2SO_4 = M \times \text{L solution}$$
$$\text{mol } Na_2SO_4 = 0.50 \text{ M} \times 0.100 \text{ L}$$
$$= 0.50 \frac{\text{mol}}{\text{L}} \times 0.100 \text{ L} = 0.050 \text{ mol}$$

The number of moles of sodium sulfate required can be converted into the mass required, using the molar mass.

$$0.050 \text{ mol} \times \frac{142 \text{ g}}{1 \text{ mol}} = 7.1 \text{ g } Na_2SO_4$$

Therefore, the required solution can be prepared by dissolving 7.1 g of sodium sulfate in enough water to produce a solution with a total volume of 100 mL.

EXAMPLE 7.12:

What total volume of a 0.20 M solution of sodium chloride in water can be prepared from 50 g of the salt? The molar mass of NaCl is 58.5 g/mol.

Solution

$$M = \frac{\text{mol NaCl}}{\text{L solution}}$$

Or,

$$\text{L solution} = \frac{\text{mol NaCl}}{M}$$
$$= \frac{50 \text{ g} \times \dfrac{1 \text{ mol}}{58.5 \text{ g}}}{0.20 \dfrac{\text{mol}}{\text{L}}}$$

$$= \frac{0.85 \text{ mol}}{0.20 \dfrac{\text{mol}}{\text{L}}} = 4.3 \text{ L}$$

The solution, therefore, could be prepared by dissolving the 50 g of sodium chloride in the amount of water required to give a total volume of 4.3 L.

7.5   Acids and Bases

GOALS: To describe the major theories to define acids and bases.
To know the experimental properties that identify acids and bases.
To describe pH, including the degree of change of [H$^+$] that accompanies a change of pH.
To appreciate the difference between acid concentration and acid strength.
To appreciate the nature of common solutions such as hydrochloric acid, ammonia water, and carbonated water.
To understand the relationship among the three species carbonic acid, bicarbonate, and carbonate, including the relative basicity of the ions.
To explain why $CaCO_3$ is a satisfactory antacid, whereas Na_2CO_3 is not.
To write equations for the reactions that lead to the formation of stalagmites and stalactites.

Acidic and basic substances are familiar to everyone. The sour taste of vinegar is due to the presence of acetic acid, and lemon juice is sour due to citric acid. Common bases include lye, lime, milk of magnesia, and baking soda. Bases are often described as bitter, and when they are dissolved in water, as slippery feeling.

Litmus paper is a porous paper impregnated with an indicator dye called litmus, which turns red in acid and blue in base.

The chemist uses the following criteria to determine the presence of acids and bases. Acids turn litmus paper from blue to red, neutralize bases, and form solutions having a pH of less than 7. Bases turn litmus paper from red to blue, neutralize acids, and form solutions that have a pH that is greater than 7. We will consider these three properties after defining acids and bases.

Definitions

Three definitions are commonly used. Svante Arrhenius, a Swedish scientist, stated that **an acid is a substance that releases hydrogen ions, H$^+$.** The common mineral acids hydrochloric acid, nitric acid, and sulfuric acid all fit this definition when they are dissolved in water.

$$HCl \xrightarrow{H_2O} H^+ (aq) + Cl^- (aq)$$
$$\text{hydrochloric acid}$$

$$HNO_3 \xrightarrow{H_2O} H^+ (aq) + NO_3^- (aq)$$
$$\text{nitric acid}$$

$$H_2SO_4 \xrightarrow{H_2O} 2 H^+ (aq) + SO_4^{2-} (aq)$$
$$\text{sulfuric acid}$$

Arrhenius defined a base as a substance that produces hydroxide ions, OH$^-$. Sodium hydroxide (Figure 7.11), potassium hydroxide, calcium hydroxide, and magnesium hydroxide (milk of magnesia) all fit this definition.

Figure 7.11 Drain cleaners often contain highly caustic bases, such as sodium hydroxide. See Chapter 22 for details.

$$\text{NaOH (s)} \xrightarrow{\text{H}_2\text{O}} \text{Na}^+ \text{(aq)} + \text{OH}^- \text{(aq)}$$
sodium
hydroxide

$$\text{KOH (s)} \xrightarrow{\text{H}_2\text{O}} \text{K}^+ \text{(aq)} + \text{OH}^- \text{(aq)}$$
potassium
hydroxide

$$\text{Ca(OH)}_2 \text{ (s)} \xrightarrow{\text{H}_2\text{O}} \text{Ca}^{2+} \text{(aq)} + 2 \text{ OH}^- \text{(aq)}$$
calcium
hydroxide

$$\text{Mg(OH)}_2 \text{ (s)} \xrightarrow{\text{H}_2\text{O}} \text{Mg}^{2+} \text{(aq)} + 2 \text{ OH}^- \text{(aq)}$$
magnesium
hydroxide

In 1923 a second theory of acids and bases was introduced as a result of independent efforts by J. N. Brönsted, a Danish chemist, and T. M. Lowry, an English chemist. According to the Brönsted-Lowry definitions, **an acid is a substance that will *donate* hydrogen ions, H$^+$,** and **a base is a substance that will *accept* hydrogen ions.** In place of the phrase *hydrogen ion,* the chemist routinely uses the term *proton,* since this is what a hydrogen ion is—a hydrogen atom minus the valence electron. An acid can therefore be described as a **proton donor.** The simplest of the common acids is hydrochloric acid, which is a solution of hydrogen chloride, HCl, in water. Pure hydrogen chloride is a gas but when dissolved in water, it ionizes to release H$^+$ and Cl$^-$.

The tendency for acids to dissociate and release hydrogen ions in water is a consequence of the ability of the water molecule to accept a proton; that is, to act like a base:

$$\underset{\text{acid}}{\text{HCl}} + \underset{\text{base}}{\text{H}_2\text{O}} \rightarrow \text{H}_3\text{O}^+ + \text{Cl}^-$$

The species, H$_3$O$^+$, is called *hydronium ion.* The ability of the water molecule to accept a proton and form a hydronium ion is easily understood from the electron dot structure, as shown in the following equation:

$$\text{H}^+ + :\ddot{\text{O}}:\text{H} \rightarrow \left[\begin{matrix} \text{H} \\ :\ddot{\text{O}}:\text{H} \\ \ddot{\text{H}} \end{matrix} \right]^+$$
$$\hspace{2.5cm} \ddot{\text{H}} \hspace{2.2cm} \text{hydronium ion}$$

Bases dissolve in water to form *basic* or *alkaline* solutions. The term *alkali* is sometimes used in place of the word base. Sodium hydroxide is a common base sometimes called *lye*. This compound provides OH⁻ (Arrhenius's definition) to accept hydrogen ions (Brönsted-Lowry definition). Since hydroxide ions will accept protons to form water molecules, any Arrhenius base also fits the Brönsted-Lowry description of a base. The latter is preferred, however, because it includes bases other than hydroxide ion.

The limitations of the Arrhenius description of bases are illustrated by ammonia, NH_3, another common base. Ammonia is a gas that is somewhat soluble in water. The resulting solution is known either as *ammonia water* or *ammonium hydroxide*. The latter name is attributable to the reversible reaction.

$$:NH_3 + H_2O \; \underset{}{\overset{\longrightarrow}{\longleftarrow}} \; NH_4^+ + OH^-$$

$$\underset{\text{ammonia}}{} \qquad\qquad \underset{\substack{\text{ammonium} \\ \text{ion}}}{}$$

This reaction proceeds more efficiently to the left, so ammonia water is largely a solution of NH_3 gas in water, containing only a small amount of ammonium hydroxide, NH_4OH.

In the equation, ammonia acts as a base by gaining a proton to form ammonium ion. The proton is provided by water, which is acting as an acid. Since water has little tendency to give up a proton, this reaction does not go to completion. On the other hand, if ammonia comes in contact with a more acidic substance, the formation of ammonium ion is complete.

The conversion of ammonia into ammonium ion by addition of H⁺ can be understood from their structures. See Section 3.4 for details.

One interesting example is the reaction of ammonia gas and hydrogen chloride gas to form ammonium chloride, NH_4Cl.

$$HCl + NH_3 \rightarrow NH_4Cl$$
$$\underset{\text{ammonium chloride}}{}$$

Since ammonia water is a solution of NH_3 gas in water, some of the gas tends to move out of solution and escape from the bottle. Similarly, a bottle of hydrochloric acid is an aqueous solution of HCl gas, which can also come out of solution and leave the bottle. The vapors from each bottle can be tested with wet litmus paper to show that HCl is acidic and NH_3 is basic. If these two bottles are placed close together, the two gases undergo an acid-base reaction to form NH_4Cl; this substance is a white powder that appears as smoke above the bottles (Figure 7.12).

Figure 7.12 The formation of ammonium chloride from NH_3 gas and HCl gas. The white smoke over the tops of the bottles is solid NH_4Cl.

Neutralization

Another important characteristic of acids and bases is their ability to *neutralize* one another. Neutralization can usually be described as the reaction of an acid and a base to form a salt plus water. The following equation can be written for the neutralization of potassium hydroxide with nitric acid.

$$K^+(aq) + OH^-(aq) + H^+(aq) + NO_3^-(aq) \rightarrow H_2O(l) + K^+(aq) + NO_3^-(aq)$$

Since the spectator ions (Section 4.1) are not changed during the reaction, the equation can be simplified to:

$$H^+(aq) + OH^-(aq) \rightarrow H_2O(l)$$

The reaction of ammonia and hydrogen chloride, described in the previous section, is an unusual neutralization reaction since no water is formed. But the acid and base are neutralized.

pH

Another property used to identify acids and bases in the laboratory is pH, a numerical scale that describes the acidity of solutions. The pH scale is related to the water ionization reaction, which can be described by either of the following equations:

$$H_2O\,(l) \rightleftharpoons H^+(aq) + OH^-(aq)$$

$$H_2O\,(l) + H_2O\,(l) \rightleftharpoons H_3O^+ + OH^-(aq)$$

The first equation represents the ionization of water without showing the location of the hydrogen ion that is released. The second equation represents the interaction of two water molecules, one acting as a Brönsted-Lowry acid and the other as a Brönsted-Lowry base. The second equation also emphasizes that the hydrated proton exists as a hydronium ion.

Both equations show that the ionization of water yields equal amounts of hydrogen ions (or hydronium ions) and hydroxide ions. In other words, the molar concentrations of the two ions that result from ionization are equal. At 25 °C, the concentration of hydrogen ion, represented by $[H^+]$ or $[H_3O^+]$, can be measured as 10^{-7} M. Similarly, the concentration of hydroxide ions, $[OH^-]$, is 10^{-7} M.

$$[H^+] = [OH^-] = 1.0 \times 10^{-7} \text{ M}$$

Such low concentrations mean that only about two in every one billion molecules is ionized. Despite the small degree of ionization, the balance of $[H^+]$ and $[OH^-]$ is the basis of the pH scale used for describing the acidity of any substance or solution.

The pH of a solution is defined as the negative logarithm of the hydrogen ion concentration in moles per liter (M):

$$pH = -\log [H^+]$$

When $[H^+] = 10^{-7}$ M:

$$pH = -\log 10^{-7} = -(-7) = 7$$

Therefore, when $[H^+] = [OH^-] = 10^{-7}$ M, the pH = 7. Since the amount of acidic and basic ions is equal under these circumstances, we describe a pH of 7 as **neutral.** When an excess of H^+ is present, $[H^+] > [OH^-]$, the solution is **acidic.** When an excess of OH^- is present, $[H^+] < [OH^-]$, the solution is **basic** or **alkaline.**

For example, if enough acid is added to water to increase the $[H^+]$ from 10^{-7} to 10^{-6}, the solution becomes acidic:

$$pH = -\log 10^{-6} = 6$$

Such a change represents a 10-fold increase in $[H^+]$. In other words, a change in pH of one unit results from a 10-fold change in hydrogen ion concentration. If the pH were to drop from 7 to 4, the change would correspond to a 1000-fold (10^3) increase in the hydrogen ion concentration.

Notice that the pH and the hydrogen ion concentration move in opposite directions: as pH increases, $[H^+]$ decreases. Furthermore, when $[H^+]$ increases, the concentration of OH^- decreases; this is due to the formation of water. When excess H^+ ions are added, they will consume hydroxide ions and cause a corresponding decrease in $[OH^-]$ so that the following equality always applies:

$$[H^+] \times [OH^-] = 10^{-14}$$

When $[H^+]$ increases to 10^{-5}, the $[OH^-]$ decreases to 10^{-9} so that this equality is maintained. Thus, a 100-fold increase of $[H^+]$ is accompanied by a 100-fold decrease of $[OH^-]$. Table 7.3 summarizes these changes for all pH values.

To summarize:
A solution is **acidic** when:

$$[H^+] > [OH^-] \text{ and pH} < 7$$

Table 7.3   pH

pH	Concentration of Hydrogen Ion (mol/L)		Concentration of Hydroxide Ion (mol/L)
1	0.1	or 10^{-1}	10^{-13}
2	0.01	or 10^{-2}	10^{-12}
3	0.001	or 10^{-3}	10^{-11}
4	0.0001	or 10^{-4}	10^{-10}
5	0.00001	or 10^{-5}	10^{-9}
6	0.000001	or 10^{-6}	10^{-8}
7	0.0000001	or 10^{-7}	10^{-7}
8	0.00000001	or 10^{-8}	10^{-6}
9		10^{-9}	10^{-5}
10		10^{-10}	10^{-4}
11		10^{-11}	10^{-3}
12		10^{-12}	10^{-2}
13		10^{-13}	10^{-1}
14		10^{-14}	10^{0}

A solution is **basic** when:

$$[H^+] < [OH^-] \text{ and pH} > 7$$

A solution is **neutral** when:

$$[H^+] = [OH^-] \text{ and pH} = 7$$

These relationships can also be seen from:

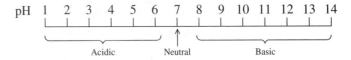

Precise measurements of pH require the use of a pH meter such as that shown in Figure 7.13. The pH values of many common substances are given in Figure 7.14.

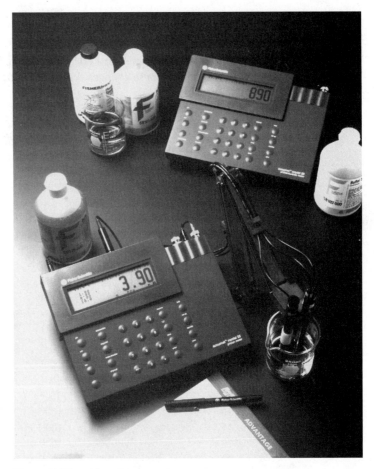

Figure 7.13   pH meters.

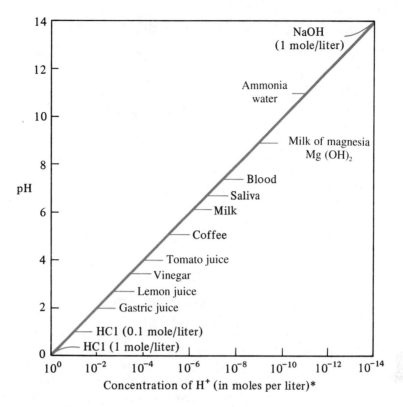

Figure 7.14 The pH of some common substances.

Concentration versus Strength

Acid *concentration* and acid *strength* are two distinctly different properties. The concentration of an acidic (or basic) solution describes the amount of acid (or base) dissolved in a specific quantity of solution. The idea of acid or base strength is more complex; it describes the relative ability of an acid to donate a proton (or a base to accept one). From arrows in the following equations, the distinction between a strong acid, such as hydrochloric acid, and a weak acid, such as acetic acid, can be easily understood.

$$\underset{\substack{\text{hydrochloric}\\\text{acid}}}{\text{HCl}} + \text{H}_2\text{O} \;\longrightarrow\; \text{H}_3\text{O}^+ + \text{Cl}^-$$

$$\underset{\text{acetic acid}}{\text{CH}_3\overset{\displaystyle \text{O}}{\overset{\|}{\text{C}}}\!-\!\text{OH}} + \text{H}_2\text{O} \;\underset{\longleftarrow}{\overset{\longrightarrow}{}}\; \text{H}_3\text{O}^+ + \text{CH}_3\overset{\displaystyle \text{O}}{\overset{\|}{\text{C}}}\!-\!\text{O}^-$$

The arrows give a qualitative indication of the tendency of each acid to *dissociate* into its respective ions. The strong acid ionizes completely in water, whereas the acetic acid ionizes only slightly, usually less than 5%. The pH of an acidic solution depends on both the concentration and the strength of an acid since both affect the amount of available hydrogen ion.

Table 7.4 Common weak acids

Weak Acid	Common Source
Carbonic	Soft drinks
Acetic	Vinegar
Citric	Citrus fruits
Tartaric	Grapes
Malic	Apples, grapes
Lactic	Sour milk, cheese
Boric	Eyewash solutions

Common Acids and Bases

The most common strong acids are hydrochloric acid, nitric acid, and sulfuric acid. They dissociate according to the reactions shown at the beginning of this section. Some important weak acids and common sources are listed in Table 7.4 and their structures are given in Figure 7.15. Notice that most of the weak acids are carboxylic acids (Section 6.3) and some have more than one proton that may be acidic. The acidic protons are highlighted.

The common strong bases are substances that contain hydroxide ion or produce it in large amounts when dissolved in water. Thus, sodium hydroxide, NaOH, potassium hydroxide, KOH, calcium hydroxide, $Ca(OH)_2$, and magnesium hydroxide, $Mg(OH)_2$, are common strong bases.

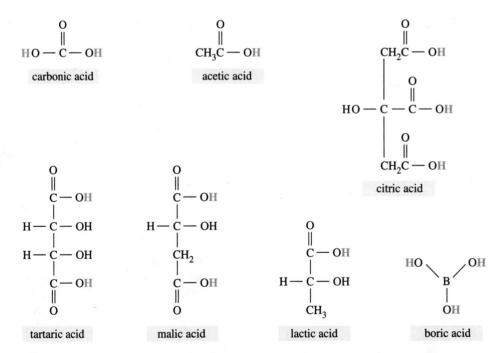

Figure 7.15 Some common weak acids. The acidic hydrogens are shown in color.

Sodium carbonate, Na_2CO_3, also is a base. It releases carbonate ions, which then form bicarbonate and hydroxide ions according to the following reactions.

$$Na_2CO_3 \rightarrow 2\,Na^+ + CO_3^{2-}$$
sodium carbonate · carbonate ion

$$CO_3^{2-} + H_2O \rightleftharpoons HCO_3 + OH$$
carbonate ion · bicarbonate ion · hydroxide ion

Bicarbonate ion from sodium bicarbonate, $NaHCO_3$, is also a base due to the following reaction:

$$HCO_3^- + H_2O \rightleftharpoons H_2CO_3 + OH^-$$
bicarbonate ion · carbonic acid · hydroxide ion

But the bicarbonate ion has much less tendency than the carbonate ion to capture a proton and cause the release of hydroxide ions. In other words, the basicity of carbonate ions is much greater than that of bicarbonate ions. These two ions are part of a series of three species that has carbonic acid as the parent compound.

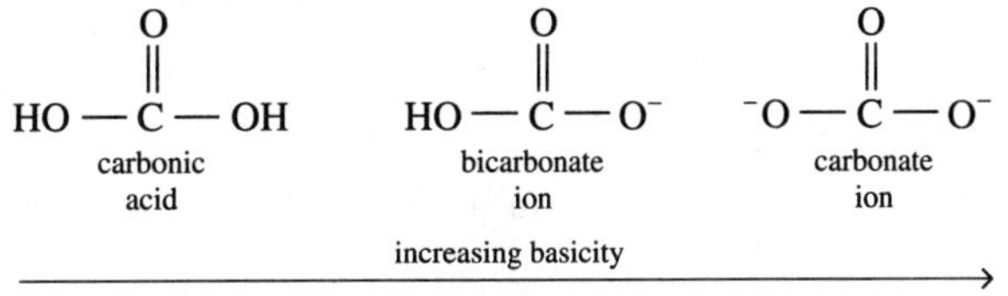

These complete structures help our understanding of the increase in basicity (or decrease in acidity) in going from left to right. The carbonate ion has the greatest affinity for hydrogen ions due to the -2 charge. After a proton is captured and bicarbonate ion forms, the decrease to a -1 charge causes the bicarbonate ion to be much less basic. As a result, sodium bicarbonate, $NaHCO_3$ (baking soda), is much less basic than sodium carbonate, Na_2CO_3 (washing soda). The latter is discussed further in Chapter 22. The carbonic acid is, of course, the least basic.

Since bicarbonate ion has an acidic proton available, it can act as an acid in the presence of a strong base. Nevertheless, bicarbonate is more often used as a weak base to neutralize acids. For example, several commercial antacids combine sodium (or potassium) bicarbonate with a weak organic acid such as citric acid. Both citric acid and sodium bicarbonate are solids and will not react with one another as long as they are kept dry. But once water is added, the compounds dissolve and the chemistry begins.

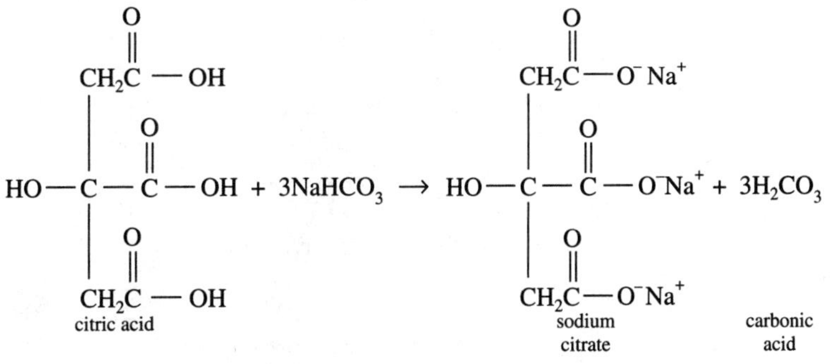

Carbonic acid is an unstable compound that decomposes according to the following reaction:

$$H_2CO_3 + H_2O \;\underset{\longleftarrow}{\longrightarrow}\; H_2O + CO_2 \text{ (gas)}$$

Thus, when carbonic acid forms in large quantity, bubbles of CO_2 (carbonation) are observed. Viewed another way, carbonic acid is merely a solution of CO_2 gas in water. When an excess is present, the gas will come out of solution; that is why there may be a "pop" when a can of carbonated soda is opened, since extra CO_2 was dissolved under pressure. Similarly, when $NaHCO_3$ is used as an antacid (for example, in Alka-Seltzer) and dry tablets containing sodium bicarbonate and citric acid are dropped into water, a fizzing action is observed. When some of the excess sodium bicarbonate in an Alka-Seltzer reaches the stomach and reacts with stomach acid, HCl, more CO_2 is formed in the reaction and a burp may result.

The sodium bicarbonate that does reach the stomach accounts for some of the antacid action of an Alka-Seltzer, but the sodium citrate released during the fizzing is also basic and neutralizes the strong HCl by forming the weak citric acid.

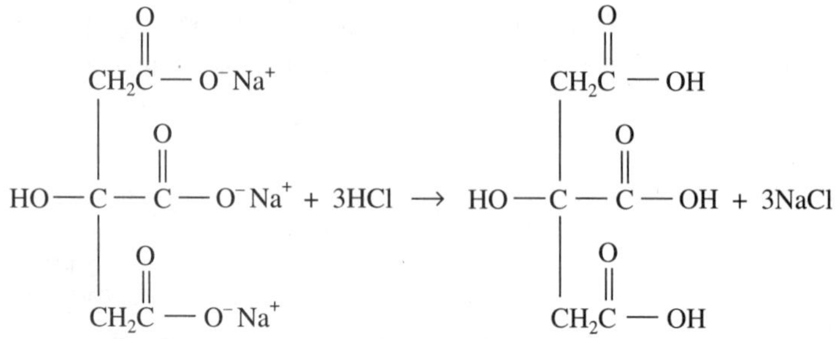

sodium citrate   hydrochloric acid   citric acid
(strong acid)   (weak acid)

Lime, also known as *quicklime,* is calcium oxide; it reacts with water to form *slaked lime,* which functions as a base by releasing hydroxide ions.

$$CaO + H_2O \rightarrow Ca(OH)_2$$

lime   slaked lime

Limestone is calcium carbonate, $CaCO_3$. It acts as a base like sodium carbonate, although calcium carbonate is practically insoluble in water if no acid is present. A solution of sodium carbonate is quite basic, whereas a solution of calcium carbonate is not very basic since little of this salt dissolves. If an acid is added, however, calcium carbonate will neutralize it.

The difference between Na_2CO_3 and $CaCO_3$ accounts for the choice of $CaCO_3$ rather than Na_2CO_3 as an antacid in such products as Tums, Alka 2, Titralac, and others. The $CaCO_3$ can be taken by mouth without any hazard; it will not react until it reaches the stomach, where it encounters HCl. On the other hand, when Na_2CO_3 comes in contact with water, it will dissolve and release hydroxide ions (see earlier equation), which may be damaging to tissues of the mouth and esophagus.

Calcium carbonate is used heavily in agriculture, where it is known as agricultural lime. Refer to Chapter 15 for details. Although it is quite insoluble in water, it can be changed by the following reaction:

$$CaCO_3 + H_2CO_3 \;\underset{\longleftarrow}{\longrightarrow}\; Ca(HCO_3)_2$$

calcium   carbonic   calcium
carbonate   acid   bicarbonate

Notice from the balanced equation that two bicarbonate ions form; both result from carbonate capturing a single proton from carbonic acid. The carbonic acid forms by the following reaction of CO_2 dissolved in water.

$$CO_2 + H_2O \xrightleftharpoons{} H_2CO_3$$

As the arrows show, this reaction is an equilibrium that can proceed in either direction, although the decomposition of carbonic acid is favored. If the concentration of carbon dioxide is high, some carbonic acid will form, which can then react with $CaCO_3$. As carbonic acid undergoes this acid-base reaction, more carbonic acid is formed from CO_2, thus allowing a large amount of $CaCO_3$ to dissolve as $Ca(HCO_3)_2$.

Calcium bicarbonate is quite soluble in water and moves readily when water flows through deposits of limestone. Cold water running through underground channels can dissolve a large amount of CO_2 and thus carry a large amount of $Ca(HCO_3)_2$. But when water carrying this salt enters an environment where evaporation can occur, the two reactions leading to the formation of $Ca(HCO_3)_2$ move in reverse, and large deposits of insoluble $CaCO_3$ result. Where underground water flows down to the ceiling of an underground cave, icicle-shaped stalactites form from above as water drips and evaporates. Water that drips to the floor of a cave, before evaporating, leads to formation of stalagmites. These columns of limestone protrude upward, usually directly beneath stalactites (Figure 7.16).

Tap water or even distilled water that has been in contact with air for some time shows a pH somewhat below 6. An acidic pH is due to the presence of CO_2, which is dissolved in the water and is undergoing the previous reaction. Although carbonic acid is unstable and is not present in large amounts, enough can exist in water to lower the pH to slightly acidic. Freshly distilled water will show a pH equal to 7 due to the loss of CO_2 gas during heating. But, when water is in contact with the air for a short time, the pH will again drop to below 6.

Figure 7.16 Stalagmites and stalactites.

SUMMARY

A solution is a homogeneous mixture in which a solute is dissolved in a solvent. Both the solute and the solvent may be gases, liquids, or solids. Air, carbonated beverages, salt water, and metal alloys are common solutions. Examples of alloys are yellow gold, sterling silver, and dental amalgams.

The solubility of most salts in water and the ability of oil and water to mix are typical solubility phenomena, easily explainable based on the relative polarity of solvents and solutes. Electronegativity is a property used for explaining the polarity of different compounds. When the electronegativity difference between two atoms is large, a bond between the atoms is highly polar. Hydrogen bonding between water molecules results from the high polarity of the oxygen-hydrogen covalent bond.

Temperature, pressure, and solute/solvent characteristics affect solubility. An increase in temperature generally increases solubility, although the solubility of gases increases as the temperature decreases. When a solution freezes, however, solubility of a gas decreases. The loss of CO_2 from carbonated beverages can cause a dangerous pressure buildup that could cause a can or bottle to explode.

The concentration of solutions may be expressed using several different units including percentage concentration as mass percent (% m/m), mass/volume percent (% m/v), and volume percent (% v/v). Very low concentrations may be given in parts per million (ppm) or parts per billion (ppb). The concentration of ethyl alcohol in alcoholic beverages is often given in proof, which is double the volume percent. Karat, not to be confused with carat (200 mg), is a unit of concentration commonly used to describe the gold content of gold alloys used in making jewelry. Molarity is the most common unit of concentration used in describing the concentration of solutions in the laboratory.

Certain properties characterize and define acids and bases, including the ability to neutralize one another. An acid, defined by the Arrhenius and Brönsted-Lowry theories, is a proton donor. According to the Arrhenius theory, a base is a source of hydroxide ions. The Brönsted-Lowry theory defines a base as a proton acceptor and includes several substances that do not supply hydroxide ions. The pH scale designates the exact level of acidity or basicity of any solution, including the level in many common fluids. A pH change of one unit corresponds to a 10-fold change in the concentration of hydrogen ions (or hydronium ions).

Acid concentration and acid strength are different properties. Concentration tells how much solute is present in a solution; strength describes the extent to which an acid releases hydrogen ions.

Among the common bases are carbonate, CO_3^{2-}, and bicarbonate, HCO_3^-, each of which may accept hydrogen ions to form carbonic acid, H_2CO_3, an unstable compound that decomposes to form water and carbon dioxide, resulting in a fizzing action. Limestone, calcium carbonate, is quite insoluble in water, but it may be made soluble by the presence of dissolved carbon dioxide, which yields carbonic acid. Reaction with the $CaCO_3$ forms soluble calcium bicarbonate. In underground caves, as a solution of calcium bicarbonate evaporates, carbon dioxide is lost and calcium carbonate forms again; this can lead to unusual limestone deposits, such as stalagmites and stalactites.

PROBLEMS

1. Give a definition or an example of each of the following.

   a. Solute
   b. Solvent
   c. Solution
   d. Aqueous
   e. Carbonation
   f. Alloy
   g. Amalgam
   h. Solvation
   i. Miscible
   j. Saturated
   k. Unsaturated
   l. Supersaturated
   m. Litmus paper
   n. Hydrochloric acid
   o. Alkaline
   p. Lye
   q. Lime
   r. Slaked lime
   s. Limestone
   t. Neutralization
   u. Baking soda
   v. Washing soda
   w. Electronegativity

2. What are the nature and a consequence of hydrogen bonding?

3. What is meant by "Like dissolves like"?

4. Describe the interaction between water molecules and sodium and chloride ions when NaCl dissolves in water.

5. Under what circumstances would a salt fail to dissolve in water?

6. Why do oil and water not mix?

7. What two properties of ethylene glycol make it so well suited for use as the primary component of antifreeze?

8. Why is it potentially dangerous to place a carbonated beverage in a freezer?

9. How do mass percent (% m/m) and mass/volume percent (% m/v) differ? When are they numerically equal?

10. When are solutions described in ppm or ppb? What is the concentration of a 1 ppm solution expressed in mass percent? In ppb?

11. What volume of ethyl alcohol is present in 1 L of 80-proof vodka?

12. What is the weight in grams of a 1-carat diamond?

13. How many grams of gold are present in 10 g of 18-karat gold?

14. Distinguish between ammonia water and ammonium hydroxide. What is the composition of each?

15. Distinguish between Arrhenius and Brönsted-Lowry acids and bases. Give an example of each type.

16. Explain the formation of ammonium chloride "smoke" when bottles of ammonium hydroxide and hydrochloric acid are opened and placed side by side.

17. Write balanced equations for the neutralization of calcium hydroxide by nitric acid; sulfuric acid by ammonium hydroxide; hydrochloric acid by sodium bicarbonate; hydrobromic acid, HBr, by calcium carbonate; limestone by carbonic acid; acetic acid by potassium hydroxide; citric acid by potassium bicarbonate; stomach acid by milk of magnesia; and boric acid by calcium hydroxide.

18. What color would a strip of litmus paper be after it had been moistened with lemon juice, milk of magnesia, vinegar, ammonia water, a solution of baking soda, a solution of washing soda, and carbonated water?

19. a. Give two examples of strong acids and explain why they are strong. b. Give two examples of strong bases and explain why they are strong.

20. Distinguish between concentration and strength as applied to acids.

21. What is the change in the concentration of hydrogen ions present when the pH changes from 2 to 4?

22. A popular shampoo was advertised as "low pH, nonalkaline." Does that make sense?

23. Explain how the ingredients of stalagmites and stalactites may be transported to the site of deposition, and what chemical change accompanies their deposition.

24. What mass of boric acid is present in 250 mL of a 4% (m/v) solution of boric acid in water?

25. What mass of sodium nitrate is required to prepare 1.5 L of a solution that is 2.0% (m/v)?

26. How would you prepare 250 mL of a 0.67% (m/v) solution of NaCl in water?

27. Calculate the percentage concentration, v/v, of a solution that contains 12.0 mL of an alcohol dissolved in 50.0 mL of solution.

28. What volume of a 0.67% (v/v) solution of ethyl alcohol in water can be prepared from 10.0 mL of the alcohol?

29. What mass of KCl is present in 1 kg of a solution that is 5% (m/m)?

30. What mass of water is required to prepare a 1% (m/m) solution of glucose using 10 g of glucose?

31. What is the concentration in ppm and ppb of a citric acid solution that is 0.000076% (m/m)?

32. What is the % (m/m) of a solution that contains 44 ppb of $NaNO_3$?

33. Calculate the molarity of each of the following solutions:

    a. A solution containing 0.36 mol Na_2SO_4 dissolved in 1.5 L of solution

    b. A solution prepared by dissolving 10.0 g of sodium hydroxide in enough water to produce 200 mL of solution (The molar mass of NaOH is 40 g/mol.)

34. What mass of NaOH would be required to prepare 250 mL of a 0.1 M solution in water?

35. What volume of 0.5 M solution of boric acid can be prepared from 25 g of the acid? The molar mass of boric acid is 62 g/mol.

Polymers

Of all the known organic compounds, polymers are among the most important to consumers. Polymers are very large molecules. They have very high molar masses and are made up of long chains of repeating units called **monomers.**

Within the last 50 years, we have been bombarded with an increasing list of products, including plastics, packaging materials, and synthetic fibers, in which organic polymers are the basic ingredient. Specific examples are described in Sections 8.2 through 8.5.

Of equal importance are the natural biopolymers, including proteins (polypeptides), nucleic acids (polynucleotides, DNA, RNA), and polymeric carbohydrates (polysaccharides). The last of these is discussed in Chapter 9; the others are described in Sections 8.6 through 8.9. Fats are natural biochemicals, and although they are very large molecules, they are not classified as polymers. They are discussed in Chapter 21.

8.1 Synthetic Polymers

GOAL: To recognize the importance of polymers in well-known consumer products.

The word polymer is derived from the Greek *poly*, "many," and *meros*, "parts."

We often encounter chemical terms such as *polyesters, polyurethanes, acrylics,* and *vinyls.* Brand names such as *Orlon, Acrilan, Dacron, Plexiglas, Lucite, Teflon, Saran, Styrofoam,* and *Formica* are well known in products such as fibers for clothing and carpets, plastic wrap, nonstick cooking utensils, adhesives, luggage, garden hoses, and insulating material, to name a few.

8.2 Addition Polymers

GOALS: To study the nature of common vinyl and acrylic polymers.
To appreciate the formation of addition polymers.
To compare the structure and properties of hard and soft contact lenses.

Polymers that form in chemical reactions in which monomers simply add together to form long chains without any side products are called **addition polymers.** Furthermore, functional groups that are present in the monomers do not participate in the formation of addition polymers, although they do influence the properties of the resulting polymers.

Polyethylene

Unsaturated compounds are monomers used to synthesize a wide range of polymers. Ethylene, for example, can be polymerized to form polyethylene by treatment with the proper catalyst. The polymer consists of the units—CH_2CH_2— repeating over and over to give giant molecules with masses in the hundreds, thousands, or even millions of atomic mass units.

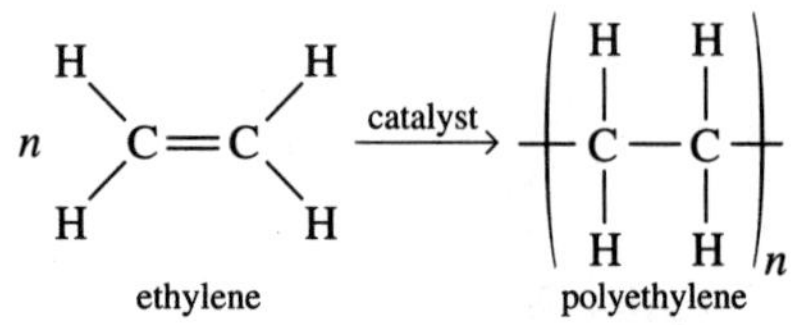

$$n \quad \underset{H}{\overset{H}{\diagdown}} C = C \underset{H}{\overset{H}{\diagup}} \xrightarrow{\text{catalyst}} \left(\begin{array}{cc} H & H \\ | & | \\ C - C \\ | & | \\ H & H \end{array} \right)_n$$

ethylene ⟶ polyethylene

A closer look at the polymerization reveals that only the carbon-carbon double bond is participating in the reaction. In fact, only one bond is affected. This is not surprising in view of the nature of a carbon-carbon double bond; although the two bonds are usually pictured as equivalent, they are different. One important difference is their strength. One of the bonds, called a sigma (σ) bond, is very strong. The other is called a pi (π) bond and is much weaker. It is the electrons of the π bond that react in polymerization.

If we recall that a covalent bond is formed by combining a pair of electrons, we can picture the ethylene molecule as follows:

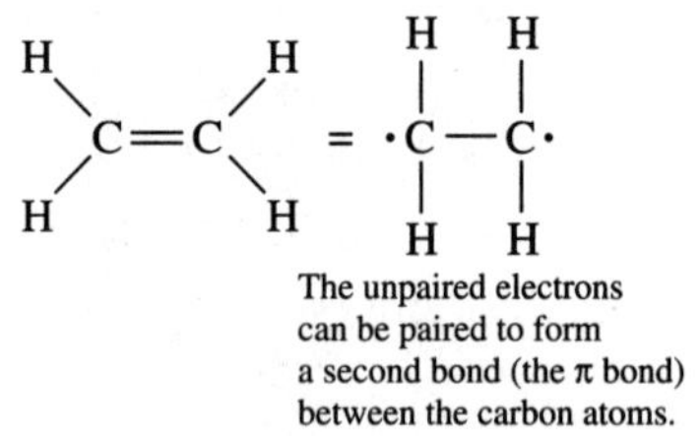

The unpaired electrons can be paired to form a second bond (the π bond) between the carbon atoms.

We can also see how two ethylene molecules might be induced to combine under the influence of a suitable catalyst.

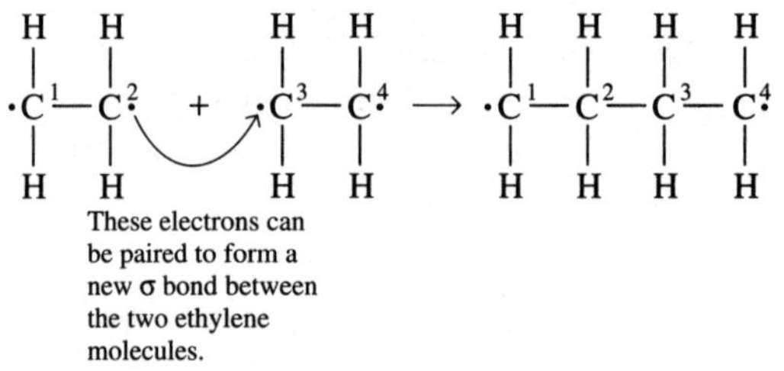

These electrons can be paired to form a new σ bond between the two ethylene molecules.

The same process can then continue on and on until all the monomer is used up or the chain stops growing.

Polyethylene is an important polymer that is used for bags, bottles, pails, pipe, electrical insulation, and toys. The method of processing it depends on its intended use.

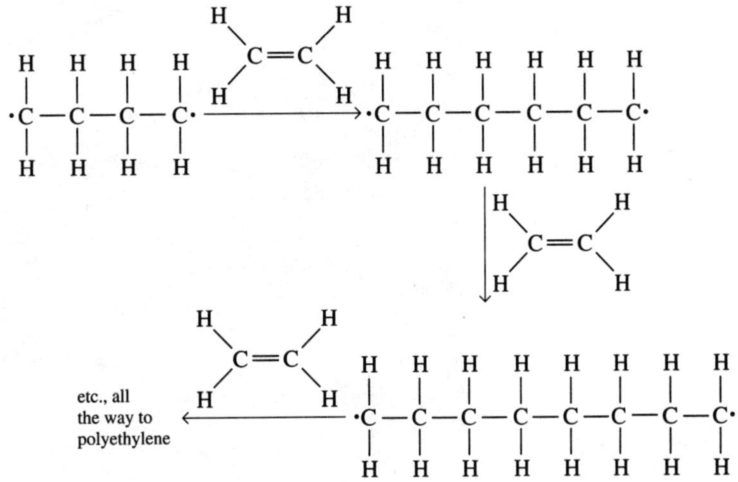

Vinyl Polymers

Now that we have seen that only the double bond is involved in polymerization, imagine what would happen if a methyl group or a functional group, such as a chlorine atom, were present in place of one (or more) of the hydrogens in the ethylene molecule.

When a methyl group is attached to the double bond, the monomer is propylene and the polymer is called polypropylene.

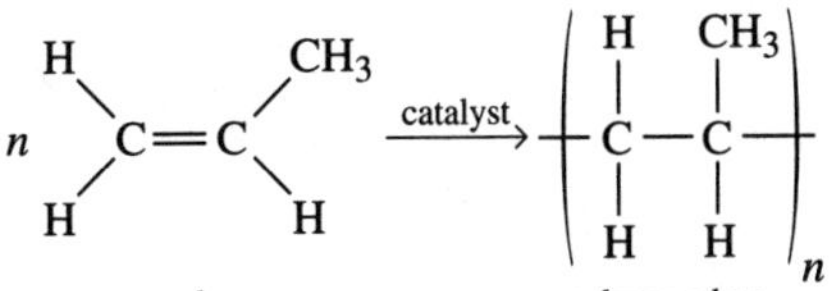

propylene   polypropylene

The coefficient n indicates that large numbers of monomer units are linked together. Polypropylene can tolerate higher temperatures than polyethylene and is used in fibers, steering wheels, pipe, indoor-outdoor carpeting, plastic bottles, and kitchenware.

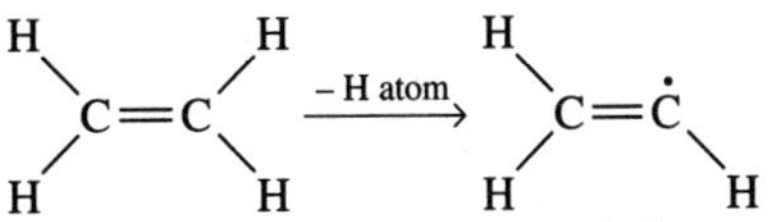

vinyl
radical

The name *vinyl* is used to identify the radical that forms when a hydrogen atom is removed from ethylene.

If a chlorine is present in place of one of the hydrogens in the ethylene molecule, the monomer becomes chloroethylene, better known as vinyl chloride. The polymer is called polyvinyl chloride, or PVC (Figure 8.1). It is used as plastic pipe, floor tiles, siding for

Figure 8.1   Polyvinyl chloride is used in many consumer applications.

houses, packaging, plastic raincoats, garden hoses, and auto seat covers. In addition, PVC replaces leather in upholstery and luggage.

$$n \quad \underset{\text{vinyl chloride}}{\overset{\displaystyle \begin{matrix} H & & Cl \\ \diagdown & & \diagup \\ & C = C & \\ \diagup & & \diagdown \\ H & & H \end{matrix}}{}} \quad \xrightarrow{\text{catalyst}} \quad \underset{\text{poly(vinyl chloride)}}{\left(\!\! \begin{matrix} H & Cl \\ | & | \\ -C-C- \\ | & | \\ H & H \end{matrix} \!\!\right)_{n}}$$

Another well-known polymer is Teflon (Figure 8.2), in which all the hydrogens of ethylene are replaced by fluorine atoms. This renders the polymer inert and stable, even

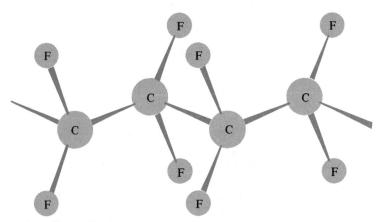

Figure 8.2   The Teflon polymer chain.

at high temperatures. Nonstick Teflon coatings on frying pans are very popular. Some other well-known vinyl polymers are shown.

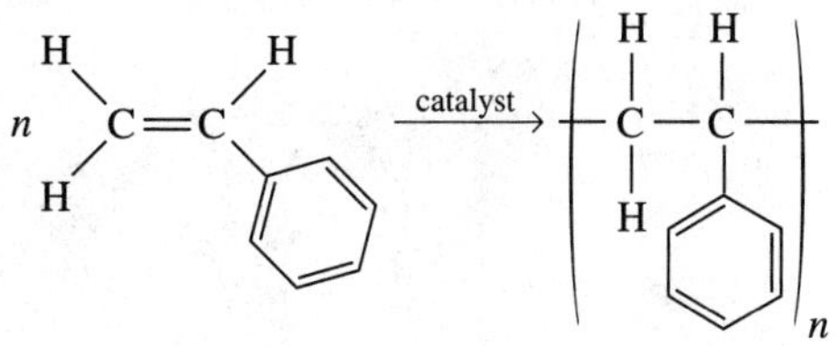

styrene

polystyrene
(Styrofoam)

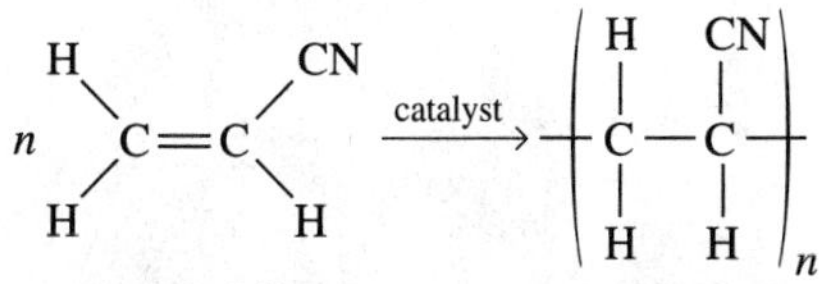

acrylonitrile

polyacrylonitrile
(fibers for rugs and
clothing; Acrilan, Orlon)

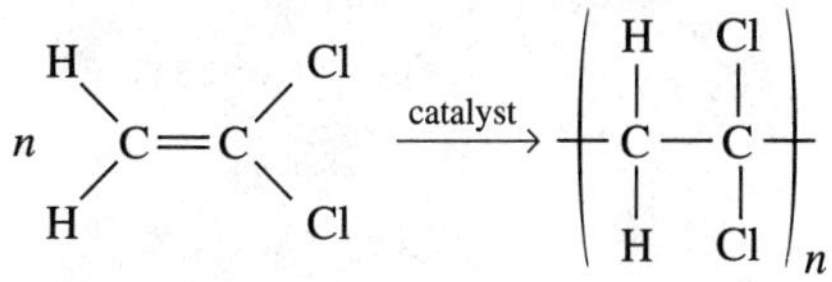

vinylidene chloride

poly(vinylidene chloride)
(Saran)

The well-known acrylics are polymers; they are formed from monomers that are variations of the compound called *acrylic acid.*

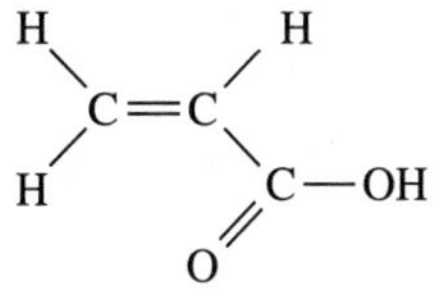

Whereas acrylic acid is a potential monomer, polyacrylic acid is not a useful polymer due to the high reactivity of the acid group. Since polymers are applied where durability and resistance to chemical attack are required, reactive functional groups cannot be present.

Poly(methyl methacrylate) and poly(2-hydroxyethyl methacrylate) are acrylic polymers formed from monomers derived from acrylic acid (Figure 8.3).

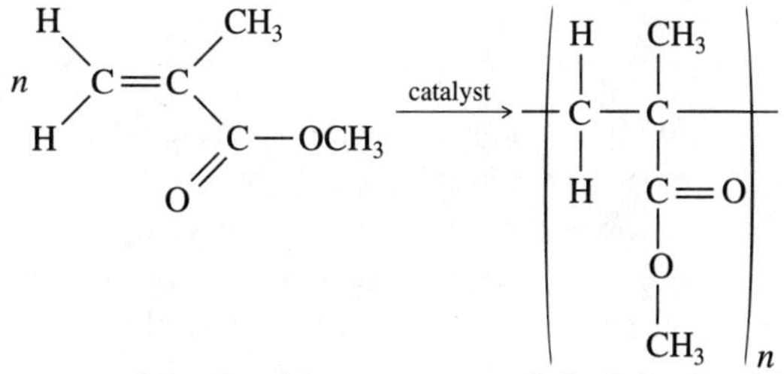

methyl methacrylate

poly (methyl methacrylate)
(hard contact lenses;
Lucite, Plexiglas)

Figure 8.3   A familiar product made of poly(methyl methacrylate), better known as Plexiglas.

In both monomers used here, the acid group (of acrylic acid) is present as an ester (Section 6.3). The methyl group attached to the carbon-carbon double bond in each monomer accounts for the name **meth**acrylate.

The two polymers provide an interesting contrast of properties. They are used in making a variety of plastics, including hard and soft contact lenses, respectively. Both are transparent plastics but have different physical properties. As the name suggests, hard contact lenses are hard, brittle, plastic lenses that transport oxygen poorly. As a result, they are often irritating to the eyes. On the other hand, the polymeric material in soft contact lenses absorbs a large quantity of water because of the added $-OH$ group, which closely resembles the water molecule. Consequently, the soft contact lens acts like a film of water resting in the eye and usually causes very little irritation.

8.3   Condensation Polymers

GOAL: To appreciate the distinction between addition and condensation polymers.

So far, we have considered polymers formed from ethylene or substituted ethylenes, that is, the vinyl and acrylic addition polymers. As noted earlier, functional groups do not participate in the formation of these polymers.

An entirely different kind of polymerization occurs during the production of **condensation polymers.** In this process, functional groups react with one another to link the monomer molecules and form polymeric chains. In addition, a small molecule (often water) is usually released when condensation polymers form.

Polyesters

An important class of condensation polymer is the *polyester*. A simple ester can be formed by combining an alcohol and an acid (Section 6.3).

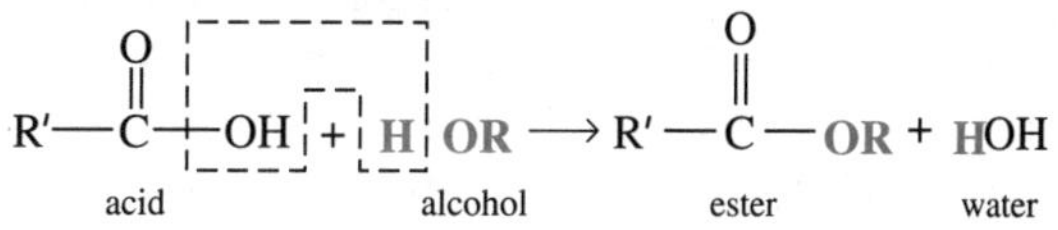

$$R'-\overset{\overset{O}{\parallel}}{C}-OH + H-OR \longrightarrow R'-\overset{\overset{O}{\parallel}}{C}-OR + HOH$$

acid   alcohol   ester   water

A polyester will form if both the alcohol and the acid are *bifunctional* monomers, as in the case of ethylene glycol (with two alcohol groups) and terephthalic acid (with two acid groups).

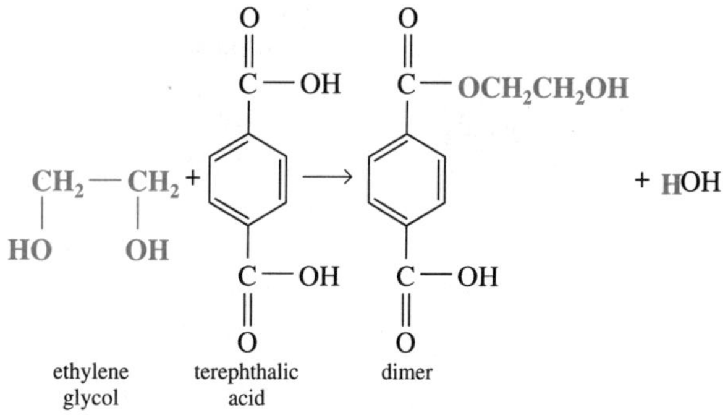

ethylene   terephthalic   dimer
glycol   acid

That a polymer forms is most easily appreciated by looking at the dimer (combination of two monomers) stage, the first step of the polymerization. Free alcohol and acid groups are still available at opposite ends of the dimer for continued growth of the polymer chain. Due to the bifunctional nature of the monomers, the growing chain always has a functional group available at each end. This particular polyester, known as *polyethylene terephthalate* (PET), is marketed under the brand name Dacron; it is used in making fibers for clothing and carpeting and in making tire cords. It is also used under the trade names Fortrel and Mylar in electrical insulation, photographic film, magnetic tape, gear wheels, and plastic soda bottles.

8.4 Thermoplastic and Thermosetting Resins

GOAL: To understand the importance of cross-linking between polymer chains.

Polymers that are processed into useful end products are usually mixed with other ingredients that impart special properties. Plasticizers are often added; these are nonvolatile organic compounds, such as dibutyl phthalate, that convert a hard, brittle plastic into a soft, pliable product.

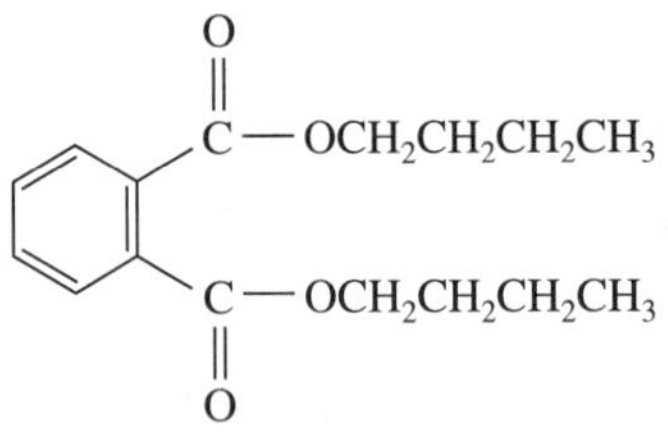

$$
\begin{array}{c}
O \\
\parallel \\
C - OCH_2CH_2CH_2CH_3 \\
\\
C - OCH_2CH_2CH_2CH_3 \\
\parallel \\
O
\end{array}
$$

dibutyl phthalate

Dyes and antibacterial agents are often added as well.

The basic polymer component is called a **resin.** The polymers considered so far are called **thermoplastic** resins, because they can be heated and melted without decomposing. For this reason they can be molded into many different shapes and processed into many useful items. Thermoplastic polymers have long linear chains of repeating units as their main feature.

Another type of resin is the **thermosetting** polymer. It is characterized by a complex, three-dimensional, weblike structure in which the individual polymer chains are interconnected. A thermosetting resin cannot be melted to allow molding into various shapes. Instead, it must be polymerized directly into its final shape.

A polyester can be a thermosetting resin if at least one of the monomers is trifunctional (Figure 8.4). In the reactions shown, glycerine combines with terephthalic acid. Two chains are growing; each consists of five monomer units. The extra −OH functional groups located on each of the chains let the diacid monomer form a cross-link between the chains, as shown. The result is a thermosetting polyester. Thermosetting resins are very hard materials. The hardness, coupled with excellent insulating properties, makes these resins ideal for radio cases, gear wheels, electrical circuit boards, and circuit board coatings.

Rubber

Vulcanization of rubber is another process that introduces cross-links between polymer chains. Natural rubber is an addition polymer with the following repeating unit:

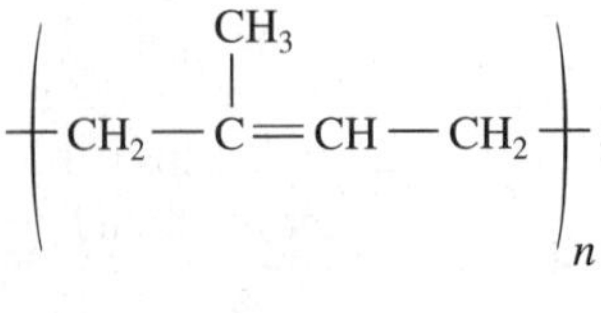

$$
\left(CH_2 - \underset{\underset{CH_3}{|}}{C} = CH - CH_2 \right)_n
$$

cis-polyisoprene (natural rubber)

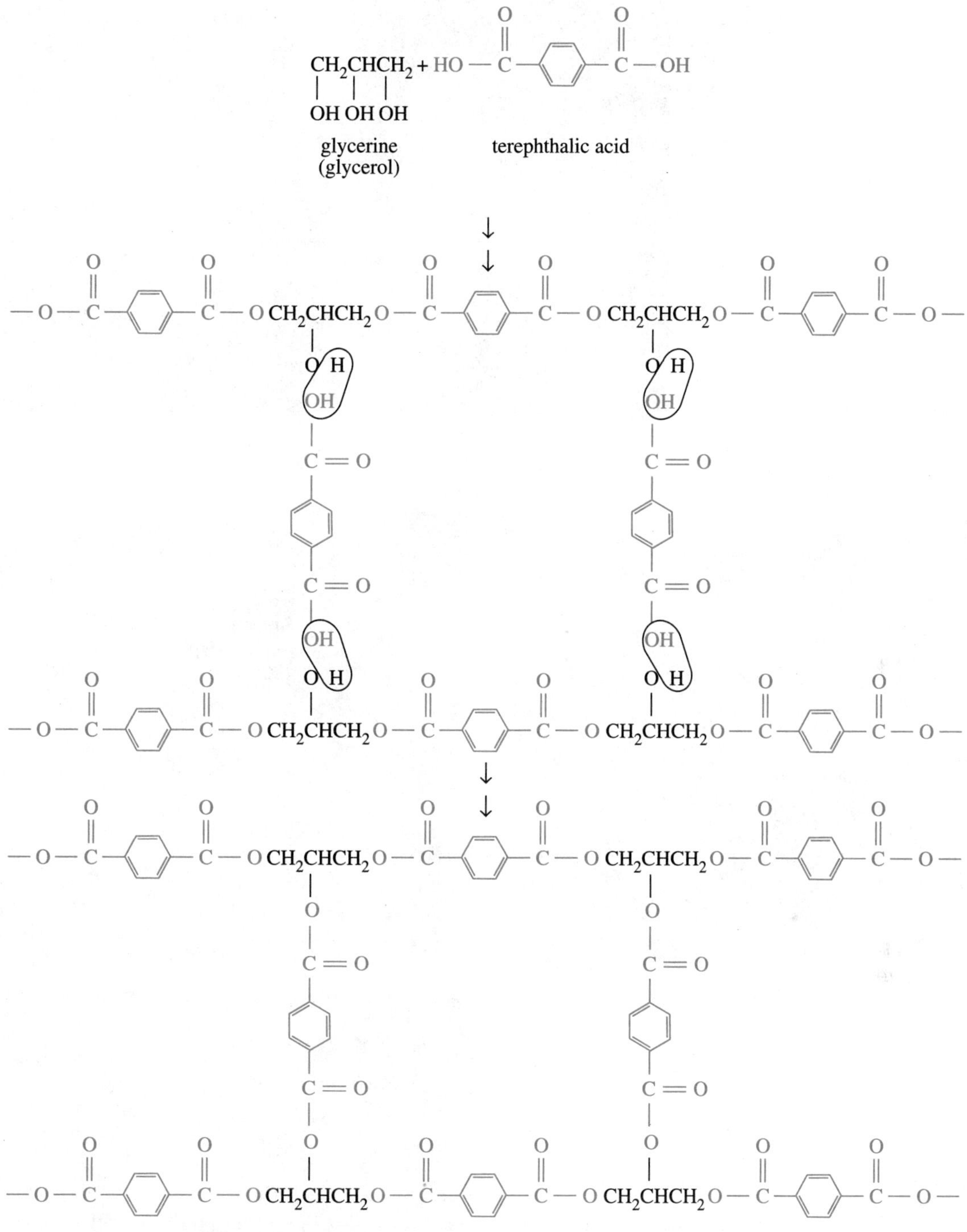

Figure 8.4 Formation of a cross-linked thermosetting polyester.

Natural rubber is an elastic substance that becomes very sticky or tacky when hot. In 1839, Charles Goodyear (Figure 8.5) perfected the process he called *vulcanization,* in which natural rubber is treated with sulfur to form a network of cross-links. This makes the rubber harder and stronger, and eliminates tackiness (Figure 8.6).

Figure 8.5 Charles Goodyear
discovers the miracle of
vulcanization.

Figure 8.6 Steps in tire production. (top) Worker
measures sulfur, which is needed in vulcanization. (top
right) The uncured tire is placed in a curing mold, and
subjected to pressure and heat for a specified period. This
causes the soft, gummy, "green" tire to be transformed
into a tough, long-wearing tire. Simultaneously, the tread
is impressed and the tire emerges ready for final
inspection. (right) Open molds show indentations that
determine the design of the tread.

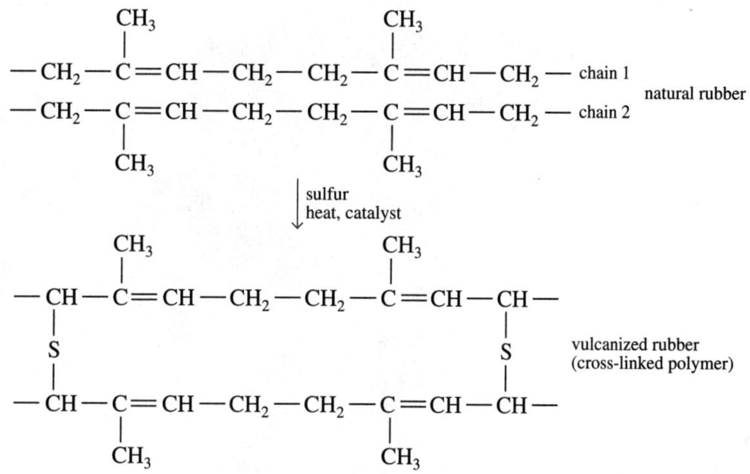

natural rubber

sulfur heat, catalyst

vulcanized rubber (cross-linked polymer)

8.5 Other Important Polymers

GOAL: To know the structure of polyamides, polyurethanes, and common formaldehyde polymers.

Although the list of possible polymeric materials is almost endless, let us conclude our look by considering three widely used classes of polymers.

Polyamides

Nylon has long been used as a fiber for clothing, stockings, carpets, and tire cords. To the chemist, nylon is not a single polymer but is any one of many polymers called **polyamides.** They are formed by condensation polymerization of a diacid and a diamine in a type of reaction that is analogous to formation of polyesters. The most successful nylon is known as nylon 66, which is formed from two 6-carbon monomers as follows:

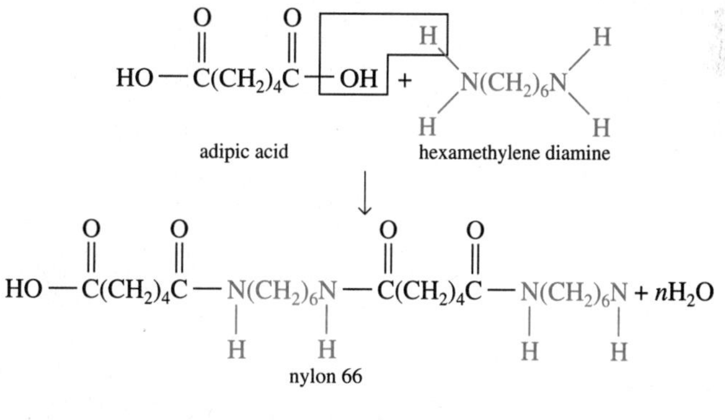

adipic acid   hexamethylene diamine

nylon 66

Proteins are also polyamides, usually called *polypeptides*. See Section 8.7.

Polyurethanes

The polyurethanes (Figure 8.7) have become an important class of polymers, particularly as foams in furniture, packing material, automobile safety padding, soles for shoes, mattresses, pillows, insulation, life preservers, and fiber for carpets. A polyurethane is a sort

Figure 8.7   Polyurethane foams have many consumer applications, including packing material (for cushioning), sponges, filters, and insulation.

of combination polyester and polyamide. Like most condensation polymers, it results from the reaction of the functional groups in a diisocyanate and a diol (or triol). Unlike other reactions forming condensation polymers, this reaction releases no water or other small molecule.

$$CH_2CH_2 + O = C = N - (CH_2)_4 - N = C - O$$

$$\underset{\substack{\text{ethylene}\\\text{glycol}}}{\overset{\displaystyle |\quad |}{OH\ OH}} \qquad \underset{\substack{\text{tetramethylene}\\\text{diisocyanate}}}{}$$

$$CH_2CH_2O - \overset{\displaystyle \overset{O}{\|}}{C} - \underset{\displaystyle H}{N}(CH_2)_4 \underset{\displaystyle H}{N} - \overset{\displaystyle \overset{O}{\|}}{C} - OCH_2CH_2 -$$

$$\underset{\substack{\text{a polyurethane}}}{\overset{\displaystyle |}{OH}}$$

Formaldehyde Polymers

The simplest aldehyde, formaldehyde, has the formula CH_2O. It can act as a monomer by forming 1-carbon bridges between units of other monomers such as phenol and melamine. The products are the well-known thermosetting resins Bakelite and Formica, respectively. Both are very hard and durable.

Bakelite was developed by Leo H. Baekeland, a Belgian-born chemist who came to the United States in 1899. In 1909 he received a patent for this process for producing a thermosetting resin from the combination of phenol and formaldehyde. It was the first commercial synthetic resin and dominated the plastics field until the 1930s.

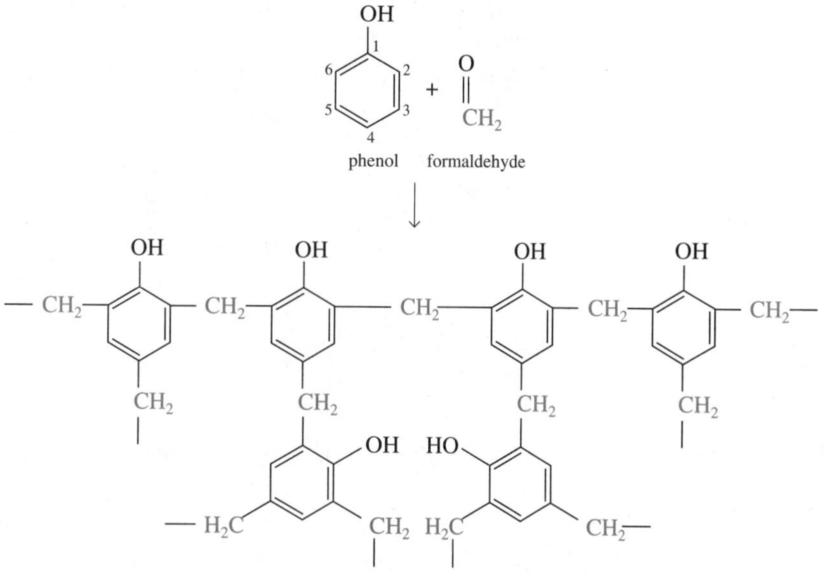

phenol   formaldehyde

a phenol-formaldehyde resin
(Bakelite)

The monomer phenol has three sites (numbered 2, 4, and 6) that can react with formaldehyde to form a complex, weblike structure. Baekeland perfected the process for interrupting the polymerization before the polymer was fully cross-linked, so that it could be molded into various shapes before the final cure (complete cross-linking), to form a hard thermosetting plastic. The Bakelite resin structure can be represented by an abbreviation with the symbol P to signify the phenol ring:

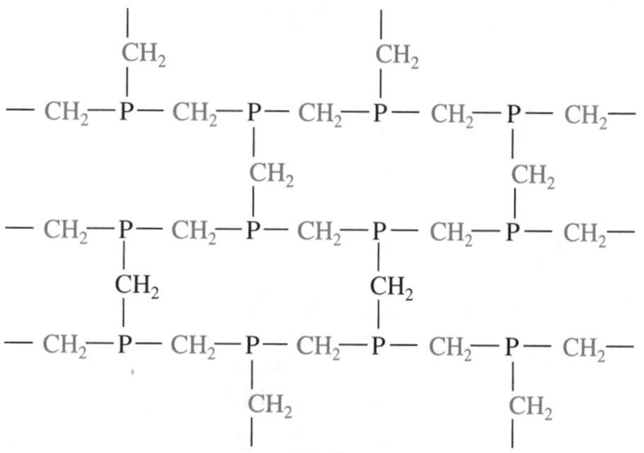

Bakelite
(P = phenol)

Bakelite resins are extremely resistant to heat and chemicals. They are used as electrical insulation, brake linings, pulleys, handles for cooking utensils, adhesives, telephone headsets, radio cases, buttons, and many more applications in which a hard, durable plastic is required.

In a similar fashion, the trifunctional melamine molecule can form a rigid thermosetting resin, with formaldehyde forming a 1-carbon bridge between the melamine monomers. The fully cross-linked structure of the melamine-formaldehyde resin is like the phenol-formaldehyde resin (shown previously), except that each phenol is replaced by melamine. Melamine-formaldehyde resins are used in making dinnerware, and under the brand name Formica, as decorative surface coatings for counter tops, tables, and wall coverings.

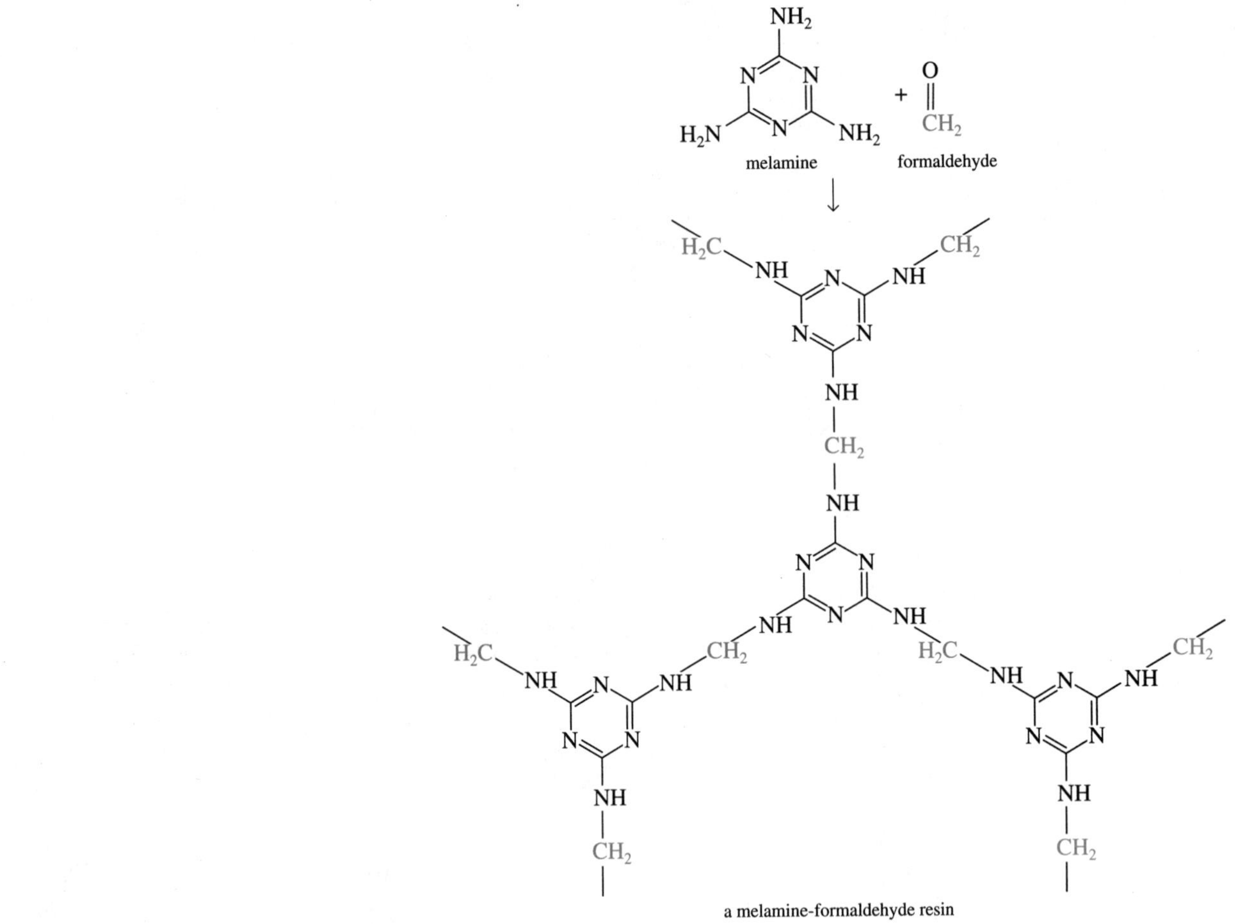

a melamine-formaldehyde resin

8.6 Natural Polymers

GOAL: To appreciate the range of biopolymers.

While synthetic polymers appear in many forms that are familiar to the consumer, *biopolymers* constitute all living organisms. Proteins, nucleic acids, and polysaccharides are the biopolymers. Along with fats (see Chapter 21), minerals, and many small organic molecules, biopolymers are responsible for the structural and functional chemistry of all plants and animals. For example, proteins are major components of skin, hair, nails, horns, feathers, wool, connective tissue, and muscle. Proteins also function as catalysts (enzymes), hormones, and carriers (such as hemoglobin, which transports oxygen throughout the body); they provide immunity (as immunoglobulins; see Chapter 23) and are the ultimate expression of genetic information.

The storehouse of genetic information is the nucleic acids (Section 8.8), another important type of biopolymer. The nucleic acids, known as DNA (deoxyribonucleic acid) and RNA (ribonucleic acid), direct the synthesis of proteins and therefore control all functions of living organisms.

The polymeric sugars, known as polysaccharides, also have both structural and functional roles in living things. Polysaccharides are an important component of cell membranes and serve as a storage form of energy. These polymers and their corresponding monomers are discussed in Chapter 9.

8.7 Proteins

GOALS: To appreciate the molecular structure of polypeptides, including the monomer units, and the importance of proper sequence.
To know the meaning of an essential amino acid.

Proteins are also polyamides (Section 8.5), usually called *polypeptides,* in which the monomers are **amino acids.** Amino acids are unusual monomers for condensation polymers, because both the acid (−COOH) and amino (−NH₂) functional groups are present in the same compound.

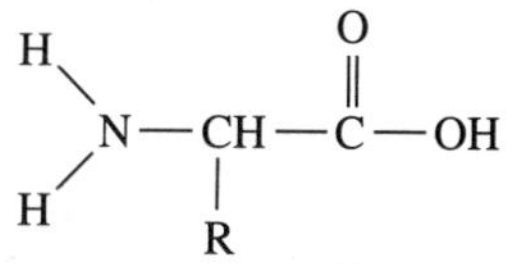

an amino acid

This would seem to make proteins simpler than other polyamides, but any of 20 different groups may be represented by R. In other words, 20 different amino acids are incorporated into polypeptides. The number of possible polypeptides is enormous. Regardless of R, the amino acids combine as shown, with the loss of a water molecule:

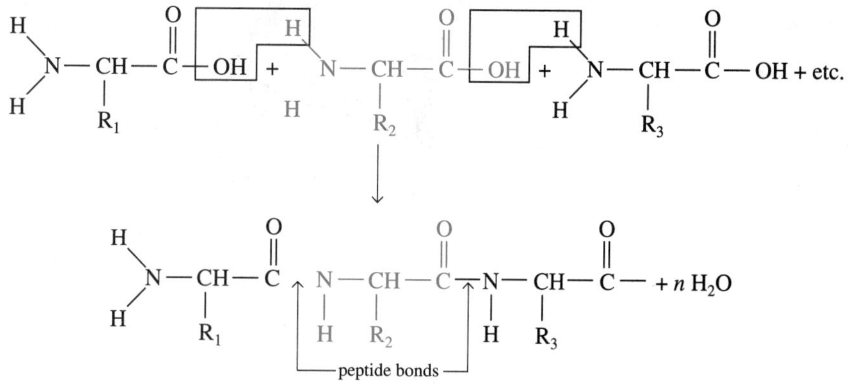

a polypeptide

Peptide bonds connect the individual amino acid units within the polypeptide chains. These bonds are easily broken in a reaction called **hydrolysis,** which is the reverse of the formation of polypeptides. Proteins consumed in the diet are broken down by enzymes in the intestinal tract into the individual amino acids. The amino acids are then absorbed and reassembled into polypeptides within the cells of the body.

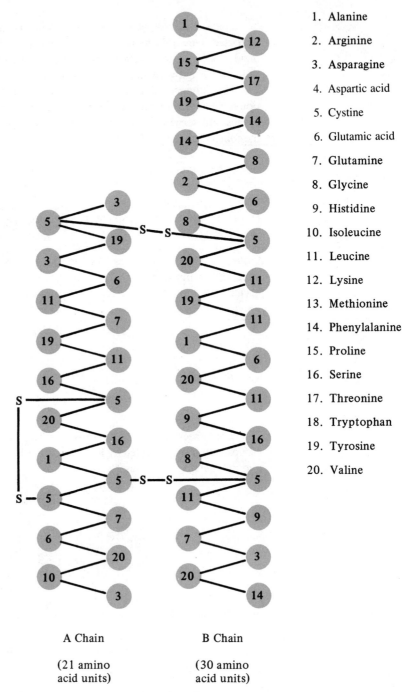

1. Alanine
2. Arginine
3. Asparagine
4. Aspartic acid
5. Cystine
6. Glutamic acid
7. Glutamine
8. Glycine
9. Histidine
10. Isoleucine
11. Leucine
12. Lysine
13. Methionine
14. Phenylalanine
15. Proline
16. Serine
17. Threonine
18. Tryptophan
19. Tyrosine
20. Valine

A Chain

(21 amino
acid units)

B Chain

(30 amino
acid units)

Figure 8.8   The amino acid sequence of beef insulin.

Of the 20 amino acids, 10 are classified as *essential amino acids* to signify that they cannot be synthesized and must be consumed in the human diet. The other 10 can be synthesized within the body from sugars and other sources of carbon, oxygen, nitrogen.

Each cellular protein of any and every organism has a specific *sequence* of amino acids that is responsible for all of the characteristics necessary for a protein to function. Outside the body, conditions for combining amino acids are the same as those used to make synthetic polyamides. A protein cannot readily be synthesized in the laboratory, however, since it is necessary to assemble the amino acid units in exactly the right sequence. Figure 8.8 indicates the sequence in beef insulin. This relatively small polypeptide (51 amino acid units) consists of two chains designated A and B. If the sequence of amino acids is altered, this polypeptide has less or no biochemical activity.

If one were to mix all the necessary amino acids in the laboratory and allow polymerization to occur, the polymer would have a random sequence of amino acid units. In 1963, R. B. Merrifeld developed an automated technique for synthesizing proteins; he received a Nobel Prize for this accomplishment in 1984. The procedure is a very tedious one in which the polymer chain is lengthened by reaction with one amino acid at a time. Many days are required to complete the synthesis of even very simple polypeptides. Although truly a monumental research accomplishment, it is impractical for synthesizing proteins on a commercial scale.

When a polypeptide is required as a hormone, antibiotic, or enzyme, it is usually isolated from an animal or microorganism. The protein insulin, for example, is routinely isolated from the pancreas of cattle or hogs, a by-product of the food industry. In recent years, using the techniques of biotechnology, microorganisms have been developed to synthesize polypeptides of commercial importance.

Biotechnology is concerned with the synthesis of natural polypeptides in living organisms. Since polypeptide synthesis is under the control of genetic information in the polymeric nucleic acids DNA and RNA, they are the topic of the following sections.

8.8   Nucleic Acids

GOALS: To understand the role of nucleic acids in living organisms.
To recognize the monomer units of DNA and RNA.
To understand the location and nature of the genetic code.
To know the meaning of a retrovirus.

The nucleic acid polymers are all very large polymer chains, in contrast to polypeptides, which range in size from small to very large. Two types of nucleic acid polymers are essential to all living organisms: **DNA** and **RNA**. The monomer units of nucleic acids are called *nucleotides;* DNA and RNA are *polynucleotides.*

The DNA molecules function as genes. Their coded information directs the synthesis of all proteins in all life forms; that is, the nucleic acids provide the information to build all structures and to carry out all functions of life. The code in the sequence of nucleotides is thus of great interest in genetic disease, for personal health, and in industry.

Long strings of DNA, or genes, supercoiled around each other and supported by proteins, are what we see as **chromosomes,** colored bodies, with an ordinary laboratory microscope. Most human cells contain 46 chromosomes, 46 enormous polymers of DNA. Even greater detail can be seen under an electron microscope.

The five nitrogen-containing compounds of nucleotides are called bases because their nitrogen atoms can accept a hydrogen ion, as can ammonia (Section 7.5).

Each nucleotide consists of three parts: a nitrogen base, a sugar called ribose (or deoxyribose), and a phosphate group. An example of each component separately and in combination is given in Figure 8.9.

The nucleotide adenosine monophosphate (AMP) is one of four nucleotides found in the group of nucleic acids known as **ribonucleic acids,** or **RNA.** In the **deoxyribonucleic acids,** or **DNA,** a typical nucleotide is **deoxy**adenosine monophosphate (**d**AMP).

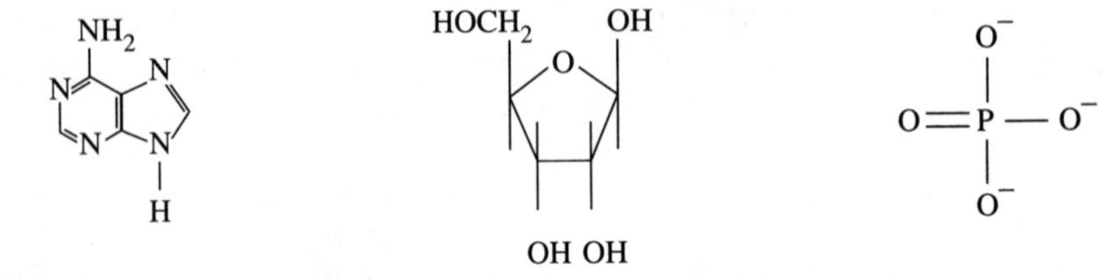

adenine (A)
(a nitrogen base)

ribose
(a sugar)

phosphate

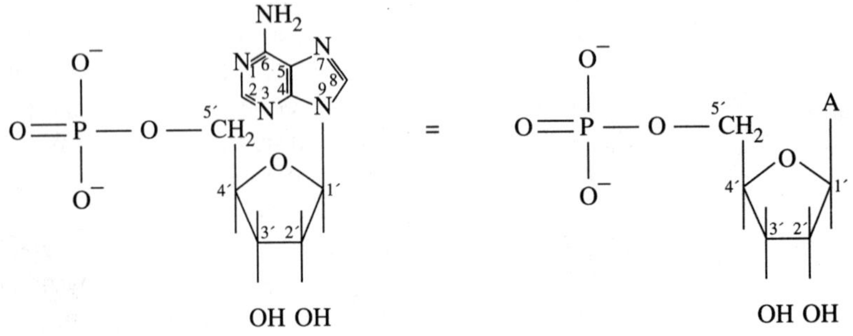

adenosine monophosphate (AMP)
(a nucleotide) A = adenine

Figure 8.9 Nucleotide molecules consist of a nitrogen base (such as adenine, A), a 5-carbon sugar, and a phosphate group. The sugar may be ribose or deoxyribose; see text. Both the ring atoms of the base and the carbon atoms of the sugar are numbered. Prime numbers are used on the sugar to avoid confusion with the numbering of the atoms of the base.

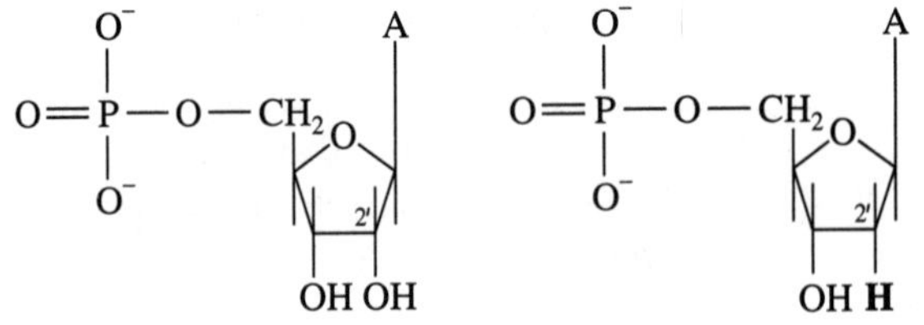

AMP (a ribonucleotide)

dAMP (a 2′-deoxyribonucleotide)

As you can see, the structures of AMP and dAMP are identical except for the absence of an −OH group in the deoxy sugar. In addition, the bases found in DNA are A, G, C, and T. In RNA, the base T is absent but U is present. The five bases in DNA and RNA are shown in Figure 8.10.

Note the numbering of the atoms in the nucleotides in Figure 8.9. Each carbon and nitrogen atom of the bases is given a number. Each carbon atom of the sugar has a prime number. A nucleotide of DNA is a 2′-deoxyribonucleotide. The nitrogen bases are attached at the 1′ positions of the ribose units.

The prime numbers are also used to describe the linkages between the nucleotide monomer units in DNA and RNA. The polymer chain includes only the sugar and the phosphate groups. The linkage joins a 5′ carbon of one sugar to a 3′ carbon of the next sugar unit with a phosphate group serving as a bridge. See Figure 8.11 for an example.

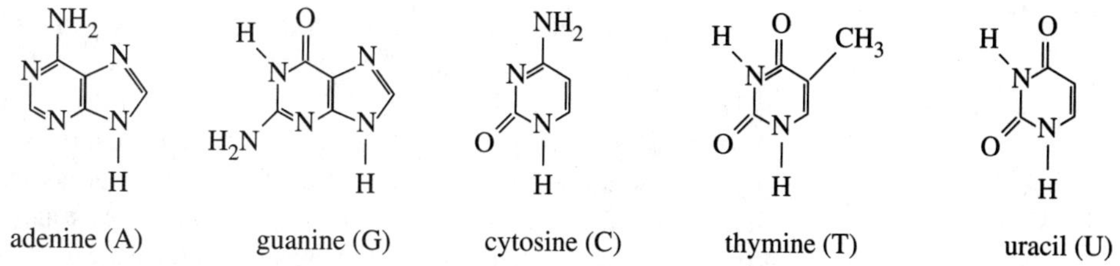

Figure 8.10 The five bases in DNA and RNA. The bases A, C, G, and T are present in DNA; A, C, G, and U are in RNA.

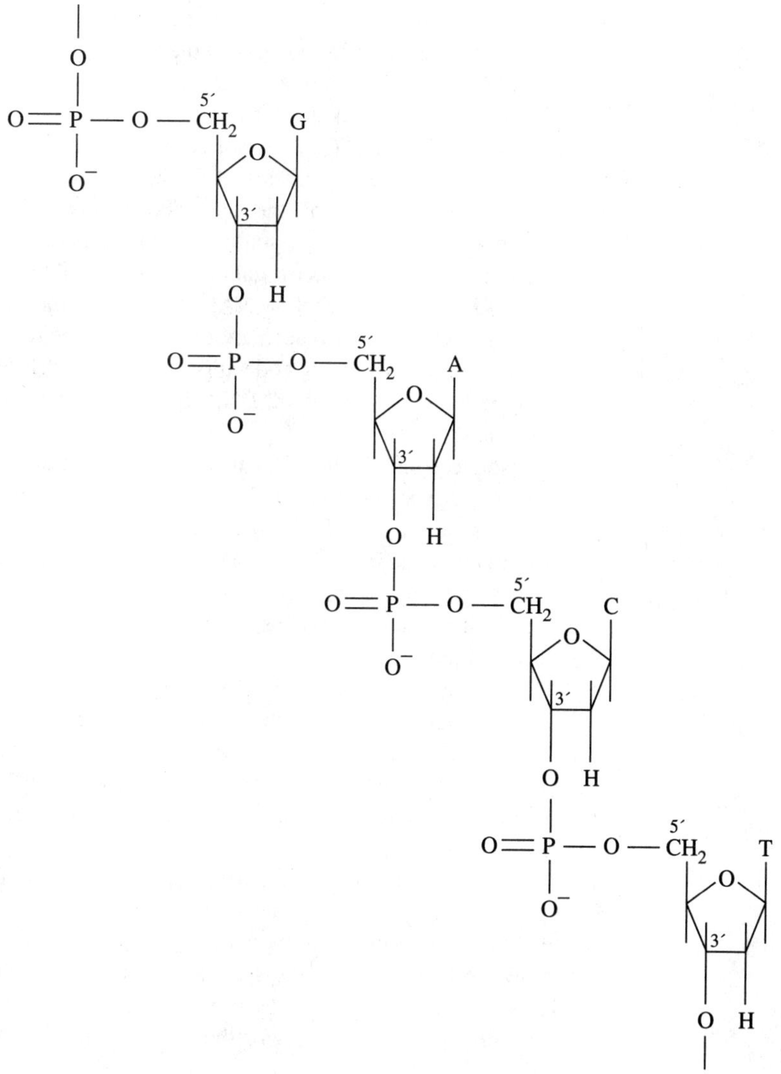

Figure 8.11 A segment of a DNA molecule with the sequence—GACT—.

Notice that the nitrogen bases, which distinguish each monomer unit from the next, are not part of the polymer chain. The sequence of bases is critical to the proper functioning of DNA and RNA molecules.

Consider the role of DNA and RNA in living organisms. DNA molecules function as *genes;* they contain coded information that directs the synthesis of proteins in cells of plants, animals, and microorganisms. RNA molecules function as carriers of the coded information and structural sites for protein synthesis. Since proteins have both structural and functional roles, genetic control over the quantity and types of proteins that are synthesized constitutes the nature of an entire organism. Genetic defects, such as the inability to synthesize a particular protein, may be lethal to an organism; or a rare genetic change may be selected as advantageous.

Each gene directs the synthesis of one protein. In other words, each gene contains information to cause the proper placement of amino acids in the correct sequence in a growing polypeptide chain. There are small genes with information to guide the synthesis of small proteins, such as insulin, and larger genes used for synthesis of larger proteins, such as enzymes.

The information in the DNA molecules, the *genetic code,* is found in the *sequence* of bases (A, G, C, and T) in each gene. Each set of three nucleotides in the sequence constitutes one bit of code. Thus, a portion of a DNA molecule with the sequence TACGCACTGAACTGT . . . is a series of coded signals TAC, GCA, CTG, AAC, and TGT, each of which specifies 1 of the 20 amino acids. However, the synthesis of proteins is not guided directly by the sequence of bases in DNA. In an intermediate step the sequence of bases in DNA guides the synthesis of a corresponding chain of RNA, called *messenger RNA,* in a process called *transcription.* Once the proper base sequence has been synthesized into messenger RNA, synthesis of the protein chain, called *translation,* can take place with proper placement of the amino acids.

The complete genetic code appears in Table 8.1. Notice the code is not given in terms of three-letter sequences of DNA bases but in terms of sequences of bases in RNA molecules.

Notice from Table 8.1 that, using three-letter sequences, 64 possible codes, called codons, are possible. Since only 20 amino acids are incorporated into proteins, the genetic code has an excess of information. Many of the amino acids are specified by several different codons. In addition, four of the codons have special significance; they are signals to start and stop the protein synthesis. The codon AUG is special for two reasons: first, because it is used to initiate synthesis, and second, because it is used to specify the amino acid methionine.

You are now in a position to understand why an alteration in genetic information can have profound effects. A change of even a single base in a DNA molecule is called a mutation. A mutation could result in synthesis of an inactive protein, in no synthesis, or in uncontrolled synthesis. The latter could cause cells to become cancerous. Mutations can be brought about by treatment with chemicals or irradiation, including irradiation from radioactive isotopes.

The normal flow of information is from DNA to RNA to protein by transcription and translation. In a rare process known as *reverse transcription,* information flows from RNA to DNA. Long thought to be a curious oddity of some RNA viruses, *retroviruses* are now known to be the viruses responsible for some tumors and for acquired immunodeficiency syndrome (AIDS). Once they infect cells, their RNA directs the synthesis of viral DNA, and normal genetic control of the cells is lost. See Section 23.5 for more details about AIDS.

Table 8.1   The genetic code found in RNA is a series of three-letter codes, each of which specifies an amino acid that is to be placed in a growing polypeptide chain

First Position (5′ end)	Second Position				Third Position (3′ end)
	U	C	A	G	
U	Phe	Ser	Tyr	Cys	U
	Phe	Ser	Tyr	Cys	C
	Phe	Ser	Stop	Stop	A
	Phe	Ser	Stop	Trp	G
C	Leu	Pro	His	Arg	U
	Leu	Pro	His	Arg	C
	Leu	Pro	Gln	Arg	A
	Leu	Pro	Gln	Arg	G
A	Ile	Thr	Asn	Ser	U
	Ile	Thr	Asn	Ser	C
	Ile	Thr	Lys	Arg	A
	Met*	Thr	Lys	Arg	G
G	Val	Ala	Asp	Gly	U
	Val	Ala	Asp	Gly	C
	Val	Ala	Glu	Gly	A
	Val	Ala	Glu	Gly	G

*Also a start signal.

8.9   Biotechnology

GOAL: To know the basic principles and potential value of recombinant DNA technology.

Armed with knowledge of the role of DNA in directing protein synthesis, scientists have labored for many years to put this information to use in producing large quantities of important proteins and small molecules, such as hormones and antibiotics. New techniques known collectively as **recombinant DNA technology** allow genes to be transferred between organisms, or *recombined*. Therefore, microorganisms may receive genes from animal sources and be *genetically engineered* to carry out the synthesis of useful molecules. Gene *cloning* is achieved when the *host* cells take up transferred DNA and reproduce it.

Abundant examples of genetic engineering are proposed in agriculture. For example, microbes capture (fix) nitrogen gas from the air and convert it to a form that is useful as a nutrient for plants (see Section 15.1). Microbial pesticides are under development using recombinant DNA technology (see Section 16.1). Microorganisms can be employed for cleaning up oil spills or other environmental disasters. New vaccines are being produced as a result of genetic alteration of disease-causing organisms (see Section 23.3). New antibiotics have been isolated from genetically engineered microbes (see Section 24.3).

Genetic engineering also can be used to develop new or improved plants as food sources. This may be particularly useful in feeding people in underdeveloped countries or in soil-depleted regions of the developed world (see Chapter 15), or it may just be used to develop a better tomato that can withstand storage and shipping from source to point of consumption.

Biotechnology is not a new field. Careful breeding and selection of food crops and microorganisms, such as yeasts used in brewing and winemaking (see Chapter 17), have been going on for centuries. Breeders were selecting for a particular genetic makeup long before genes or the existence of DNA was proposed. Now that recombinant DNA technology has entered the picture, genetic engineering has much greater potential.

SUMMARY

A primary application of organic chemistry is polymer chemistry. Addition polymers are formed by polymerization of unsaturated monomers. A variety of functional groups may be present in the monomer but do not participate in polymerization; they do, however, strongly influence the properties of the finished product. Addition polymers are generally thermoplastic polymers or resins that can be molded into a variety of shapes.

Condensation polymers are formed when monomers contain two or more functional groups and the groups react with one another. Polyesters, polyamides, and polyurethanes may be thermoplastic (if each monomer has only two functional groups), or thermosetting (if additional functional groups are available for cross-linking of the polymer chains). Vulcanized rubber is harder and sturdier than natural rubber due to cross-linking of the polymer chains. Formica and other melamine-based polymers are very durable, rigid, thermosetting plastics that have many uses.

Proteins (polypeptides) and nucleic acids (polynucleotides) are important classes of biopolymers. Proteins serve both structural and functional roles. The synthesis of polypeptides occurs under the control of information present in DNA. The monomer units of polypeptides are 20 amino acids, which are joined as are polyamides by formation of peptide bonds. Hydrolysis is the process in which polypeptides are broken down into amino acids; in this way, dietary proteins become a source of amino acids for incorporation into human polypeptides.

The synthesis of polypeptides occurs under the control of information in DNA in the form of the genetic code. Chromosomes are DNA molecules, smaller segments of which are genes. Each gene has the code to direct the synthesis of one polypeptide. The coded information is transcribed into the form of messenger RNA, after which protein synthesis, called translation, uses the code to assemble polypeptide chains. Genetic information almost always flows from DNA to RNA, although some retroviruses, such as the AIDS virus, have RNA genes that direct the synthesis of DNA in infected cells.

By the techniques of recombinant DNA technology, genes may be synthesized and transferred between organisms. Thus organisms may be genetically engineered into forms with useful properties. Some genetically engineered plants may prove to be useful food sources. Microorganisms have been engineered to synthesize proteins such as hormones and antibiotics, insecticides, and other chemicals.

PROBLEMS

1. Give a definition or example for each of the following:
   a. Monomers
   b. Polymers
   c. Bifunctional monomer
   d. Resin
   e. Thermoplastic resin
   f. Thermosetting resin
   g. Vulcanization
   h. Orlon
   i. Acrilan
   j. Dacron
   k. Plexiglas
   l. Lucite
   m. Teflon
   n. Saran
   o. Nylon
   p. Bakelite
   q. Formica
   r. Essential amino acid
   s. Polypeptide
   t. Hydrolysis
   u. Nitrogen base

v. Nucleotide
w. Polynucleotide
x. Gene
y. Chromosome
z. 2′-deoxy
aa. Mutation
bb. mRNA
cc. Recombinant DNA technology
dd. Retrovirus

2. What are the main differences between addition polymers and condensation polymers?

3. What is the repeating unit in polyvinylchloride?

4. What is the repeating unit in a common acrylic polymer?

5. What is the monomer used to make soft contact lenses?

6. What are some of the uses of polyurethanes?

7. What are the monomers used in making nylon 66? Explain this name.

8. Why is it difficult to synthesize functional proteins in the laboratory?

9. What is the meaning of the symbols A, C, G, T, and U in describing nucleic acids?

10. What is the possible consequence of an error in achieving the proper sequence of amino acids in a polypeptide? Nucleotides in DNA?

11. Figure 8.11 shows a segment of a DNA chain. How might you change the molecular structure to represent an RNA chain?

12. What is the genetic code? What is it used for in the cell?

13. How does the number of monomer units compare in a gene and its corresponding protein?

14. Distinguish between transcription and translation. What is reverse transcription?

15. List some possible products of biotechnology.

Carbohydrates

9

The important class of compounds called *carbohydrates* will be encountered repeatedly in later chapters. Carbohydrates are biochemical compounds that appear in many forms, ranging from simple sugars to polymers such as starch and cellulose. Many carbohydrates are edible but some cannot be metabolized within the human body. Some that cannot be metabolized by humans can be broken down by other animals such as termites and cattle.

9.1 Monosaccharides and Disaccharides

Goals: To understand the structure of glucose, including α and β forms.
To consider the disaccharides sucrose, maltose, and lactose.

The simplest carbohydrates are called *sugars*, or **monosaccharides.** Of these, the most common contain six carbons and have the formula $C_6H_{12}O_6$, or $C_6(H_2O)_6$, which accounts for the use of the word *carbohydrate* (hydrated carbon). Two important 6-carbon monosaccharides are **glucose** (Glu) and **fructose** (Fru). In Section 8.8 we saw the importance of the 5-carbon sugars ribose and deoxyribose as building blocks of RNA and DNA. Monosaccharides chemically combine to form **disaccharides.** Glucose and fructose bond to form a disaccharide called sucrose (Glu-Fru), or common table sugar.

The notation used for carbohydrates is illustrated for glucose in two ways, in detail and in the common abbreviated form; the latter will be used throughout the text.

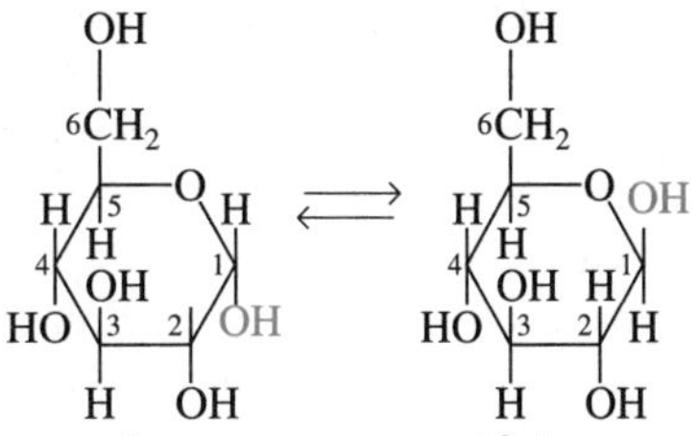

α-glucose   β-glucose

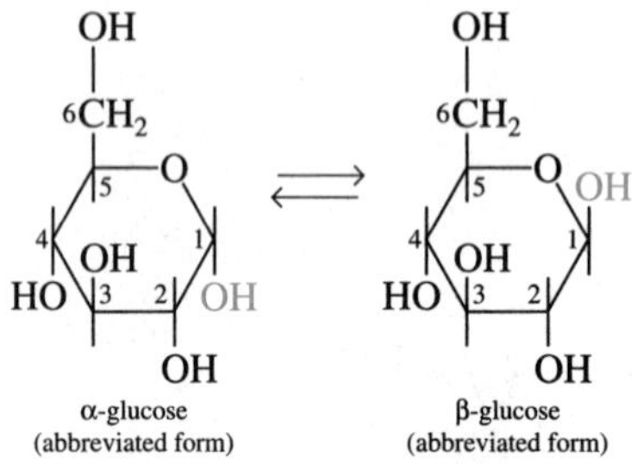

α-glucose
(abbreviated form)

β-glucose
(abbreviated form)

Both forms have a six-membered ring consisting of five carbons and one oxygen. Imagine that the ring is lying in the plane of the page. The lines pointing from the ring toward the top of the page represent bonds to groups or atoms lying above the ring, and lines pointing toward the bottom of the page represent bonds below the ring. In the abbreviation, all of the hydrogens attached directly to the ring are not shown.

Glucose can exist in two slightly different isomeric forms, symbolized by the Greek alpha (α) and beta (β). These isomers rapidly interconvert when glucose is dissolved in water; they differ only in the position of the −OH at C-1. In all the monosaccharides that we will study, the α isomer will have the −OH at C-1 below the ring, and the β isomer will have the −OH at C-1 above the ring.

None of the other −OH groups of glucose can change position, although other isomers do exist. For example, in galactose, the −OH at C-4 points up above the ring rather than below, as in glucose. But glucose and galactose do not normally interconvert.

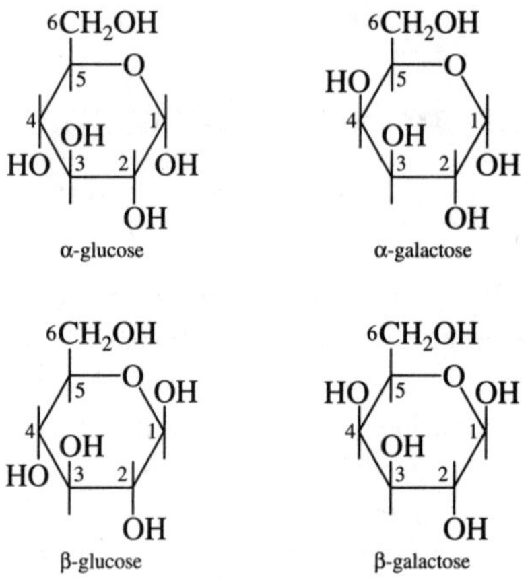

α-glucose

α-galactose

β-glucose

β-galactose

It is the α form of glucose that bonds to the fructose unit to form sucrose (common sugar) as shown. Fructose also exists in α and β forms, but the β configuration is incorporated into sucrose. Since fructose is the only sugar we will study that has a five-member ring, we need not worry about the difference between its α and β forms. When considering sucrose and other disaccharides, only the α, β isomerism of the monosaccharide ring on the left is important. The arrangement of atoms at the point of attachment to the ring on the left determines if a carbohydrate is a useful food source. Consequently, representing sucrose with the fructose ring abbreviated as R is enough to show that the −OR (R = fructose) is attached to the glucose ring in the same way as the −OH at C-1 in α-glucose. The bond between the two monosaccharide units in sucrose is α.

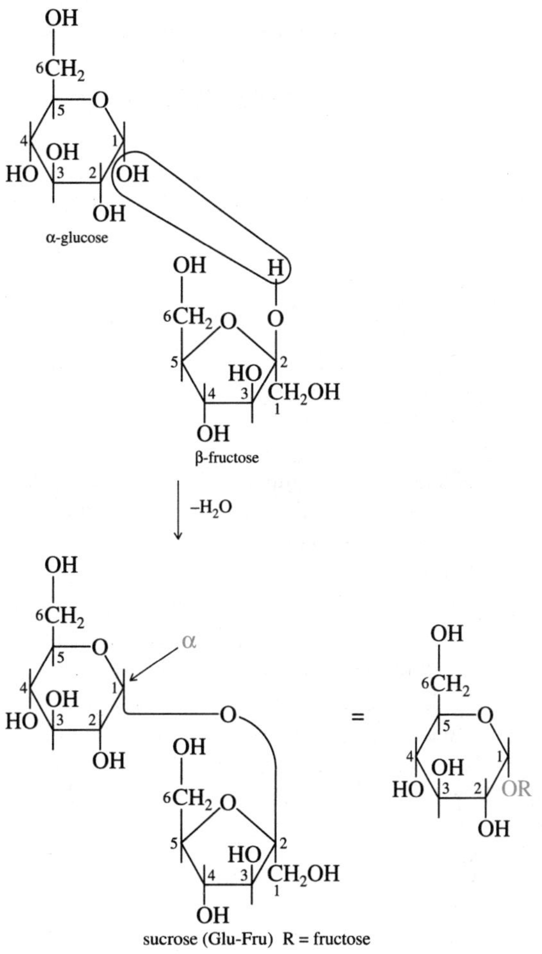

α-glucose

β-fructose

−H₂O

sucrose (Glu-Fru) R = fructose

As a result of the α linkage, sucrose can be digested by all animals, including humans. We have the necessary enzymes to cleave an α linkage and split sucrose into glucose and fructose, both of which can be used directly as a source of fuel. Humans can metabolize any monosaccharide, either directly or by conversion to glucose, but can use a disaccharide or polysaccharide only if it can be broken down into monosaccharides. The enzymes required to catalyze the cleavage of an α linkage are normally available; but the enzymes required for cleavage of a β linkage often are missing.

When sucrose is cleaved, the reaction is the reverse of that shown for formation of sucrose from glucose and fructose. The equal mixture of glucose and fructose that is produced in this reaction is called *invert sugar.* It is the main sugar in honey, since honeybees have enzymes that cleave sucrose in flower nectar. Glucose is abundant in nature. Fructose occurs primarily in honey and fruit. It is the sweetest of the carbohydrates, followed in decreasing order by sucrose, glucose, and the disaccharides maltose and lactose.

Sucrose, which we usually call *sugar,* is obtained for commercial use from sugarcane and sugar beets. The raw sugar is 96% to 98% sucrose, and is brown due to the presence of molasses. Brown sugar ranges from light to dark brown, containing some molasses. The lighter, more purified brands have a milder flavor.

Molasses is the sweet, sugary syrup remaining after all the sugar that can be economically recovered is crystallized out of sugarcane or sugar beet juice. It is used primarily as animal feed.

Two other important disaccharides are **maltose** (malt sugar) and **lactose** (milk sugar). Maltose consists of two glucose units, Glu-Glu, whereas lactose is a combination of a galactose and a glucose unit, Gal-Glu.

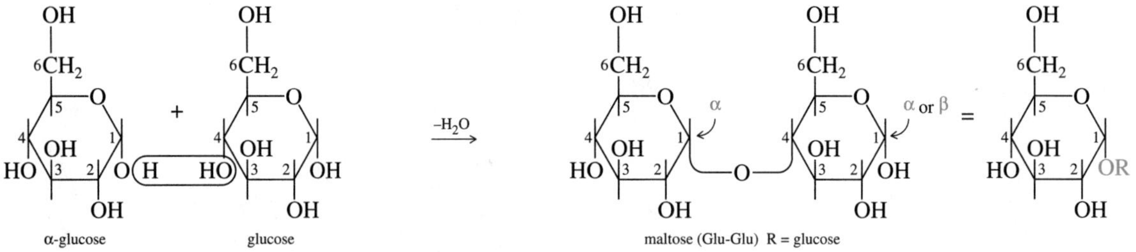

α-glucose   glucose   maltose (Glu-Glu) R = glucose

Once again, we must take account of α and β isomeric forms. The carbons that are linked in the disaccharides are also important. Sucrose is distinctive because the oxygen bridge between the two rings links C-1 of glucose to C-2 of fructose. For most other carbohydrates, including maltose and lactose, the linkages are 1,4; C-1 of the ring on the left is linked to C-4 of the ring on the right by an oxygen bridge.

Maltose is readily metabolized by humans. Yeast enzymes also cleave maltose during the production of alcoholic beverages by fermentation (see Chapter 17). In both cases, the α linkage to the left-hand ring makes the disaccharide susceptible to attack by enzymes that humans and yeasts have available.

Lactose (milk sugar) differs from maltose in two ways: each lactose molecule has one glucose and one galactose, and the monosaccharide units are joined by a β linkage.

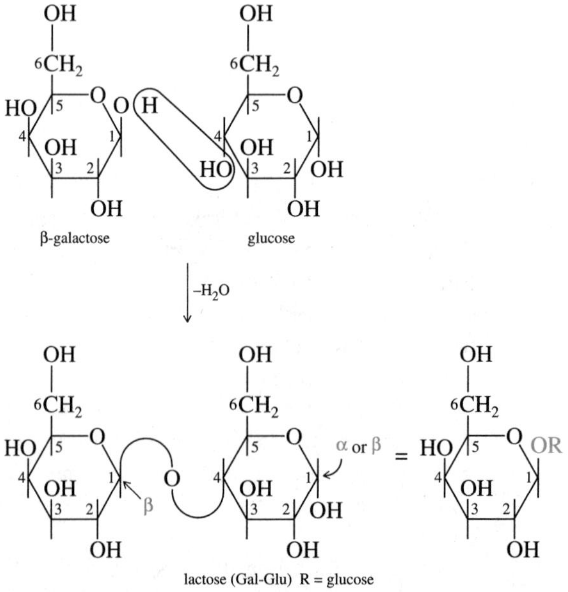

The importance of barley malt in brewing is discussed in Chapter 17.

β-galactose   glucose

lactose (Gal-Glu) R = glucose

Recall that galactose differs from glucose only in the location of the −OH at C-4, and humans are able to metabolize galactose once it is cleaved from the disaccharide. But because of the β linkage, some people are unable to metabolize lactose. They lack an enzyme known as β-galactosidase (lactase), which catalyzes the cleavage of the β linkage in lactose. About 70% of the world's adult population—including people of African, Asiatic, and South American Indian origins—cannot digest lactose. As a result, they suffer from a condition known as *lactose intolerance.* The importance of this disorder is discussed in detail in connection with dairy products in Chapter 19.

9.2 Polysaccharides

Goals: To study the structure of some important polysaccharides.
To appreciate the difference between amylose and amylopectin.
To appreciate some chemistry of starch.

Common commercial sources of sucrose (sugarcane and sugar beets) store most of their carbohydrates as sucrose. The flavor of fresh, sweet corn is due to sucrose, which is converted to less sweet starch as the corn ages.

When animals consume carbohydrates, they burn (metabolize) some and store the rest. Plants are able to produce carbohydrate (glucose) by the process known as *photosynthesis.* Once a plant has acquired the carbohydrate, the plant can metabolize it or store it, just as humans and other life forms do. The two main storage forms of carbohydrate in plants are **cellulose** and **starch.** Animals store carbohydrate primarily in the form of **glycogen,** discussed later.

cellulose (Glu)$_n$ β linkages

Cellulose

Cellulose is a polysaccharide consisting of glucose units (usually more than 1500) joined by β linkages. Cellulose is an important structural component of most plants. Its high content in wood and cotton contributes to their rigid structures.

Because of the β linkages between glucose units, cellulose is indigestible by humans, since we lack the necessary enzymes, called *cellulases.* Termites, cows, and other ruminants (cud-chewing animals) are able to digest grasses and other feed high in cellulose, even though they also lack the necessary enzymes. The difference is the multicompartment stomach of cows. In the section known as the *rumen,* microorganisms thrive and supply the necessary enzymes for breaking down cellulose. The attack does not simply cleave the cellulose to glucose. Instead, cellulose is converted into 2-, 3-, and 4-carbon fragments, which are stored as carbohydrate and fat. Methane and hydrogen are also formed in the breakdown of cellulose, so cows are sometimes accused of contributing to environmental pollution.

Vegetables with a high cellulose content increase the bulk in the human diet and stimulate contractions of the intestines; they have little caloric value, however, despite the high content of carbohydrate, because humans cannot metabolize cellulose. Dietary fiber is recommended because it improves the flow of waste materials through the large intestine. Vegetables with a coarse texture (lettuce, cabbage, peas, beans, broccoli, etc.), fruits, and cereal bran have a high fiber content.

Starch

Starch is a polysaccharide of repeating glucose units, but it differs from cellulose in two ways. First, the linkages between the monosaccharide units are α. This means starch is readily digested by animals, since the enzymes required for cleavage of α linkages are ubiquitous. Second, there are two forms of starch. One form is linear and is called **amylose.** The other is branched and is called **amylopectin.**

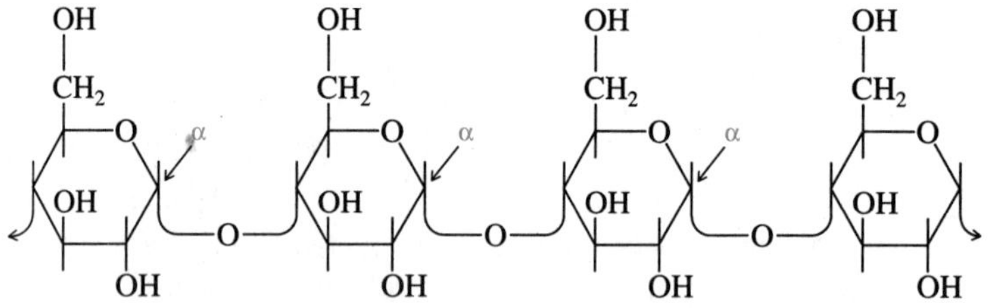

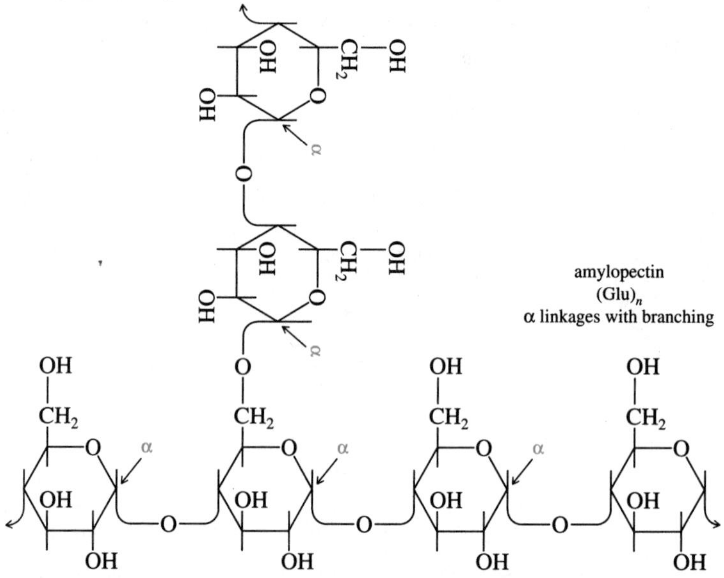

amylose
$(Glu)_n$
α linkages

amylopectin
$(Glu)_n$
α linkages with branching

Note the structures of amylose and amylopectin. Amylose is identical to cellulose except for the α linkages (rather than β) between C-1 and C-4 of adjacent glucose units. In amylopectin, most linkages are again α C-1 to C-4, but there is an occasional link between C-1 and C-6. This gives amylopectin a branched structure and different properties. (One consequence of the difference in structure is important in brewing; see Chapter 17.)

Amylose chains are usually 1000 to 4000 glucose units long, whereas amylopectin often has more than 5000 units. Natural starch (corn, potato, wheat, tapioca) consists of 20% to 30% amylose and 70% to 80% amylopectin, but some hybrids (primarily corn) may have different compositions. A hybrid known as waxy corn contains 100% amylopectin. High-amylose starches with contents typically 50% to 70% are also available.

The proportion of the two components determines some properties that influence the use of different starches. When amylose is heated and later cooled, it forms an opaque, rigid gel that makes it applicable for use in puddings and gum candies. Heated amylopectin cools to a clear nongelling paste with a waxy consistency, suitable for thickening gravies and sauces.

The different behavior of amylose and amylopectin is attributed to differences in hydrogen bonding (Chapter 7) between −OH groups of separate chains or within the same chain. The highly branched structure of amylopectin keeps the chains from approaching one another so water penetrates the solid starch granules more easily. In contrast, the unbranched amylose structure allows chains to lie close together with tight hydrogen bonding. Lower penetration of water molecules during cooking results in a more rigid and opaque gel.

Modification of Starches

The physical characteristics of a starch, such as shelf life, resistance to physical change due to freezing and thawing, and viscosity, can be chemically modified and improved. A number of chemical reagents have been used for several decades to alter starch. Two important ones are phosphoryl chloride, $POCl_3$, and acetic anhydride. The $POCl_3$ is a cross-linking agent that stabilizes swollen starch granules for use in applications where starch is exposed to high temperatures. Acetic anhydride replaces some of the −OH groups with acetate (ester) groups, resulting in fewer hydrogen bonds between starch molecules.

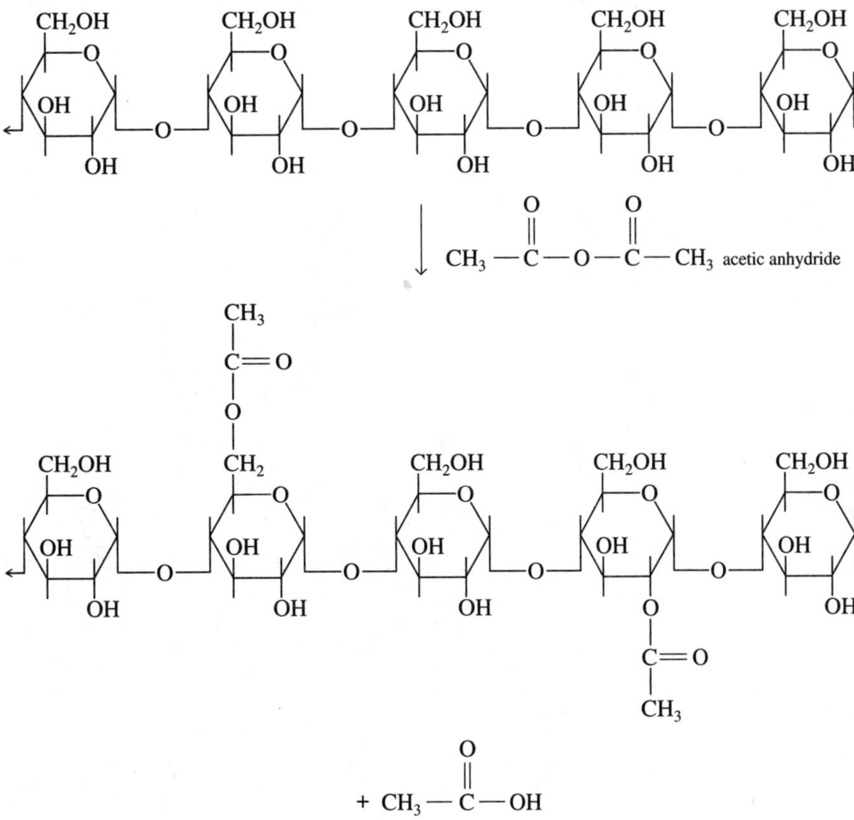

This increases the clarity of the starch and improves other properties. Modified starches are used in many prepared foods.

Starches and dextrins are used in numerous adhesive applications, such as bottle labeling, carton sealing, envelope sealing, and gummed tape manufacture. Dextrins are produced from all sources of starch; they are formed by cooking starch with a catalyst, usually

HCl, followed by neutralization of the acid. This treatment causes hydrolysis (depolymerization) and recombination, called *condensation,* of starch fragments. During the process, the starch chains become shorter and more highly branched, which increases the tackiness of the resulting starch gel. Some moisture is required to assist in cleaving starch chains, but condensation reactions are encouraged by low moisture content (less than 3%)(*1*). If an abundant supply of water were present, the starch would be cleaved all the way to glucose.

Glycogen

Another polysaccharide of glucose is *glycogen,* the storage form of carbohydrate in animals. Glycogen has a structure very similar to amylopectin, except for shorter chains, typically 12 to 18 units long, and more frequent branches. As in starch, there are α linkages between all glucose units, so that glycogen is readily metabolized.

A common representation of glycogen is given in Figure 9.1. The drawing is a cross section showing the treelike structure created by branched chains. The short inner segments are parts of branches extending above and below the cross section. Every circle represents a glucose unit; only one (arrow) does not have its C-1 attached to another monosaccharide unit.

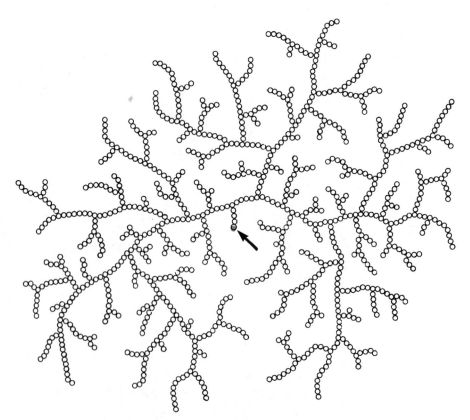

Figure 9.1   The structure of glycogen. (From McGilvery, R. W. *Biochemistry: A Functional Approach,* 3rd ed; W. B. Saunders: Philadelphia, 1983.)

Other Polysaccharides

Many natural polysaccharides have sugars other than glucose as the repeating units. One important example is hyaluronic acid, as shown.

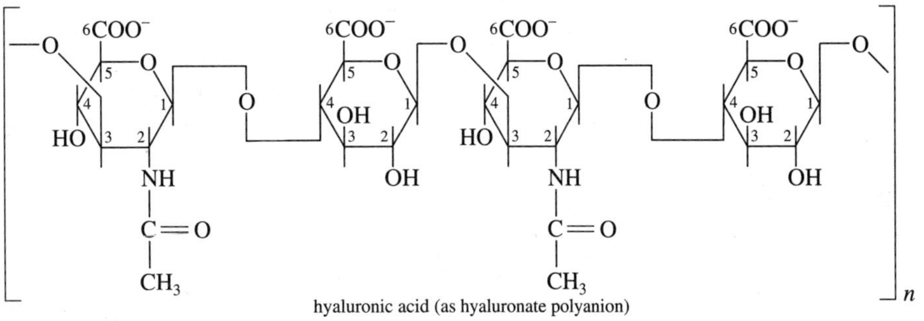

hyaluronic acid (as hyaluronate polyanion)

A close inspection of this structure shows a polysaccharide of two monosaccharide units alternating within the chain. Two other unusual features distinguish it from starch: the presence of an acid group (in ionic form) as carbon 6 of every unit, and a 1,3 linkage alternating with the more common 1,4 linkage.

Called hyaluronate in the polyanionic form, it acts as a lubricant and buffer against mechanical damage in joints throughout the body. Because of the repulsion between charges on adjacent chains, the chains readily slide past one another, rather than sticking together. This accounts for the lubricating action.

SUMMARY

Monosaccharides, disaccharides, and polysaccharides are the common forms of carbohydrates. Glucose is the most abundant monosaccharide. It can exist alone as a monosaccharide, as the disaccharide maltose, or as the repeating unit in the polysaccharides starch, glycogen, and cellulose. Fructose and galactose are the other principal monosaccharides. Sucrose and lactose are other common disaccharides.

The α and β configurations of the linkages between monosaccharide units are important in the digestibility of disaccharides and polysaccharides. The α arrangement, found in sucrose, maltose, and starch, makes these carbohydrates readily digestible. A large portion of the world's adult population lacks the enzyme lactase necessary to break the β linkage in lactose. These people are unable to tolerate milk because they cannot metabolize the milk sugar.

Only ruminants and termites can metabolize cellulose, with its β linkages between glucose units. In ruminants, microorganisms thriving in the intestinal tract provide the enzymes required to break down the polysaccharide.

Amylose and amylopectin are the components of starch. Amylose is a linear polysaccharide and amylopectin is branched. Branching greatly affects the properties of different starches and determines their suitability for use in various food products. Starches can be modified chemically to improve their properties. Glycogen is the storage form of carbohydrate in animals. It has a structure similar to that of amylopectin.

PROBLEMS

1. Give definitions or examples of each of the following.

   a. α and β linkages
   b. Monosaccharide
   c. Disaccharide
   d. Polysaccharide
   e. Starch
   f. Glycogen
   g. Cellulose
   h. Ruminant
   i. Brown sugar
   j. Amylopectin

2. Identify the following disaccharides or polysaccharides.

   a. Gal-Glu
   b. Glu-Fru
   c. Glu-Glu
   d. -Glu-Glu-Glu-Glu-Glu-Glu-Glu-Glu-
   e. -Glu-Glu-Glu-Glu-Glu-Glu-Glu-Glu-Glu-Glu-Glu-

3. Draw a complete formula for maltose. How would you modify your drawing to symbolize the amylose component of starch? How would you modify your drawing of amylose to symbolize cellulose?

4. Why are humans unable to use cotton as a dietary source of carbohydrate? Are any animals able to do so?

5. What are the two ways in which maltose and lactose differ in structure?

6. Identify some foods, including sauces, gravies, thickeners, or salad dressings, that contain modified starch.

Reference

1. Whistler, R. L.; Bemiller, J. N.; Paschall, E. F. *Starch: Chemistry and Technology.* Academic: New York, 1984; Chapters 10 and 20.

Nuclear Chemistry I: Spontaneous Radioactive Decay

10

I n this and the following chapter we will examine the controversial subject of nuclear science. This topic spans the traditional scientific disciplines of physics, biology, and chemistry, as well as certain areas of engineering and medicine.

Spontaneous radioactive decay will be considered first so that you can appreciate the nature of radiation and how it is detected. We will also consider radiation exposure from natural sources, such as cosmic radiation and radon, and from human-made sources. The application of radioactive isotopes in archeology and medicine receives special attention.

Chapter 11, the first chapter in the section titled Energy and Environment, introduces the topic of induced radioactive decay. This leads to a discussion of transmutation and to the use of nuclear processes for generating energy.

10.1 Isotopes

GOAL: To understand isotopes.

As discussed in Sections 2.2 and 2.3, protons and neutrons are located in the central portion of the atom known as the nucleus. Because of their location, these particles are also called **nucleons.** Recall that the number of protons, given by the atomic number, determines what element is present. Atoms of many elements can exist in different forms, called **isotopes** or **nuclides,** that differ in the number of neutrons present in the nucleus.

For the element carbon, the isotopes $^{12}_{6}C$, $^{13}_{6}C$, and $^{14}_{6}C$, are known. In this designation, the subscript numeral represents the number of protons (atomic number). The use of the subscript is optional and will be used hereafter only for emphasis, since all isotopes of any atom have the same atomic number.

The superscript is the **mass number** of the nuclide; it is defined as the sum of the number of protons plus the number of neutrons. (The mass of an electron is negligible compared with the mass of a proton or a neutron.) Thus, ^{12}C, the most abundant isotope of the element carbon, has six protons and six neutrons in the nucleus; these are balanced by six electrons in the orbital system, to give a net charge of zero. The isotope ^{13}C has seven neutrons, and ^{14}C has eight. The isotope ^{13}C makes a small but significant contribution to all naturally occurring carbon. In contrast, ^{14}C is unstable and undergoes spontaneous radioactive decay. It is therefore present only in trace amounts in carbon-containing compounds.

Recall that the average atomic mass of carbon is 12.011 amu (Section 2.3). This value takes account of the predominance of ^{12}C (about 99% natural abundance), but also acknowledges the small contribution of ^{13}C (about 1% natural abundance). Similarly, there are three isotopes of hydrogen, ^1H (protium), ^2H (deuterium), and ^3H (tritium), with an average atomic mass of 1.008 amu. Deuterium makes up about 0.02% of natural hydrogen. Tritium is present in only trace amounts since, like ^{14}C, it is unstable and undergoes radioactive decay. The isotope ^1H makes up about 99.98% of naturally occurring hydrogen.

10.2 Radioactivity

GOALS: To appreciate the importance of the neutron to proton ratio (n/p).
To understand how n/p ratio can be altered by spontaneous radioactive decay.
To recognize the significance of a drip line.
To understand the nature of α, β, and γ radioactive decay.
To see the variety and sources of ionizing radiation.

An unstable isotope undergoes radioactive decay by liberating energy and forming a more stable isotope. To understand radioactive decay of isotopes such as ^{14}C and ^3H, consider the role of neutrons in the atomic nucleus. The number of neutrons strongly influences the stability of the nucleus. A stable atom is a delicately balanced combination of attractive and repulsive forces. Particles with opposite charges, such as protons and electrons, attract one another. On the other hand, particles with the same charge repel one another; electrons repel one another, and protons repel other protons. Neutrons affect the balance in a more subtle way, and they are important in determining whether an atomic nucleus is stable or will undergo radioactive decay.

The neutron-proton (n/p) ratio has a great effect on stability. A plot of this ratio in Figure 10.1 shows that the stable isotopes lie in a narrow, gently curving band (dark squares), with little variation in ratios. The current nuclear chart contains a full 2200 entries, each representing an isotope for which at least one property has been measured. Of the 2200 known isotopes, synthetic as well as natural, only 271 are stable, that is, nonradioactive. When the n/p ratio is either too high or too low, an isotope is unstable and radioactive decay occurs. Isotopes that undergo radioactive decay are known as **radionuclides.**

Those isotopes with more than the number of neutrons required for stability appear to the right of the band of stable isotopes; they can be characterized as having a *neutron excess.* The *neutron drip line* shown in Figure 10.1 identifies the limit on the number of neutrons that can be accommodated by a particular element. Therefore, the drip line also gives the maximum atomic mass that is possible among the isotopes of any element. To the left of the band of stable isotopes, the *proton drip line* describes the lower limit of atomic mass for isotopes that can exist long enough to be observed. To the left of the proton drip line, too many protons are present compared with the number of neutrons, and the isotopes are unknown.

As an isotope decays, the n/p ratio moves closer to a stable value. Even a more favorable ratio, however, does not stabilize any nuclei with a mass greater than that of bismuth-209 due to the large number of protons. Consequently, all elements above $^{209}_{83}Bi$ are unstable, and therefore all of the isotopes of these elements undergo some form of radioactive decay. (Notice where the dark squares end in Figure 10.1.)

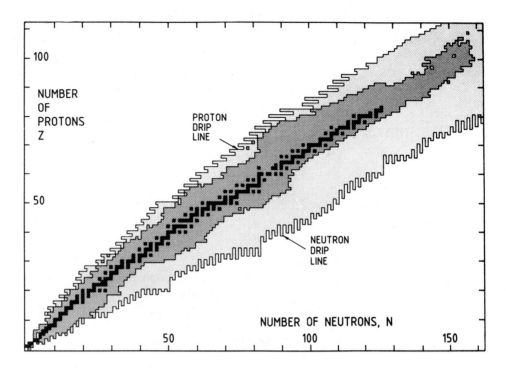

Figure 10.1   Chart of the nuclides. Stable isotopes are represented by black squares. Moving out from the stable isotopes are areas (dark gray) representing isotopes that have already been synthesized and identified. Moving farther out are isotopes that theory indicates could exist for an observable period of time, but have not yet been produced. Finally, beyond the drip lines, no combinations of neutrons and protons are expected to yield observable isotopes *(1)*.

When a radionuclide decays, a new nucleus forms. If that nucleus is stable, no further decay occurs. If it is not stable, decay will continue until a stable nuclide is finally produced. Radioactive decay may take several forms; the most common types are alpha (α), beta (β), and gamma (γ) decay, all of which are forms of **spontaneous radioactive decay.**

Alpha Decay

An α particle (helium nucleus), symbolized α, $^4_2\alpha$, or ^4_2He, can be emitted by a nucleus, which gains stability in the process. This type of decay is particularly common to elements with high atomic numbers (above 83), since the loss of an α particle causes a decrease in the number of protons in the nucleus. Since the α particle consists of two protons and two neutrons, its n/p ratio is 1. The emission of an α particle from a nucleus causes the affected atom to proceed to a *higher* n/p (since n/p is always greater than 1 for the starting isotope). A typical reaction is

$$^{235}_{92}\text{U} \rightarrow {}^4_2\alpha + {}^{231}_{90}\text{Th} \quad \text{(\alpha decay)}$$

$$\text{n/p} = \frac{235 - 92}{92} = \frac{143}{92} = 1.554 \qquad \text{n/p} = \frac{141}{90} = 1.567 \quad \text{(n/p increases)}$$

Remember that the number of neutrons is equal to the mass number minus the number of protons (atomic number).

Beta Decay

The conversion of a neutron to a proton can occur within a nucleus, resulting in a *lower* n/p ratio. A neutron can be formed by combining a proton and an electron. Therefore, when a neutron converts to a proton inside the nucleus, the process must be accompanied by the release of a negative electron, also known as a *negatron,* from the nucleus. When a negatron is emitted from a nucleus, the electron is called a β^- particle to denote its origin. The decay process is alternately known as **negatron emission** or **β^- decay.** An example is the decay of indium-116, in which an indium atom changes into a tin atom due to the gain of a proton:

$$^{116}_{49}\text{In} \rightarrow {}^{0}_{-1}\beta + {}^{116}_{50}\text{Sn} \qquad \text{(β^- decay, electron emission)}$$

$$\text{n/p} = \frac{67}{49} = 1.367 \qquad \text{n/p} = \frac{66}{50} = 1.320 \qquad \text{(n/p decreases)}$$

Regardless of the source of an electron, it has a charge of minus one (represented by the subscript) and no mass (the superscript).

Two ways other than α decay for *increasing* n/p are **β^+ decay**, also called **positron emission,** and **electron capture.** These two processes are the opposite of β^- decay and occur when the n/p is too low. For instance,

$$^{116}_{51}\text{Sb} \rightarrow {}^{0}_{+1}\beta + {}^{116}_{50}\text{Sn} \qquad \text{(β^+ decay, positron emission)}$$

$$\text{n/p} = \frac{65}{51} = 1.275 \qquad \text{n/p} = \frac{66}{50} = 1.320 \qquad \text{(n/p increases)}$$

Positron emission tomography (PET) offers a means for studying normal and abnormal body functions. See Section 10.8.

In this example, a proton converts into a neutron and a positron. The change of atomic number from the starting nuclide to the product shows the loss of the proton. However, the mass number does not change; this indicates that the total number of nucleons (protons plus neutrons) remains the same.

In the electron-capture process, an electron enters the nucleus and combines with a proton to form a neutron, as illustrated.

$$^{195}_{79}\text{Au} + {}^{0}_{-1}\text{e} \rightarrow {}^{195}_{78}\text{Pt} \qquad \text{(electron capture)}$$

$$\text{n/p} = \frac{116}{79} = 1.468 \qquad \text{n/p} = \frac{117}{78} = 1.500 \qquad \text{(n/p increases)}$$

Gamma Decay

Many nuclei exist in unstable forms; some can relieve part or all of the instability by emitting energy in the form of γ rays. Unlike α and β particles, γ rays are not particles at all. They are hard to visualize and to describe. They are a type of electromagnetic radiation, like light. Light is described in two ways: as wavelike and as small bundles of energy called *quanta* or *photons.*

In most cases, γ rays are given off only when other types of decay occur. One common example is observed with cobalt-60, which undergoes β^- decay to nickel-60 and emits two γ rays in the process.

$$^{60}_{27}\text{Co} \rightarrow {}^{60}_{28}\text{Ni} + {}^{0}_{-1}\beta + 2\gamma$$

In a few cases, it is possible to observe the release of γ radiation when there is no α or β decay. One important example is the decay of technetium-99m as follows:

$$^{99m}\text{Tc} \rightarrow {}^{99}\text{Tc} + \gamma$$

This process is commonly known as an **isomeric transition,** symbolized IT, in which an unstable form of the isotope becomes more stable by emitting radiation. The m (for meta-stable) in ^{99m}Tc signifies that the nucleus is unstable and will become stable by γ emission. Such a process in which only γ emission occurs is uncommon, because the isotopes that undergo such changes are very unstable. Some metastable isotopes remain active for only a fraction of a second. Technetium-99m loses half its radioactivity in only 6 hours. Isomeric transition presumably occurs in the decay of ^{60}Co (see previous equation), but ^{60m}Ni, which is initially formed, is so unstable that it immediately decays by emitting radiation to form stable ^{60}Ni, the observed product.

Other Types of Ionizing Radiation

When particles or radiation emitted during radioactive decay come into contact with surrounding matter, some atoms of target material are converted to ions. Thus, radioactive emissions (α, β, and γ) are collectively known as **ionizing radiation.** Table 10.1 lists all of the common types of ionizing radiation.

X-rays are given off after electron capture. An outer-orbital electron drops into the space vacated by the captured electron, and energy is released in the form of x-rays (Figure 10.2). Electron capture by the radioactive beryllium nucleus leaves an unstable arrangement of the orbital electrons. The system stabilizes when an electron moves from the outer shell to the inner shell and emits energy in the form of an x-ray.

Table 10.1   Types of ionizing radiation

Type of Radiation	Symbol	Mass	Charge	Description	Source
α	$^4_2\alpha$ or ^4_2He	4	+2	Helium nucleus	Spontaneous radioactive decay (primarily from heavy atoms)
β (negatron)	$^0_{-1}\text{e}$ or $^0_{-1}\beta$	0	−1	Negative electron	Spontaneous radioactive decay
β (positron)	$^0_{+1}\text{e}$ or $^0_{+1}\beta$	0	+1	Positive electron	Spontaneous radioactive decay
Proton	^1_1H	1	+1	Positive particle	Artifically produced by nuclear reactors (accelerators)
Neutron	^1_0n	1	0	Neutral particle	Artifically produced by nuclear reactors (accelerators)
γ	γ	0	0	Electromagnetic radiation	Spontaneous radioactive decay
X-ray		0	0	Electromagnetic radiation	X-ray machines, rearrangement of orbital electrons

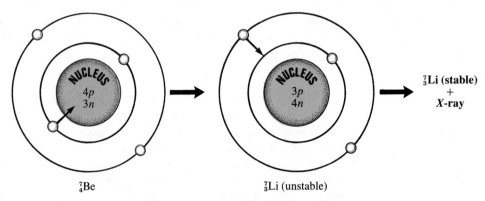

Figure 10.2   An example of x-ray emission after electron capture.

Large quantities of x-rays are generated by machines for clinical, industrial, and research uses. The method of generation is similar to throwing a rock against a wall. In the process, the wall takes on the energy of the moving rock, and emits it as sound waves and heat. Similarly, in an x-ray machine, a target material is bombarded by high-speed (high-energy) electrons. These electrons interact strongly with electrons of the target, knocking out some of the target electrons and causing other orbital electrons to relocate. The result is the release of energy as x-rays.

10.3   Nuclear Reactions

GOALS: To know the technique and importance of balancing nuclear equations.
To appreciate the value of the periodic table in predicting the products obtained by radioactive decay.
To understand the phenomenon of half-life and some of its consequences.
To appreciate factors that influence the ability of radiation to penetrate matter.

Each radioactive decay process has been described by writing a nuclear reaction. To write these complete and balanced equations accurately, adhere to the following rules:

1. The sums of the atomic numbers on each side of the equation must be equal.

2. The sums of the mass numbers on each side of the equation must be equal.

Review the nuclear equations written thus far. Then study the following examples, each describing a radioactive decay process. Properly balance each equation to identify the missing product.

EXAMPLE 10.1:

Tritium (^{3}H) undergoes radioactive decay, emitting an electron (β^- particle). Write the complete reaction.

Solution:

The problem should be approached stepwise. First write the portion of the equation that is known.

$$^3\text{H} \rightarrow\ _{-1}^{0}\beta + \ ?$$

Next, include the atomic number for hydrogen. This gives the sum of the atomic numbers on the left side of the equation and allows us to write the atomic number and mass number for the unknown product nucleus.

$$_1^3\text{H} \rightarrow \ _{-1}^0\beta + \ _2^3?$$

Consult a table of elements to find the symbol and name of the element that has an atomic number of 2. It is helium, and the complete and balanced equation for β^- decay of tritium is:

$$_1^3\text{H} \rightarrow \ _{-1}^0\beta + \ _2^3\text{He}$$

This equation confirms that β^- decay results in an increase of one proton and a decrease of one neutron.

EXAMPLE 10.2:

Aluminum-25 undergoes radioactive decay by electron capture. Write the complete reaction.

Solution:

Once again, write as much of the equation as possible from the information given in the problem. Find the atomic number of Al in a table of the elements. This gives:

$$_{13}^{25}\text{Al} + \ _{-1}^0\text{e} \rightarrow \ _{12}^{25}?$$

A table of the elements shows that magnesium is the element with atomic number 12. Therefore, the final equation is:

$$^{25}\text{Al} + \ _{-1}^0\text{e} \rightarrow \ _{12}^{25}\text{Mg}$$

This equation confirms that electron capture results in a decrease of one proton and an increase of one neutron.

EXAMPLE 10.3:

Thorium-232 undergoes radioactive decay yielding radium-228. Write the complete equation.

Solution:

This time, the product is known, but not the mode of decay. The problem can be approached in the same way, however. If necessary, consult Table 10.1.

$$_{90}^{232}\text{Th} \rightarrow \ _{88}^{228}\text{Ra} + \ _2^4?$$

$$_{90}^{232}\text{Th} \rightarrow \ _{88}^{228}\text{Ra} + \ _2^4\alpha$$

Radioactivity and the Periodic Table

In α and β radioactive decay processes, one element is converted to another, and the principal nuclear changes are:

1. For α decay, the number of protons decreases by two.
2. For β decay, (a) electron emission—the number of protons increases by one; (b) positron emission—the number of protons decreases by one; and (c) electron capture—the same as positron emission.

Positron emission and electron capture have the same net effect on a nucleus: both yield a product element that precedes the starting element in the periodic table.

In Example 10.1, electron emission increases the atomic number to yield the next element in the periodic table (Figure 10.3). Thus, hydrogen (atomic number 1) was converted to helium (atomic number 2). In Example 10.2, the opposite occurred; aluminum (atomic number 13) was converted to magnesium (atomic number 12), because a proton is converted to a neutron when the electron is captured by the nucleus.

In α decay, two protons (and two neutrons) are emitted, so the atomic number decreases by two. An illustration appeared in Example 1.3, in which thorium (atomic number 90) was converted to radium (atomic number 88).

Data on Radioactive Isotopes

Since type and energy of radiation vary greatly, it is necessary to have tables of data to assess the dangers of any radioactive isotope. Reference sources, including handbooks of physics and chemistry, list the kind of information given in Table 10.2.

atomic number / symbol of the element / average atomic mass: 1 **H** 1.008

	1 IA	2 IIA	3 IIIB	4 IVB	5 VB	6 VIB	7 VIIB	8	9 VIIIB	10	11 IB	12 IIB	13 IIIA	14 IVA	15 VA	16 VIA	17 VIIA	18 VIIIA
1																		2 **He** 4.003
2	3 **Li** 6.941	4 **Be** 9.012											5 **B** 10.81	6 **C** 12.01	7 **N** 14.01	8 **O** 16.00	9 **F** 19.00	10 **Ne** 20.18
3	11 **Na** 22.99	12 **Mg** 24.31											13 **Al** 26.98	14 **Si** 28.09	15 **P** 30.97	16 **S** 32.07	17 **Cl** 35.45	18 **Ar** 39.95
4	19 **K** 39.10	20 **Ca** 40.08	21 **Sc** 44.96	22 **Ti** 47.88	23 **V** 50.94	24 **Cr** 52.00	25 **Mn** 54.94	26 **Fe** 55.85	27 **Co** 58.93	28 **Ni** 58.69	29 **Cu** 63.55	30 **Zn** 65.38	31 **Ga** 69.72	32 **Ge** 72.59	33 **As** 74.92	34 **Se** 78.96	35 **Br** 79.90	36 **Kr** 83.80
5	37 **Rb** 85.47	38 **Sr** 87.62	39 **Y** 88.91	40 **Zr** 91.22	41 **Nb** 92.91	42 **Mo** 95.94	43 **Tc*** (98)	44 **Ru** 101.1	45 **Rh** 102.9	46 **Pd** 106.4	47 **Ag** 107.9	48 **Cd** 112.4	49 **In** 114.8	50 **Sn** 118.7	51 **Sb** 121.8	52 **Te** 127.6	53 **I** 126.9	54 **Xe** 131.3
6	55 **Cs** 132.9	56 **Ba** 137.3	57 **La** 138.9	72 **Hf** 178.5	73 **Ta** 180.9	74 **W** 183.9	75 **Re** 186.2	76 **Os** 190.2	77 **Ir** 192.2	78 **Pt** 195.1	79 **Au** 197.0	80 **Hg** 200.6	81 **Tl** 204.4	82 **Pb** 207.2	83 **Bi** 209.0	84 **Po*** (209)	85 **At*** (210)	86 **Rn*** (222)
7	87 **Fr*** [223]	88 **Ra*** 226.0	89 **Ac*** 227.0	104 **Unq*** [261]	105 **Unp*** [262]	106 **Unh*** [263]	107 **Uns*** [262]	108 **Uno*** 265	109 **Une*** [266]									

Lanthanides	58 **Ce** 140.1	59 **Pr** 140.9	60 **Nd** 144.2	61 **Pm*** (145)	62 **Sm** (145)	63 **Eu** 152.0	64 **Gd** 157.3	65 **Tb** 158.9	66 **Dy** 162.5	67 **Ho** 164.9	68 **Er** 167.3	69 **Tm** 168.9	70 **Yb** 173	71 **Lu** 175.0
Actinides	90 **Th*** 232.0	91 **Pa*** [231]	92 **U*** 238.0	93 **Np*** [237]	94 **Pu*** [244]	95 **Am*** [243]	96 **Cm*** [247]	97 **Bk*** [247]	98 **Cf*** [251]	99 **Es*** [252]	100 **Fm*** [257]	101 **Md*** [258]	102 **No*** [259]	103 **Lr*** [260]

*All isotopes are radioactive.
[ ] Indicates mass number of the isotope of longest known half-life.

Group numbers 1–18 represent the system recommended by the International Union of Pure and Applied Chemistry.

Figure 10.3 Periodic table of the elements.

Table 10.2   Isotope information for ^{14}C and ^{203}Hg

Isotope Information	^{14}C	^{203}Hg
Natural abundance (%)	Near zero	Zero
Atomic mass	14	203
Half-life	5568 yrs	47 days
Mode of decay	β^-	β^-, γ
Energy of decay	0.155 MeV	0.21, 0.279 MeV

Half-Life

The **half-life** is the length of time required for the radioactivity of an isotope to diminish by one-half. This can best be understood using ^{203}Hg as an example. Table 10.3 shows that half of the radioactivity of ^{203}Hg is lost every 47 days. The significance of this property is even clearer in the graph in Figure 10.4. It shows that decay is most rapid when the amount of radioactive material is greatest, although the half-life remains the same at all levels of radioactivity. The half-life of an isotope at any level of activity can be determined by making a series of measurements over a suitable period of time. The first measurement is taken as time = 0; the activity level then is taken as 100% relative to all later measurements. This does not mean that no decay occurred before the first measurement. In fact, the isotope may have only a small fraction of its original activity, but the half-life does not change with time.

Half-lives vary greatly, ranging from fractions of a second to thousands of years. Carbon-14 dating depends on half-life (Section 10.7). Iodine-131 is sometimes used to treat an overactive thyroid gland (Section 10.8). The length of time that radioactive emissions will continue from this isotope, its shelf life or useful life, can be calculated accurately. Such information is valuable for knowing how long a patient will be exposed to the radioactivity from a given dose of ^{131}I.

Knowing half-lives of radionuclides is also important in planning safe disposal. An isotope with a short half-life, such as ^{203}Hg ($t_{1/2}$ = 47 days), may simply be allowed to decay in a properly shielded container. Isotopes with very long half-lives may have to be buried in a suitable location under the direction of the Nuclear Regulatory Commission.

Table 10.3   Time and percentage activity for ^{203}Hg

Time (Number of Half-Lives)	Activity (%)
0	100
1 × 47 days (1)	50
2 × 47 days (2)	25
3 × 47 days (3)	12.5
4 × 47 days (4)	6.25
5 × 47 days (5)	3.125
10 × 47 days (10)	0.00098

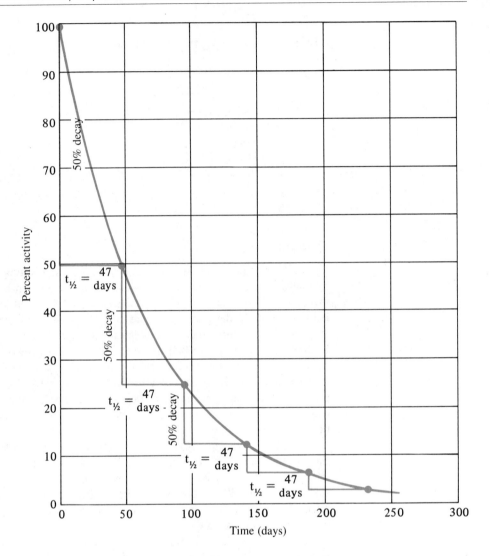

Figure 10.4 Radioactive decay of mercury-203.

Decay Energy

Decay energy is the energy of the emitted particle or ray (see Table 10.2). The situation is complicated for β particles, which are emitted with a range of energies; the energy value listed describes the maximum energy observed. A particle called a **neutrino** is emitted along with the β⁺ particle; an **antineutrino** is emitted with a β⁻ particle. The neutrino (or antineutrino) carries some of the energy released during radioactive decay. The tabulated decay energy for β particles is the total amount of energy released in β decay, distributed between the β particle and the neutrino (or antineutrino).

Decay energy is important for predicting the ability of radiation to penetrate matter. High-energy radiation has great penetrating power. Practical consequences include knowing whether radiation will penetrate the body during treatment of certain diseases (Section 10.8). Safe storage conditions can be determined only by knowing how effectively the radiation will penetrate a potential container. Some isotopes can be handled safely in glass

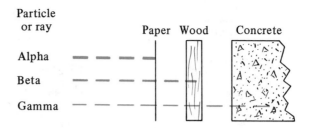

Figure 10.5 Relative penetration of α, β, and γ radiation.

and may not even penetrate air to any extent. Others must be stored in thick-walled lead containers to prevent escape of radiation.

Mass and Size

The size (mass) of different kinds of radiation also determines their ease of penetration. The heavy α particle interacts strongly with anything that it strikes and therefore has very little ability to penetrate any matter, even air. When the particle mass is smaller, as it is for β particles and γ radiation, the degree of penetration is greater (Figure 10.5). For this reason, isotopes that emit α particles are safest to handle, since the particles cannot pass through air or body tissues. If an α emitter is acquired orally or by inhalation, however, the worst possible effect from radiation exposure exists. Tissues in the immediate vicinity of the isotope will receive the full dose of radiation.

10.4 Radiation Detection and Exposure

GOALS: To understand the methods used to detect radioactive materials.
   To be aware of ways in which people are subjected to radioactivity.
   To be aware of the large difference between the amount of radiation to which the population is normally exposed and the amount required to produce obvious clinical symptoms.

As noted earlier, radioactive emissions are collectively known as ionizing radiation since they cause the formation of ions when they come in contact with matter. This explains radiation damage, and it also provides a means for detecting radiation.

Counters

When ionization occurs in a chamber filled with a suitable gas, the amount of radiation causing the ionization can be measured. Figure 10.6 gives a schematic diagram of a typical radiation counter, a **Geiger-Müller** counter.

   The counter also contains a negatively charged wire (collecting electrode) and a positively charged outer wall (electrode). The ions formed in the gas by the ionizing radiation are attracted to the electrodes and produce a signal that flows to the meter, when they make contact. Such a device can be calibrated so that the signal is a direct measure of the amount of the radiation.

   Radiation is described as *soft* if it has difficulty penetrating any matter, including the window of a counter (Figure 10.7). Soft radiation is often counted by *liquid scintillation*. In this technique, a special chemical compound is added to the sample to absorb the radiation energy. This compound then reemits a portion of the energy

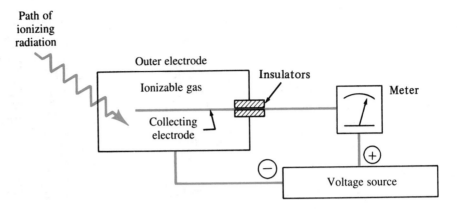

Figure 10.6 Schematic diagram of a typical radiation counter *(2)*.

as light that can be measured in order to describe the amount of radioactivity. Carbon-14, one of the most widely used isotopes in research, is often monitored by liquid scintillation. It decays by emission of a low-energy β^- particle. The reaction is:

$$^{14}_{6}\text{C} \rightarrow {}^{14}_{7}\text{N} + {}^{0}_{-1}\beta$$

In Section 10.5 we will consider the hazards of β and other kinds of radiation. For now, you should be aware that the radiation from ^{14}C is very soft, which means that the isotope is safe to handle, although care must be taken to avoid any ingestion; and the radiation is stopped (absorbed) before it can pass through the window of a radiation counter, so liquid scintillation is usually required to detect it.

It is also important to monitor the amount of radiation to which a person may be exposed. Radiation detection badges or **dosimeters** are worn at all times by researchers

Figure 10.7 A radiation counter.

Figure 10.8
A radiation dosimeter.

who handle radioactive materials and by workers at nuclear power plants. The dosimeter pictured in Figure 10.8 can be carried in the pocket. It operates on the same principle as the Geiger-Müller counter except that it works on a low voltage and contains a device, an electrometer, that provides a record of accumulated dose.

Many badges are made of photographic film that is shielded against exposure to light rays but is susceptible to the more penetrating types of ionizing radiation. Badges made to detect different types of radiation consist of several layers of photographic film separated by filters (absorbers) of increasing absorbing power. Thus, the top layers of film are exposed to all radiation. The type of radiation can be determined by developing and examining the different layers of film. The amount of radiation is calculated from the degree of darkening of each layer of film.

Units of Activity and Dose

To understand the potential hazards of radioactive materials, one must be familiar with some of the ways of describing radiation. The following are conventional units.

Curie (Ci)   Unit that describes the amount of radioactive material in a sample by the number of radioactive disintegrations per second. One Curie is equal to 37 billion (3.7×10^{10}) radioactive disintegrations per second, which is approximately equal to the activity of 1 g of radium-226. The unit is named after Marie Curie (Figure 10.9), who discovered the isotope ^{226}Ra. The **picocurie (pCi)** is 10^{-12} Ci; it is a common unit used to describe many environmental samples. The Curie has been superseded by the Becquerel in modern usage.

Becquerel (Bq)   SI (System International) unit that describes the amount of radioactive material in a sample by the number of radio-

Figure 10.9

Marie Sklodowska Curie (1867–1934). A native of Poland, Marie Curie studied under Henri Becquerel in the 1890s in France. In 1902 Madame Curie and her husband Pierre reported the isolation of radium, a new radioactive element. Becquerel and the Curies shared the Nobel Prize in 1903 for their work. Marie Curie received a second Nobel Prize in 1911 for further work. Her son-in-law and daughter, Irene Joliot-Curie, received yet another Nobel Prize in 1935 for work on artificial radioactivity.

	active disintegrations per second. A 1-Bq sample undergoes one disintegration per second (dps). Since 1 Bq is 1 dps and 1 Ci is 3.7×10^{10} dps, the Bq is a much smaller quantity.
Working level (WL)	Unit that describes the level of radon gas in an air sample in terms of its radioactive decay products (see Section 10.11 for details); 200 pCi/L is equal to 1 WL.
Roentgen (R)	Unit that describes the quantity of radiation given off by a sample, applicable only to x- or γ radiation.
rad	Unit of radiation absorbed dose.
Gray (Gy)	SI unit of absorbed dose; 1 Gy equals 100 rad.
rem	Unit of absorbed dose expressed in terms of biological effect of radiation; the effect varies from one tissue to another and with different types of radiation. The unit is derived from the phrase roentgen-equivalent-man.
Sievert (Sv)	SI unit of biological effect; 1 Sv equals 100 rem.

Levels of Radiation Exposure

It is useful to put some statistics into proper perspective when trying to understand the hazards of radiation. The data in Tables 10.4 and 10.5 provide a striking contrast between usual exposure and the levels of radiation that bring about rapid and sometimes acute symptoms. As you examine the tables, it will be apparent that the normal level of exposure is comparatively low.

The data in Table 10.5 are somewhat variable, affected by individual lifestyle and place of residence. Exposure to cosmic radiation is dependent on altitude. At 1 mile altitude, such as in Denver, exposure is 3 to 4 times higher than at sea level. Airplane flights increase exposure. A one-hour flight at 39,000 feet results in an additional dose of approximately 0.5 mrem (5 μSv). Although this is an insignificant dose, repeated flights can lead to a much higher annual dose. Most of the radiation exposure by inhalation is due to radon (Section 10.6); the level of exposure is quite variable. Exposure to medical and dental x-rays also varies widely for each individual.

The recommended maximum annual exposure from human-made sources is 0.170 rem (170 mrem), about 3 times the 63 mrem in Table 10.5, so the normal exposure is well within the suggested limit. In comparison with the levels in Table 10.4, there seems to be no cause for concern even if the 170-mrem level is exceeded. This does not imply that it is appropriate to ignore the recommended maximum level. Its existence alone indicates that low levels of radiation are cause for concern.

Staunch nuclear advocates are quick to point out that the average radiation exposure to residents living near Three Mile Island during the 1979 accident was only 1.2 mrem, despite the fact that this was the most serious nuclear accident in United States history. Compared with the average annual dose of over 300 mrem from natural sources alone, this level of exposure is clearly insignificant. Statistical analyses suggest that this extra exposure will have the effect of reducing life expectancy by 2 minutes *(5, 6)*. The details of the Three Mile Island and Chernobyl accidents are presented in Chapter 11.

Table 10.4   Probable effects of whole-body short-term radiation doses *(3, 4)*

Dose		*Probable Clinical Effect*
(rem)	*(Sv)*	
0–25	0–0.25	No observable effects.
25–100	0.25–1	Slight blood changes but no other observable effects.
100–200	1–2	Vomiting in 5–50% of exposed people within 3 hrs, with fatigue and loss of appetite. Moderate blood changes. Except for the blood-forming system, recovery will be complete in all cases within a few wks.
200–600	2–6	For doses of 300 rem and more, all exposed individuals will vomit within a few hrs. Severe blood changes, accompanied by hemorrhage and infections. Loss of hair after 2 wks for doses over 300 rem. Recovery in 20–100% of people within 1 mo–1 yr.
600–1000	6–10	Vomiting within 1 hr, severe blood changes, hemorrhage, infection, and loss of hair. From 80–100% of exposed individuals will die within 2 mos; those who survive will be convalescent over a long period.

Table 10.5 Average annual radiation exposure in the United States (5, 6)

Source	Dose (mrem)	Dose (μSv)
Natural sources		
External to the body		
From cosmic radiation	27	270
From the earth	28	280
From building materials	10	100
Inside the body		
Inhalation of air	200	2000
Elements found naturally in human tissues	41	410
Total natural sources	**306**	**3060**
Human-made sources		
Medical procedures		
Diagnostic x-rays	39	390
Nuclear medicine	14	140
Subtotal	53	530
Atomic energy industry, laboratories	0.05	0.5
Luminous watch dials, television tubes, radioactive industrial waste, glass and ceramics, tobacco products, etc.	10	100
Radioactive fallout	Negligible	
Total human-made sources	**63**	**630**
Overall total	**369**	**3690**

10.5 Biological Effects of Radiation

GOALS: To distinguish between internal and external exposure to radiation.
To consider some ways in which exposure to radiation could occur.
To appreciate the distinction between somatic and genetic effects.

The desirability of using nuclear power to generate electricity is controversial, but there is no question that radioactive materials are potentially hazardous. We have a natural fear of radiation, associated with cancer and nuclear bombs. Although sophisticated devices can detect radiation, we cannot see it or feel it, and this adds to the mystery and the fear. Data like those in Table 10.4 make us anxious, whereas those in Table 10.5 make us feel more comfortable about the low levels of radiation that we experience daily.

Many volumes have been written trying to evaluate the data gathered in controlled laboratory experiments, and learned after historical events that became unplanned experiments. These include bombs dropped in Japan at the end of World War II and the Chernobyl nuclear plant accident in 1986. Cancer treatment with radiation provides additional information (Section 10.8), as have many experimental tests using laboratory animals.

The General Picture

This section will acquaint you with the language used to describe radiation effects and give you a sense of the complexity of the topic. First we must distinguish between exposure to radiation externally and internally.

Exposure to α or β particles from outside the body is normally of little concern because of the low ability of these particles to penetrate matter. But if a radioactive isotope is ingested or inhaled and remains in the body, the full dose of radioactivity is delivered to the tissues as long as the isotope continues to decay or until it is eliminated from the body. Furthermore, if an isotope is not eliminated by excretion or exhalation, its effects may persist even after decay because the decay products of the original isotope may also be radioactive.

Soft radiation (Section 10.4) is exemplified by the weakly penetrating low-energy β radiation emitted by carbon-14, safe to work with unless it is taken internally. But, consider the classic case of workers who painted radium coatings on watch dials to make them glow in the dark. Radium-226 emits weakly penetrating α particles and therefore seems quite safe to handle. Unfortunately, many of the painters were found to have repeatedly moistened the brush tips with their lips to get a fine point on the brushes. Thus, they acquired a large amount of radioactive material internally, and many developed mouth and throat cancer and leukemia.

The long half-life of ^{226}Ra (1622 yrs) plus its tendency to accumulate in bone make it a serious health hazard. The tendency to accumulate is expected, since calcium, a primary component of bones, and radium have similar properties. Both are in group IIA in the periodic table.

Small exposure occurs continually as radioactive isotopes enter the body because of their natural presence on earth. Potassium is an abundant element in all living things, and a small percentage of its atoms are radioactive potassium-40; this isotope accounts for most of the exposure to radiation from within human tissues. Most of the radiation exposure that is attributed to inhalation is due to radon gas (Section 10.6).

Radioactive isotopes such as thallium-201 and technetium-99m are given internally in nuclear medicine procedures. They can be taken with little risk because they have very short half-lives, their decay products are not radioactive, and the dose required for diagnosis is very small.

If levels of exposure were greatly increased due a nuclear disaster, any danger would result from the presence of radioactive isotopes in the environment as well as the possibility of ingesting or inhaling them. Only those persons directly in the path of a nuclear bomb would be exposed to radiation released in the explosion. But if a full-scale explosion were to occur, radioactive isotopes would be released into the atmosphere and could travel over long distances.

Just as important as the radioactive fallout from a nuclear bomb is that from an explosion such as occurred at Chernobyl, or from processing of ores, fuels, and radioactive wastes. If radioactive materials get into water or food, the chance exists that they may be ingested, with serious consequences. For example, fallout on land used for grazing may cause radioactive isotopes to be consumed by sheep or cattle. These could be transferred to meat or milk consumed by humans.

Radioactive isotopes of the noble, or inert, gases, such as radon, argon, and krypton, may be released into the atmosphere at various stages in the handling of radioactive materials, such as in electric power production. For example, several isotopes of krypton are produced by the breakdown (fission) of uranium in nuclear reactors. Most of the krypton is retained in the fuel rods in the reactors and decays later when the fuel rods are in storage. Any leakage in the fuel rods permits escape of radioactive gases into the coolant water in the reactors. By analyzing these gases it is possible to monitor the integrity of the reactor cores.

Although the level of radiation from natural sources is regarded as low, it is likely that continual dosing contributes to aging. Nevertheless, good evidence suggests that cells of the body are able to repair radiation damage. When high doses of radiation are received, however, this ability may be overwhelmed. Acute exposure, such as some of the higher doses listed in Table 10.4, can result in blood changes due to destruction of bone marrow cells, which are required for production of blood cells. Diminished ability to produce white blood cells reduces the capacity to combat infection. The effect can be permanent or transient, depending on the time and level of exposure. The kinds of effects listed in Table 10.4 for very large doses (greater than 25 rem) occur rapidly in the short term. Survivors may experience long-term effects such as the development of leukemia, a condition common to many Japanese after World War II.

Specific Concerns

Finally, we must make the distinction between genetic and somatic effects that may result from exposure to radiation. **Somatic** effects are those that lead to either short- or long-term symptoms, but are not passed on to future generations.

If cells of the reproductive system are exposed to radiation, **genetic** effects may appear in offspring. If reproductive cells are killed by radiation, those cells cannot reproduce and the negative effects are lost. If the DNA content of cells is damaged and can be repaired, the damage is not carried over into future generations. But if DNA is altered directly by radiation or if repair is not totally accurate, the effects are carried into eggs and sperm to affect later generations. When DNA molecules are changed, we say that they have undergone **mutation.**

Genetic mutation occurs constantly, but the probability of serious consequences is controversial. In humans, the spontaneous rate is about 12 mutations per 1,000 live births for persons with ordinary background exposure.

Data derived from survivors of bombings in Hiroshima and Nagasaki suggest that the radiation dose required to double the mutation rate is at least 100 rem, about 300 times the yearly dose described in Table 10.5. At such a high dose, much of the ability to repair damaged DNA is also lost. Therefore, when this information is extrapolated down to evaluate the dangers of very low doses of radiation, even without considering the benefits of repair processes, the increased dangers become statistically insignificant. For example, the 1-mrem dose received by the population within 50 miles of Three Mile Island is expected to cause 0.00012 mutations per 1,000 live births. That this number is so far below 1 indicates a low probability that *any* radiation-induced genetic effects will ever occur due to the accident *(7, 8).*

Rapidly multiplying cells are particularly susceptible to the effects of radiation. Skin, digestive lining, bone marrow cells, and reproductive cells are so sensitive because they divide often. This is the basis for using radiation to kill cancer cells, which tend to multiply more rapidly than other cell types (Section 10.8). Similarly, the cells of a growing fetus are particularly susceptible, so x-rays and other sources of radiation must be avoided during pregnancy.

10.6   Radon

GOALS: To be familiar with the origin and fate of radon and radon daughters.
To appreciate the variables that influence radon levels in the home.
To consider common methods used to measure radon levels.

On December 19, 1984, Stanley J. Watras went to work at the Limerick nuclear power station in Pottstown, Pennsylvania, where he was employed as an engineer. When he arrived, he set off radiation detector alarms because he was carrying an unusually high level of radon decay products on his clothes and in his hair. The source was traced to his home, where a radon level of 2800 picocuries per liter (pCi/L), or 14 Working Levels (WL), the highest environmental level of radon ever recorded, was observed in his basement. This discovery triggered a flood of concern and investigation into the nature and extent of the problem.

Radon is a colorless, odorless, chemically inert gas with atomic number 86. The danger of exposure to humans is not directly due to radon. The gas is breathed in and out and does not accumulate in the lungs. However, the products formed from the radioactive decay of radon, known as **radon daughters,** are solids that may be inhaled as dust particles. Since radon daughters emit α and β particles, they are very hazardous within the respiratory system. Their damaging effects concentrate in the trachea and the lungs since they penetrate no farther through body tissues. It is believed that 20% to 50% of lung cancer deaths among nonsmokers are caused by radon.

Most radon daughters also emit γ radiation, which is used to detect these isotopes. However, the α and β particles are considered to be potentially far more damaging to tissues than γ rays.

Origin and Decay of Radon

Most of the radon in nature is ^{222}Rn. It is formed by a series of radioactive decay steps beginning with ^{238}U, the most abundant isotope of uranium. The ^{222}Rn is formed directly from radium-226 (^{226}Ra). The decay process is shown in Figure 10.10, beginning with ^{238}U and ending with ^{206}Pb, a stable (nonradioactive) isotope of lead. It can be seen from the figure that eight α particles and six β particles (negatrons) are emitted. Along the way, the dangerous radon daughter isotopes include three different radioactive isotopes of polonium (Po), two of bismuth (Bi), and two of lead (Pb) that form prior to the stable ^{206}Pb.

Uranium and its decay products, which are widely distributed throughout the earth, account for a significant portion of the natural environmental radiation. All rocks and soils contain at least traces of radon, and in some regions, concentrations are high. Underground miners, particularly uranium miners, are generally exposed to higher than normal amounts because of uranium in soils, and because radon gas tends to collect in low places since it is more dense than air. Since coal and uranium are not often found together, coal miners are not usually exposed to radon.

A region extending through southeastern Pennsylvania from near Reading well north into New Jersey to near Trenton is known as the Reading Prong. It is a mass of black shale that runs parallel to the Appalachian Mountains. Black shales, phosphate rock, granites, and carbonate rocks normally contain high concentrations of uranium. The map in Figure 10.11 shows that radon is a problem throughout the United States.

Since most house foundations are built partially underground, radon may diffuse out of the soil and into the basement level. Newer homes, built since energy conservation

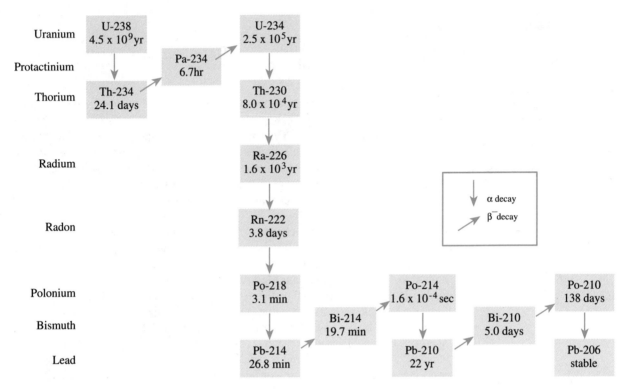

Figure 10.10   The uranium-238 decay series *(9)*.

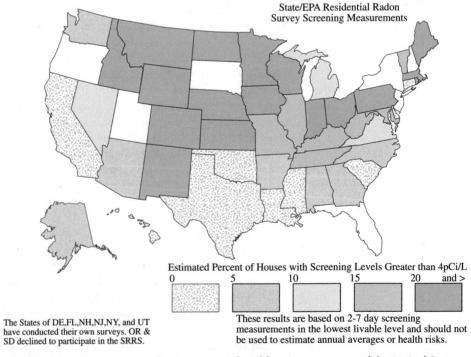

Figure 10.11   Exposure to radon is a potential problem in many parts of the United States.

became a goal, tend to be more airtight. This condition may prevent entry in the first place or trap more radon. Since radon can dissolve in water and accumulate in wells, water for bathing, laundry, and dishwashing may sometimes be an important source.

The U.S. Environmental Protection Agency (EPA) has targeted a level of 4 pCi/L (0.02 WL) as the so-called *action level* for radon. That is, if a home is found to have this level or more, it is recommended that some corrective action be taken to alleviate the problem. When high levels are discovered, ventilation is often the simplest remedy. Other remedies include pressurization (to prevent radon from flowing in) by air conditioning or forced-air heating, capping sumps, sealing cracks, using traps to prevent gas from flowing up into floor drains, and waterproofing exterior basement walls. It has become commonplace to require radon testing as a condition of purchasing a property.

Radon Testing

Basement levels should always be tested. Not only is radon most likely to enter through the basement, but several factors make concentration there likely. Radon is more dense than air, which would cause it to remain on a lower level. Furthermore, ventilation is usually more efficient on floors above ground where windows and doors are open at least some of the time.

Seasonal variations should also be taken into account. Radon levels are normally greatest in winter, partially because windows and doors are likely to be closed, but also because higher temperatures inside the house tend to create a chimney effect (or stack effect), which draws the gas from the ground into the house.

Two types of detectors are used to measure radon levels in the home. The *track-etch detector* filters out α-emitting materials such as solid radon daughters, thus allowing only gases to enter, including radon. When decay occurs inside the detector over a period of several weeks, the emitted α particles produce visible tracks on a special plastic film. This type of detector is used for long-term measurements of radon levels.

The *diffusion-barrier charcoal-absorption (DBCA) collector* or *charcoal canister* samples the air continuously. Charcoal absorbs radon, and a diffusion barrier prevents its escape. At a testing laboratory, γ radiation from radon daughters, primarily ^{214}Pb and ^{214}Bi, is counted. This detection device is normally used for short-term (typically 2–4 days) measurement.

The Extent of the Radon Problem

Although the area along the Reading Prong has received the greatest notoriety, high levels of radon have been found in virtually every state and worldwide. The EPA has estimated that more than 1 million homes in the United States have levels in excess of the action level.

It is estimated that 5,000 to 20,000 lung cancer deaths occur each year in the United States due to radon. This compares to 85,000 each year due to smoking and 120,000 lung cancer deaths per year total *(10, 11)*. At the action level of 4 pCi/L, radon is thought to impose an incremental risk—in addition to the risk from other sources—of 1% to 3%, assuming continuous exposure over a period of 70 years. This level of risk is equal to 10,000 to 30,000 cases of lung cancer per million people. This level of risk is far greater than the one per million that often triggers action to deal with other environmental pollutants. Acceptable risk is not decided by scientific principles, but by regulatory agency decisions.

Since radon is a natural pollutant, rather than a human-made problem, it is not always taken seriously. We cannot see or smell it; we cannot feel its effects; homes are generally regarded as safe; since some forms of remediation are potentially difficult and costly, we have even more excuses not to take the problem seriously. Persons who expect to sell a home within a few years may not want to know if they have a radon problem. If no data are available, a seller may avoid a moral crisis when selling a home. That is why many home buyers should have the protection of prepurchase radon testing.

The Watras home at 14 WL (2800 pCi/L) of radon was as hazardous as smoking 140 packs of cigarettes per day. The risk of 200 pCi/L (1 WL) is regarded as the cancer-causing equivalent of 10 packs per day *(9)*. The risks associated with lower levels of radon, such as the action level, are more controversial *(12, 13)*.

10.7   Applications of Radiochemistry in Archeology

GOALS: To understand the basis, technique, and limitations of carbon-14 dating, including the phenomenon of the steady state.
To appreciate the value of the rubidium-strontium clock as an alternative to carbon-14 dating.

Radiochemistry has many applications in science and medicine. The remainder of this chapter addresses some of these in archeology, research, and medicine.

Carbon-14 Dating

The general population receives a continual dose of cosmic radiation. It is mostly γ radiation at the surface of earth, but in space it is a mixture of many types. The solar wind, for example, contains a high level of protons from the sun. Cosmic rays can interact with gases in the upper atmosphere to produce neutrons. These neutrons can then bombard nitrogen atoms, present as N_2 gas in the atmosphere, forming radioactive carbon according to the following reaction.

$$^{14}_{7}N + {}^{1}_{0}n \rightarrow {}^{14}_{6}C + {}^{1}_{1}H$$

The shorthand notation $^{14}N(n,p)^{14}C$ also describes this process. Thus, the interaction of solar protons causes the production of neutrons, which lead to the formation of ^{14}C and more protons.

The ^{14}C then finds its way into earth's atmosphere and becomes a part of the biological carbon pool (the biosphere) as it is incorporated into live plants and animals. Any living material continually takes up and gives off ^{14}C until equilibrium is reached and the level remains constant.

On the average, approximately two ^{14}C atoms are deposited every second on every square centimeter of earth's surface. It is readily detected because of the β^- emission that accompanies its decay:

$$^{14}_{6}C \rightarrow {}^{14}_{7}N + {}^{0}_{-1}\beta$$

This reaction also accounts for the loss of ^{14}C from the biosphere. Thus an equation for formation and another for destruction of ^{14}C describe a situation in which the total concentration reaches a **steady state,** or constant concentration. The steady state is easily understood by noting that a greater buildup of ^{14}C would lead to faster decay, since the rate of decay is proportional to the concentration of material that is decaying, as shown in Figure 10.12.

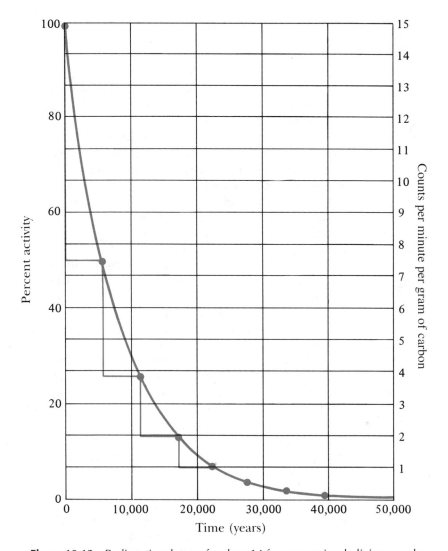

Figure 10.12   Radioactive decay of carbon-14 from a previously living sample.

All living things have the same steady-state level of ^{14}C, since they are continually exchanging ^{14}C with the biosphere. When death occurs, however, the intake of ^{14}C ceases, whereas radioactive decay continues at a rate governed by the half-life of 5568 years. After 5568 years, half the normal level of ^{14}C present in the organic material will be gone. The normal steady-state level of ^{14}C in living organisms undergoes approximately 15 radioactive disintegrations per gram of carbon per minute, or 15 counts per minute (cpm) per gram of carbon. Any carbon-containing sample that displays only 7.5 cpm per gram of carbon must be about 5568 years old.

To determine the ^{14}C level, the carbon content of a sample is burned to release CO_2, and the radioactivity of the CO_2 is monitored to determine the amount of ^{14}C isotope. In all cases, the ^{12}C isotope is the one present in greatest amount; it is not radioactive.

Coal is an example of a material that has been out of contact with the biosphere for so long (more than 50,000 yrs) that virtually all its ^{14}C has decayed. Thus, the very long half-life of ^{14}C is much too short to allow any accurate determination of coal's age. Research scientists normally place the outer limit for the carbon-14 technique at about

50,000 years, since by then the level of activity due to ^{14}C is so low (Figure 10.12). The Dead Sea scrolls and wood found in Egyptian pyramids have ages within the range of ^{14}C dating.

The steady-state level of ^{14}C has been upset by many factors over the centuries. The most dramatic example is the burning of fossil fuels, such as coal and oil, which have not been part of the biosphere for thousands of years. Since the ^{14}C level of these fuels is therefore unusually low (near zero), burning predominantly changes ^{12}C to $^{12}CO_2$. This release of nonradioactive carbon causes dilution of the normal steady-state carbon pool. Some researchers have predicted as much as a 12% dilution effect, which would then increase the apparent age of any materials exposed. However, some plants that died in 1950 showed only a 2% dilution. The difference between predicted and observed levels has been attributed to absorption of the CO_2 by the oceans.

Perfection of the technique for counting old ^{14}C-containing samples was an important technological advance in itself. The difficulty lies in distinguishing the very low radiation levels that may be a small fraction of the normal background radiation (mostly cosmic rays). Cosmic rays are very penetrating (mostly γ radiation) and thus easily detectable by any radiation counter. *Anticoincidence counting* was developed by W.F. Libby to counteract the problem. In this technique, a container is surrounded by efficient shielding, so that all external radiation except cosmic rays will be filtered out. The system has two counters. One counter is exposed to both cosmic radiation and the ^{14}C (β emitter) sample; the other is shielded from the weak emission and detects only cosmic radiation. The difference between the observation of the first counter (cosmic plus β) and the reading of the second counter (cosmic only) gives the intensity of radiation due to the ^{14}C content.

A technique called accelerator mass spectrometry (AMS) allows direct detection of ^{14}C at concentrations of 3 atoms of ^{14}C in 10^{16} atoms of ^{12}C. This corresponds to a radiocarbon age of about 70,000 years. Thus, the traditional method, which measures the radioactive decay of ^{14}C, can be used for samples up to about 40,000 years old, and the newer AMS method can be used for samples up to 100,000 years old *(14, 15)*.

Carbon-14 dating has also been used to estimate the age of certain iron tools and utensils obtained from archeological sites. The three major forms of processed iron are wrought iron, cast iron, and steel, all of which are alloys (mixtures) of iron and carbon. It is possible to determine the age of many iron alloys provided that the carbon source had not been dead for long before being alloyed. For example, if coal were the source of the carbon, the iron sample could not be dated. Wood and charcoal were common fuels in iron smelting in early times and thus became the usual sources of the carbon content of the alloys. Detailed information on the age of certain iron samples helps piece together the historical development of the Iron Age, which began about 6000 B.C.

The Rubidium-Strontium Clock

Several other examples of archeological dating are based on slow (long half-life) radioactive decay. One popular method is the rubidium-strontium clock, which depends on the following radioactive decay:

$$^{87}Rb \rightarrow {}^{87}Sr + {}_{-1}^{0}\beta \qquad (t_{1/2} = 4.7 \times 10^{10} \text{ yr})$$

Since ^{87}Sr is not radioactive, it remains unchanged once it is formed from radioactive Rb. From the ^{87}Rb:^{87}Sr ratio, the age of the mineral can be calculated. Since ^{87}Rb has a half-life of 47 billion (47×10^9) years, the rubidium-strontium clock can be used even to determine the age of the Earth, which is estimated at 4.5 billion years. Also the age of some lunar samples brought back during the Apollo 15 mission was determined as 3.3×10^9 years.

Since the rubidium-strontium clock method of dating is complicated, the ^{14}C method is used the most, even though it is limited to a maximum of 50,000 years.

10.8   Applications in Research and Medicine

GOALS: To appreciate the use of radioactive tracers in research.
To appreciate the technique of dialysis.
To see the variety of applications of radionuclides in diagnosis and therapy.
To understand the strategy in the use of radiochemical cows.

Isotopes are frequently used as tracers to detect the amount and location of compounds or ions in samples, living or dead. Applications are found in research, and medical diagnosis and treatment. Compounds labeled with radioisotopes can be used to monitor chemical changes in animals, plants, and microorganisms.

Isotopic Tracers in Research

A particular example of the use of tracers is in studying the extent to which toxic mercury ions bind to human serum albumin, a protein in blood. The technique is called *equilibrium dialysis*. A schematic diagram of the apparatus is shown in Figure 10.13, in which radioactive $^{203}Hg^{2+}$ ions are present.

An important component of the dialysis experiment is the semipermeable membrane that separates the two compartments of the system. This membrane is permeable to small molecules, such as H_2O and Hg^{2+} salts but not to large protein molecules, such as albumin. Thus, all the chemicals, including the $^{203}Hg^{2+}$ ions, can move freely throughout the system, except protein, which remains fixed.

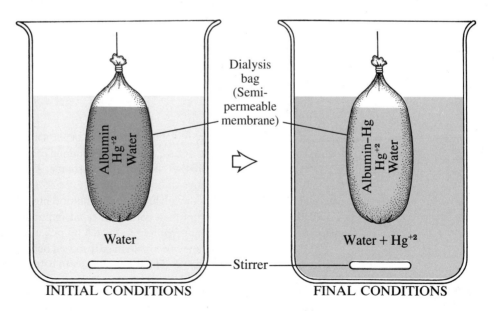

Figure 10.13   An equilibrium dialysis study of the binding of mercury to serum albumin. See text for details.

The experiment begins as shown on the left side of Figure 10.13 with all of the $^{203}Hg^{2+}$ ions placed inside the dialysis bag. After sufficient time has passed, a sample can be removed from each compartment, and the radioactive ^{203}Hg (β and γ emission) counted. The mercury will exist as:

Outside the dialysis bag:	free Hg^{2+} only
Inside the dialysis bag:	free Hg^{2+}
	Hg^{2+} bound to albumin

Since the membrane allows the unbound mercury ions (Hg^{2+}) to pass freely in both directions, the concentration of free Hg^{2+} will become the same both inside and out. The excess count inside the bag can be attributed to the mercury-203 that is bound to the albumin.

Another use of radioactive isotopes is in the many phases of drug testing. For example, some drugs are designed to be applied to an area of the skin, but absorption through the skin and into the blood may be both unnecessary and undesirable. Toxicity and side effects could make the drug dangerous. In other cases, it may be desirable for a drug to be absorbed through the skin. For instance, if an oral drug is irritating to the stomach, if it is degraded in the stomach, or if it is not absorbed easily into the bloodstream when taken orally, a skin application could be an alternative to injection.

Analysis of blood samples, urine samples, and expired air for the presence of a radioactively labeled drug or its breakdown products reveals the degree of absorption of the drug through the skin. Even if the drug itself is metabolized in the body, the radio-nuclide will still be present in some form and its location can be determined easily in test animals *(16)*.

Clinical Applications of X-rays and Radioisotopes

Nuclear medicine procedures have been developed and tested for diagnosis and treatment since 1950. Their main advantage is that they leave body tissues intact, and when they are effective, they are preferable to invasive surgical procedures.

When a radionuclide is chemically or physically bound to a substance that can be introduced into the body, the resultant product is termed a **radiopharmaceutical** agent. Therapeutic applications can be divided into two types: **teletherapy,** or external exposure, and internal radiation.

External Radiation Therapy

High-energy electromagnetic radiation can readily penetrate body tissues when externally focused on a particular organ or tissue. The most common examples are machine-generated x-rays and γ rays from a cobalt-60 or cesium-137 source. The ideal radiation treatment is one that preferentially destroys tumor cells and leaves normal cells unaffected. Unfortunately, such an ideal is not yet available. Rapidly multiplying cells are much more sensitive to radiation than relatively dormant cells, and rapid multiplication is a characteristic of many tumor cells, so they are often more affected by radiation than normal cells.

The α particles have a range of less than 0.01 mm in body tissues, so penetration from an external source is not possible. The β particles seem better suited for external radiation therapy but must be accelerated to extremely high energies.

Internal Radiation Therapy

Internal radiation therapy, using radioactive isotopes, is less common than external therapy, because an isotope must localize at the site of the tumor growth. Unfortunately, only a few radiopharmaceuticals show high selectivity; in some cases the target organ or tissue may be too sensitive to permit such treatment.

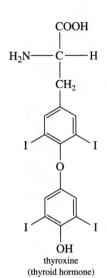

thyroxine
(thyroid hormone)

Iodine-131   Hyperthyroidism (Graves' disease) often is treated with ^{131}I. It kills specific numbers of thyroid cells, reducing excess thyroid hormone production down to a normal level. The isotope works well for this purpose for two reasons. First, it bonds selectively to the thyroid gland, since iodine atoms are incorporated into the thyroid hormones. Second, its half-life is 8.06 days. This short lifetime allows good control of the dose. A large dose can be administered to achieve the desired effect, with prompt subsequent loss of radioactivity. A long half-life would require much lower doses to avoid adverse long-term effects of large amounts of radiation. Iodine is normally administered as a sodium iodide pill, NaI.

Phosphorus-32   Sodium phosphate labeled with radioactive phosphorus-32, ^{32}P, is non-specific. Phosphate is a common chemical component of all living things; it freely circulates throughout the body in the blood. This feature has been turned to advantage in the sometimes successful treatment of leukemia (overproduction of white blood cells) and polycythemia vera (overproduction of red blood cells).

Brachytherapy   Surgical implantation of a radiopharmaceutical agent in or near a tumor achieves a high degree of localization of radiation. This is known as **brachytherapy** (short-distance therapy); it has been used successfully in treatment of cancers of the prostate, uterine cervix, endometrium, vagina, brain, breast, head, and neck.

The most common isotope used in brachytherapy is ^{137}Cs, although ^{192}Ir, ^{125}I, and ^{103}Pd are also incorporated into temporary or permanent implants in the form of needles or seeds. Since the implantation is invasive, it is often desirable for the material to remain permanently so that additional surgery for retrieval is unnecessary. This means a short-half-life isotope is preferred.

Boron-10 Neutron-Capture Therapy   Perhaps the most imaginative form of radiation therapy for cancer is boron-10 neutron-capture therapy. The method includes both internal and external procedures. The nuclear equation is

$$^{10}_{5}B + {}^{1}_{0}n \rightarrow {}^{7}_{3}Li + {}^{4}_{2}\alpha$$

Administration of a dose of some boron-containing substance is followed by external irradiation with neutrons. Because neutrons are uncharged, they readily penetrate tissue and interact with boron atoms. When this happens, lithium ions and α particles are produced at the site of the boron. Both ^{10}B and ^{7}Li are stable isotopes, nontoxic in the quantities used. Since α particles are the most massive of the common forms of radiation, they are biologically the most destructive. In neutron-capture therapy, they can theoretically be produced directly at the site of the tumor; the specificity can be greatly enhanced by incorporating boron-10 atoms into substances that localize in tumors. One of the positive features of this approach is the absence of any dose of radioactive isotopes. Even if some of the boron goes to normal tissues, these tissues will be unaffected unless neutrons are beamed directly on them. The radiation dose can be well controlled, since α particles are produced only when the external source of neutrons is operating *(17, 18)*.

Diagnosis

Diagnostic applications of radiochemistry combine external and internal methods. In a typical procedure, an isotope is administered orally or by injection, and must concentrate

Figure 10.14   A scintillation camera.

to some extent in the organ to be studied. After a period that allows the isotope to become concentrated, γ radiation is checked (scanned) by placing a suitable detector near the place on the skin closest to the organ being investigated.

The *scintillation camera* measures radioactive emissions from inside the body *imaging* an entire organ in which an isotope concentrates (Figure 10.14). This permits detection of abnormalities such as blockage or leakage.

Imaging can also detect tumors as "cold spots," or as areas of unusually low concentration of the isotope when these areas are avoided by the radiopharmaceutical. Since some radiopharmaceuticals preferentially localize in tumors, "hot spots" of radioactivity may also indicate tumors. Specific examples of diagnostic radiochemistry follow.

Thyroid Function

One common test monitors thyroid function by monitoring the uptake of radioactive ^{131}I. Ingested iodine is rapidly absorbed from the intestine as iodide ion, I^-, and is circulated through the bloodstream. The iodide moves to the thyroid gland where it is concentrated and incorporated into two thyroid hormones that regulate the general rate of metabolism.

Iodine-131 is commercially available in capsules of sodium iodide, NaI. The patient swallows the pill, and measurements of the radioactivity level of the thyroid are made at 6- and 24-hour periods by placing an external radiation detector at the neck. Normal thyroid accumulates 7.5% to 25% of the oral dose in 6 hours and 12% to 46.5% in 24 hours. An absorption about half of normal is typical of congestive heart failure. A high reading suggests hyperthyroidism.

Technetium-99m

Of particular interest in nuclear medicine is the radioisotope technetium-99m, ^{99m}Tc, which is used in about 85% of all diagnostic procedures involving radionuclides. This isotope decays by isomeric transition—emission of γ radiation. Since no charged particles are emitted, little tissue damage occurs when the isotope enters the body. Equally important, the amount of the radiation needed is unusually low, since most of the emitted radiation has an energy that is ideal for normal detection cameras. With many other isotopes, much of the radiation is emitted with such high energies that it cannot be detected easily. The dose must then be increased to get enough radiation for detection; the result is greater damage to the tissues.

Heart Function by Cardiac Imaging

A number of isotopes have been used over the years to monitor the flow of blood through the heart and other parts of the body. By far the most popular is thallium, as ^{201}Tl, which decays by γ emission with a conveniently low half-life (74 hrs). The γ rays from ^{201}Tl are emitted as a result of electron capture and have energies that are easily detected, thus allowing for the use of very small doses.

Thallous chloride, $TlCl$, given by intravenous injection, accumulates in healthy heart muscle tissue due to the similarity of thallium and potassium ions, which are abundant there. In clinical studies, areas of heart muscle that do not acquire $^{201}Tl^+$ indicate an obstruction of blood vessels (an area of dead or diseased tissue). Although thallium is a very toxic substance, adverse effects are not observed due to the low dose of radioactive isotope required for testing.

Another cardiac imaging agent is Cardiolite, a ^{99m}Tc radiopharmaceutical. The imaging properties of ^{99m}Tc and its long half-life are advantageous, and the characteristics of the remainder of the molecule ensure binding to heart tissue *(19)*.

Bone Scanning

Bone can be examined using radiopharmaceuticals. Either ^{85}Sr or ^{99m}Tc may be injected intravenously, and scans taken at a specified later time (usually 72 hrs). The ^{85}Sr is taken up by bone (primarily calcium phosphate), since strontium is chemically similar to calcium. Hot spots of radioactivity indicate rapid bone growth. This may indicate normal bone development in children or the presence of tumor cells, since both are often characterized by rapid growth. The scan in Figure 10.15 shows normal uptake of ^{99m}Tc in a 15-year-old in whom the high levels (dark areas) indicate new bone growth.

Positron Emission Tomography

Isotopes that emit positrons offer an interesting approach to diagnosis of certain diseases and to better understanding of normal body functions. The key to appreciating the special value of a positron emitter is the mode of destruction of a positron. Wherever a positron is released, it collides with an electron in matter around it to produce two γ rays.

$$^{0}_{+1}\beta + ^{0}_{-1}\beta \rightarrow 2\gamma$$

This reaction explains why positrons have been characterized as *antimatter*. They cause the annihilation of matter, in this case electrons, and release energy in the form of γ radiation.

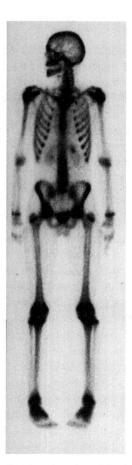

Figure 10.15 A technetium-99m bone scan.

Furthermore, the two γ rays released from each annihilation reaction travel in exactly opposite directions. Therefore, if a positron-emitting isotope is deposited in a particular tissue or organ, it is possible to produce an image of that part of the organ where the isotope is located.

Even metabolic activity, both normal and abnormal, within the brain can be monitored using positron emitters. For example, ^{18}F can be incorporated into compounds such as glucose and dopa, which are known to localize in the brain. The technique is known as positron emission tomography (PET). A ring of sensors is placed around a patient's head to detect γ rays. A computer processes the signals and produces an image of a cross-sectional slice of the organ. After a series of tomographic slices has been analyzed, a three-dimensional computer image becomes available for the location of the radio-pharmaceutical.

When ^{18}F-containing glucose is administered, it enters the bloodstream and moves to the brain. Whenever any portion of the brain is active, increased use of glucose supports the activity. The tomographic images locate increased activity, either normal or abnormal, providing the opportunity to study brain function associated with disorders such as epi-

The word tomography is derived from the Greek *tomos,* meaning ''a piece cut out, or a section.''

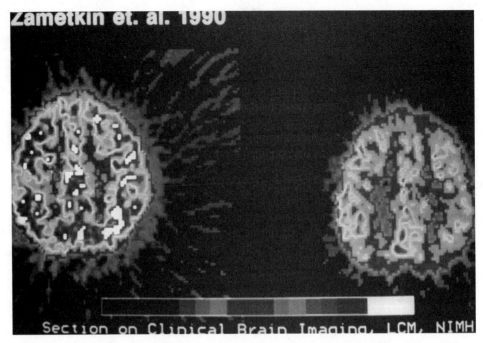

Figure 10.16   A PET study of hyperactivity. The hyperactive patient (right) has decreased activity in areas of the brain that control movement and attention. The scans suggest that hyperactivity may result from a neurological disorder rather than from psychosocial problems *(20)*.

lepsy, schizophrenia, Alzheimer's disease, cerebral palsy, and stroke. Figure 10.16 compares the degree of brain activity in a normal person and a hyperactive person *(20, 21)*.

Radiochemical Cows

Milk that is made by cows can be removed by milking. Radioactive isotopes decay, producing other isotopes that can be separated by "milking." Here, milking involves the separation of two elements that differ in some physical property, such as the solubility of their ions in water. One example is the "cow" that serves as a source of ^{99m}Tc.

$$\underset{t_{1/2} = 67 \text{ hrs}}{^{99}_{42}\text{Mo}} \xrightarrow{\beta^- \text{ decay}} \underset{t_{1/2} = 6 \text{ hrs}}{^{99m}_{43}\text{Tc}}$$

Technetium-99m is used extensively in nuclear medicine procedures. Its short half-life $(t_{1/2})$ makes transport and storage impractical; but the molybdenum parent, ^{99}Mo, can be made (Section 11.2), shipped, and used at a convenient pace as a source of ^{99m}Tc.

A typical cow system has the parent isotope embedded in a medium like an ion-exchange resin. As decay occurs, some of the daughter isotope takes the place of the parent. When a supply of daughter isotope is required, the chemist washes the resin to remove the daughter isotope from the resin while the parent remains behind. An example is shown in Figure 10.17.

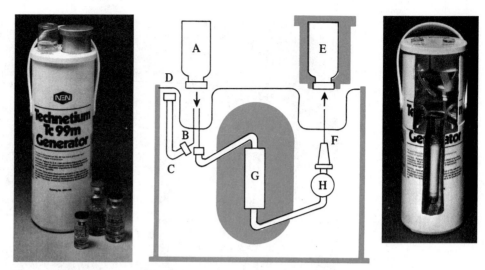

Figure 10.17   A "cow." Photographs of a generator for producing samples of technetium-99m. As the drawing shows, a charge vial A containing salt solution is placed onto the double needle B and vented through tube C fitted with a cotton filter D. The salt solution moves from vial A and through the system when an evacuated, shielded (to prevent escape of radiation) collection vial E is placed onto needle F. As the salt solution moves from A to E, it passes through shielded column G loaded with parent molybdenum-99. Technetium-99m is selectively dissolved into the salt solution from G, after which it passes through filter H and needle F into collection vial E.

SUMMARY

Most elements exist in several forms, called isotopes, that differ in the number of neutrons found in the nucleus. Some of these isotopes are stable, whereas others are radioactive and undergo α, β, or γ decay. X-rays may be emitted as a result of electron capture or by machines. The products of nuclear decay can be predicted by properly balancing the equations that describe these processes.

Emitted radiation is accurately detected by sophisticated counting devices, including Geiger-Müller counters, dosimeters, and film badges. Liquid scintillation is used for counting soft radiation.

A comparison of the radiation doses that cause dramatic clinical symptoms and the average annual dose levels to which the population is exposed suggests that there is no cause for concern over current or somewhat increased dose levels.

To determine the hazards and proper techniques for handling a particular radioisotope, we must be able to interpret published data on the isotope. The half-life determines the useful lifetime of an isotope and how long it is a potential hazard. The mode of decay and the decay energy determine the ability of the decay particle or γ ray to penetrate matter and the means for detecting the isotope.

Isotopes that emit α and β particles are harmless when kept outside the body, whereas γ rays are able to penetrate through tissues. If ingested or inhaled, α- and β-emitting isotopes are hazardous. Both somatic and genetic effects are possible due to exposure to radiation. Most tissues are capable of repairing some radiation damage, as long as the damage is not too severe. Rapidly dividing cells are most susceptible.

Since more than a million homes in the United States may have radon levels greater than the action level specified by the EPA, it is important to understand how radon is detected and what variables may affect measurements that are made. These include seasonal variation, duration of the analysis, and location tested. Most radon in the form of ^{222}Rn originates from ^{238}U by a series of 14 radioactive decay reactions—8 α decays, and 6 β decays. It is the radon daughters that present the major health hazard, because they are α and β emitters that can be inhaled as dust particles and contribute to lung cancer.

Radon may enter the home in the water supply and from the soil or rocks below or around the foundation of a home. It may be removed, or its entry may be prevented, in several ways. Radon levels may be tested in several ways, including the use of a track-etch detector for long-term monitoring or by using a diffusion-barrier charcoal-absorption detector for a period of a few days.

Carbon-14 dating is an important technique for determining the age of samples derived from sources that were once alive. Cosmic radiation interacts with nitrogen-14 in the atmosphere to produce carbon-14. The carbon can then be incorporated into living organisms, and the supply is continuous. The radioactive decay of ^{14}C is also continuous, and the two processes tend to balance

in a condition called a steady state. When an organism dies, incorporation ceases and only decay continues, so that the level of ^{14}C decreases at a rate determined by its half-life (5568 yrs). It is possible to determine the age of a sample by measuring the level of ^{14}C that remains. One hundred thousand years is the upper limit for this dating technique, since the ^{14}C content becomes too low for accurate measurement beyond that time. Fossil fuels are too old to be dated by this technique. The rubidium-strontium clock permits dating of older samples, for example, materials from the moon, since rubidium-87 has a very long half-life.

Radioactive isotopes are important as tracers in many fields of research. They make it possible to monitor the movement of isotopes in systems undergoing change.

Nuclear medicine is a versatile field. Both internal and external radiation therapies have been used for many years. Radionuclides are used in a variety of diagnostic procedures, including sophisticated procedures such as cardiac imaging and positron emission tomography.

PROBLEMS

1. Give a definition or example of each of the following.

   a. Isotopes
   b. Atomic number
   c. Atomic mass
   d. Average atomic mass
   e. α particle
   f. β particle
   g. n/p ratio
   h. X-ray
   i. Isomeric transition
   j. Soft radiation
   k. Radionuclide
   l. Nuclide
   m. Nucleon
   n. Drip line
   o. Negatron
   p. Positron
   q. Radon daughters
   r. Action level
   s. Track-etch detector
   t. Reading prong
   u. Chimney effect
   v. Anticoincidence counting
   w. Rb-Sr clock
   x. Cosmic radiation
   y. Cast iron
   z. Teletherapy

   aa. Radiopharmaceuticals
   bb. Brachytherapy
   cc. Radiochemical cow
   dd. m (as in ^{99m}Tc)
   ee. Tracer

2. Give the number of protons, neutrons, and electrons in each of the following atoms or ions. Also give the name of each and indicate whether the symbol describes a cation, anion, or neutral atom. Refer to the inside front cover for a table of atomic symbols, numbers, and average atomic masses.

   Example: $^{59}Fe^{3+}$
   Answer: $^{59}_{26}Fe^{3+}$ has   26 protons
                                           33 neutrons
                                           23 electrons
                          and is a   cation of iron

   a. $^{64}Cu^{2+}$
   b. ^{198}Au
   c. ^{235}U
   d. $^{131}I^-$
   e. ^{113}Sn
   f. $^{203}Hg^{2+}$
   g. ^{74}As
   h. $^{68}Ga^{3+}$
   i. $^{90}Sr^{2+}$
   j. $^{19}F^-$

3. Write complete nuclear equations to describe each of the processes listed.

Example: β^- decay of ^{203}Hg

Answer: $^{203}_{80}\text{Hg} \rightarrow {}^{203}_{81}\text{Tl} + {}^{0}_{-1}\beta$

   a. Electron capture by tin-113 accompanied by γ emission
   b. α decay of bismuth-214
   c. Positron emission from indium-106
   d. γ emission by technetium-99m
   e. β^- decay of krypton-87 accompanied by γ emission
   f. α decay of cerium-142
   g. Positron emission by cesium-125
   h. γ emission by yttrium-91m
   i. Electron capture by platinum-186
   j. β^- decay of europium-158
   k. $^{14}\text{N} + n \rightarrow {}^{14}\text{C}$
   l. Decay of $^{14}\text{CO}_2$
   m. Isomeric transition of $^{99\text{m}}$Tc

4. Write a complete nuclear equation to describe each of the processes listed. See footnote *after* solving the problems.*

   a. Decay of plutonium-241 to produce americium-241
   b. Decay of oxygen-15 to produce nitrogen-15
   c. Decay of thorium-232 to produce radium-228
   d. Decay of nickel-59 to produce cobalt-59
   e. Decay of bromine-74 to produce selenium-74
   f. Decay of iron-59 to produce cobalt-59
   g. Decay of lead-205 to produce thallium-205

5. Determine whether the n/p ratio increases, decreases, or remains the same for the isotopes in each of the following processes.

   a. Problem 3e
   b. Problem 3c
   c. Problem 3h
   d. Problem 3a
   e. Problem 3b

6. Explain the advantage of liquid scintillation counting compared with other techniques.

7. Which is the larger quantity in each of the following pairs?

   a. 1 rem or 1 Sv
   b. 10 Sv or 600 rem
   c. 10 μSv or 10 mrem
   d. 10 rad or 1 Gy
   e. 1 Bq or 1 Ci
   f. 1 pCi or 1 μCi

8. How many half-lives would be required for an isotope to decay to below 1% of its original activity?

9. What isotope contributes most of the radiation dose received from within human tissues?

10. It has been said that Stanley Watras set off an alarm heard around the world. What does this mean?

11. Why are radon daughters regarded as more dangerous than radon?

12. Write the nuclear equations for the following.

   a. The first step in the decay chain leaHding to ^{222}Rn
   b. The step leading to formation of radon from radium
   c. The decay of ^{222}Rn
   d. The final step in the decay chain that begins with uranium and ends with a stable isotope of lead

13. Referring to a periodic table and to Figure 10.10, track the decay series detailed in the figure.

14. Write nuclear equations to account for the formation of isotopes of three different elements that are radon daughters.

15. Each α decay causes a loss of two protons and two neutrons, whereas each β^- decay causes the conversion of a neutron into a proton. It is possible to account for the loss of four α particles (8 protons lost, 8 neutrons lost) and two β^- particles (2 neutrons lost, 2 protons gained) in the decay of uranium-238 to radon-222.

   a. Confirm that the gain and loss of nuclear particles in the uranium decay series do really account for the formation of radon-222.
   b. Thorium-232 decays in several steps to form thoron (radon-220). How many protons and neutrons are lost between the ^{232}Th and the ^{220}Rn? What is a possible combination of α and β^- decays that could account for the overall conversion?

16. What risks are associated with the action level of 4 pCi/L set by the EPA?

17. What are some possible solutions for curing a radon problem in the home?

18. Why should basement levels be checked for radon? How can the season of the year influence radon levels?

19. Each of the following questions pertains to the radioactive decay of carbon-14-labeled CO_2 obtained from living systems. Consult Figure 10.12 as necessary.

   a. What is meant by the steady state?
   b. How are the readings obtained for comparison with data plotted along the vertical axis?
   c. Why is a value of 15 cpm plotted on the vertical axis?
   d. Why is 100,000 years considered the upper limit for using ^{14}C dating?

---

*Parts (b), (d), (e), and (g) appear to be the same process. However, oxygen-15 decays by positron emission, nickel-59 by electron capture, bromine-74 by electron capture, and lead-205 by positron emission.

e. What is the apparent age of a wood sample, found in an Egyptian pyramid, that was burned to produce CO_2 with a radioactivity level of 6.7 cpm/g carbon?

f. What is the apparent age of a cast iron sample that was similarly treated to produce CO_2, which has a radioactivity level of 11 cpm/g carbon?

20. What is the advantage of the newer technique of accelerator mass spectrometry over the traditional method of monitoring radioactive decay for measuring the ^{14}C content of a sample?

21. How might a radioactively labeled drug be used to monitor uptake through the skin?

22. What type of radiation is most damaging to tumors when administered outside the body? Inside the body? Explain.

23. What is the reaction for boron-10 neutron-capture therapy?

24. Describe a diagnostic application of radiochemistry.

References

1. Hardy, J. C. "Exotic Nuclei and Their Decay." *Science* **1985,** *227,* 993.

2. Kastner, J. *The Natural Radiation Environment;* U.S. Atomic Energy Commission: Washington, DC, 1968; p. 28.

3. Glasstone, S.; Sesonske, A. *Nuclear Reactor Engineering;* Van Nostrand: Princeton, NJ, 1963; p. 532.

4. Mann, W. B. J. *Radioactivity Measurements: Principle and Practice;* Pergamon: New York, 1988; pp. 34–40.

5. Cohen, B. L. *The Nuclear Energy Option;* Plenum: New York, 1990; Chapter 5.

6. Knoche, H. *Radioisotopic Methods for Biological and Medical Research;* Oxford University: New York, 1991; pp. 329–332, 353.

7. Wolfson, R. J. *Nuclear Choices;* MIT: Cambridge, 1991; Chapter 4.

8. Cohen, B. L. *The Nuclear Energy Option;* Plenum: New York, 1990; Chapter 5, Appendix.

9. Makofske, W. J.; Edelstein, M. R. *Radon and the Environment;* Noyes: Park Ridge, NJ, 1988, pp. 49, 367.

10. "Radon Tagged as Cancer Hazard by Most Studies, Researchers." *Chemical and Engineering News,* February 6, 1989, pp. 7–13.

11. Wagner H. N.; Ketchum, L. E. *Living with Radiation: The Risk, the Promise;* Johns Hopkins University: Baltimore; 1989; pp. 78–87, 136–143.

12. Abate, T. "Radon Research Typifies Challenges Facing Risk Assessment." *The Scientist,* February 17, 1992; pp. 13–14.

13. Luckenbaugh, R. W. "Radon and Lung Cancer." *Chemical and Engineering News,* August 19, 1991, pp. 4–5.

14. Hedges, R. E. M.; Gowlett J. A. J. "Radiocarbon Dating by Acceleration Mass Spectrometry." *Scientific American* **1986,** *254,* 100.

15. Roth, E.; Poty, B. *Nuclear Methods of Dating;* Kluwer Academic: Boston, 1989; Chapter 16.

16. Evans, E. A.; Oldham, K. G. *Radiochemicals in Biomedical Research;* John Wiley & Sons: New York; 1988; pp. 26–27.

17. Barth, R. F.; Soloway, A. H.; Fairchild, R. G. "Boron Neutron Capture Therapy for Cancer." *Scientific American* **1990,** *263,* 100.

18. Harling, O. K.; Bernard, J. A.; Zamenhof, R. G., *Neutron Beam Design, Development, and Performance for Neutron Capture Therapy;* Plenum: New York, 1990.

19. Dagani, R. "^{99m}Tc Compounds Approved for Cardiac Imaging." *Chemical and Engineering News,* January 14, 1991; pp. 24–25.

20. *1992 Science Year, the Worldbook Annual Science Supplement;* World Book: Chicago, 1991; p. 333.

Energy and Environment

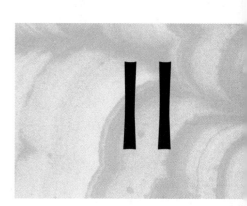

II

Nuclear Chemistry II: Induced Radioactive Decay

11

In this chapter we begin the subject of energy by considering the science of nuclear energy. Having addressed spontaneous radioactive decay in Chapter 10, we turn to ways in which isotopes, both radioactive and nonradioactive, can be *induced* to undergo various kinds of nuclear reactions. The reaction types include fission, fusion, and conversion to other isotopes, known as *transmutation.* All are discussed in the context of the generation of energy.

Since nuclear energy is very controversial, this chapter also describes the two best-known nuclear power accidents, Three Mile Island and Chernobyl. These two events taught us many lessons, and they will continue to have an impact on the nature of nuclear power.

The disposal of radioactive waste and nuclear arsenals is a nagging problem for the nuclear power industry. Their potential solutions are considered, and the chapter ends with a discussion of the design of nuclear bombs to help you appreciate why a nuclear power plant is not a potential nuclear bomb.

Chapters 12 and 13 continue the topic with a discussion of other sources of energy. Once you are aware of the potential uses and difficulties associated with each one, you will more fully appreciate how nuclear energy might play a role in filling worldwide demands for electricity both in the short and the long term.

11.1   Production of Other Particles: Induced Radioactive Decay

GOALS: To be aware of the generation of certain particles from nuclear processes.
To understand the use of accelerators and moderators.

The phenomena known as *transmutation, nuclear fission,* and *nuclear fusion* are related. They all occur when certain nuclei are bombarded with high-energy radiation such as alpha (α) and beta (β) particles. Other particles are available from nuclear reactions such as the two that follow. Here, the nuclei are bombarded with α particles, causing production of neutrons and protons:

$$^{9}_{4}\text{Be} + ^{4}_{2}\alpha \rightarrow ^{12}_{6}\text{C} + ^{1}_{0}\text{n}$$

shorthand notation: $^{9}_{4}\text{Be}(\alpha, \text{n})^{12}_{6}\text{C}$

$$^{14}_{7}\text{N} + ^{4}_{2}\alpha \rightarrow ^{17}_{8}\text{O} + ^{1}_{1}\text{H} \text{ (or } ^{1}_{1}\text{p)}$$

shorthand notation: $^{14}_{7}\text{N}(\alpha, \text{p})^{17}_{8}\text{O}$

The shorthand notations indicate the starting and product nuclei, and the bombarding and product particles (in parentheses).

We are now aware of possible sources of five different types of radiation—α, β, γ, n, and p (Table 10.1)—that can be used to bombard other nuclei and induce other nuclear reactions. However, not all particles are emitted with energies that permit them to interact with other nuclei productively. Some particles are emitted with energies that are too low to penetrate other nuclei, whereas other particles have such high energies that they pass right through the nuclei without causing any change. Charged particles, both positive and negative, tend to be repelled by electrons surrounding the nuclei or by the nuclei themselves. Thus a **linear** or **cyclic accelerator** *(cyclotron)* is used to increase the kinetic energy (the energy of motion) of the charged particles and overcome the repulsive forces.

In contrast, neutrons, which have a zero charge, may penetrate a nucleus without any effect. Carbon in the form of graphite is often used as a **moderator** to slow neutrons down so that they can come under the influence of nuclear forces and cause reactions.

11.2   Transmutation of Elements

GOAL: To appreciate some of the major applications of transmutation.

Eighty-one elements, from hydrogen to bismuth, are found naturally in stable forms on earth. Two other elements—thorium and uranium—are unstable, but have such long radioactive lifetimes that they are believed to have been present when the planet was formed. An additional 25 elements have been produced artificially or have been observed as natural decay products of other elements.

New elements may be produced from known elements by **transmutation;** this is accomplished by bombarding the nuclei of an element with α particles, β particles, protons, or neutrons. Accelerators or moderators are needed to allow the target nuclei to capture the particles. The following equations provide four examples of reactions that have produced previously unknown isotopes:

$$^{96}_{42}\text{Mo} + ^{2}_{1}\text{H} \rightarrow ^{1}_{0}\text{n} + ^{97}_{43}\text{Tc} \qquad \text{technetium-97}$$
$$^{142}_{60}\text{Nd} + ^{1}_{0}\text{n} \rightarrow ^{0}_{-1}\beta + ^{143}_{61}\text{Pm} \qquad \text{promethium-143}$$
$$^{209}_{83}\text{Bi} + ^{4}_{2}\alpha \rightarrow 3\,^{1}_{0}\text{n} + ^{210}_{85}\text{At} \qquad \text{astitine-210}$$
$$^{230}_{90}\text{Th} + ^{1}_{1}\text{p} \rightarrow 2\,^{4}_{2}\alpha + ^{223}_{87}\text{Fr} \qquad \text{francium-223}$$

Section 10.8 (Radiochemical Cows), discussed the production of technetium-99m from molybdenum-99. This is an example of transmutation:

$$^{99}_{42}\text{Mo} \xrightarrow{\beta^- \text{ decay}} ^{99\text{m}}_{43}\text{Tc}$$
$$t_{1/2} = 67 \text{ hrs} \qquad t_{1/2} = 6 \text{ hrs}$$

Because of the extensive use of ^{99m}Tc and its short half-life ($t_{1/2}$), it is necessary to have a continual source of the isotope available. The ^{99}Mo is the source, but it is also radioactive with a limited half-life. It is not present in nature, but is a synthetic isotope that can be produced in a nuclear reactor by neutron bombardment, according to the following transmutation reaction:

$$^{98}_{42}\text{Mo} + {}^{1}_{0}\text{n} \rightarrow {}^{99}_{42}\text{Mo} + \gamma$$

shorthand notation: $^{98}\text{Mo}(\text{n}, \gamma)^{99}\text{Mo}$

The ^{98}Mo source of ^{99}Mo is a stable isotope that makes up about 24% of molybdenum present in nature *(1)*.

Some of the most ambitious examples of transmutation are the synthese

s of the **transuranium elements.** These elements, which follow uranium (atomic number 92) in the periodic table, may have been present in earth's crust when it formed about 4.5 billion years ago, but the half-lives of the isotopes are too short to permit survival to present times.

The first successful synthesis was reported in 1940, when elements 93 (neptunium) and 94 (plutonium) were produced according to the following reactions:

Names were not given to elements 104 through 109 because United States and Russian scientists could not agree on discovery rights and names.

$$^{238}_{92}\text{U} + {}^{2}_{1}\text{H} \rightarrow {}^{239}_{92}\text{U} + {}^{1}_{1}\text{H} \qquad \text{deuteron bombardment}$$
$$^{239}_{92}\text{U} \rightarrow {}^{239}_{93}\text{Np} + {}^{0}_{-1}\beta \qquad \text{spontaneous decay, } t_{1/2} = 23.5 \text{ min}$$
$$^{239}_{93}\text{Np} \rightarrow {}^{239}_{94}\text{Pu} + {}^{0}_{-1}\beta \qquad \text{spontaneous decay, } t_{1/2} = 2.35 \text{ days}$$

The complete list of known transuranics appears in Table 11.1.

Table 11.1   The transuranium elements *(2)*

Element	Symbol	Atomic Number	Discovery Year
Neptunium	Np	93	1940
Plutonium	Pu	94	1940
Americium	Am	95	1944
Curium	Cm	96	1944
Berkelium	Bk	97	1949
Californium	Cf	98	1950
Einsteinium	Es	99	1952
Fermium	Fm	100	1953
Mendelevium	Md	101	1955
Nobelium	No	102	1958
Lawrencium	Lr	103	1961
Unnilquadium	Unq	104	1964
Unnilpentium	Unp	105	1970
Unnilhexium	Unh	106	1974
Unnilseptium	Uns	107	1981
Unniloctium	Uno	108	1984
Unnilennium	Une	109	1982

11.3 **Nuclear Fission**

GOALS: To understand the process of nuclear fission.
To appreciate chain reactions.
To understand what makes a nuclear reactor subcritical, critical, or supercritical.
To consider the importance of breeding.

Fission is the splitting of certain large atoms into smaller atoms after the absorption of bombarding neutrons. Due to the difference in size of the nuclear forces of the parent nucleus and the smaller fission product nuclei, the entire process is accompanied by the liberation of huge amounts of energy. Mass is lost during fission and is converted to energy according to $E = mc^2$ (c = speed of light). Atomic bombs used in World War II were fission bombs. Nuclear fission is also the process that is used to generate electricity in nuclear power plants.

One common example is the fission of uranium-235, which results when uranium is bombarded with neutrons, as indicated by the following four equations. Each reaction is accompanied by liberation of energy.

$$\begin{aligned}
^{235}_{92}U + {}^{1}_{0}n &\rightarrow {}^{135}_{53}I + {}^{97}_{39}Y + 4\,{}^{1}_{0}n \\
&\rightarrow {}^{139}_{56}Ba + {}^{94}_{36}Kr + 3\,{}^{1}_{0}n \\
&\rightarrow {}^{103}_{42}Mo + {}^{131}_{50}Sn + 2\,{}^{1}_{0}n \\
&\rightarrow {}^{139}_{54}Xe + {}^{95}_{38}Sr + 2\,{}^{1}_{0}n
\end{aligned}$$

All of these fission reactions are characterized by production of more neutrons than are consumed. The process can thus become a self-perpetuating *chain reaction,* in which the product neutrons become bombarding particles to cause fission of more ^{235}U.

Fortunately, it is possible to control the rate of these reactions by carrying them out under **critical** or **subcritical** conditions, in which an average of one or less product neutron becomes a bombarding neutron for continuation of the chain (Figure 11.1). Such conditions can be readily achieved by introducing an efficient neutron-absorbing material, such as cadmium or boron, that can be added in any amount desired to maintain the critical or subcritical condition. Cadmium metal and boron carbide are commonly incorporated into **control rods** in nuclear reactors used to generate electricity. If enough control rods are inserted into a reactor, fission can be suppressed entirely. As more and more rods are removed, fission can reach critical.

If provisions are not made to capture some neutrons, and the fission process is allowed to take place with an average of more than one product neutron bombarding other ^{235}U nuclei, the system is termed **supercritical;** this condition may result in loss of control of the fission process. Since the level of neutron bombardment, fission, and production of more neutrons can rapidly accelerate when conditions are supercritical, an atomic explosion may result (Section 11.8).

Breeding

Uranium-235 is in limited supply and does not constitute a long-range energy resource. The isotope ^{238}U is much more prevalent in nature, but it is not fissionable. Breeder reactors provide an important means of carrying out fission of more abundant starting materials such as ^{238}U. The following sequence of reactions converts nonfissionable ^{238}U into fissionable plutonium-239, or *breeds* ^{239}Pu.

$$\begin{aligned}
^{238}_{92}U + {}^{1}_{0}n &\rightarrow {}^{239}_{92}U \qquad &&\text{neutron bombardment} \\
^{239}_{92}U &\rightarrow {}^{239}_{93}Np + {}^{0}_{-1}\beta \qquad &&\text{spontaneous decay, } t_{1/2} = 23.5 \text{ min} \\
^{239}_{93}Np &\rightarrow {}^{239}_{94}Pu + {}^{0}_{-1}\beta \qquad &&\text{spontaneous decay, } t_{1/2} = 2.35 \text{ days}
\end{aligned}$$

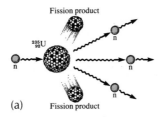

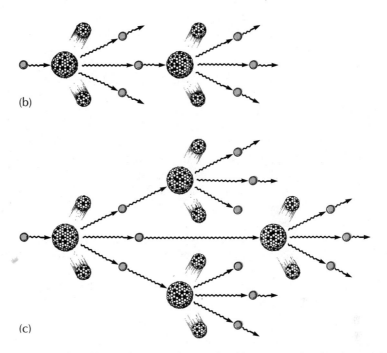

Figure 11.1   A nuclear fission chain reaction initiated by neutron bombardment of uranium-235. After fission, the neutrons that are released may be lost or captured by other nuclei, which also undergo fission. The fraction of product neutrons that bombard other ^{235}U nuclei determines the state of a nuclear reactor. (a) If an average of *less than one* of the product neutrons becomes a bombarding neutron, the chain reaction will be *subcritical* and not sustainable. (b) If an average of *one* of the product neutrons becomes a bombarding neutron, the chain reaction will be *critical* and self-sustaining. (c) If an average of *more than one* of the product neutrons becomes a bombarding neutron, the chain reaction becomes *supercritical* and can lead to an explosion.

Similarly, the isotope thorium-232 is abundant in nature and nonfissionable, but it can be converted to a fissionable isotope, ^{233}U, by the following sequence:

$$^{232}_{90}\text{Th} + {}^{1}_{0}\text{n} \rightarrow {}^{233}_{90}\text{Th} + \gamma \qquad \text{neutron bombardment}$$

$$^{233}_{90}\text{Th} \rightarrow {}^{233}_{91}\text{Pa} + {}^{0}_{-1}\beta \qquad \text{spontaneous decay, } t_{1/2} = 22 \text{ min}$$

$$^{233}_{91}\text{Pa} \rightarrow {}^{233}_{92}\text{U} + {}^{0}_{-1}\beta \qquad \text{spontaneous decay, } t_{1/2} = 27 \text{ days}$$

The overall breeding reactions are:

$$^{238}_{92}\text{U} \rightarrow {}^{239}_{94}\text{Pu}$$
$$^{232}_{90}\text{Th} \rightarrow {}^{233}_{92}\text{U}$$

A nonfissionable isotope that can be used to breed a fissionable product isotope is said to be *fertile*.

11.4 Nuclear Fusion

GOALS: To be aware of the fusion process and the challenge of containment.
To recognize the form of energy released by fusion.

Fusion is the process in which small nuclei are combined to form heavier nuclei, in contrast to nuclear fission in which large nuclei split into smaller fragments. Surprisingly, both fission and fusion liberate large amounts of energy, even though it seems an impossible contradiction that two opposite processes can both give off energy. However, just as we know that isotopes above ^{209}Bi lack stability, we also know that maximum stability characterizes isotopes at about mass number 60; this is consistent with the high abundance in nature of elements such as iron and nickel. Therefore, fusion of small nuclei, below mass number 50, releases energy to build larger, more stable nuclei, just as fission of large nuclei releases energy to form more stable, smaller nuclei.

Fusion requires the combination of nuclei that exhibit strong repulsions for one another, so extremely high energy must be supplied to initiate the process. Although it holds great promise as an energy source for the future, great technological problems lie along the way. Extremely high temperatures (75–100 million °C) are required, which makes containment impossible in any known vessel. Much effort has gone into developing some sort of magnetic bottle to contain the reaction. In addition, reaching such temperatures is a major obstacle; lasers may offer a solution by providing a source of intense energy.

Once fusion is initiated, it is self-sustaining. According to the following reactions, it is thought to be responsible for the energy output of the sun by the following reactions:

$$2\,_{1}^{1}H \rightarrow \,_{1}^{2}H + \,_{1}^{0}\beta$$

$$_{1}^{1}H + \,_{1}^{2}H \rightarrow \,_{2}^{3}He$$

Many engineering designs are possible for exploiting fusion as a commercially important source of energy. The most studied system in the United States to date is a design reported by the Russians in 1968, known as the Tokamak (Figure 11.2). The reactive materials are heated to such high temperatures that they exist in the form of a plasma as positively charged nuclei (ions) and electrons. Tokamak uses the same fusion process as that of the hydrogen bomb:

Tokamak is derived from the Russian for "toroidal magnetic chamber." A toroid is a doughnut shape.

$$_{1}^{2}H + \,_{1}^{3}H \rightarrow \,_{2}^{4}He + \,_{0}^{1}n$$

$$_{3}^{6}Li + \,_{0}^{1}n \rightarrow \,_{2}^{4}He + \,_{1}^{3}H$$

$$_{1}^{2}H + \,_{1}^{3}H \rightarrow \,_{2}^{4}He + \,_{0}^{1}n \text{ etc. (chain continues)}$$

Source of starting materials:

deuterium (^2H):   from D_2O (^2H_2O) of seawater

tritium (^3H):   from second reaction (above)

neutrons: from nuclear reactors and reactions above

lithium: from ores

The fusion of deuterium, ^2H, and tritium, ^3H, is the step that releases a large amount of energy. About 80% is carried by the high-energy, fast-moving neutron product. As the neutron slows down, it releases its energy as heat in the *blanket*, a region of the reactor adjacent to the plasma. The heat can be used to produce steam to generate electricity.

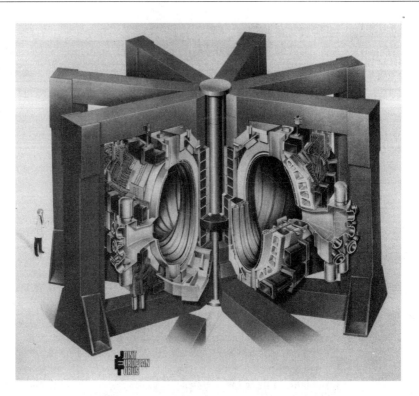

Figure 11.2   Tokamak-type fusion devices use strong magnetic fields to confine dense plasma within a toroidal vacuum chamber, and prevent contact with the vessel wall. The Joint European Torus (JET, depicted here in cross section) is the largest now under construction. In the JET apparatus, the fusion reaction will produce less energy than the electrical power necessary to operate the electromagnets and heat the plasma. It will provide essential operating data for the next generation of fusion reactors capable of net power production.

The α particle product carries the remaining 20% of the energy from fusion. It remains in the reactor, and the heat generated sustains the plasma temperature. Thus, fusion, like fission, can be self-sustaining once it has been initiated.

Up to now, when fusion has been achieved, the energy released has been less than the amount introduced into the system to initiate the process. Commercial use is unlikely in this century and could be delayed well beyond 2000, if fusion ever becomes economically competitive.

11.5 Nuclear Power Generation

GOALS: To appreciate the importance of nuclear power for generating electricity.
To be aware of the various designs of nuclear power plants, including the advantages and disadvantages associated with the use of different coolants and moderators.
To appreciate the possible role of the breeder reactor.

The use of nuclear fission to generate electric power is controversial. Nuclear power has had a stormy history, and projections of its future are difficult. On one hand, we can see

Table 11.2   Trends in energy sources used to generate electricity in the United States, 1970–1990

Energy Resource	1970 (%)	1980 (%)	1990 (%)
Nuclear	1	11	20
Coal	46	51	57
Oil	13	11	6
Hydroelectric	16	12	8
Gas	24	15	9
Totals	100	100	100

Source: U.S. Department of Energy

from Table 11.2 that the U.S. electric utility industry has become quite dependent on nuclear fission. On the other hand, events at Three Mile Island (March 1979) and Chernobyl (April 1986) emphasized concerns about the safety of nuclear power. Although many observers predicted an end to the use of nuclear power after the Three Mile Island (TMI) accident, the data in Table 11.2 do not substantiate this claim. Of course, construction on many nuclear power plants was well under way in March 1979. Thus, most of the increased generation has been unaffected by TMI. Planning for further construction of nuclear power plants in the United States has virtually ceased, and construction on others has been halted. New construction and percentage use in European countries, however, continue to increase.

Nuclear power is not just an issue of concern to the United States. Many countries lack natural supplies of oil, gas, or coal, but dependence on imported oil or coal has economic and political consequences. As a result, over 400 nuclear reactors are in operation in 26 countries. That figure is projected to increase to about 600 by the year 2000. In France, nearly 70% of electricity is generated from nuclear energy, as is 60% in Belgium and 50% in Sweden (Figure 11.3).

Figure 11.3
Pennsylvania Power and Light Company's Susquehanna steam electric station with twin BWR generating units.

Thus we have to understand nuclear power in its various forms, its advantages, and its hazards and limitations. We must appreciate what happened at TMI and at Chernobyl to determine how power plants can be designed to prevent any repeats of these accidents. On the positive side, we have to understand why nuclear power plants are not potential atomic bombs. On the negative side, we must consider the problems of radioactive waste that comes from the plants in the form of spent fuel.

All of these topics constitute the remainder of this chapter. The role of nuclear power in the context of the overall need for energy, electric and other forms, is discussed further in Chapters 12 and 13.

Types of Nuclear Power Plants

The three nuclear processes that could be used for electric power generation are ^{235}U-induced fission, ^{239}Pu or ^{233}U breeding followed by induced fission, and fusion. Supplies of ^{235}U are dwindling. Even long before they are depleted, the supply-and-demand picture could price ^{235}U up to uneconomical levels. As a replacement, the all but inexhaustible nonfissionable ^{238}U may become prominent in energy production. As noted earlier, commercial fusion is unlikely in the near future. The plutonium breeder system in the form of the *liquid-metal fast breeder reactor* (LMFBR) seems to offer the most likely long-range answer to energy production from nuclear sources.

The system known as the *light water reactor* (LWR) uses ^{235}U fission, and H_2O acts as the coolant. **Heavy water,** ^2H_2O or **D_2O,** called *deuterium,* is water in which the element hydrogen is present as the isotope that has one neutron in the nucleus. Light water contains only the isotope ^1H. The fuel for the LWR contains about 3% ^{235}U. Since uranium ore contains only about 0.7% ^{235}U (and 99.3% ^{238}U), the ore must be *enriched* in ^{235}U from 0.7% to 3.0% before use in power plants. Nuclear weapons use uranium containing about 97% ^{235}U, requiring several thousand additional stages of enrichment.

Other nuclear reactors for electric power production are listed in Table 11.3. Each name indicates something about its mode of operation. The possible variables include the type and design of the coolant system, the moderator, whether the process is simple fission or breeding, and whether the reactor is a fast or a thermal neutron reactor system.

First, consider the distinction between a **fast reactor** and a thermal reactor. The words *fast* and *thermal* describe the energy of the neutrons used to bring about fission or to breed a fissionable isotope. Fast neutrons are high-energy neutrons, generally 1 million electron-volts (MeV) or higher, whereas thermal neutrons are slow, low-energy neutrons, less than 1 MeV. In the jargon of the nuclear scientist, the **cross section** (σ) of a nucleus describes

Table 11.3   Nuclear power plant designs

Abbreviation	Reactor Type
LWR	Light water reactor
HWR	Heavy water reactor
PWR	Pressurized water reactor
BWR	Boiling water reactor
HTGCR	High temperature gas-cooled reactor
LMFBR	Liquid metal fast breeder reactor

the probability that it will capture a neutron. A large cross section signifies a high probability of capture (and possibly fission).

Some isotopes will not capture fast neutrons, so a moderator is required to decrease the energy of neutrons and promote capture. An example of an isotope with a high-fission cross section for thermal neutrons but a low cross section for fast neutrons is ^{235}U. Therefore, the fission of ^{235}U can be initiated only with neutrons that have been properly moderated.

On the other hand, ^{238}U requires fast-bombarding neutrons for capture and breeding ^{239}Pu (Section 11.4). In other words, ^{238}U has a zero cross section for capturing thermal neutrons but a high cross section for capturing fast neutrons.

There are many consequences of this difference between ^{235}U and ^{238}U. Since ^{235}U makes up 0.715% of natural uranium, there is one ^{235}U atom for every 139 ^{238}U atoms. In this proportion, ^{235}U cannot be made to undergo sustained fission to power a nuclear reactor. The reason is that the ^{238}U nucleus can cause an inelastic scattering of the thermal neutron (even though it cannot capture the neutron). This lowers the energy of the neutron to a level below even the moderated level required for ^{235}U fission. However, when the uranium ore is enriched to about 3% ^{235}U, fission becomes a significant event and can be built up and maintained at critical, self-sustaining levels (Figure 11.4). Fission reactions produce more neutrons than they consume, so even though ^{238}U captures some neutrons, at least one product neutron from each fission can be moderated and used again as a bombarding particle.

Figure 11.4 Measuring a uranium oxide fuel element is a routine quality control procedure. In a nuclear power plant, each pellet can provide about as much energy as a ton of coal.

Table 11.4   Coolants and moderators, advantages and disadvantages

Coolants	Advantages	Disadvantages
Gas (He, CO_2)	Efficient heat transfer. Chemically inert.	Expensive.
Light water (H_2O)	Inexpensive. Captures many neutrons.[1] Can also serve as moderator.	Requires high pressure containment.
Heavy water (D_2O)	Captures very few neutrons. Can also serve as moderator.	Expensive. Requires high pressure containment.
Liquid metal (sodium)	Can tolerate high temperature without pressure buildup. Efficient heat transfer.	Reacts violently with air and water. Poor moderator.[2]

Moderators	Advantages	Disadvantages
Light water	Inexpensive. Captures many neutrons.[1] Good moderator.	
Heavy water	Good moderator. Captures very few neutrons.	Expensive.
Graphite (carbon)	Inexpensive. Fair moderator. Captures few neutrons.	
Beryllium (Be)	Captures few neutrons.	Toxic. Expensive. Fair moderator.

[1]Requires enriched fuel.
[2]An advantage for fast reactors.

There has logically been concern over the speculation that a runaway nuclear power plant could become a nuclear bomb. Such an event is not possible for several reasons. First, the fuel composition is important. Just as uranium ore has too little ^{235}U to give the *critical* condition required to sustain fission, even uranium enriched to 3% ^{235}U is not sufficient to produce the conditions necessary for a nuclear explosion. Samples enriched to about 97% ^{235}U are common for nuclear weapons.

Second, the reactor design cannot contain the nuclear reactions that would allow the system to reach bomb intensity. If control were lost, the heat would cause the core of the reactor to come apart (melt). Third, a bomb requires that the components be combined in about 10^{-6} seconds, which is impossible within the design of a power-generating plant. The design of nuclear bombs is discussed in Section 11.8.

Table 11.4 lists the most important examples of coolants and moderators used in nuclear power plants around the world. Quite clearly, they all have both advantages and disadvantages.

The *boiling water reactor* (BWR) is a system in which either light water or heavy water is used as both coolant and moderator. A schematic diagram appears in Figure 11.5. The system performs in the following way:

1. Heat is generated by nuclear fission in the core of the reactor.

2. Water flows through the core and acts as moderator and coolant.

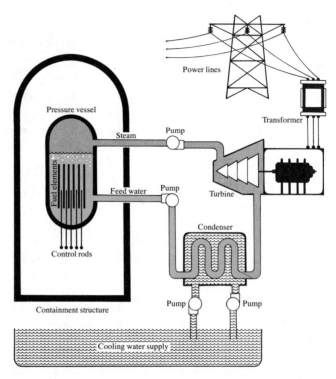

Figure 11.5 A BWR nuclear power plant.

3. As a coolant, the water extracts heat from the core and is converted to steam.

4. The steam is used to drive a turbine, which generates electricity.

In a *pressurized-water reactor,* PWR (Figures 11.6 and 11.7), the water again acts as coolant and moderator but is kept under pressure so that it does not boil as it circulates past the core. Instead, the steam generator is separated from the core. With the coolant water kept pressurized, the operating temperatures can be much higher and the efficiency of the unit is improved. This advantage is largely offset by the expense of the additional component of the system (compare Figures 11.5 and 11.7). An important application of the PWR is the nuclear submarine, which benefits from the compact size of the unit. Equally important is the absence of the large demand for oxygen, which is needed to burn fossil fuels. Conventional submarines must surface frequently to take on a supply of air.

A *heavy-water reactor* (HWR) is simply one that uses heavy water, D_2O, but may actually have either a BWR or a PWR design. Compared with ordinary (light) water, heavy water absorbs few neutrons, so it is not necessary to use enriched uranium to sustain the fission process. This cost saving is offset by the increased cost of the heavy water. The HWR also can be one in which heavy water is the moderator and light water is the coolant. The only commercial HWR is the CANDU (<u>Can</u>adian <u>d</u>euterium-<u>u</u>ranium) design. The CANDU reactors are more expensive to build, but they have been proven to have the highest reliability of all nuclear power reactors in terms of possible operating time.

The setup of the *high-temperature gas-cooled reactor* (HTGCR) is like that of the PWR, except that the gas acts only as a coolant. Therefore, a moderator (usually graphite) must be included in the core (Figure 11.8).

The *liquid-metal fast breeder reactor* (LMFBR), also simply known as the *liquid-metal reactor* (LMR) or the *fast breeder reactor* (FBR), is the most sophisticated design

Figure 11.6   Pressure vessel being lowered into the chamber at Brunswick nuclear power plant (Carolina Power & Light Company). The 821-megawatt plant unit 2 is located in Southport, North Carolina.

considered here (Figure 11.9). The coolant is liquid sodium. Since fast neutrons are required for the productive bombardment of ^{238}U to breed fissionable ^{239}Pu (Section 11.3), this design has no moderator. Liquid sodium is pumped through the left-hand loop (heat exchanger) and picks up the heat from the core. Liquid sodium also flows through the intermediate loop, picks up heat from the sodium that moves through the core, and then flows to the steam generator and transfers heat there. Since sodium reacts violently with water, the sodium flowing through the core is isolated from the water in the steam generator by the intermediate sodium loop. If the system does spring a leak, the core and the steam generator are protected from the site of the problem. Liquid sodium, which flows through the core, becomes intensely radioactive through neutron bombardment. The intermediate loop isolates the radioactivity from the water supply passing through the steam generator, so that any water-sodium reaction could occur only with nonradioactive sodium.

In each type of system, the steam drives turbines and generates electricity as outlined in Figure 11.5. Power plants that are run on coal, oil, or gas all use steam in essentially the same way (Figure 11.10).

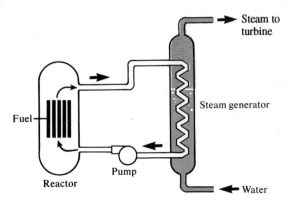

Figure 11.7 Nuclear steam-supply components in a PWR.

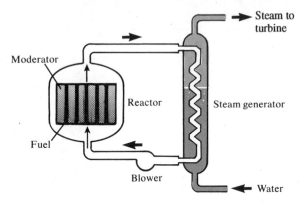

Figure 11.8 Nuclear steam-supply components in a GCR.

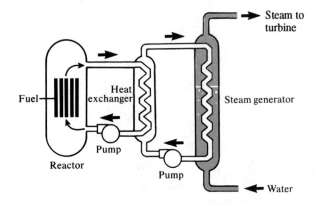

Figure 11.9 Nuclear steam-supply components in an LMFBR. Liquid sodium flows through the core and the intermediate loop that carries heat to the steam generator.

Figure 11.10 Steam turbine for generating electricity at the Carolina Power & Light Company's Brunswick plant. The 821-megawatt turbine uses steam from a BWR to drive a generator that is located beyond the wall at the top of the photograph. The turbine section shown is 117 feet long. The generator is on the same shaft and extends for another 59 feet. The S-shaped structures on either side of the turbine are insulated steam pipes and valves.

NUCLEAR GENERATING UNIT CAPACITY

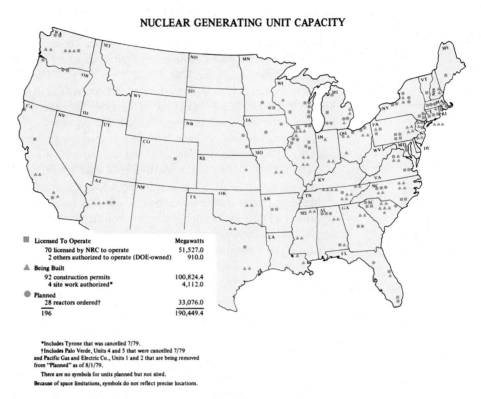

	Megawatts
■ **Licensed To Operate**	
70 licensed by NRC to operate	51,527.0
2 others authorized to operate (DOE-owned)	910.0
▲ **Being Built**	
92 construction permits	100,824.4
4 site work authorized*	4,112.0
● **Planned**	
28 reactors ordered†	33,076.0
196	190,449.4

*Includes Tyrone that was cancelled 7/79.
†Includes Palo Verde, Units 4 and 5 that were cancelled 7/79
and Pacific Gas and Electric Co., Units 1 and 2 that are being removed
from "Planned" as of 8/1/79.
There are no symbols for units planned but not sited.
Because of space limitations, symbols do not reflect precise locations.

Figure 11.11   Status of nuclear generating units, December 31, 1991. Due to space
limitations, symbols do not represent exact locations.

11.6   Two Nuclear Accidents: Chernobyl and Three Mile Island

GOALS: To understand some of the safety hazards associated with nuclear power plants.
To understand the weakness of the RBMK reactor design and the important
operator errors made at Chernobyl.
To understand the events that occurred at Three Mile Island and the influence of
operator errors.

Two historic events shook the nuclear industry: Chernobyl and Three Mile Island (TMI).
The accident at TMI was effectively contained. It had only minor consequences to humans,
although the effects on the electric utility that owns the power plant and on the U.S. nuclear
power industry have been profound. In contrast, the Chernobyl accident killed 31 people
and devastated the lives of many thousands more. The events that took place at Chernobyl
may actually prove beneficial to the nuclear power industry, due to lessons learned about
the unique design of the plant and the incredible series of errors that were made by the
operators.

Chernobyl

On April 26, 1986, a series of events took place at Chernobyl, 120 km northwest of Kiev
in the Ukrainian Soviet Socialist Republic. The type of reactor in use at Chernobyl is the
RBMK, a Russian acronym for "water-cooled and graphite-moderated." An uncontrolled
power surge overheated the reactor and expelled the upper shield, causing complete loss
of cooling. After the explosive release, heat from the decay of the fission products caused

the temperature to become hot enough to allow fission products to distill out of the reactor. Nine days after the initial accident, the daily release of radioactive materials was nearly as high as it was at the time of the initial release. About 50 million curies (Ci) of the noble gases ^{133}Xe and ^{85}Kr were released, as well as about 50 million Ci of ^{131}I, ^{134}Cs, and ^{137}Cs, plus small amounts of ^{89}Sr, ^{90}Sr, ^{141}Ce, ^{144}Ce, ^{238}Pu, ^{239}Pu, and ^{240}Pu *(3, 4)*.

When water acts as both moderator and coolant (as in the LWRs in use in the United States), if coolant is lost, so is the moderator. Thus the fission process stops if water is lost, since the low cross section ^{235}U will not capture fast neutrons (Section 11.5). This reactor design is said to have a **negative void coefficient**—if the coolant is lost, the intensity of the fission decreases and the reactor shuts down automatically.

The RBMK design has a **positive void coefficient,** which contributed to the development of the accident. If coolant water is lost, the graphite moderator allows fission to continue and even accelerate. Besides acting as coolant, the water also captures some neutrons and thus lowers the power level of the reactor. At Chernobyl, the reactor power surged causing the reactor to overheat. This heating caused too much water to be converted to steam, and control was lost in a few seconds.

Since the RBMK reactor has this design flaw, it includes sophisticated cooling mechanisms that were supposed to deal with the positive void coefficient. Among these is an emergency core cooling system (ECCS). At the time of the accident, the reactor was scheduled to be shut down for maintenance. In anticipation of this, the operators were testing the feasibility of using electric power from the reactor, as it was being shut down, to sustain the flow of coolant water to the reactor. This was done to mimic a situation in which the reactor might be cut off from external electric power during a shutdown *(5, 6)*.

In their overzealous attempt to proceed with the experiment, the operators inactivated several control systems, each of which could have prevented the catastrophe that followed. The ECCS was disconnected, an automatic reactor trip system was disabled, and most of the reactor's 211 control rods (Section 11.3) were withdrawn from the core, even though operating regulations prohibited all of these actions. The total effect was to turn a reactor core that had too much water into one that had too little water within a period of less than a minute. Due to the positive void coefficient, the reactor power surged, and with all the emergency systems deactived, there was no way to control the rapid increase. Computer models of the accident indicate that the power went from 200 megawatts (MW) to 3800 MW in 2.5 seconds. In another 1.5 seconds it rose to at least 120 times the normal power output of the reactor *(6, 7)*.

Ensuing fires carried the fuel and fission products high into the air, spreading the contamination over a wide area. Counting firefighters and plant operators, 31 people died, many from radiation doses from being close to the site. Radioactive fallout moved mostly in a northerly direction. It was first detected in Sweden and Finland, after which the whole world was alerted to the catastrophe *(8)*.

Many analyses of the design problems, the actions of the operators, and consequences of the accident have appeared. The RBMK is a derivative of the reactors built to produce plutonium for the first Soviet atomic bombs *(7)*. When the Soviets adopted the design in the 1950s, they had little choice since they lacked the technology to construct some of the major components of LWRs. However, they had begun a move away from the RBMKs to PWRs prior to this disaster, although several RBMKs were still under construction.

Three Mile Island

The accident at Three Mile Island (TMI) was minor compared with the one at Chernobyl. Nevertheless, it was the most serious nuclear accident that has yet happened in the United States.

The events began on March 28, 1979. The unit 2 reactor was the PWR type. The PWR has a steam generator that is separated from the reactor core (Section 11.5, Figure 11.7). Thus water flows through the core, picking up heat generated by fission. This component of the system is known as the *primary loop.* Water also flows through the steam generator or *secondary loop;* it comes in contact with the primary loop, allowing heat to pass from the primary to the secondary loop to be carried to the turbine generator.

A series of problems occurred at TMI beginning with a failure of water pumps in the secondary loop. Since the power plant was designed to provide backups in the event of malfunctions, additional pumps were activated in the secondary loop. Unknown to the operators, valves had been closed in the secondary loop, preventing the backup pumps from delivering water to the secondary loop. Since heat was still being transferred from the primary loop, the secondary loop went dry—all of the water was converted to steam. This automatically caused the turbine generator to trip (shut down) and led to the insertion of some control rods into the reactor. Since the secondary loop was no longer able to draw heat away from the primary loop, the pressure in the primary loop began to rise above normal. At an elevated pressure, a pilot-operated relief valve (PORV) opened to allow steam to escape and to prevent the pressure from rising too high. Shortly after, the reactor tripped—control rods were inserted and the fission was stopped.

At that point, the reactor was under control, but although fission had ceased, generation of heat in the core had not. Up to 7% of the heat released in a nuclear reactor is from the decay of radioactive fission products, particularly γ emitters. The amount of heat from this source depends on the length of time the fuel has been in the reactor. When the fuel is fresh, few fission products are present, and radioactive decay contributes little heat. But when the fuel has been in the reactor for many months, fission products make their greatest contribution to the heat liberated within the core. It so happened that the fuel in the TMI reactor was near the end of its lifetime, and thus the level of fission products was near a maximum. This made the reactor particularly vulnerable at the time of the accident.

Recall that the transfer of heat from the primary loop was still not taking place since no water was in the secondary loop. Therefore, the temperature in the primary loop continued to rise. This caused the ECCS in the primary loop to activate and dump more water into the core. However, unknown to the operators, another problem had developed. As the core cooled, the pressure relief valve was supposed to close, and indications inside the control room were that it had. But the PORV had actually become stuck in an open position and remained that way for more than 2 hours. Consequently, a large amount of water flowed out of the reactor onto the floor of the containment building, and the pressure became lower than normal in the primary loop. When the operators discovered the low pressure, they deactivated the ECCS. About 10 minutes after the start of the accident, the operators recognized that the valves in the secondary loop were closed and they restored the flow of water to the secondary loop. Although several minutes passed between the time the pumps were activated and water actually began to flow, this malfunction did not lead to any significant damage to the plant. It did add to the confusion, however, and provided a distraction from the serious problem of the stuck-open PORV.

With the PORV still open and indications of low pressure in the primary loop, the operators had a condition that was considered impossible: high temperature and low pressure. Therefore, they decreased the flow of cooling water to the core to allow some pressure to develop; this caused a portion of the core, which is normally completely covered with water, to be uncovered because of formation of steam bubbles. This allowed the core to heat up to temperatures in excess of 2200 °C and caused approximately half of the core to melt. This type of "core meltdown" is commonly known as the "China principle" in which it is imagined that the nuclear core would melt its way through the center of the earth and come out in China.

Approximately 4 hours into the accident the stuck PORV was discovered and cooling was restored to the core. With the whole world watching events unfold, the accident was a continuing concern for another 48 hours until it was decided that no further danger existed.

Comparing TMI and Chernobyl

We have seen that operator errors contributed heavily to each accident. At Chernobyl, the fission process went totally out of control, causing a massive power surge resulting in an explosion and virtually total destruction of the plant. At TMI, heat from the fission products, rather than the fission process itself, led to overheating and melting of a substantial portion of the core.

At TMI, 17 Ci of radioiodine was released because the containment building effectively trapped the liquid and solid radioactive materials that escaped from the reactor. Greater amounts, thought to be a few million curies of noble gases, ^{133}Xe and ^{85}Kr, escaped and were dispersed into the atmosphere. Whereas radiation detectors some distance away detected the noble gases, no people were exposed to any significant dose of radiation.

In contrast, massive amounts of radioactive elements of several kinds were released at Chernobyl and spread over a wide area, partly due to the fires that followed the explosion and burned for many hours. Many individuals received large doses of radiation. Some died within a short time after the accident; for others the effects will be observed in future decades and even in future generations. A study supported by the U.S. Department of Energy estimates that 21,000 Europeans may die of cancer in the next 50 years because of exposure to radiation from the accident. Potassium iodide tablets were distributed to inhabitants of the area around Chernobyl to try to protect them from the effects of ^{131}I, which was released in large amounts from the reactor. Since iodine is efficiently captured by the thyroid gland for incorporation into thyroid hormones, it was hoped that nonradioactive potassium iodide would be able to saturate each person's thyroid and prevent uptake of the radioactive iodine (9).

Many lessons were learned from these two accidents. Designs that are under consideration for future construction rely heavily on passive components, such as systems that do not require operator intervention, to bring reactors under control in the event of malfunctions. Smaller, modularized units are also being contemplated to permit less costly construction and greater ease of operation. Above all, designs like the RBMK, with a positive void coefficient, are no longer being considered. Through proper design of reactor core and moderators, it is possible to ensure that a reactor will always have a negative void coefficient so that the fission process cannot run out of control. More efficient passive cooling systems seem to be the way to prevent meltdowns. Meanwhile, for existing plants, more safety measures have been included, such as more extensive training of operators.

Other than the $1 billion required to clean up the TMI plant and the financial devastation of the utility that owns it, no great harm has resulted from that accident. The consequences of Chernobyl will be studied and discussed for many decades, just as are the effects of nuclear bombs on the Japanese citizens in World War II.

11.7   Safety Concerns

GOALS: To consider the conclusions of the Inhaber report.
To appreciate the challenges of disposal of spent nuclear fuel.
To understand the strategy of accelerator transmutation of waste.

When we consider the hazards of any technology, we would prefer to use a process that is trouble-free. But in the real world, all technologies have risk, and it becomes necessary to assess the relative risk of each.

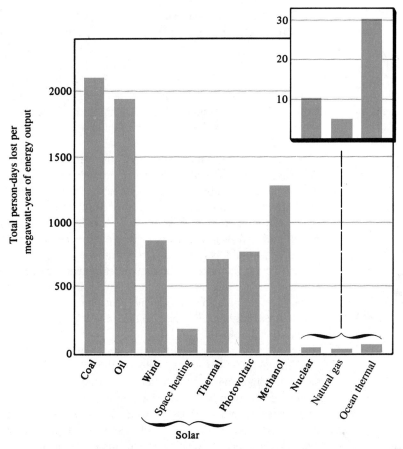

Figure 11.12 Total risk per unit energy output (1 megawatt-year, or MW-yr) for 10 energy systems. Only the maximum risk is shown for each energy form. Because natural gas, nuclear, and ocean thermal energy conversion have low risk, their bars are shown on an expanded scale. For comparison, modern nuclear reactors that produce electric power are rated at about 1000 MW, or slightly higher. A unit with a capacity of 1000 MW operating without interruption for 1 year would deliver 1000 MW-yr of electrical energy.

One death is taken as equivalent to 6000 person-days lost. Therefore, using the figures 2100 person-days lost per MW-yr (from the graph) for coal-fired plants and 10 person-days lost per MW-yr for nuclear plants, the Inhaber report predicts for 100 equivalent 1000-MW coal-fired and nuclear plants, respectively, a maximum of 35,000 and 170 deaths per year of operation. Both figures are said to include catastrophic accidents (10).

In judging the relative risk of nuclear power, we must consider the risk of not using nuclear power. None of the alternatives has the catastrophic potential of nuclear energy, but a detailed analysis of all the human damage that results (or would result) from many energy technologies, a study commissioned by the Canadian Atomic Energy Control Board, concluded that only natural gas is safer (10).

According to the Inhaber report, when the production and construction of materials, emissions caused by production of materials, operation, maintenance, energy backup, energy storage, and transportation are all considered, the relative hazards can be evaluated; Figure 11.12 diagrams this evaluation for both catastrophic and noncatastrophic risks. For comparison, the risks are plotted as person-days lost for a given amount of energy. (The energy forms included in Figure 11.12 are discussed in Chapters 12 and 13.)

Radioactive Waste Disposal

One of the most controversial aspects of nuclear power is the disposal of spent fuel (long-half-life isotopes and fission products) from the reactors. The fuel consists of about 3% ^{235}U and 97% ^{238}U. During fission, ^{235}U falls to a level below 1%, and can no longer sustain fission.

Most of the ^{238}U remains unchanged; however, some is attacked by neutrons, followed by β decay, to form neptunium and plutonium (Section 11.2). Both α and β decay, and neutron capture by many of these heavy isotopes form a whole series of isotopes of the actinide elements, actinium, thorium, americium, curium, and so on (see periodic table). They are all toxic, and some have very long half-lives. For example, plutonium-239, which forms in both light-water reactors and breeder reactors, has a half-life of 25,000 years; thus, waste disposal has repercussions for the very long term.

Fission products make up the remainder of spent fuel. For these products, γ radiation from ^{90}Sr and ^{137}Cs presents the greatest problem. In fact, after 5 years the decay of these isotopes is responsible for most of the heat released by radioactive waste. The isotopes have half-lives of 29 and 30 years, so that in 700 years less than one ten-millionth of their radioactivity will remain.

It has even been suggested that the burden of radioactive waste will actually become an asset in about 500 years. The γ emissions will be mostly gone then, and the remaining hazard will come from easily shielded α particle emissions from the actinide elements. The actinides can then be separated and used as reactor fuels or for other applications *(8)*.

Of course, such a view assumes that the waste material can be safely stored for 500 years. Critics question the certainty of guaranteed continuous safe isolation from the environment. We appear to be placing a tremendous burden on future generations. The thought of safe storage for tens of thousands of years, while even the actinides decay, is hard to imagine. The danger of leaching into water supplies makes even the α emitters a cause for great concern, since they are biologically more damaging than γ emitters if they are ingested (Section 10.5).

Nevertheless, much research is geared toward developing procedures for safe, permanent storage of nuclear waste. A multibarrier approach is favored, applying several techniques simultaneously so that if one component of the system fails, others will serve as backups. The most often mentioned candidate is deep burial in remote and geologically safe locations such as salt beds, domes, granite, and shale. Both remoteness and geology are barriers, however.

Burial in salt deposits seems ideal. The presence of solid salt deposits at locations in New Mexico (near Carlsbad) and elsewhere suggests that no water has been present in these places for some time, if ever. If it had, the salt would likely have washed away. Thus, the potential for leaching seems nonexistent. Salt also exhibits plastic flow under pressure, so that back-filling burial sites with crushed salt could result in self-sealing as the formation returns to a solid mass. Critics argue that the possibility of volcanic or seismic activity precludes the certainty that the geological environment will remain intact for the long time required for actinides to decay.

A third barrier might be a canister that could withstand even severe environmental conditions. A fourth barrier might be achieved by converting nuclear waste to a solid form that would itself be resistant to leaching. Almost 20 years of research has provided information on *vitrification,* which is conversion to glass. Conversion to a ceramic material is also under investigation. It has been suggested that radioactive waste could be propelled into outer space and forgotten forever. But the risks of getting the material safely off the earth are regarded as unacceptable.

Critics often argue that we do not know how to dispose of nuclear waste safely. A supportive view is that we have several choices and plenty of time to determine which approach is best and to develop better technologies. Meanwhile, spent fuel rods from nuclear reactors are merely stored in large pools of water, and the water absorbs the radiation and heat as the waste steadily decays.

Until the late 1970s it was assumed by many that spent fuel would be reprocessed. This means that the components of the spent fuel would be separated so that all fissionable isotopes (for example, ^{235}U and ^{239}Pu) and fertile isotopes (for example, ^{238}U) could be recovered and recycled into a reactor. One relevant sentiment has been phrased as, ''Plutonium was born in a reactor; it has to die in a reactor.'' This statement argues for recovering the plutonium and returning it to a reactor to be disposed of by fission *(11)*. In addition, there might be uses for some of the actinides, and the pure waste material that has no value could be concentrated and stored.

Various factors have prevented reprocessing from being developed for commercial use. The low price of uranium is one; it has made reprocessing economically uncompetitive. In addition, a serious concern always is the possible diversion of plutonium to unscrupulous individuals or governments. Plutonium is highly toxic and can be used to produce nuclear weapons.

Whether reprocessing ever becomes significant commercially in the United States remains to be seen. Political and environmental pressures against it make it highly unlikely for many years to come. Meanwhile, another approach may be developed for disposal of some dangerous isotopes. A technique known as accelerator transmutation of waste (ATW) is being studied. By this method, long-lived radioactive isotopes are bombarded by neutrons and undergo transmutation to form nonradioactive isotopes or short-lived radioisotopes that decay rapidly into a harmless form. The process is illustrated by the conversion of the fission product technetium-99, which has a half-life of 250,000 years, into nonradioactive ruthenium-100 *(12)*.

$$^{99}_{43}\text{Tc} + ^{1}_{0}\text{n} \rightarrow ^{100}_{43}\text{Tc}$$
$$t_{1/2} = 21{,}000 \text{ yrs}$$

$$^{100}_{43}\text{Tc} \rightarrow ^{100}_{44}\text{Ru} + ^{0}_{-1}\beta$$
$$t_{1/2} = 15.8 \text{ sec} \quad \text{nonradioactive}$$

11.8   Nuclear Bombs

GOALS: To appreciate the three designs of nuclear bombs.
To understand why a nuclear power plant is not a potential nuclear bomb.

Three types of nuclear bombs have been devised. The best known is a simple fission device that is commonly called the *atomic bomb,* which uses either ^{235}U or ^{239}Pu as fuel. The other two are the *hydrogen bomb* and the *neutron bomb,* which are fusion bombs.

Let us first consider the fission bomb. Two characteristics of this weapon best illustrate the design that allows release of such violent energy. One property is the purity of the fissionable material, ^{235}U or ^{239}Pu. Highly concentrated ^{235}U (approximately 97%) is necessary to avoid the absorption of neutrons by ^{238}U once the device is triggered. This ensures that repeated bombardment of fissionable isotopes will be an efficient, rapidly self-accelerating, supercritical process that quickly builds to produce a tremendous explosive yield.

The bomb dropped on Hiroshima was a uranium bomb, whereas a plutonium bomb was used at Nagasaki.

The other principal feature of the fission bomb is its **critical mass,** the amount of isotope required for the reactor to achieve the supercritical condition in which more than one product neutron becomes a bombarding neutron (Figure 11.1). The concept of critical mass can be understood by considering a hypothetical spherical sample of fissionable material. By the well-known relations

$$\text{Volume of a sphere} = \frac{4}{3}\,\pi r^3$$

$$\text{Surface area of a sphere} = 4\pi r^2$$

as r increases (corresponding to a greater mass), volume will increase faster than surface area. Therefore, the ratio volume/surface area increases as sample size increases. When the surface area is large, product neutrons can easily escape from it, and they do not again become involved in the reaction. However, as the volume of the sphere becomes larger, a greater percentage of the product neutrons cannot escape, and therefore, they become bombarding neutrons. When a certain critical mass is reached, the reaction can become rapidly self-accelerating, leading to a nuclear explosion.

It is necessary in designing a bomb to arrange that the fuel forms a critical mass, but not until exactly the right instant—the moment of detonation. For this purpose, there are two designs, illustrated in Figure 11.13. In the first design, a propellant (small explosive) displaces the subcritical mass at one end of the system very rapidly into the subcritical mass on the other end, producing a larger critical mass that explodes.

In the second design, a nonnuclear explosive is detonated, causing an implosion of the subcritical mass in the fissionable core (^{235}U or ^{239}Pu). Although the mass is not changed, the fissionable material is compressed. The compression results in a decrease in the surface area of the core and thus reduces the probability that a neutron will escape

DESIGN I

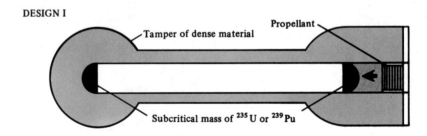

DESIGN II

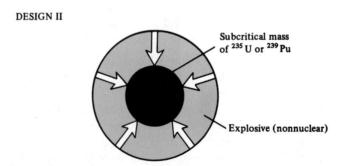

Figure 11.13   Designs of nuclear bombs *(13)*. (From Behrens, C. F. et al. *Atomic Medicine,* 5th ed.; Williams & Wilkins: Baltimore, 1969.)

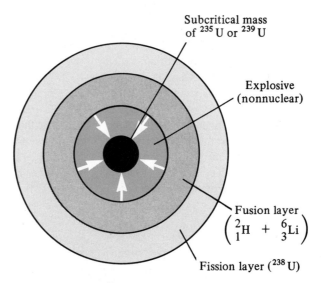

Subcritical mass
of ^{235}U or ^{239}U

Explosive
(nonnuclear)

Fusion layer
$\left(\begin{matrix} {}^{2}_{1}H & + & {}^{6}_{3}Li \end{matrix}\right)$

Fission layer (^{238}U)

Figure 11.14   The hydrogen bomb and the neutron bomb. The fission of ^{235}U or ^{239}Pu is triggered by a nonnuclear explosive that compresses the core of fissionable material. The energy and the neutrons released by the fission then initiate fusion involving deuterium, D, lithium, Li, and tritium, T, in the layer surrounding the core. Tritium is formed from the reaction of lithium with neutrons (Section 11.4). The blanket of ^{238}U increases the explosive yield as the uranium undergoes fission on being bombarded by neutrons released during fusion. Uranium-238 is normally regarded as a nonfissionable isotope. It does undergo fission, however, when bombarded by the extremely high-energy neutrons released during fusion. The neutrons released in an LWR or even the LMFBR do not have enough energy to cause fission of ^{238}U.

The neutron bomb (also known as the enhanced-radiation weapon) has the same basic design as the hydrogen bomb except that the ^{238}U blanket is omitted. Although this causes the explosive yield to be less, the release of neutrons is greatly enhanced. Neutrons are extremely damaging to all living things *(14)*.

from the core. The probability then becomes greater that each product neutron will become a bombarding neutron. Thus, the implosion of the fissionable core creates the conditions required for a supercritical process, as a subcritical mass is made critical.

The hydrogen bomb and the neutron bomb are pictured and explained in Figure 11.14. Both devices are triggered as in design II of the fission bomb (Figure 11.13). Fission results and triggers fusion. The hydrogen bomb is more damaging to the physical surroundings, whereas the neutron bomb (also called the enhanced-radiation weapon) is more damaging to living things.

SUMMARY

Transmutation is an important application of radiochemistry. It has led to the synthesis of many new elements and is a common way of producing radioactive isotopes for use in medicine and research. Transmutation involves bombarding a target nucleus with some sort of particle that is usually either accelerated or moderated to achieve reaction. Transuranium elements and other synthetic isotopes have been produced by transmutation.

Fission is the principle on which nuclear power plants operate. Breeder reactors function by producing fissionable fuel from fertile, nonfissionable isotopes. Nuclear fusion is the ultimate goal for nuclear power production, but many technological problems are yet to be solved. The energy output of the sun and the energy released by hydrogen bombs are the results of nuclear fusion.

The designs of nuclear power plants are quite varied. The light-water reactor (LWR) and the liquid-metal fast-breeder reactor (LMFBR) are the most popular. The probability (cross section) that a nucleus will capture a neutron determines whether an unmoderated (fast) or a moderated (thermal) reactor design is used. Other major variables are the coolant and moderator.

The problems of safety to the population and the environment are serious concerns in nuclear power generation. A nuclear power plant is clearly not a potential bomb, but the release of radiation due to loss of coolant is an ever-present danger that must be avoided by proper operation and design.

The disposal of radioactive wastes may be the most nagging problem that the nuclear industry has to deal with. Recycling (reprocessing), actinide burning, and accelerator transmutation of waste (ATW) are alternatives to burial in safe repositories.

The history-making accidents that occurred at TMI in 1979 and Chernobyl in 1986 have been studied thoroughly. Heat from the decay of fission products was the cause of the partial meltdown of the core at TMI. The event was triggered by faulty pumps and compounded by operator actions. Although the reactor was a total loss, no humans have or will likely ever suffer any ill effects from TMI. At Chernobyl, the fission process was allowed to run out of control, in part because of the reactor design that has a positive void coefficient. Operator errors also contributed greatly. The heat of decay of fission products caused radioactive materials to be liberated from the site for many days after the initial explosion. The accident cost many lives at the time and is expected to cause many cases of cancer in years to come.

PROBLEMS

1. Give a definition or example of each of the following.
   a. Coolant
   b. Moderator
   c. Transmutation
   d. Fission
   e. Fusion
   f. Breeders
   g. Plasma
   h. Tokamak
   i. Critical mass
   j. Supercritical
   k. Boiling water reactor
   l. Pressurized water reactor
   m. Heavy water reactor
   n. Cross section
   o. CANDU
   p. Negative void coefficient
   q. RBMK

2. Complete the equation for each of these nuclear processes.
   a. $^2H + {}^3H \rightarrow \alpha + ?$
   b. $^6Li\ (n, \alpha)?$
   c. $^{235}U + n \rightarrow {}^{94}Rb + {}^{140}Cs + ?$
   d. $^{27}_{13}Al\ (^2H, \alpha)?$
   e. $^{235}U + n \rightarrow ? + {}^{95}Kr + 3\,n$
   f. $^{238}U + n \rightarrow$
   g. Isomeric transition of ^{99m}Tc

   h. $^{99}Mo \rightarrow {}^{99m}Tc$
   i. $^{98}Mo \rightarrow {}^{99}Mo$
   j. $^{108}In\ (e, \gamma)?$
   k. $^{197}Au\ (n,\gamma)\ {}^{198}Au$
   l. $^{27}Al\ (n,p)\ {}^{27}Mg$
   m. Transmutation of ^{99}Tc to ^{100}Ru

3. A mixture of 1 g of ^{226}Ra (an α emitter) and a few grams of ^9Be makes an excellent source of neutrons (about 10^7 neutrons/sec). Write the nuclear reactions that account for this fact.

4. Why are moderators necessary for neutrons but not for electrons or other charged particles?

5. For sustained fission of ^{235}U, a moderator must be present. Why?

6. Compare the relative advantages of light water and heavy water as coolants. As moderators.

7. The letters LMFBR could become increasingly important when you turn on lights, appliances, and the like. What do the letters stand for and how does each component of the system, described in the name, differ from alternative systems?

8. Write an example of a fission reaction of ^{235}U leading to formation of:
   a. 2 neutrons + iodine-131
   b. 3 neutrons + krypton-85

9. What are the purpose and nature of control rods in nuclear reactors?

10. What is the source of the heat energy that could lead to the meltdown of the core of a nuclear reactor?

11. Compare the nuclear reactions that were used to produce ^{239}Pu by transmutation (Section 11.2) and by breeding (Section 11.3).

12. Explain how control rods would be handled to take a fission reactor from subcritical to critical.

13. What is the form of the potentially useful energy that is released by fusion?

14. Initiation of fusion requires a large input of energy. How does fusion become self-sustaining?

15. Compare the events at Chernobyl and TMI in terms of:
    a. Operator error
    b. The short- and long-term consequences
    c. The extent of damage to the reactors
    d. The amount and type of radioactive materials that escaped
    e. The nuclear processes that caused loss of control

16. Why was the TMI plant especially vulnerable to a meltdown at the time of the accident? How did the core become uncovered?

17. What is the purpose of giving potassium iodide tablets in the event of a nuclear disaster?

18. What is the major design flaw in the RBMK? Why is this so critical?

19. What is currently being done with spent fuel from U.S. nuclear power plant reactors?

20. Suppose you were the ruler of a country with little wealth and only one energy resource—a mountain of uranium. How would you handle the problem of disposal of radioactive waste from the nuclear power plant that you are forced to operate? Is this a ridiculous question or does it have some relevance in the real world?

21. What are the conclusions of the Inhaber report?

22. Explain the two designs of atomic bombs shown in Figure 11.13.

23. Compare the hydrogen and neutron bombs.

24. Two samples of ^{239}Pu have radii of 5 cm and 25 cm. Calculate volume/surface area for each sample, and explain why one is more likely to be suitable for use in an atom bomb.

References

1. Keller, C. *Radiochemistry;* John Wiley & Sons: New York, 1988; Chapter 7.

2. Kauffman, G. B. "Beyond Uranium." *Chemical and Engineering News,* November 19, 1990, pp. 18–29.

3. Anspaugh, L. R.; Catlin, R. J.; Goldman, G. "The Global Impact of the Chernobyl Reactor Accident." *Science* **1988,** *242,* 1513–1519.

4. Ahearne, J. H. "Nuclear Power After Chernobyl." *Science* **1987,** *236,* 673–679.

5. Marples, D. R. *Chernobyl and Nuclear Power in the USSR;* St. Martin's: New York, 1986; pp. 181–183.

6. Atwood, C. H. "Chernobyl—What Happened." *Journal of Chemical Education* **1988,** *65* (12), 1037–1041.

7. Sweet, W. "Chernobyl—What Really Happened." *MIT Technology Review,* July 1989, pp. 43–52.

8. Devell, L.; Tovedal, H.; Bergstrom, U.; Appelgren, A.; Chyssler, J.; Andersson, L. "Initial Observations from the Reactor Accident at Chernobyl." *Nature* **1986,** *321,* 192–193.

9. Marshall, E. "Recalculating the Cost of Chernobyl." *Science* **1987,** *236,* 658–659.

10. Inhaber, H. *Risk of Energy Production.* Atomic Energy Control Board: Ottawa, 1978.

11. Steinberg, M. "Nuclear Fuel Cycle and Special Problems in Nuclear Energy." *Abstracts, American Chemical Society, 4th Chemical Congress;* New York, 1991.

12. Amato, I. "Aid to Alchemy." *Science* **1992,** *256,* 443.

13. Behrens, C. H.; King, E. R.; Carpender, J.; Waste. J. *Atomic Medicine,* 5th ed.; Williams & Wilkins: Baltimore, 1969; Chapter 2.

14. Kaplan, F. M. "Enhanced-Radiation Weapons." *Scientific American* **1978,** *238* (5), 44.

Energy: The Problems and Some Conventional Solutions

12

Our most common energy resources are oil, coal, and natural gas, which are used by industry and consumers. Even for countries with adequate supplies of coal or natural gas, the reality is that we cannot run our cars—or buses, planes, trains, and trucks—on these fuels. Many modes of transportation, and many heating systems, use liquid fuels, which thus play a critical role in world economies.

For some countries, the highest priority in energy planning is energy independence in order to retain economic and political stability. For others, the energy crisis is primarily one in which we try to develop technologies using alternatives to oil without destroying the environment.

The United States was involved in a war in the Middle East after Iraq overran Kuwait. The war was explained as a humanitarian effort, but it was also a war for oil for the West. The consumer tends to judge the cost of energy when handing over cash at the gas pump, but what about the cost of the war effort? Should we pay for that when we fill our cars? Gasoline prices soared when there was a possibility that oil supplies might be threatened. When the crisis was resolved, the prices returned to more normal levels. The cost of the war is, in reality, a part of the cost of being able to buy gasoline at a reasonable price. As one author described the matter, "Using the most conservative estimate, the U.S. Department of Defense spends about $23.50 on each barrel of imported oil to safeguard oil supplies from the Persian Gulf" (1). Thus it is clear that petroleum is a precious commodity for countries that need it.

Natural gas is a desirable alternative source of energy, a truly premier fuel for uses such as generating electricity, heating homes, and supplying energy for industry. It has the advantage over other resources of ease of distribution (by pipeline). It has

247

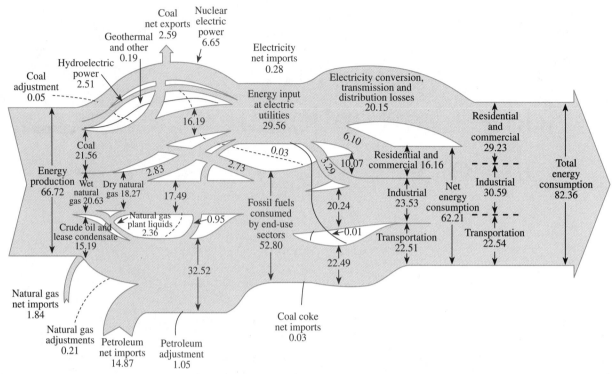

Figure 12.1   Total energy flow for 1992. All numbers are given in quadrillion Btu (Q). See Section 12.3 for a discussion of units. Note: both "wet" and dry natural gas are predominantly methane, but wet natural gas also contains ethane, propane, butanes, and pentanes. (From *Annual Energy Review 1992*; Department of Energy/Energy Information Administration: Washington, DC, 1993.)

a relatively low environmental impact except for the production of carbon dioxide, which contributes to the greenhouse effect (Section 14.4). But burning of all fossil fuels releases carbon dioxide.

Coal is abundant in some parts of the world, including the United States, but it has some serious drawbacks as a source of energy. Mining is a dirty and hazardous process, so coal can have a major environmental impact even before it is used. Burning coal makes its environmental problems most evident. But use of this fuel is likely to rise for reasons to be considered in this chapter.

Electricity (Section 13.12) appears to be a very clean form of energy until we recall that it is derived mostly from coal and nuclear fission. Each has environmental consequences that are regarded by many as undesirable. Solar energy, geothermal energy, hydroelectric power, wind power, and tidal power are other alternatives for supplying energy; and "negawatts," or energy conservation, is the cry most often heard from many of the staunchest environmental protectionists *(2)*. All of these topics are discussed in this and the following chapter. The sources and uses of the various energy forms are shown in Figure 12.1, including the energy used and lost in producing electricity (Section 12.2).

12.1   Primary and Secondary Energy Sources

GOAL: To appreciate the distinction between primary energy resources and secondary energy forms.

The fossil fuels (oil, gas, coal), nuclear energy, falling water, geothermal, and solar energy are classified all as *primary* energy resources. We must recognize that electricity is a *secondary* form of energy, derived from primary sources of energy. Fuels derived from coal, such as synthetic natural gas (syngas) and synthetic gasoline, as well as alcohol fuels from plant materials or from coal or natural gas, also are secondary energy forms.

12.2   Energy Conservation and Conversion

GOALS: To be aware of the unavoidable inefficiency of the generation of electric power. To see how cogeneration can be used to improve efficiency.

One way to deal with an energy shortage is to use less energy; that is, practice conservation. Much energy is wasted due to the day-to-day habits of individuals, and to industrial and commercial practices, particularly when energy costs are low. The shortages and high prices of the 1970s and early 1980s caused everyone to try to conserve energy. The success of conservation, despite continued economic growth, is indicated by the pattern of energy use in Figure 12.2. The units on the vertical axis of this figure will be explained later in

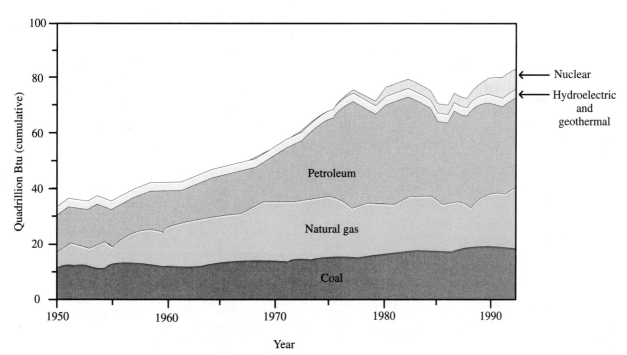

Figure 12.2   United States consumption of energy by source, 1949–1992. See section 12.3 for an explanation of the units of energy appearing on the vertical axis. The peaks at 1973 and 1979 correspond to times of high demand, which led to the oil shortages; this caused fuel prices to soar, resulting in conservation. In 1986, crude oil prices dropped sharply, causing another trend of rising energy consumption. (From *Annual Energy Review 1992;* Department of Energy/ Energy Information Administration: Washington, DC, 1993.)

Table 12.1   Efficiency of intercity freight transport

Form	Energy Consumption Rate (Btu/ton-mile)	Efficiency (ton-mile/gal)
Pipeline (liquid fuels)	450	280
Railroad	670	188
Waterway	680	185
Truck	2,800	45
Airplane	42,000	3

this section. In 1986, crude oil prices dropped sharply and have remained relatively low. This led to increased usage, which has leveled off in recent years.

The well-known conservation methods include insulation, improved efficiency of indoor air cooling and heating, more efficient appliances, a shift to more efficient forms of transportation, and other strategies that simply translate into making do with less. Table 12.1 gives some facts on energy efficiencies in transportation. By far the most frequent means of transport, generally used for convenience or speed, are the least efficient. Figure 12.3 shows the impact of smaller, more efficient automobiles on the consumption of gas-

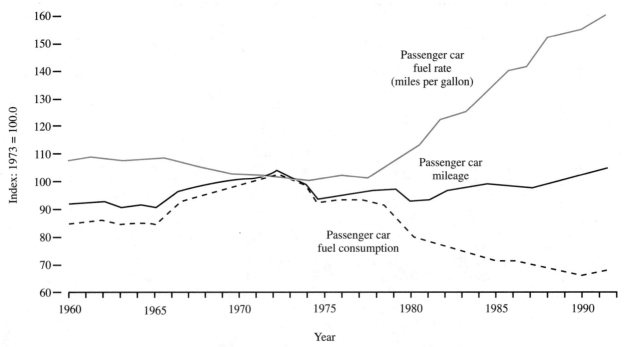

Figure 12.3   Passenger car efficiencies, 1960–1991. The fuel rates, total mileage, and fuel consumption are all expressed as percentages relative to the levels in 1973, the year of the first oil shortage. (From *Annual Energy Review 1992;* Department of Energy/Energy Information Administration: Washington, DC., 1993.)

oline beginning in the early 1970s. Note how the most significant changes began as a consequence of the 1979 oil shortage and subsequent price rise.

Conservation means more than just doing with less, however. It also includes recycling products made from aluminum, steel, glass, paper, and plastics. Each of the processes involved in making such products is very energy intensive; that is, they consume large amounts of energy. Recycling the materials is generally less energy intensive.

Conservation can also mean more efficient use of energy resources by applying new technologies. Many of these developments are not really new, but they have not been used commercially while oil has been cheap and abundant. Fortunately, they can be taken off the shelf and applied as necessary.

One new area of technology is the use of urban and industrial waste as a source of energy. Tremendous quantities of waste materials are carted off to landfills for burial. Some of them have a significant fuel value that can be released by burning or other processes.

Cogeneration

Cogeneration includes methods in which energy from coal, oil, or any other primary source supplies both heat and electricity. In such a system, the energy available in the fuel can be more completely extracted. Techniques called *topping, bottoming,* and *magnetohydrodynamics* (MHD) use fuels more efficiently and thus conserve them.

To understand the strategy of cogeneration, we have to acknowledge the importance of the science of thermodynamics and its relation to generating electricity.

The term *thermodynamics* comes from the Greek words for heat and power. The science involves heat, work, and energy, and the changes they produce.

A steam turbine to generate electricity can be driven by steam from the core of a nuclear reactor or by steam from burning coal, oil, or any other fuel. When a turbine is turned, the process is a form of work (mechanical energy). This mechanical energy is converted to electrical energy, which can be converted to heat (say for space heating) or back to mechanical energy (to drive electrical appliances or machinery, for example). Similarly, in a steam engine, expanding steam causes a piston to move, thereby doing mechanical work.

The laws of thermodynamics tell us two things about converting heat energy from steam to work:

1. The conversion of heat to work cannot be 100% efficient. A sizable portion of the original heat energy that flows to the turbine or engine is wasted.

2. The efficiency of converting heat to work increases as the heat temperature increases. In other words, the hotter the steam, the more efficient the conversion.

The quad (Q) is short for 1 quadrillion Btu or 10^{15} Btu. See Section 12.3 for details.

In converting mechanical energy to electricity in the steam turbine, some of the energy in the steam will be wasted. Unfortunately, the amount wasted exceeds the amount converted to electricity. Only the most efficient electrical generators have efficiencies of even 35% to 38%. Most have efficiencies of one-third or less, so that two-thirds or more of the coal, oil, or any other primary energy source burned to generate electricity is lost as waste heat. According to the laws of thermodynamics, most of the waste is unavoidable. Of the more than 80 Q of energy consumed in the United States, approximately 24% is lost in conversion to electric power and 1% in transmission of electricity (Figure 12.1). A more detailed discussion of electricity can be found beginning in Section 13.6.

Compare the efficiency of several forms of energy conversion in Table 12.2. When there is a need to conserve energy and to try to use a resource as efficiently as possible, the data are not encouraging. That is why scientists have suggested that efforts should be

Table 12.2   Efficiency of some energy conversion devices

Device	Efficiency (%)
Dry-cell flashlight battery	90
Home gas furnace	85
Storage battery	70
Home oil furnace	65
Small electric motor	62
Fuel cell	55
Steam power plant	38
Diesel engine	38
High-intensity lamp	32
Automobile engine	25
Fluorescent lamp	22
Incandescent lamp	4

directed toward conserving some of the energy wasted when generating electricity. To this end, topping, bottoming, and MHD are receiving much attention.

Topping and Bottoming

Topping is a form of cogeneration in which a primary energy resource is first applied to generating electricity and then the waste heat is used directly. This approach allows for efficiency as high as 80%. Recognizing that the generation of electricity is only about one-third efficient, much of the other two-thirds of the heat energy is used directly for such things as space heating. Topping is practical for district heating in apartments, shopping centers, factories, universities, greenhouses, and other facilities where the heat can be transported short distances and used directly.

Bottoming is a variation of the same general procedure. The primary energy source is burned to produce high-temperature heat energy for an industrial process. Hot exhaust gases still contain enough energy to make steam for generating electricity.

Magnetohydrodynamics

Another promising technology is still in development. Magnetohydrodynamics uses two processes for generating electricity. It burns fuel (most likely coal) and passes hot gases produced in combustion through a magnetic field. If the temperature of the combustion gases is maintained above 2000 °C, the gases will ionize. If a small amount of potassium (as potassium carbonate) or other alkali metal salt is added, the concentration of ions in the gas stream will be great enough that the movement of these ions through the magnetic field will generate an electric current (Figure 12.4). This technique can be adapted for use in conjunction with a second device for generating electricity, namely, a steam or gas turbine.

The chief advantage of MHD is increased efficiency in generating electricity. The efficiency of converting heat energy to the work of driving a turbine is low. In MHD, the energy released by burning a fuel is converted directly to electricity. Since the hot gases are not used to turn the blades of a steam or a gas turbine, much less energy is lost. The hot exhaust gases can then be sent to a conventional turbine generator to produce additional electric power.

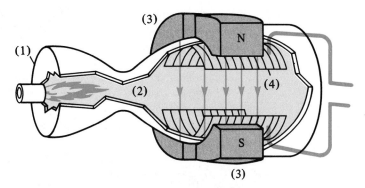

Figure 12.4   A schematic of an MHD generator. Powdered coal burns fiercely in a combustion chamber (1). Squirting through a nozzle (2) at supersonic speed, combustion gases stream between magnets (3), yielding an electric current taken off by electrodes (4). (Reprinted with the permission of Macmillan College Publishing Company from ENERGY by Gerard M. Crawley. Copyright © 1975 by Gerard M. Crawley.)

The MHD could be described as a topping device, since it is first used to generate electricity. But since the second component is also used to generate electricity, the entire system is described as a *combined cycle.* It is expected that total efficiency will approach 60%. Several nagging technological problems still prevent the commercial use of MHD, but experts believe these could be overcome by the end of the 20th century.

12.3   Energy Units and Use

GOALS: To consider some of the common units of energy and power.
        To learn how to make unit conversions.
        To understand the importance of peaking.

To discuss energy, we must be familiar with the common ways of describing its quantities, including the energy content of various fuels. One common standard is the *British thermal unit,* abbreviated **Btu,** which is defined as the amount of energy required to raise the temperature of 1 lb of water by 1 °F. The Btu describes the energy that furnaces and air conditioners deliver.

Another unit of energy is the *calorie,* abbreviated cal. One calorie is the amount of energy required to raise the temperature of 1 g of water by 1 °C. Although scientists commonly use calories as a unit of energy, nutritionists and average consumers normally express the energy content of foods in Calories (note the capitalization). One Calorie, sometimes called the large calorie, equals 1000 calories. Scientists commonly use the term *kilocalorie* (kcal) instead of Calorie.

The relationship between calories and British thermal units is

$$1 \text{ Btu} = 252 \text{ calories} = 0.252 \text{ kcal}$$
$$4(1 \text{ Btu}) = 4(0.252 \text{ kcal}) = 1.008 \text{ kcal}$$

In other words, 1 kcal equals approximately 4 Btu.

The *joule* (J) is related to the Btu and the calorie as shown:

$$1 \text{ Btu} = 1055 \text{ J} = 1.055 \text{ kilojoules (kJ)}$$
$$1 \text{ cal} = 4.184 \text{ J}$$

Many scientists consider the joule to be the only internationally acceptable unit of energy; even the British have abandoned the Btu *(3).* But most of the literature (books, articles, etc.) still use other units, particularly the Btu. Therefore, so that you can more easily read current literature and become conversant with the topic, we will use the familiar terms and not the joule. At the end of this section, we relate the joule to all of the common units of energy so that you have a convenient reference if you have to convert from one unit of energy to another.

Two other units often appear in discussions of energy: the *horsepower* and the *watt.* These are not units of energy. They are units of *power,* and describe how fast energy can be delivered and used up. One **watt (W)** equals 3.412 Btu per hour (Btu h^{-1}). One **horsepower (hp)** equals 746 W.

Almost all electrical appliances and light bulbs are labeled to show how much power they use. If a 100-W light bulb is burned for 1 hour, it consumes 100 watt-hours (W-h) of electricity. The **watt-hour** is thus another unit of energy, used only to describe electrical energy; however, since it is such a small amount of energy, we more often use the unit *kilowatt (kW)* in descrbing electrical *power,* and the *kilowatt-hour (kW-h)* to describe electrical *energy.* Thus, a 100-W light bulb would consume 0.1 kW-h in 1 hour, or 1 kW-h if left on for 10 hours.

By comparison, a typical room air conditioner might have a power demand of 1500 W, which means that there would be a flow of 1.5 kW-h of electricity for every hour the unit is cooling. Notice that this corresponds to consumption 15 times as great as what a 100-W light bulb uses during each hour of operation.

An additional unit is useful for describing very large quantities of energy. The *quad (Q)* is short for one quadrillion Btu or 10^{15} Btu. It is commonly used to describe amounts of energy consumed by large segments of the population. For example, the total consumption of primary energy in the United States in 1958 was about 40 Q. Current levels exceed 80 Q (Figure 12.1).

One final unit that often enters into discussions of energy is the *barrel* (bbl). One barrel equals 42 gal. The energy content of a barrel of crude oil is approximately 5.8 million Btu.

Interconverting Energy Units

For some perspective on the relative magnitude of the various energy units, examine the following list. The quad is the largest unit of energy that you are likely to confront. The other units, listed in order of decreasing size, are compared with the quad.

$$
\begin{aligned}
1 \text{ quadrillion Btu} &= 8.79 \times 10^{11} \text{ kW-h thermal (kW-h}_t) \\
&= 2.93 \times 10^{11} \text{ kW-h electric (kW-h}_e) \\
&= 2.52 \times 10^{14} \text{ kcal} \\
&= 1.00 \times 10^{15} \text{ Btu} \\
&= 1.06 \times 10^{15} \text{ kJ} \\
&= 2.52 \times 10^{17} \text{ cal} \\
&= 1.06 \times 10^{18} \text{ J}
\end{aligned}
$$

Although the kilowatt-hour is the common unit of electrical energy, it can also be used to measure heat (thermal energy). Due to the inefficiency of converting heat to electricity (Section 12.2), it is sometimes desirable to emphasize the inefficiency by acknowledging that only about 1 kW-h of electrical energy (1 kW-h$_e$) can be obtained from 3 kW-h of thermal energy (3 kW-h$_t$), assuming an efficiency of 33% for converting heat energy to electrical energy.

Using the numerical relations between the quad and the other units, you can convert units back and forth. Since quantities that equal the same quantity are equal to each other, any of the expressions of energy equal any of the others. An example follows in which the mathematical relation between electric kilowatt-hours and kilojoules is solved in two ways. The appropriate equalities can then be used to convert units as required.

$$2.93 \times 10^{11} \text{ kW-h}_e = 1.06 \times 10^{15} \text{ kJ}$$

$$1 \text{ kW-h}_e = \frac{1.06 \times 10^{15}}{2.93 \times 10^{11}} \text{ kJ} = 3.62 \times 10^3 \text{ kJ}$$

$$1 \text{ kJ} = \frac{2.93 \times 10^{11}}{1.06 \times 10^{15}} \text{ kW-h}_e = 2.76 \times 10^{-4} \text{ kW-h}_e$$

Peak Demand

At some times of the day and in some seasons, demands for energy peak at unusually high levels. Thus it is important in evaluating alternative forms of energy, particularly as sources of electrical energy, to know how they might help to meet peak demands. Such demands occur only a few weeks of the year, and are concentrated in a few hours of the day. Because of air conditioning, most peak demand periods are during the summer months, but utilities serving the far northern states and those that aggressively promote electric heating often have winter peaks.

Electric utilities must have the generating capacity to meet the peak demand for a few hundred hours a year, even though much of the capacity is dormant most of the time. They use a variety of methods to generate electric power during greatest use. Diesel engines and jet engines may be used to drive generators. They are relatively expensive to operate, however, and their liquid fuels are derived from precious petroleum.

Utilities must maintain large overcapacities for several reasons: peaking, the anticipation of increased future demand, and the long time required to plan and build power plants (more than 10 yrs for a nuclear power plant). When alternative energy sources are considered as replacements for nuclear and fossil fuels in generating electricity, all these factors are considered. One of the big obstacles to solar energy is the unpredictability of the weather; cloudy weather during peak demand times would only aggravate matters. This problem and some possible solutions are discussed in Section 13.11.

12.4   The Use of Fossil Fuels

GOAL: To understand why energy crises occurred in the past and may again.

The entire world has come to depend heavily on three fossil fuels: coal, natural gas, and oil (petroleum). Coal was available and abundant before the other two came into prominence, but it has a serious environmental impact. Coupled with the great convenience and low cost of oil and gas, coal lost favor in many applications. This section looks at the facts regarding the supplies of oil and gas and shows why increased dependence on coal is likely (Section 12.5).

Petroleum

Figures 12.5 and 12.6 trace the components of the oil picture that led to shortages, triggered serious attempts at conservation, and will surely provide concerns for the future.

Figure 12.5 shows the steady rise in production and imports of petroleum. The rise actually began in about 1910 and continued without a significant break until the 1973 Arab

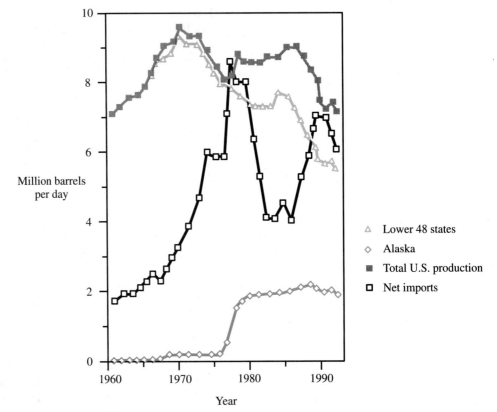

Figure 12.5 Crude oil production and imports, 1961–1992. (From the U.S. Department of Energy.) Note the increasing importance of oil production in Alaska, as less is produced in the lower 48 states. But far less is produced in Alaska than is imported from other countries.

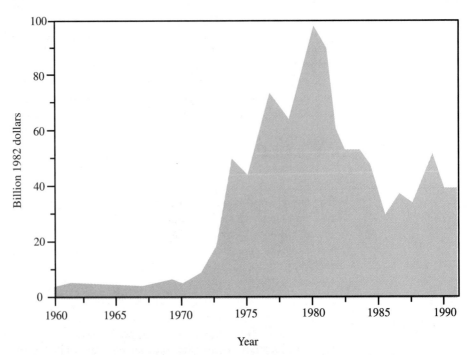

Figure 12.6 Value of fossil fuel net imports, 1959–1992. (From *Annual Energy Review 1992;* Department of Energy/Energy Information Administration: Washington, DC, 1993.)

oil embargo. After a brief downturn of demand, imports again sharply increased and led to an even more severe crunch in 1979, resulting from political turmoil in Iran. By 1982, dependence on imported oil had decreased to below 1972 levels as costs remained high. Imports remained at low levels until 1986 when the price of crude oil decreased sharply. Saudi Arabia increased oil production in 1985, and from that point on, the dependence on imported oil again rose. Largely due to the availability of cheaper foreign oil, the supply of domestic oil decreased in the late 1980s. Oil wells that were only marginally profitable were shut down, and exploration for new domestic oil became a lower priority. In the mid 1990s, the balance between imported and domestic crude oil is close to what it was in 1979. Thus the United States has again become quite vulnerable to an oil crisis *(4)*.

It is also increasingly difficult to recover domestic oil reserves. It is not that U.S. reserves are about to run out, but that much of the easily recoverable oil in the lower 48 states is now gone. Thus the increased cost of producing domestic oil makes it difficult to compete with the large-volume, low-cost production that is possible elsewhere, especially in the Middle East. To illustrate the magnitude of the problem, consider that about 100 oils wells in Saudi Arabia produce more oil than 180,000 wells in the United States *(5)*. The high level of imports, which has a negative balance of export-import payments for the United States, is seen most clearly in Figure 12.6. The contribution of oil, including imports, and its place in the overall scheme of energy resources is shown in Figures 12.1 and 12.2.

Natural Gas

Natural gas is an important component of energy supply and consumption. When oil prices soared in the early 1980s, the price of natural gas followed as demand increased. In the past, natural gas provided 20% of the primary energy for generation of electricity. By 1986 it accounted for only 9%, and the figure has remained fairly constant *(6)*.

Natural gas is collected from subterranean deposits, put under pressure, and piped nationwide for use as home heating fuel and in industrial applications (Section 6.1).

Natural gas is predominantly methane (CH_4). It offers some distinct advantages over other fuels, although wide price swings make some users (both home and industrial) uneasy about developing renewed heavy dependence on this fuel. Nevertheless, supplies are abundant, its environmental impact is lower than that of most other fuels, and new technologies have become available for more efficient use.

Of these factors, the environmental impact is the most compelling. Natural gas contains virtually no sulfur or sulfur compounds, which contribute to acid rain (Section 14.3). Coal and some oils are high in sulfur. Natural gas has the lowest percentage of carbon of the fossil fuels and therefore contributes the least amount of carbon dioxide per unit of energy. This is important, since CO_2 is regarded as the major contributor to the greenhouse effect (Section 14.4). Gasoline and coal produce 40% and 85% more CO_2, respectively, for the same amount of thermal energy delivered *(7)*. Coal burning also releases ash, contributing to smog, eye irritation, and other environmental effects.

12.5   Coal: Supply and Demand

GOAL: To understand the origin and ranks of coal.

The primary energy resource most often mentioned as the solution to our long-range energy problems is coal. Of the 21.55 Q of coal used to produce energy in the United States in 1991, 16.06 Q produced electric power. The electric utilities also consumed 6.54 Q of nuclear energy. The mix of nuclear and coal makes up about 76% of the energy, with the rest coming from hydroelectric power, geothermal, natural gas, and petroleum (Figure 12.1). As demands for electric power rise and pressures mount against nuclear power, coal is likely to gain increased importance.

Coal was once largely avoided as dirty, undesirable, and inconvenient. In many ways, it still is, but our awareness and concern for the environment virtually ensure that we will use it in the cleanest ways possible. Problems such as mine subsidence, often happening years after underground mining, acid mine drainage, which pollutes rivers and streams (Section 14.7), abandoned mine fires (which cost Pennsylvania $60 million in 1960 alone), soil erosion, reclamation costs, and black lung disease for miners are old, well-known problems that remain from years past.

Most estimates indicate that the world's supply of coal will last through as much as 600 years of demand. Fortunately, technologies are available that lessen its negative environmental consequences. We will consider the history and nature of coal, its uses, the sulfur problem, and the availability of gases and liquids derived from coal.

Coal was the fuel that bridged the gap between the period of wood as the major fuel source and the heavy use of petroleum and natural gas beginning in the 1940s. It may even be appropriate to regard petroleum and natural gas as the energy sources that are bridging the gap between two periods of coal as the predominant fuel—the past (1890–1940) and the future.

Origin of Coal

A chemical formula cannot be written to describe coal. Coal actually exists in many forms, and even two samples from a single location may not have the same chemical composition.

Like crude oil and natural gas, coal is a fossil fuel. This means that it is largely derived from remains of living things, mostly plant material, that underwent chemical and geological alterations over tens or hundreds of millions of years. The geochemical events, known as *coalification,* occurred in a series of three steps. (1) After plants died they underwent chemical decay to form a product known as *peat.* (2) Over many years, thick layers of peat formed, particularly in swampy, wooded areas where underbrush and trees died and decayed under the influence of bacterial microorganisms and fungi. (3) Peat was converted to coal by geological phenomena such as land subsidence, flooding, and mountain building. These earth disturbances often buried the peat to great depths, so that (4) it was subjected to great pressures and temperatures. As a consequence of the changing geological environment, it underwent gradual chemical changes that characterize the transition to the various types of coal.

Since coal was formed from plants that grew many millions of years ago, and since these plants grew by using the energy of sunlight in photosynthesis, it is reasonable to say that coal burning releases solar energy that was locked away long ago.

Ranks of Coal

The main types of coal are *anthracite* and *bituminous.* These and other substances like them are ranked according to various properties including fuel value. The most mature and highest-ranking is anthracite, often called *hard coal.* It has an energy content in the range of 14,000 to 16,000 Btu/lb. These and all other energy content values are for ash-free coal. The average coal used in the United States is at least 15% ash. Bituminous coal, sometimes called *soft coal,* may have an energy content as low as 10,500 Btu/lb, although much of it has an energy content in excess of 15,000 Btu/lb, higher than some anthracite coals.

The distinction between anthracite and bituminous coals is not a simple difference in energy content, however. Consider how coal behaves when it is heated to high temperatures, typically 900 °F in the absence of air, without burning. This process is called *destructive distillation*. By excluding oxygen and providing heat from an external source, a series of gases and liquids is released from the coal. These substances are collectively known as *volatile matter*. They are derived from resins and tarry materials that were present in trees and other plant material that were converted to coal. In older, harder, anthracite coal, these tars have matured into coal due to exposure to pressure and heat in the earth.

In bituminous and lower-ranking coals, the tars were never exposed to conditions that permitted the complete maturing, or coalification. Thus, bituminous coals differ from anthracite coals primarily in the amount of volatile matter; this volatile matter has a range of 0% to 14% for anthracite and may exceed 40% for some bituminous coals. The residual that remains when the volatile matter is removed is called *char*.

Table 12.3 shows the categories common in classifying coal. The column headed Fixed Carbon really describes the amount of char, since char is nearly pure carbon. The percentages of fixed carbon plus volatile matter total 100% for dry, ash-free coal. As you can see, the table gives no energy content (tabulated as calorific values) for anthracite and for the highest-ranking bituminous coals. These coals are differentiated on the basis of volatile matter and fixed carbon.

When coal is burned, volatile matter is released and burned as well. In fact, it ignites readily, so that softer coals are easier to burn than anthracite. Since the combustion of the volatile matter releases energy, the energy content of bituminous coal is often as high as that of some anthracite coals or even higher. Beginning with some of the middle-ranking bituminous coals and going through the lower ranks to *subbituminous* coal and low-energy *lignite,* distinction among them is made on the basis of energy content instead of volatile matter.

Of the almost 1 billion tons of coal produced in the United States, less than 1% is anthracite, 67% is bituminous, 23% is subbituminous, and 9% is lignite *(6)*. Worldwide, Russia, the United States, and China together have 87% of the world's coal deposits *(8)*. Most other countries do not have abundant supplies of coal.

The four ranks of coal have their principal deposits in different locations in the United States (Figure 12.7). Figure 12.8 gives more detailed information on the amounts available east and west of the Mississippi River (more about this later). The only significant deposits of anthracite coal are in northeastern Pennsylvania. They are considered part of the large bituminous deposits that extend the length of the Appalachian Mountains where the extreme geological environment necessary for formation of anthracite existed. In fact, it is currently believed that the anthracite coal was formed during a geological period when the land mass that is now Africa collided with North America. See Figure 12.9 (p. 262) for details. Recovering this coal usually requires deep mining.

Ash and Particulate Removal

All the data in Table 12.3 are given on a dry, mineral-matter-free basis. Different coals have different amounts of moisture and minerals trapped within them. The moisture vaporizes on burning, but minerals do not burn, so they are left behind as ash, and ash constitutes an environmental problem. Some *fly ash* is released into the atmosphere when combustion gases (mostly carbon dioxide and water vapor) carry it up a smokestack. Also called *particulate matter* or simply *particulates,* it can be a serious health hazard when inhaled. Electric utilities routinely remove the particulates in equipment called *baghouses,* which function much like giant vacuum cleaner bags. *Electrostatic precipitators,* which are used more than baghouses, remove particulates by attracting them to charged electrodes. Moving particles carry static electricity, so they collect on the electrodes and can be removed.

Table 12.3   Classification of coals by rank

Class	Group	Fixed Carbon Limits, % (dry, mineral-matter-free basis)		Volatile Matter Limits, % (dry, mineral-matter-free basis)		Calorific Value Limits, Btu/lb (moist, mineral-matter-free basis)[1]	
		Equal to or greater than	Less than	Greater than	Equal to or less than	Equal to or greater than	Less than
I. Anthracitic	1. Meta-anthracite	98	...	...	2	...	...
	2. Anthracite	92	98	2	8	...	...
	3. Semianthracite	86	92	8	14	...	...
II. Bituminous	1. Low-volatile bituminous coal	78	86	14	22	...	...
	2. Medium-volatile bituminous coal	69	78	22	31	...	...
	3. High-volatile A bituminous coal	...	69	31	...	14,000[2]	...
	4. High-volatile B bituminous coal	...	...	...	...	13,000[2]	14,000
	5. High-volatile C bituminous coal	...	...	...	...	11,500	13,000
						10,500	11,500
III. Subbituminous	1. Subbituminous A coal	...	...	...	...	10,500	11,500
	2. Subbituminous B coal	...	...	...	...	9,500	10,500
	3. Subbituminous C coal	...	...	...	...	8,300	9,500
IV. Lignitic	1. Lignite A	...	...	...	...	6,300	8,300
	2. Lignite B	...	...	...	...	...	6,300

[1] *Moist* refers to coal containing natural, inherent moisture; it does not include visible water on the surface of the coal.

[2] Coals having 69% or more fixed carbon on the dry, mineral-matter-free basis are classified according to fixed carbon, regardless of calorific value.

Reprinted with permission from the *Annual Book of ASTM Standards*, Part No. 260, p. 388. Copyright, American Society for Testing and Materials, 1916 Race St., Philadelphia, PA 19103.

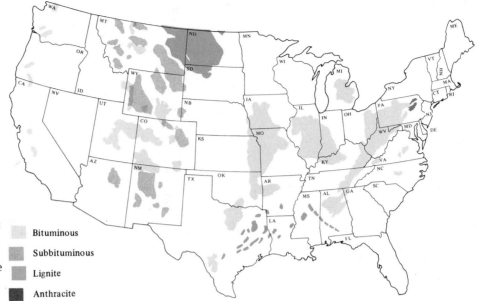

Figure 12.7   Main coal deposits of the United States, outside of Alaska and Hawaii, showing the kinds of coal. (Adapted courtesy of the National Coal Association.)

- Bituminous
- Subbituminous
- Lignite
- Anthracite

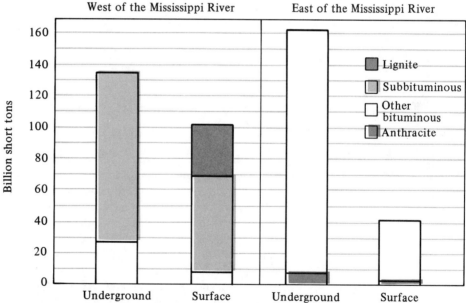

Figure 12.8   Demonstrated coal reserve base by rank, potential method of mining, and region. Approximately 52% of the demonstrated coal reserve base is high-grade bituminous, 38% is subbituminous, 8% lignite, and 2% anthracite. About 85% of high-grade bituminous coal and virtually all anthracite coal exist east of the Mississippi River, whereas most of the subbituminous coal and lignite is in the West. One-third of the demonstrated reserve base, or 141 billion tons, is in beds so close to the surface or so thick that underground mining is considered impractical. Most of this reserve base (nearly three-fourths) is in states west of the Mississippi River, but over half the base that can be mined by underground methods is east of the Mississippi River. (From the U.S. Department of Energy.)

Figure 12.9 At the end of a geological period known as the Carboniferous Period, it is theorized that the continents formed one supercontinent known as Pangaea. On a geological time scale, the continents slammed together with such force that the rocks folded and buckled. The patterns of folding can still be seen in rocks cut to build highways crossing the Anthracite region of eastern Pennsylvania. The Appalachian mountains, which now extend from Pennsylvania southwestward into northern Alabama, formed during this period. The heat and pressure generated by the folding produced anthracite. The subsequent fragmentation and drift of Pangaea resulted in the continents as we know them today; the precursors of North America, South America, and Africa are evident from the drawing *(8)*.

12.6 Carbonization and Coal By-Products

GOAL: To know how coke is made and how it is used in producing iron and steel.

Heating coal without enough oxygen to permit burning is called *carbonization,* since the residue or char is nearly pure carbon. The volatile matter that is released consists of water, coal tar, light oil, ammonia (NH_3), and coal gas. Both coal tar and light oil are good sources of chemicals for making drugs, cosmetics, detergents, plastics, fibers, and additives for motor fuels. Ammonia is also used to make chemicals such as nitric acid, HNO_3, ammonium sulfate, $(NH_4)_2SO_4$, and diammonium phosphate, $(NH_4)_2HPO_4$. Ammonium salts and ammonia itself are important fertilizers (see Chapter 15). Coal gas can be used for heating and for making chemicals.

Char also is valuable. When bituminous coal with a volatile matter content of 23% to 32% is heated without oxygen, it softens and the volatile matter bubbles out as gases. The product resolidifies as a shiny, porous, hard, black solid called *coke,* which is used in processing metals, or metallurgy. The volatile by-products of coke making are a rich variety of organic chemicals that were the basis of the organic chemical industry for many years. Recently, crude oil has been a cheaper and more convenient source for some of these chemicals. In years to come, as oil supplies dwindle, carbonization of coal may once again become the predominant source.

In recent years, about 86% of the coal mined in the United States has been used as fuel by the electric utilities, 5% is used to make coke, and the rest is distributed in various applications such as industrial steam generation. A small amount goes into space heating.

Coke in Metallurgy

The production of iron and steel from iron ore depends heavily on carbon monoxide, CO, formed by partial burning of coke in blast furnaces. Much of the iron in iron ore exists as iron oxides, particularly magnetite, Fe_3O_4, and hematite, Fe_2O_3. At high temperatures in a blast furnace in an atmosphere of CO, the oxygen in the iron ore is burned off as carbon dioxide, CO_2, and free iron forms:

$$3\ Fe_2O_3 + CO \rightarrow 2\ Fe_3O_4 + CO_2$$
$$Fe_3O_4 + CO \rightarrow 3\ FeO + CO_2$$
$$FeO + CO \rightarrow Fe + CO_2$$

The burning coke supplies heat for the blast furnace. It also forms CO by reacting with oxygen in the air blast, according to the following reactions:

$$C + O_2 \rightarrow CO_2$$
$$CO_2 + C \rightarrow 2\ CO$$

The coal used to produce coke for making iron and steel is called metallurgical coal. It has a low sulfur content, so contamination of iron and steel by sulfur is minimized.

12.7 Sulfur in Coal

GOAL: To learn the forms of sulfur in coal and the means for their removal.

Supplies of coal are plentiful, leading to optimism about increased use as a major energy resource. Nevertheless, one particularly severe problem is the sulfur content of many coals (Figure 12.10).

Sulfur dioxide is a major air pollution problem, and also contributes to acid rain. See Sections 14.2 and 14.3 for details.

When coal is burned, sulfur is released primarily as sulfur dioxide, SO_2, which is a serious pollutant. The Environmental Protection Agency (EPA) has some complex regulations that restrict the SO_2 released from coal-burning plants. For plants that were operating before June 1979, the regulations allow that no more than 4 lb of SO_2 may be released per 1 million Btu of energy consumed. For plants for which construction began after September 1978, the regulations became more restrictive at 1.2 lb of SO_2 released per 1 million Btu *(9)*.

Consider a typical bituminous coal with an energy value of 12,500 Btu/lb. If it contains about 2.5% sulfur, burning will release about 4 lb of SO_2 per 1 million Btu. This is within regulations for older plants, but not for newer ones. For subbituminous coal or lignite, with a much lower energy range, the sulfur content must be lower than 2.5% to meet the EPA standards for older power plants. Fortunately, this is often the case, so subbituminous and lignite coals may be used in facilities far from where they are mined. For example, Commonwealth Edison Company provides electric power for much of northern Illinois, including Chicago. Illinois has a wealth of coal (Figure 12.7), but the sulfur content is high, typically 3.5%. Therefore, Commonwealth Edison burns low-sulfur, subbituminous, western coal (Figures 12.8 and 12.10).

The difficulty in meeting the standards is reflected in Figures 12.8 and 12.10, which show that most of the low-sulfur western coal is the relatively low-energy (8300–11,500 Btu/lb) subbituminous variety, for which the sulfur content must be well below 0.75% to meet the older regulation. The reserves of low-sulfur coal that can meet that standard are

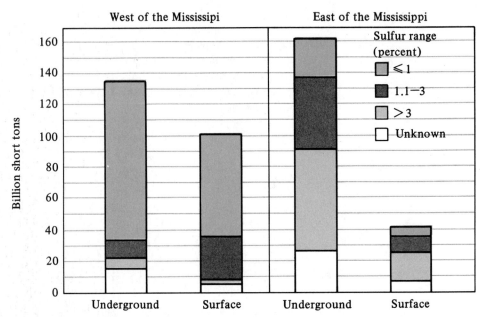

Figure 12.10   Demonstrated coal reserve base by sulfur content, potential method of mining, and region. The coal with an intermediate sulfur content (1.1–3%) composes approximately 21% of the total reserve base, of which approximately 40% is west of the Mississippi River. Approximately 46% of the base has a sulfur content of 1% or less. Of this, approximately 84% is west of the Mississippi River. (From the U.S. Department of Energy.)

still considerable, although that coal is in greatest demand and has the highest price. Since most eastern coal and even some of the western coal cannot meet the new standard (1.2 lb of SO_2/million Btu), sulfur has to be removed. The new regulations also stipulate that SO_2 emissions must be reduced by at least 70% on coals that would release less than 1.2 lb of SO_2 per million Btu, and by at least 90% on coals that would release more than that.

Of interest, anthracite coal has been exempted from the regulation. It normally is very low in sulfur (typically 0.5%) and has a high energy content (usually greater than 14,000 Btu/lb). The exemption was created to spur development of this prime form of coal, which must be deep-mined and is located in economically depressed areas of the country.

Coal Cleaning

Several methods of removing sulfur are possible and are in various stages of development. Methods applied before burning include cleaning, solvent refining, gasification, and liquefaction. Scrubbers are used to trap SO_2 when coal is burned. To understand these techniques, you must first know that sulfur exists in two forms, inorganic and organic. The chief form of inorganic sulfur is *pyritic sulfur,* or *iron pyrites,* FeS_2, also known as fool's gold. A small amount is sulfates, mostly calcium sulfate, $CaSO_4$, and some iron, copper, and magnesium sulfates. Organic sulfur is sulfur bound to carbon in complex coal molecules derived from plants. No simple chemical formula can be drawn to symbolize this form of sulfur.

Some pyritic sulfur can be separated from coal with water or other liquid (Figure 12.11). Since pure bituminous coal has a much lower density than iron pyrites, the coarse

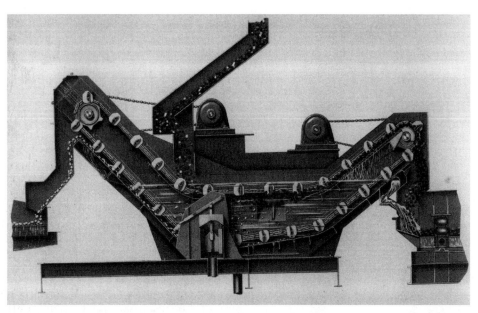

Figure 12.11   The McNally Lo-Flo Dense Media Vessel can be used to separate coal from iron pyrites because of the large difference in density of the two substances. Coal (black) contaminated with iron pyrites (white) enters from the top and passes into the liquid medium. The density of the medium is controlled so that the low-density coal floats near the top and the heavier iron pyrites sink. A conveyor moving in a clockwise direction skims off the floating solid material and deposits it at the right. The same conveyor then picks up the heavier material and deposits it at the left.

particles or clumps of pyrites settle out of a liquid much faster than the coal. Finely divided pyritic sulfur does not settle well and is not so easily removed. Were the procedure modified to remove almost all the pyritic sulfur, much coal would be lost in the process. None of the organic sulfur is affected by the procedure.

Coal Gasification

One of the most intriguing methods for using coal in a clean form is gasification, in which coal is converted to synthetic gas. Natural gas is mostly methane, CH_4, with small amounts of ethane and other alkanes (Sections 6.1 and 12.4). Natural gas, with an energy content of approximately 1000 Btu per standard cubic foot (scf), is routinely transported long distances by pipeline. Any synthetic substitute must have a comparable Btu content to be economically competitive. Synthetic gas with a Btu content near 1000 Btu/scf is known as high-Btu gas. Medium-Btu gas, at 320 Btu/scf, can be economically transported short distances (up to 20–30 miles), but low-Btu gas is normally used only near the site of production.

Several processes have been developed for coal gasification; all use much the same chemical reactions, with variations aimed at increasing efficiency. The Lurgi gasifier, described in Figure 12.12, is one of the most successful.

Gasification is carried out in four steps: devolatilization, steam-carbon reaction, CO-shift reaction, and catalytic methanation. First, coal is exposed to high temperatures (range

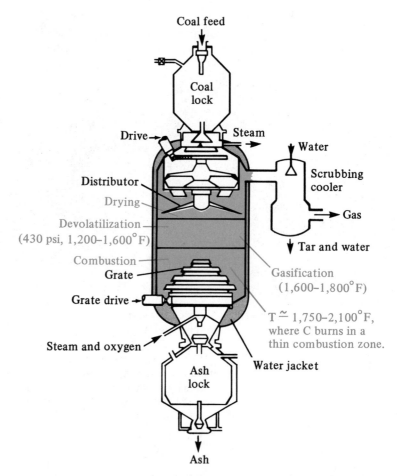

Figure 12.12 The Lurgi gasifier is currently the most widely used commercial system for gasifying coal. Coal is put into the Lurgi at the top, and a mixture of steam and oxygen is put in near the bottom. The colored labels describe chemical changes. A medium-Btu gas exits from the gasifier and may be upgraded by catalytic methanation if desired. (From Penner, S. S., Icerman, L. *Energy;* Vol. 2, p. 105, figure 10.3-7. Copyright 1975 by Addison-Wesley: Reading, MA. Reprinted with permission.)

1100–1500 °F). This step is called **devolatilization.** The volatile matter is released and decomposes to methane and a char residue.

$$\text{Coal volatile matter} \quad \xrightarrow{\text{1100–1500 °F}} \quad CH_4 + char$$

A key step in coal gasification is the addition of hydrogen. Since hydrogen is expensive and its production requires energy, hydrogen is supplied as water, H_2O, in the form of steam. The steam is caused to react with the char in the so-called *steam-carbon reaction,* according to the following equation:

$$\underset{\text{char}}{C} + \underset{\text{steam}}{H_2O} \rightarrow \underset{\substack{\text{carbon} \\ \text{monoxide}}}{CO} + \underset{\text{hydrogen}}{H_2}$$

Both gases have an energy content of about 320 Btu/scf, so the mixture is classified as a medium-Btu gas. Such a mixture is an excellent fuel for nearby use, but if it has to be transported a long distance, it must be upgraded to a high Btu first. Two additional steps are required.

Some (but not all) of the carbon monoxide is reacted with more steam, to form more hydrogen in the *CO-shift reaction:*

$$\underset{\substack{\text{carbon}\\\text{monoxide}}}{CO} + \underset{\text{steam}}{H_2O} \rightarrow \underset{\substack{\text{carbon}\\\text{dioxide}}}{CO_2} + \underset{\text{hydrogen}}{H_2}$$

This additional hydrogen is then caused to react with the remaining carbon monoxide, in a process called *catalytic methanation:*

$$CO + 3\,H_2 \rightarrow H_2O + CH_4$$

Once the methane is purified, it qualifies as a high-Btu, or *substitute natural gas* (SNG), and can be transported economically long distances throughout the country in the natural gas pipeline system.

During gasification, energy must be provided. Some oxygen gas must be introduced to provide the energy by burning some of the coal. Providing energy is significant for two reasons. For one thing, it limits the energy efficiency of gasification to 55% to 67%. The low efficiency is simply regarded as part of the price paid for converting a dirty fuel (coal) into a clean fuel (SNG).

The other reason for being concerned about the use of oxygen is its high cost. Moreover, energy is consumed in producing oxygen (Figure 4.8). Therefore, air (21% oxygen, 78% nitrogen) is used as the source of oxygen. Nitrogen dilutes the CO, H_2, and CH_4, and greatly lowers the Btu content of the fuel since it is noncombustible. Removing the nitrogen is not practical, since it has a low chemical activity. Nevertheless, a low-Btu gas (80–170 Btu/scf) containing CO, H_2, and N_2 is an acceptable substitute for natural gas at sites close to the gasification system.

Once coal is gasified into a clean fuel, ash and sulfur remain. The ash is trapped inside the gasification unit and later removed. Sulfur leaves the system as gaseous hydrogen sulfide, H_2S. The *Claus process* converts the H_2S to free sulfur, S, which can be sold as a by-product. The three-step process uses iron oxide, FeO, which is regenerated:

(1) $H_2S + FeO \rightarrow FeS + H_2O$
(2) $2\,FeS + 3\,O_2 \rightarrow 2\,FeO + 2\,SO_2$
(3) $2\,H_2S + SO_2 \rightarrow 2\,H_2O + 3\,S$
net: $2\,H_2S + O_2 \rightarrow 2\,H_2O + 2\,S$

In 1985 the first commercial coal gasification plant was completed by the Great Plains Gasification Associates near Beulah, North Dakota. Although the plant, which uses the Lurgi gasifier, is regarded as a technical success, the price that must be charged for the gas to make the project an economic success is currently about 3 times the cost of natural gas that is available in distribution pipelines *(8)*. Thus this technological progress can be judged only in terms of economic and political factors, such as the possibility of having to deal with future energy emergencies.

Nitrogen is usually not burned at all during combustion except under extreme conditions. One notable example is the internal combustion engine of the automobile, in which some nitrogen is burned to form nitrogen oxides (Section 14.2) due to the high temperatures and pressures that are encountered.

See problem 19.

Liquefaction

The strategy for liquefying coal begins with the gasification products CO and H_2, which are subjected to catalytic reactions known as the Fischer-Tropsch process. Fischer and Tropsch studied how iron and cobalt could catalyze the reaction of carbon monoxide with hydrogen to form a variety of liquid hydrocarbons. Their work led to large-scale production of gasoline, diesel fuels, and aviation fuels, which were used by Germany during World War II, with peak production of 200 million gallons per year in 1943.

By varying the conditions of the reaction, hydrogen and carbon monoxide can be made to combine to form simple molecules such as formaldehyde and methanol, or higher-molar-mass alcohols and hydrocarbons. Thus, all of the coal (except the mineral matter), not just the small percentage of volatile matter, can be converted to H_2 and CO and then to liquids.

In the reactions to form formaldehyde and methanol, notice that the amount of hydrogen introduced determines whether the aldehyde or the alcohol will be the final product.

$$CO + H_2 \rightarrow H-\overset{\overset{\displaystyle O}{\|}}{C}-H \quad \text{formaldehyde}$$

$$H-\overset{\overset{\displaystyle O}{\|}}{C}-H + H_2 \rightarrow CH_3OH \quad \text{methanol}$$

As in simple gasification, all sulfur is removed. In fact, it has to be removed just to prevent contamination of the catalysts used in the Fischer-Tropsch process. In effect, complex molecules in coal are taken apart by gasification and put back together by the Fischer-Tropsch synthesis.

Scrubbers

When coal itself is burned, and as much sulfur has been removed as possible, some sulfur dioxide in the combustion gases may still have to be removed to conform to environmental standards. Equipment known as a *flue-gas desulfurization device,* or more commonly a *scrubber,* removes SO_2 from combustion gases.

There are several variations of scrubbing devices. All are expensive to install and maintain, and they present many technical problems. So far, the electrical utilities, which are the greatest consumers of coal, have mostly avoided scrubbers and simply burn low-sulfur coal in the effort to meet EPA standards. But as new coal-fired units go into operation and more stringent regulations must be adhered to, scrubbers are becoming more common.

One important and relatively simple scrubbing device operates according to the following chemical reactions; limestone, which is alkaline, reacts with SO_2, which is acidic:

$$\underset{\substack{\text{calcium}\\\text{carbonate}\\\text{(limestone)}}}{CaCO_3} + SO_2 \rightarrow \underset{\substack{\text{calcium}\\\text{sulfite}}}{CaSO_3} + CO_2$$

Limestone is often used as a slurry in water. In the presence of water, SO_2 exists as H_2SO_3 due to the reaction

$$SO_2 + H_2O \rightarrow H_2SO_3$$

So the chemistry of SO_2 removal is actually

$$CaCO_3 + H_2SO_3 \rightarrow CaSO_3 + H_2CO_3$$

The carbonic acid, H_2CO_3, then decomposes to form CO_2 and H_2O *(10)*.

12.8   Conclusions About Coal

GOAL: To draw a time line showing the brief era of fossil fuel use.

Thus far in the 1990s, the low prices of crude oil and natural gas do not make either gasification or liquefaction of coal economically competitive. But the desire for energy independence is a political factor that may justify subsidies to make these industries economically viable.

During the time of the oil shortages in the 1970s, the title of an article in *Smithsonian* characterized the potential for greater reliance on coal as follows: ''Coal Is Cheap, Hated, Abundant, Filthy, Needed'' *(11)*. The author suggested that ''the second reign of King Coal is not likely to be a tranquil one.''

There may even be a third reign of King Coal far in the future if new technologies allow us to conserve coal and use it primarily as a source of chemicals, called petrochemicals, now obtained from petroleum. If not, the coal reserves will be depleted just as surely as oil supplies will. It has been estimated that world oil production will peak by 2000, and that 90% of the oil originally present on earth will be gone by 2030 or before. The rate of consumption depends on conservation and success in developing alternative sources of energy. It is even harder to estimate how long coal will be available, since projections must reach much farther into the future.

One interesting way of viewing the history of fossil fuels, including coal, is the graph in Figure 12.13, prepared by Hubbert who said: ''It is difficult . . . to realize how transitory the fossil-fuel epoch will eventually prove to be when it is viewed over a longer span of human history. The situation can better be seen in the perspective of some 10,000 years,

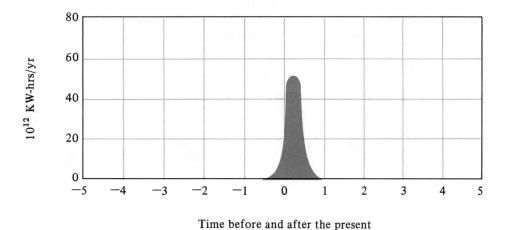

Figure 12.13   Total world production of fossil fuels in time perspective. (From Hubbert, M. K. ''Energy Resources: A Report to the Committee on Natural Resources of the National Academy of Sciences–National Research Council.'' Publication 1000-D. NAS-NRC: Washington, DC, 1962).

half before the present and half afterward. On such a scale, the complete cycle of the exploitation of the world's fossil fuels will be seen to encompass perhaps 1300 years, with the principal segment of the cycle (defined as the period during which all but the first 10 percent and the last 10 percent of the fuels are extracted and burned) covering only about 300 years'' *(12)*.

SUMMARY

The fossil fuels (oil, gas, coal), nuclear fission, falling water, geothermal heat, and the sun are primary sources of energy and can be used to produce secondary forms of energy such as electricity. Energy conservation, which includes insulation and energy-efficient transportation for both private and commercial use, is important in meeting the demands for energy.

Some energy conversion processes are inefficient and are accompanied by the waste of a large portion of the primary energy input. The inefficiency of the conversion of heat energy to work (such as turning a turbine to generate electricity) is a consequence of the laws of thermodynamics and is unavoidable. Some waste heat can be used directly for space heating and for industrial processes. This and other forms of cogeneration allow more efficient use of primary energy resources. Magnetohydrodynamics (MHD) offers another approach to conserving energy, since it allows direct conversion of some of the heat to electricity. This permits greater efficiency of fuel use by skipping the intermediate conversion of heat to work, but commercial-scale use of MHD is not expected soon.

The kilowatt-hour (kW-h), calorie, joule, and British thermal unit (Btu) are common units of energy. Conversion between them is often necessary. The quad (Q) is short for quadrillion Btu and is often used to describe very large amounts of energy. The watt (and kilowatt) and horsepower are common units of power, describing the rate at which energy can be delivered.

Peaking is a problem that electric utilities must contend with, and they often do so by using precious fossil fuels. Alternative primary energy resources (such as solar energy) will be most effective if they can provide energy continuously, particularly at times of peak demand.

Coal is an attractive alternative to oil, particularly in the United States, but presents many environmental challenges. The production of SO_2 during coal combustion is a severe problem.

The four ranks of coal are anthracite, bituminous, subbituminous, and lignite; they are ranked according to volatile matter and energy content. The western United States has abundant low-sulfur subbituminous coal that is often shipped great distances for direct use by electric utilities. Anthracite coal is also a low-sulfur fuel; but most bituminous coal, which is abundant in the United States, has a high sulfur content and must somehow be processed to allow environmental protection. Coal cleaning, gasification, liquefaction, and scrubbing of stack gases are techniques that are becoming increasingly important. Nevertheless, even though coal is available in clean forms, it is also a fossil fuel and the supply is limited. In a chart covering thousands of years, the use of all fossil fuels will likely show up as a small blip representing about 300 years. The world's population will subsequently have to rely totally on other primary energy sources.

PROBLEMS

1. Give a definition or an example of each of the following.

   a. Cogeneration
   b. Topping
   c. Bottoming
   d. Hard coal
   e. Soft coal
   f. Anthracite

   g. Bituminous
   h. Volatile matter
   i. Char
   j. Fixed carbon
   k. Coke
   l. Particulates
   m. Baghouses

n. Electrostatic precipitators
o. Carbonization
p. Iron pyrites
q. High-Btu gas
r. Medium-Btu gas
s. Low-Btu gas
t. Substitute natural gas
u. Scrubber
v. Fischer-Tropsch process

2. Give a complete list of the primary energy resources. What are the prevalent secondary energy forms?

3. What advantage is available from magnetohydro-dynamics?

4. Distinguish between the kilowatt and the kilowatt-hour. Why is there a difference between kilowatt-hours thermal and kilowatt-hours electric?

5. Express 4.2×10^6 kJ in kW-h_e. Express 1.7×10^2 kcal in Btu.

6. Identify the larger quantity of energy from each of the following pairs. Consult Section 12.3 for help.
   a. 1 Q or 10^{12} Btu
   b. 1 kW-h_e or 1 kW-h_t
   c. 2.93×10^{11} kW-h_e or 2.52×10^{17} cal
   d. 80 Q or 2.52×10^{19} cal
   e. 1000 kW-h_e or 1000 Btu
   f. 10^6 kJ or 10^6 cal

7. If automobiles in the United States consumed fuel at the rate of 20.0 miles per gallon in 1988, what was the rate of consumption in 1973? Refer to Figure 12.3.

8. Referring to Figure 12.2, determine the number of years that it took to double the consumption of energy in the United States beginning in 1950. How many years will it take to double consumption beginning in 1979?

9. Knowing that a barrel (42 gal) of oil has an energy content of 5.8 million Btu, how many million barrels of oil were consumed in the peak year of United States oil consumption? See Figure 12.2.

10. From Figure 12.1, calculate the percentage of energy consumed for generating electricity that was lost in conversion and transmission.

11. What are the four ranks of coal and how do they differ?

12. Where are the principal deposits of each of the four classes of coal in the United States? Why is anthracite coal found in only one location?

13. Describe how coke is used in producing iron and steel.

14. What are the EPA regulations on SO_2 emissions from coal-fired power plants?

15. What four methods are available to keep coal from polluting the environment with SO_2?

16. What advantage does high-Btu gas from coal have compared with medium- or low-Btu gas?

17. Coal gasification can be carried out using either oxygen or air. What difference does it make?

18. How is hydrogen supplied for gasification of coal?

19. Refer to the equations for the Claus process for removing H_2S on page 267. Confirm that the net equation is the result of the three individual reactions. Hint: Keep each equation balanced, but multiply exponents so that FeO and SO_2 will cancel out of the sequence.

20. Write a reaction for removing SO_2 from flue gases.

References

1. Hubbard, H. M. "The Real Cost of Energy." *Scientific American* **1991,** *264* (4), 36–42.

2. Lovins, A. B. "The Electricity Industry." *Science* **1985,** *229,* 914.

3. Barrow, B. B. "Units of Energy." *Science* **1979,** *179,* 1181.

4. Hirsch, R. L. "Impending United States Energy Crisis." *Science* **1987,** *235,* 1467–1473.

5. Ecklund, E. E.; Mills, G. A. "Alternative Fuels: Progress and Prospects, Part 1." *Chemtech* **1989,** *19* (9), 549–556.

6. *Annual Energy Review 1989;* Department of Energy/Energy Information Administration: Washington, DC, 1990; pp. 183, 203.

7. Burnett, W. M.; Ban, S. D. "Changing Prospects for Natural Gas in the United States." *Science* **1989,** *244,* 305–310.

8. Schobert, H. H. *Coal: The Energy Source of the Past and Future;* American Chemical Society: Washington, DC, 1987; pp. 36–44, 228–235.

9. 40 Code of Federal Regulations, Part 43a; U. S. Government Printing Office: Washington, DC, 1989.

10. Corcoran, E. "Cleaning Up Coal." *Scientific American* **1991,** *264* (5), 106–116.

11. Stein, J. "Coal Is Cheap, Hated, Abundant, Filthy, Needed." *Smithsonian* **1973,** *3* (11), 18.

12. Hubbert, M. K. "Energy Resources of the Earth." *Scientific American* **1971,** *225* (3), 60–70.

Energy: Alternatives for the Future

13

Besides the conventional fossil fuels, and the gases and liquids that can be derived from them, other energy resources will likely achieve increased prominence in the future. Synthetic fuels, such as shale oil, alcohols, and hydrogen, as well as solar energy, are the most likely candidates.

The demand for fuels for transportation, home and commercial heating, and electric power generation will remain high. Almost all forms of transportation use liquid fuels, so new energy resources must be evaluated as possible replacements. In view of the heavy demand for electricity, energy resources such as solar energy must also be considered as possible replacements for coal, oil, and nuclear power for generating electricity. The need for large-scale energy storage capacity is also a key component of the energy picture.

Developments in the energy field were spectacular in the early 1980s after the oil crisis in 1979. For the latter part of the 1980s, progress slowed due to abundant cheap oil and lack of government incentives. Nevertheless, progress in solar energy has been dramatic, and other energy sources, such as methanol, await further development when economics or environmental pressures make them competitive. Automobile fuels in the year 2000 will surely not be the same as they were in the 1980s, and even the electric or solar-powered car may be a significant item.

As for electric power, economic growth makes it likely that the demand will increase greatly despite dramatic conservation efforts. If nuclear energy regains acceptance, it and coal will dominate development by electric utilities, although nuclear energy cannot make a long-term contribution without development of the breeder reactor or fusion. Enhanced storage facilities can draw from electric energy at off-peak times for delivery at times of peak demand.

Many energy resources have the disadvantage of being *nonrenewable*. Some are regarded as *renewable* because they come from biomass, plants, and animal products that can be restored or regrown; or they result from capturing wind, or the energy of flowing water, or solar radiation. Interest in them dropped during the later 1980s, but research and development are rising again. This chapter will describe the potential role that may be played by each form of energy.

13.1 Synthetic Fuels

GOAL: To understand the meaning of synthetic fuels.

Some of the most prominently mentioned alternatives to fossil fuels are synthetic fuels, or *synfuels*. The word *synthetic* means that these fuels are processed or manufactured in some way from natural raw materials. In effect, this means that only coal and natural gas are natural fuels, since they can be burned in their natural state to release their energy content. Gaseous and liquid fuels that are produced from coal are prime examples of synthetic fuels.

See Petroleum (Section 6.1) for a discussion of cracking and reforming.

Petroleum provides an example of the confusion that sometimes accompanies the phrase *synthetic fuels*. Gasoline, kerosene, heating oil, and several other fuels are the result of extensive processing (including cracking and reforming) of petroleum. The chemical changes are as drastic as those involved in gasification and liquefaction of coal, since in all these processes, molecules are literally taken apart and put back together in other forms. Still, the fuels derived from crude oil are not generally classified as synthetic fuels.

We already know some of the complex chemistry of synthetic fuels from having considered coal gasification and liquefaction. The other contestants in the development of synthetic fuels are oil from shale and tar sands, ethyl alcohol and methyl alcohol, methane, and hydrogen.

13.2 Oil from Shales and Tar Sands

GOAL: To consider the availability and recovery of fuels from shale and tar sands.

Oil shales constitute the second largest fossil fuel resource in the United States, behind only coal. Indeed, the oil in shale is not really oil; it is a waxy organic material called *kerogen*. But when the shale is mined, crushed, and heated to about 500 °C, gaseous and liquid hydrocarbons form. The heating process is called *retorting*. The liquids that are released by retorting can be hydrogenated and refined into liquid fuels for use as gasoline and heating oil. Like coal liquids, shale oils are deficient in hydrogen and must be treated to adapt them to conventional refinery processes.

The yield of gases and liquids from oil shale is commonly in the range of 10 to 60 gal/ton of shale. At least 25 to 30 gal/ton is necessary if the economics are to be favorable. Almost all the deposits of oil shale that yield more than 15 gal/ton are found in geological formations on the western slopes of the Rocky Mountains in the Green River Basin in Colorado, Wyoming, and Utah (Figure 13.1).

Spent shale presents a disposal problem. Some can be returned to the mines, but since processing increases the volume of shale, only part of the residue can be returned.

The United States also has some deposits of tar sands, but they are not very important. Large deposits in the Canadian province of Alberta are known as the Athabasca sands and are already being converted to synthetic crude oil, called *syncrude*, on a commercial scale. Current Canadian output is approximately 50,000 barrels per day of crude oil from about 120,000 tons of rock. This yield is somewhat lower than that expected from the highest-grade Colorado shales.

The tar sands are heated with hot water and steam to release a gooey material known as *bitumen*, which is upgraded by refining. (Cracking and distillation temperatures are

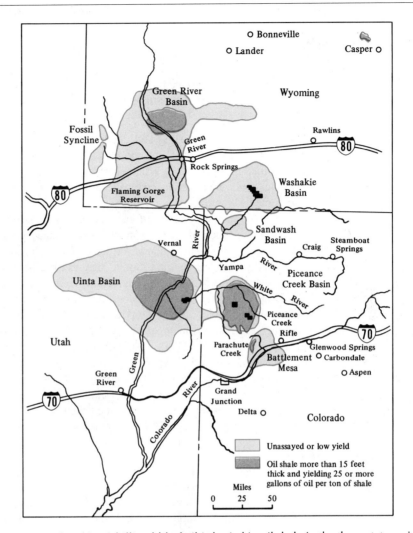

Figure 13.1   About 1800 billion bbl of oil is buried in oil shale in the three-state region of Utah, Colorado, and Wyoming. Almost all the prime shales, with at least 30 gal/ton in seams 30 feet thick or more, are in the Piceance Creek Basin in Colorado. Much high-quality shale is in a formation called the Mahogany Zone; this deposit is almost 70 feet thick, is visible on the exposed surfaces of some canyon walls, and originated from the sedimentation of Lake Uinta, a freshwater lake that existed in an old geological period. Black areas indicate tracts of federal land leased for private development *(1)*. (Copyright 1974 by the American Association for the Advancement of Science.)

above 500 °C.) The average bitumen content of Canadian tar sands averages 12%, and ranges to 18%. It has been estimated that 275 billion barrels of this oil is recoverable. This corresponds to about 80% of reserves of crude oil in the Middle East.

Syncrude from oil shale and tar sands generally costs much more than natural crude oil, because of the complexities and costs of recovering the synthetics. In fact, some experts regard the cost of syncrude (and other synfuels) as the true value of crude oil, since it represents the total cost of replacing crude oil as an important energy resource.

13.3 Liquid Fuels: Gasoline

GOALS: To consider the challenge of maintaining adequate octane ratings.
To understand the problems of fuel volatility.
To appreciate the current use and future potential of oxygenated fuels.
To understand the conversion of methanol to gasoline.

One of the most interesting and challenging aspects of the energy picture is the large demand for gasoline to fuel automobiles. The burning of gasoline in the automobile engine is described in Figure 13.2. When energy shortages occurred in the past, many economic consequences were experienced by all. Two obvious ones were increased prices of gasoline and home heating oil.

A preliminary discussion of gasoline can be found in Section 6.1 (see Petroleum). A complete discussion of gasoline must consider many variables and problems—chemical, technological, and environmental. Primarily for environmental reasons, gasoline has been a constantly changing product. Some changes have been subtle and have gone unnoticed by the average consumer. The phase-out of leaded gasoline was more obvious.

Despite pollution-control devices, air pollution from cars burning gasoline remains a serious problem, particularly in urban areas. The most critical forms are ozone and smog. Although essential in the upper atmosphere, ozone at the surface of earth adversely affects plants and animals. It is both a component of and a contributor to the formation of smog. Gasoline engines produce varying amounts of oxides of nitrogen, symbolized as NO_x to represent both NO and NO_2. These compounds participate in the formation of smog and are irritating to the lungs and eyes. Detailed discussion is in Chapter 14. These serious environmental challenges and economic problems require strategies for formulating gasoline or substitutes. They will be discussed in the following sections.

Octane Ratings

Lead additives were used for many years as octane enhancers to control the burning of gasoline and give a smoother-running engine (Section 6.1). In the late 1970s, before significant pressures to eliminate lead, the amount present was typically 3 g of Pb/gal *(2)*. By January 1986, the permissible level had dropped to only 0.1 g/gal. At this level, octane improvement was so slight that petroleum refiners totally eliminated it and began to rely on other means of improving octane ratings. By then, newer cars required unleaded fuels to prevent contamination of catalytic converters installed to control pollution, so suppliers already had many years of experience providing unleaded gasoline.

Aside from any additive, fuel molecules influence the octane rating. In general, fewer carbons in a hydrocarbon molecule give a higher rating, as do branching of hydrocarbon chains and the presence of cyclic structures, particularly aromatic compounds.

Branched, unbranched, cyclic, and aromatic hydrocarbons are discussed in Chapter 6.

The lead used to improve octane ratings of motor fuels was added in the form of tetramethyllead, $(CH_3)_4Pb$, or tetraethyllead, $(CH_3CH_2)_4Pb$, or a combination of the two. Engine knock or ping results from uncontrolled detonation of fuel; it is most likely to be noticed when an engine is operating under loads, such as accelerating uphill at low speed.

During the 1950s there were extraordinary efforts to develop more powerful, more efficient automobile engines that required compression of the air-fuel mixture to a fraction of its original volume (Figure 13.2). High *compression ratios* (9.0–11.0) of uncompressed to compressed fuel mixtures spurred the use of high levels of lead antiknock additives as a means of raising octane levels high enough to allow the engines to operate under the designed loads. In the early 1970s, engines operating at lower compression ratios (8.0–9.0) became more common as cars were designed to run on unleaded gasoline out of

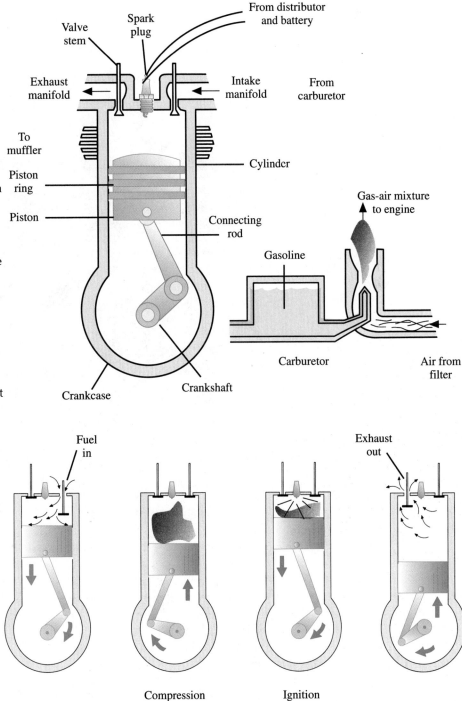

Figure 13.2
The automobile engine. The engine gets its power when a fuel-air mixture is ignited. Gasoline is vaporized in a carburetor and carried into the cylinder in a stream of filtered air through the intake manifold. The piston in the cylinder is shown in four positions. In (a) the piston is moving downward and draws in the fuel-air mixture with the inlet valve in the open position. In (b) the piston is moving upward with both the inlet and outlet valves closed so that the contents are compressed. In (c) the fuel-air is fully compressed, and the spark plug receives a flow of electric current from the distributor, causing the fuel to ignite and forcing the piston downward. Ignition causes the temperature to rise dramatically and causes the pressure in the cylinder to increase at least 10-fold; this pressure forces the piston downward. The downward movement also turns the crankshaft, which delivers power to propel the automobile and to continue the motion of the piston upward, so that in (d) the outlet valve opens and the exhaust gases are expelled.

Figure 13.3   An instrument for measuring octane rating.

concern for the environment *(2)*. Newer ways to improve octane rating are discussed in the following sections.

Octane ratings are normally expressed in the form of the RON (research octane number) or the MON (motor octane number), both of which are determined by laboratory measurements of fuel samples (Figure 13.3). A typical motor fuel might have a RON of 94 and a MON of 86, so a RdON (road octane number) of 90 is the calculated average of the RON and the MON. The RdON, also known as the **antiknock index** or **AKI,** is the number displayed on the gas pump.

The Fuel Volatility Problem

See Properties and Uses of Hydrocarbons in Chapter 6 for information on the effects of branching of hydrocarbons.

Small hydrocarbon molecules, such as butanes and pentanes, have good antiknock properties and offer a relatively inexpensive way to solve the lead problem. This has resulted in formulation of gasolines with increasing amounts of butanes over the last two decades, particularly as lead was phased out. Unbranched butane has a boiling point of 0 °C, so it and branched butane (isobutane) are gases at normal temperatures.

Although they dissolve to a significant extent in gasoline, butanes and pentane isomers are very volatile; evaporation of unburned hydrocarbons into the atmosphere, called

Figure 13.4 July gasoline volatility levels. Vapor pressures in psi: A = 9.0; B = 10.0; C = 11.5. See text for discussion. (Reprinted with permission from *Chemical and Engineering News,* July 6, 1987. Copyright © 1987 American Chemical Society.)

evaporative emissions, makes a major contribution to smog formation (see Chapter 14). Thus, significant pressure has been exerted to find chemical solutions to correct the problem.

The volatility of gasoline has a major influence on automobile performance. If fuel is too volatile, vapor lock may occur in an engine, particularly in warmer climates. If volatility is too low, engines may be hard to start, particularly in colder climates. In fact, for each month of the year, as average daily temperatures vary, refiners alter fuel volatility in response (Figure 13.4). Thus during summer months, in states with soaring temperatures, refiners maintain fuel volatility at relatively low levels. For July, the volatility levels range from A (least volatile) in the desert southwest to C in many of the far northern states. During the coldest winter months, all volatility levels are higher, ranging to E (most volatile) in the north.

The pattern of severe evaporative emissions during the warmer months coincides with the highest ozone levels in many cities. Hydrocarbon emissions come from the fuel tank and the carburetor. Newer cars contain charcoal canisters in the fuel tanks to trap hydrocarbon vapors when the cars are not in use. When the engine is operating, the canister is flushed with air to carry the trapped hydrocarbons into the engine. Fuel-injection engines, designed to replace engines with carburetors, decrease evaporative emissions that are released from a hot carburetor immediately after an engine is turned off.

The Oxygenates

Several compounds collectively known as **oxygenates** offer interesting ways of improving octane ratings. The most prominent (Table 13.1) are methyl alcohol (methanol), ethyl alcohol (ethanol), methyl-t-butyl ether (MTBE), and ethyl-t-butyl ether (ETBE).

Table 13.1   Properties of some fuel oxygenates

Compound	Antiknock Index[1]	Heat of Combustion (Btu/gal)	Boiling Point (°C)
Methanol	101	64,500	64.6
Ethanol	101	76,500	76.5
tert-Butyl alcohol	100	101,100	82.6
MTBE	108	108,500	55.4
ETBE	111	116,500	72.8
Gasoline	87	124,800	32–205

[1]AKI = ½ (RON + MON)

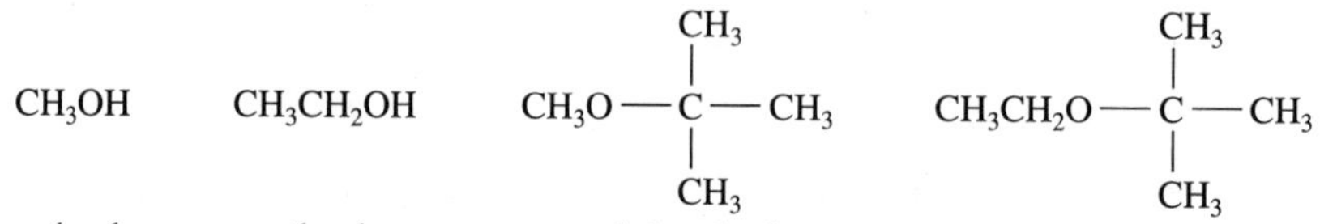

CH$_3$OH	CH$_3$CH$_2$OH	CH$_3$O—C(CH$_3$)$_2$—CH$_3$	CH$_3$CH$_2$O—C(CH$_3$)$_2$—CH$_3$
methanol	ethanol	methyl-*tert*-butyl ether (MTBE)	ethyl-*tert*-butyl ether (ETBE)

Methanol and ethanol are frequently promoted to be used in place of gasoline or in blends with gasoline. The value of the alcohols, as well as that of MTBE and ETBE, as octane enhancers, is obvious from the data in Table 13.1. The ethers burn well in automobile engines and they are completely miscible with gasoline in all proportions. Of the two, MTBE is far less expensive and already is used as an octane enhancer; it has been used in motor fuels in Europe since about 1975. Levels as high as 15% MTBE blended with gasoline give a high-octane fuel and replace a relatively large amount of gasoline. This reduces dependence on foreign oil.

The synthesis of MTBE occurs by the following reaction:

$$\text{CH}_3\text{OH} + (\text{CH}_3)_2\text{C}=\text{CH}_2 \rightarrow \text{CH}_3\text{O}-\text{C}(\text{CH}_3)_2-\text{CH}_3$$

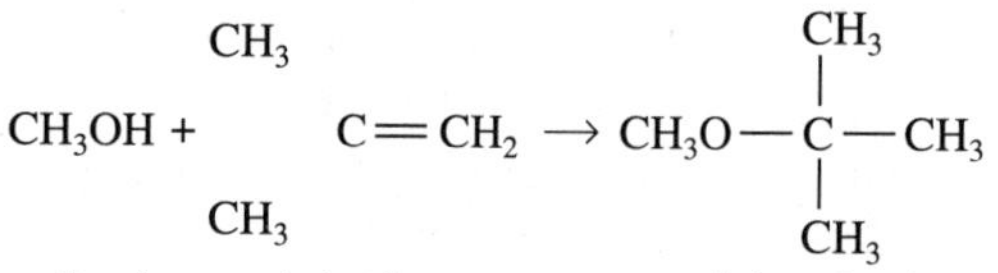

methanol   isobutylene   methyl-*tert*-butyl ether (MTBE)

As for the starting materials that are needed to synthesize MTBE, the methanol can be obtained from *synthesis gas* (CO + H$_2$), which can be produced from coal or natural gas. The isobutylene can be formed from *tert*-butyl alcohol according to the following reaction. The *tert*-butyl alcohol is currently produced as a by-product of the synthesis of propylene oxide (an important monomer). It can also be made from synthesis gas.

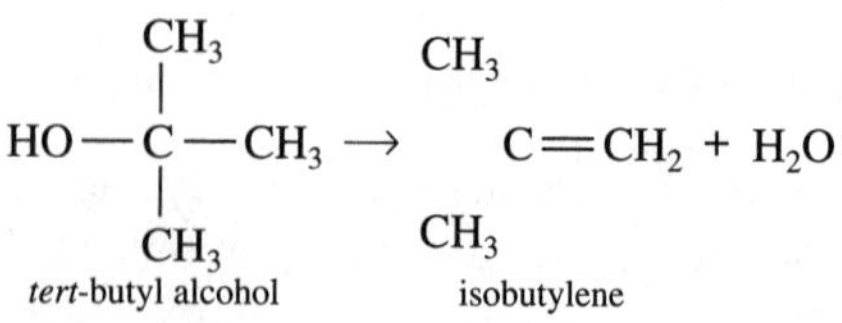

$$\text{HO}-\text{C}(\text{CH}_3)_2-\text{CH}_3 \rightarrow (\text{CH}_3)_2\text{C}=\text{CH}_2 + \text{H}_2\text{O}$$

tert-butyl alcohol   isobutylene

Ethyl Alcohol: Gasohol

Ethyl alcohol (ethanol) may be obtained by a purely chemical synthesis. If it is to become competitive as a motor fuel, the bulk will be produced biochemically by the action of yeast enzymes on sugars. This is the process, called fermentation, used for production of alcoholic beverages. (Details are in Chapter 17.)

In Brazil, ethyl alcohol is heavily used as a motor fuel, accounting for about one-third of the total consumption by automobiles. A blend of 20% ethanol and 80% gasoline is common, but much of the use is in the form of 96% ethyl alcohol (symbolized E96), called hydrated ethanol, which contains 4% water. Blends of ethanol and gasoline are commonly called *gasohol* (or gasahol).

Several circumstances in Brazil led to the use of ethanol. The country lacks crude oil and is therefore totally dependent on foreign imports for its gasoline. Equally important, a major crop is sugarcane, so abundant raw material is available for the fermentation process. Nevertheless, it has taken a major political commitment to subsidize the production of ethyl alcohol fuels.

The need to subsidize ethanol in Brazil or elsewhere can be understood by considering the energy balance sheet associated with the production of the alcohol: the energy input exceeds the output. The energy available from burning the end product is 76,500 Btu/gal, as shown in Table 13.1. The energy input required to produce ethyl alcohol from carbohydrates has been variably estimated to be about 150,000 Btu/gal. Two-thirds of this is consumed by distillation and other steps. One-third is consumed in growing the crop, particularly by fertilizer production and crop cultivation. Therefore, some form of government subsidization is necessary to ensure a reasonable price at the gas pump. This, of course, means that all taxpayers help to pay the bill, as they do to maintain oil supplies.

The economic viability of ethanol production varies with different crop sources and depends on the economic status of each source. For example, in 1990, high world sugar prices led sugarcane processors in Brazil to divert more of their industrial output to sugar and away from ethanol. This created a severe shortage of ethanol and triggered imports of methanol to use in a blend consisting of 33% methanol, 60% ethanol, and 7% gasoline as motor fuel *(3)*.

In the United States, gasohol has been vigorously promoted in the midwestern farm states known as the corn belt. Vast amounts of corn are grown as food for humans and as a commodity, particularly as feed for beef cattle. Corn cobs, husks, and stalks, collectively known as *corn stover,* have no food value for humans because we cannot metabolize their large amount of cellulose and other polysaccharides. But cellulose is a polymer of glucose, which can be broken down with enzymes and fermented into alcohol.

The structure of cellulose and its resistance to breakdown are discussed in Section 9.2.

Even when crop wastes are used as raw material, the large energy input required to produce the alcohol remains the major drawback. Only tax incentives, some as high as 20¢ per gallon (on 10% ethanol–90% gasoline), and/or lower state and federal gasoline taxes make ethanol-gasoline blends competitive in this country. Proponents point to increased tax revenues and higher levels of employment as offsetting the tax subsidies. What is clear is that the decision to use gasohol is less a scientific issue than it is an economic and political one.

Aside from the unfavorable energy balance, ethanol is an attractive fuel source. Since it is derived from plant matter, commonly referred to as *biomass,* it can be regarded as a renewable resource, unlike fossil fuels. Whether it is used alone or in blends with gasoline, it will stretch supplies of gasoline, which will likely become increasingly precious.

At the 10% level of ethanol, gasohol is interchangeable with straight gasoline and requires no engine modifications. Since pure ethanol has a RON of 110 (AKI = 101), the

RON of gasohol (10%) is 94 (Table 13.1). At this level, no additional octane enhancement is required for most cars.

Compared with gasoline (124,800 Btu/gal), a significant drawback is the lower energy content of ethanol (76,500 Btu/gal). This means that as pure alcohol, the energy content of ethanol is about 60% that of gasoline, or in a 13% blend, gasohol would have about 96% the energy content of gasoline. Either way, more frequent trips to the gas pump would be necessary.

Methanol

Methanol is important as a solvent in many applications. As a fuel, it has been used in race cars for many years, primarily because its high octane rating (Table 13.1) allows cars to be run at very high compression ratios for greater power.

Methanol is sometimes known as *wood alcohol* because it was originally recovered as a by-product in the production of charcoal from wood by *destructive distillation* (the burning of wood with an insufficient air supply). Nowadays, it is obtained mostly from synthesis gas ($CO + H_2$), which comes from either natural gas or coal. The abundant high-sulfur coal in the United States may become an appealing source, preserving low-sulfur coal for direct burning. For now, natural gas remains as the primary source of synthesis gas because of its relatively low cost and because of the lower capital costs of its conversion technology. The reactions for synthesis of methanol from methane follow:

$$CH_4 + H_2O \rightarrow CO + 3\ H_2$$
$$CO + 2\ H_2 \rightarrow CH_3OH$$

Both the economic and energy balance sheets are quite favorable for methanol due to its likely sources. It seems to have a distinct advantage over ethanol as a major motor fuel for the future.

As a fuel, methanol could be used as a minor component of a blend with gasoline or, at the other extreme, "neat" or pure. A fuel known as M85, a blend of 85% methanol and 15% gasoline, is considered a major prospect for future use. Blends have several advantages over neat methanol. The presence of some volatile hydrocarbons assists in cold starting, and the gasoline discourages ingestion of the toxic methanol. Furthermore, although methanol burns with a cooler flame than gasoline, a fact that adds to its popularity as a racing fuel, the methanol flame is clear blue and almost invisible; thus someone could approach a methanol fire without recognizing it. Gasoline adds color and smoke to the flame so that it can be readily detected. Methanol can also be burned in a leaner mixture with air. This will make incomplete combustion less of a problem and decrease releases of carbon monoxide.

Methanol burns cooler than gasoline, so fewer oxides of nitrogen, NO_x, are formed. This provides a double benefit. Not only are the NO_x noxious compounds, but they also participate in the formation of smog (Figure 14.6). Government initiatives to deal with smog might well make methanol an important motor fuel even before shortages of crude oil do. Replacement of gasoline by methanol would decrease hydrocarbon release.

Methanol has several potential disadvantages. It is a toxic substance that must be handled with great care. It is well known for its ability to cause permanent blindness when ingested orally; it can also be inhaled and absorbed through the skin. These are grave concerns, although methanol may not be any more hazardous than gasoline.

Any incomplete combustion (oxidation) of methanol could lead to releases of toxic formaldehyde by the following equation. However, the amounts are regarded as small and could be eliminated using exhaust catalysts.

$$CH_3OH + \frac{1}{2}O_2 \rightarrow H-\overset{\overset{O}{\|}}{C}-H + H_2O$$
$$\text{formaldehyde}$$

$$HO-\overset{\overset{\displaystyle CH_3}{|}}{\underset{\underset{\displaystyle CH_3}{|}}{C}}-CH_3$$
tert-butyl alcohol

See Factors Affecting Solubility (Section 7.3) for a discussion of the interaction between alcohols and water.

Blends with gasoline present some potential problems. Even a very small amount of water destabilizes methanol-gasoline mixtures. If any water gets into a storage or fuel tank, it will extract the alcohol and form a separate layer at the bottom of the tank. As water accumulates, contamination of the fuel and corrosion of the fuel system could result. Therefore, such blends are certain to contain additives called **cosolvents** that inhibit corrosion and stabilize the mixtures. Blends of 10% methanol and 90% gasoline would require 5% to 10% of a higher alcohol such as *tert*-butyl alcohol (TBA) to stabilize them. Stability is much greater for ethanol-gasoline mixtures.

Methanol releases only 48% as much energy as gasoline (Table 13.1); thus fuel tanks would have to be filled twice as often. This disadvantage is offset somewhat by the higher efficiency possible in burning methanol due to the higher octane rating. This allows for higher compression ratios (up to 13:1) and greater horsepower, and increases the available energy content by about 10% so that 1.8 gal of pure methanol (M100) would replace 1 gal of pure gasoline.

Blends of gasoline with 5% or 10% methanol are likely to be prominent in the early stages because they can be used without significant engine modifications. They would have enhanced octane numbers and would stretch gasoline supplies.

Methanol to Gasoline

The Mobil methanol-to-gasoline (MTG) process provides a route to high-quality, high octane gasoline from nonpetroleum sources, coal, or natural gas (Section 13.8). The process is relatively simple, and high energy efficiency is associated with the conversion. Like other sources of synthetic motor fuels, the MTG process is not yet economically competitive with gasoline produced from cheap crude oil.

The MTG process achieved commercial success in New Zealand in 1986 under the combined sponsorship of the government of New Zealand and Mobil. Forces that drove it to commercialization were the total dependence on imported liquid fuels and an abundant supply of natural gas. The country's economy was devastated by the oil shocks of 1973 and 1979. In the 1990s, with the MTG process in place, New Zealand is importing only about one-half the amount of crude oil as in 1973 *(4)*.

The overall reaction for the MTG process is represented by the following equation, where $(CH_2)_n$ is the average formula of the alkane-aromatic hydrocarbon product mixture:

$$n\ CH_3OH \rightarrow (CH_2)_n + n\ H_2O$$

However, this simple equation does not reflect the complex chemistry that occurs. The MTG process depends on a highly selective catalyst, known as ZSM-5 zeolite, developed by Mobil and disclosed in 1976. The term **zeolite** signifies a group of **aluminosilicates.** These natural or synthetic materials consist of silicon and oxygen represented by the

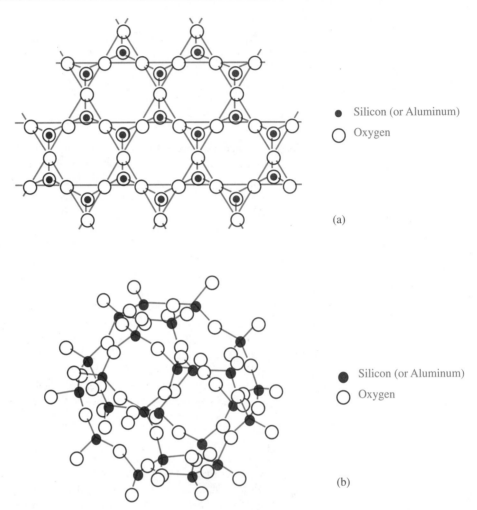

Figure 13.5 Either silica or an aluminosilicate may be represented by the two views of portions of the structure. Within the structure, each silicon (or aluminum) is bonded to four oxygen atoms and each oxygen is bonded to two silicon atoms. Part (a) is a cross section of the structure. Some oxygen atoms are located above or below the silicon atoms and are not shown. Part (b) shows the atoms in three dimensions. When some of the silicon atoms are replaced by aluminum atoms, the structure is called an aluminosilicate. Some crystalline forms of aluminosilicate are called zeolites. (From Cotton, F. A., Wilkinson, G. *Advanced Inorganic Chemistry,* 5th ed. Copyright 1988 by John Wiley & Sons. Reprinted by permission of John Wiley & Sons.)

formula $(SiO_2)_n$ with varying amounts of aluminum replacing some silicon atoms. Figure 13.5 shows two representations of a zeolite.

Since silicon is tetravalent, the formula $(SiO_2)_n$ denotes a complex three-dimensional structure in which each silicon atom is surrounded by four oxygen atoms and each oxygen acts as a bridge between two silicon atoms. The zeolites occur in a variety of forms depending on the amount of aluminum present and on the methods used to produce them.

The molecular-size spaces known as cavities or channels within the zeolite crystals can have various sizes and shapes. These channels restrict the size of reactants that can gain access to the interior and be attacked within the zeolite. The channels also restrict the size and shape of the molecules that can escape without modification.

Since methanol molecules are quite small, they can readily move into the three-dimensional framework of the zeolite and become exposed to chemical attack. The atoms of the zeolite catalyze the chemical reactions. In the pure $(SiO_2)_n$ structure, the silicon atoms are uncharged. But each aluminum atom is negatively charged and requires a cation (for instance Na^+). These sites within the zeolite are locations of high chemical reactivity.

The overall process can be represented by the following set of reactions:

$$2\ CH_3OH \longrightarrow H_2O\ +\ CH_3OCH_3\ \text{(dimethyl ether)}$$
$$\downarrow$$
$$\text{low-molecular-weight alkenes} + H_2O$$
$$\downarrow$$
$$C_5 \text{ or greater alkenes}$$
$$\downarrow$$
$$\text{alkanes} + \text{cycloalkanes} + \text{aromatic hydrocarbons}$$

Notice the formation of water in two steps of the scheme, accounting for the loss of the oxygen of methanol. The resulting mixture of alkanes, cycloalkanes, and aromatic hydrocarbons is similar to that obtained by refining petroleum. The channels in the zeolite have diameters just large enough to release hydrocarbons (mostly C_5 to C_{10}) that boil in the same range as gasoline (Table 6.3). In addition, the aromatic hydrocarbons, which make up about 40% of the mixture, have very high octane numbers *(5)*. The cycloalkanes also have good octane ratings. If any linear alkanes, with poor octane numbers, are produced and escape from the zeolite structure, their shape permits easy reentry and further alteration by the catalyst to cause formation of more highly branched alkanes or alkenes *(6)*. The final product mixture typically has a RON of 92 to 94.

It will be interesting to see how economic factors influence the competitiveness of the MTG process in relation to the use of gasoline from crude or shale oil, or blends of gasoline and methanol (or ethanol) as alternative motor fuels.

13.4 Methane: Fuels from Biomass

GOAL: To appreciate the potential for generating methane from waste.

Besides natural gas and substitute natural gas made from coal, another possible source of methane is the anaerobic treatment of sewage and garbage. Sewage contains an abundant supply of **anaerobic** microorganisms, particularly bacteria. In the absence of oxygen, the bacteria spontaneously cause the decay or fermentation of the sewage, producing methane gas. If the methane is collected, it can be burned as a fuel in place of natural gas. Peat bogs and rice paddies, too, generate methane, and beef cattle feeding lots are another prime location. It has been estimated that 100,000 cattle produce enough waste to yield methane to supply the needs of 30,000 people. Since the waste material used to generate methane comes from plant matter consumed by animals, the resource is described as renewable biomass derived from solar energy, which was originally required for plant growth.

The fossil fuels also could be regarded as derived from solar energy since they too were formed primarily from plant matter. But fossil fuels cannot be regarded as renewable because of the long time and special geological conditions required for them to form.

13.5 Hydrogen

GOALS: To study various sources of hydrogen.
        To consider the difficulty of storing hydrogen.

Last on this list of important synthetic fuel candidates for future use is hydrogen (H_2). Although the current expense of producing it, and the potential difficulties of transporting and storing it, have prevented widespread use, it is an excellent fuel. The potential for its extensive use has led many to refer to a "hydrogen economy." Proponents of hydrogen point to its low environmental effect. The only combustion product is water. There is no release of carbon dioxide, sulfur dioxide, particulates, or unburned hydrocarbons that plague the use of fossil fuels.

Sources and Production

Hydrogen does not exist as such in nature; it must be released from its combined forms. It can be generated by gasification of coal (Section 12.7) or by steam reforming of methane from natural gas. Natural gas is currently the primary source:

$$CH_4 + H_2O \rightarrow CO + 3\ H_2$$

This is the reverse of the methanation process in which methane is made from the CO and H_2 that have been formed from coal.

Another source of hydrogen is water, through the process of electrolysis (Section 4.2). If an electric current is passed through water in a device such as shown in Figure 13.6, oxygen is produced at one electrode (anode) and hydrogen is produced at the other electrode (cathode):

$$2\ H_2O \rightarrow O_2 + 2\ H_2$$

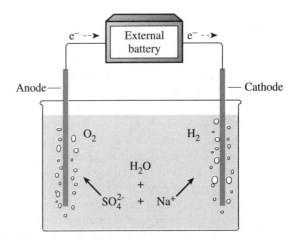

Figure 13.6 Electrolysis of water. Electric current from a battery or other source can be passed through water to form oxygen gas and hydrogen gas.

This method is most convenient, since it provides oxygen at the same time that it provides hydrogen. The drawback is economic, since the process requires electrical energy. In other words, hydrogen from this source (as from natural gas and coal) is a secondary energy resource. Even electricity is a secondary energy resource, formed by burning natural gas, coal, oil, or nuclear fuel.

One interesting aspect of this method is that hydrogen would be part of a cycle; the cycle would begin with water as the source of hydrogen and end with water as the product of combustion of the hydrogen:

$$2 \text{ H}_2 + \text{O}_2 \rightarrow 2 \text{ H}_2\text{O}$$

Transport and Storage

Hydrogen has an energy content of 325 Btu/scf. Although this is not competitive with natural gas for long-distance transport through pipelines (Section 12.7), the importance of hydrogen will increase as supplies of natural gas diminish.

Hydrogen is a flammable substance that forms explosive mixtures with oxygen in a wide range of concentrations. Some skeptics are quick to remind us of the burning of the German blimp *Hindenburg,* which was filled with hydrogen when it burst into flames in New Jersey in 1937. Some critics may even claim that there is some connection with the hydrogen bomb. We have already seen (Section 11.8), however, that the hydrogen bomb is a nuclear fusion device and has nothing whatsoever to do with hydrogen as a fuel.

A significant challenge to the use of hydrogen comes from its very low density and very low boiling point, −253 °C. Although it can be distributed as a gaseous fuel, it can be liquefied, gaining the advantages of transport and storage that are available with other liquid fuels.

Sophisticated and energy-consuming refrigeration equipment and insulated storage containers are required to liquefy hydrogen and keep it in liquid form. Such *cryogenic* storage involves a tank with a cold inner compartment for the liquid, surrounded by a second compartment that is a vacuum chamber that keeps environmental heat from reaching the chamber containing the hydrogen (Figure 13.7). This is the principle of the Thermos bottle and the Dewar flask for keeping contents either hot or cold.

Liquid hydrogen thus does not offer the ease of use and storage that we normally associate with other liquid fuels. In fact, it may never be used as a fuel for surface transportation vehicles.

The energy content of liquid hydrogen is 52,000 Btu/lb, compared with 19,000 Btu/lb for gasoline. This higher *energy per pound* is considered an important potential advantage for hydrogen as fuel for airplanes, for which weight is an important factor. On the other hand, liquid hydrogen has about one-tenth the density of gasoline, so that its *energy per gallon* is about 3.5 times less than that of gasoline.

In other words, the *energy density* of hydrogen is high on a weight basis but low on a volume basis. To fuel an automobile, a 20-gal gasoline tank would have to be converted to a 78-gal hydrogen tank to provide a comparable mileage range between fill-ups. Also, the weight of 20 gal of gasoline is about 120 lb; 78 gal of liquid hydrogen would weigh only about 47 lb, but the large *insulated* tank for hydrogen might easily weigh 200 lb or more and cost over $2000, which would become part of the purchase price of a car.

Furthermore, some hydrogen would boil off and escape from even the best-insulated tanks, so provisions would have to be made for venting. Therefore, long-term storage

Figure 13.7 The cryogenic storage and transport of liquid hydrogen. The tank of a liquid hydrogen transport trailer is wrapped with alternating layers of aluminum foil and fiberglass to a thickness of about 2 in. Once the tank is in place, the insulation space is maintained in a high vacuum to minimize heat transfer to the stored liquid hydrogen, so that evaporation is minimized.

would not be possible, and frequent fill-ups, probably by exchanging the empty tank for a prefilled one, would be necessary.

Substances called *metal hydrides,* which may offer an alternative means for storing hydrogen in a compact form, are being actively investigated. Some metals (for example, magnesium) and some metal alloys (for example, Mg_2Ni) are known to bind hydrogen in compounds such as MgH_2 and Mg_2NiH_4. These compounds are stable at normal temperatures but dissociate to release hydrogen at high temperatures; such high temperatures could be provided by hot exhaust gases. Thus, a tank of powdered or porous metal or metal alloy could be charged with hydrogen gas. On a volume basis, the concentration of hydrogen is greater in some hydrides than in liquid hydrogen.

Proponents of hydrogen tend to compare it with another secondary form of energy, electricity. Both can be generated from a variety of sources, whatever is economically most feasible. The relationship between the two is interesting, particularly when one considers that hydrogen can be made from water. Electricity cannot currently be stored in any significant quantity; it must be generated exactly when it is needed.

Some experts envision that cheap electricity produced during off-peak periods could be used for electrolysis of water to produce hydrogen. Since hydrogen can be stored as a liquid or as a gas (in underground caverns, for example), electrolysis could be carried out during off-peak times and hydrogen could be available as fuel when energy demands are high.

In a comparison of hydrogen and electricity, equivalent amounts of energy are much cheaper to transport as hydrogen. Some experts point to this as the greatest advantage offered by hydrogen and describe hydrogen as a ''relay device'' by which energy is moved from one form to another and from one location to another.

13.6 Electricity

GOAL: To consider the contributions of various primary energy resources for generating electric power.

From the first occasion of electric power generation in Chicago in 1887, and the first transmission over a distance of 4 miles carried out by engineers working for George Westinghouse in 1893, the importance of electric power has steadily increased, and it is expected to be even more important in the future. Figure 13.8 shows the use of the primary energy resources for generation of electricity over the last 4 decades, not including losses due to low efficiency of energy conversion (discussed in Section 12.2). Figure 13.9 shows the same information in more detail for 1991, but also reflects the low conversion efficiency associated with generating electricity.

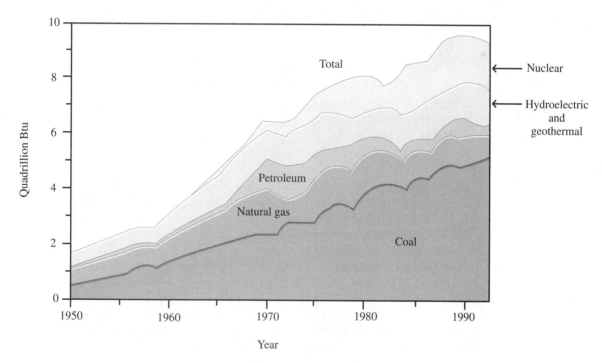

Figure 13.8 Net generation (not including losses due to low conversion efficiencies) of electricity by electric utilities by energy source, 1950–1992 (From *Annual Energy Review 1992;* Department of Energy/Energy Information Administration: Washington, DC, 1993.)

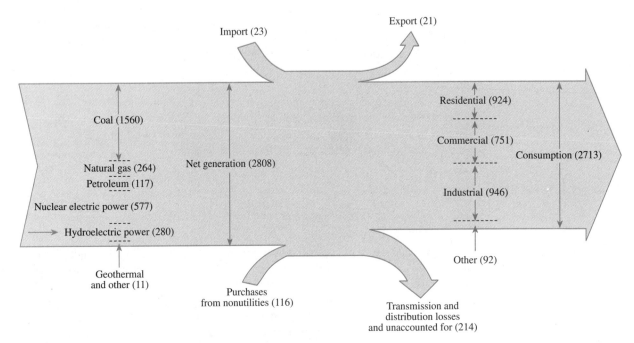

Figure 13.9 Electric utility electricity flow, 1991 (quadrillion Btu). Conversion losses account for about 66% of the total energy consumed. Power plants' use of electricity accounts for about 5% of the gross generation. Transmission and distribution (T&D) consume about 9% of the gross generation. (From *Annual Energy Review 1992*; Department of Energy/Energy Information Administration: Washington, DC, 1993.)

The following sections discuss other aspects of electrical energy, including some of the more esoteric means of generating and storing it, and the projected role of the electric vehicle.

13.7 Hydroelectric Power

GOAL: To see the potential for generating electric power from falling water.

Falling water is a renewable energy resource that is used to produce hydroelectric power. The largest single project in the United States is the 4000-MW Grand Coulee Dam on the Colorado River in Washington state. Hydroelectric dams are now operating in 47 states. New York has an extensive network of dams, including one at Niagara Falls.

As of 1990, 275 billion kW-h$_e$ of electric power were generated in the United States from hydroelectric power, accounting for 9.8% of the total. This is a decrease from the peak year 1983 when 332 kW-h$_e$ (14.3%) came from hydroelectric power *(7)*. Thus a resource of about 1 Q makes only a small contribution to the almost 10 Q of electric energy produced. Remember, however, that almost 3 Q of nuclear or fossil fuel energy would be required to replace renewable hydroelectric power due to the low (about 33%) efficiency of conversion to electric energy (Section 12.2).

The quad (Q) is short for 1 quadrillion Btu or 10^{15} Btu. See Section 12.3 for details.

13.8 Geothermal Energy

GOAL: To see the potential for generating electric power from geothermal energy.

Geothermal energy is heat energy found in large deposits that emerges as hot springs from just below earth's surface. It has been used for thousands of years for bathing and cooking.

Figure 13.10   Valley of steam. The photograph shows part of the geothermal steam field located in Sonoma and Lake counties about 90 miles northeast of San Francisco, where Pacific Gas and Electric Company generates 908,000 kW of electricity from underground steam. On top of the hill are units 7 and 8, which went into operation in 1972, and which each generate 53,000 kW. In the lower left corner are units 3 and 4, which became operational in 1967 and 1968 and were retired in 1992. Steam at the right and center comes from active geothermal wells. The electric power generated here is enough to supply almost 1 million residential customers.

Now, electric power is produced from it at several places around the world. Steam from the region around Larderello, Italy, has been used to generate electricity since 1904.

In the United States, the only geothermal resource in commercial use is the Geysers' steam field north of San Francisco, used by the Pacific Gas and Electric Company to generate much of the electricity for the city of San Francisco (Figure 13.10). The generating capacity at the Geysers' site is being continually expanded. Current capacity is approximately 1000 MW (1 gigawatt, GW), with estimates of the ultimate capacity ranging to 4 GW. The Geysers' accounted for about 0.1% of the production by electric utilities in 1990. We classify geothermal energy as another important renewable energy resource, but the total capacity is limited by the number of available sites.

The steam released from geothermal sources has a lower temperature than the steam generated in nuclear and fossil-fueled power plants. Therefore, the efficiency of generating electricity from geothermal energy is only in the range of 20% to 25%. Some of the heat is not even hot enough to generate electricity.

On the other hand, much of that heat can be used efficiently for space heating in home, commercial (particularly greenhouse), and industrial applications. About two-thirds of the space-heating requirements of Iceland are supplied by geothermal energy. Some estimates suggest that 1.5 to 2.0 Q could be used as direct heat in the United States by the end of the century.

Geothermal energy is a primary energy resource, even though the heat comes mostly from the decay of radioactive materials in the earth's interior. Pollution problems include the release of carbon dioxide, sulfur, arsenic, hydrogen sulfide, some radioactive materials, and high levels of water vapor. Although it is generally believed that the most serious pollutants can be contained, these concerns have stimulated considerable discussion.

13.9 Wind Power

GOAL: To see the potential for generating electric power from wind.

A primary renewable resource that has been used for centuries is the wind. Sailing ships once accounted for travel throughout the world, and windmills have been used to pump water and to generate electricity for many years. But only recently has serious attention been paid to the possibility of wind power for current energy needs. Unfortunately, it is unreliable and intermittent; therefore, its useful application is primarily as a supplemental source of electric power. For electric power generation, wind could help meet peak demands or be used with storage devices so that energy can be available whenever it is needed. Northern electric utilities with peak demands during the winter could well use wind power, since wind velocities tend to be higher at that time of the year.

Possibly wind power could be used in conjunction with hydroelectric power. When reservoir water levels were low and wind velocities were high, wind power could be used to pump water uphill into a reservoir. Electricity could then be generated using the hydroelectric capacity.

The uncertanties involved in wind power technology are so great that estimates of use by the end of the century range from 0.05 Q to as high as 1.25 Q.

13.10 Tidal Power

GOAL: To see the potential for generating electric power from changing tides.

Variations in water levels due to tides offer an alternative to hydroelectric power generation. In a few places in the world, tidal variations amount to 40 feet or more twice daily. A few locations have large amounts of water that are easily dammed during high tide. During low tide, this water can be released by emptying the basin through turbine generators.

The first tidal power plant to be built is across the estuary of the Rance River between Saint Malo and Dinard, France, where tides often exceed 40 feet. As the tide drops, the trapped water flows through 24 turbines, each of 13-MW capacity, for a total power output of 312 MW. This capacity compares favorably with nuclear power reactors and coal-fired plants, which usually have capacities of 500 to 1150 MW. But the nuclear and coal-fired units can provide an uninterrupted flow of electric power. Electricity from tidal power is intermittent, typically available for two periods a day, each lasting about 5 hours.

13.11 Solar Energy

GOAL: To consider various means of using solar energy.

The ultimate energy resource is the sun. Nuclear fusion processes in the sun release large amounts of energy, some of which penetrates earth's atmosphere. The amount of solar energy falling on the United States annually is well over 600 times the amount that we currently consume. This fact is potentially misleading, since a 10% efficiency of conversion to other energy forms is usually considered respectable. Therefore, the figure is reduced to 60 times the amount of required energy, and decreasing due to increasing demand. This has led some experts to visualize our turning two-thirds of New Mexico or Arizona into a huge solar collector, since these are the areas of greatest *insolation* or solar intensity (Figure 13.11). Realistically, solar energy might develop to the point of providing 5 to 10 Q of the 100 Q of total energy likely to be required by the end of the century.

Problems include the intermittency of sunlight, available at high intensity for only 6 to 8 hours a day, and only on sunny days. At other times, backup systems and/or energy storage capacity would be necessary for large power plants and small home solar systems.

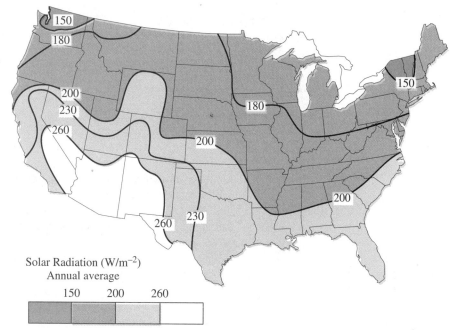

Figure 13.11   Annual solar intensity in the United States *(13)*. (Copyright 1974 by the American Association for the Advancement of Science.)

In evaluating the potential contribution from solar energy, we consider solar thermal energy, photovoltaic conversion, and passive solar heating. We may also think of more exotic forms, such as ocean thermal energy conversion and solar power satellites. Wind power has already been considered. You may be surprised that wind power is put under the heading of solar energy, but the heating effect of the sun on the earth's atmosphere causes most of the air flow. Also, fuels from biomass, including ethyl alcohol, are logically included under solar energy. Energy from the sun fuels the growth of plant material through photosynthesis.

Solar Thermal Energy

Solar energy is already being converted to thermal energy for space heating and water heating. The strategy is to collect the rays of the sun on a solar collector, a black absorbing surface, where it is transformed into heat energy, or infrared radiation. If air or water is circulated through the solar collectors, the heat can be carried to where it is needed or to storage. Large water tanks, rock piles, or other devices can be used to store the heat during daylight hours for release when it is required. A backup system, based either on electricity or fossil fuels, is usually required to meet the demands on cloudy days. High cost and questionable reliability of solar units for heating homes have limited their use.

Solar thermal energy can be used to generate electricity. From massive arrays of collectors, the collected heat is conducted to one central unit, where it is used to generate electricity by conventional means.

Another approach would place a central collector, called a power tower, so that sunlight would be reflected toward it by a large array of mirrors that rotate, tracking the sun during the day. Much higher temperatures can be reached than with other thermal energy methods, so the efficiency of generating electricity is greatly increased. This technique requires direct sunlight and would probably be practical only in the southwestern United States, where cloudy days are infrequent. An array of collectors can function in diffuse light, but the lower temperatures mean lower efficiency.

Photovoltaic Energy Conversion

Photovoltaic (PV) cells, also called solar cells, convert solar energy directly to electricity without intermediate conversion to heat. The cells' chief components are semiconductors, which are wafers of silicon treated with small amounts of other elements such as gallium or arsenic.

Although a seemingly more direct way of generating electricity, the chief drawbacks to PV cells are high cost and low efficiency. Considerable progress has been made in overcoming both problems. Crystalline silicon and gallium arsenide (GaAs) solar cells have been developed with efficiencies near 30%, dropping costs from $15 to about $0.30 per kW-h without storage. At this price, the PV power systems are within the range currently paid by some utilities for peaking power on hot summer days. To be competitive with other future electric generation options, photovoltaics must operate with total costs in the range of $0.06 to $0.12 per kW-h *(9, 10)*.

As noted earlier, the intermittent nature of solar energy is a major limitation. Variations in cloud cover, both short- and long-term, must be anticipated (Figure 13.12). Without significant storage capacity, PV power production may be limited to daytime energy replacement. However, the best times for PV power output match well to utility peak demands due to air conditioning on hot summer days.

It is common to describe some energy sources as *dispatchable* to signify that they can be called on at any time to deliver energy to meet a particular demand. Solar energy must be categorized as *nondispatchable* unless massive storage is available as well *(11)*. Thus

Electric utilities often use diesel or jet engines to drive generators to meet peak demands for electric power. These devices are relatively expensive to operate (Section 12.3).

Figure 13.12   500-kW photovoltaic collectors, installed in 1990, are linked to the local electric utility grid in Mt. Soliel, Switzerland.

its future as a significant source of electric power may well be determined by associated storage capacity.

Passive Solar Heating

Space heating using solar collectors is commonly called *active* solar heating. South-facing windows and skylights are two strategies that fall under the heading of *passive* solar heating. By emphasizing these designs in construction of new homes, shopping centers, and other buildings, the capture of solar energy can be maximized without great extra expense.

More elaborate versions of passive solar heating appear in Figure 13.13. In many instances, natural convection within the home can be used to circulate the heat, avoiding costly plumbing or ductwork required for active solar heating systems. Nevertheless, both passive and active systems require heat storage capacity.

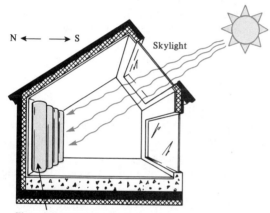

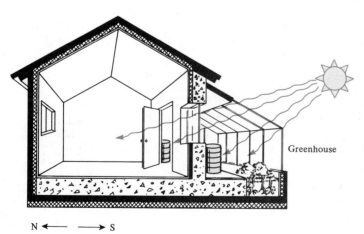

Figure 13.13 Passive solar heating using a traditional design (a) or a solar greenhouse (b). In (a) the south-facing skylight permits direct sunlight to strike north wall of room. Water-filled tubes or drums are placed along north wall to store heat. In (b) heat enters through door and windows. The south-facing greenhouse gains heat directly from the sun. This energy is stored in water-filled drums and in the massive floor. After the sun goes down and the greenhouse cools, heat is radiated from storage to supply the greenhouse and the living area. With proper design, plants and vegetables thrive in the greenhouse all year.

Sometimes large (typically 55-gal) metal drums or steel tubes filled with water are used for heat storage. They are painted black, so solar rays are more efficiently absorbed. In addition, as the space around the containers heats up, the water is warmed indirectly. When the sun is no longer shining and the temperature of the space decreases, heat is radiated from the water-filled containers into the room so that energy is available at night and on sunless days.

Two chemicals offer an interesting alternative means of storing heat in a passive system. Sodium sulfate decahydrate, $Na_2SO_4 \cdot 10\ H_2O$ (also known as Glauber's salt) and calcium chloride hexahydrate, $CaCl_2 \cdot 6\ H_2O$, have melting points slightly above normal room temperature (20–22 °C). Glauber's salt melts at 32 °C (90 °F) and hydrated calcium chloride at 27 °C (81 °F). When one of these salts is exposed directly or indirectly to the heat of the sun, it changes from solid to liquid. The melting process takes up a large amount of heat energy, called the *heat of fusion,* so that a large amount of energy can be stored in a relatively small volume. This energy is later released into a cooler room when the temperature drops, and the sample solidifies, liberating the heat of fusion.

Hydrated calcium chloride is commercially available as a product known as Thermol 81 from PSI, Inc. It is encased in polyethylene tubes 6 feet long and 3.5 inches in diameter. The tubes are expensive compared with barrels of water, but their advantage is that they can store almost 5 times as much heat energy in the same volume.

Ocean Thermal Energy Conversion

The sun causes water near the surface of any body of water to be warmer than the water at greater depths, and it may become possible to take advantage of the temperature difference to generate electricity. In *ocean thermal energy conversion* (OTEC), a fluid circulating within the conversion system can be alternately vaporized and condensed. For example, liquid ammonia or propane would vaporize by absorbing heat from warmer water near the surface of the ocean, and this vapor (like steam) could turn a turbine to generate electricity. Then the vapor would be circulated to lower depths, where the colder water would condense it back to a liquid. The cycle could be repeated, as shown in Figure 13.14. If such a system became practical, it is possible that the electrical energy could be used to generate hydrogen, which would be the means of transporting the energy to land.

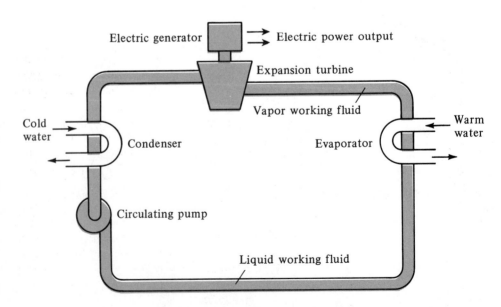

Figure 13.14
A schematic diagram of an ocean thermal energy conversion (OTEC) system *(11)*. See text for discussion. (Reprinted with permission of the copyright owner, the American Chemical Society, from *Chemical and Engineering News*.)

Solar Power Satellite

One scheme involves a satellite in space that would collect solar energy. The collecting could be continuous, without interference from clouds, and at much higher intensity, since earth's atmosphere absorbs much energy from sunlight. Solar energy would be converted to electrical energy and then to microwaves on the satellite. The microwaves would be beamed to a receiver on earth for conversion back into electrical energy.

13.12   Energy Storage

GOALS: To consider the need for large-scale energy storage, including pumped-storage and batteries.
To study some of the batteries used by consumers.
To appreciate the need for and the challenge of the electric vehicle.

In looking to the future and contemplating increased dependence on other primary energy resources, we must give energy storage much attention. Much present research and planning are focused on ways of storing electrical energy, since electric power can be produced so easily from various primary and secondary sources. Hydrogen storage has already been mentioned (Section 13.5). The intermittent nature of wind, sun, and tides, and the practice of using costly fossil fuels to generate electric power at times of peak demand also increase the need for storage capacity.

One convenience of petroleum that is often overlooked is the ease of storage. This is illustrated by the following comparison of alternatives (12). Each of the following items releases the same amount of energy:

7 gallons of petroleum

200 metric tons traveling at 100 m/s (224 mph)

250 charged lead-acid car batteries

1000 tons of water falling through 100 meters vertical drop

70 kg (154 lb) of wood

5 tons of rock, 250 °C above the surrounding temperature

2.5 tons of water, 100 °C above the surrounding temperature

1000 jelly doughnuts

Pumped Storage

At the present time, only one storage mechanism is commercially significant. **Pumped storage** uses off-peak electric power for pumping water uphill into a reservoir. During peak demand, the water is allowed to flow downhill and generate hydroelectric power.

The largest pumped-storage facility, nearly 1900 MW at maximum capacity, is operating at Ludington, Michigan (Figure 13.15). The plant uses Lake Michigan as the lower reservoir; the upper reservoir is a human-made lake more than 3 km long and 1.5 km wide. The facility was built at a cost of $340 million, which points out one of the obstacles to large-scale pumped storage. Few locations have suitable elevation differences for such a facility, and even at these the environmental effects are drastic.

Pumped storage also has an efficiency problem, as do other forms of energy conversion. For every three units of electric power needed for pumping, only two are recovered. This is not so bad compared with some inefficient steam generators that run on fossil fuels and are often pressed into service to meet peak demands.

Figure 13.15

The pumped-storage hydroelectric facility at Ludington, Michigan, is the largest in the world. During off-peak hours, water from Lake Michigan is pumped into the upper reservoir (top right), from which it flows back downhill to generate hydroelectric power during hours of peak demand.

Batteries

Some of the most intense research is directed toward developing new storage batteries. Since batteries produce electrical energy by chemical reactions within the batteries, they are called *electrochemical cells* (see Section 4.2 for details). Two factors are stimulating interest in batteries. One is electric utility peaking; the other is the push toward development of the electric vehicle.

Consumers are familiar with the *dry cell* used to power flashlights, transistor radios, and so on. Reactions at the negative electrode (the *anode*) and the positive electrode (the *cathode*) are:

$$Zn \rightarrow Zn^{2+} + 2\ e^- \qquad \text{at the anode}$$

$$2\ MnO_2 + 2\ NH_4^+ + 2\ e^- \rightarrow Mn_2O_3 + 2\ NH_3 + H_2O \qquad \text{at the cathode}$$

When the circuit is closed, as when a radio is turned on, electrons are produced in the

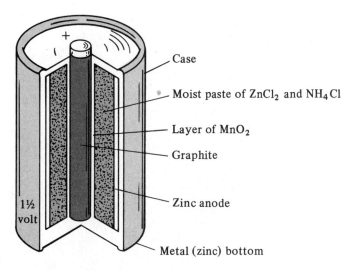

Figure 13.16 Dry-cell battery.

reaction at the anode and taken up in the reaction at the cathode. This flow of electrons is the electric current that powers the radio.

The dry cell pictured in Figure 13.16 has manganese dioxide, MnO_2, and ammonium ions, as NH_4Cl, contained in a moist paste between the anode and the cathode. The anode is the zinc can that acts as the container for the battery. A central graphite (carbon) rod acts as the cathode and conducts electrons into the battery to drive the cathode reaction. Since the anode reaction causes the zinc to dissolve (and convert to $ZnCl_2$), dry-cell batteries are prone to leak when they are worn out. Ammonium chloride, NH_4Cl, acts as an *electrolyte* to conduct current (electrons) between the electrodes in the battery.

As electrons are flowing from anode to cathode, the battery is said to be *discharging*. Some batteries can be *recharged* by attaching them to a device that forces electrons to flow from cathode to anode. If electrons are delivered to a zinc anode, zinc ions can be converted back to zinc metal. But the dry-cell battery is not rechargeable since the ammonia, NH_3, released at the cathode is tied up by the zinc chloride, $ZnCl_2$, in the form $Zn(NH_3)_2Cl_2$. Therefore, the cathode reaction cannot be forced in the reverse direction.

The dry-cell battery is commonly considered a general-purpose battery; it is also known as the zinc-carbon battery to identify the electrodes.

The major competitor for the dry-cell battery is the zinc-alkaline-manganese dioxide battery, known as the *alkaline battery*. Similar in appearance to the dry-cell battery, it also has a zinc anode and uses MnO_2 as the cathode-reactive material. But zinc is present as a powder, which gives the alkaline battery far more activity so that it will deliver more power than the dry cell. The container is nickel-plated steel rather than zinc, so it does not participate in the electrochemical reactions as in the dry cell. Potassium hydroxide, KOH, replaces ammonium chloride as the electrolyte; this accounts for the name *alkaline cell*.

The MnO_2-graphite blend of the alkaline battery is highly conductive in direct contact with the inside of the can. This arrangement provides direct electrical contact with the button cap, so a graphite rod to conduct electrons into the battery is unnecessary. Because of the greater energy output of the alkaline cell, the cost is normally higher than for the dry-cell battery *(13)*.

The zinc-air cell is used primarily in button form in hearing aids. It has the highest energy density of any commercially available battery, because atmospheric oxygen is the

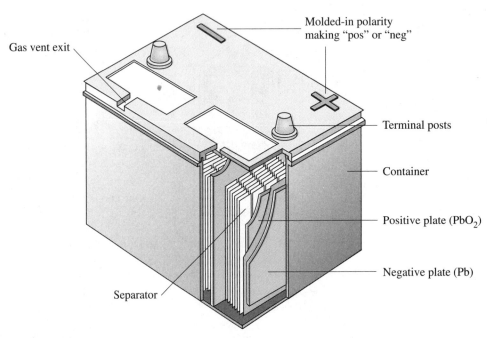

Gas vent exit

Molded-in polarity making "pos" or "neg"

Terminal posts

Container

Positive plate (PbO_2)

Negative plate (Pb)

Separator

Figure 13.17 The lead-acid battery, also called the lead-storage battery, consists of a set of lead, Pb, and lead oxide, PbO_2, plates immersed in a sulfuric acid solution. As the battery discharges, electrons flow from the Pb anode to the PbO_2 cathode due to the reactions described in the text. During recharging, the opposite occurs.

cathode material. The electrode itself can be very thin, allowing more space for the zinc anode material, and giving the battery greater capacity. Air holes on one side are sealed until the battery is installed, to exclude entry of oxygen and increase shelf life.

Batteries that cannot be recharged are called *primary cells. Secondary cells,* also called *storage cells,* are able to go through a series of discharge-recharge cycles.

The best-known storage battery, devised in 1859, is the lead-acid battery (also called the lead-storage battery) used in automobiles. It uses a lead, Pb, anode and a lead dioxide, PbO_2, cathode. The electrode reactions are as follows:

$$Pb + SO_4^{2-} \rightarrow PbSO_4 + 2\ e^- \quad \text{at the anode}$$
$$PbO_2 + SO_4^{2-} + 4\ H^+ + 2\ e^- \rightarrow PbSO_4 + 2\ H_2O \quad \text{at the cathode}$$

Sulfuric acid, H_2SO_4, acts as a medium between the electrodes, as shown in Figure 13.17.

The Electric Vehicle

One of the most intriguing uses of batteries under development throughout the world is the electric vehicle (EV). It is intended as a nonpolluting form of transportation primarily within cities over short distances, with a targeted range of 80 to 120 miles.

Electric cars are often seen as the ultimate answer to the emissions pouring out of the exhaust pipes of the world's 500 million vehicles. In smoggy cities like Los Angeles and Tokyo, vehicle emissions of unburned hydrocarbons, carbon monoxide, and nitrogen oxides are serious problems (Section 14.6). Attempts to make cars cleaner by using catalytic converters in exhaust systems have been disappointing. Battery-driven electric cars, which release no exhaust at all, seem to offer a solution.

In California, a market for EVs is virtually guaranteed by a mandate from the state Air Resources Board. Beginning in 1998, 2% of all new cars sold in the state must be zero-emission vehicles; 10% are required by 2003. So far, only the electric car appears likely to meet the 1998 deadline. Some 170,000 cars are currently sold in California each year. Assuming 200,000 will be sold in 2003, the 10% rule would require 20,000 "volts-wagons" to be sold annually by that year *(14–16)*.

It has been suggested that electric utilities might go into the business of supplying owners with both off-peak-rate electricity and batteries for the life of the vehicle. Perhaps a flat-rate annual "power fee" would include financing charges for replacement battery packs as needed *(17)*. The electric companies are eager to provide the energy for electric cars, since their off-peak capacity in the evenings is readily available. As a result, the Electric Power Research Institute of Palo Alto, California, a nonprofit corporation that represents 680 utilities, joined with the United States Advanced Battery Consortium, a partnership made up of the big three U.S. automobile makers, to develop new-generation batteries for EVs *(14, 18)*.

Although the well-developed lead-acid battery delivers the small bursts of energy necessary to start cars and power some industrial and recreational vehicles, it is not well suited for the EV. The *specific energy* of the battery, which is the amount of energy that can be stored per kilogram of battery weight, is low. A high specific energy translates into a longer driving range for the EV.

Since lead is a very heavy metal, the specific energy of the lead-acid battery is only about 35 W-h/kg. The significance of this is apparent from simple calculation. A typical goal for energy storage capacity for batteries that can power EVs is about 60 kW-h. Thus, a lead-storage battery at 35 W-h/kg that could provide 60 kW-h would weigh more than 1700 kg (almost 3800 lb).

$$60 \text{ kW-h} \times \frac{1 \text{ kg}}{35 \text{ W-h}} \times \frac{1000 \text{ W-h}}{1 \text{ kW-h}} = 1714 \text{ kg}$$

From Table 13.1 and Section 12.3, we know that 1×10^{15} Btu equals 2.93×10^{11} kW-h, and that a gallon of gasoline has an energy content of 124,800 Btu/gal. This allows us to calculate the energy content of a gallon of gasoline in kW-h as:

$$124,800 \text{ Btu/gal} \times \frac{2.93 \times 10^{11} \text{ kW-h}}{1 \times 10^{15} \text{Btu}} = 36.6 \text{ kW-h/gal}$$

Therefore, the energy stored in a 1700-kg set of batteries that can deliver 60 kW-h is equivalent to less than 2 gal of gasoline. This means that the range of an EV powered by lead-acid batteries is fairly limited. That is why improvements in specific energy are so important. In addition, the lead-acid battery does not stand up well to many full discharge-recharge cycles. The upper limit is normally 300 cycles.

General Motors will probably be first to mass-produce an electric car, currently known as the Impact (Figures 13.18 and 13.19). The design of the vehicle in its current stage of development reveals the promise and the challenges that lie ahead. The Impact is a two-seat, 95-inch-wheelbase coupe, and weighs only 2,200 lb, including an 870-lb battery pack of 32 ten-volt lead-acid batteries housed in a central tunnel *(19)*.

The Impact benefits from a lightweight aerodynamic body and advanced electronics that allow the use of lighter and more efficient electric motors. In spite of this, when battery

Figure 13.18 The Impact from General Motors. See also Figure 13.19.

Figure 13.19 The components of the Impact showing the battery pack, DC to AC converter, and port for charging.

replacement costs and the price of electricity are combined, GM estimates the Impact's energy costs at twice those of a similar-size gasoline-powered car.

The design incorporates two 57-hp AC induction motors, each driving one front wheel to produce an 8-second 0- to 60-mph time, with a 120-mile projected range at 55 mph. Adding lights and air conditioning will decrease the range. The carrying capacity of the Impact is only 350 lb, hinting at how difficult it is to minimize the total weight. The battery pack can be recharged overnight at a rate of up to 50 amperes at 400 volts through a built-in charger. Thus, instead of taking 3 minutes to fill a car at the gas pump, recharging the depleted battery will take 6 to 8 hours.

A critical factor is the cost of battery replacement. Because the life span of batteries is limited, replacing them on a regular basis should be the main cost of operating the car. Estimates vary, but lead-acid batteries are typically projected to last about 2 years or 20,000 to 30,000 miles in average commuter use.

It is clear that GM adopted the lead-acid battery because the technology is so well-developed. Other companies developing EVs have focused on alternative batteries. The nickel-zinc storage battery (Edison cell), invented by Thomas Edison in 1910, has a short cycle life (limited number of recharges), but it has the highest specific energy compared with lead-acid and nickel-iron batteries. The specific energy values are 50 W-h/kg for the nickel-iron battery and 70 W-h/kg for the nickel-zinc battery. Although both of them seem to be superior to the lead-acid battery, neither is under serious consideration for incorporation into the EV. The high cost of nickel and the need for frequent replacement make them uneconomical.

The most interesting of the systems being considered is the sodium-sulfur battery. Its specific energy rating is estimated to be as high as 200 W-h/kg (not including battery housing, insulation, and hardware)—about 4 times as high as the lead-acid battery's. Since it would weigh much less, either range or payload (weight-carrying capacity) or some combination can be improved. Recall the small payload anticipated for the Impact. For most developers, this limitation is regarded as more important than the limited range (Figure 13.20).

Some interesting drawbacks are associated with the sodium-sulfur battery. Most batteries are constructed with metal plates in contact with a liquid electrolyte. In contrast, the

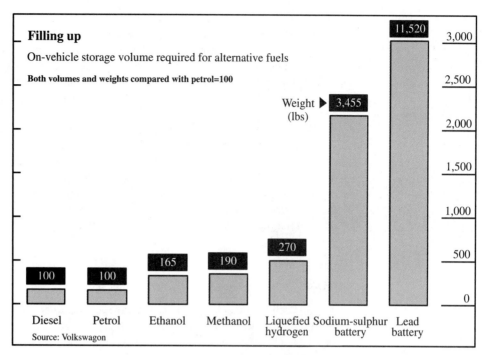

Figure 13.20 On-vehicle storage volume and weights required for alternative fuels. (From *The Economist,* October 13, 1990.) *(16)*

sodium-sulfur battery has two liquid electrodes separated by a solid electrolyte. Molten sodium (anode) and molten sulfur (cathode) are separated by a high-temperature ceramic. The heat generated by the battery and trapped within a heavily insulated capsule keeps the electrodes molten unless the car is shut down for several days. Then a warm-up stage requires heating by an internal unit plugged into an outside power supply.

At the present time, the life of sodium-sulfur batteries is relatively short, but improved technology, driven by the changing marketplace, could easily lengthen it. In addition, a potential hazard must be solved. Since sodium metal reacts violently with water, release of sodium in the event of an accident must be prevented, and all such batteries must be recycled safely.

One final potential car battery is the zinc-air battery. As noted earlier for hearing aids, it uses zinc as the anode and oxygen (from air) as the cathode leading to formation of zinc oxide (ZnO). When it is recharged, the zinc is reconstituted and oxygen is given off. Still in the developmental stages, this battery has a potential energy density of about 200 W-h/kg; energy densities of 160 W-h/kg have been observed in an EV driving on city streets *(18)*.

Convenience and cost will have to appeal to the consumer for the EV to become attractive. For convenience, a reasonable range, access to electricity, and ease and safety of recharging batteries will be the main considerations. Although the cost of electricity will be quite low compared with gasoline, the replacement cost of battery packs will undoubtedly determine the economic appeal of the EV. Civic incentives may also be important. Los Angeles is considering a range of tax credits, reduced electric rates, special driving and parking privileges, direct price subsidies ($2,000–$5,000 per car), requiring charging stations in new construction and corporate parking lots, as well as buying a large number of EVs outright *(15)*.

At some point, hybrid vehicles will be developed. A combustion engine will drive the wheels outside environmentally sensitive areas, switching to battery power in restricted zones. Audi has demonstrated the viability of this concept with its Duo Hybrid system, which combines front-wheel drive by a gasoline engine for long-distance cruising with an electric motor powering the rear wheels as a nonpolluting alternative. In fact, the dual ability also provides the potential to recharge the batteries using combustion to generate electrical energy.

But even when the ideal battery system comes along for use in the EV and has a significant impact on urban pollution, we must recall that electrical energy is a secondary energy resource. It must be produced from a primary resource, usually either coal or nuclear energy. Although the polluting effects of nuclear energy are controversial, the negative environmental impacts of coal are well known. Therefore, the environmental impact of the EV is merely transferred to the electric power-generating station. Above all, this kind of dilemma illustrates the interplay between energy and environment.

13.13 Fuel Cells

GOAL: To see the chemistry and uses of fuel cells.

The fuel cell has been left as the last storage method to be considered because it combines aspects of three topics discussed in this chapter: hydrogen, electricity, and electrochemical cells. Although fuel cells may take many forms, the simplest uses oxygen and hydrogen in an electrochemical cell to generate electricity. It resembles a battery in that there are electrode reactions in which electrons released at an anode are taken up at a cathode.

For the hydrogen-oxygen fuel cell, the following reactions apply when an alkaline medium is used:

$$2\ H_2 + 4\ OH^- \rightarrow 4\ H_2O + 4\ e^- \quad \text{at the anode}$$
$$O_2 + 2\ H_2O + 4\ e^- \rightarrow 4\ OH^- \quad \text{at the cathode}$$

Although this is formally similar to the reactions in dry-cell batteries and storage batteries, the distinction between a fuel cell and a battery is that the fuel cell does not have storage capacity for either fuel. In a lead-storage battery, Pb and PbO_2 electrodes react electrochemically to produce electricity. In an oxygen-hydrogen fuel cell, both hydrogen and oxygen are supplied from an outside source, as indicated in the schematic diagram in Figure 13.21.

Adding the two electrode reactions for the oxygen-hydrogen fuel cell, the net reaction is:

$$2\ H_2 + O_2 \rightarrow 2\ H_2O$$

This reaction also occurs when hydrogen is burned in the presence of oxygen. In fact, the fuel cell provides a means for the two to react without combustion and without even coming into contact with each other.

If the hydrogen and oxygen were allowed to burn, the heat could be used to boil water and generate electricity with a steam turbine. As we have seen before, this process is inefficient (typically 60–70% of the heat energy is wasted) due to the step in which heat is transformed to the mechanical energy of the steam turbine. But in the fuel cell, the energy released by the chemical reactions is converted directly to electricity, so that efficiencies up to 55% are possible. Until now, the best-known application of the fuel cell has been in the space program, where cost has not been a limiting factor.

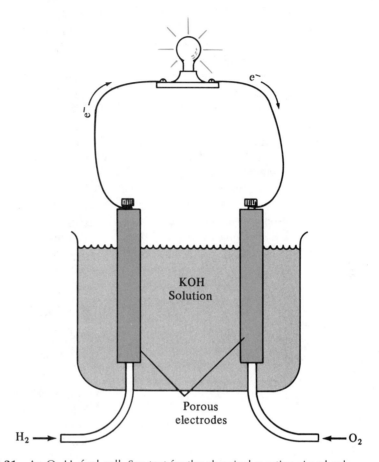

Figure 13.21 An O_2-H_2 fuel cell. See text for the chemical reactions involved.

SUMMARY

Synthetic fuels will likely become increasingly important as sources of energy to meet needs that are now met by the traditional fossil fuels. Gaseous and liquid fuels from coal, oil from shale and tar sands, methane and ethyl alcohol obtained from plant and animal wastes, methyl alcohol, and hydrogen are the major possibilities.

The most concentrated source of oil shale extends through Colorado, Wyoming, and Utah. The fuel is recovered from shale by retorting to release kerogen, which is hydrogenated and refined into liquid fuels. The biggest deposits of tar sands are in Alberta, Canada.

Anaerobic decay of sewage and garbage that have been acted on by bacteria provides a source of methane. Yeast cells can ferment sugars to form ethyl alcohol, which can be used alone or in a combination with gas-oline, called gasohol, as a motor fuel. The sugars can be derived from plant wastes by chemical breakdown of complex polysaccharides that are not fermentable. Sewage, garbage, plant waste, and other such sources of methane or ethyl alcohol are called biomass.

The economics of alcohol production, together with the energy efficiency of the process, argue against its large-scale use. Nevertheless, the waste, expense, and complications of using alcohol as a fuel may have to be accepted, considering the demand for liquid fuels. Methyl alcohol, produced from coal as methyl fuel, could be combined with gasoline.

Hydrogen is an attractive (although challenging) alternative synthetic fuel. It is excellent for many applications, although the economics of producing it and the methods required to store it are drawbacks. It may find its greatest use as a relay device or a storage device for other forms of energy.

Electric power may assume increasing importance for space heating and powering electric vehicles as petroleum supplies dwindle. Dependence on nuclear power, which, as an energy source, is useful only for generating electricity, is likely to increase, although the breeder or fusion may be necessary to meet the long-term demand.

Hydroelectric power, geothermal energy, wind power, and tidal power are alternatives to traditional steam-electric systems for providing electricity. Each is limited in total capacity, but their chief advantage is that they are renewable.

The ultimate energy resource is the sun. Wind power and fuels from biomass are sometimes grouped under solar energy, but solar thermal energy conversion, photovoltaic conversion, passive solar heating, ocean thermal energy conversion (OTEC), and solar-powered satellites involve more direct use of solar energy.

Energy storage is undergoing intense research since it is a necessary feature when replacing petroleum as a fuel. The convenience of petroleum as a storage form is often overlooked; it contains energy that can be used to produce heat or electric power at any moment. Solar energy, wind power, and similar alternatives do not offer this convenience; therefore, backup systems and storage devices become essential for energy independence.

Pumped storage and batteries (electrochemical cells) are two storage devices. Batteries for peaking and electric vehicles are under investigation. If hydrogen is developed as a means of storing energy, the hydrogen-oxygen fuel cell may become important to efficient use of hydrogen.

PROBLEMS

1. Give a definition or example of each of the following.

   a. Kerogen
   b. Retorting
   c. Syncrude
   d. Bitumen
   e. Stover
   f. Biomass
   g. Antiknock index
   h. Oxygenates
   i. MTBE
   j. E96
   k. M85
   l. Neat
   m. Formaldehyde
   n. Cosolvents
   o. MTG
   p. Zeolite
   q. Aluminosilicates
   r. Wood alcohol
   s. Cryogenic
   t. Heat of fusion
   u. Glauber's salt
   v. OTEC
   w. Photovoltaics

2. What are the candidates for synthetic fuels as liquids? As gases?

3. What is gasohol? What advantages and disadvantages would be associated with its use?

4. What compounds are being used to maintain octane ratings of gasoline as replacements for lead?

5. Why is fuel volatility a problem?

6. Why would cosolvents be used in blends of 10% methanol and 90% gasoline?

7. What special circumstances exist in Brazil and New Zealand, and how have synthetic fuels come into prominence there?

8. Why would M85 be preferred over M100 as a motor fuel?

9. Describe the ZSM-5 zeolite in terms of chemical structure, catalytic effects, and selectivity of attack on hydrocarbons and other compounds.

10. What are the principal sources of hydrogen?

11. What advantages and disadvantages are associated with hydrogen as a fuel?

12. What is a potential role for metal hydrides?

13. Where is geothermal energy now being used?

14. Where is tidal power now being used?

15. Seven forms of energy can be placed under the heading of solar energy. What are they? Which of these are now being used?

16. What is a dry-cell battery? Where do the electrons come from as a dry-cell battery is discharged to deliver electric current? Answer the same question for the lead-storage battery.

17. Compare the dry-cell battery and the alkaline cell battery.

18. Distinguish between primary cells and secondary cells.

19. What is the significance of specific energy?

20. Why might a dry-cell battery leak when it is fully discharged?

21. Would you like to put in an order for an Impact from General Motors? Why or why not? Is the EV a solution to pollution?

22. Gasoline has an energy density of about 400 W-h/kg. How does this compare with the lead-acid battery, and what conclusions do you draw from the comparison?

References

1. Metz, W. D. "Oil Shale: A Huge Resource of Low-Grade Fuel." *Science* **1974,** *184,* 1271.

2. Benson, J. "What Good Are Octanes?" *Chemtech* **1976,** *6* (1), 16–23.

3. Anderson, E. "Brazil's Fuel Ethanol Program Sputters." *Chemical and Engineering News,* January 15, 1990, p. 6.

4. Maiden, C. J. "A Project Overview." *Chemtech* **1988,** *18* (1), 38–41.

5. Meyers, R. A. *Handbook of Synfuels Technology.* McGraw-Hill: New York, 1984; pp. 2–79.

6. Meisel, S. L. "Catalysis Research Bears Fruit." *Chemtech* **1988,** *18* (1), 32–37.

7. *Annual Energy Review 1991.* Department of Energy/ Energy Information Administration: Washington, DC, 1992; p. 211.

8. Calvin, M. "Solar Energy by Photosynthesis." *Science* **1974,** *184,* 375–381.

9. Post, H. N.; Thomas, M. G. "Photovoltaic Systems for Current and Future Applications." *Solar Energy* **1988,** *41* (5), 465–473.

10. Hubbard, H. M. "Photovoltaics Today and Tomorrow." *Science* **1989,** *244,* 297–304.

11. Henrie, J. O.; Beck, E. J.; Zener, C.; Fetkovich, J. "Ocean Thermal Gradients—A Practical Source of Energy?" *Science* **1977,** *195,* 206.

12. Rose, D. J. *Learning About Energy.* Plenum: New York, 1986; pp. 445–446.

13. "Batteries: Disposable or Rechargeable?" *Consumer Reports* **1991,** *56* (11), 720–723.

14. Welter, T. R. "GM Makes an Impact: Environmental Regulations and Technologic Advances Propel the Electric Car." *Industry Week* **1991,** *240* (2), 40.

15. Frank, L.; McCosh, D.; Normile, D. "Electric Vehicles Only." *Popular Science* **1991,** *239* (4), 39.

16. "How Clean Is the Plug-In Car?" *Economist* **1991,** *317,* 76.

17. Brown, S. F. "California Dreaming." *Popular Science* **1992,** *240* (4), 39.

18. Fischetti, M. "Electric-Car Start-Ups." *Popular Science* **1991,** *239* (4), 39.

19. Fischetti, M. "Here Comes the Electric Car—It's Sporty, Aggressive and Clean." *Smithsonian* **1992,** *23* (1), 34–43.

Air and Water Pollution

14

One of our most abundant consumer products is pollution. It is the consequence of virtually all human activity. The food we eat; the energy we consume by lighting, heating, and cooling our homes and businesses; and the plastics and other consumer products that we use all have profound effects on the environment. Pollution is inevitable, but once recognized and understood, it can often be controlled.

In many instances, the well-intended decision to avoid some environmental damage is made without realizing the consequences of the alternatives. Would you vote for construction of a nuclear power plant, with its potential hazards and disposal problems? Or would you rather get your electricity from a coal-fired plant that releases sulfur dioxide, contributes to acid rain, and releases carbon dioxide that contributes to the greenhouse effect? Or would you prefer a modern coal-fired power plant with much higher electric bills to support the technology required to prevent the release of sulfur dioxide?

Of course solar energy may be less polluting than fossil fuels, but energy derived from the sun is not always available when and where we need it. Conservation is always desirable, but what are you willing to give up to make your contribution? We all enjoy our creature comforts and conveniences. We appreciate the supermarkets with vast arrays of foods that are produced using chemical fertilizers and pesticides, and that are transported to us by burning fossil fuels to give us the quality and prices that we expect. We prefer the convenience of driving to the supermarket rather than using public transportation, which would burn less fossil fuel and help protect the environment. We appreciate the convenience of liquid and solid detergents for washing clothes, dishes, and the like. We like to have our clothes come out clean, bright, soft, and without static cling, even if the chemicals that drain out of washing machines may contribute to pollution of water supplies. In other words, we often ignore the consequences that our actions have on the environment.

But then we come across some troubling information about the ozone layer, the greenhouse effect, or acid rain, and this captures our interest, at least until we hear that scientists do not all agree on the seriousness of a problem. Such controversies are both good and bad. The good news is that scientists are often stimulated to acquire reliable data to support their opinions, even though it often takes many years. The bad news is that some individuals may argue without the benefit of sufficient facts, causing the public to become confused and tune out the issue altogether. Most questions must be answered before testing has determined the environmental impact of each pollutant.

This chapter is meant to acquaint you with several environmental issues, including what is known about them and what controversies exist. In some cases, you will discover that information is unknown or incomplete, so you can listen with an educated ear for further developments. In the meantime, you will be able to filter extreme comments on all sides of some issues.

Many environmental issues are addressed elsewhere in the text, including nuclear energy, radon, generation of electricity, motor fuels, pesticides, and detergents. In this chapter we will consider the information necessary to understand acid rain, the greenhouse effect, the effect that chlorofluorocarbons (CFCs) have on the ozone layer, smog, and other aspects of air and water pollution.

14.1   Clean Air

GOAL: To appreciate common units used in describing the composition of air.

Air is a mixture of gases. The definition of clean air must include a judgment on what air is without any undesirable substances added to it. Some normal components are regarded as pollutants because more of them are present than the definition of clean air allows. Some components of clean air are listed in Table 14.1; the list is not exhaustive, however.

The components in the table are listed according to the *percentage by volume;* the other common percentage is weight percentage. Volume percentage is calculated as follows:

$$\text{Volume \%} = \frac{\text{volume of component}}{\text{total volume of air}} \times 100$$

Various concentration terms, such as mass percent, volume percent, and parts per million, are discussed in Section 7.4.

Air contains many components in minute amounts. When the volume percentage is very small, it becomes an awkward way to describe concentration and we change to *parts per million (ppm)*, which is calculated as follows:

$$\text{parts per million (ppm)} = \frac{\text{volume of component}}{\text{total volume of air}} \times 10^6$$

Table 14.1   Components of clean dry air (1–3)

Component	Volume Percent	ppm by Volume
Nitrogen, N_2	78.084	
Oxygen, O_2	20.946	
Argon, Ar	0.934	
Carbon dioxide, CO_2	0.035	
Neon, Ne	0.001 818	18.18
Helium, He		5.24
Methane, CH_4		1.7
Krypton, Kr		1.14
Hydrogen, H_2		0.5
Nitrous oxide, N_2O		0.3
Carbon monoxide, CO		0.1
Xenon, Xe		0.08
Ozone, O_3		0.02 to 10
Ammonia		0.006
Nitrogen dioxide, NO_2		0.006
Nitric oxide, NO		0.0006
Sulfur dioxide, SO_2		0.0002
Hydrogen sulfide, H_2S		0.0002

Thus, the concentration in parts per million is essentially the number of parts of a component in 1 million parts of air, where the number of parts is determined by volume measurements.

14.2   Polluted Air

GOAL: To understand the specific problems associated with many of the major pollutants—carbon monoxide, sulfur oxides, nitrogen oxides, ozone, carbon dioxide, particulates, and smog.

Air pollutants are both solid particulates and gases. Of the gases listed in Table 14.1, the following are regarded as air pollutants: carbon monoxide, sulfur dioxide, sulfur trioxide, nitrous oxide, nitric oxide, nitrogen dioxide, ozone, carbon dioxide, methane, ammonia, and hydrogen sulfide. The two major chlorofluorocarbons, CFC-11, and CFC-12, are not tabulated; they are present in the atmosphere at concentrations of approximately 0.00023% and 0.00038% by volume, respectively (Section 14.5). Radon is not listed because all of its isotopes are radioactive (unstable) (Section 10.6).

The impact of some pollutants is not always easy to judge. Levels of carbon dioxide or the CFCs may not be directly harmful, for example, but may contribute to problems such as global warming or depletion of the stratospheric ozone layer.

Carbon Monoxide

Carbon monoxide is a well-known poison for living systems as well as an air pollutant. Three special dangers associated with it are that it is colorless, odorless, and tasteless, so exposure is not readily noticed. If the concentration of carbon monoxide is 400 to 500 ppm and healthy human beings inhale it for 1 hour, it has no appreciable effect. If the concentration is 600 to 700 ppm, inhalation for 1 hour gives barely detectable effects such as shortness of breath. A concentration of 1000 to 2000 ppm is classified as dangerous, and 4000 ppm or higher is fatal in less than 1 hour *(4)*. The data do not include different tolerances for individuals who may be elderly or have respiratory disease. Neither do they suggest the effects on living systems besides humans. For example, what is the effect of CO on growing crops, trees, cattle, hogs, or other species essential to human life?

Sulfur Oxides

Burning coal releases one of the more highly publicized air pollutants, sulfur dioxide, SO_2 (Section 12.7). Sulfur dioxide in contact with atmospheric oxygen is slowly oxidized to sulfur trioxide, SO_3:

$$2\ SO_2\ (gas)\ +\ O_2\ (gas) \rightarrow 2\ SO_3\ (gas)$$

Sulfur trioxide reacts quickly with water to produce sulfuric acid:

$$SO_3\ (gas)\ +\ H_2O\ (liquid) \rightarrow H_2SO_4\ (liquid)$$

See Common Acids and Bases (Section 7.5) for a discussion of the action of calcium carbonate as a base.

Large amounts of sulfuric acid close to the earth's surface can cause extensive damage to buildings and statues. Marble is rock originally deposited as limestone, $CaCO_3$. Heat and pressure change the size of the calcium carbonate crystals. Since $CaCO_3$ is a base, we expect that items made of limestone (Figure 14.1), including marble, will be susceptible

Figure 14.1
The damaging effects of acidic air pollution can be seen by comparing these photos of a decorative statue on the Field Museum in Chicago. The left photo was taken c. 1920, the right in 1990.

to the effects of sulfuric acid resulting from SO_2 pollution. The effect is explained by the following chemical reaction:

$$CaCO_3 + H_2SO_4 \longrightarrow CaCO_4 + H_2CO_3$$
$$\text{marble} \qquad\qquad\qquad \text{unstable}$$
$$\downarrow$$
$$H_2O + CO_2$$

When its surface is exposed to sulfuric acid, marble is changed to calcium sulfate, which is more soluble in water than $CaCO_3$. The surface coating is washed away more readily by rainfall, and a fresh surface of $CaCO_3$ is exposed. Even $CaSO_4$ is not very soluble, so the effects take place slowly, although the greater the SO_2 pollution, the faster the erosion of the marble. Priceless statuary and buildings of Venice, Italy, are being severely eroded by sulfuric acid, due to SO_2 produced by extensive industry along the coast. The famous lions outside the New York City Public Library are suffering the same fate.

Unlike carbon monoxide, the presence of sulfur dioxide is immediately noticeable, irritating the throat and lungs. Inadvertent poisonings are not possible. Since sulfur dioxide, sulfur trioxide, and sulfuric acid are so irritating, it has become commonplace for power plants that produce sulfur dioxide to have very tall smokestacks. This may effectively spare residents of nearby areas by delivering the pollutants into the wind to be dispersed over a much wider area. This strategy is a throwback to the old saying, ''The solution to pollution is dilution.'' But, all too often, the result is *acid rain* (Section 14.3).

Nitrogen Oxides

Although there are eight different nitrogen oxides, the only three that are normally detectable in air are nitrous oxide, N_2O, nitric oxide, NO, and nitrogen dioxide, NO_2. Of these, only NO and NO_2 are normally found in significant amounts in air; they are collectively symbolized as NO_x.

Nitrous oxide, often known as laughing gas, is a colorless, nonflammable, nontoxic gas with a slightly sweet taste. Only a small amount is present in the atmosphere as a result of natural processes carried out by soil microorganisms. It is used in dentistry and in surgery (Section 25.2) as an anesthetic.

Nitric oxide and nitrogen dioxide are normally regarded as air pollutants. Nitrogen, N_2, and oxygen, O_2, in the air have very little tendency to combine chemically. Combustion of fuels such as coal and gasoline causes very high temperatures at which nitrogen and oxygen may react to a small extent as follows:

$$N_2 \text{ (gas)} + O_2 \text{ (gas)} \rightarrow 2NO \text{ (gas)}$$

This reaction also takes place in the atmosphere when lightning strikes.

Once NO has formed, it oxidizes in air to form NO_2.

$$2 \text{ NO (gas)} + O_2 \text{ (gas)} \rightarrow 2 \text{ NO}_2 \text{ (gas)}$$
$$\text{nitric} \qquad\qquad\qquad \text{nitrogen}$$
$$\text{oxide} \qquad\qquad\qquad \text{dioxide}$$

Since NO_2 has a strong choking odor, its presence is easily detected. In the atmosphere, it undergoes a variety of chemical reactions, including the formation of nitric acid, HNO_3. This contributes to both acid rain and smog (Section 14.6).

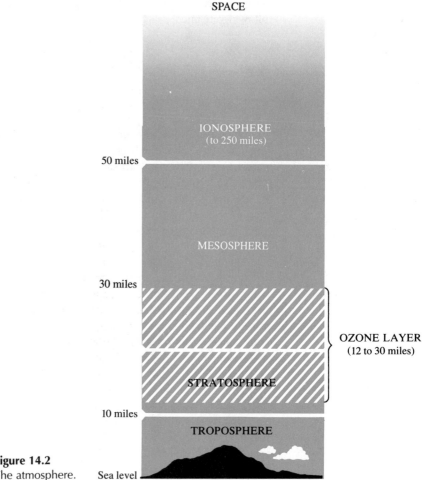

Figure 14.2
The atmosphere.

Ozone

Ozone, O_3, is most often regarded as a beneficial compound that provides protection from the harmful ultraviolet rays of the sun. Simultaneously, it is a toxic gas with a chlorine-like odor that causes severe respiratory problems. Ozone is a highly reactive substance that reacts with many organic materials and makes a significant contribution to the formation of smog. But in the region of the atmosphere known as the stratosphere (Figure 14.2), away from any living organisms, ozone is indeed beneficial (Section 14.5).

Particulates

Thus far we have dealt with gaseous pollutants, but the atmosphere carries vast amounts of *particulate matter.* These particles, which are quite variable in size and composition, may be solids or liquids, organic or inorganic.

The smallest particles, called *aerosols,* are smaller than one micron (1 μ, 10^{-6} m), too small to be settled by gravity. Particulates over 10 μ settle rapidly to the earth, usually within a day or two. The smaller particles usually combine to form larger aggregates, and then settle to the ground.

Particulates in the atmosphere include viruses, smoke, automobile exhaust, bacteria, fly ash, coal dust, cement dust, and pollen, to name a few. The particle size that is most dangerous, in the range 0.1 to 3 μ, can cause scar tissue in the lungs, eventually leading to emphysema and death. Small particles of asbestos are especially likely to cause this scar tissue.

14.3   Acid Rain: Air Pollution Sometimes Equals Water Pollution

GOAL: To understand the origin and consequences of acid rain.

Rainfall in clean air is slightly acidic due to the presence of carbon dioxide in the air. The CO_2 dissolves in water and then reacts with water to produce carbonic acid, a weak acid (Section 7.5). The carbonic acid dissociates slightly to release hydrogen ions.

$$CO_2 \text{ (gas)} + H_2O \rightleftharpoons H_2CO_3$$
$$H_2CO_3 \rightleftharpoons H^+ + HCO_3^-$$

The pH of rain may drop as low as 5.6, depending on the amount of CO_2 it contains. The pH of water that has been freshly distilled is 7, but it drops quickly to 5.6 on standing in contact with air.

Acid rain means precipitation with a pH below 5.6. The major contributors are oxides of nitrogen and sulfur, nitric acid, HNO_3, and sulfuric acid, H_2SO_4. When spewed from tall smokestacks, these compounds may travel long distances before settling to the ground. In areas where there is very little precipitation, acid rain is considered a misnomer since acidic deposition occurs in the form of dry, microscopic particles (aerosols).

In the western United States, nitrogen oxides from automobiles make the greatest contribution to acid rain. Elsewhere in this country, combustion of fossil fuels, particularly coal, releases large quantities of sulfur dioxide, which accounts for most of the acid rain. Like the loss of stratospheric ozone and the buildup of atmospheric carbon dioxide, acid rain has no respect for state or national boundaries.

During the 1980s, coal-fired utilities accounted for a dramatic rise in the levels of sulfur dioxide emissions in the United States. The increased use of coal was the result of the oil price crunch of the 1970s and a reluctance to move toward nuclear power.

Some fish are harmed at a pH of 5.5, but a pH of 5.0 or below is lethal to most. In some locations, acid rain has not yet proved damaging because alkaline mineral deposits, such as limestone, $CaCO_3$, may have an opportunity to neutralize acidic water before it flows into lakes and rivers. Nevertheless, many lakes have become devoid of fish due to lowered pH.

Low pH conditions in the soil also can increase the solubility of aluminum, which is quite insoluble at neutral or alkaline pH. Aluminum concentrations in some Canadian lakes have been measured as high as 372 ppb, while a concentration of 100 ppb is toxic to fish. Thus, much of the toxicity attributed to acidic water may actually be a consequence of the increased aluminum concentration *(5)*.

14.4   Carbon Dioxide: The Greenhouse Effect

GOAL: To understand the nature of the greenhouse effect.

Gases that can absorb heat (infrared radiation) and prevent its escape from earth's surface are known as *greenhouse gases*. Water vapor, as clouds, minimizes radiational cooling, which takes place during the nighttime hours. A clear sky at night allows for more rapid cooling at earth's surface. The excessive trapping of infrared radiation because of increasing concentrations of certain atmospheric gases is known as the *greenhouse effect*.

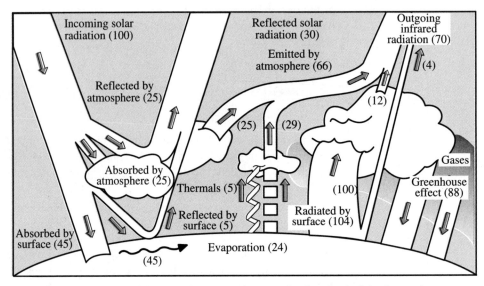

Figure 14.3

The flow of energy to and from earth's atmosphere. In the first third of the figure, for every 100 units of incoming solar radiation, 25 units (25%) are reflected by the atmosphere before reaching earth; 25 units are absorbed by the atmosphere; 5 units are reflected by earth's surface; and 45 units are absorbed. Of the 45 units of solar radiation absorbed by earth, 5 units are lost to the atmosphere as heat (thermals), and 24 units are consumed in evaporation of water from earth's surface. The remaining 16 units become part of the 104 units of energy "radiated by surface."

The greenhouse effect appears in the right third of the figure where 88 units of energy captured by greenhouse gases and clouds are transferred back to the earth from the atmosphere. This means 104 (16 + 88) units can be radiated from the surface, of which 4 units pass through the atmosphere unaffected. These 4 units combined with 54 (25 + 29) units from the atmosphere plus 12 units of energy radiated upward by the greenhouse gases, combine for a total of 70 units of heat (infrared radiation) that is lost to space. The 70 units of heat and 30 units of reflected solar radiation total to match the 100 units of incoming solar radiation (8). All numbers are approximate. (Source: Stephen H. Schneider.)

The concern associated with the greenhouse effect is global warming. The extreme is illustrated by the planet Venus where the temperature is reportedly about 500 °C. This is believed to be due to the dense atmosphere, including carbon dioxide, which reduces the loss of heat from the surface of the planet.

Solar radiation has its maximum intensity in the visible region, between 400 and 700 nm of the spectrum. Earth re-emits heat energy at wavelengths about 20 times longer than solar radiation. This energy is found as infrared radiation, mostly between 8 and 14 μm. Radiation at these wavelengths corresponds to the energies of internal vibrations of poly-atomic molecules, such as carbon dioxide, methane, water, and ozone. This correspondence allows for strong absorption of infrared radiation by these molecules in the atmosphere so that heat is retained near the surface. A detailed description of the flow of energy to and from earth appears in Figure 14.3.

Since infrared radiation only partially escapes from the lower atmosphere, the result is comparable to the capture of energy in a greenhouse. Visible radiation energy can pass through the windows and enter the greenhouse, but infrared radiation (heat) cannot move out through glass.

It is generally agreed that the last 100 years underwent an average air temperature

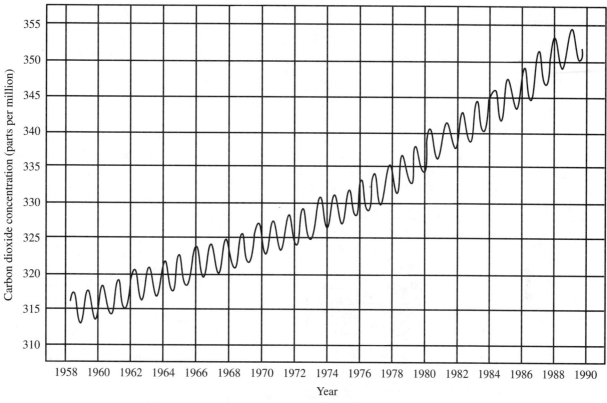

Figure 14.4 Atmospheric carbon dioxide can be seen increasing at a steady rate. The regular oscillation is attributable to seasonal changes in absorption of carbon dioxide by growing plants. Measurements made at Mauna Loa, Hawaii. (Source: Charles D. Keeting, Scripps Institution of Oceanography; National Oceanic and Atmospheric Administration.)

warming of about 0.5 °C. The concentration of carbon dioxide in earth's atmosphere increased from about 280 ppm in 1950 to about 315 ppm in 1957, and in recent years, to about 350 ppm (Figure 14.4). Over the past 30 years carbon dioxide emissions have more than tripled worldwide. One expectation is that these figures will continue to increase due to the increased use of fossil fuels by developing countries *(3, 6–8)*.

No proof is available that the buildup of greenhouse gases has been responsible for the increase in global temperatures. Natural cycles may play a part. However, the fear is that additional increases in the amounts of greenhouse gases could lead to significant changes in climate and greater melting of the polar ice caps. These, in turn, would produce substantial changes in the location of agricultural zones and coastlines. Severe periods of drought could result in loss of some productive food-growing regions.

Although scientists agree on the potential for the greenhouse effect to become serious, they do not concur on the timing or severity of global warming, or the necessary remedies. Warming might be minimized by increasing the efficiency of energy consumption, thus reducing carbon dioxide emissions. Greater emphasis on alternative energy sources, such as solar energy and nuclear power, also would lessen the release of carbon dioxide. In addition, aggressive reforestation efforts and preservation of tropical forests could have the benefit of removing carbon dioxide by enhancing its consumption in photosynthesis. The so-called Global ReLeaf effort calls for planting 100 million trees aimed at reducing atmospheric carbon dioxide *(9, 10)*.

14.5   Stratospheric Ozone Loss

GOALS: To understand why ozone can be dangerous in the troposphere, and important to preserve in the stratosphere.
To describe the codes used to identify CFCs, HFCs, HCFCs, and halons.
To appreciate how CFCs contribute to ozone destruction in the stratosphere.
To account for the description of chlorine monoxide as a smoking gun.
To be aware of the role of reservoir molecules in removing destructive forms of chlorine.

Both ozone, O_3, and ordinary oxygen, O_2, absorb high-energy ultraviolet light in the stratosphere and prevent this radiation from reaching the earth where it can harm animals and plants. In 1974, F. Sherwood Rowland and Mario J. Molina wrote a historic paper on the contribution of a group of compounds known as the **chlorofluorocarbons,** or **CFCs,** to the destruction of the stratospheric ozone layer *(11)*.

CFCs

Two of the best-known and commercially most important CFCs are trichlorofluoromethane, $CFCl_3$, and dichlorodifluoromethane, CF_2Cl_2.

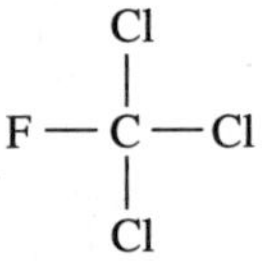

trichlorofluoromethane
(CFC-11)

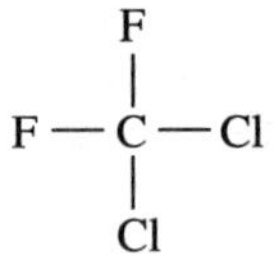

dichlorodifluoromethane
(CFC-12)

Each CFC has a code name used in most chemical and industrial literature. We will use another important CFC, CFC-113, to illustrate:

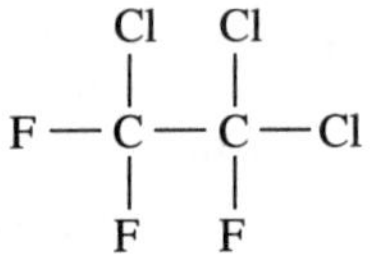

CFC-113

the units digit = number of F atoms
the tens digit = number of H atoms +1
the hundreds digit = number of C atoms −1

Since all CFCs contain at least one fluorine atom, the units digit is always greater than zero. The tens digit is also greater than zero in all cases due to the +1, even if no hydrogen atoms are present. But for CFCs with only one carbon, such as CF_2Cl_2, CFC-12, the hundreds digit is equal to zero and is therefore omitted *(12)*. The number of chlorine atoms is not specified in this symbolism, but can be calculated by remembering that carbon is always tetravalent and assuming that any missing atoms are chlorine atoms.

The term CFC is commonly used only for compounds lacking hydrogen atoms; these are the ones most likely to cause damage to the ozone layer. Therefore, we must be aware of the hydrofluorocarbons (HFCs) and hydrochlorofluorocarbons (HCFCs), since some of them are likely to emerge as replacements for the CFCs in certain applications. They are described in a later section *(13–18)*.

The CFCs became popular partly because of their chemical stability; they tend to be inert, nontoxic, and resistant to chemical breakdown or reaction. Also known as Freon-12 and Genetron-12, CFC-12 has been used as a cooling fluid in refrigerators and air conditioners since the early 1930s. Trichlorofluoromethane, $CFCl_3$ or CFC-11, was the most popular propellant in aerosol spray cans until increasing concern about the threat to the ozone layer discontinued its use in the United States in 1978.

In the troposphere (Figure 14.2), CFCs are unusually stable. This inertness is only part of the problem, as they also are unaffected by sunlight; that is, they are *photochemically inert* to the wavelengths of sunlight that pass through the stratosphere and reach the surface of earth. The CFCs move slowly upward on air currents into the stratosphere, a surprising movement because they are heavier than air.

Once in the stratosphere, CFCs contribute to the destruction of ozone. Approximately 80% of the total atmospheric ozone is in the stratosphere, where it acts as a filter for high energy radiation, allowing only wavelengths greater than 290 nm to reach earth's surface.

Concern about the loss of ozone comes from estimates that for every 1% decrease in ozone concentration in the stratosphere, a 2% increase in ultraviolet radiation reaching earth is expected. A 2% increase of ultraviolet radiation could bring a 4% to 6% increase in some types of skin cancers. Estimates of the loss of ozone that is likely to occur range from as high as 15% to 18% by the end of the century to as little as 2% to 4% by the end of the next century *(19, 20)*.

The CFCs are susceptible to photochemical attack in the stratosphere by high-energy ultraviolet light with wavelengths below 290 nm. The C—Cl bond is the weakest in the molecule, so it undergoes photochemical cleavage, called **photolysis,** resulting in the loss of a chlorine atom (Cl•):

$$F - \underset{\underset{Cl}{|}}{\overset{\overset{F}{|}}{C}} - Cl + h\upsilon \text{ (light)} \longrightarrow F - \underset{\underset{Cl}{|}}{\overset{\overset{F}{|}}{C}}\bullet + Cl\bullet \qquad \textbf{(I)}$$

Both products carry one of the electrons from the bond that is cleaved. Such particles, with unpaired electrons, are called **radicals,** or free radicals. Other particles involved in reactions to be described later are also radicals, but the odd electron of the chlorine atom and other radicals will not be shown subsequently.

The chlorine atoms released in reaction **I** react with ozone to produce chlorine monoxide:

$$Cl + O_3 \longrightarrow \underset{\substack{\text{chlorine} \\ \text{monoxide}}}{ClO} + O_2 \qquad \textbf{(II)}$$

Chlorine monoxide also is quite reactive, particularly toward oxygen atoms:

$$ClO + O \rightarrow Cl + O_2 \qquad \textbf{(III)}$$

Notice that this reaction forms more chlorine atoms, which can attack more ozone according to reaction **II.** Therefore the initial formation of chlorine atoms in reaction **I** leads to reactions **II** and **III,** after which reactions **II** and **III** are repeated over and over, a pattern known as a **chain reaction.** Thus a single CFC molecule may ultimately lead to the destruction of 100,000 ozone molecules in the stratosphere *(12)*.

The oxygen atoms in reaction **I** come from the photolysis of oxygen in the stratosphere:

$$O_2 + \text{light} \rightarrow 2\ O$$

Each oxygen atom is then able to react with chlorine monoxide (reaction **III**) or with more O_2 to form ozone:

$$O_2 + O \rightarrow O_3$$

Thus, the formation of oxygen atoms is necessary for ozone to form, but the process contributes to the destruction of ozone as well, due to reaction **III.**

Despite the 1974 paper by Rowlands and Molina, it was not until late 1987 that measurements showed *all* ozone disappeared for a few weeks from some regions of the stratosphere over Antarctica. Each September and October, beginning in the late 1970s, just as winter is coming to an end in the Southern Hemisphere, stratospheric ozone levels shrink drastically and then return to normal by late November. The complete absence of ozone, termed an *ozone hole,* jolted even the most skeptical observers. This finding was coupled with the observation of a concentration of approximately 1.2 ppb of ClO measured at 18 km (11 miles) above Antarctica. This chlorine monoxide, produced in reaction **II,** has become known as a *smoking gun* because its presence indicates prior damage to ozone *(21, 22).*

Chlorine atoms and chlorine monoxide may be removed from the atmosphere by reaction with methane and nitrogen dioxide, which are also present in the stratosphere:

$$Cl + CH_4 \rightarrow \underset{\substack{\text{hydrogen} \\ \text{chloride}}}{HCl} + \underset{\substack{\text{methyl} \\ \text{radical}}}{CH_3}$$

$$ClO + \underset{\substack{\text{nitrogen} \\ \text{dioxide}}}{NO_2} \rightarrow \underset{\substack{\text{chlorine} \\ \text{nitrate}}}{ClONO_2}$$

Hydrogen chloride and chlorine nitrate are **reservoir forms** of chlorine not destructive to ozone.

The conditions in Antarctica are thought to be especially conducive to ozone depletion, even more than in the arctic regions of the Northern Hemisphere. It is now known that damaging forms of chlorine may be formed from hydrogen chloride and chlorine nitrate under certain circumstances, such as those encountered in the Antarctic winter. During the polar night, stratospheric temperatures drop below -90 °C, cold enough to form clouds even in the very dry conditions there. Ice crystals in the clouds provide surfaces on which the chlorine reservoir compounds react:

$$HCl + ClONO_2 \rightarrow \underset{\text{chlorine}}{Cl_2} + \underset{\substack{\text{nitric} \\ \text{acid}}}{HONO_2}$$

$$H_2O + ClONO_2 \rightarrow HOCl + HONO_2$$

Nitric acid then sticks to the cloud surfaces. The gaseous HOCl and Cl_2 remain to be photolyzed when sunlight returns in September at the end of the polar winter:

$$Cl_2 + light \rightarrow 2\ Cl$$
$$HOCl + light \rightarrow Cl + OH$$

The chlorine atoms from these reactions attack ozone so that its level drops sharply. After about a month of sunlight, the polar stratosphere warms, chlorine atoms are again converted to reservoir forms, and ozone levels rise *(23).*

Winter temperatures in the Arctic only drop to -80 °C compared to -90 °C in the Antarctic. This difference is significant since the type of ice clouds that assist the release of Cl_2 and HOCl do not form until temperatures drop to about $-85\ °C$. Nevertheless, the presence of chlorine monoxide in the Arctic region (at concentrations 100 times less than in the Antarctic) does indicate the likelihood of loss of ozone due to CFCs *(23)*.

It has been observed that the stratosphere over the latitude band that includes Dublin, Moscow, and Anchorage had about 8% less ozone in January 1986 than it had in January 1969 *(24)*. Even damage as far away as Antarctica should not be dismissed as insignificant. The absorption of ultraviolet radiation by ozone in the stratosphere over the Antarctic is partly responsible for maintaining normal temperatures in that region.

Because of their low chemical reactivity, CFCs have lifetimes of a century or more. It takes 50 to 150 years for them to move from earth's surface into the stratosphere. Even with an immediate ban on CFCs, it would take centuries for the atmosphere to escape their effects *(12, 24–26)*. Work on substitutes is under way, and there is increasing emphasis on developing technology for recovering and recycling CFCs to minimize their effects on the atmosphere.

Alternatives to CFCs

The CFCs became popular as aerosol propellants, as blowing agents in the formation of polymer foams (for example, polyurethane foams), as solvents for cleaning sensitive electronic components, as refrigerants (including air conditioners), and as fire extinguishers. Besides low toxicity and lack of chemical reactivity, they are nonflammable, highly miscible with other organic materials, and have low thermal conductivity. Although the use of CFC-11, $CFCl_3$, in aerosols was discontinued in the United States in 1978, heavy use of this and other CFCs has continued worldwide. International agreements will phase out CFC use by 1996 in industrialized countries and by 2006 in developing countries.

A related class of compounds containing bromine atoms and known as **halons** are popular in liquid fire extinguishers:

$$
\begin{array}{ccc}
\ \ \ \ \ F & \ \ \ \ \ F & \ \ \ \ F\ \ \ \ F \\
\ \ \ \ \ | & \ \ \ \ \ | & \ \ \ \ | \ \ \ \ | \\
F - C - Br & F - C - Br & F - C - C - F \\
\ \ \ \ \ | & \ \ \ \ \ | & \ \ \ \ | \ \ \ \ | \\
\ \ \ \ Cl & \ \ \ \ F & \ \ \ \ Br\ \ Br \\
\text{halon-1211} & \text{halon-1301} & \text{halon-2402}
\end{array}
$$

They are identified by a code similar to that used for the CFCs except that the numbers are simpler and more logical:

$$
\begin{array}{l}
\ \ \ \ F \\
\ \ \ \ | \\
F - C - Br \ \ \ \ \text{halon-1301} \\
\ \ \ \ | \\
\ \ \ \ F
\end{array}
$$

the units digit = number of Br atoms
the tens digit = number of Cl atoms
the hundreds digit = number of F atoms
the thousands digit = number of C atoms

In other words, reading from left to right, the digits represent the number of carbons, fluorines, chlorines, and bromines.

Halons are far more effective in putting out fires than other chemicals. Being nontoxic and chemically inert makes them superior for many uses where people or delicate equipment could be harmed. Halon-1301 and halon-1211 are installed in aircraft and military tanks to extinguish gasoline fires with great efficiency.

Since halons are potentially damaging to stratospheric ozone, international agreements specified their production would cease on January 1, 1994, but existing supplies may be used as long as they are available. By eliminating testing and training exercises, the supply may be adequate for many years. This practice will greatly curb releases of halons into the atmosphere even while they remain available *(27)*.

It has become a priority to identify suitable alternatives for these compounds, but full-scale testing of toxicities and of economical methods for synthesis of alternatives will likely require several more years of research. Much of the emphasis has been directed to the development of **hydro**chlorofluorocarbons (**HCFCs**) and **hydro**fluorocarbons (**HFCs**). Three compounds from these classes are shown:

$$
\begin{array}{ccc}
\underset{\text{HFC-134a}}{\text{F}-\overset{\displaystyle \overset{\text{F}}{|}}{\underset{\displaystyle \underset{\text{F}}{|}}{\text{C}}}-\overset{\displaystyle \overset{\text{H}}{|}}{\underset{\displaystyle \underset{\text{H}}{|}}{\text{C}}}-\text{F}
&
\underset{\text{HCFC-22}}{\text{H}-\overset{\displaystyle \overset{\text{F}}{|}}{\underset{\displaystyle \underset{\text{F}}{|}}{\text{C}}}-\text{Cl}
&
\underset{\text{HCFC-141b}}{\text{H}-\overset{\displaystyle \overset{\text{H}}{|}}{\underset{\displaystyle \underset{\text{H}}{|}}{\text{C}}}-\overset{\displaystyle \overset{\text{Cl}}{|}}{\underset{\displaystyle \underset{\text{Cl}}{|}}{\text{C}}}-\text{F}
\end{array}
$$

The code numbers are the same as for the CFCs. The a and b acknowledge the possible existence of constitutional isomers.

Consult Section 6.1 for a discussion of constitutional isomers.

The significant feature is the presence of hydrogen atoms, which are susceptible to cleavage by hydroxide radicals (• OH), which are abundant in the troposphere. The hope is these compounds will be attacked and decomposed as they move upward through the troposphere many years before reaching the stratosphere.

Most alternatives are expensive. Since more than 10^9 kg per year of CFCs were produced worldwide in recent years, the economic consequences are apparent. More subtle economic factors may arise when new compounds are targeted to particular applications. For example, CFCs have been used heavily as blowing agents in insulating foams, where they remain trapped inside the foams. Since they have unusually low thermal conductivities, that is, they conduct heat very poorly, newer insulating materials may be much less efficient, resulting in greater energy consumption *(13–18)*.

14.6 Smog

GOAL: To describe the formation of photochemical smog.

One of the more complex forms of air pollution is smog. The term, which combines the words smoke and fog, was first applied to London-type fogs. High concentrations of smoke particles and sulfur dioxide were produced by coal burning for home heating and for industrial processes.

The hazy smog associated with Los Angeles is known as *photochemical smog* since sunlight is needed for its formation. Its ingredients are nitrogen oxides, NO_x, and *volatile organic compounds* (VOCs), both of which come largely from automobiles. Once formed, the major components are ozone and oxidized organic compounds. Summertime provides the optimum conditions for photochemical smog. A temperature inversion layer (Figure 14.5) traps smog precursors near the ground on more than 80% of the days in Los Angeles. Intense sunlight permits the photochemical reactions.

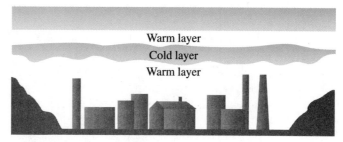

Figure 14.5   A temperature inversion occurs when a layer of warm air is trapped below a layer of cooler air, so air pollutants at the surface are unable to rise and escape into the atmosphere.

The NO_x form in a series of reactions. Ordinarily, nitrogen, N_2, is regarded as non-flammable. But as the major component of air, large amounts of nitrogen enter automobile cylinders where fuel combustion takes place. Under such extreme conditions, some nitrogen undergoes combustion to form nitric oxide:

$$N_2 + O_2 \xrightarrow{\text{combustion}} 2\ NO$$

Once NO exits in exhaust gases and enters the atmosphere, it is quickly oxidized to NO_2.

$$2\ NO + O_2 \rightarrow 2\ NO_2$$

Once the sun comes up during a morning rush hour, the following reaction triggers a series of steps. The first step is the photolysis (photochemical cleavage) of NO_2.

$$NO_2 + \text{light} \rightarrow NO + O \tag{I}$$

The oxygen atoms released may react in one of two ways:

$$O + O_2 \rightarrow O_3 \tag{IIa}$$
$$O + H_2O \rightarrow 2\ OH \tag{IIb}$$

Changes in concentrations of NO_x and ozone are plotted in Figure 14.6 for a typical day in Los Angeles.

In addition, the OH radicals formed in reaction **IIb** initiate a complex series of reactions involving oxygen, NO, and volatile hydrocarbons. The overall process is illustrated for butane *(28)*.

$$CH_3CH_2CH_2CH_3 + OH + O_2 + NO \tag{III}$$

$$\downarrow 9 \text{ steps}$$

$$\underset{\substack{\text{methyl ethyl ketone}\\\text{(MEK)}}}{CH_3\overset{\overset{\displaystyle O}{\|}}{C}CH_2CH_3} + \underset{\text{acetaldehyde}}{CH_3\overset{\overset{\displaystyle O}{\|}}{C}H} + NO_2 + H_2O$$

Further reactions of the acetaldehyde lead to the formation of an important component of

Figure 14.6
The levels of several compounds involved in smog formation. Nitric oxide (NO) is released from motor vehicles during rush hour; it is quickly oxidized to NO_2. The NO_2 then participates in the formation of ozone and other components of photochemical smog. As the intensity of the sunlight diminishes in late afternoon and early evening, the formation of ozone ceases even though some NO_2 remains *(28)*.

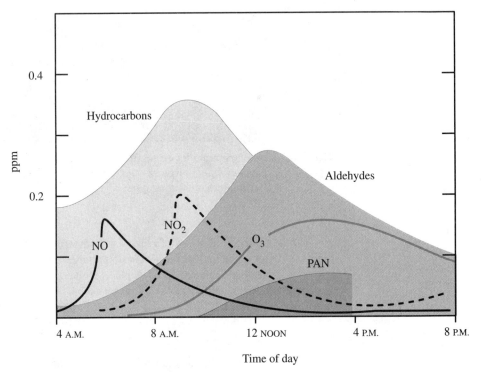

smog known as peroxyacetyl nitrate (PAN):

$$\underset{\text{O}}{\overset{\text{O}}{CH_3\overset{\|}{C}H}} + NO_2 + O_2 + OH \qquad \textbf{(IV)}$$

$$\downarrow \text{3 steps}$$

$$\underset{\substack{\text{peroxyacetyl nitrate} \\ \text{(PAN)}}}{CH_3\overset{\overset{\text{O}}{\|}}{C}-O-O-NO_2} + H_2O$$

Ozone, PAN, aldehydes, ketones, and nitric acid (from NO_x) are responsible for the extreme eye and lung irritation caused by photochemical smog.

14.7   Water Pollution

GOALS: To appreciate the essentials of the hydrologic cycle.
To understand the chemistry of hard water, the difference between temporary and permanent hardness, and the chemistry of softening water in the home.
To appreciate the common methods for treating drinking water and the associated problems.
To consider the subject of biochemical oxygen demand.
To appreciate the magnitude of the problem of disposing of hazardous chemical wastes.

Components of air often end up in water. The presence of dissolved CO_2 is well understood, as is the phenomenon of acid rain and its effects on natural lakes and rivers (Section 14.3). In this section we will consider the meaning of clean water, and study the nature and problems of other substances that are often found in water.

Clean Water

Defining clean water is even more difficult than defining clean air. Clean water from the ocean, for instance, is not the same as clean water from a freshwater lake. Ocean water contains dissolved salts such as sodium chloride and magnesium chloride in high concentrations.

Pure water is easy to define: it contains *only* H_2O. Natural water is never pure, however. Running along or through earth's surface provides ample opportunity for water to dissolve impurities. Not all these impurities are objectionable, and some are even beneficial. We can classify various types of natural water by the amount of dissolved minerals, as shown in Table 14.2. The table does not contain the category hard water, which indicates the presence of specific cations, such as calcium and magnesium, in the range of 120 to 250 ppm.

Water from the oceans averages 35,000 ppm of dissolved minerals. Some of the elements that are components of minerals dissolved in the salty oceans are shown in Table

Table 14.2   Classifications of natural waters *(1)*

Natural Water	Dissolved Minerals (ppm)
Fresh	0–1,000
Brackish	1,000–10,000
Salty	10,000–100,000
Brine	>100,000

Table 14.3   Elements present in seawater *(29)*

Element	Concentration (ppm)
Chlorine	19,000
Sodium	10,000
Magnesium	1,300
Sulfur	900
Calcium	400
Potassium	380
Bromine	65
Carbon	28
Oxygen	8
Strontium	8
Boron	4.8
Silicon	3.0
Fluorine	1.3
Nitrogen	0.8
Argon	0.6
Lithium	0.2
Rubidium	0.12
Phosphorus	0.07
Iodine	0.05

14.3. Magnesium, bromine, and iodine, all in ionic form, have been mined from the sea; and sodium chloride for use as common table salt is derived from seawater. There has been discussion of extracting gold, at 4×10^{-6} ppm, and silver, at 3×10^{-4} ppm, from the oceans. The low concentrations of these precious metals mask the vast absolute amounts actually present.

The Hydrologic Cycle

Water on earth is distributed in various places, as shown in Table 14.4. It does not exist forever in only one form; it moves through the *hydrologic cycle* (Figure 14.7). Within this cycle, water is carried from the oceans into the atmosphere by evaporation and transpiration (from plants), and is deposited on land as rain or snow. The remainder falls back into the oceans. Some water on land runs off into the oceans.

All the water on earth is not in motion at any one time, but all the water is in the cycle. This is what allows renewal of lakes and rivers from rain, even as fresh water runs off into the oceans. In effect, this natural cycle provides fresh water for the earth on a continual basis. Every day, the sun removes well over 1 trillion metric tons of water by evaporation from the oceans. Some of it falls on the land masses as rain. From this rainfall, the United States has almost 5 billion tons of water flowing through its rivers each day.

Table 14.4   Distribution of water on earth *(30)*

Source of Water	Percentage Contained
Oceans	94
Ice and snow	4.2
Underground	1.2
Surface and soil	0.4
Atmosphere	0.001
Living species	0.00003

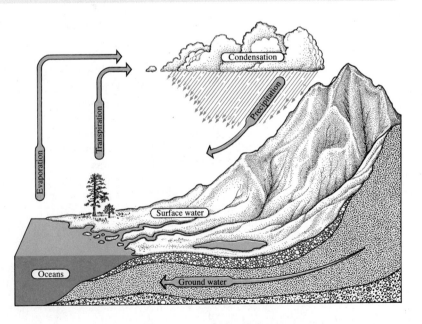

Figure 14.7
The hydrologic cycle.

Unfortunately, the demands for fresh water are great and are increasing at an accelerating rate. It has been estimated that the demand will equal the supply shortly after the year 2000. This means that water must be recycled and/or the demand decreased. Only an extremely small percentage of the demand is for drinking; agriculture and industry use vast amounts. Where the demand cannot be decreased, used water will have to be recycled to a condition fit for reuse.

Hard Water

Hard water is characterized by the presence of soap-precipitating agents, the deposition of scale in boilers and kettles, the inability to clean clothes properly even though the soap is adequate, and the inability to raise suds from soap. All these conditions are related to the presence of cations such as calcium and magnesium. These metal ions form scums, or insoluble compounds, with soaps (Section 22.3). The source of the hardness is minerals in land over which the water flows. Limestone, $CaCO_3$, and dolomite, $CaCO_3$ plus $MgCO_3$, formations provide excellent sources of Ca^{2+} and Mg^{2+} (See Figure 14.8).

The solubility of calcium carbonate in water is only about 9 ppm. Since we characterize hard water as having dissolved minerals in the range 120 to 250 ppm, most of which is $CaCO_3$, some mechanism must increase the concentration of $CaCO_3$ in water. This mechanism is related to the presence of CO_2 in the atmosphere. Carbon dioxide dissolves in water to form carbonic acid, and carbonic acid reacts with $CaCO_3$:

$$CO_2 + H_2O \rightarrow H_2CO_3 \text{ (carbonic acid)}$$
$$CaCO_3 + H_2CO_3 \rightarrow Ca^{2+} + 2\ HCO_3^-$$

The interconversion between $CaCO_3$ and $Ca(HCO_3)_2$ accounts for the formation of stalagmites and stalactites (Section 7.5).

The second reaction is an acid-base (neutralization) reaction, in which the carbonic acid reacts with the base, CO_3^{2-}.

The $Ca(HCO_3)_2$ is easily decomposed if water is boiled or even heated. The reaction is reversed, CO_2 is driven off, and $CaCO_3$ precipitates out of the water. This is the reason a hard crust forms in boilers and steam irons: the heat drives off the carbon dioxide and precipitates $CaCO_3$. A parallel reaction sequence is used for $MgCO_3$. This condition in water is known as *temporary hardness,* because the hardness cations are so easily removed from solution by precipitation.

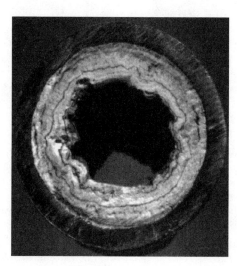

Figure 14.8
Actual pipe sections filled with scale from continued use of hard water.

If $CaCO_3$ and $MgCO_3$ were the only compounds responsible for water hardness, the problem would be solved easily, as simple heating would precipitate most of these materials. Then treatment with strong acid would remove the precipitate:

$$CaCO_3 + 2\ H^+ \rightarrow Ca^{2+} + CO_2 + H_2O$$

The $CaCO_3$ does indeed precipitate, and if acid is added to the crust, it does bubble and give off CO_2 gas; but not all the scum reacts, nor does all the hardness precipitate on heating. In addition to carbonates, natural waters pick up calcium sulfate, $CaSO_4$, an abundant species that also causes hardness. Unlike $CaCO_3$, $CaSO_4$ does not precipitate on heating, nor does it dissolve in strong acid easily. The $CaSO_4$ is a primary cause of so-called *permanent hardness.*

Washing machines clog and clothing appears dingy due to the precipitation of soap by hardness cations, Ca^{2+} and Mg^{2+}. This has led to the use of water-softening equipment. Water softeners act by binding the hardness cations onto another material, called an *ion-exchange resin,* and exchanging them for sodium ions, Na^+ (Figures 14.9 and 14.10).

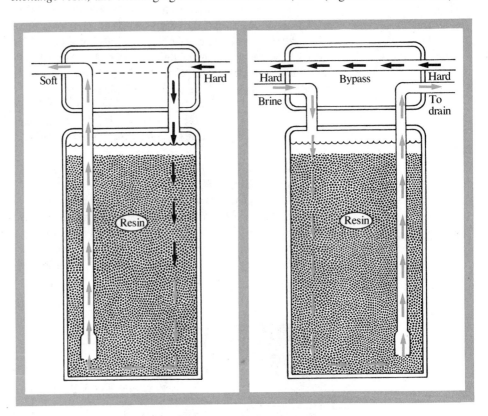

Figure 14.9

A home water softener in schematic form. Water may be softened by ion-exchange resins, which remove calcium and magnesium ions from the water. The tank contains the ion-exchange resin that removes the hardness cations from the water as it passes through. A second tank, not shown, contains brine, NaCl, which periodically is required for recharging the resin. The schematic diagram shows the operation of the system (a) during the softening cycle and (b) the recharging cycle. During the recharge, water is passed through the tank of brine, picks up a high concentration of NaCl, and passes through the ion-exchange resin. The excess NaCl is washed off the resin, and the softening cycle is resumed. The details of the chemistry are discussed in Figure 14.10. (Courtesy of Culligan Water Institute.)

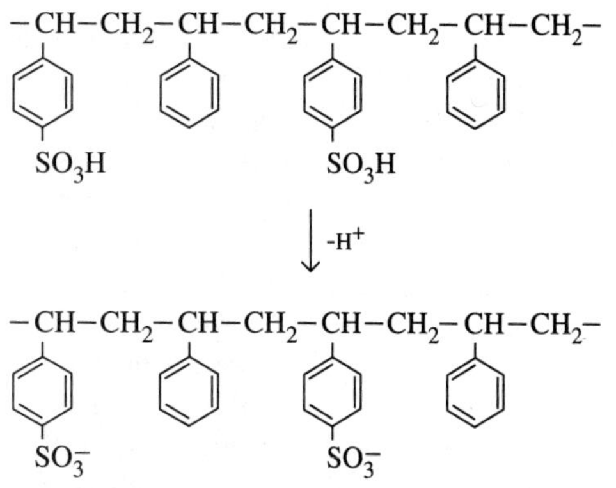

$$-CH-CH_2-CH-CH_2-CH-CH_2-CH-CH_2-$$

SO$_3$H   SO$_3$H

$\downarrow$ -H$^+$

$$-CH-CH_2-CH-CH_2-CH-CH_2-CH-CH_2-$$

Figure 14.10   SO$_3^-$ SO$_3^-$

Polystyrene is described in Section 8.2.

The chemistry of water softening. Ion-exchange resins are complex substances with a variety of chemical structures. Water softening uses an ion-exchange resin called a sulfonated polystyrene. A portion of a polystyrene resin is represented with sulfonic acid groups ($-SO_3H$) attached, pictured in both neutral and ionic forms.

Not every ring contains a sulfonic acid group, but each negative site on the resin can bind a cation. When the ion-exchange resin is ready for softening, sodium ions, Na$^+$, are bound to the anionic sites. As water containing hardness cations flows through the resin, Ca^{2+} and Mg^{2+} displace Na$^+$ and bind to the anionic sites. This is the phenomenon of ion exchange. The Ca^{2+} and Mg^{2+} replace Na$^+$ due to their greater positive charges causing tighter binding to the anionic sites. When all the Na$^+$ have been replaced by hardness ions, the resin is recharged by stripping off the hardness cations. This is done by flushing the tank with concentrated NaCl solution, obtained by directing a flow of water through a second tank containing NaCl (brine) and then through the resin. Although Ca^{2+} and Mg^{2+} do bind more tightly than Na$^+$ to the resin, when the concentration of Na$^+$ in the flow is high enough, Na$^+$ will successfully compete for the negative sites on the polystyrene; then the resin will be restored to the sodium form and become an effective water-softening agent.

The impact of synthetic detergents on water supplies is discussed in Section 22.3.

Synthetic detergents were developed to eliminate the difficulties of using soap in hard water. The chemistry of soaps and detergents is examined in Sections 22.2 and 22.3.

Drinking Water

The water we drink must have one overriding characteristic—it must be safe! Human history is replete with plagues and epidemics due to unsafe drinking water. For example, cholera and hepatitis are spread when drinking water is too close to sewage runoff. Water contains disease-carrying or disease-producing species when waste water and drinking sources are mixed. Today, drinking water is tested and often sterilized by chlorination or oxygenation. Chlorine, Cl$_2$, destroys germs and disease-producing species. It also may generate some dangerous by-products if the water contains organic compounds.

In oxygenation, water is sprayed into the air to allow maximum opportunity for atmospheric oxygen to dissolve in the water. Many undesirable organisms are anaerobic and cannot tolerate oxygen. The process also improves the taste of the water.

Dissolved Oxygen

Oxygen gas reaches saturation in water at a concentration of 10 ppm. This small amount is critical for sustaining life in natural waters. Fish, for example, require 4 to 5 ppm of oxygen for life.

Oxygen is depleted from bodies of water when an overabundance of oxidizable material is present. This material can be anything from mine runoff to sewage or industrial wastes. Most of the material with which oxygen reacts is biochemical. The *biochemical oxygen demand,* or BOD, is formally defined as the amount of dissolved oxygen consumed by contents of the water during a 5-day period at 20 °C. The paper used to make this book generates organic waste with a BOD of 16,000 to 25,000 ppm, whereas brewing beer generates waste with a BOD of 500 to 1200 ppm. Certainly these BODs far exceed the oxygen available at saturation.

Swiftly moving streams with rapids and falls expose more water to the air and replenish their oxygen faster. Slow-moving streams and lakes replenish oxygen far more slowly. A lake or stream is considered dead if the BOD is so great that no life can be supported except organisms involved in oxidizing waste. In deep lakes into which waste is dumped, the oxygen at the bottom is used up quickly and replaced slowly. During the present century, the amount of oxygen near the bottom of the northern end of the Baltic Sea has decreased from approximately 3 ppm to virtually zero because of waste disposal.

Dissolved oxygen can cause objectionable materials in water. An example is its reaction with iron pyrites, FeS_2, which is frequently exposed by mining. Sulfuric acid and insoluble iron compounds are produced according to the following sequence of reactions.

$$4 \ FeS_2 + 14 \ O_2 + 4 \ H_2O \rightarrow 4 \ Fe^{2+} + 8 \ SO_4^{2-} + 8 \ H^+$$
$$4 \ Fe^{2+} + O_2 + 4 \ H^+ \rightarrow 4 \ Fe^{3+} + 2 \ H_2O$$
$$4 \ Fe^{3+} + 12 \ H_2O \rightarrow 4 \ Fe(OH)_3 + 12 \ H^+$$
$$\text{``yellow boy''}$$

The result is acid mine drainage, which pollutes streams due to acidity and deposits of "yellow boy," the insoluble iron compound.

Hazardous Waste Disposal

Water supplies in some locations are particularly in danger of possible contamination by hazardous waste materials. Industries involved in the production of textiles, paper, metals and metal products, refined petroleum products, and the manufacture of other chemical products (drugs, plastics, etc.) generate a huge amount of waste, some of which is quite hazardous. As consumers, we purchase many items without being aware of the waste problems associated with their production.

Love Canal is one of the best known examples of damage caused by improper disposal of hazardous wastes. A portion of the southeast corner of the city of Niagara Falls was a partially dug section of a canal that William T. Love had begun to construct in 1894 as a means of joining the upper and lower Niagara Rivers. Soon after construction had begun, Love's financial backers deserted him and the project was abandoned. For several years, the partially dug canal served as a swimming hole for many local children. But starting in 1920, the location was used as a dumping site for several chemical companies and the city of Niagara Falls.

The dumping continued until 1953 when the site was covered with earth. Within 10 years, about 100 homes and an elementary school were built on this 16-acre rectangular piece of land. During the middle and late 1960s, 6 years of unusually heavy rainfall caused a rise in the water table in the area, and more than 75 chemicals, including several suspected carcinogens, began to show up in the basements of homes. A multitude of health problems, such as miscarriages, birth defects, cancer, and nervous system disorders, also occurred. In 1978 the forced evacuation of 255 homes caused President Carter to declare the site a federal disaster area. Although declarations of disaster are not unusual, this was the first issued for a human-made disaster. The estimate is that some 80,000 55-gallon drums of chemical wastes were deposited in the Love Canal by Hooker Chemical Company over 10 years.

Another well-publicized dumping ground has become known as the "Valley of the Drums." Some 100,000 drums of waste chemicals were dumped in a site near West Point, Kentucky, where they rusted, corroded, and broke open to release the toxic contents.

About 300,000 drums were buried at a location near Toone, Tennessee, where hazardous pesticides were disposed of for many years. Pollution of drinking water resulted in a wave of sickness and birth defects.

Unfortunately, these are not the only examples, although they are among the most publicized. In 1980 the United States Congress passed legislation known as Superfund to provide support to finance the proper cleanup of many waste sites. Over the years, more and more locations have been identified as requiring attention. Current estimates are that many thousands of sites may require as much as $100 billion from the Superfund. Other legislation has been adopted to require industries to track the flow of chemicals through their processes so that additional contributions to the waste supply will be minimal, although still far from zero. Meanwhile, the cleanup process required as a result of totally unregulated waste disposal in the past will continue for many years.

SUMMARY

The oxides of carbon, CO and CO_2, the oxides of nitrogen, NO_x, the oxides of sulfur, SO_2 and SO_3, and ozone, O_3, are frequent air pollutants. Ozone serves a desirable function in the stratosphere, however, where it absorbs harmful, high-energy ultraviolet radiation, thus protecting inhabitants of Earth from effects such as skin cancer. Unburned hydrocarbons or volatile organic carbons (VOCs), along with the NO_x compounds, are responsible for formation of ozone and photochemical smog in cities like Los Angeles.

The oxides of sulfur and nitrogen are dangerous pollutants both directly (discussed earlier) and indirectly. The indirect effect comes from the formation of sulfuric acid and nitric acid, which cause acidic deposition, including acid rain. Acid rain is responsible for damage to marble statuary and to fish populations in rivers and lakes.

The greenhouse effect links the buildup of carbon dioxide to global warming. Although global carbon dioxide concentrations and temperatures have risen over many decades, there is as yet no proof of a connection between the two.

The loss of stratospheric ozone over Antarctica observed in recent years increased concern over the effect of chlorofluorocarbons (CFCs) on the stratosphere. Although CFCs were banned from aerosol sprays in the United States in 1978, their use continues even in aerosols in other countries and in other applications everywhere. Since their destructiveness is now generally accepted, active development of replacement compounds and efforts at recovery and recycling are under way.

The CFCs remain chemically unaffected for decades in the troposphere until they reach the stratosphere, where ultraviolet radiation initiates a chain reaction that begins by formation of chlorine atoms. The chlorine atoms directly destroy ozone molecules and form chlorine monoxide in the process. Since the presence of chlorine monoxide in the stratosphere indicates destruction of ozone molecules already completed, ClO is characterized as a smoking gun.

Reservoir molecules, such as HCl and the chlorine nitrate, convert to chlorine (Cl_2) and HOCl, respectively, both of which are subject to photolysis to form chlorine atoms. The conversion of the reservoir molecules to chlorine and HOCl is assisted by interaction with ice crystals in the stratosphere over Antarctica during the very cold winter when sunshine is absent. As winter ends in September and October, the photolysis reactions take place and massive destruction of ozone follows. Although the same conditions may not exist over the Arctic regions, chlorine monoxide has also been found there, although in far smaller amounts than in the Antarctic stratosphere. Probable alternatives to CFCs and halons are similar compounds containing hydrogens and known as hydrofluorocarbons (HFCs) and hydrochlorofluorocarbons (HCFCs). They are susceptible to attack by hydroxide radicals in the troposphere.

It is easy to define pure water, but not so easy to define clean or polluted water. Salt water from the ocean may not be polluted, but neither is it fit to drink. Hard water is categorized according to degree of temporary or permanent hardness, distinguished by the ability of CO_2 to affect the solubility of the dissolved ions Ca^{2+} and

Mg^{2+}. Water softeners can be used to remove these cations, but they do not remove the sulfate ions present in permanently hard water.

Chlorine and oxygen are used routinely to treat drinking water to remove biological pollutants. Chlorination can affect certain dissolved pollutants and render them more dangerous. Even dissolved oxygen can contribute to pollution in reacting with iron pyrites, FeS_2, released during mining. The biochemical oxygen demand, BOD, can be used to evaluate the extent of some pollution problems.

PROBLEMS

1. Give a definition or example of each of the following.

   a. Percent by volume
   b. Percent by weight
   c. Parts per million (ppm)
   d. Stratosphere
   e. Troposphere
   f. Ozone
   g. Ionosphere
   h. Mesosphere
   i. Particulates
   j. London smog
   k. Greenhouse gas
   l. Photolysis
   m. Radical
   n. Chain reaction
   o. Ozone hole
   p. Reservoir molecule
   q. Halons

2. What is the origin of sulfur oxides, and how are they damaging to the environment? Include the reactions for the deterioration of marble statuary.

3. Why would an immediate ban on CFCs not yield immediate preservation of stratospheric ozone?

4. What are the four major applications of CFCs and halons?

5. Draw structures to fit the following codes:

   a. CFC-13
   b. CFC-114
   c. HFC-125
   d. HCFC-131
   e. Halon-2222

6. How is it possible for 1 CFC molecule to cause the destruction of 100,000 ozone molecules?

7. Certain chemical reactions take place during the Antarctic winter and after the sun returns following a long period of darkness. Describe these reactions and explain how they contribute to formation of an ozone hole in the Antarctic region.

8. Explain, with equations, why ClO has been characterized as a smoking gun.

9. Regarding photochemical smog:

   a. What ingredients and atmospheric conditions are required to initiate its development?
   b. How do you explain the rise and fall of NO in Los Angeles air (Figure 14.6)? Consider the source and the fate of NO.
   c. How do you explain the rise and fall of NO_2 in Los Angeles air (Figure 14.6)? Consider the source and the fate of NO_2.
   d. What are the source and role of volatile organic compounds?
   e. Why is strong sunlight required for this type of smog to develop?

10. Why is it difficult to define unpolluted water?

11. What is the hydrologic cycle?

12. What causes temporary hardness, and what effect does it have?

13. What causes permanent hardness, and how can it be removed?

14. How can water be made safe for drinking, and what are the drawbacks of these methods?

15. What is acid mine drainage, and how does it contribute to water pollution?

16. How does oxygen contribute to acid mine drainage?

17. What is a dead lake or stream?

References

1. Giddings, J. C. *Chemistry, Man, and Environmental Change;* Canfield: San Francisco, 1973; pp. 199, 229.

2. Weast, R. C., Ed. *Handbook of Chemistry and Physics,* 70th ed.; CRC: Boca Raton, FL, 1989; p. F178.

3. Hileman, B. "Global Warming." *Chemical and Engineering News,* March 13, 1989, p. 25.

4. Sax, N. I. *Dangerous Properties of Industrial Materials;* Van Nostrand Reinhold: New York, 1968; pp. 533–534.

5. Maugh, T. H. "Acid Rain's Effects on People Assessed." *Science* **1984,** *226,* 1409.

6. White, R. M. "The Great Climate Debate." *Scientific American* **1990,** *263* (1), 36.

7. Jones, P.D.; Wigley, T. M. L. "Global Warming Trends." *Scientific American* **1990,** *263* (2), 84.

8. Byrne, G. "The Greenhouse Effect: Science and Policy." *Science* **1989,** *243,* 771.

9. Swartz, J. "CO_2 Reduction and Reforestation." *Science* **1988,** *242,* 1494.

10. Byrne, G. "Let 100 Million Trees Bloom." *Science* **1988,** *242,* 371.

11. Molina, M. J.; Rowland, F. S. "Stratospheric Sink for Chlorofluoromethanes: Chlorine-atom Catalyzed Destruction of Ozone." *Nature* **1974,** *249,* 810.

12. Rowland, F. S. "Chlorofluorocarbons and the Depletion of Stratospheric Ozone." *American Scientist* **1989,** *77* (1), 36.

13. Manzer, L. E. "The CFC-Ozone Issue: Progress on the Development of Alternatives to CFCs." *Science* **1990,** *249,* 31.

14. Zurer, P. S. "Industry, Consumers Prepare for Compliance with Pending CFC Ban." *Chemical and Engineering News,* June 22, 1992, pp. 7–13.

15. Ember, L. "EPA Proposes Nationwide Recycling Program for Ozone-Depleting CFCs." *Chemical and Engineering News,* May 7, 1990, p. 44.

16. Zurer, P. S. "Ozone-Safe Technology Fund Likely Vital to Montreal Treaty Compliance." *Chemical and Engineering News,* August 6, 1990, p. 19.

17. Ember, L. "Lack of Market Stalls New Chlorofluorocarbons." *Chemical and Engineering News,* May 25, 1987, p. 18.

18. Crawford, M. "EPA: Ozone Treaty Weak." *Science* **1988,** *242,* 25.

19. Hively, W. "How Bleak Is the Outlook for Ozone?" *American Scientist* **1989,** 77 (3), 219.

20. Wong, J. L. "Controlling Environmental Carcinogens." *Chemtech* **1987,** *17,* 46.

21. Zurer, P. S. "Ozone Hole's Hidden Chlorine Chemistry Explained." *Chemical and Engineering News,* May 21, 1990, p. 43.

22. Kerr, R. A. "Stratosphere Ozone Is Decreasing." *Science* **1988,** *239,* 1489.

23. Zurer, P. S. "Chlorine Eroding Arctic as Well as Antarctic Ozone, Scientists Confirm." *Chemical and Engineering News,* March 19, 1990, p. 22.

24. Zurer, P. S. "Ozone Layer: Study Finds Alarming Global Losses." *Chemical and Engineering News,* March 21, 1988, p. 6.

25. O'Sullivan, D; Zurer, P. S. "Saving the Ozone Layer: Key Issues Face Tough Negotiations." *Chemical and Engineering News,* March 13, 1989, p. 4.

26. Finlayson-Pitts, B. J. *McGraw-Hill Yearbook of Science and Technology 1989;* McGraw-Hill: New York, 1990; pp. 41–45.

27. Browne, M. W. "As Halon Ban Nears, Researchers Seek a New Miracle Firefighter." *New York Times,* December 15, 1992, p. C4.

28. Warneck, P. *Chemistry of The Natural World;* Academic: New York, 1988; Chapter 5.

29. Pryde, L. T. *Environmental Chemistry;* Benjamin/Cummings: Menlo Park, CA, 1973; p. 213.

30. Ward, R. C. *Principles of Hydrology;* McGraw-Hill: New York, 1975; p. 5.

Agricultural Chemistry

III

How Does Your Garden Grow?

15

In this section we will consider some of the ways in which chemistry plays a role in agriculture. The *agrichemicals,* which include fertilizers and pesticides, are used in huge amounts worldwide, a practice called conventional farming. Pesticides are commonly subdivided into herbicides (weed killers) and insecticides. The latter are discussed in Chapter 16. Agrichemicals are important not only to commercial farmers, who grow vast amounts of food for direct consumption and as feed for animals, but also to the home gardener, who deals with the same issues on a smaller scale.

Nitrogen, phosphorus, and potassium are required in large amounts by all plants and must be continually replenished. Therefore, fertilizers or suitable alternatives must be provided for plants to grow. Chemical fertilizers are used most often. For many years, however, the negative consequences of these products have become increasingly apparent. Heavy use of synthetic chemicals has often led to declining soil productivity, deteriorating environmental quality, and lower profits. Some problems arise from the decreased ability of soils to bind the chemicals due to the loss of organic matter. When chemicals leach out of the soil readily, it becomes necessary to increase their use, which increases the burden on the environment.

This has led to increased awareness of the need to practice *sustainable agriculture,* in which the emphasis is on rotating crops to build up soil, and on controlling pests naturally. Many farmers grow crops without chemicals, instead preferring natural organic products.

15.1 Fertilization

GOALS: To understand the classifications of soils and the importance of organic matter in soils.
To understand the classification of nutrients as macronutrients, secondary nutrients, and micronutrients, so as to appreciate the strategies of fertilization.
To consider the importance of soil pH and learn some ways to adjust it.
To examine the major sources of the macronutrients (N, P, K) and the use of mixed fertilizers.
To appreciate the importance of nitrogen fixation by legumes.

All living things, plants as well as animals, must have an adequate supply of nutrients for growth. Successful fertilization requires an understanding of two closely related subjects, soils and fertilizers, since soil structure has a great influence on fertilizer requirements.

Soils

Soil is a loose mass of broken and chemically weathered rock mixed with organic matter. Both the organic and the inorganic mineral components influence the texture and the capacity for binding chemical nutrients. Organic matter, known as **humus,** is found mostly near the surface. It consists of partially decomposed residues of plants, animals, and microorganisms.

The inorganic portion of the soil is mostly oxides (compounds containing oxygen) of silicon (Si), aluminum (Al), and iron (Fe), and combinations of these elements. The oxides combine in complex units of variable size. Soils are classified as clay, silt, sand, or a combination, depending on particle size. Clay soils have the smallest particle size, below 0.002 mm in diameter. Silt particles range from 0.002-0.05 mm and sand from 0.05-2 mm in diameter. In some cases, the classifications are further divided (for example, coarse and fine sand). **Loam** is a mixture of clay, silt, and sand that shows some of the characteristics of each.

Soil texture usually is described in terms of the relative percentage of clay, silt, and sand. The most common designations are listed in the margin in order of increasing fineness.

Sand   Coarse
Loamy sand   ↑
Sandy loam
Loam
Silt loam
Silt
Sandy clay loam
Clay loam
Silty clay loam
Sandy clay
Silty clay
Clay   Fine

Loamy soil is usually the best for growing crops. The two extremes, clay and sand, have properties that are less conducive to good growing, although recognizing their associated difficulties suggests ways to compensate.

Since clay has a small particle size, it has a high surface area. It also has surface charges that give it an unusually high capacity for holding nutrients. Thus, the loss of nutrients from clay soils is negligible compared with that from sandy soils. But this seeming advantage means that nutrients may not be released to plants.

Clay soils also hold much more water than does sand, so leaching of nutrients is low. Too much water may cause a deficiency of air, however, and good aeration is required for root growth and for high levels of soil microorganisms. Clay soils are heavy and sticky. They tend to crack when dry, although expansion and contraction help aeration, since air is alternately forced out of and drawn into the soil.

Sandy soils have some characteristics opposite to clay. Sand has little capacity for water and little attraction for nutrients. Water tends to pass through it freely, often leaching out nutrients. Failure of sand particles to bind nutrients tightly is an advantage, since the nutrients are more available to plants as long as they remain in the soil. But sandy soils require more frequent fertilization.

Loamy soils are well aerated for three reasons: their constituent material is loosely packed; they have a low capacity for water retention; and the water held by the clay particles changes often. Loamy soils bind nutrients well, due to the clay component, but not excessively, due to the sand.

Some of the chemicals required for plant growth are obtained from minerals in the soil, but much of the nutrient supply comes from fertilizers or decomposed organic matter. Organic matter is an essential component of fertile soil. Besides providing a supply of nutrients, it tends to improve soil texture by countering the overcapacity for water retention of clay soils and by moderating the opposite effect of sandy soils. Microorganisms play an active role in determining the chemical composition of soil; they decompose and dispose of plant and animal residues, releasing chemicals that are directly usable by plants.

Some bacteria even fix nitrogen, one of the chief nutrients required for plant growth. **Nitrogen fixation** is the process whereby nitrogen as N_2 is removed from the air and converted to a useful ionic form as NO_3^-. Most microorganisms, however, simply feed on plant and animals residues, later releasing the nutrients back into the soil. The process whereby organic matter is decomposed and inorganic ions are released is **mineralization.**

Most of the nitrogen in organic matter is present as protein, unavailable for use by growing plants. It is released from the protein as ammonia, NH_3, which remains trapped in the soil as ammonium ions, NH_4^+. The ammonium ions are then oxidized by microorganisms in two steps to form nitrate ions. This process of **nitrification** releases the nitrogen in the form most useful to growing plants.

$$NH_4^+ \quad \rightarrow \quad NO_2^- \quad \rightarrow \quad NO_3^-$$
$$\text{ammonium ion} \qquad \text{nitrite ion} \qquad \text{nitrate ion}$$

Fertilizers

The primary challenge in agriculture is to supply all the necessary nutrients to growing plants. Just as the human diet must contain certain essential components (amino acids, vitamins, minerals, etc.), plants too require nutrients they cannot store or make. Plants and animals have sophisticated chemical reaction pathways (metabolic pathways) for using nutrients, either directly or by conversion to forms suitable for both function and structure.

One interesting phenomenon is the way in which plants and animals complement one another. Each can do things that the other cannot. For example, humans can produce, transport, and apply external nutrients to plants. Plants can use certain mineral nutrients, for example, nitrogen as nitrate ion, NO_3^-, to produce amino acids, which they incorporate into protein. We cannot use nitrate as a source of nitrogen, and depend on plants and on animals that feed on plants to get protein with a full complement of the 10 essential amino acids, plus dietary requirements such as carbohydrates, fats, vitamins, and minerals.

Some 16 nutrients are commonly considered essential for proper plant growth. For emphasis, they are often divided into three groups: macronutrients, secondary nutrients, and micronutrients (Table 15.1).

Table 15.2 shows that plants contain large, medium, or trace amounts of the three levels of nutrients. Carbon, hydrogen, and oxygen are not tabulated since they are acquired from water absorbed from the soil and from photosynthesis as follows:

$$\overset{\text{light}}{CO_2 + H_2O \rightarrow O_2 + \text{carbohydrate}}$$

Monosaccharides, disaccharides, and polysaccharides are discussed in Chapter 9.

The carbohydrate produced by photosynthesis appears in plants as starch, cellulose, and simple sugars (monosaccharides and disaccharides). For most plants, 60-95% of the dry weight is carbohydrate, so photosynthesis can be described as **carbon fixation.**

Table 15.1   Plant nutrients

Macronutrients	Secondary Nutrients	Micronutrients
Carbon (C)	Calcium (Ca)	Boron (B)
Hydrogen (H)	Magnesium (Mg)	Copper (Cu)
Oxygen (O)	Sulfur (S)	Iron (Fe)
Nitrogen (N)		Manganese (Mn)
Phosphorus (P)		Zinc (Zn)
Potassium (K)		Molybdenum (Mo)
		Chlorine (Cl)
		Sodium (Na)

Table 15.2   Nutrient content of various foods (1)

Crop	Yield/Acre	Nutrients in Crop (lb/acre)									
		N	P	K	Ca	Mg	S	Cu	Mn	Zn	B
Corn	100 bu	90	15	21	6	6	7	0.04	0.06	0.10	$\cdots$
Apples	500 bu	30	5	37	8	5	10	0.03	0.03	0.03	0.01
Potatoes	400 bu	80	13	120	3	6	6	0.04	0.09	0.05	0.05
Tomatoes	15 ton	90	13	11	5	8	10	0.05	0.10	0.12	0.14
Soybeans	40 bu	150	15	43	7	7	4	0.04	0.05	0.04	0.01

Table 15.3   Nutrients present in the upper seven inches of undisturbed, unfertilized soil (1)

Nutrients	Pounds/Acre
Nitrogen	1000–6000
Phosphorus	800–2000
Potassium	To 49,000
Calcium	To 500,000
Magnesium	8000–26,000
Sulfur	To 3000
Copper	4–400
Manganese	To 200,000
Zinc	20–600
Boron	40–400
Chlorine	200–18,000
Iron	400–over 200,000
Molybdenum	About 4

The remaining nutrients are provided from the soil; nitrogen is one of the most important. It also can be obtained directly from the air by bacteria associated with certain plants called *legumes,* discussed later in this chapter. Table 15.3 gives the content of the remaining 13 nutrients found in the uppermost 7 inches, the soil layer in which most plants grow. In effect, Table 15.2 shows what plants take from the soil, whereas Table 15.3 shows what is present in the soil. A quick comparison suggests that soil is in good shape and does not really seem to need fertilizer at all. Unfortunately, whereas nutrients may well be there, they may be unavailable to the plants. Therefore, fertilizers provide them in forms that plants can use.

pH

Acidity and pH were considered in Chapter 7.

In some cases, adequate plant nutrition may be provided by preparing the soil to release the available nutrients. One method is controlling soil pH. Different types of soils have different pH ranges. As a rule, clay soils are acid, limestone soils are basic or alkaline, and sandy soils are neutral.

At one time farmers checked pH simply by tasting soil. If it tasted sweet, the ground was alkaline. The generally desirable pH is slightly alkaline. If soil is very alkaline, the required metal nutrients become soluble enough to wash away with rain. This preference is only a general rule, however; there are dramatic exceptions. Blueberry plants grow well in many soils, but they bear fruit only if the soil is acid. The best pH for them is about 4.5. Rhododendrons flower best in a sour ground. The ideal pH range for general gardening is 6-8. As a general guide, consider the information given in Table 15.4. Plants not listed usually prefer a neutral soil.

High Alkalinity

When **alkalinity is too high,** acidity can be increased by adding certain substances.

The formula $KAl(SO_4)_2 \cdot$ 12 H_2O indicates that 12 molecules of water are present for every 1 potassium, 1 aluminum, and 2 sulfate ions in a crystal of alum.

Alum   Alum, $KAl(SO_4)_2 \cdot 12\ H_2O$, reacts in water to release hydrogen ions and neutralize excess hydroxide ions. It is soluble in water but does not wash away easily in rain, since the ions bind to the soil at the point of application.

Table 15.4   pH requirements for plants and vegetables *(2)*

Acid-Loving Plants (pH 4-6)	Neutral (pH 7)	Alkaline Soil Plants (pH > 7)
Azalea	Apple	Asparagus
Blackberry	Cornflower	Bean
Blueberry	Gardenia	Beet
Chrysanthemum	Pansy	Cabbage
Cranberry	Pumpkin	Cantaloupe
Huckleberry	Rice	Carnation
Lily	Turnip	Cauliflower
Marigold		Celery
Oak		Cucumber
Peanut		Lettuce
Radish		Nasturtium
Raspberry		Onion
Rhododendron		Pea
Spruce		Rhubarb
Strawberry		Squash
Sweet potato		
Watermelon		

The reactions of alum are as follows; we assume that the reaction is only with OH^- and not with any other alkaline component of the soil. The reacting species of alum is hydrated aluminum Al(III) ion, $Al(H_2O)_6^{3+}$, in which six water molecules are bound to aluminum and supply hydrogen ions:

$$Al(H_2O)_6^{3+} \quad\quad + OH^- \rightarrow Al(H_2O)_5(OH)^{2+} + H_2O$$
$$Al(H_2O)_5(OH)^{2+} + OH^- \rightarrow Al(H_2O)_4(OH)_2^+ + H_2O$$
$$Al(H_2O)_4(OH)_2^+ + OH^- \rightarrow Al(H_2O)_3(OH)_3 \quad + H_2O$$
$$Al(H_2O)_3(OH)_3 \quad + OH^- \rightarrow Al(H_2O)_2(OH)_4^- + H_2O$$

If the medium is strongly basic, an additional conversion can occur.

$$Al(H_2O)_2 (OH)_4^- + 2\ OH^- \rightarrow Al(OH)_6^{3-} + 2\ H_2O$$

According to these reactions, the water molecules are able to release protons to neutralize OH^- present in the soil. This can be rewritten as follows for the release of the first proton:

$$Al(H_2O)_6^{3+} \rightarrow H^+ + Al(H_2O)_5(OH)^{2+}$$

The reactions for aluminum sulfate, $Al_2(SO_4)_3$, are the same as those given for alum. It also can be used to decrease soil pH.

Natural Sources   Leaves (particularly oak leaves) and pine needles liberate acid when they decompose. Citrus peels contain citric acid. Nut shells and barks from most trees contain tannic acid. All can be introduced into the soil to release acid.

High Acidity

When **acidity is too high,** soils can be corrected by adding any of several substances.

Lime   To a chemist, lime (also known as quicklime) is calcium oxide, CaO, which is extremely alkaline and will burn plants. Slaked lime is calcium hydroxide, $Ca(OH)_2$, which is equally harmful. Lime is converted to slaked lime when it comes in contact with water, according to the following equation:

$$\underset{\text{lime}}{CaO} + H_2O \rightarrow \underset{\text{slaked lime}}{Ca(OH)_2}$$

See Common Acids and Bases in Section 7.5 for a discussion of carbonate and bicarbonate ions.

To the farmer, lime is calcium carbonate, $CaCO_3$, better known as limestone. Thus, "agricultural lime" is the same as crushed limestone. It provides calcium, in addition to correcting acid pH. The neutralizing action of $CaCO_3$ is described by the following, where carbonate and bicarbonate ions act as bases:

$$CaCO_3 = Ca^{2+} + CO_3^{2-}$$
$$\underset{\substack{\text{carbonate}\\\text{ion}}}{CO_3^{2-}} + H^+ \rightarrow \underset{\substack{\text{bicarbonate}\\\text{ion}}}{HCO_3^-}$$
$$HCO_3^- + H^+ \rightarrow \underset{\substack{\text{carbonic}\\\text{acid}}}{H_2CO_3}\ \text{(unstable)}$$
$$H_2CO_3 \rightarrow H_2O + CO_2$$

Dolomite Dolomite is a mixture of magnesium carbonate, $MgCO_3$, and $CaCO_3$. Limestone usually contains some $MgCO_3$. Both $CaCO_3$ and $MgCO_3$ dissolve slowly in the soil; therefore, one application provides acid-neutralizing materials for several years. Calcium and magnesium are secondary plant nutrients.

Wood Ashes When wood is burned, potassium is retained in ashes as potassium carbonate, K_2CO_3, also called *potash*. It is a soluble material that leaches out of soil easily. Like other sources of carbonate ion, K_2CO_3 is highly alkaline and can neutralize an acid soil. Because of the rapid release of carbonate, wood ashes are so basic that they are usually not allowed to come directly in contact with plants and leaves, but are applied to the soil surrounding them. The word potash has another meaning, discussed in the section on potassium fertilizers.

Nitrogen

Nitrogen is often considered the most important plant nutrient since it is required for the synthesis of proteins. Organic and conventional (chemical) farmers use different sources of nitrogen.

In organic gardening the most concentrated source is blood meal (15% nitrogen). Others are hoof meal and horn dust (12.5% nitrogen), cottonseed meal (7% nitrogen), and manure. Fresh grass clippings have about 1% nitrogen. Compost has a good nitrogen content if properly made (Section 15.3). Manure is a frequent but not very concentrated source. Raw garbage normally contains 1-3% nitrogen.

Sewage that has been processed to remove potentially harmful bacteria and heavy metals is called sludge. It contains about 2% nitrogen, a higher concentration than in raw sewage. Processing involves bubbling air through raw sewage, which promotes the growth of aerobic bacteria that consume some of the carbohydrate content. This leaves nitrogen in a more concentrated form that coagulates and settles. Dried, heat-treated sludge has a 5-6% nitrogen content and 3-6% phosphorus content. The common chemical sources of nitrogen are described in the following paragraphs.

Ammonia, NH_3, 82% Nitrogen

Conventional farmers find the best source of nitrogen is ammonia, since it is a highly concentrated source of nitrogen (82% m/m), is easy to apply to the soil, and is very inexpensive. Ammonia (bp -33 °C) is soluble in water up to 28%. Used as a water solution or pumped into the ground as a gas (Figure 15.1), it binds readily to many components of the soil and is readily and quickly converted to usable plant food. H^+ reacts with gaseous NH_3 to form soluble NH_4^+ so that it is no longer a gas. Nitrogen is taken up by plants as nitrate, NO_3^-, which is produced from NH_3 by bacteria in the soil by nitrification, described earlier in this chapter.

Ammonia is produced on a commercial scale by the Haber process, described by the equation

$$3 H_2 + N_2 \xrightarrow[\substack{500 °C \\ 500-1000 \text{ atmospheres pressure}}]{\text{catalyst}} 2 NH_3$$

It is very alkaline, but the alkalinity disappears as the ammonia is converted to nitrate in the soil. Ammonia is toxic and injurious to living tissues, and is difficult to handle. It adds

Mass percent (m/m) is discussed in Section 7.4.

German chemist Fritz Haber (1868–1934) received the Nobel Prize in 1918 for his work on the synthesis of ammonia.

Figure 15.1

Pumping ammonia into the soil.

no organic matter to soil. In fact, its nitrogen content is either used quickly by plants or washed away, but the increased growth of plants due to ammonia is spectacular. Consequently, it is the largest source of nitrogen for agricultural use despite its disadvantages.

Ammonium Sulfate, $(NH_4)_2SO_4$, 21% Nitrogen

Ammonium sulfate is the most important solid nitrogen fertilizer used throughout the world. It contains readily available nitrogen as NH_4^+, is soluble in water, and supplies sulfur, a secondary nutrient. Because of its high water solubility, it must be applied frequently, especially where rainfall is heavy.

Urea, $(NH_3)_2CO$, 47% Nitrogen

Urea, a metabolic waste product of animals, is another good source of nitrogen. It is easily synthesized on a commercial scale, and is soluble, relatively nontoxic, and easily handled. Urea reacts slowly to release ammonia over time, thus providing a longer-term supply of NO_3^- by nitrification.

Biuret and triuret are slight modifications of the urea molecule that also are convertible to nitrate by the action of bacteria in the soil. Since the water solubility decreases:

$$urea > biuret > triuret$$

prolonged release of the nitrogen results, more in line with the growth pattern of many plants. Specifically, triuret nitrogen becomes completely available in 6-12 weeks *(3)*.

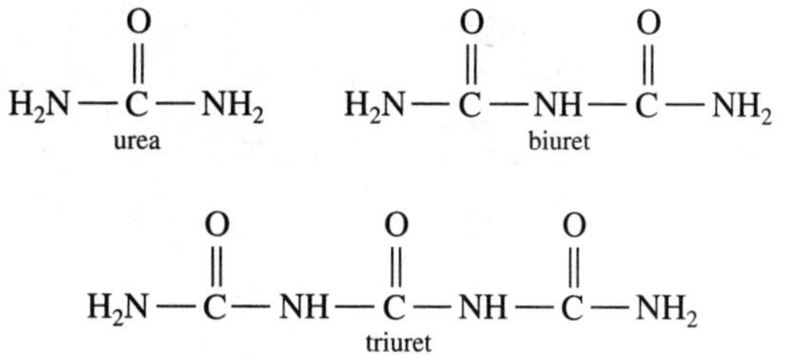

$$
\begin{array}{ccc}
& O & & O & & O \\
& \parallel & & \parallel & & \parallel \\
H_2N-C-NH_2 & & H_2N-C-NH-C-NH_2 \\
\text{urea} & & \text{biuret}
\end{array}
$$

$$
\begin{array}{ccccc}
& O & & O & & O \\
& \parallel & & \parallel & & \parallel \\
H_2N-C-NH-C-NH-C-NH_2 \\
& & & \text{triuret}
\end{array}
$$

Ammonium Nitrate, NH_4NO_3, 35% Nitrogen

Ammonium nitrate is a solid, granular material that is soluble in water. It has a major problem associated with it: it is a powerful explosive, reacting according to the equation

$$NH_4NO_3 \rightarrow N_2O \text{ (g)} + 2\ H_2O \text{ (g)}$$

This reaction needs only to be initiated for an explosion to occur. Ammonium nitrate will burn without exploding and it requires reasonably serious conditions to explode, but the precise nature of these conditions is not known, a fact that makes it difficult to ensure safe handling. One of the most serious explosions was a shipload that destroyed a major part of the city of Texas City, Texas, in 1947. More than 400 people were killed in the incident, which touched off explosions and fires in nearby petroleum plants *(4, 5)*.

Other Ammonium Salts

Almost any ammonium salt with a beneficial or at least an innocuous anion can be used as a nitrogen fertilizer. One of the most important ones is ammonium phosphate, the main multinutrient (N plus P) fertilizer in use today. See further discussion under phosphorus fertilizers.

Many ammonium salts (including ammonium sulfate and nitrate) have an acidifying effect that is usually countered by adding limestone. For every 100 lb of ammonium nitrate, 59 lb of calcium carbonate (limestone) are required to neutralize the acid *(6)*. The acidity is due to the reaction

$$\underset{\substack{\text{ammonium} \\ \text{ion}}}{NH_4^+} \rightarrow \underset{\text{ammonia}}{NH_3} + H^+$$

Figure 15.2

Nitrogen-fixing nodules on soybean roots.

Nitrogen Fixation for Sustainable Agriculture

Plants known as legumes can fix nitrogen, N_2, from the air directly to form NH_3. The fixation is carried out by bacteria that reside in nodules in the roots of leguminous plants, as shown in Figure 15.2. Alfalfa is the most potent nitrogen fixer, followed by clovers, soybeans, other beans, peas, and peanuts.

The nodular nitrogen-fixing bacteria exist in a symbiotic (mutually beneficial) relation with these plants. The plants contain chlorophyll and produce carbohydrates by photosynthesis, while the microorganisms use carbohydrates as fuel for fixing nitrogen, which ultimately ends up as plant protein.

If crops are rotated from year to year and include legumes in some years, the soil nitrogen content may be increased to a level requiring only minimal chemical nitrogen sources. Usually, alfalfa is grown and plowed back into the soil so that both nitrogen content and organic matter are increased. It is common knowledge that legumes reduce fertilizer costs, soil erosion, and problems with diseases and pests in subsequent plantings *(7–9)*.

The value of nitrogen fixation is so great that researchers have tried to induce other nonnodular plants to carry out the process. By genetic engineering, the genes for nitrogen fixation are introduced directly into plant cells. So far, the key enzyme *nitrogenase* is rapidly inactivated by abundant oxygen in plants *(8, 10)*.

Phosphorus

$$\% \text{ P} \times 2.29 = \% \text{ P}_2\text{O}_5$$
$$\% \text{ P}_2\text{O}_5 \times 0.437 = \% \text{ P}$$

Phosphorus is usually found as some form of phosphate, PO_4^{3-}, HPO_4^{2-}, or $H_2PO_4^-$. The amount of phosphorus is commonly quoted as percentage P_2O_5 even though the fertilizer contains no P_2O_5. This practice is related to the conversion of P_2O_5 to phosphoric acid, H_3PO_4:

$$P_2O_5 + 3 \text{ H}_2O \rightarrow 2 \text{ H}_3PO_4$$

Table 15.5   Organic sources of phosphates

Source	% as P_2O_5
Bone meal	22–25
Dried shrimp wastes	10
Raw sugar wastes	8
Dried ground fish	7
Activated sludge	3–6
Dried blood	1–5
Wool wastes	2–4
Nutshells	1
Manure	0.2–1.4
Wood ashes	1.5
Tankage (from slaughter houses)	Up to 3

Organic Sources, Variable P_2O_5

Organic sources of phosphates are listed in Table 15.5. Until the beginning of this century, these were the primary sources.

Phosphate Rock, Variable P_2O_5

During this century, calcium phosphate ores known as *phosphate rock* have become the primary source of phosphates. Extensive deposits of these ores are found in beds that were originally ocean floor. In general, they are a combined phosphate fluoride associated with $CaCO_3$ and symbolized $Ca_{10}F_2(PO_4)_6$, although the composition varies with the source. The principal deposits are in North Africa, the United States, and Russia. These mineral ores are the primary source of all phosphate fertilizers. Organic gardeners use phosphate rock directly without any chemical modification.

The need for increased efficiency of transport and application for large-scale farming led to the production of materials from phosphate rock of higher phosphate content. These are described in the following paragraphs.

Normal Superphosphate, 16–22% P_2O_5

Superphosphate is produced from phosphate rock by the action of sulfuric acid, as illustrated in the equation:

$$Ca_{10}F_2(PO_4)_6 + 7\ H_2SO_4 + 3\ H_2O \rightarrow 3\ Ca(H_2PO_4)_2 \cdot H_2O + 7\ CaSO_4 + 2\ HF$$

The $CaSO_4$ is not removed, and the total product has 16–22% P_2O_5. This is normal superphosphate.

Triple Superphosphate, 44–47% P_2O_5

If phosphate rock is treated with phosphoric acid instead of sulfuric acid, triple superphosphate is formed:

$$Ca_{10}F_2(PO_4)_6 + 14\ H_3PO_4 + 10\ H_2O \rightarrow 10\ Ca(H_2PO_4)_2 \cdot H_2O + 2\ HF$$

It has a P_2O_5 content about 3 times that of normal superphosphate, since no $CaSO_4$ is present in the final product.

Ammonium Phosphate, Variable P_2O_5

Phosphoric acid also can be combined with ammonia to yield ammonium phosphate, a mixed (N + P) fertilizer mentioned with the nitrogen fertilizers. The nitrogen and phosphorus contents are variable, depending on the production process. As the following reactions show, the relative amounts of both substances determine the final product:

$$H_3PO_4 + \quad NH_3 \rightarrow NH_4^+ \, H_2PO_4^-$$
$$H_3PO_4 + 2\, NH_3 \rightarrow (NH_4^+)_2 \, HPO_4^{2-}$$
$$H_3PO_4 + 3\, NH_3 \rightarrow (NH_4^+)_3 \, PO_4^{3-}$$

The second product, diammonium phosphate, is a common fertilizer.

The solubility of phosphate fertilizers can be a problem. **Phosphate fixation,** unlike nitrogen fixation, is the precipitation of phosphates (by iron and aluminum in the soil), which makes them inaccessible. Liming is often important to adjust the pH for efficient use of phosphates, which are most available near pH 7. At pH 6 or above, iron oxides and aluminum oxides in the soil are largely precipitated, so iron and aluminum cannot precipitate the phosphates.

Potassium

% K × 1.2 = % K_2O
% K_2O × 0.83 = % K

Just as the word "lime" has different meanings to a chemist and a farmer, the term "potash" also has different meanings depending on where it is used. To the chemist, potash is potassium carbonate (see Wood Ashes). The percentage of potassium in fertilizers is normally reported as % K_2O, known as *percent potash,* even if no K_2O is actually present.

Wood ashes are an excellent source of potash (Figure 15.3), but the strong alkalinity of K_2CO_3 makes them less useful than most other potassium fertilizers. All the common potassium fertilizers are acceptable to both organic and nonorganic farmers since they are all natural materials.

Common Potassium Fertilizers

Granite Dust   Granite dust is an excellent source of slow-working potash. The potash content varies from 3-5%, although one Massachusetts quarry yields a granite dust with 11% potash. Granite also supplies nutrients such as phosphorus, calcium, magnesium, iron, and manganese.

Greensand   Also known as greensand marl, this is a natural marine deposit made up of iron, potassium, silicon, and oxygen. It contains 6-7% potash.

Manures   Manures are excellent sources of potassium. Since potassium cannot decompose, a livestock farm on which manure is handled properly requires little or no potassium fertilization except to make up for losses due to leaching.

KCl   Deposits of KCl are found in many parts of the world, the largest in central Europe, the western United States, and Canada. Other common ores are potassium sulfate, K_2SO_4, and potassium nitrate, KNO_3. The sulfate is also produced commercially by combining

The United States.

To all to whom these Presents shall come. Greeting.

Whereas Samuel Hopkins of the city of Philadelphia and State of Pennsylvania hath discovered an Improvement, not known or used before such Discovery, in the making of Pot ash and Pearl ash by a new Apparatus and Process, that is to say, in the making of Pearl ash 1st by burning the raw Ashes in a Furnace, 2nd by dissolving and boiling them when so burnt in Water, 3rd by drawing off and settling the ley, and 4th by boiling the ley into Salts which then are the true Pearl ash; and also in the making of Pot ash by fluxing the Pearl ash so made as aforesaid; which Operation of burning the raw Ashes in a Furnace, preparatory to their Dissolution and boiling in Water, is new, leaves little Residuum, and produces a much greater Quantity of Salt: These are therefore in pursuance of the Act, entitled "An Act to promote the Progress of useful Arts", to grant to the said Samuel Hopkins, his Heirs, Administrators and Assigns, for the Term of fourteen Years, the sole and exclusive Right and Liberty of using, and vending to others the said Discovery, of burning the raw Ashes previous to their being dissolved and boiled in Water, according to the true Intent and Meaning of the Act aforesaid. In Testimony whereof I have caused these Letters to be made patent, and the seal of the United States to be hereunto affixed Given under my Hand at the City of New York this thirty first Day of July in the Year of our Lord one thousand seven hundred & Ninety.

G Washington

City of New York July 31st 1790.—

I do hereby certify that the foregoing Letters patent were delivered to me in pursuance of the Act, entitled "An Act to promote the Progress of useful Arts"; that I have examined the same, and find them conformable to the said Act.

Edm: Randolph Attorney General for the United States.

(Endorsement on back of grant)

Delivered to the within named Samuel Hopkins this fourth day of August 1790.

Th Jefferson

Figure 15.3

The first United States patent grant, July 31, 1790. Note the signatures of Washington and Jefferson. The patent was issued to Samuel Hopkins for his process for making potash and pearl ash (purified potash).

KCl with magnesium sulfate, $MgSO_4$. This expensive process, which limits the demand for K_2SO_4, is primarily used for making tobacco fertilizer.

Mixed Fertilizers

Mixtures of the three macronutrients are commonly used. It is generally less expensive and safer to combine these chemicals at the production level than on the farm, so that all three can be applied at once. The familiar formulation 5:10:5 means 5% N, 10% P_2O_5, and 5% K_2O. This is actually 5% N, 4.4% P, and 4.2% K.

Other Nutrients

Of the other nutrients (secondary and micro) listed in Table 15.1, several accompany the important nutrients or are introduced when pH is being modified. For example, liming with dolomite introduces the secondary nutrients calcium and magnesium. Potassium and ammonium sulfate, as well as superphosphate, supply sulfur.

Micronutrients are normally present in sufficient amounts in mineral deposits, and their chemical modifications are used for fertilization or pH adjustment. Phosphate rock contains molybdenum, and granite dust contains iron and manganese, for example. These micronutrients are integral parts of plant enzymes in nitrogen fixation (Mo), photosynthesis (Zn), and respiration (Fe).

15.2 Herbicides

GOALS: To understand the selectivity of some herbicides.
       To consider the use of TCDD in Agent Orange.

Weeds are very costly to agriculture. In addition to lowering the quality and quantity of crops, they may poison livestock, induce off-flavors in milk, reduce the flow of irrigation waters, and hinder mechanization of crop production. Chemical weed killers or herbicides serve a useful purpose, but are often regarded as unacceptable due to their detrimental effect on the environment.

The selectivity of weed killers is critical. The best-known are the chlorophenoxys 2,4-D and 2,4,5-T, and some related compounds such as silvex and MCPA. The *phenoxys* are growth regulators with hormone-like activity. Whereas 2,4-D selectively attacks broad-leaf weeds, it is harmless to cereals and other grassy crops. The resistance of grasses is attributed to their high levels of detoxifying enzymes.

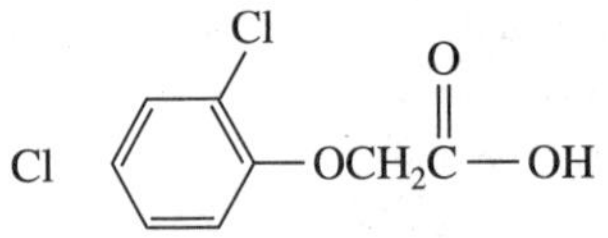

2,4-D
2,4-dichlorophenoxyacetic acid

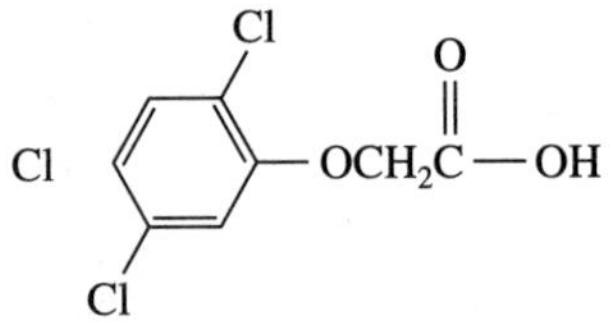

2,4,5-T
2,4,5-trichlorophenoxyacetic acid

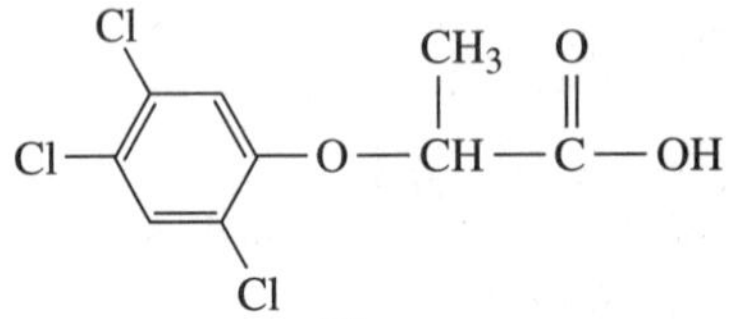

Silvex
2-(2,4,5-trichlorophenoxy)propionic acid

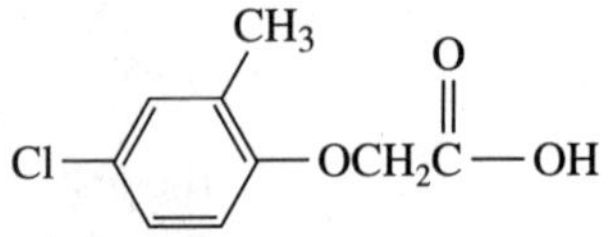

MCPA
2-methyl-4-chlorophenoxyacetic acid

The hormonal effects of the phenoxys lead to rapid growth of the stems of a weed: the spindly tall growth forms weak tubes that may get crushed, blocking the flow of nutrients and water up the stem.

A ban on the use of 2,4,5-T has been in effect for many years. A compound commonly known as TCDD (2,3,7,8-tetrachloro-dibenzo-p-dioxi) stirred much controversy with regard to some phenoxy herbicides. It is a highly toxic carcinogen and teratogen (causes birth defects) in laboratory animals, and causes the severe skin disease chloracne in humans. Similar effects were once attributed to 2,4,5-T until it was recognized that TCDD, a by-product of the synthesis of 2,4,5-T, contaminated many samples of this herbicide. Careful control of temperature and alkalinity can prevent formation of TCDD.

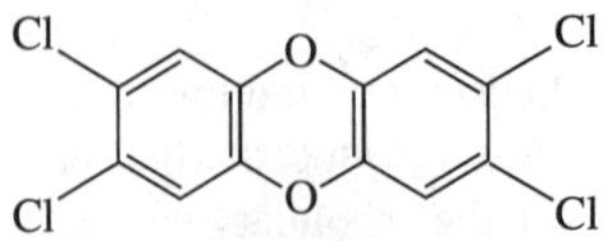

TCDD

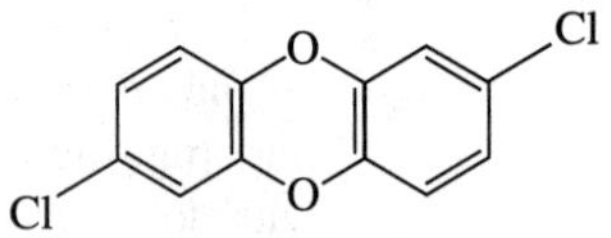

DCDD

During the Vietnam War, a 1:1 mixture of 2,4-D and 2,4,5-T was used as a chemical defoliant known as Agent Orange. Many veterans and Vietnamese citizens attribute health problems to exposure to this defoliant, another lasting controversy of that conflict *(11)*.

Several industrial accidents exposed large numbers of humans to TCDD. The evidence accumulated from these incidents indicates that TCDD is less toxic to humans than to laboratory animals. It is neither a carcinogen nor a teratogen in humans, although severe chloracne was often observed *(12, 13)*.

The corresponding dioxin DCDD could potentially form in the synthesis of 2,4-D. This is unlikely, but if it does occur, DCDD is not regarded as hazardous even to laboratory animals.

15.3 The Organic Way

GOAL: To understand the various approaches to gardening that are used by organic farmers, and the chemical changes that occur during composting.

Organic farmers take a different approach to satisfy the requirements for fertilization and pH control. Table 15.6 lists most of the acceptable and unacceptable components of the organic farmer's tools. Chemicals applied directly as minerals have already been mentioned.

First and foremost is **compost.** Scientists have been studying composting for as long as ecology has existed. Essentially, it is new topsoil produced from a wide variety of organic matter. The starting materials are weeds, grass clippings, kitchen garbage, garden residues, manure, sewage sludge, bone meal, blood meal, dried fish, leaves (oak leaves for acidity), and lime (for alkalinity). These contain many minerals and can be decomposed by aerobic and anaerobic bacteria and fungi. As these microorganisms digest the organic matter, they reduce the carbon content, thus increasing the percentage of nitrogen and other nutrients.

Compost is acclaimed as the perfect soil conditioner. The five methods listed by the editors of *Organic Gardening and Farming Magazine (14)* differ primarily in the techniques for encouraging the bacteria. These methods of composting are:

1. The Indore method

2. The 14-day method

3. The anaerobic method

4. The sheet composting method

5. The earthworm method

Table 15.6   The organic approach

Acceptable		Unacceptable
Phosphate rock	Compost	Insecticides
Dolomite	Earthworms	Herbicides
Granite dusts	Mulching	Commercial fertilizers
Greensand marl	Oil sprays (for insect control)	
Pulverized limestone	Plant extractives	
Ashes	Crop rotation	
	Manures	

The Indore method was first proposed by Sir Albert Howard due to his work in Indore, India in the early 1930s.

The Indore method is carried out by layering different organic materials; this causes rapid decomposition to usable soil material. The pile consists of a 5-6-in. layer of green material (plant residues, grass clippings, etc.) and then a 2-in. layer of manure, blood meal, sewage sludge, or other high-protein material. This is followed by a layer of topsoil, limestone, and phosphate rock. The layers are repeated until the pile reaches a height of 5 ft and a width of 10 ft. Stakes that are positioned during pile formation are removed to provide ventilation to the interior of the pile. The pile is watered plentifully, but not so it becomes soggy. In 3 months the pile converts to a sweet-smelling, dark, crumbly soil.

Within 24 hours after this kind of compost pile is constructed, the temperature of the interior begins to rise due to bacterial action, a sign of organic decomposition in the pile. The temperature approaches 82 °C within a few days, will stay in this range for several weeks, and will then begin to decrease. Ventilation is important because the bacteria and fungi are aerobic. If they cannot get enough air, anaerobic putrefaction may set in. As the pile cools, it is turned to allow entry of more air, and the temperature begins to rise again. The higher the temperature, the shorter the time required to finish the decomposition, as long as it does not rise so high that the bacterial activity is inhibited.

During decomposition, many changes occur in the pile. A compost heap constructed by the Indore method goes through five distinct changes to convert the ingredients to finished compost (15).*

First change: Chemical oxidation produces heat in the same manner as burning. In order for plant components to be oxidized, their protective outer layering must be removed. This can be accomplished by shredding the components. During the oxidation, the pile decreases in height from 5 feet to 3 1/2 feet.

Second change: Aerobic fungi penetrate the plant material and cover it with white growth, which are new generations of fungi. Heat, moisture, and air facilitate their growth. This phase is complete in about 3 weeks. It is very similar to the rotting of fruit once the skin is broken. If the skin is intact, the fruit keeps much longer.

Third change: Aerobic bacteria follow the path of the dead fungi and take over the decomposition of the plant material, further breaking down the plant cells. The bacteria work for about 3 weeks, at the end of which time most of the heat is lost and the heap is a brown-colored compost.

Fourth change: By this phase the compost heap excludes most atmospheric oxygen. Anaerobic bacteria take over the decomposition. The need for some oxygen is provided by slow diffusion through the pile, as long as it has not been tamped down or become sodden. If oxygen is excluded, some bacteria will break down nitrate to obtain oxygen, resulting in the formation of ammonia. The odor of ammonia gas may be detected as it escapes. The loss decreases the amount of nitrogen present in the compost. This phase is complete in 4 to 8 weeks. Cold weather will retard this process.

Fifth change: If enough lime has been added to prevent the heap from becoming too acid, and if air is allowed to penetrate into the pile, nitrogen-fixing bacteria will begin to fix appreciable quantities of nitrogen into the heap.

Data on the overall changes during composting are illustrated in Table 15.7. The chief change is decreased carbon by oxidation to CO_2, which is lost from the pile. This loss is favorable, since it causes an increase in the concentration of other plant nutrients that must

Table 15.7   Typical changes during composting *(15)*

Component (%)	Beginning	End
Ash	7.6	17.4
C	41.7	34.5
N	1.66	2.45
P	0.13	0.27
K	0.80	1.65
C/N ratio	24.9	14.1

be obtained from the soil. Atmospheric CO_2 is a ready source of carbon during photosynthesis.

Compost can be used as a fertilizer, side dressing, and general soil conditioner. No single chemical analysis can be given for compost because of the wide variety of starting materials. In general, it is a rich and versatile fertilizer.

The 14-day method is a modification of the Indore method in which all components are ground or pulverized to speed decomposition. The main effect is an increased surface area, so microorganisms can multiply at a greatly increased rate.

The heap is not layered, but is turned every 2-3 days. Figure 15.4 describes the progress of a compost pile constructed by the 14-day method. As the graph shows, in this particular case the temperature reached peaks of 81 °C and 74 °C, and cooled in 12-14 days. The compost was already sufficiently decayed for garden use.

The anaerobic method requires the elimination of air from the pile. The simplest approach is to cover the heap with plastic, but it can be difficult for the home gardener to handle a large pile in this way. Some European cities compost trash and garbage anaerobically in large bins. The method includes inoculating the pile with anaerobic bacteria.

The sheet composting method introduces manure and other organic matter directly into the soil during nongrowth periods, where it proceeds to decay. This is a slower process

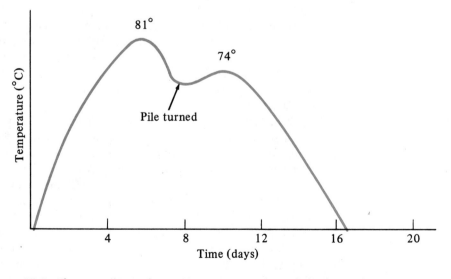

Figure 15.4   The temperature changes in a compost pile made by the 14-day method *(15)*.

but it avoids loss of nutrients (mostly ammonia) of other methods caused by the high temperatures.

The earthworm method is interesting. Worms eat and digest all kinds of organic matter and leave behind their waste matter, called *castings*. These castings are very rich in nitrogen and phosphorus. Worms also work on compost piles.

Worms can be bought by the thousands from seed companies and other garden suppliers. They multiply rapidly, so only a small fraction of the total required have to be introduced into the pile. Their castings replace the manure normally added in other methods. They can be used as a follow-up to the Indore method after the heat has dissipated, or they can be used on shallow piles (2 ft maximum) that do not have a heat buildup.

Mulching is another technique used by the organic farmer. A layer of material is laid down on the soil; it may be straw, plastic, rocks, leaves, compost, wood chips, gravel, bark chips, or other. Its advantages include discouraging weeds, preserving moisture, keeping low-growing plants clean and dry, and insulating by keeping the soil warmer in winter and cooler in summer. Mulch may also fertilize. Of these advantages, the most notable may be the blockage of weeds; thus chemical weed killers can be avoided.

Oil sprays and certain acceptable **plant extractives,** for example, rotenone and pyrethrum dust, are used to a limited extent for insect control. More often, companion crops are planted with garlic cloves or marigolds, which repel many insects. Other approaches to insect control, some of which are acceptable to organic farmers, are discussed in Chapter 16.

By avoiding chemical additives, the organic farmer ensures that such materials will not be present in foods. The organic farmer also argues that chemicals not only destroy harmful insects, microorganisms, and weeds, but also may destroy beneficial species. For example, microorganisms known to produce antibiotics that may defend plants from disease can be destroyed by insecticides. Also, the destruction of worms, which provide both fertilization and better aeration of the soil, reduces natural soil fertility.

Composting Municipal Wastes

Municipalities in the United States generate 180 million tons of solid waste each year, of which 65% is reusable organic material, mostly consisting of paper, paperboard, and yard and food wastes. Experts envision a national composting effort that goes well beyond households creating backyard piles of garden wastes and food. They see cities creating mountains of saleable compost from solid municipal waste using the best technology available. Interest in composting municipal waste waned in the 1980s, but reawakened in the 1990s largely out of concerns about water quality, population growth, and diminishing landfill space.

Nationally, the number of private and municipal solid waste composting ventures has increased dramatically in recent years. Now 19 plants have a combined capacity of 3770 tons per day, and several dozen more are being planned. Much of the impetus comes from landfill legislation. Fourteen states have banned yard, food, and paper waste from landfills, forcing cities and towns to find a use for it. Some states have standards for municipal compost. In Minnesota, it is judged on the following criteria: the feedstock source and sanitation (pathogen levels); inert materials (glass and other foreign matter); chemicals such as PCBs and heavy metals; and maturity (the age of the compost heap).

Most studies show that with the use of compost from municipal waste, crop yields dramatically increased and plant diseases decreased. In addition, the use of chemical fertilizers, with all their attendant environmental effects, was greatly decreased or eliminated.

Possible toxicity due to trace metals in solid waste compost is a concern leading to required chemical analysis and federal guidelines. Nickel, zinc, and copper are injurious to plants when present in high concentrations. Cadmium and lead cause human health problems. Two factors suggest that metals will not be a problem. Very little inorganic material originates in food and paper waste to make its way into the compost, and metal ions that do end up in the compost appear to be tightly bound to the compost and the soil to which they are applied *(16)*.

SUMMARY

Soils are primarily composed of oxides of silicon, aluminum, and iron. Common designations are clay, silt, and sand, in order of increasing particle size. Clay has the greatest affinity for nutrients and water; sand has the least. Loam is the easiest soil to work with. Organic matter, largely made up of microorganisms, contributes to soil fertility through mineralization.

Some crops grow very well in alkaline soils, others in an acidic environment. Also, some nutrients are more available within certain pH ranges. The pH of the soil can be altered in a variety of ways.

Plant nutrients are commonly categorized as macronutrients, secondary nutrients, and micronutrients according to their relative abundance. The macronutrients are carbon, hydrogen, oxygen, nitrogen, phosphorus, and potassium. The first three are provided by moisture and photosynthesis. The last three must be provided externally and are available in a wide variety of forms.

Mixed (N-P-K) fertilizers are often used by the home gardener; they contain varying amounts of nitrogen, phosphorus, and potassium, described in terms of the percentages of N, P_2O_5, and K_2O (potash). Secondary and micronutrients are usually also present.

Sustainable agriculture emphasizes crop rotation and nitrogen fixation by legumes. Organic farmers use methods other than commercial fertilizers for conditioning soil. Compost is important for them and for many cities. Unprocessed minerals such as dolomite and phosphate rock are considered acceptable for organic gardening. Most insecticides and herbicides are avoided.

Composting is a complex series of aerobic and anaerobic processes carried out by microorganisms that thrive on the organic matter and convert it into fertile topsoil. The carbon content is reduced, with a resultant increase in the percentage of other useful nutrients. The Indore method is the best-known for making compost. Others are the 14-day, anaerobic, sheet composting, and earthworm methods. Composting municipal organic wastes is under development in many areas, primarily due to problems encountered in landfills.

PROBLEMS

1. Give a definition or example of each of the following.

   a. Photosynthesis
   b. Mineralization
   c. Lime (2 meanings)
   d. Limestone
   e. Agricultural lime
   f. Dolomite
   g. Legumes
   h. Nitrification
   i. Phosphate rock
   j. Carbon fixation
   k. Nitrogen fixation
   l. Phosphate fixation
   m. Potash (2 meanings)
   n. Greensand
   o. TCDD
   p. Compost

2. Write the reaction for photosynthesis.

3. What are the main components of soil?

4. What are the main differences among clay, silt, sand, and loam?

5. What are the advantages and disadvantages of a clay soil?

6. What is humus? What purpose does it serve?

7. What are the major strategies of sustainable agriculture?

8. What are the advantages and disadvantages of ammonia as a fertilizer?

9. What does it mean if a fertilizer is labeled 5-10-5?

10. How might Agent Orange have caused problems for some Vietnam veterans?

11. What is the main difference between the Indore and the 14-day methods of making compost?

12. Which of the following happen during composting, and why are these changes beneficial?

a. The carbon content increases.
b. The nitrogen content increases.
c. The phosphorus content increases.
d. The potash content increases.

References

1. Slack, A. V. *Defense Against Famine;* Doubleday: Garden City, NY, 1970; p. 40.

2. Rodale, J. I. *Encyclopedia of Organic Gardening;* Rodale: Emmaus, PA, 1959; p. 8.

3. Hays, J. T.; Hewson, W. B. "Controlled Release by Chemical Modification of Urea: Triuret." *Journal of Agricultural and Food Chemistry* **1973,** *21,* 498.

4. "If You Heat NH_4NO_3, Will You Laugh or Cry?" *Chemistry* **1965,** *38* (4), 33.

5. "The Biggest Day in Texas City History." *Chemical and Engineering News,* September 24, 1979, p. 15.

6. Bear, F. E. *Soils and Agriculture;* John Wiley & Sons: New York, 1953; p. 222.

7. Reganold, J. P.; Papendick, R. I.; Parr, J. F. "Sustainable Agriculture." *Scientific American,* June **1990,** 112.

8. Moffat, A. S. "Nitrogen-Fixing Bacteria Find New Partners." *Science* **1990,** *250,* 910.

9. *McGraw-Hill Yearbook of Science and Technology 1989;* McGraw-Hill: New York, 1990; p. 6.

10. Moffat, A. S. "Nitrogenase Structure Revealed." *Science* **1990;** *250,* 1513.

11. *McGraw-Hill Yearbook of Science and Technology 1990;* McGraw-Hill: New York, 1991; p. 179.

12. Tschirley, F. H. "Dioxin." *Scientific American,* February **1986,** 29.

13. Hanson, D. "No Rise in Birth Defects After Italy's Dioxin Leak." *Chemical and Engineering News,* March 28, 1988, p. 5.

14. Gerras, C. *300 of the Most Asked Questions about Organic Gardening;* Bantam Books: New York, 1974; Chapter 4.

15. Rodale, J. I. *The Complete Book of Composting;* Rodale: Emmaus, PA, 1960; pp. 54–56, 87, 115.

16. Gillis, A. M. "Shrinking the Trash Heap: Scientists Are Evaluating Composted Municipal Solid Waste to Determine Whether the Nation's Garbage Can Be Put to Better Use." *BioScience* **1992,** *42* (2), 90.

Insect Control and Chemical Communication

16

Insect damage has always been one of the greatest problems for farmers and home gardeners. More than one-third of everything grown or stored is consumed by 10,000 species of damaging insects. The boll weevil alone causes losses of $200–300 million annually. It is the most significant farm pest.

Half of all human deaths and deformities due to disease are traceable to insects. Malaria is the biggest killer of humans worldwide, even though only a small percentage of those who contract the disease die of it. The *Anopheles* mosquito, which transmits malaria, is the most harmful disease-carrying pest.

Insect control is a multidisciplinary subject. Scientists from several branches of biology and chemistry actively engage in research, often combining efforts. New techniques and instrumentation have made it possible to isolate, purify, and identify complex mixtures of chemicals obtained from insects. The most obvious and usual approach to insect control is with insecticides, a purely chemical approach. Biological approaches, and several combined ones, depend on studies of insect behavior and the effect of chemicals on behavior.

CHAPTER PREVIEW
16.1 Insecticides: The Bulldozer Approach

16.2 Chemical Communication

16.3 Modern Methods of Insect Control

16.1 Insecticides: The Bulldozer Approach

GOALS: To recognize the magnitude of the problem of insect control.
To appreciate the effectiveness and limitations of the use of insecticides, including resistance and environmental problems.

The strategies in using insecticides vary depending on the behavior of the target insects. Throughout this section we will use dichlorodiphenyltrichloroethane, or DDT, as an example. Although first synthesized in 1873, its insecticidal properties were not discovered until 1939. J. R. Geigy received the Nobel Prize in 1948 for the discovery. This best known, most effective insecticide is only one of many in a versatile arsenal, and has both advantages and problems.

The *Anopheles* mosquito must have a blood meal before it lays eggs. If it bites a person (or animal) infected with malaria, it may carry off the malaria parasite in the blood sample (Figure 16.1), and 10–12 days later it can transmit the parasite through another bite. During this time, the mosquito habitually rests on interior walls. That was the key to controlling this insect and thus combating the spread of malaria. Interior wall surfaces were routinely sprayed with DDT, which kills by attacking the mosquito's central nervous system.

Although DDT is inexpensive and easily applied, and its effectiveness in combating diseases has been dramatic, the Environmental Protection Agency (EPA) banned its use in most applications in 1973. Despite its effectiveness in killing mosquitos, adverse environmental consequences were thought to be too great. One problem is its lack of specificity in killing insects. Most insects are either helpful or have no apparent value, although they all contribute to the natural biological balance of nature; only a small percentage are harmful to humans. The use of DDT and other insecticides has been described as "dropping nuclear bombs on pickpockets" *(1)*. Two serious problems are associated with insecticides. The first is the development of resistance; the second is environmental.

Resistance

An insect population may become resistant to the effects of an insecticide for several reasons, some of which are known, while others remain a mystery. In those for which the cause of resistance is known, insects are able to alter chemically the insecticide molecule so that the insecticidal properties are lost. The original insecticide remains lethal, but it is modified (detoxified) by the insect before it reaches the target organ, which is the central nervous system for DDT. The enzyme DDT-dehydrochlorinase, or simply DDTase, catalyzes the conversion of DDT to DDE. The DDE molecule is not insecticidal.

A phenyl radical forms when a hydrogen atom is removed from benzene.

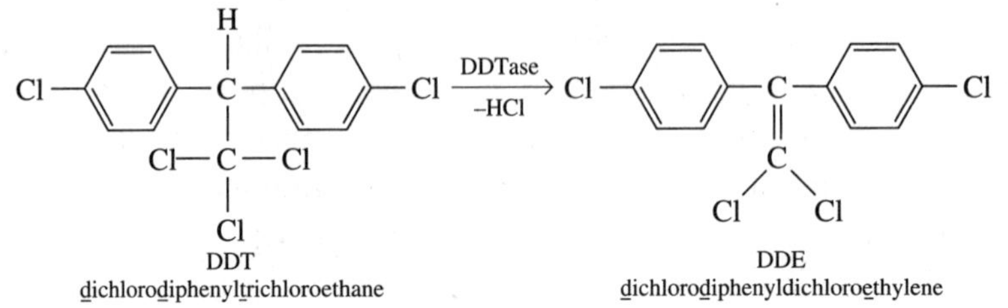

Some DDT-resistant insects can be made vulnerable to DDT by adding a synergist that enhances the action of the DDT. Two such synergists are similar to DDT in structure:

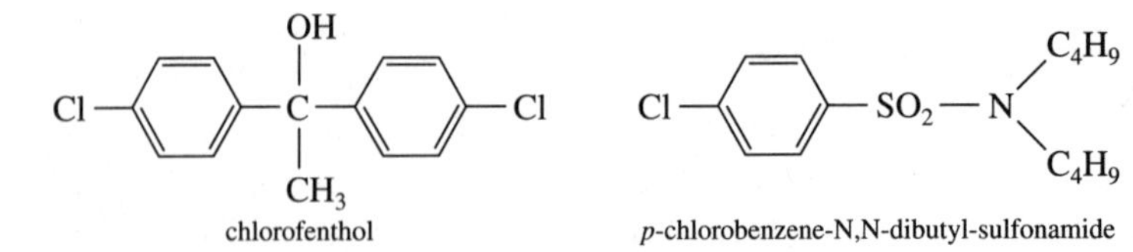

Either molecule is recognized by DDTase and can become bound to the enzyme as if it were DDT. Binding inhibits the enzyme and prevents it from attacking DDT, leaving DDT intact and lethal to insects.

Not only may insecticides become ineffective due to resistance, they may even cause the development of resistant insects. In other words, the *population of resistant insects*

Figure 16.1 A mosquito drawing blood.

increases as a result of using insecticides. We might speculate that insects adapt by developing the ability to synthesize DDTase, but this is not how resistance develops. Instead, a portion of an insect population that is resistant to an insecticide survives and passes on the resistance to its next generation. This trait may become common in later generations.

The population of a particular insect may be sharply decreased in the short term by applying an insecticide. But when subsequent generations appear in later months or years, resistant insects may far outnumber those that remain vulnerable. This occurs because of the prior use of an insecticide that killed off susceptible insects. The survivors are better able to multiply because the susceptible insects no longer compete for food supplies.

Three strategies for managing resistance are moderation, saturation, and multiple attack *(2)*. **The moderation strategy** uses less insecticide than is necessary to eradicate an insect population completely. A more tolerable number of nonresistant insects survive so that resistant ones do not become dominant. If insects target high-priced food commodities, however, this tactic may not be acceptable.

In California, the Mediterranean fruit fly, or Medfly, has had an enormous economic impact on fruit crops in some years. Sometimes quarantines have been imposed forbidding the export of certain fruits into other states. Finding even a single Medfly is feared, so heavy doses of insecticide may be applied to protect a crop even though it encourages resistance.

Saturation uses very high doses of insecticide so that even the resistant insects are not able to modify all of it and are killed. Nowadays, such an approach is carried out only as part of several simultaneous steps called integrated pest management (IPM) to minimize development of resistance among the surviving insects (see later discussion).

In **multiple attack,** several insecticides are applied in rotation, so that no insect with a particular resistance could multiply out of control. In other words, an insect may carry the gene for resistance to one insecticide, but may be susceptible to other chemicals. Unfortunately, none of these approaches can guarantee success in preventing the development of resistance, and each one places other burdens on the environment.

Environmental Problems

Among the many desirable insects are those that act as predators on the undesirable insects. Killing most insects, including helpful predators, in an effort to eliminate a harmful one may leave the latter in a *biological vacuum*. It might then multiply far beyond its original population when the application of insecticides is decreased or when resistance develops. Many insects became serious problems only after insecticides were used. One author described insecticides as "ecological narcotics." Increasingly higher doses and frequency of use, plus the variety required for control as resistance develops, is much like what drug addicts experience *(3)*.

Several chlorinated hydrocarbon pesticides (Figure 16.2) came into prominence in the late 1940s. The first to be banned in this country was DDT, followed by aldrin, dieldrin, and others *(4)*.

Besides encouraging resistance, DDT is a serious pollutant, and was banned for that reason. Its environmental effects are largely attributable to **biomagnification;** this phenomenon is a result of the solubility properties of DDT in water and fatty solvents or tissues. The solubility of DDT is about 100,000 ppm in fat but only 0.0012 ppm in water at 20 °C (68 °F). In other words, if DDT were placed in a container and mixed thoroughly with equal amounts of liquid fat and water, 10^8 molecules of DDT would dissolve in the fat for every 1 that dissolved in the water.

In a living organism, this means that DDT tends to build up in fatty deposits rather than in blood (mostly water). Elimination from stored fat is much slower than from blood.

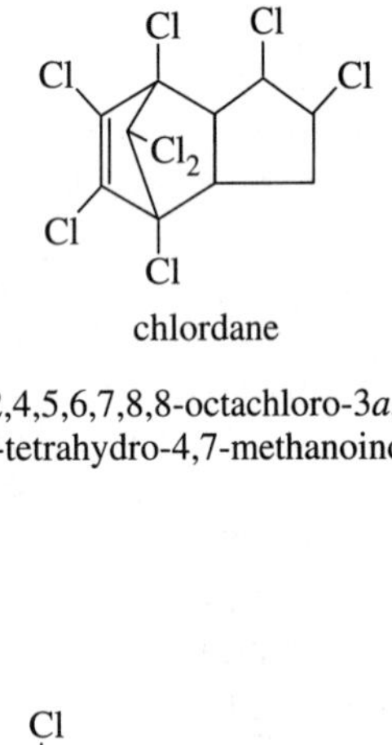

chlordane

1,2,4,5,6,7,8,8-octachloro-3a,4,7,
7a-tetrahydro-4,7-methanoindane

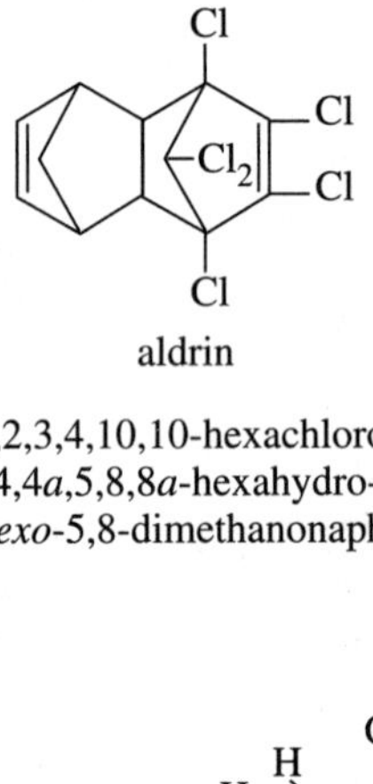

aldrin

1,2,3,4,10,10-hexachloro-
1,4,4a,5,8,8a-hexahydro-1,
4-*endo-exo*-5,8-dimethanonaphthalene

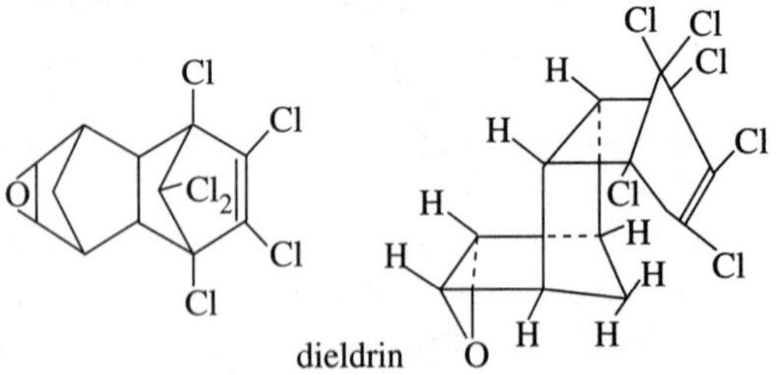

dieldrin

1,2,3,4,10,10-hexachloro-6,7-epoxy-
1,4,4a,5,6,7,8,8a-octahydro-1,4-*endo-
exo*-5,8-dimethanonaphthalene

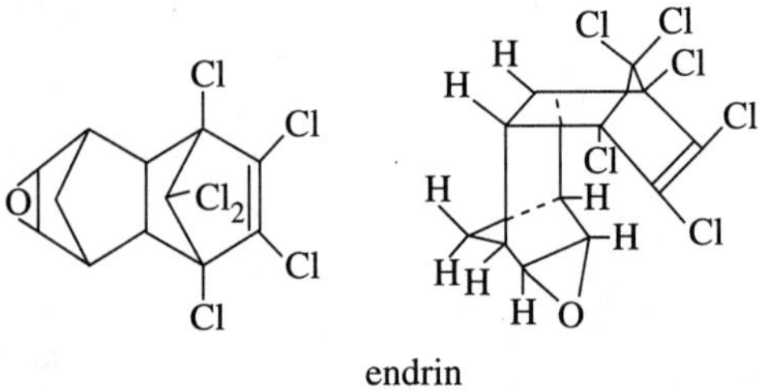

endrin

1,2,3,4,10,10-hexachloro-6,7-epoxy-
1,4,4a,5,6,7,8,8a-octahydro-1,4-*endo-
endo*-5,8-dimethanonaphthalene

Figure 16.2 Some chlorinated hydrocarbon insecticides now banned by the Environmental Protection Agency.

Table 16.1   Food chain showing biomagnification of DDT *(5)*

Location of DDT	Concentration (ppm)
Lake Michigan	0.000 002
Amphipods	0.410
Fish	3–6
Gulls	99

This inability to eliminate DDT causes a large magnification of the concentration, in the direction from lake water up the predatory food chain to gulls at the top, as shown in Table 16.1. That is, as fish eat amphipods (small, shrimplike, water-dwelling insects) and as gulls eat fish, the concentration of DDT in each is increased. In this example, the gulls have a DDT concentration *in their fatty tissue* about 50 million times the concentration in the lake water.

In addition, DDT affects the reproduction of many species of birds. Several predatory birds, high on a food chain, have vanished from areas where they were once prominent. The most common explanation for the loss comes from the observation of thin eggshells in some species. Eggshells are largely crystalline calcium carbonate, $CaCO_3$, and an enzyme called *carbonic anhydrase* catalyzes the conversion $CO_2 \rightarrow HCO_3^-$ (bicarbonate ion), which makes carbonate available for eggshell formation. This enzyme is inactivated by DDT. Fortunately, since the ban on the use of DDT and other insecticides, there is evidence of the recovery of populations of peregrine falcons, bald eagles, and other birds *(6)*.

Further condemnation of the use of these products comes from estimates that less than 0.1% of all the pesticides (herbicides plus insecticides) applied to crops actually reaches the target pests. The rest may become an environmental burden *(7)*. On the other hand, a balanced view must consider their advantages. Many insects spread human diseases that have caused millions of deaths (Table 16.2). Insecticides have reduced the annual death

Table 16.2   Some of the most common diseases known to be transmitted to humans by insects, ticks, or mites *(8)*

Disease	Vectors
African sleeping sickness	Tsetse flies
Anthrax	Horse flies
Bubonic plague	A rat flea
Chagas' disease	Assassin bugs
Dengue fever	Two mosquitoes
Dysenteries	Several flies
Encephalitis	Several mosquitoes
Endemic typhus	Oriental rat flea
Epidemic typhus	Human louse
Filariasis	Several mosquitoes
Hemorrhagic fevers	Several mites and ticks
Leishmaniases	Psychodid flies

Disease	Vectors
Louping ill	Castor bean tick
Malaria	*Anopheles* mosquitoes
Onchocerciasis	Several black flies
Pappataci fever	A psychodid fly
Q fever	Ticks
Relapsing fevers	Several ticks
Rocky Mountain spotted fever	Two ticks
Scrub typhus	Chigger mites
Trypanosomiasis	Several flies
Tularemia	Several flies, fleas, lice, ticks
Yaws	Several flies
Yellow fever	Several mosquitoes

rate due to malaria alone from 6 million people in 1939 to less than 1 million today. Many other outbreaks of serious diseases were controlled or prevented using these agents.

A variety of farming practices, including the use of pesticides, have increased the output of many important crops by more than 5-fold. The United States grows 48% of the world's corn and 63% of its soybeans. With earth's population expected to increase from 4.4 billion in 1980 to 6.4 billion by the year 2000, damage to food crops by pests will become even more intolerable.

Other Insecticides

The most successful and the most disappointing results have been associated with chlorinated hydrocarbon insecticides, such as DDT, dieldrin, and chlordane. But other products have been used both before and since the chlorinated hydrocarbons came and went.

The Organophosphates

One of the largest and best known classes of insecticides is the organophosphates (OPs), including such compounds as diazinon, chlorpyrifos (Dursban), methyl parathion, and malathion.

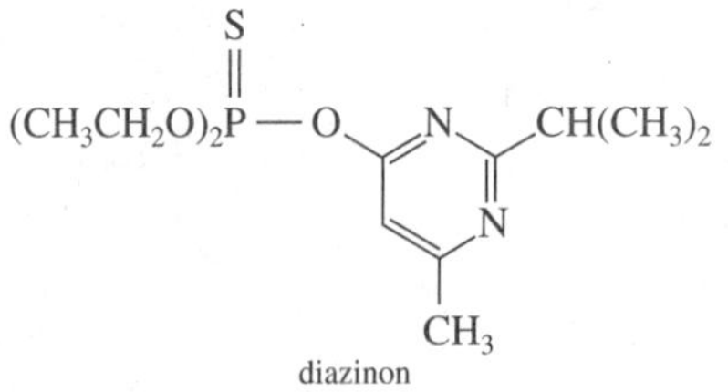

diazinon

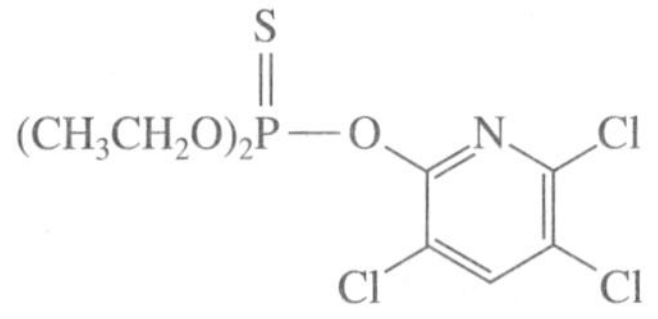

chlorpyrifos (Dursban)

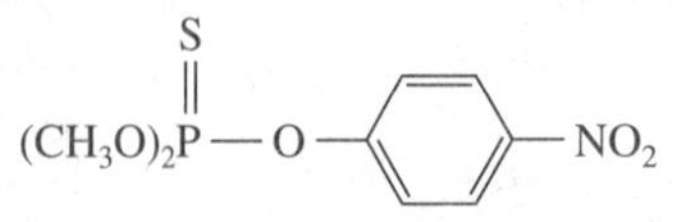

methyl parathion

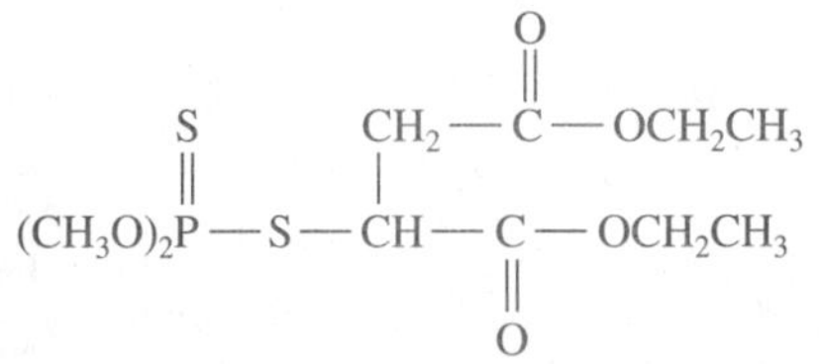

malathion

$$\underset{\text{dichlorvos}}{(CH_3O)_2 \overset{\displaystyle \overset{O}{\|}}{P} - O - CH = CCl_2}$$

The OPs have one important advantage and one important disadvantage compared with the chlorinated hydrocarbons. Due to greater water solubility and chemical instability, the OPs are readily degraded in the environment. Thus, they do not persist as a long-term environmental burden as do the chlorinated hydrocarbons. However, some of the OPs are less potent insecticides and are more toxic to animals and humans than the chlorinated hydrocarbons.

Malathion is the oldest and most heavily used of the OPs. Introduced in 1950, it was liberally applied to most vegetables, fruits, and forage crops. Since its toxicity is very low, it is safe for use in the home. It is a common component of flea powders for dogs, cats, and other animals, and is even prescribed by physicians to control head and body lice on humans. In the early 1980s, after the ban on the chlorinated hydrocarbons, malathion became the insecticide of choice in controlling the Medfly that invaded fruit crops in California. It was mixed with a protein bait of molasses and yeast, and sprayed from the ground and from the air over infested areas *(4)*.

Dichlorvos is an OP that is more volatile than most insecticides, making it suitable for use as a fumigant. The vapor spreads within a closed space to provide insecticidal activity. Therefore dichlorvos has been incorporated into pet collars and in ''no-pest strips,'' from which its vapors are released slowly over several months. Dichlorvos was generally regarded as safe but is now classified as a possible human carcinogen *(8)*.

Biopesticides

Certain plants produce chemicals that give them natural resistance to insects. Genetic engineering strategies are being aimed at making specific crops toxic to insect pests. The general approach takes various forms. Most often, the toxic substance is a protein.

The use of chemical insecticides has been termed the *bulldozer approach,* for reasons cited above. The next question is, What are the alternatives to this purely chemical approach to insect control? The possibilities range from purely biological control to bio-chemical approaches. Some methods depend on how chemicals influence the behavior of insects, and on how insects use chemicals to communicate; this is the subject of the following sections.

16.2 Chemical Communication

GOAL: To develop an understanding of the many forms of chemical communication among insects, and thus to appreciate the potential for turning natural communication mechanisms into methods of insect control.

Figure 16.3 summarizes the roles of chemicals in interactions between individuals within a particular species (pheromones) and between different species (allelochemicals). Although the flow sheet is applicable to all animals, this discussion focuses on insects.

Allelochemicals

The balance of nature can be more fully appreciated by considering the **allelochemicals** or **allelochemics.** They are defined as nonnutrient substances originating in one organism but affecting the behavior and welfare of organisms of other species. The organisms range

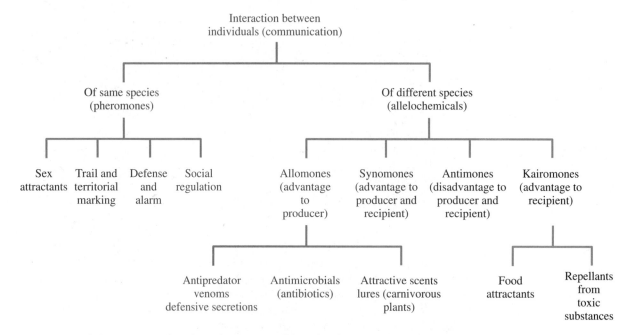

Figure 16.3   A flowsheet description of the role of chemicals in insects and other organisms. Examples are discussed in the text.

from plants, which may be consumed or pollinated by insects, to higher animals, which may be predators of insects. Even microorganisms that may be present on plants, insects, or other animals may be affected by allelochemicals. Allelochemicals help to organize and structure a community by controlling the interactions among all organisms present.

For example, pine trees produce chemicals known as terpenoids that interfere with the growth of microorganisms and plants on the surface of the trees. Thus terpenoids act as **allomones,** chemicals advantageous to the producer but not to the recipient predator. At the same time, terpenoids are attractive to bark beetles, which can damage the trees by boring into the bark. This gives the terpenoids **kairomone** activity since the emitter (the tree) suffers and the receivers (the beetles) benefit. The terpenoids also attract predators of the bark beetles, a **synomone** (mutually beneficial) activity since both the tree and the predators of the beetles benefit *(9)*.

Similarly, weeds are not totally negative when they exude allelochemicals that repel insects or attract insect predators. This can protect the primary crop that might otherwise be damaged by an insect.

Allelochemicals also include **antimones,** substances that are deleterious to both producer (a plant) and recipient (an insect). Consider that a plant may release an allomone to protect itself against a damaging insect. Suppose the same chemical repels another insect that is a predator of the damaging insect. Thus both the plant and the predator are harmed by the antimone (mutually harmful) activity.

Defense

Chemical defense has been studied in many species of insects and other animals. In insects, long-range and short-range sprays or surface secretions (Figure 16.4) are well known. There are also the alarm pheromones and sting venom of the honeybee discussed later in this chapter.

Figure 16.4   Grasshopper with defensive froth.

Some species of millipedes store a compound known as *mandelonitrile,* which is a source of *hydrogen cyanide,* according to the reaction shown.

$$C_6H_5-\underset{\underset{CN}{|}}{\overset{\overset{OH}{|}}{C}}-H \rightarrow C_6H_5-\overset{\overset{O}{\|}}{C}-H + HCN$$

mandelonitrile   benzaldehyde   hydrogen cyanide

The defensive secretion given off by an irritated skunk consists of the following three compounds in the ratio 4:3:3.

$$\underset{H}{\overset{H_3C}{>}}C=C\underset{CH_2SH}{\overset{H}{<}}$$

trans-2-butene-1-thiol

$$CH_3CHCH_2CH_2SH$$
$$|$$
$$CH_3$$

3-methyl-1-butanethiol

$$\underset{H}{\overset{H_3C}{>}}C=C\underset{CH_2SSCH_3}{\overset{H}{<}}$$

methyl-1-(*trans*-2-butenyl) disulfide

A millipede attacked by a predator carries out this reaction and discharges the toxic repellent. Other millipedes ooze a mixture of compounds when irritated. The chief repulsive components are known as *quinones.*

The whip scorpion, which is only a few centimeters long, can accurately direct a spray many times that distance. A would-be predator receives a mixture of acetic acid, caprylic acid (also known as octanoic acid), and water. The acetic acid is the irritant, but its effectiveness is enhanced by the fat-soluble caprylic acid, which permits penetration of the waxy coating on many recipients.

$$H-\overset{\overset{O}{\|}}{C}-OH \qquad CH_3\overset{\overset{O}{\|}}{C}-OH \qquad CH_3(CH_2)_6\overset{\overset{O}{\|}}{C}-OH$$

formic acid   acetic acid   caprylic acid

Some ants defend themselves by discharging a spray of formic acid as far as 20 cm. Other ants use acetic acid.

Figure 16.5 Monarch butterflies.

Monarch butterfly larvae acquire an effective chemical substance by feeding on milkweed plants, and the chemical is carried through to the adult butterfly. Predatory birds vomit violently after consuming a monarch and quickly learn not to repeat their mistake. The monarch adult is brightly colored (Figure 16.5), which helps potential predators to identify clearly and avoid this species. Yet monarchs raised on cabbage are safely consumed. In fact, if birds are conditioned to eat these monarchs, they will also attack the unsafe butterflies and suffer the consequences *(9)*.

The bombardier beetle produces a quinone secretion when attacked. In Figure 16.6(a), a predatory toad is about to consume a tasty morsel. Unfortunately for the toad, the defensive secretion is detected when the toad strikes the prey (b), and the beetle is rejected (c) *(10)*. The defensive chemical spray of a bombardier beetle is also shown in Figure 16.7.

Defense Mechanisms of Plants

A subtle balance of nature also controls the interaction of many plants with insects or animals. Pyrethrin is a natural insecticide present in chrysanthemums, which makes it acceptable to organic gardeners. It is a potent toxin for many insects; others are unaffected by it. Those not affected have detoxification enzymes capable of metabolizing this and other plant toxins. (This situation is analogous to DDT resistance due to DDTase.) The plants themselves may produce a natural synergist to block detoxification. *Sesamin* is a known synergist of pyrethrin in chrysanthemums.

Catnip is an intriguing substance produced by certain plants in the mint family. Its effect on cats is interesting, but its purpose is to repel insects *(11)*.

Questions have even been raised about whether some plants practice chemopsychological warfare as a defense mechanism *(12)*. For example, Does an insect that has fed on a fungus containing LSD mistake a spider for its mate, or does a zebra that has eaten a plant rich in alkaloids become so intoxicated that it loses its fear of lions?

(a)

(b)

(c)

Figure 16.6
Toad being repelled by
a bombardier beetle.
The toad (a) eyes the
beetle, (b) strikes at it
with its sticky tongue,
and (c) rejects it.

Figure 16.7
A bombardier beetle is shown as it fires a defensive spray
forward in response to having a front leg pinched with
forceps. For purposes of photography, the beetle has been
attached to a hook with wax.

Pheromones

Allelochemicals control interactions between different species; pheromones are used to regulate many activities by providing communication between insects of the same species. Affected insects include the honeybee, gypsy moth, boll weevil, and bark beetle.

The Honeybee

Honeybees are used extensively for making honey and, because of their pollen-gathering activity, for pollinating some 90 food crops worth more than $20 billion annually. Some 1600 United States beekeepers rent out honeybee hives for this purpose *(13)*.

The honeybee *(Apis mellifera)* is one of the best understood and most striking examples of complex chemical communication among insects. They also use vision, hearing, and touch to communicate among themselves and with other insects; we will consider only the chemistry related to some of the categories in Figure 16.3.

Bees have pheromones that affect reproduction, food seeking, and defense. The social structure of their colony consists of only one developed female, the *queen.* The other inhabitants of the hive are many thousands of *workers* (Figure 16.8) and the males, or *drones,* whose sole purpose is to mate with the queen. Drones are absent from the beehive most of the time. The workers, which are undeveloped females, do not have a reproductive role unless a queen becomes lost.

The most thoroughly studied pheromones of the honeybee are the first two shown in Figure 16.9. The 9-oxo compound, known as *queen substance,* is a multipurpose substance produced by the mandibular gland of the queen bee. The 9-hydroxy compound is known as *queen scent.* Queen substance acts with queen scent, HVA, and HOB to signal the presence of a dominant reproductive queen. Another effect of this group of pheromones is the inhibition of ovary development in the workers. If a queen is removed from the colony and pheromone output ceases, the workers will build queen cells to rear new queens from eggs previously laid by the queen. In addition, the ovaries of some of the workers ripen, and they begin to lay eggs that develop into drones. If a queen is removed from a hive but queen substance is supplied, ovary development is significantly inhibited.

Figure 16.8   Queen honeybee on comb with workers.

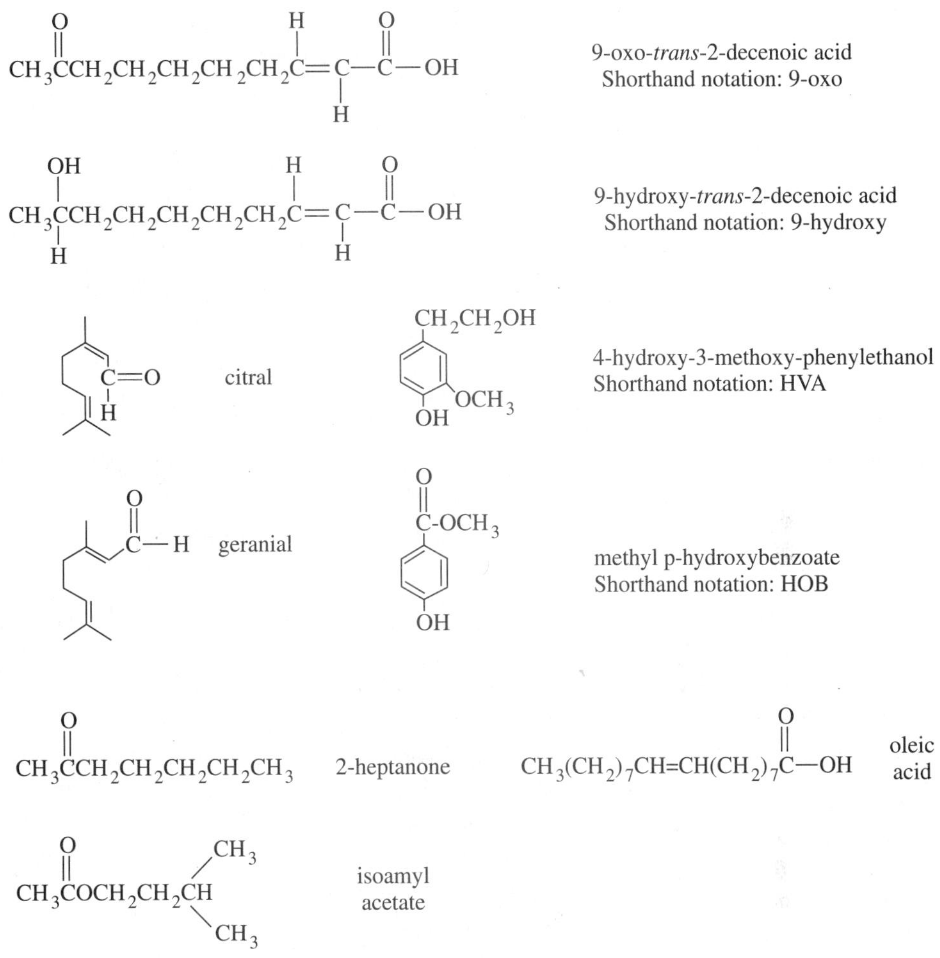

Figure 16.9 Pheromones of the honeybee *(14).*

Queen substance also serves as a sex attractant used by the queen to attract drones when she leaves the nest to go on her nuptial flight. Although honeybees can see a queen only within 1 meter (m) or less, drones can be lured from as much as 60 m downwind.

Citral and geranial (Figure 16.9) are substances released by workers when they locate a food source. They are attractive to other foraging bees and increase the efficiency of locating food for an entire colony.

Pheromones also have a role in defense of the hive. A single sting increases the likelihood of attack by more workers. This is due to the release of isoamyl acetate (Figure 16.9), a sweet-smelling component of banana oil that elicits aggressive behavior in other workers and directs them to sting in the same location. Isoamyl acetate is released only during stinging, whereas 2-heptanone, an **alarm pheromone,** is released whenever workers are excited. This organizes defense against robber bees or other attack. Conversely, some robber bees in search of food exude citral, which disorganizes host workers when they are under attack. Some colonies are not affected by citral and are not vulnerable to attack by robbers.

Different honeybees show different sensitivities to alarm pheromone. For example, older bees produce more of it and are also more sensitive to it. High sensitivity to alarm

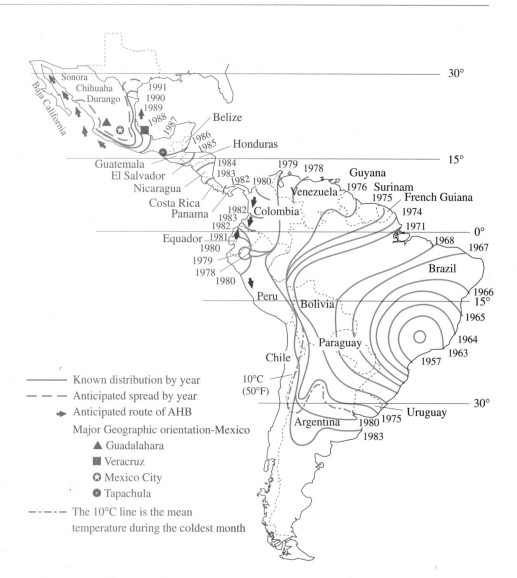

Figure 16.10
Movement of African
honeybees in the
Americas since their
accidental release in
Brazil in 1957.
(Courtesy of John G.
Thomas.)

pheromone is also associated with the species *Apis mellifera scutella,* usually known as
the African honeybee, or ''killer bee'' *(15).* It has been the subject of many sensational
articles and several horror movies.

In 1956, 47 honeybee queens were imported from Africa to Brazil as part of an
experiment to breed a sturdier bee. Unfortunately, in 1957, many swarms headed by African queens accidentally escaped. In the years since, the African bees took over the bee
populations of South America by hybridizing existing colonies. The spread of these populations is shown in Figure 16.10, including entry into the United States in the fall of 1990.

For many years, beekeeping, called *apiculture,* declined dramatically in South America due to the African bee. But as people learned to control the aggressive behavior of the
bees, it again became a thriving industry even in areas where the traditional European bees
once barely survived. Among the common strategies are the use of large smokers, which
reduce honeybees' response to alarm pheromone *(16).*

The venom released by the sting of a honeybee is the ultimate form of defense. The
venom contains certain proteolytic (protein-cleaving) enzymes that attack proteins
involved in nerve impulses, and may cause paralysis and death in some recipients. Hem-

orrhaging and hemolysis (disintegration cleavage) of red blood cells are other toxic effects of bee venom.

Honeybees also have a chemical that signals the death of a member of the colony and stimulates workers to remove the dead bee from the hive. Workers will carry even a live queen from the hive if she has been brushed with a dab of oleic acid (Figure 16.9). However, after the acid evaporates, the ''undertaking activity'' dissipates and the queen is accepted back into the hive *(17)*.

The Gypsy Moth

The gypsy moth *(Porthetria dispar)* causes heavy losses of forest timber. Up to now, the destruction has been primarily in the northeastern United States, but recent evidence shows infestation as far south as Alabama and as far west as Wisconsin. It first became established in the northeast by importation in 1869 by a Massachusetts scientist who hoped to breed a sturdier silkworm. Unfortunately, some moths escaped, and the species flourished in the absence of its natural enemies. The larvae and eggs can be carried long distances on cars and other vehicles, which accounts for its designation as a gypsy.

In the spring, the larvae emerge from eggs laid the previous year. After the stages of larval development and a final change into the pupal stage, the adult moth emerges. The larvae, or caterpillars, are what cause the damage. A single 5-cm caterpillar eats about 0.1 sq m (1 sq ft) of leaf surface a day. Some trees are killed by even a single defoliation and many more are killed by two successive defoliations. Adult moths do not feed and live only a short time, but they mate and propagate the species. Only the male gypsy moth flies (Figure 16.11). The larger female, loaded down with eggs and unable to fly, locates

Figure 16.11   (top) Male gypsy moth; (above) female gypsy moth; (right) larvae.

on tree trunks. She releases a potent attractant pheromone, which the male detects, flying upwind to find her. The sex attractant is more dense than air, so the male moth flies close to the ground.

Three compounds have been associated with the attraction by the female: gyptol, gyplure, and disparlure. Gyptol isolated from the female gypsy moth seemed, in most experiments, to be attractive to males. Synthesized gyptol was needed for further studies, but it was difficult to make in the laboratory. A similar compound, gyplure, was more easily synthesized from a component of castor oil, but was unattractive to males. The confusion over the attractiveness of gyptol is now attributed to trace contamination of the isolated gyptol with the extremely potent real attractant *disparlure.*

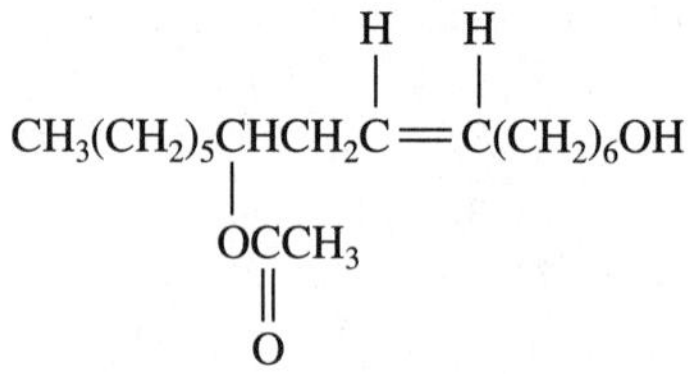

$$CH_3(CH_2)_5\overset{\displaystyle H}{\underset{\displaystyle \underset{\displaystyle O}{\overset{\displaystyle \|}{OCCH_3}}}{C}}HCH_2\overset{\displaystyle H}{C}=\overset{\displaystyle H}{C}(CH_2)_6OH$$

10-acetoxy-*cis*-7-hexadecen-1-ol
gyptol

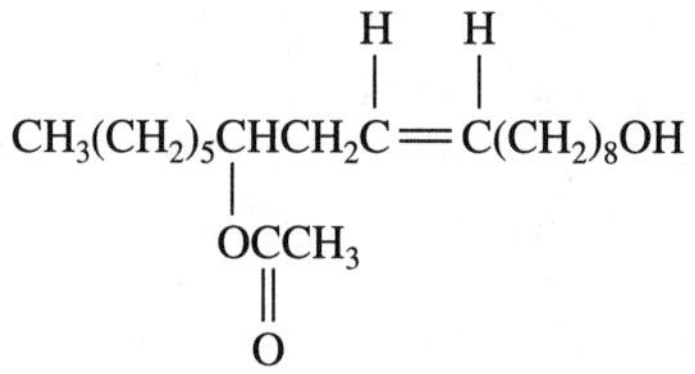

$$CH_3(CH_2)_5\overset{\displaystyle H}{\underset{\displaystyle \underset{\displaystyle O}{\overset{\displaystyle \|}{OCCH_3}}}{C}}HCH_2\overset{\displaystyle H}{C}=\overset{\displaystyle H}{C}(CH_2)_8OH$$

12-acetoxy-*cis*-9-octadecen-1-ol
gyplure

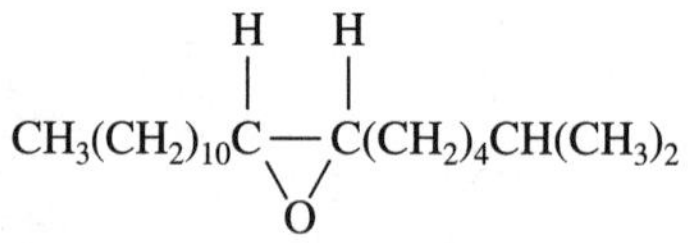

$$CH_3(CH_2)_{10}\overset{\displaystyle H}{\underset{\displaystyle O}{C}}\diagdown\overset{\displaystyle H}{\diagup}C(CH_2)_4CH(CH_3)_2$$

cis-7,8-epoxy-2-methyloctadecane
disparlure

The bioassays necessary to determine attractiveness frequently are the most difficult part of the research, and results are often conflicting or confusing. In some cases, differences in results have been traced to the daily behavioral patterns of certain insects. For instance, some insects mate only at certain times of the day or night, so a sex attractant may be effective for only a few hours.

The Boll Weevil

The boll weevil *(Anthronomus grandis)* is the most destructive agricultural pest (see Figure 16.12). The males produce an attractant that is best described as **an aggregation phero-mone,** since it attracts both males and females equally. The males are attractive to females only when their distance from each other is a few inches—presumably a visual attraction; when the males produce the pheromone, they are attractive within 30 ft.

The pheromone, known as *grandlure,* is a combination of the four compounds below.

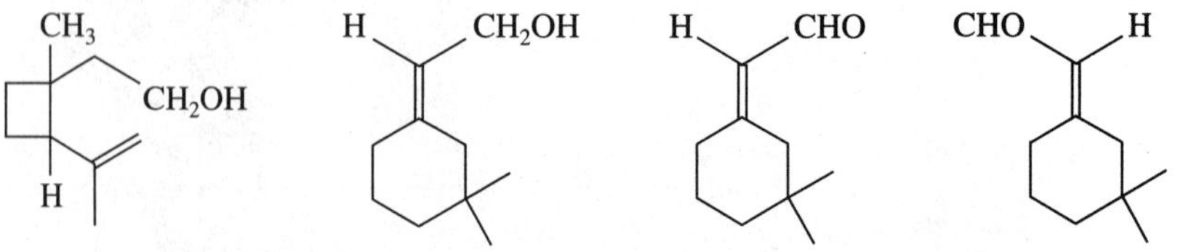

Figure 16.12
The boll weevil.

A cotton diet leads to greater attractiveness of the males than an artificial diet. It is thought that some constituent of cotton is converted to one or more components of the pheromone mixture.

Bark Beetles

Dutch elm disease is caused by a fungus that is transmitted by several species of bark beetles. Well over 50% of the elm trees in the United States have been killed by the disease, more than 40 million trees destroyed. By 1977 the costs of control programs totaled about $30 million per year covering all but six states *(18)*. In Norway, during 1978–1980 about 10 million trees died after attack by bark beetles *(19, 20, 21)*.

As one illustration, Douglas fir beetles have been studied extensively. In a single day, a 300-foot-tall Douglas fir can be overwhelmed by thousands of beetles, each about a quarter of an inch long, boring into its bark. The beetle is helped in killing the tree by a fungus that it carries.

The fir beetle attracts other beetles to the site of attack using an aggregation pheromone, which allows them to mount a coordinated attack. Douglas fir seedlings need direct sunlight, and they are typically harvested in solid blocks called clear cuts. Spraying the aggregation pheromone on an area scheduled for cutting attracts large numbers of beetles. If the trees were cut and removed within a short time, the density of beetles remaining in nearby areas would be decreased.

Some beetles are well adapted to attacking trees. Those that successfully attack healthy trees usually release their pheromone before tunneling into the bark. The tree's resin may kill an initial invader, but continued attack by more insects eventually causes the flow of resin to diminish, and the tree is killed.

Other species normally attack only trees weakened by adverse weather or disease. They usually release aggregation pheromone after attack. If such a beetle attacks a healthy tree, it probably will be killed off by resins before it signals other beetles to continue. Further attack would not be likely to succeed and might result in the death of many beetles.

Pheromone Duplication in Different Insects

One intriguing aspect of pheromones is the finding that several different insects may use the same one for the same purpose. This seems reasonable except when sex attractants are involved, and confusion could have interesting consequences. There are several reasons why confusion normally is minimal. Insects may have different daily or seasonal cycles and thus mate at different times of the day or year. Different geographical location and the attraction of different species for different plants are other obvious ways that food-seeking and aggregation confusion are avoided. Morphological (form) incompatibility for copulation and genetic incompatibility also could account for lack of productive mating.

The cabbage looper (Figure 16.13) and the alfalfa looper use the same sex attractant, *looplure,* but they have a subtle mechanism that provides separation. The amount of pheromone released by the female and the sensitivity of the corresponding male seem to be attuned. The male cabbage looper is less sensitive than the alfalfa looper, that is, requires more of the pheromone for attraction *(22, 23).* This observation nicely accounts for the low attraction between the male cabbage looper and the female alfalfa looper (which secretes very little sex attractant). It does not, however, account for the failure of the female cabbage looper (which puts out a higher level of pheromone) to attract the male alfalfa looper. It is not total avoidance of one species by the other, but strong preference between the males and females of each species. But even the strong preference leads to the conclusion that the detection system of the very sensitive male alfalfa looper may become saturated by a high concentration of female cabbage looper pheromone when the male is far downwind from the female. Travel farther upwind would increase the amount of pheromone reaching the male, but if the detection apparatus were already saturated, he could not recognize any progress in reaching the female and would not continue.

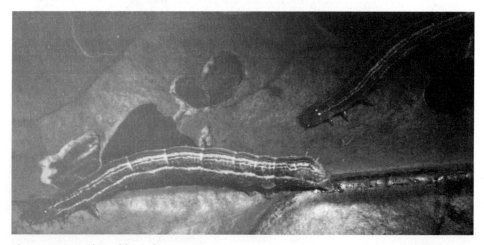

Figure 16.13   The cabbage looper.

Another subtle means of separation is used by the Indian meal moth and the almond moth, which also have the same sex attractant pheromone:

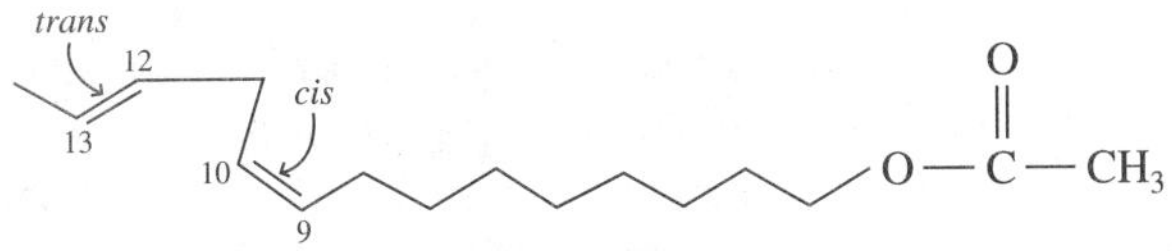

cis-9-*trans*-12-tetradecadien-1-ol acetate

The female almond moth also puts out *cis*-9-tetradecen-1-ol acetate, however, which stimulates almond male moths and inhibits Indian meal moths. The female Indian meal moth produces *cis*-9-*trans*-12-tetradecadien-1-ol, which inhibits the male almond moths. Thus, both synergism and masking help prevent confusion *(22)*.

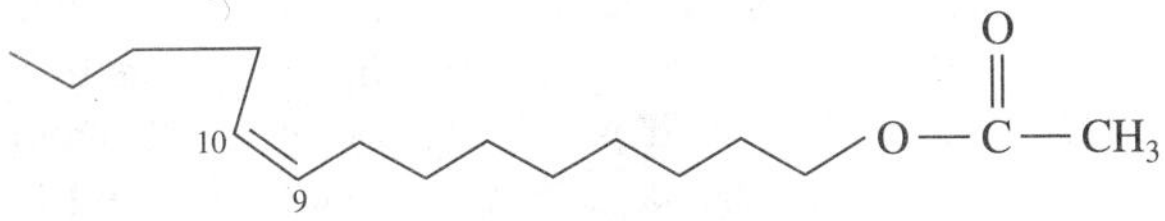

cis-9-tetradecen-1-ol acetate

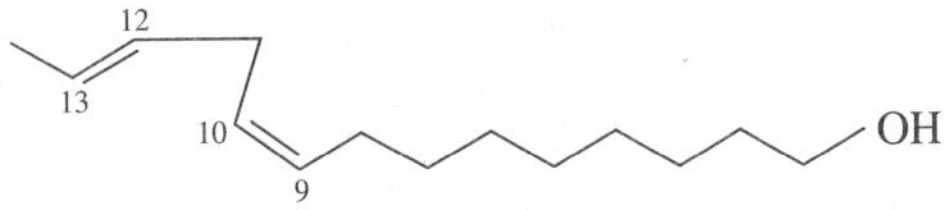

cis-9-*trans*-12-tetradecadien-1-ol

16.3   Modern Methods of Insect Control

GOAL: To understand integrated pest management for controlling insects.

An expert in **entomology** (the branch of zoology that deals with insects) might disagree with the title of this section, since several of the methods are by no means modern. Many of the ones now recognized as effective and environmentally acceptable were shelved when the easier and less expensive DDT came into prominence in the 1940s. Now that the hazards of insecticides are known, **integrated pest management (IPM)** is applied whenever possible. This involves a variety of methods to control, but not eradicate, an insect pest. In other words, the plan is to maintain an acceptable level of both pests and predators.

For large infestations of insects, chemical pesticides will undoubtedly continue to be used. In some cases, a two-phase attack starts with insecticides to kill off a large percentage of an insect population, followed by another, more ecologically oriented approach to maintain control. Even heavy use of insecticides can be kept under greater control than in the past. Biodegradable agents are preferred as a means of minimizing pollution. Controlled release or encapsulated products should permit a lower level and lower frequency of application.

Pheromones provide different means of combating insects and all have one thing in common. Each takes advantage of natural communication mechanisms essential to survival and propagation of a species. The methods either interfere with the mechanisms or turn them against the insect. Even though pheromones are chemicals, they usually have a negligible impact on the environment. They have low toxicity, can be used in minute amounts, and are biodegradable.

Attractant pheromones, such as looplure, grandlure, and disparlure, are used to lure unwanted insects into traps. Such traps could also contain insecticides or a sticky substance to capture insects without environmental contamination. It seems improbable that insects would develop resistance to pheromones. If an insect developed the ability to metabolize its own sex attractant and resist its effect, the propagation of the species would be greatly hampered.

Heavy infestations may not be controllable using pheromone-baited traps. When the population density is high, vision is more significant in mating. This is why insecticides may sometimes be necessary before using other methods.

Confusion and masking are two strategies in using pheromones. Confusion is illustrated by a study in which small pieces of paper treated with disparlure (gypsy moth attractant) were air-dropped over a wide area at the rate of 20 mg per acre. Baited traps containing either unmated female gypsy moths or disparlure were placed in both treated and untreated areas. Far more males were trapped in control areas that were not treated. Apparently, they had become confused by the abundance of sex attractant and were unable to locate either the females or the baited traps.

Masking introduces another substance that eliminates the attractiveness of the normal sex attractant. It is patterned after the behavior of the almond and Indian meal moths, which use the same sex attractant but avoid confusion due to the presence of a masking agent that inhibits the attractiveness of the pheromone.

The pink bollworm moth provides another illustration of the potential for masking. The male of the species is attracted to a substance known as propylure, which is the *trans* isomer of the following structure:

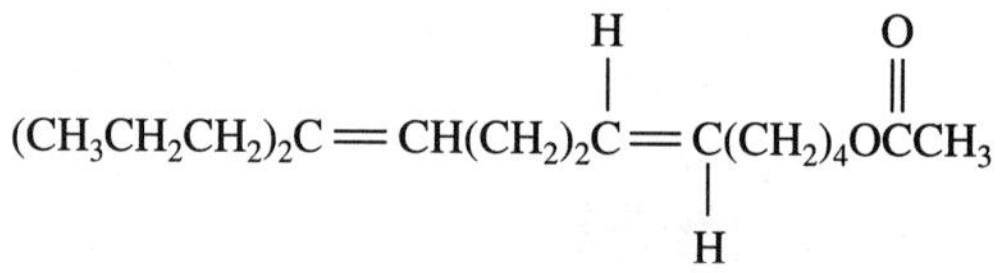

$$(CH_3CH_2CH_2)_2C = CH(CH_2)_2\overset{\overset{\displaystyle H}{|}}{C} = \underset{\underset{\displaystyle H}{|}}{C}(CH_2)_4O\overset{\overset{\displaystyle O}{||}}{C}CH_3 \qquad \text{propylure}$$

trans

It has been shown that the *cis* isomer inhibits the attractiveness of the *trans* and could therefore provide a means of masking the normal attractant. If mating is prevented, insects do not produce another generation.

Sterilization of insects has been termed *autocidal control.* Trapping males and sterilizing them with chemosterilants or ionizing radiation (usually γ radiation from a cobalt-60 source) is often effective. Sterilization of females would have little more effect than killing them, whereas a sterile male that may mate many times could prevent fertilization in many females, which usually mate only once.

The use of sterile males was an important component of the effort to eradicate the Medfly in California in the 1980s. Over 4 billion sterile flies were introduced in a multi-pronged attack that included insecticides, baited traps, stripping of fruit, and even inspection of cars at roadblocks to prevent inadvertent entry of the Medfly, all at a cost in excess of $160 million.

A substance known as *juvenile hormone* (JH), or certain other chemicals that mimic JH or other insect hormones, seem to hold a key to success in controlling some insects. The hormone controls progressive phases of development from larva to adult. Larval stages require JH; but for a mature larva to change into a sexually mature adult, the supply must be cut off. Humans can interfere at this stage. In the presence of an external source of JH, a susceptible insect continues to develop as a larva. Or it may change into an immature adult that retains many larval characteristics and cannot reproduce.

Compared with an insecticide, the primary advantage of the use of hormones is specificity—JH is not likely to kill off beneficial soil microorganisms, although it could eliminate some beneficial insects. Development of resistance is unlikely, since JH is essential to the insect in some stages of development. Finally, it is a natural substance, and should be biodegradable and therefore nonpolluting.

There is one potential problem with using JH: larvae are big eaters and do great damage to crops. Thus, if the larval stages are prolonged, crop loss could temporarily increase until the total insect population decreases.

SUMMARY

Insects can cause tremendous damage by spreading disease and by consuming crops. Insecticides, particularly DDT, were the main defense against insects but have fallen into disfavor due to adverse environmental effects. Biomagnification may cause insecticides to become serious hazards, killing birds and helpful insects. In some cases, insecticides actually promote the establishment of harmful insects by killing off harmless predators and selecting for resistance. Enzymes that convert an insecticide to a harmless substance, as when DDT is converted to DDE, are normally responsible for resistance.

Pheromones and allelochemicals (allomones, kairomones, synomones, antimones) influence individuals of the same and other species, respectively. Pheromones cause honeybees, gypsy moths, boll weevils, bark bee-

tles, and most other insects to mate, feed, and defend themselves. It is possible to use this information to turn insects against themselves. An example is the use of sex attractants to trap or confuse insects. Allelochemicals provide benefits and disadvantages to producer and recipient insects.

In integrated pest management (IPM), pheromones can be used in combination with insecticides in methods that avoid environmental pollution. Masking can be used to interfere with normal attraction. Sterilization of trapped males may follow in some cases and be used to interfere with normal species propagation. Juvenile hormone (JH) can be used to block development of reproductive adult insects.

PROBLEMS

1. Give a definition or example of each of the following.
   a. Biomagnification
   b. Biological vacuum
   c. Allelochemical
   d. Pheromone
   e. Allomone
   f. Kairomone
   g. Synomone
   h. Antimone
   i. Synergist
   j. Aggregation pheromone
   k. Biopesticide
   l. Hemolysis
   m. Entomology
   n. Biodegradability
   o. Disparlure
   p. Grandlure
   q. Integrated pest management
2. Name some diseases that are transmitted by insects.
3. How does an insect population develop resistance to DDT?
4. How can DDT kill an insect that is resistant to DDT?
5. Why was DDT banned by the Environmental Protection Agency?
6. What are the three strategies for managing resistance?
7. What are the advantages and disadvantages of organophosphates over chlorinated hydrocarbons?

8. How has dichlorvos been used? Why is it well suited for this application?
9. Why is a person who receives one bee sting likely to get several more from the same hive?
10. What is an important strategy used to control the African honeybee?
11. In what stage of development does the gypsy moth do its greatest damage?
12. What is the most significant agricultural pest? Why?
13. What is the function of queen substance in a honeybee hive?
14. What is an example of an insect that derives an important allomone from its normal food source?
15. How do cabbage loopers and alfalfa loopers differentiate between themselves for mating purposes?
16. Confusion and masking are two approaches to using pheromones for insect control. What is meant by each?
17. How does biological control of insect populations differ from other types of control?
18. What is one important technique for sterilizing insects? What is the purpose? Why is it done to males rather than females?

References

1. Shapley, D. "Mirex and the Fire Ant: Decline of Fortunes of a 'Perfect' Pesticide." *Science* **1971,** *172,* 358.

2. Matsumura, F. *Toxicology of Insecticides,* 2nd ed.; Plenum: New York, 1985; p. 339.

3. Debach, P. *Biological Control by Natural Enemies;* Cambridge University: London, 1974; pp. 11–19.

4. Ware, G. W. *Pesticides: Theory and Application;* W. H. Freeman: San Francisco, 1983; pp. 13, 41–45, 192.

5. Metcalf, R. L. "Pests and Pollution." In *Wednesday Night at the Lab;* Rinehart, K. L., Jr.; McClure, W. O.; Brown, T. L., Eds.; Harper & Row: New York, 1973; pp. 102–103.

6. Stock, T. "A Stunning Comeback." *Chemical and Engineering News,* December 22, 1986, p. 3.

7. Pimental, D.; Levitan, L. "Pesticides: Amounts Applied and Amounts Reaching Pests." *Bioscience* **1986,** *36* (2), 86–91.

8. "Hazardous Pest Strips Sneak Back on the Market." *Consumer Reports* **1990,** *55* (7), 445.

9. Whitman, D. W. "Allelochemic Interactions Among Plants, Herbivores, and Their Predators." In *Novel Aspects of Insect-Plant Interactions;* Barbosa, P.; Letourneau, D. K., Eds.; John Wiley & Sons: New York, 1988; Chapter 1.

10. Eisner, T.; Meinwald, J. "Defensive Secretions of Arthropids." *Science* **1966,** *153,* 1341–1350.

11. Seltzer, R. J. "Role of Insect Sex Lure Isomers Probed." *Chemical and Engineering News,* August 20, 1973, p. 19.

12. Ehrlich, P. R.; Rowen, P. H. "Butterflies and Plants." *Scientific American* **1967,** *216,* 104.

13. Silva, J. A. M. "Invasion of the 'Killer' Bees." In *Science Year 1991: The World Book Annual Science Supplement;* World Book: Chicago, 1992; pp. 26–39.

14. Slessor, K. M.; Kaminski, L.; King, G. G. S.; Borden, J. H.; Winston, M. L. "Semiochemical Basis of the Retinue Response to Queen Honey Bees." *Nature* **1988,** *332,* 354–356.

15. Collins, A. M. "Genetics of Honey Bee Colony Defense." In *Africanized Honey Bees and Bee Mites;* Needham, G. R., et al., Eds.; John Wiley & Sons: New York, 1988; Chapter 13.

16. Camazine, S.; Morse, R. A. "The Africanized Honeybee." *American Scientist* **1988,** 76 (5), 465–471.

17. Gould, J. L.; Gould, C. G. *The Honeybee;* Scientific American: New York, 1988; pp. 50–51.

18. Owen, D. R.; Lownsbery, J. W. "Dutch Elm Disease." In *Eradication of Exotic Pests;* Dahlsten, D. L.; Garcia, R.; Lorraine, H., Eds.; Yale University Press: New Haven, 1989; Chapter 9.

19. Bakke, A. "The Utilization of Aggregation Pheromone for the Control of the Spruce Bark Beetle." In *Insect Pheromone Technology: Chemistry and Applications;* Leonhardt, B. A. and Beroza, M., Eds.; ACS Symposium No. 190; American Chemical Society: Washington, DC, 1982; pp. 219–229.

20. Bakke, A. "The Pheromone of the Spruce Bark Beetle *Ips typographus* and Its Potential Use in the Suppression of Beetle Populations." In *Insect Suppression with Controlled Release Pheromone Systems;* Kydonieus, A. F.; Beroza, M.; and Zweig, G., Eds.; Chemical Rubber Co.: Boca Raton, FL, 1982; Volume II, Chapter 1.

21. Bakke, A. "Mass Trapping of the Spruce Bark Beetle *Ips typographus* in Norway as a Part of an Integrated Control Program." In *Insect Suppression with Controlled Release Pheromone Systems;* Kydonieus, A. F.; Beroza, M.; and Zweig, G., Eds.; Chemical Rubber Co.: Boca Raton, FL, 1982; Volume II, Chapter 2.

22. Tumlinson, J. H., et al. "Sex Pheromones and Reproductive Isolation of the Lesser Peachtree Borer and Peachtree Borer." *Science* **1974,** *185,* 614.

23. Kaal, R. S., et al. "Pheromone Concentration as a Mechanism for Reproductive Isolation Between Two Lepidopterous Species." *Science* **1973,** *179,* 487.

Food

Alcoholic Beverages

17

The chemistry underlying the production of alcoholic beverages (beer, wine, distilled beverages) is similar to the chemical processes involved in producing some dairy products, baked goods, and fermented foods, in the spoilage of foods, and in the production of antibiotics.

17.1 Brewing

GOALS: To understand the procedures and purpose of malting and mashing.
To consider the chemistry of fermentation, and the consequences of maintaining either aerobic or anaerobic conditions.
To understand how the progress of fermentation can be monitored.
To understand the strategies used in making light beers.

The production of beer is a complex chemical process of importance throughout the world. Although some of the minute details of the chemistry are unknown, many of the phenomena are well understood. We begin by considering the malting of barley, the principal grain used in brewing. The purpose of malting is to develop enzymes that convert starch into sugars. The malt is then subjected to mashing, addition of hops, and fermentation. Yeast is used during fermentation to provide enzymes that convert sugars into alcohol. Lagering and packaging complete the production of beer *(1)*.

Malting

Barley (Figure 17.1) is the grain of choice for beer production because it is relatively easy to malt; barley malt provides the enzymes α-amylase and β-amylase, which are required for efficient conversion of the plant starches into fermentable sugars; and barley hulls remain solidly attached to the kernel even during threshing, providing a good, sturdy compartment for some of the stages of malting.

After the kernels are cleaned and sized, malting is a three-step process: steeping, germination, and kilning. *Steeping* consists of soaking the barley kernels in water at about 15 °C for 48–72 hours. A water content of 45-50% is achieved, which provides conditions suitable for germination.

Figure 17.1   Barley, a field of beer in the making.

Germination is encouraged by exposing the barley to high humidity and controlled temperature, typically 15–25 °C. In about a week, much of the rigid structural starches and proteins break down, and enzymes develop. The formation of *green malt* is accompanied by respiration and root development. Respiration provides energy for synthesis of the enzymatic proteins and other changes that occur. Root development, the biological result of germination, is wasteful for brewing, but the roots are high-protein material used as livestock feed.

Kilning (drying) reduces the water content of the barley to about 10%. Temperatures of 65–80 °C for 1–3 days yield light-colored malt. Temperatures up to 100 °C produce dark malts, which are used for porters, stouts, ales, and other dark beers. The lower temperatures favor retention of maximum enzyme activity developed during germination. High temperatures destroy more of the enzyme activity but cause development of characteristic colors and flavors.

After kilning, the roots are removed, and the clean malt is either stored or used directly. Malt contains less moisture than barley grain, making it more suitable for storage and grinding. Malt provides several components required and desirable for beer production: fermentable sugars, starch, amylase activity, other proteins that contribute to body and foam, amino acids, and flavor ingredients.

Other Ingredients

Adjunct unmalted starches are added to provide sufficient fermentable sugar. Corn starch is often used, and rice occasionally.

Hops are another important component of beer. The blossoms of the hop plant *(Humulus lupulus)* add a bitter character, countering the sweetness of the carbohydrate. This is due to chemical compounds such as humulone and lupulone, which also have an antiseptic effect. Approximately 1 g of hops per liter of beer is average, although some beers are more heavily hopped.

$(CH_3)_2C=CHCH_2$ — OH — $COCH_2CH(CH_3)_2$

$(CH_3)_2C=CHCH_2$ OH OH
humulone

$(CH_3)_2C=CHCH_2$ — OH — $COCH_2CH(CH_3)_2$

$(CH_3)_2C=CHCH_2$ OH $CH_2CH=C(CH_3)_2$
lupulone

Brewer's yeast supplies the many enzymes necessary for fermentation. It causes vigorous fermentation, then settles well, permitting easy clarification. Water is also important. Mineral composition varies greatly, so some waters are better suited for brewing than others.

Mashing

Review the structure of starches, maltose, glucose, and other carbohydrates in Chapter 9.

The actual brewing process begins with preparation of the mash, by crushing the malt to expose all its starches and proteins to the enzymes and water. Much of the barley starch was already broken down to monosaccharides and disaccharides during malting. The mash is gradually heated to about 70 °C over a period of about 2 hours (Figure 17.2). During this time, the enzymatic attack by amylase enzymes on adjunct starches is completed.

Recall that starch consists of both amylose and amylopectin. A linear amylose molecule has only 1,4 linkages between the glucose units. The more complex amylopectin has some branching of the main chains. On the average, the branches are about 12 glucose

Figure 17.2 Brewing kettles.

units long and are attached to the main chain by 1,6 linkages at about every twelfth unit of the chain.

The cleavage of amylose by the amylase enzymes forms maltose (a disaccharide of glucose, Glu-Glu) plus glucose. The β-amylase specifically and successively cleaves off maltose units from one end of the chain to the other.

$$— Glu — Glu + Glu — Glu + Glu — Glu + Glu — Glu + Glu — Glu — \quad \text{amylose}$$

$$\downarrow \text{β-amylase}$$

$$n \ \text{Glu—Glu} \quad \text{maltose}$$

The α-amylase is a less specific enzyme that randomly cleaves amylose chains. This forms many fragments with varying numbers of glucose units, which eventually are cleaved into both maltose and glucose.

Neither the α- nor the β-amylase completely breaks down amylopectin, since these enzymes do not cleave the 1,6 linkages at the branches. The resulting starch fragments are known as **limit dextrins.** Produced when amylases have acted to near the limit of their ability, limit dextrins commonly range from disaccharides to hexasaccharides. They are not fermentable and therefore remain in the finished beer, adding to the carbohydrate content and contributing to the stability of the foam.

The product of the mashing step is known as a **wort.** Its solids are allowed to settle and are removed. Then the wort is boiled with the hops for 2–3 hours, filtered, and placed in a fermentation tank. Yeast is added, a step described as *pitching,* or *injection,* to initiate fermentation.

Fermentation

Two categories of yeast are in common use: *top-fermenting* and *bottom-fermenting.* During the vigorous stages of fermentation, when large amounts of CO_2 are being produced, both yeasts tend to remain suspended, but as the fermentation subsides, a top-fermenting yeast rises to the top and a bottom-fermenting yeast settles.

Typical conditions for fermentation are 4–10 °C for 7–11 days when using a bottom-fermenting yeast, and 10–16 °C for 6 days or less for a top-fermenting yeast. The bottom-fermenting yeast is used in the production of the lager beer that predominates in the United States.

The overall chemical change during fermentation is described by the equation that follows:

$$C_6H_{12}O_6 \rightarrow 2 \ CH_3CH_2OH + 2 \ CO_2$$
$$\text{glucose} \qquad \text{ethyl} \qquad \text{carbon}$$
$$\text{alcohol} \qquad \text{dioxide}$$

This is a tremendous oversimplification of fermentation (also called **glycolysis**), also used in baking, cheese making, and curing foods. Note the important features of the process, highlighted in Figure 17.3.

Each of the 12 steps leading to ethanol (ethyl alcohol) requires a separate enzyme provided by the yeast. Since yeast is living microorganisms, the reaction pathways represent chemical reactions, that is, *metabolism,* necessary for the survival of its cells.

Most important to the yeast is the production of *adenosine triphosphate, ATP,* the most important source of chemical energy for all living organisms. It is the ''common currency'' of energy in biological systems. (Your brain is using the energy of ATP as you

The word enzyme originates from the Greek *en,* "in," and *zume,* "yeast."

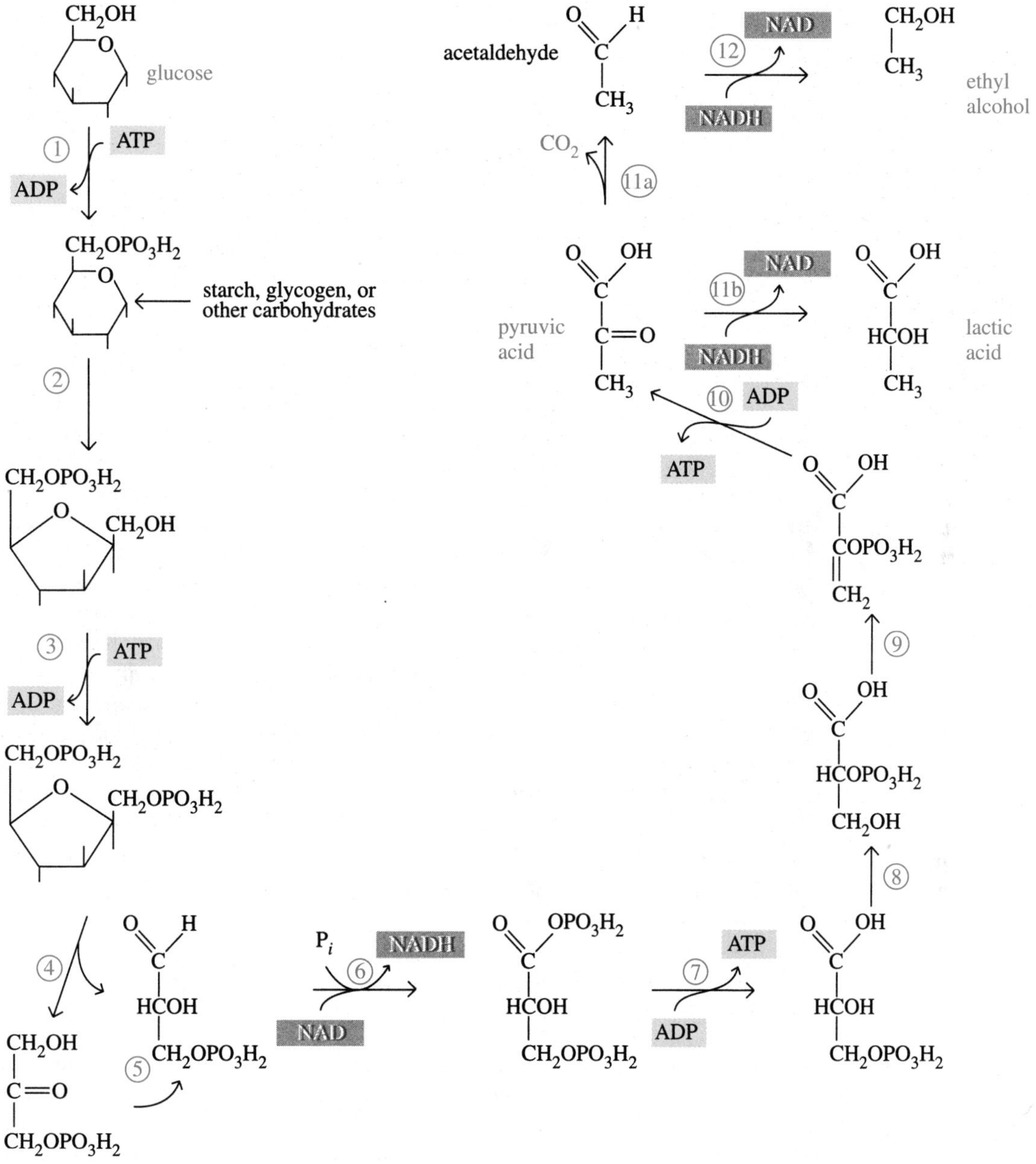

Figure 17.3 Fermentation (glycolysis). Only the key compounds in the sequence are named. P_i is shorthand for inorganic phosphate ions.

read this page.) The ATP is converted to *adenosine diphosphate, ADP,* by removing the terminal phosphate group (shown on p. 386 in color) and by release of its energy for biochemical work. It is convenient to think of ATP as the charged form of the battery used to power living systems. The ADP is the discharged form; recharging converts it back to ATP. A related high-energy compound is *guanosine triphosphate, GTP;* the low-energy form is *guanosine diphosphate, GDP.* Collectively, ATP and GTP are known as *high-energy phosphate compounds.*

adenosine triphosphate (ATP)

In the fermentation sequence, the conversion of glucose to ethyl alcohol leads to a net production of two ATP; this explains why yeast and other organisms (including humans) carry out all or part of this sequence as they metabolize glucose. Steps 1 and 3 each consume one ATP molecule. In step 4, the 6-carbon sugar is converted to two 3-carbon fragments, after which step 5 converts one fragment into the other. In other words, the starting compound for step 6 is formed twice during this sequence. Therefore, all steps beginning with step 6 occur twice for every one glucose that is metabolized. *One ATP is produced for each 3-carbon fragment that passes through steps 7 and 10;* thus a net of two ATP is produced per molecule of 6-carbon glucose that is metabolized.

One special quality of yeast is its ability to survive either aerobically or anaerobically. The difference begins with *pyruvic acid,* the product of step 10. The presence or absence of oxygen has no effect on steps 1–10, but does determine the fate of the pyruvic acid. If no oxygen is available, pyruvic acid is converted in two steps to ethyl alcohol plus CO_2 (steps 11a and 12). If oxygen is available, pyruvic acid is completely metabolized to 3 CO_2 by the eight steps in the series of reactions known as the *Krebs cycle,* or *citric acid cycle,* shown in Figure 17.4. When this happens, no ethyl alcohol forms. To understand why yeast cells dispose of pyruvic acid in different ways under different conditions, we must consider the role of *nicotinamide adenine dinucleotide,* NAD, in the metabolic sequences.

There are four ways to dispose of pyruvic acid: steps 11a, 11b, 11c, and 11d. Step 11a leads to formation of ethyl alcohol. Step 11b forms lactic acid. Steps 11c and 11d provide compounds necessary for the citric acid cycle.

Regenerating NAD

Notice in Figure 17.3 that NAD is consumed (converted to NADH) at step 6. For the production of alcohol to continue, NAD must be regenerated. There are three ways to do this. Two possibilities are shown in Figure 17.3: steps 12 and 11b. Step 12 forms ethyl alcohol. Step 11b, an alternative anaerobic method, regenerates NAD by converting pyruvic acid to lactic acid. This result, **lactic acid fermentation,** is the most common pathway used for metabolism of glucose by anaerobic microorganisms other than yeast.

If oxygen is present, pyruvic acid enters the Krebs cycle by step 11c. Within the cycle, even more NAD is consumed; it is regenerated later in a process that requires oxygen (discussed in the following section).

The Role of B Vitamins

The B vitamins, present in most every commercially available vitamin supplement, are important in metabolism. Both NAD and NADH are forms of **niacin,** vitamin B_1, referred to as **active niacin** and **inactive niacin,** respectively. Active niacin is required for survival of all living things.

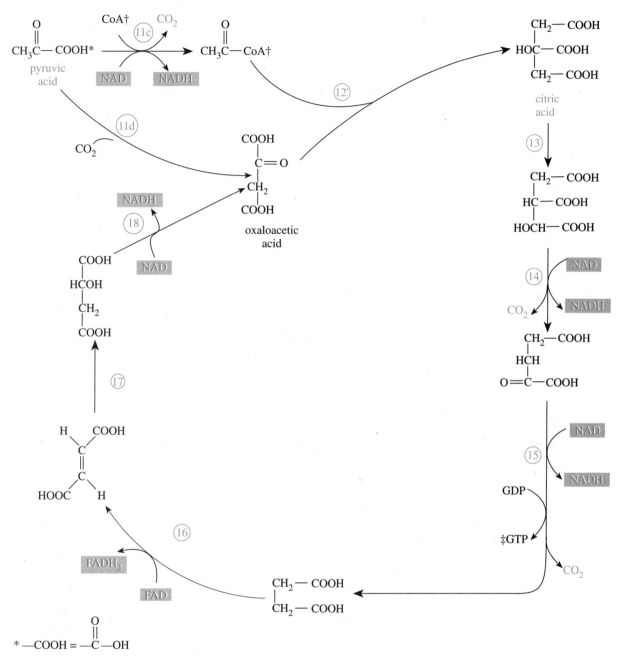

* —COOH = —C—OH
          ‖
          O

† CoA = coenzyme A, which is derived from one of the B vitamins.

‡ GTP and GDP are high- and low-energy compounds like ATP and ADP.

Figure 17.4   Metabolism of pyruvic acid by the citric acid cycle (Krebs cycle). Step numbers continue from the glycolysis scheme, Figure 17.3.

Under aerobic conditions, active niacin is regenerated by the following, a shortened version of seven steps:

$$NADH + O_2 \xrightarrow[\substack{3\ ADP \qquad 3\ ATP}]{7\ steps} NAD + H_2O$$

Yeast can regenerate active niacin under both aerobic and anaerobic conditions, but yeast grows more vigorously under aerobic conditions due to the high yield of ATP during regeneration. However, the growth occurs at the expense of ethyl alcohol production since steps 11a and 12 are bypassed. Consequently, the brewer or wine maker routinely limits the supply of oxygen during fermentation and receives two ethyl alcohol molecules for every glucose that enters the scheme.

Both FAD and $FADH_2$, which also appear in the Krebs cycle, are two forms of vitamin B_2, called **riboflavin.** The FAD is called **active riboflavin,** and $FADH_2$ **inactive riboflavin.** The former can be regenerated only under aerobic conditions, according to the following process in six steps:

$$FADH_2 + O_2 \xrightarrow[\text{2 ADP} \quad \text{2 ATP}]{\text{6 steps}} FAD + H_2O$$

The Yield of High-Energy Phosphate

The metabolism of glucose to form pyruvic acid yields 2 ATP, as already discussed. Since the regeneration of each NAD yields 3 ATP and each FAD yields 2 ATP, we can locate five steps (6, 11c, 14, 15, and 18) in which NAD is consumed, and one step (16) in which FAD is consumed. Remembering that all of these steps occur twice, we can calculate a total of 34 ATP produced. Adding the 2 ATP from the formation of pyruvic acid and the formation of 2 GTP due to step 16, we arrive at a total of 38 high-energy phosphate per 1 glucose.

Review of Fermentation

The highlights of the metabolic pathways in Figures 17.3 and 17.4 are summarized in this section by considering a series of examples.

EXAMPLE 17.1:

All reactions beginning with step 6 are used twice per glucose.

How many steps are involved in the metabolism of glucose?

Solution:

1. Anaerobically:

    12 steps from glucose to ethyl alcohol (19, counting repeated steps)

    11 steps from glucose to lactic acid (17, counting repeated steps)

2. Aerobically:

    15 steps from glucose to pyruvic acid (including repeated steps)

    2 steps (11c) to the Krebs cycle (including repeated steps)

    14 steps in the Krebs cycle (including repeated steps)

    7 steps to regenerate each NAD

    6 steps to regenerate each FAD

    3 steps for production of ATP from each NADH

    2 steps for production of ATP from each $FADH_2$

Step 11d is not counted, since it occurs only occasionally. This is because the product of step 11d is continually regenerated at step 18 of the cycle.

Since steps 6–18 all occur twice and NAD is consumed in steps 6, 11c, 14, 15, and 18, for every glucose, 70 (5 x 7 x 2) steps are required to regenerate all of the NAD. Similarly, step 16 occurs twice, so that 12 (1 x 6 x 2) steps are required to restore the FAD.

Since 10 NADH are formed in the sequence from glucose through the Krebs cycle, 30 steps are required to generate 30 ATP. Similarly, 2 $FADH_2$ requires 4 steps to form 4 ATP.

All of the above total up to 147 steps per glucose metabolized.

EXAMPLE 17.2:

What is the purpose of these pathways?

Solution:

The process generates useful energy (stored in the ATP molecules) that is required by the yeast and other organisms.

1. Anaerobically:
   Two ATP are produced per glucose metabolized.
2. Aerobically:
   By the Krebs cycle and the subsequent steps, an additional 34 ATP and 2 GTP are produced. The aerobic process is much more efficient in fulfilling the purpose, but if a supply of O_2 is not available, or if an organism lacks the necessary enzymes, one of the anaerobic pathways must be used.

EXAMPLE 17.3:

Why does each pathway end where it does?

Solution:

1. Anaerobically:
   Both anaerobic pathways proceed until NAD is regenerated. Yeasts carry out alcoholic fermentation. Other organisms and animals carry out lactic acid fermentation under anaerobic conditions. The choice of pathway depends on what enzymes are available for catalysis.
2. Aerobically:
   The aerobic pathway ends when the fuel (glucose) is completely oxidized to CO_2 and water, all the NAD and FAD are regenerated, and all the available energy is stored as ATP and GTP.

The ethyl alcohol and lactic acid produced in the anaerobic pathways represent incomplete use of energy. Each compound has energy content that cannot be used by anaerobic organisms. This is indicated by the production of only 2 ATP per glucose anaerobically, compared with 36 ATP and 2 GTP per glucose aerobically. When alcohol is consumed, the remaining energy of the ethyl alcohol molecule is released due to complete metabolism in the body. Similarly, lactic acid is fully metabolized when consumed.

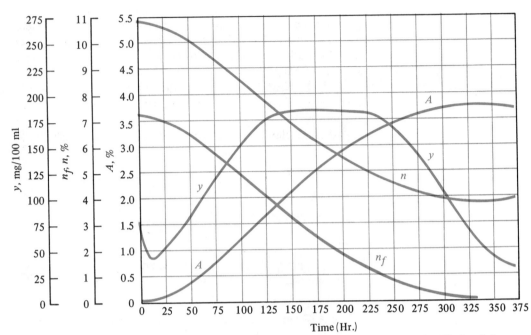

Figure 17.5   Changes that occur during fermentation. The curves are identified as follows: *A*, alcohol; *n*, extract; n_f, fermentable extract; *y*, suspended yeast solids. (From Kirk, R. E.; Othmer, D. F. *Encyclopedia of Chemical Technology;* John Wiley & Sons: New York. Reprinted by permission of John Wiley & Sons.)

The Course of Fermentation

The curves in Figure 17.5 provide additional information on the changes during fermentation, as the wort is changed to beer. The alcohol content is determined by the amount of fermentable sugar contributed by malt extract and adjuncts. That part of the extract that is fermentable (n_f) decreases to zero coincident with the alcohol content reaching a plateau. The release of CO_2 gas (step 11a, Figure 17.3) causes foaming that keeps the yeast suspended, but as fermentation subsides, the yeast settles. The unfermentable extract (*n*) that remains in the beer consists mainly of malt proteins and limit dextrins.

Section 1.5 contains a discussion of density and specific gravity.

The progress of fermentation is routinely monitored by a hydrometer, sometimes called a saccharometer, that measures the specific gravity of the wort. During fermentation, high-specific-gravity glucose is replaced by low-specific-gravity ethanol. The result is illustrated by Figure 17.6, which shows the hydrometer sinking farther down into the liquid when most of the sugar has been replaced by alcohol. The effect resembles the experience of a swimmer in fresh versus salt water. A person is buoyed up much more by the high-specific-gravity salt water.

Lagering and Packaging

Lagering (storage) follows the fermentation. Storing beer, typically for 2–6 months, allows for some additional fermentation. The beer clarifies, and some flavor characteristics develop. After lagering, proteolytic (protein-cleaving) enzymes may be added to digest and solubilize any proteins that might precipitate when the beer is cooled. This step is called chill-proofing.

Figure 17.6 Hydrometer for monitoring fermentation. (a) Example of hydrometer reading of 1.042. This is the average specific gravity of beer before yeast is added. (b) Hydrometer reading is 1.010, the typical level near the end of the fermentation. (Figure 6 from BREW IT YOURSELF by Leigh Beadle. Copyright © 1971 by Leigh Beadle. Reprinted by permission of Farrar, Straus & Giroux, Inc.)

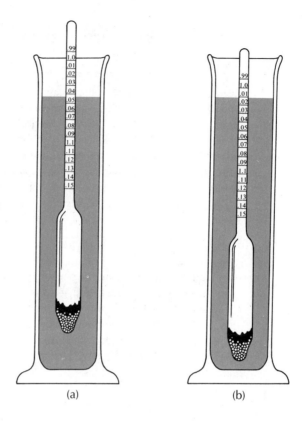

(a)   (b)

Finally the beer is packaged. This includes adding CO_2 under pressure and pasteurization (heating to deactivate microbes). Later, when the beer is opened and poured, the CO_2 pressure is released, causing foam to build. Unlike carbonated soft drinks, beer retains this foam as a head with the help of limit dextrins and malt proteins.

Specialty Beers

Many types of beer are in production worldwide. Table 17.1 lists some of their important characteristics. Barley malt not only supplies some fermentable carbohydrate, but is also the source of the protein in beer. The amylases in malt are important as the starch-breaking enzymes, but other malt proteins stabilize beer foam. When the wort is boiled before

humulone   →   isohumulone

fermentation, the compound humulone extracted from hops is converted to isohumulone. This is particularly important, since isohumulone interacts with the protein in the beer to

Table 17.1   Characteristics of various types of beer

Type of Beer	Characteristics (Alcohol Content %v/v)
Lager	Light; bottom yeast; 3–4% alcohol; variable hops content
Munich	Dark brown (from dark malt); aromatic, sweet; high hops; up to 5% alcohol
Bock	Heavy; very dark; variable alcohol
Porter	Dark; low hops; 4–6% alcohol; heavy, rich, malty flavor
Stout	Very dark; heavy, strong malt taste; 5–7% alcohol; high hops
Ale	Heavy; slightly dark; top yeast; up to 6% alcohol; high hops
Cream ale	Light color; usually top yeast; low hops; 4–6% alcohol
Malt liquor	Heavy; slightly dark; low hops; 4–5% alcohol
Light beer	Light color; 2.5–3.3% alcohol; high hops

stabilize the foam and cause a phenomenon called *lacing* (or *cling*) in which the foam tends to cling to the sides of a glass after it has subsided. This gives beer a creamy appearance that is very important. Limit dextrins also assist foam formation and retention.

Since both the barley proteins and isohumulone are required for stable foam formation and lacing, rich beers made with a large amount of barley malt protein, and those that are heavily hopped tend to have the best appearance. The heavy, dark, German beers, porter, and stout are typical examples.

Light Beers

A recent addition to the list of beers is the low-calorie ''light'' beer, which has become increasingly popular in this country. Its chemistry is interesting. Since the calories in beer come from the carbohydrate, protein, and alcohol contents, some or all of these must be lower than normal.

The carbohydrate content of beer is due to the limit dextrins. The amount of limit dextrins can be minimized in several ways. Less barley malt as a source of branched-chain starches (amylopectin) can be used. Barley malt also provides protein, which is needed for foam and lacing. To make up the reduced barley malt protein, more hops are used, causing light beers to tend to taste more bitter. Adjunct starches can be added in lesser amounts, but this reduces the amount of fermentable carbohydrate. In fact, light beers do contain less alcohol and fewer calories (Table 17.2). At least one major American brewer uses glucose as an adjunct, which can be fermented completely.

Another strategy is to use the enzyme **amyloglucosidase,** which cleaves the limit dextrins into fermentable sugars. This has led to an advertising controversy, since one brewery touts its light beer as a ''natural,'' rather than a ''laboratory,'' light beer. The word *natural* means that the amyloglucosidase that cleaves the limit dextrins occurs naturally in their barley malt due to the method for preparing the malt. Other brewers who want to use amyloglucosidase add commercially available enzyme extract. Either way, the enzyme does the same thing. Most light beers are made using a combination of less starch and added amyloglucosidase.

Such beers naturally have less flavor. To make up for this, some other tricks are used. High-temperature fermentation causes the formation of more yeast metabolism products that contribute to beer flavor. ''Heavy brewing'' is another technique for increasing beer flavor. This method uses a more concentrated wort during fermentation so that more yeast

Section 6.3 contains a discussion of alcohols, acids, and esters.

Table 17.2   Calorie count of typical regular and light beers

Ingredient	Composition/12 oz. Regular Beer		Composition/12 oz. Light Beer	
	Grams	Calories	Grams	Calories
Protein @ 4 Cal/g	1.08	4.35	0.85	3.40
Carbohydrate @ 4 Cal/g	13.58	54.32	5.08	20.32
Alcohol @ 6.9 Cal/g	13.48	93.01	10.53	72.66
Total Calories		151.65		96.38

Information courtesy of Schlitz Brewing Co.

metabolism products are formed. Afterward, water is added to decrease the concentration to the desired level.

Among the yeast metabolism products are esters that contribute both flavor and aroma to beer. Esters form when alcohols react with acids, as illustrated in this equation.

$$CH_3CH_2OH + CH_3-\overset{O}{\overset{||}{C}}-\overset{O}{\overset{||}{C}}-OH \rightarrow CH_3-\overset{O}{\overset{||}{C}}-\overset{O}{\overset{||}{C}}-OCH_2CH_3$$

ethyl alcohol / pyruvic acid / ethyl pyruvate (an ester)

Esters and other flavor ingredients that form as a result of fermentation are sometimes called *congeners* (Section 17.3).

17.2 Wine Making

GOALS: To realize the relationship between sugar content of grapes and the alcohol content of wines.
To appreciate the chemical significance of a good year, and to see how high-acid, low-sugar grapes can be used to make a good wine.
To understand the importance of sulfur dioxide in making wine.
To understand the two types of carbonated wines.

Wine is fermented grape juice. When the juice of other fruits is fermented, the source is identified, such as peach wine or apple wine. The grapes are crushed (Figure 17.7) to produce a **must** in which yeast enzymes ferment the available sugar.

Climate is the main factor that influences the development of sugar in grapes. Certain areas of the world are well known for their distinctive wines. Even in the United States, California and eastern wines are quite different. The more temperate climate of California produces better natural (unfortified) wines.

In a good year the climate is warm enough to permit full ripening of grapes to give a high level of fermentable sugar. If the climate is too warm, the wine may have low acidity. Acidity is an important factor. The main acids are tartaric and malic (Latin *malum,* ''apple''). Even though they are present in small amounts, 0.1–0.8%, they greatly influence a wine's character. In overly warm years, the sugar content may be high enough but the acidity too low for a good dry (not sweet) wine. The high-sugar, low-acid grapes may be used to produce sweet dessert wines.

Figure 17.7

The Wilmes press. Inside the cylinder there is an inflatable rubber sleeve (bladder). In operation, the cylinder is filled with grape pulp, its covers are fastened in place, and the cylinder is revolved. A trough beneath the press collects the juice. When the free-run juice ceases to flow, the bladder is inflated with compressed air, which presses the pulp against the inside of the cylinder. This pressing action expresses the remainder of the juice. Solids (pomace) are discharged by removing the cylinders and rotating the cylinder. A second trough containing a screw conveyor is moved into place beneath the press. This trough conveys the pomace away. The Wilmes press can hold about four tons of pulp. The complete pressing cycle takes about thirty minutes.

Organic acids are discussed in Section 7.5.

Table 17.3 shows a typical pattern of change in the sugar level, pH, and contents of three acids in grapes during the last 6 weeks of a growing season *(2)*. The concentration of sugars more than doubles, whereas the level of acidity, particularly malic acid and tartaric acid, decreases dramatically. Malic acid is translocated from the grapes into the

Table 17.3   Changes in levels of sugar and acidity of Cabernet Sauvignon grapes during the last 6 weeks of a typical growing season (2)

Date	Sugars (g)	pH	Acidity	Tartaric Acid	Malic Acid	Citric Acid
18 August	84	2.75	356	171	200	2.7
24 August	108	2.85	264	152	140	2.4
31 August	128	3.00	184	118	105	2.0
7 September	156	3.17	152	119	75	2.2
14 September	156	3.17	136	114	65	2.0
21 September	178	3.30	128	124	50	2.4
28 September	188	3.35	120	124	43	2.1
5 October	184	3.35	108	106	45	3.6

Concentrations of acids are in mEq/L. One mEq/L equals 1 mmol/L for a monoacid; 2 mEq/L equals 1 mmol/L for a diacid; 3 mEq/L equals 1 mmol/L for a triacid.

leaves. Therefore, in a growing season shortened by poor weather, the composition of the grapes may be quite different.

Grapes grown in the eastern United States tend to be high in acid and low in sugar. This low-sugar defect is routinely corrected by adding sugar, a technique known as **fortification,** but it does not introduce any flavor components. High acidity can be corrected by encouraging a secondary fermentation known as the malo-lactic fermentation.

Malo-Lactic Fermentation

In this conversion, malic acid with two acid groups is converted to lactic acid with only one acid group. This means a decrease in acidity and a mellower wine.

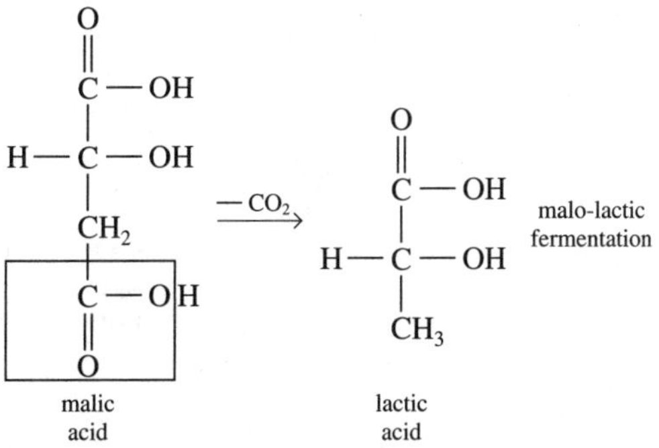

Malo-lactic fermentation may be either encouraged or discouraged, depending on the method of handling the wine after the primary fermentation. High-acid eastern wines (like Concord and Catawba) are invariably made to undergo this secondary fermentation. California wines usually require malo-lactic fermentation only in cool years, when acidity is relatively high. High SO_2 concentration can be used to discourage the process by inhibiting the growth of microorganisms that carry it out.

Technical Variations

Alcohol Content

The sugar content is expressed as a percent by weight (grams of sugar per 100 g of must). The alcohol content is expressed in percentage by volume (liters of alcohol per 100 L of wine).

Alcohol content is a serious concern in wine making. As shown in Table 17.4, wines are categorized and taxed accordingly. Most have an alcohol content somewhat less than 14%. They are called table wines. So-called dessert wines have an alcohol content above 14%. Most are fortified by addition of alcohol, usually brandy. For some dessert wines, fermentation is stopped by addition of alcohol while some sugar remains (Figure 17.8). Examples are muscatel (10–15% sugar, 16% alcohol) and port (9–14% sugar, 18–20% alcohol).

Vermouths contain 16–18% alcohol, along with 3–4% sugar in dry vermouth or 12–20% sugar in sweet vermouth. Vermouth is an unusual wine, produced in all areas of the world, with spices and herbs added to give it a special aroma and bitter taste.

Because of taxation, producers of table wine do not want to exceed 14% alcohol. Approximately 9% is necessary for stability; this amount will form during fermentation if the must contains 16.4% sugar. In general terms, each 1% sugar ferments to 0.55% alcohol. For example, 25% sugar can be achieved in good years and yields a wine with 13.75% alcohol. In some parts of Europe, high humidity and a long, warm growing season encourage the growth of a grape-dehydrating fungus, *Botrytis cinerea.* As the grapes overripen, water evaporates, and the berries shrink and become raisins. The fungus also consumes some of the acid in the grapes, so the overall effect gives a high-quality fruit for wine making. The loss of water from botrysised grapes may increase the sugar content to well over 30% *(3).*

Sulfur Dioxide

Fermentation in wine making differs only slightly from the method used in brewing. The chemical changes are the same, but a few technical details vary. For one thing, wild yeasts routinely grow on the skins of the grapes. Since these wild varieties may provide a different and perhaps less desirable group of enzymes, sulfur dioxide, SO_2, is commonly introduced to the must to kill the microorganisms. The common wine yeasts are not sensitive. The SO_2 acts as a general antiseptic by inhibiting the growth of any bacteria that might cause spoilage even after fermentation.

Sulfur dioxide is a gas that can be pumped directly into the must. It can also be introduced in the form of Campden tablets, which are commonly used by the home wine maker. The tablets are sodium metabisulfite, $Na_2S_2O_5$, which reacts in water according to the following reaction:

$$Na_2^+S_2O_5^{2-} + H_2O \rightarrow 2\ Na^+HSO_3^-$$

sodium   sodium
metabisulfate   bisulfite

Table 17.4 Wines and taxation

Type of Wine	Percentage Alcohol	Tax (per gal) (2)
Still (nonsparkling)	Up to 14	$1.07
Still	14–21	1.57
Still	21–24	3.15
Champagne or sparkling wine		3.40
Artificially carbonated wine		3.30

Figure 17.8
Fermentation tanks.

Sodium bisulfite is the half-neutralized form of sulfurous acid, H_2SO_3, which is formed when SO_2 is dissolved in water.

$$SO_2 + H_2O \rightarrow H_2SO_3$$
$$\text{sulfurous acid}$$

$$H_2SO_3 \xrightarrow{-H^+} HSO_3^- \xrightarrow{-H^+} SO_3^{2-}$$
$$\text{sulfurous acid} \qquad \text{bisulfite ion} \qquad \text{sulfite ion}$$

Sulfur dioxide also combines with any unconverted acetaldehyde; this substance would give the wine an undesirable character. Acetaldehyde is the immediate precursor of ethyl alcohol in the fermentation scheme. Sulfur dioxide is also an antioxidant; it prevents oxidation. Oxidation may take many forms. For alcoholic beverages, the one that is of concern is the conversion of ethyl alcohol in two steps to acetic acid, as shown.

Acetic acid is the ingredient of vinegar that gives it a sour taste; such a taste may be fine for vinegar but not for wine *(4)*. For NAD to oxidize ethyl alcohol or acetaldehyde, oxygen must be available to regenerate NAD, as discussed earlier. To protect wine, it is necessary either to exclude oxygen totally, which is difficult, or to add an antioxidant such as SO_2. The SO_2 functions by intercepting oxygen and being oxidized itself according to the following equation:

$$2 \ SO_2 + O_2 \rightarrow 2 \ SO_3$$

Color

White wines are fermented free of grape skins, whereas red wines derive their color from pigments extracted from the skins.

Carbonation

Sparkling wines and *bubbling wines* are effervescent, with a high content of carbon dioxide that is released when the wine is opened. Bubbling wines, also called *crackling wines,* are produced by saturating a wine with CO_2 gas under pressure, as with a carbonated soft drink. The gas is released rapidly and only lasts for a short time.

Sparkling wines are produced by a secondary alcoholic fermentation. (The gas from the initial fermentation escapes.) After the wine is produced and bottled, a specific amount of beet sugar solution is added and the bottle is corked. If too little sugar is added, the sparkling and foaming will be insufficient. If too much sugar is added, excessive pressure will cause a high percentage of cracked bottles. The best known sparkling wine is champagne, first produced in Champagne, France, by the monk Dom Perignon near the end of the seventeenth century.

In sparkling wine formed by secondary fermentation, the sparkling lasts for a long time as the bubbles form slowly and from the bottom of the glass. This longer duration is explained by the fact that the CO_2 is present in three forms. Some CO_2 is free gas, some is dissolved in the wine, and some is chemically bound but released slowly when the wine is opened. Exactly how CO_2 is bound is not known, but various methods of handling wine, such as heating, are known to have significant effects on the ability of the wine to bind CO_2. In crackling wines, no CO_2 exists in bound form and the free and dissolved CO_2 escapes rapidly.

During aging, the bottles are stored with the corks down so that the yeast sediment settles on top of the corks. After fermentation is complete and the wine is sufficiently aged, the bottles must be opened to remove the sediment. Special equipment is used to freeze the wine and sediment in the necks of the bottles, so the corks can be extracted and the sediment removed with little loss of wine or CO_2. The bottles are recorked and fastened with a small wire basket to prevent premature blowing out of the cork due to the pressure of CO_2 gas *(3, 5)*.

Newer technology makes it possible to do the secondary fermentation in tanks rather than in the bottles, although the quantity of CO_2 present is usually much less than when fermentation is done in the bottle. The process carried out in tanks is less expensive, however, and lends itself better to large-scale production *(5)*.

17.3 Distilled Alcoholic Beverages

GOAL: To appreciate the sources of fermentable carbohydrate in distilled alcoholic beverages.

When producing distilled alcoholic beverages, the mash is heated to boiling and distilled. All solids and other high-boiling materials are left behind as a residue, while the volatile components, including alcohol, are boiled off and collected.

Table 17.5   **Congeneric content of principal types of distilled alcoholic beverages**(6)

Component	American Blended Whiskey	Canadian Blended Whiskey	Scotch Blended Whiskey	Straight Bourbon Whiskey	Bonded Bourbon Whiskey	Cognac Brandy
Fusel oil	83	58	143	203	195	193
Total acids (as acetic acid)	30	20	15	69	63	36
Esters (as ethyl acetate)	17	14	17	56	43	41
Aldehydes (as acetaldehyde)	2.7	2.9	4.5	6.8	5.4	7.6
Furfural	0.33	0.11	0.11	0.45	0.90	0.67
Total solids	112	97	127	180	159	698
Tannins	21	18	8	52	48	25
Total congeners (% m/v)	0.116	0.085	0.160	0.292	0.309	0.239

Numbers are in g/100 L at 50% alcohol.

Source: Kirk, R. E.; Othmer, D. F. *Encyclopedia of Chemical Technology;* John Wiley & Sons: New York, 1993. Reprinted by permission of John Wiley & Sons.

Various grain starches may be the source of alcohol and are handled in different ways. In making Scotch whiskey, malted barley is dried over peat fires. This gives the malt a smoky flavor that carries through into the final product. Other grains are generally blended into the mash for both Scotch and other whiskeys. Canadian whiskey is made from a blend of corn, rye, and barley malt. Rye whiskey is made from a mash that is predominantly rye grain. Gin is usually a blend of various grains plus certain berries, most notably juniper, that provide flavor constituents. Bourbon is a blend in which corn predominates, although real corn whiskey must be made from a minimum of 80% corn.

Brandy is distilled from fermented grapes. Other brandies normally specify the source, as in apricot brandy. Brandy is often added to make high-alcohol dessert wines. Rum is a distillate derived from fermented juice of sugar cane, sugar syrup, and molasses. Vodka also is distilled from fermented grain mash. When the distillate is charcoal filtered, all pigments are removed.

Congeners (Table 17.5) are side products of fermentation to ethyl alcohol. Aging contributes to full development of congeners; these substances may even include materials extracted from storage containers, such as oak barrels.

SUMMARY

Brewing is a worldwide industry. Before the actual brewing, barley is malted in a three-step process of steeping, germination, and kilning. During germination, structural starches and proteins are broken down and complex enzyme systems are developed. These enzymes break down barley starches and adjuncts into fermentable sugars. The green malt that results from this process is kiln dried. High-temperature kilning yields dark malts that are used in dark beers, ales, and other products.

Hops are added to give beer a bitter character that counteracts the sweetness of the carbohydrates. Complete breakdown of starches by the action of α- and β-amylases is accomplished during mashing. The product, a wort, is fermented.

Fermentation is catalyzed by yeast enzymes that use fermentable sugars as food, and convert them to ethyl alcohol and carbon dioxide under anaerobic conditions. (Under aerobic conditions, the sugars are largely oxidized to carbon dioxide and water.) Fermentation can be monitored with a hydrometer that measures the specific gravity of the wort as it becomes a beer. The beer is then lagered, chill-proofed, and packaged with carbonation.

Wines are made by fermenting sugars in fruits, usually grapes. Although some are fortified to increase the alcohol content, better wines result when the climate during the growing season is moderate and the season long so grapes mature naturally to produce a satisfactory balance of sugar and acidity. Low sugar and high acidity are common characteristics of eastern United States wines, which can be modified by addition of sugar, and by encouraging malo-lactic fermentation to lower the acidity. When the alcohol content exceeds 14%, taxation increases dramatically, as in the case of dessert wines.

Sulfur dioxide is often added to wine. It acts as an antioxidant, suppresses wild yeasts and other microorganisms, and combines with and removes acetaldehyde.

Carbonated wines are classified as either sparkling or bubbling (crackling) wines depending on the means of introducing excess carbon dioxide. Sparkling wine is produced using secondary fermentation. Bubbling wine has carbon dioxide added under pressure.

Distilled alcoholic beverages are usually made from grains, except for brandies, which are distilled from fermented fruits. Congeners are side products of fermentation, which greatly influence the flavor and aroma.

PROBLEMS

1. Give a definition or example of each of the following.
   a. Anaerobic
   b. Hydrometer
   c. Congeners
   d. Malting
   e. Steeping
   f. Kilning
   g. Hops
   h. Limit dextrins
   i. Wort
   j. Enzyme
   k. ATP
   l. NAD
   m. Specific gravity
   n. Lagering
   o. Proteolytic
   p. Humulone
   q. Isohumulone
   r. Lacing
   s. Cling
   t. Esters
   u. Must

2. What is the chief grain used in brewing? Why?

3. Why does yeast ferment glucose? Why does the fermentation go all the way to ethyl alcohol instead of stopping at pyruvic acid?

4. Give one reaction that occurs during the fermentation of glucose. How is it catalyzed?

5. What are adjuncts? What is their purpose?

6. What is the alcohol content of the typical American beer?

7. What step in the production of beer determines whether the final product will be light or dark?

8. What is the source of limit dextrins in beer?

9. Describe three methods that are used to decrease the carbohydrate content of a light beer.

10. What are the technique and strategy of heavy brewing?

11. Advertising for certain light beers commonly indicates a one-third reduction in calorie content compared with regular beer. For the regular and light beers described in Table 17.2, what percentage reduction in calories is attributable to each component and how is each achieved?

12. What is the source of fermentable sugar for wines?

13. Why is sulfur dioxide added to wine?

14. What are Campden tablets? Why are they used?

15. What is the main difference between the procedure used in making white wines and that in making red wines?

16. What do wine makers mean by a good year? What can they do in a bad year to make a good wine?

17. Why is the 14% alcohol level important in wine making?

18. Describe and explain the superiority of sparkling wines over crackling wines.

19. What is the primary source of fermentable carbohydrate used in making Scotch whiskey? Rye whiskey? Gin? Bourbon? Brandy? Rum?

20. What is the origin of the smoky flavor of Scotch?

21. Confirm the figure 147 for the number of steps in the metabolism of glucose under aerobic conditions.

References

1. Briggs, D. E.; Hough, J. S.; Stevens, R.; Young, T. W. *Malting and Brewing Science;* Chapman & Hall: New York, 1981; Vol. I and II.

2. Farkas, J. *Technology and Biochemistry of Wine;* Gordon and Breach: New York, 1988; Vol. 1, p. 94 and chapter 20.

3. Farkas, J. *Technology and Biochemistry of Wine;* Gordon and Breach: New York, 1988; Vol. 2, pp. 421–426.

4. Potter, N. N. *Food Science,* 4th ed.; Van Nostrand Reinhold: New York, 1986; pp. 545–546.

5. Ough, C. S. ''Chemicals Used in Making Wine.'' *Chemical and Engineering News,* January 5, 1987, pp. 19–28.

6. Kirk, R. E.; Othmer, D. F. *Encyclopedia of Chemical Technology,* 2nd ed.; John Wiley & Sons: New York, 1993; Vol. 1, p. 510.

Baking

18

In many ways, the chemistry of baking is similar to that involved in making beer, wine, and distilled products. Yeast enzymes catalyze the fermentation of sugars to produce ethyl alcohol plus carbon dioxide in alcoholic beverages. In baking, it is the CO_2 that is the important product because of its role in leavening (raising) many baked goods.

> **CHAPTER PREVIEW**
> **18.1 Flour**
> **18.2 Leavening**
> **18.3 Changes During Baking of Bread**
> **18.4 Other Baked Goods**

18.1 Flour

GOALS: To understand the distinction between hard and soft wheats, and their relationship to strong and weak flour.
To appreciate the role of gluten in leavened baked goods.
To consider the role of dough ingredients other than flour.

To appreciate the chemistry of baking, it is necessary to consider the characteristics of the grains involved. Whereas barley is well suited for brewing and grapes are ideal for making wines, flour made from wheat is by far the most important grain for making baked goods. Rye flour is also used, usually in combination with wheat flour.

Common wheat accounts for more than 90% of the wheat grown in the United States; its classifications are in Table 18.1, in approximate order of decreasing importance. Winter wheats are planted in the fall and harvested in the spring. They are generally grown in areas with less harsh winters. Spring wheats are planted in the spring and harvested in the fall. They are generally grown in the cooler northern areas (Figures 18.1 and 18.2).

The distinction between *hard* and *soft* (see Table 18.1) relates to the texture of the endosperm, which is the chief location of the starchy material used for making flour (Figure 18.3). White wheats may be either the spring or winter variety; those grown in the United States are mostly soft. Durum wheat varieties are used for making pastas.

Hard and soft wheats have different amounts of protein, a big factor in deciding their suitability for different baked goods. Among the many kinds of wheat proteins, two of the most important are *gliadin* and *glutenin*. Combined in the presence of water, they take on water and form a complex mass called *gluten*. Gluten, with other components of the dough embedded in it, provides the structural framework for trapping gases in leavened baked goods.

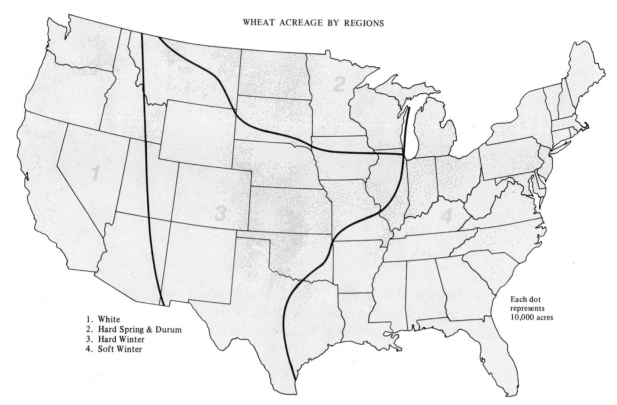

WHEAT ACREAGE BY REGIONS

1. White
2. Hard Spring & Durum
3. Hard Winter
4. Soft Winter

Each dot represents 10,000 acres

Figure 18.1 Wheat acreage by regions. (Courtesy of the U.S. Department of Agriculture.)

a. b. c.

Figure 18.2 Types of wheat: (a) Soft red winter wheat. (b) Hard red spring wheat. (c) Durum wheat.

Table 18.1 Some types and characteristics of common wheat

Classification	No. on Map Fig. 18.1	Main Growing Areas	Percentage Protein	End Products
Hard red winter	3	Central and southern Great Plains	12–16	Bread, rolls
Soft red winter	4	Ohio, Indiana	7–9	Cakes, cookies, pastries, biscuits
Hard red spring	2	Northern Great Plains	12–16	Bread, rolls
White	1	Pacific Coast, New York, Michigan	<7	Cakes, cookies, pastries, biscuits
Durum	2	North Dakota, South Dakota		Pasta products

Gliadin provides elasticity. Glutenin provides strength to the dough mixture (1). The two properties are well illustrated by pizza dough; it undergoes extensive kneading and shaping, and exhibits considerable strength and elasticity. The hard wheats with their high protein content yield flour that produces a high-gluten dough. The high gluten content makes the dough suitable for well-leavened breads and rolls. Low-gluten doughs do not have enough strength to allow them to trap the quantity of gas necessary to produce what we consider normal bread. Flat breads made from soft wheat flours are common in some parts of the world. More frequently, soft flours are used in making cakes, pie crusts and other pastries, and crackers, all of which require little leavening. Some of the strong wheats are usually blended with the weaker varieties. Common white flour is a relatively weak blend.

After the addition of shortening, water, sugar, and yeast, dough is developed fully by mechanical or hand kneading. The process involves a series of complex chemical changes, during which the arrangement of the fibers in the dough proteins is modified. The chief physical changes are improvement of the gas-retaining properties and formation of many

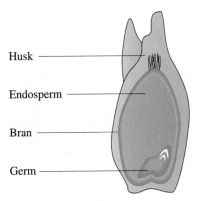

Husk

Endosperm

Bran

Germ

Figure 18.3 A cross section of a kernel of wheat. The starchy endosperm, used for making white flour, makes up more than 80% of its mass. The object of flour milling is to separate the endosperm. Germ and bran are included in whole-wheat flour.

minute air cells. These can later expand when the dough rises during fermentation or chemical leavening.

After bread dough is given an opportunity to rise, it is generally punched down to eliminate large pockets of gas that might appear as large holes in the finished product. At the same time, the yeast cells acquire a fresh supply of oxygen. Home baking generally involves milder handling, so softer flours are generally suitable.

Flour can be modified in several ways before use. Besides blending strong and weak flours, aging improves the baking properties of flour. During aging, a bleaching process occurs due to the oxidation of highly colored pigments called *xanthophylls*. Bleaching and other changes can be accomplished by chemical treatment, which reduces the expense of storing flour to permit natural aging. Chemical changes that improve the dough are complex, but apparently involve the S-H functional groups of certain amino acids in the flour protein.

Chemical aging has often been criticized. It destroys many of the natural nutrients of wheat flour, and even though enriching nutrients may be added, nutritional value is decreased.

Whole-wheat flour includes bran and germ. It is generally considered the most nutritious of all flours; it has been argued, however, that the high cellulose content of bran reduces its nutritional value. Cellulose is not digestible, but it provides bulk, or roughage, that stimulates contraction of the intestines and speeds the passage of food through the system. But too much bulk can carry digestible food out with it.

Other Dough Ingredients

Sugar is added to the dough to provide sweetness and to feed the yeast. During baking above 135 °C, caramelization of the sugar on the surface of the bread is partly responsible for browning of the crust. An interaction between the starch and protein also contributes to browning.

Shortening or cooking oil is another important ingredient in baking dough. Shortness, or richness, of baked products is characterized by a crumbly texture. Oil contributes flavor and assists in the leavening by helping to trap expanding gas bubbles. Wheat germ has a high oil content, although it constitutes only about 3% of the flour. Since oil is prone to spoilage, the germ decreases the shelf life of flour.

Sometimes wheat flour is enriched. Wheat protein frequently lacks certain essential amino acids such as lysine. This is a minor problem when other dietary components make up for the deficiency, but in some areas of the world other protein may be scarce. In these areas, hybrid grains are a partial solution, but enrichment with either amino acids or soybean protein (up to 8% soy flour) may be more practical. Certain vitamins are commonly used to enrich flour even in the United States. They are thiamine (vitamin B_1), riboflavin (vitamin B_2), and niacin (vitamin B_3).

Blends of rye and wheat flour are common in rye and pumpernickel breads. Rye flour alone has too little gluten to permit the normally desired amount of leavening. A typical formula consists of one-third rye flour and two-thirds hard spring wheat flour. Caramel color is often added, especially to commercial pumpernickel breads. Westphalian pumpernickel bread is made from 100% rye flour. Leavening is poor and the bread is very dense, dark, and chewy *(2)*.

Sourdough is used for rye and pumpernickel breads. A *sour* is a culture of old dough containing wild yeasts and other bacteria that provide flavor, aroma, and texture to the finished product.

18.2   Leavening

GOAL: To appreciate the four basic processes that may be involved in leavening.

Leavening action is caused by one or a combination of the following: yeast, chemical leavening, air expansion, and steam production during baking.

Yeast

In metabolizing glucose, yeast cells can function either aerobically or anaerobically (Chapter 17). For maximum production of ethyl alcohol, the anaerobic condition is encouraged, since each glucose molecule yields two molecules of ethyl alcohol and two molecules of CO_2 when oxygen is excluded.

Under aerobic conditions, each glucose molecule can be converted to six molecules of CO_2. For leavening this would be ideal, but it is not attainable since the supply of oxygen in the interior of a bread dough is inadequate to maintain aerobic conditions. Nevertheless, some air is trapped in the interior of bread dough, so metabolism is partly aerobic and partly anaerobic. Either way, much CO_2 is liberated. Even the alcohol formed during anaerobic fermentation can be vaporized during baking, thus contributing to leavening and to the aroma of fresh-baked bread.

Chemical Leavening

Chemical leavening is used in making many baked goods, including cakes, cookies, pies and other pastries, pancakes, and pizza. Baking soda, $NaHCO_3$, is the most important chemical leavening agent. It releases carbon dioxide according to the following reaction with acid.

$$Na^+HCO_3^- + H^+ \rightarrow Na^+ + H_2CO_3 \quad \text{(unstable)}$$
$$\downarrow$$
$$CO_2 + H_2O$$

Baking powder (Figure 18.4) is a mixture of several components. Baking soda is the active ingredient. The other important ingredients are an acid (the source of H^+ for the previous reaction) and inert ingredients (to prevent premature reaction between the acid and the baking soda). Calcium sulfate, calcium lactate, and starch (mostly corn starch) are common inert ingredients.

By law, a baking powder formulation must provide an amount of CO_2 equal to at least 12% of the weight of baking powder, which means that at least 23% $NaHCO_3$ must be present. Most baking powders contain 26–30% $NaHCO_3$.

Several acids are used in baking powder formulations. Combinations of fast- and slow-acting acids can provide leavening action throughout the baking period. The difference is usually due to solubility, since some of the acids are quite insoluble, and therefore inactive, at low temperatures, but become soluble, and therefore active, as the temperature rises.

One of the acids often used in baking powder formulations is sodium aluminum sulfate, which may be written $Na_2SO_4-Al_2(SO_4)_3$ or $NaAl(SO_4)_2$. It is acidic by virtue of the chemical reactions of $Al(H_2O)_6^{3+}$ described in Section 15.1 (see margin). It is relatively slow acting and is added to many baking powders to provide slow but continuous leavening action.

A series of protons released by $Al(H_2O)_6^{3+}$ can be used in a variety of applications, such as lowering soil pH. Details are found in Section 15.1. The release of one proton occurs by the following reaction:

$$Al(H_2O)_6^{3+} \rightarrow H^+ + Al(H_2O)_5(OH)^{2+}$$

Figure 18.4 Baking soda and baking powder.

Several phosphate salts and their analogs are also used as acids. Both $H_2PO_4^-$ and HPO_4^{2-} are weakly acidic ions. Salts of pyrophosphoric acid and tartaric acid also are frequently used to provide acidity in baking powder.

$Na_2H_2P_2O_7$
sodium acid pyrophosphate

$KHC_4H_4O_6$
potassium acid tartrate
(cream of tartar)

Finally, ammonium bicarbonate is a good chemical leavening agent because it releases NH_3 and CO_2 when heated:

$$NH_4^+ HCO_3^- \xrightarrow{\Delta} NH_3 + H_2CO_3 \quad \text{(unstable)}$$
$$\downarrow$$
$$CO_2 + H_2O$$

Both ammonia gas and carbon dioxide provide leavening action, but since ammonia cannot be tolerated in the final product, ammonium bicarbonate is used only in cookies, eclair and cream puff shells, and crackers. Each of these items is baked to such a dry consistency that water-soluble ammonia is completely expelled *(3)*.

Air Expansion and Steam Production

Air trapped in bread dough provides significant leavening as it expands in response to heating. The inside of the dough can reach a temperature of 100 °C (212 °F). It does not exceed 100 °C because steam forms from trapped water at that temperature, and vaporization has a cooling effect.

The Ideal Gas Law

The leavening contribution of the air can be estimated by calculating the volume change predicted by the equation for the ideal gas law, which describes the influence of pressure *(P)*, temperature *(T)*, and the amount of gas *(n*, the number of moles) on the volume of a gas. In this equation, *R* is a constant with the value of 0.082 L-atmosphere/mole-degree.

$$PV = nRT$$

The complex units of R are explained by cross-multiplying the equation to give

$$R = \frac{PV}{nT}\left\{\frac{atm \times L}{mol \times deg.}\right\}$$

Interconversion between Fahrenheit and Celsius can be carried out by the following relation.

$$°F = \frac{9}{5}°C + 32 \text{ or}$$

$$°C = \frac{5}{9}(°F - 32)$$

Interconversion between Celsius and Kelvin can be carried out by the following relation.

K = °C + 273

Consult Section 1.6 for details.

in which *P* has the unit atmospheres, *V* is expressed in liters, and temperature is expressed on the Kelvin (K) temperature scale. For example, 20 °C (68 °F) would equal 293 K.

It is often stated that 1 mol of any gas occupies 22.4 L at standard temperature and pressure (STP). This can be confirmed by using the ideal gas equation; at STP:

$P = 1.00$ atm
$T = 0$ °C $= 273$ K
$R = 0.082$ L-atm/mol-deg (the ideal gas constant)
$n = 1$ mol

The ideal gas equation is given by:

$$PV = nRT$$

so

$$V = \frac{nRT}{P} = \frac{(1)(0.082)(273)}{1.00} = 22.4L$$

Let us use the same relationship to calculate the percentage change in the volume of the air trapped in bread dough when it is heated from room temperature, 20 °C (68 °F) to 100 °C (212 °F). The latter temperature corresponds to steam that would be formed inside the moist dough as it is baked. Since *P, n,* and *R* remain constant, they can be left in the equation as symbols. Writing the ideal gas equation in the appropriate form, we get

$$V = \frac{nRT}{P}$$

Therefore,

$$V_{20°} = \frac{nR}{P} \times 293$$

whereas

$$V_{100°} = \frac{nR}{P} \times 373$$

and therefore

$$\frac{V_{100°}}{V_{20°}} = \frac{373°}{293°} = \frac{1.27}{1} \text{ or a 27\% increase}$$

Notice that it makes no difference what gas is trapped in the dough. Air, composed mostly of N_2 and O_2, and also any gases released by leavening agents are affected in the same way.

The generation of steam from liquid water is also important in leavening. Let us calculate the change in the volume when 1 mol of liquid water at 20 °C is heated to 100 °C to produce steam during baking. The density of liquid water at 20 °C is about 1 g/mL.

For *liquid* water at 20 °C:

$$\text{molar mass of water} = 18$$
$$\text{volume in mL} = \text{mass} \div 1 \text{ g/mL}$$
$$= 18 \text{ mL}$$

For water *vapor* at 100 °C, with P assumed to equal 1.00 atm:

$$V = \frac{nRT}{P} = \frac{(1)(0.082)(373)}{1.00} = 30.6 \text{ L}$$

Using these two volumes, the expansion due to conversion from liquid water to water vapor (including the increase in temperature) is given by the following ratio:

$$\frac{V_{100}(\text{vapor})}{V_{20}(\text{liquid})} = \frac{30.6 \text{ L}}{18 \text{ mL}} = \frac{30,600 \text{ mL}}{18 \text{ mL}} = \frac{1700}{1}$$

This is just one of many illustrations of the tremendous expansion possible when a liquid (or solid) is vaporized. The result is somewhat of an exaggeration in trying to evaluate the importance of steam formation in leavening, since all the water may not vaporize and some may escape. However, even if a small portion is trapped in the dough during baking, it can have a dramatic effect.

18.3 Changes During Baking of Bread

GOAL: To understand oven spring and other events that occur during baking.

After the final fermentation, which is done at room temperature or slightly above, the raised dough is placed in an oven (Figure 18.5). Different recipes specify different oven temperatures, mostly in the range 375–450 °F (190–230 °C).

The first event during baking is a phenomenon known as oven spring, in which the dough rises sharply due to the expansion of gases (CO_2 and air) already present in response to oven heat. As the temperature of the dough increases, the yeast activity increases; the activity finally stops as the yeast organisms are killed.

As the temperature inside the dough reaches 175 °F (80 °C), the ethyl alcohol produced during fermentation vaporizes and aids in the leavening. The inside finally reaches 212 °F (100 °C), which converts some of the moisture into water vapor and aids in leavening.

When the yeast is killed, its enzymes are inactivated, and fermentation stops. This is followed by coagulation of all proteins and gelatinization of starch, which give the expanded air cells a rigid structure and cause the bread to remain leavened even when cooled.

The high temperatures at the surface (crust) of the bread cause browning. Browning is attributed to two processes: caramelization of sugar, and Maillard browning, a complex interaction between protein and starch (4).

(a)

(b)

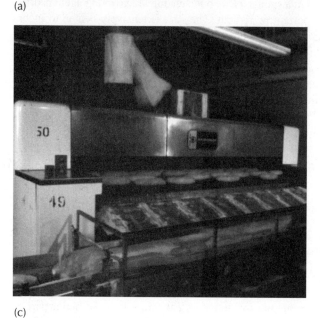

(c)

Figure 18.5   Steps in breadmaking. (a) Complete conditioning and full development of the gluten is assured by tumbling the dough in a heavy-duty mixer. (b) The dough mass is cut, shaped, and "proofed" (allowed to rise) in a steam atmosphere at about 100 °F for about 45 minutes and then (c) transferred to an oven for baking for 15–20 minutes at 400–425 °F.

Throughout baking, a wide variety of complex chemical reactions occur and cause development of many flavor and aroma ingredients, much like the congeners formed in ethyl alcohol fermentation and subsequent aging of alcoholic beverages.

18.4   Other Baked Goods

GOAL: To see some of the strategies involved in the production of baked goods other than bread.

Yeast has little use for leavening other than bread making. Chemical leavening, air expansion, and steam formation are used for other items. A mixture of ammonium bicarbonate (fast-acting) and baking powder is generally the leavening agent for eclair shells, cream puff shells, and popovers. In addition, much of the leavening action results from steam formation. Very hot ovens (450–475 °F) cause rapid formation of steam, ensuring expansion before the crust is set. After expansion is complete, the oven temperature can be reduced to prevent burning. Pie crusts and matzos are two unleavened products.

Cakes

Cakes contain several ingredients, mainly flour, water, shortening, whole eggs, egg whites, flavorings, sugar, and a leavening agent (for example, baking powder). Different types of cakes have widely differing amounts of these ingredients. Cake differs from bread in needing soft, low-gluten flours in relatively small amounts. This causes their soft, crumbly texture.

Pound, sponge, chiffon, and angel food cakes illustrate the diversity available by suitable modification of the recipe. Pound cake is traditionally based on a formulation consisting of 1 lb (454 g) each of flour, butter, whole eggs, and sugar. It has a crumbly texture that results partly from the absence of a leavening agent. The leavening that does occur is due to expansion of steam and air trapped by the shortening and eggs, particularly in beaten egg whites.

Very little shortening is used in a sponge cake, but a chemical leavening agent is often added and egg whites are an important ingredient. They are extensively beaten to incorporate large amounts of air, which contributes to leavening. Even greater amounts of egg whites are normally used in chiffon cakes. Shortening is also present, along with some baking powder for oven spring. Angel food cakes also have a high content of egg whites and contain no shortening or chemical leavening agent. Here again, steam and air expansion are responsible for leavening.

SUMMARY

Wheat flour is the chief ingredient used for baked goods. The most important types of wheat are hard and soft, reflecting a difference in the protein content. When the proteins gliadin and glutenin are combined in the presence of water, a structural network called gluten is formed. It gives dough a strong, elastic quality required for making breads and rolls, which are highly leavened. Hard wheats yield high gluten, strong doughs. Soft wheats are used for making baked goods with little leavening and a more crumbly texture.

During baking, many complex changes occur. The most important is leavening, which can be attributed to

production of gas (yeast, baking soda, steam) and expansion of gases (air, CO_2) on heating. Baking powder is a complex mixture containing baking soda, an acid, and other ingredients. Proper selection of the acidic ingredient permits a controlled production of carbon dioxide.

Steam is often an important cause of leavening. The conversion of liquid water into water vapor has the potential for dramatic leavening. It is the major part of the leavening action in eclairs and cream puff shells, as well as certain types of cakes that depend on the action of egg whites for trapping expanded gases (air and steam).

PROBLEMS

1. Give a definition or example of each of the following.
   a. Gluten
   b. Leavening
   c. Oven spring
   d. Maillard browning
2. What grain is the most important source of flour for baking?
3. What are the differences between hard and soft wheats?
4. What function does gluten perform in baking?
5. What is the purpose of egg whites in many cake recipes?
6. Give three common agents used for leavening baked goods.

7. What are the main ingredients in baking powder? How does baking powder differ from baking soda?

8. When can ammonium bicarbonate be used as a leavening agent?

9. How is leavening achieved in some baked goods even without adding a leavening agent?

10. Why is sugar added to bread dough?

11. Why does bread brown on the outside?

12. What is the function of shortening in bread dough?

13. What is sourdough bread?

References

1. Blanshard, J. M. V.; Frazier, P. J.; Galliard, T. *Chemistry and Physics of Baking;* American Chemical Society: London, 1986; Chapter 2.

2. Matz, S. A. "Modern Baking Technology." *Scientific American* **1984;** *251* (11), 122–134.

3. Matz, S. A. *Bakery Technology and Engineering,* Van Nostrand Reinhold/AVI: New York, 1992; Chapter 3.

4. Shallenberger, R. S. "Browning Reactions, Nonenzymic." In *Encyclopedia of Food Technology;* Johnson, A. H.; Peterson, M. S., Eds.; AVI: Westport, CT, 1974, Vol. 2, pp. 136–139.

Dairy Products

19

Milk and its several by-products are important foods. Not only whole milk, but skim milk, half-and-half, casein in creamer powders, yogurt, and cheese are common consumer products. The composition of milk and the chemistry of processing it are considered in this chapter.

19.1 Milk

GOALS: To appreciate the importance of the various components of milk and how they vary in content from one milk product to another.
To appreciate the special role of the milk proteins, which are of prime importance in the chemistry of cheese making.

Whole cow's milk contains approximately 87% water, 5% carbohydrate, 4% fat (butterfat), 3% protein, and 1% minerals.

Carbohydrate

The carbohydrate in the milk of humans, cows, and other mammals is lactose. It is a disaccharide with a β linkage between the monosaccharide units glucose and galactose. It is about one-sixth as sweet as common sugar (sucrose).

Human milk contains about 7.5% (m/v) lactose, whereas less-sweet cow's milk contains 4.5% (m/v). Certain mammals totally lack carbohydrate in their milk. Seals, sea lions, and walruses have milk with a high fat content, 35% (m/v) or higher. Whales and polar bears also have a low-lactose, high-fat milk. It is very much like heavy cream and permits these animals to tolerate very cold temperatures.

Many people are unable to digest milk properly because of lactose. The enzyme β-*galactosidase*, or simply *lactase*, is necessary for the digestion of milk sugar. Babies and young children have an active gene that allows them to synthesize lactase, but the gene

Humans cannot metabolize cellulose due to lack of the enzyme cellulase, which cleaves the β linkages between the glucose units of the polysaccharide. See Chapter 9 for details.

415

often ''turns off'' by adulthood. In fact, tolerance seems to be the exception rather than the rule for all but northern Europeans and their descendants, including most white Americans.

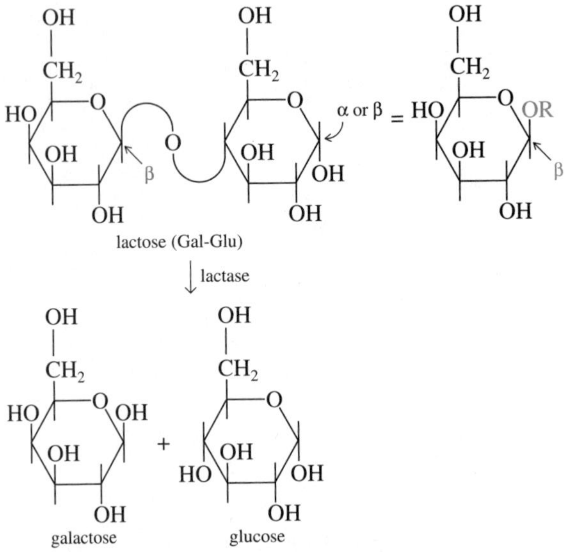

lactose (Gal-Glu)

↓ lactase

galactose   +   glucose

A 7.5% (m/v) concentration corresponds to 7.5 g of lactose per 100 mL. See Section 7.4 for a discussion of percentage as a unit of concentration.

When the enzyme is present, it is in the small intestine. When absent, animals show intolerance to lactose. Specifically milk, the main source of lactose, causes digestive tract distress. The common symptom is diarrhea. It may be due to a net flow of fluid into the intestine to dilute the lactose as it builds up, or to fermentation by the enzymes of digestive tract bacteria. Production of CO_2 and organic acids may cause bloating, cramps, and diarrhea.

Powdered milk has been shipped worldwide to help deal with malnutrition. Although it is good for small children, adult intolerance is being recognized. A separate and complicating problem is that unsanitary water may be used to reconstitute the milk.

Clinically, intolerance is readily diagnosed by administering an oral dose of lactose and testing the blood glucose level after about 15 min. The blood glucose level rises in response to lactase-catalyzed breakdown of lactose to produce glucose. No rise of glucose level indicates that the lactose has not been digested.

The origin of lactose intolerance is interesting. Many bacteria develop lactase only when grown in the presence of lactose. Studies have shown that genetic information in DNA provides these bacteria with the ability to adapt and synthesize lactase when lactose is provided, or to be energetically efficient by not synthesizing lactase when it is not needed.

Human lactase is not adaptive like the bacterial enzyme. In fact, one interesting study was conducted on six Nigerian medical students who had no tolerance for lactose. They were fed increasing amounts of lactose over 6 months, and developed tolerance. They no longer suffered from diarrhea; but their blood glucose level did not increase after a dose of lactose. Their tolerance was attributed to the development of bacteria in the gastrointestinal tract *(1)*.

One way to overcome lactose intolerance is to treat milk with lactase during commercial processing. Another is to add the enzyme to milk in the home and refrigerate the

Figure 19.1

Lactose intolerance can be dealt with using such products as Dairy Ease and Lactaid. The first of these is available in both tablet form, to be taken along with or before eating a product containing lactose, and solution form. Several drops of the solution form are added to a container of milk so that lactose is broken down before consumption.

milk for a prescribed time to give the enzyme time to work. Lactase is available commercially in tablet and solution forms (Figure 19.1). The tablets are taken when or before lactose is consumed. The solution is added to a container of milk so that much of the lactose is broken down before it is consumed.

Fat

The more general term *lipids* includes both fats and fatlike substances—for example, steroids—all of which are predominantly hydrocarbon. See Chapters 21 and 26 for details.

Centrifuges are used as commercial cream separators.

The fat, or lipid, content of milk is commonly referred to as butterfat. It exists as a suspension of minute droplets or globules that rise to the top of the liquid to form a cream layer, because it has a low density. When the cream layer separates, lactose and most of the minerals remain in the milk layer. The protein content of milk is distributed between the two layers.

Homogenization is the process whereby fat globules are reduced in size so they remain suspended. Milk is forced through a nozzle at high pressure (Figure 19.2). Churning (rapid stirring) has the opposite effect. When cream is skimmed off the milk and churned, globules aggregate into large particles that coagulate to form butter, which contains about 80% fat.

Figure 19.2   A milk homogenizer.

The health aspects of
dietary fats are discussed in
Section 26.2.

The amount of fat in milk and milk by-products determines the physical and flavor characteristics. Heavy whipping cream is at one extreme, with a fat content of about 35%, light cream usually contains at least 18%, and the product known as half-and-half contains 10–12%. Low-fat milk may contain up to 2% butterfat, whereas skim milk contains less than 0.5%. The fat content of cheese varies, often exceeding even that of heavy cream.

Minerals

Various metal salts are present in trace amounts in milk. Calcium is the most important one, not only because of its nutritional value, but also because it has a part in the production of other dairy products, such as cheese. Most of the calcium present in milk is tightly bound to protein.

Protein

Milk protein is often called *casein,* but only about 80% actually is casein. The rest is a mixture of soluble proteins that include albumins and globulins, collectively known as the *whey proteins.*

Casein is actually four different proteins, referred to as α, β, γ, and κ (kappa) caseins. Of these, α, β, and κ are the most abundant, composing roughly 50%, 33%, and 15% of casein, respectively. The α and β caseins have similar and unusual properties, being insoluble in the presence of calcium ions at the concentration of Ca^{2+} normally found in milk. The κ casein is soluble, and it also keeps the α and β caseins in solution.

19.2   Cheese

GOAL: To understand some of the variables that are possible in the production of cheeses.

We now come to the section that "Little Miss Muffet" has been waiting for. Her snack of "curds and whey" is well known. What was she eating? Probably it was not curdled milk, which is recognized by anyone who has handled any badly spoiled milk or cream. The lumpy material is called the *curd.* The curd forms by *coagulation* (also called *clotting*) of α and β caseins. In spoiled milk, lactic acid causes the clotting. Lactic acid is formed by bacterial fermentation of the monosaccharides released from lactose. Lactic acid or any other acid causes κ casein to dissociate from the other caseins. Consequently, α and β caseins become insoluble and precipitate as a curd.

During clotting, some of the lactose and most of the butterfat are trapped in the curd. The liquid that remains is the *whey.* It contains much of the lactose, plus minerals, plus the soluble noncasein proteins known as *whey proteins.* Most of the calcium is retained in the curd.

Clotting is the basic chemistry involved in making cheese. Cheese is merely curd formed and then treated under controlled conditions. Milk could be aged long enough to give bacterial enzymes time to develop and function to produce lactic acid, which would cause the milk to curdle. But this approach is time-consuming and can encourage other undesirable changes. Instead, it is common practice to add rennet to the milk. Rennet is a crude extract of calf gastric juices and contains the enzyme *chymosin,* more commonly known as *rennin.* (Rennin can also be obtained from several microorganisms.) Rennin is a proteolytic (protein-cleaving) enzyme that attacks κ casein. This cleavage of κ casein causes it to dissociate from the complex with α and β caseins, and clotting occurs *(2).*

A legend regarding the origin of cheese making tells of an Arabian merchant who traveled across the desert on a hot day. He carried a supply of fresh milk in a pouch that was made from the lining of a sheep's stomach. When he stopped to drink some of the milk, he found that it now consisted of a thin watery liquid and a soft curd with a pleasing taste. It is supposed that some rennin present in the sheep stomach had caused the curd formation.

Adding lactic acid or other acids to promote curd formation also causes a loss of certain salts. The calcium in milk is mostly fixed in a calcium caseinate-calcium phosphate complex that becomes trapped in the curd. When the pH is acidic, much of this calcium becomes soluble and ends up in the whey, making the cheese less nutritious. Also, the acidic pH restricts growth of microorganisms essential for proper *ripening* of the cheese to achieve the desired flavor and texture.

Cheese takes many forms: hard cheese, soft cheese, cottage cheese, cream cheese, Roquefort cheese, sharp cheese, Cheddar cheese, American cheese, processed cheese, and many more. Each one has particular characteristics of texture, fat content, flavor, and aroma determined by chemical differences. Rennin alone cannot provide such a wide range of end products. The varieties owe their properties to differences in production methods. The steps in cheese making may include some or all of the following, with major alternatives and variables available at every step.

Step 1.   Coagulation with acid or rennin or both

Variables

a. Type of milk
   (1) Cow, sheep, other
   (2) Whole or skimmed milk
   (3) Raw or pasteurized milk

Step 2.   Cutting and pressing curd

Variables

b. Temperature

c. Rennet addition after some lactic acid fermentation

a. Higher pressure for harder varieties

b. Amount of whey removed

c. May follow inoculation with microorganisms

Step 3.   Adding bacteria or mold culture and ripening

Variables

a. Type of microorganism

b. Culture may be added before coagulation

c. Culture on surface or injected

d. Time ranges from zero to many months

e. Addition of salt to inhibit growth of undesirable microorganisms

f. Texture (softening during long-term ripening due to extensive cleavage of proteins)

One of the first chemical changes before and during ripening is fermentation of any trapped lactose. Most of the microorganisms involved have the necessary enzymes. This is important not only for lactose-intolerant individuals but also for people on low-carbo-hydrate diets, who can use cheese as an important component of the diet.

After and accompanying lactic acid fermentation, many other complex chemical changes occur. They make each cheese different, since each microorganism provides its own complement of enzymes that catalyze a multitude of chemical reactions. Some interesting studies have been done on the chemical changes induced when curd is inoculated with *Penicillium roquefortii*. In Roquefort and most other cheeses, some of the fat is broken down to fatty acids such as octanoic acid (see Chapter 21). Many of these acids have distinctive flavors and aromas, and are convertible to other substances that also contribute flavor and aroma. The following reactions show the formation of 2-heptanone, an easily recognizable flavor and odor ingredient in blue and Roquefort cheeses.

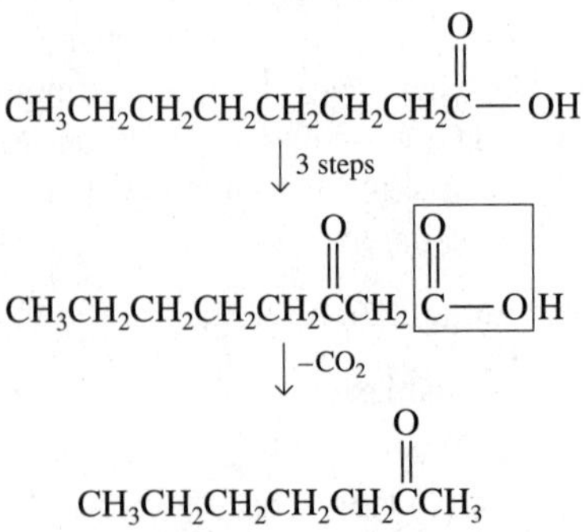

$$CH_3CH_2CH_2CH_2CH_2CH_2CH_2\overset{\displaystyle O}{\overset{\|}{C}}\!-\!OH \quad \text{octanoic acid}$$

$$\downarrow \text{3 steps}$$

$$CH_3CH_2CH_2CH_2CH_2\overset{\displaystyle O}{\overset{\|}{C}}CH_2\boxed{\overset{\displaystyle O}{\overset{\|}{C}}\!-\!O}H \quad \text{3-oxo-octanoic acid}$$

$$\downarrow -CO_2$$

$$CH_3CH_2CH_2CH_2CH_2\overset{\displaystyle O}{\overset{\|}{C}}CH_3 \quad \text{2-heptanone}$$

Types of Cheeses

Cottage cheese is the unripened curd from skimmed milk. It is produced by acid or by the combined action of acid and rennet. Since most of the fat is removed before clotting, cottage cheese is high in protein but low (0.5%) in fat content. A somewhat tastier product, creamed cottage cheese, is made by adding cream to bring the fat content to 2–6%.

Cream cheese is a soft, unripened cheese. Unlike cottage cheese, it is made from a mixture of milk and cream. The final product generally contains more than 30% fat.

Cheddar cheese is a hard cheese ripened by bacteria. Like all hard cheeses, it is curdled at high temperature and the whey is pressed out of the curd under high pressure. This treatment lowers the moisture content to 30–40%, which is typical of hard cheeses and improves the keeping properties. In contrast, soft cheeses may contain as much as 75% water. Coagulation temperatures are generally lower for soft cheeses and the product is more perishable. Cottage cheese is one soft cheese whose shelf life is particularly low because it is also unripened. The length of the ripening is quite variable for Cheddar cheeses. Short ripening periods result in formation of mild cheeses. Sharp cheeses often require more than a year to develop their full flavor.

Roquefort cheese is made from sheep's milk by using the mold *P. roquefortii.* **Blue cheese** is made from cow's milk.

Swiss cheese, such as Emmenthaler, is a hard cheese that has been ripened by bacteria; these bacteria generate CO_2 and cause the formation of eyes (holes) in the cheese (Figure 19.3).

Processed cheese is a blend of cheeses, mostly Cheddar, that results in a more consistent product. The components are finely ground, melted together, and reformed (Figure 19.4). Coloring, seasoning, and other additives may be introduced in the processing.

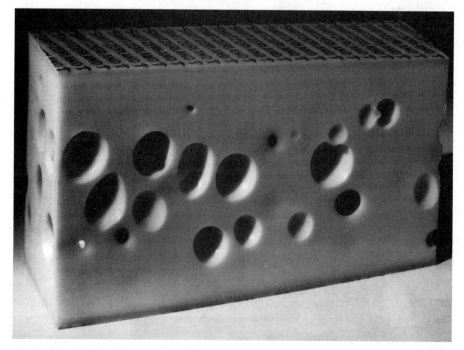

Figure 19.3   Swiss cheese.

(1)   (2)   (3)

(4)   (5)   (6)

Figure 19.4

Making processed cheese. (1) The process begins with fresh milk to which a starter (microorganism) and rennet are added. The milk is allowed to curdle in a vat until it is the consistency of custard and is ready to cut. The curd is then stirred and heated, and the whey is allowed to drain (2) while the curd is simultaneously packed against the sides of the vat. The curd is later shredded and placed in 500-lb barrels for curing and aging, after which the cheese is cut into workable pieces. (3) Cheeses of various stages of flavor development are selected and blended to produce the desired flavor and texture in the finished product. (4) The mixture then moves into a steam cooker, where it is heated to 165 °F (75 °C) or higher, and other ingredients are added. (5) The hot cheese then moves over a quick-chill roll. (6) Narrow ribbons of cheese flow from the roll onto a belt to be cut into slices.

19.3 Other Milk Products

GOAL: To consider some other products that are made from milk, and the use of sugar as a preservative.

Many other dairy products are consumed throughout the world. Yogurt, koumiss, kefir, sour cream, and cultured buttermilk are examples of products made by fermentation.

Yogurt

To make yogurt, skimmed milk is generally fermented by two microorganisms, *Lactobacillus bulgaris* and *Streptococcus thermophilus*. First the milk is heated to boiling to kill all microorganisms; inoculation then provides only the particular set of enzymes of the desired microorganisms. Yogurt itself is often used for inoculation.

Since fermentation removes lactose from milk, yogurt can be consumed by lactose-intolerant persons.

Koumiss and Kefir

The fluid milk products koumiss and kefir are actually alcoholic beverages. Both are made by double fermentation consisting of the combined action of a lactic acid bacterium and an alcohol-fermenting yeast. Consequently, these products are both somewhat sour and slightly carbonated. Both drinks are consumed heavily in Russia. Very little is consumed elsewhere.

Koumiss is made from mare's milk, which has a very low casein content and does not tend to clot. Cow's milk is not suitable since it clots easily. Acidity from lactic acid may run as high as 1.2% and alcohol as high as 2.5% in koumiss. Kefir is a slightly effervescent beverage made by fermentation of milk from goats, sheep, or cows. It contains less fat than koumiss since a soft curd that forms is removed before bottling.

Sour Cream and Cultured Buttermilk

Sour cream is produced by the action of a bacterial culture on pasteurized cream. The final product normally contains about 18% butterfat. Buttermilk is the fluid left after churning milk or cream to produce butter. It has a variable but normally low fat content. Commercial buttermilk is frequently produced by culturing with a lactic acid bacterium.

Evaporated and Condensed Milk

Unfermented milk products also are sold commercially. Both evaporated milk and condensed milk are produced by evaporation of some of the water of whole milk. To meet federal standards, evaporated milk must contain at least 7.9% butterfat and 25.9% total solids. It is sterilized by heating. Condensed milk must contain not less than 8.5% fat and 28% total milk solids, but is unsterilized. It would be quite perishable, so "sweetened condensed milk" is made by adding sucrose to give about 18% sugar content before evaporation and about 42% after evaporation. The high sugar content acts as a preservative (Section 20.2).

SUMMARY

Milk is a versatile food that can be consumed directly or converted to a wide variety of products. Its main components are water, lactose, butterfat, protein, and minerals (mostly calcium). The majority of the world's adult population cannot tolerate lactose, as they lack the enzyme lactase.

The butterfat content of milk can be handled in several ways. Milk may be homogenized by breaking up the fat droplets to keep them suspended, or the fat may be allowed to rise and form a cream layer that is skimmed off. Cream may be marketed in a variety of forms, ranging from heavy cream (35% fat) to half-and-half (10–

12% fat). Skim milk contains less than 0.5% fat. Cream may also be churned into butter.

The proteins in milk consist of 80% casein and 20% whey proteins. Curd can be concentrated to make cheese or semisolid products. Caseins separate out in the curd and trap most of the calcium and fat. Rennin can be used to initiate clotting by attack on the κ casein, which keeps α and β caseins soluble. Lactic acid, from fermentation of the lactose, also can cause curd formation.

Curd formation is the initial step in making cheese. Many variations in technique during and after this step are responsible for the wide variety of cheeses available. Among the most important variables are the choice of milk (whole, skim), the pressing of the curd (high pressure yields hard cheese), the ripening agents (bacteria, molds, or none), and the time of ripening.

Other important by-products of milk are yogurt, koumiss, kefir, sour cream, and buttermilk, all formed by fermentation with microorganisms. Evaporated and condensed milk are unfermented products made by evaporation of milk. Condensed milk is usually preserved by adding sugar.

PROBLEMS

1. Give a definition or example of each of the following.

   a. Lactose
   b. Lactase
   c. Butterfat
   d. Curd
   e. Whey
   f. Rennet
   g. Cottage cheese
   h. Cheddar cheese
   i. Processed cheese
   j. Cream cheese
   k. Yogurt
   l. Sour cream
   m. Koumiss
   n. Kefir
   o. Buttermilk
   p. Homogenization

2. What is the composition of whole milk?

3. Why are many people unable to digest milk?

4. What is the major difference between lactose and maltose?

5. What is the role of rennin in cheese formation?

6. Why is lactic acid fermentation in cheese making important to people with intolerance to milk?

7. After the clotting of whole milk, where is butterfat found?

8. How does cottage cheese differ from Cheddar cheese?

9. Why is sugar added to condensed milk?

10. Cheese has been described as a "cured curd." What is wrong with that description?

11. What do you suppose Little Miss Muffet was eating?

References

1. Kretchmer, N. "Lactose and Lactase." *Scientific American* **1972,** 227 (4), 70.

2. Kosikowski, F. V. "Cheese." *Scientific American* **1985,** 252 (5), 88.

Food Preservation

20

Part Four studies food processing in various forms. In some cases, processing means converting an already appetizing food into another popular form, such as conversion of milk to cheese, which increases the versatility of the original food. In other cases, an unpalatable food can be processed into an appetizing one, such as converting wheat into bread. Processing milk into cheese includes the important element of food preservation; several cheeses have a very long shelf life, far longer than milk. In many techniques, such as curing and pickling, chemical alterations are carried out primarily for preservation.

Controlling or modifying biochemical processes that cause spoilage of foods is a serious health and economic concern. We have seen how microorganisms can be beneficial. Now, in considering food spoilage, we see how to prevent some of their harmful effects.

> **CHAPTER PREVIEW**
> **20.1 Food Microbiology**
> **20.2 Techniques of Food Preservation**
> **20.3 Chemical Alteration**
> **20.4 The Use of Additives**

20.1 Food Microbiology

GOALS: To appreciate the nature of some microorganisms associated with foods and food poisoning.
To understand the use of pressure cookers in combating spore-forming microbes.

Many volumes have been written about food microbiology. Biology itself has been described as "a kind of superchemistry" *(1)*. This suggests the complexity of the chemistry involved in food microbiology and the difficulty in trying to comprehend the many effects, favorable and unfavorable, that microorganisms have on food.

Some microorganisms make their presence known in obvious ways. Examples are the molds *Aspergillus* and *Neurospora crassa,* which grow on bread. The lactic acid bacteria *Lactobacillus* and *Streptococcus* may cause milk to sour and eventually curdle. Other microorganisms may be more subtle and also much more dangerous. The most important example is *Clostridium botulinum,* which produces the dangerous condition called *botulism.* The anaerobic microbe thrives inside sealed containers. It is not harmful itself, and does not multiply if it is consumed; its danger lies in its ability to produce a toxin, *botulin,* under anaerobic conditions before it is eaten. The toxin, which is a protein, causes paralysis, often resulting in death. It interferes with transmission of nerve impulses, like the action of certain snake venoms. It has been said that 14 oz. of botulin would be enough to kill the entire world population *(2).*

Clostridium botulinum is abundant in soils, which accounts for its relative significance as a cause of food poisoning. Its danger is increased by its tendency to form spores. A spore is a dormant state of a microorganism, formed in response to conditions that are unsuitable for growth. When suitable growth conditions are restored, these spores are able to germinate, multiply, and produce toxins.

The *C. botulinum* spores are particularly resistant to conditions that are lethal to spores of most other microorganisms. Most microbes are destroyed by heating and maintaining boiling temperatures for a few minutes. But due to the heat resistance of *C. botulinum* spores, it is common practice to cook certain foods at temperatures above boiling in a pressure cooker (Figure 20.1) before canning. The spores are completely destroyed in 10 minutes at 250 °F (121 °C) (Table 20.1). The toxin is destroyed by even a few minutes of treatment at normal boiling temperatures, but many canned foods are simply warmed before eating. This mild treatment may allow the toxin to survive. Thorough cooking would prevent exposure, but proper canning technique using high-temperature (pressure) cooking is the best way to prevent botulism for foods that are susceptible to *C. botulinum.*

Figure 20.1 A pressure canner.

Table 20.1   Temperature and pressure in a pressure cooker

Temperature		Actual Pressure		Pressure in Excess of Normal	
°C	°F	Pounds per sq in.	Atm	Pounds per sq in.	Atm
100	212	15	1.0	0	0
109	227	20	1.35	5*	0.35
116	240	25	1.70	10*	0.70
121	250	30	2.05	15*	1.05

Note: A common unit of pressure is the atmosphere (atm). Normal atmospheric pressure is approximately 15 lb per square inch.

* Typical settings used in pressure cookers.

Salmonella is another bacterium often associated with food. Its many species are readily destroyed by heating at 140 °F (60 °C) for 15–20 min. Salmonellae are aerobic organisms that cannot multiply in a closed container (such as a can). Since they do not discolor or add a bad flavor to food, their presence is usually not obvious and they can multiply after being eaten. The bacteria produce a toxin only after the microorganism has infected the body. For this reason, exposure to *Salmonella* is known as *food infection.* Botulism is regarded as *food poisoning* since the toxin is present in the food. Although not as hazardous as botulin, the *Salmonella* toxin causes unpleasant gastrointestinal symptoms for about a week. Many mild cases are diagnosed as a viral attack. Mortality is very low, less than 1%.

Staphylococcus organisms are also frequent causes of food poisoning. Like *C. botulinum,* the danger is due to production of a toxin before the food is eaten. Cream-filled pastries are probably the best-known food that is subject to *Staphylococcus* infection. This microorganism does not form spores, so it is readily destroyed at boiling temperatures; but cream pastries and several other foods require gentle cooking and are never heated to such high temperatures.

The toxin of the strain most often associated with food poisoning is unusually heat-stable, so a food that is highly contaminated may cause poisoning even after thorough cooking. Again, the primary symptom is intestinal upset, unpleasant but rarely fatal. Proper sanitation and refrigeration are the best means of avoiding both *Salmonella* and *Staphylococcus* infections. Refrigeration slows the growth of *Salmonella* and prevents development of toxin by *Staphylococcus.*

Many disease-carrying microorganisms are less common but nevertheless serious causes of infection that can be transmitted in food or water. *Trichinosis* is a food-borne disease that is not categorized as food poisoning. *Trichinella spiralis,* a parasite often present in raw pork, is killed by thorough cooking, but if it is ingested intact, it can establish itself in the intestinal tract and cause severe symptoms.

Ptomaine poisoning is an old-fashioned and incorrect term for food poisoning. Ptomaines are diamino compounds formed during the decay of dead, protein-containing substances. Although they are very foul-smelling, they are not toxic and are not associated with any common form of food poisoning. The best known ptomaines, *cadaverine* and *putrescine,* are formed during putrefactive decay in which certain amino acids (from proteins) suffer loss of CO_2 (decarboxylation).

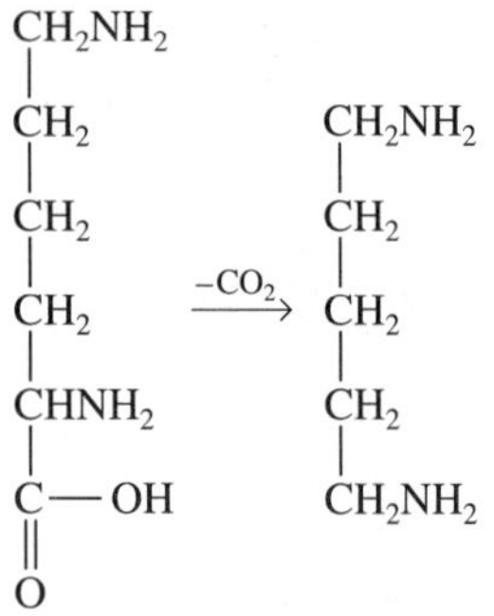

$$
\begin{array}{ccc}
\text{CH}_2\text{NH}_2 & & \\
| & & \text{CH}_2\text{NH}_2 \\
\text{CH}_2 & & | \\
| & & \text{CH}_2 \\
\text{CH}_2 & \xrightarrow{-\text{CO}_2} & | \\
| & & \text{CH}_2 \\
\text{CH}_2 & & | \\
| & & \text{CH}_2 \\
\text{CHNH}_2 & & | \\
| & & \text{CH}_2\text{NH}_2 \\
\text{C}\!-\!\text{OH} & & \\
\| & & \\
\text{O} & & \\
\text{lysine} & & \text{cadaverine}
\end{array}
$$

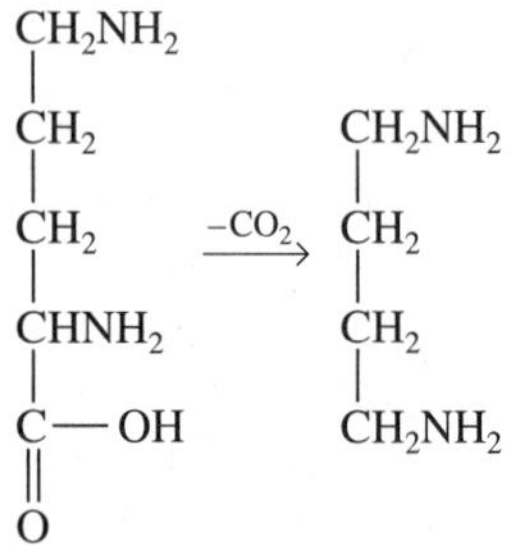

$$
\begin{array}{ccc}
\text{CH}_2\text{NH}_2 & & \\
| & & \text{CH}_2\text{NH}_2 \\
\text{CH}_2 & & | \\
| & & \text{CH}_2 \\
\text{CH}_2 & \xrightarrow{-\text{CO}_2} & | \\
| & & \text{CH}_2 \\
\text{CHNH}_2 & & | \\
| & & \text{CH}_2\text{NH}_2 \\
\text{C}\!-\!\text{OH} & & \\
\| & & \\
\text{O} & & \\
\text{ornithine} & & \text{putrescine}
\end{array}
$$

20.2   Techniques of Food Preservation

GOAL: To appreciate the strategies of several food-preservation techniques.

The major preservation techniques are shown in the following list; some are combined in handling specific foods.

- Cold storage
- Freeze-drying
- Heat processing: canning
- Irradiation
- Chemical alteration: fermentation, pickling
- Additives

Cold Storage

Foods are commonly preserved by storing them at low temperatures to suppress the growth of decay microorganisms. The low temperature slows down chemical processes, and when frozen, the water present in food is unavailable as a growth medium for microorganisms. It is often desirable to inactivate the enzymes present in microorganisms and the food itself before freezing, by heating briefly.

Freezing should be performed properly for best results. If the process is slow, large ice crystals form and break many of the cells of the food. On thawing, the tissue collapses and the texture of the food is mushy. If food is quick-frozen, the ice crystals are smaller and tissue disruption is reduced. For example, fruits must be quick-frozen to retain their natural texture. More often, they are canned.

Ice cream is greatly affected by the rate of freezing. When originally frozen, it has a smooth, creamy texture that is gradually lost after repeated opening and closing of a container. Each exposure to room temperature causes it to undergo some melting, and when it is refrozen, the ice crystals grow larger.

Finally, cooked food must be refrigerated. If food is kept cold, *Salmonella* cannot multiply. The danger of infection and multiplication arises when cooked food, for example, meat, is allowed to remain at room temperature for several hours. These conditions are ideal for multiplicaton of bacteria.

Freeze-Drying

The idea behind dehydration and freeze-drying is the same as that for freezing. Microorganisms cannot function without water. Rapid freezing and exposure to high vacuum remove all the water. Texture, appearance, palatability, and nutritional value are changed very little by this form of dehydration. The cost of operating the vacuum equipment that removes water is reflected in the cost to the consumer. Freeze-dried food is more expensive than frozen or fresh food.

Heat Processing

Commercial and home canning techniques differ somewhat, but the principle is the same: to control the growth of spoilage organisms. Commercial canning dates back to the early 1800s. As food spoilage was understood better, new techniques were developed. A substantial improvement is shorter cooking times at higher temperatures to retain fresher flavor, firmer texture, and higher nutrition.

Home canning may involve pressure packing (above boiling temperatures), hot packing (in boiling water), or cold packing. In commercial hot packing, a microorganism known as PA (putrefactive anaerobe) 3679 is used as a test of the effectiveness of heat processing. This organism is an extremely resistant spore-forming microbe. Even very resistant *C. botulinum* is more susceptible to heat than PA 3679, so safety is ensured if conditions are adequate to destroy the latter *(3)*.

In general, the less acid the food, the longer the heating time and the higher the temperature required for safe canning. Acid is a natural preservative in food. In general, fruits are high in acid and can often be cold-packed safely. Vegetables contain relatively little acid and are hot-packed or pressure-packed (Table 20.1). Meats are preserved by pressure packing, freezing, or curing.

Low pH (high acidity) reduces survival time of spores. Spores of *C. botulinum* in fruit can be completely destroyed in about 1.5 hours at normal boiling temperatures. Unfortunately, this is longer than generally called for in home canning recipes *(4)*.

Tomatoes are a particular concern since they are frequently processed by home canners. Tomatoes are normally thought to be very acidic, which would allow mild treatment for killing harmful bacteria and their spores. Unfortunately, however, many of the new hybrid varieties have low acidities, making them more susceptible to spore formation and stability. Consumers Union recommends that they be pressure packed at 250 °F at 15 lb per square inch (psi) for 15 minutes to ensure safety *(4)*.

Jams and jellies are cold-packed. Their stability is due to natural acidity in the fruit or fruit juice used. High levels of sugar also act as a preservative. Canned fruit, for example, is often packed in a heavy sugar syrup.

Sweetened condensed milk also has a high sugar content (Section 19.3). It is a paradox that foods can be protected from spoilage by an excellent food source for microorganisms, sugar. This phenomenon can almost be equated to drying the food. At very high sugar concentrations, water is drawn out of the microbe by the sugar and is unavailable as a growth medium.

Irradiation

Irradiation has been termed ''cold sterilization.'' It can be used to kill pathogenic microorganisms such as *Salmonella, Staphylococcus,* and *Clostridium.* Organisms that do not form spores are very susceptible to radiation. Low doses also control *Trichinella* and tapeworms.

Low doses (up to 100 kilorads) of radiation have been used to stop sprouting of root crops such as potatoes, carrots, onions, and garlic (Figure 20.2). Dividing cells in the sprouts are very sensitive to radiation and are killed. Low doses also delay ripening of some fruits. Higher doses (100–1000 kilorads) have been used to delay spoilage of meats, poultry, and fish. Radiation also prevents insects from reproducing in grains, fruits, and vegetables after harvest.

Some factors limit the usefulness of radiation. Spores are not destroyed easily. The chemical alterations induced may cause undesirable changes in a food. In fact, when the radiation dose is high enough to destroy all spores, the odor and flavor of the food are often altered. High-protein foods are particularly susceptible. The texture of meats and vegetables may be altered. Irradiation of milk reportedly causes changes similar to those of heating *(5).* An off flavor develops in both cases.

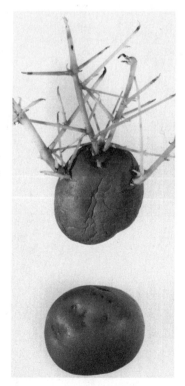

Figure 20.2 Prevention of sprouting using radiation. The potato at the top is 8 months old and was untreated. The potato at the bottom is the same age, but was irradiated with 20,000 rad of γ radiation.

Public concern is often expressed over the danger of inducing radioactivity in foods. Neutrons can sometimes induce radioactivity, but γ, x$-$, and β radiation do not. Only γ rays are used in food preservation.

20.3 Chemical Alteration

GOALS: To learn the chemistry of vinegar making.
To appreciate some chemical changes in the aging and curing of meat.
To understand the techniques used to make pickles and sauerkraut.

The production of vinegar, the curing of meats, and pickling are important examples of chemical alterations to preserve food.

Vinegar

The production of vinegar was considered something to be avoided in our study of alcoholic beverages (Chapter 17). The formation of acetic acid causes wine to taste sour.

$$CH_3CH_2OH \xrightarrow[\text{NAD} \quad \text{NADH}]{} CH_3\overset{\displaystyle O}{\overset{\displaystyle \|}{C}}{-}H \xrightarrow[\text{NAD} \quad \text{NADH}]{} CH_3\overset{\displaystyle O}{\overset{\displaystyle \|}{C}}{-}OH$$

ethyl alcohol   acetaldehyde   acetic acid

Under aerobic conditions, the regeneration of active niacin occurs by the following equation (See also Section 17.1)

$$NADH + O_2 \xrightarrow[\text{3 ADP} \quad \text{3 ATP}]{} NAD + H_2O$$

On the other hand, vinegar, even wine vinegar, is required by law to contain at least 4% acetic acid. Making vinegar is a two-step process. The first is anaerobic, alcoholic fermentation (Section 17.1) carried out by yeast. The second step is then carried out under aerobic conditions by *Acetobacter* to ensure an adequate supply of NAD, active niacin.

The acetic acid in vinegar is used as a preservative against the growth of microorganisms in foods. Both sour foods (for example, sour pickles and sauerkraut) and those with sweetening added to counter the sour taste (sweet pickles or watermelon rind) contain vinegar.

Meat

Some interesting chemical reactions go on in processing and preserving meats for both short-term consumption and long-range storage. Fresh meat undergoes some chemical changes that, if properly controlled, may greatly affect its characteristics and keeping quality. Cold storage and freezing are obvious ways of preventing spoilage. Pickling, curing, and corning all control changes that could occur, in essentially the same way.

The first event after slaughter is the onset of lactic acid fermentation in the meat. Before the kill, the circulatory system of an animal continually carries oxygen (bound to the protein hemoglobin) from the lungs to the tissues. Once the animal is killed, the oxygen supply is cut off and the tissues become anaerobic. For a time, all the natural enzymes necessary for either aerobic or anaerobic breakdown of glucose remain active.

In animal cells, glucose is stored as the polysaccharide glycogen (Section 9.2). The amount of glycogen present determines how much lactic acid can be formed. This is particularly significant, since lactic acid, like acetic acid in vinegar, acts as a preservative. It suppresses the growth of microorganisms, thus slowing spoilage and increasing the meat's keeping quality. The condition of the animal determines the amount of glycogen present in the tissues, particularly muscle and liver. Unlike fat, glycogen can be stored in only a limited quantity. Consequently, a well-fed, rested animal will have a higher gly-

Figure 20.3
Products containing papain.

cogen level than one that is hungry or has been exercised before slaughter. Meat from the first will keep much better due to the higher lactic acid content that develops. The pH will drop from about 7 to 5.4–5.7, at which point fermentation will stop even if some glycogen remains.

An animal killed in a hunt, a fish caught after a long fight, and a bull killed in a bullfight have low glycogen levels and produce very little lactic acid. Conversely, feeding before slaughter effectively builds up the glycogen supply, so beef cattle are kept in feed lots and fed corn in the weeks before slaughter.

The second big change after slaughter occurs over several days. The meat becomes more tender. Immediately after death, it is soft and hard to work with. Rigor mortis soon sets in. At this point the meat is tough and remains so for a day or two. If it is allowed to hang somewhat longer, natural proteolytic (protein-cleaving) enzymes break down some of the protein and cause it to become tender. The efficiency of the process may be increased and the required time decreased by injecting enzymes before slaughter. This provides uniform distribution of the enzymes throughout the bloodstream in all of the tissues, with resultant tenderization.

In the home, tenderizing can be mimicked by using a marinade that contains *papain,* a proteolytic enzyme obtained from the unripe fruit of papaya (Figure 20.3). Papain cleaves some of the protein fibers that can cause toughness. Most recipes call for a brief treatment, usually about 15 min before cooking, to allow the papain to penetrate the meat. Much of the tenderization occurs in the early stages of cooking, when the higher temperatures increase the activity of papain, and before the enzyme is inactivated.

Color is an important characteristic of meat. The familiar red color is often taken as an indication of freshness. Some darkening may be attributed to water loss at the surface, which causes the pigments to become more concentrated. More complex chemical changes are responsible for browning meat. The protein myoglobin, Mb, acts as a storage depot for oxygen in muscle tissue. It is a dark protein because iron, as Fe^{2+}, is tightly bound to it and is also bound to a highly colored compound called *heme.* Myoglobin can exist as simple Mb, oxymyoglobin, MbO_2, or metmyoglobin, Mb^+, as shown in the accompanying sequence of reactions. When a large piece of beef is sliced, the exposed surface is initially purple due to Mb. This quickly changes to bright red due to oxygenation to form MbO_2.

$$Mb(Fe^{2+}) \xrightarrow{\text{O}_2} MbO_2(Fe^{2+})$$

myoglobin (purple)   oxymyoglobin (bright red)

$$\downarrow \text{oxidation}$$

$$Mb^+(Fe^{3+})$$

metmyoglobin (brown)

Preserving the red color is a goal in handling meats. Success seems to depend on keeping the myoglobin as oxymyoglobin, which prevents oxidation (conversion of Fe^{2+} to Fe^{3+}) to the brown metmyoglobin. This presents a strange situation in which it is necessary to provide an adequate supply of oxygen to prevent oxidation.

A problem arises when meat is cut and packaged. Aerobic bacteria on the meat will consume some of the oxygen, causing the surface oxymyoglobin to become deoxygenated and thus susceptible to oxidation to metmyoglobin. The solution is to wrap meat in a breathing film that allows oxygen to enter and CO_2 to leave *(3)*.

Meat is sometimes cured to preserve it and to produce desired new forms, such as corned beef. A principal ingredient in curing meats is sodium nitrite, $NaNO_2$. In the presence of acid (usually either citric or acetic), the nitrite, NO_2^-, is converted to nitrous acid, HNO_2, which reacts with myoglobin to produce nitrosomyoglobin, MbNO; this substance contributes the pink color normally associated with cured meat.

The main ingredient in curing meats is salt, NaCl. Its primary function is to control the type of fermentation by determining which microorganisms can grow and cause the many changes in curing. Lactic acid bacteria are more tolerant of salt, so that lactic acid and other acids act as preservatives.

Hams are usually cured after pumping salt solution (NaCl and $NaNO_2$) into the meat through a main artery so that salt is distributed evenly throughout (Figure 20.4). Quick

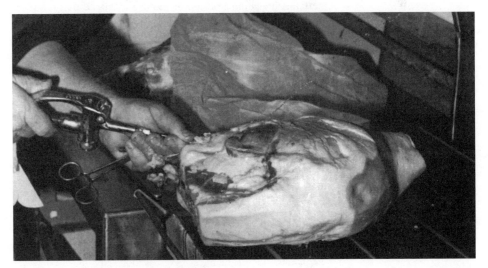

Figure 20.4   A worker is shown pumping a ham with a solution of brine (NaCl) containing other ingredients (including sodium nitrite for color) that are necessary to ensure that the ham will cure properly. The solution is pumped into the main artery through a needle so that it is distributed throughout the entire ham, ensuring a uniform cure. The ham is pumped until it increases in weight by a specific amount. Notice the salt solution spraying out of many cut blood vessels at the surface of the ham.

curing is carried out commercially by pumping salt solution into the meat through 100 or more hollow needles.

The final product of the cure is relatively dry and dense and has a high salt concentration. To be made appetizing, it must be reconstituted or freshened before eating by repeated soaking in warm water; this removes much of the salt and makes the meat juicier. Sugar and other spices are also present to control the flavor of the final product. Pork shoulders, hams, and briskets of beef are often cured in this way. For beef, the curing, or *corning,* normally takes 4–5 weeks.

Meats are commonly smoked after curing, both for flavor and as a means of preservation. It is believed that the smoke dehydrates the surface of the meat and also deposits a surface coating that consists of formaldehyde and other chemical compounds that prevent the growth of microorganisms.

Sodium nitrite has another function besides providing a familiar and pleasing color. Like sodium chloride, it is a preservative, particularly against the development of the botulism microbe in unrefrigerated canned meats and fish. There is also a negative aspect to using nitrite. Nitrite combines with certain nitrogen-containing organic compounds to form N-nitroso compounds called **nitrosamines,** many of which are known to cause tumors. The simplest possible nitrosamine forms from dimethylamine, as follows:

$$\underset{\text{dimethylamine}}{CH_3 - \underset{\underset{H}{|}}{N} - CH_3} + \underset{\underset{\text{(from NaNO}_2\text{)}}{}}{HNO_2} \longrightarrow \underset{\underset{\text{(a nitrosamine)}}{\text{N-nitrosodimethylamine}}}{CH_3 - \underset{\underset{N=O}{|}}{N} - CH_3}$$

No direct evidence links nitrosamines to human cancer, but laboratory animals are consistently susceptible to their effects. Our principal dietary sources of nitrosamines from nitrites are cured meat, sausage, frankfurters, smoked fish, canned fish, and cooked bacon. Americans ingest an average of 2.6 mg of nitrite daily, but the significance of that figure is not at all clear. It also has been estimated that 8.6 mg of nitrite is ingested daily from one's own saliva. The source of this is bacterial action that converts dietary nitrate, NO_3^-, to nitrite, NO_2^-. Thus the total human exposure to nitrites is apparently far greater than the dietary intake. Despite this load, not all persons develop cancer, so some genetic and/or environmental factors are yet to be uncovered.

Pickles and Sauerkraut

An important use of fermentation and preservation is in making pickles from cucumbers. As in meat curing, salt is the major ingredient used to control fermentation in pickling. The concentration of salt used is expressed in *degrees salometer.* These units are based on the concentration of NaCl in water at room temperature, where saturated NaCl, which is 25% NaCl, is defined as 100° salometer. Therefore, 1% NaCl equals 4° salometer.

In making pickles, an 8–10% NaCl (32°–40° salometer) solution is added initially to the cucumbers and fermentation is allowed to proceed for about a week. Dialysis occurs, in which water flows out and salt enters. Once again, lactic acid fermentation is a main pathway. After the first week, the salt concentration is increased about 1% each week for 4–6 weeks, up to 16% salt (64° salometer). When fermentation is complete, the salted, fermented cucumber is called a *salt stock,* and the pickles are well preserved. Before the pickles are packaged for eating, the salt stock is freshened by repeated washing with water to leach (dialyze) out most of the salt. If the product is to be sour pickles, vinegar is added to the freshened salt stock to give at least 2.5% acid. For sweet pickles, a sweet, spiced

Figure 20.5
Sweet pickles and dill pickles. After 4-6 weeks of fermentation, salt is washed out. Vinegar is added for sour pickles, or sweet vinegar plus spices for sweet pickles. Dill spices are added when making dill pickles.

vinegar solution is added. For dill pickles, dill seed and leaves are introduced during the fermentation or to the freshened salt stock of sour pickles. (See Figure 20.5.)

Sauerkraut is fermented cabbage. Salt (2–2.5 lb per 100 lb of cabbage) is added to shredded cabbage, and fermentation is allowed to go on for several weeks. The pickling process for sauerkraut has been thoroughly studied and found to go through three distinct stages of fermentation, carried out by three microorganisms that work in succession.

20.4   The Use of Additives

GOALS: To appreciate the range of additives for maintaining the appeal and safety of foods.
To recognize the history and present status of sweeteners.

A wide variety of additives are used to preserve foods. Many years ago, sugar, salt, and smoke were all that were available. But modern times have brought modern tastes and modern problems.

First is the practical problem that most food is grown or produced far from where it is consumed. Therefore, preservation is necessary and often achieved using additives. Second, in all advanced countries, prepared foods and convenience foods are abundant and preferred. A can of peas or soup, a box of flour or cake mix, or a bottle of salad oil might not be thought of as convenience food. But as you consider the effort that goes into creating each product, you quickly realize the convenience of time saved by the consumer, who need only visit a supermarket and fill a shopping cart. All of these convenience foods are dependent on additives to make them safe and appetizing.

Additives also can be used to cover up defects in food products. In some instances, this could increase profits for food processors and chemical companies that produce the additives. But even with the legitimate use of additives, concern exists about possible hazards.

Much testing is required to determine the safety of food additives. Often its results are controversial and open to interpretation depending on the perspectives of the interested parties. Such controversies rage regardless of whether testing shows an additive to be safe or hazardous. If an additive is concluded safe based on animal tests, critics question if the results are valid for humans. It also could be argued that adverse effects on humans might result from dietary interactions not part of an animal study. Long-term effects may not be obvious for many years; such effects are not normally studied in animals.

Then there is the case of an additive that appears to be safe in normal use but is banned due to the results of animal studies using extremely heavy doses of the chemical. The reasoning behind large doses in animals relates to their relatively short life. It is presumed that if large doses cause tumors in a significant number of animals during their short lifetime, a smaller dose may be expected to do the same over the much longer human lifetime. In addition, it is practical to study only a small number of test animals, whereas a large percentage of humans could be adversely affected. The classic example of a high-dose study was that used to test the safety of cyclamate as an artificial sweetener (see section on Sweeteners).

The following list illustrates the many classes of additives.

- Anticaking agents
- Clarifying agents
- Coloring or bleaching agents
- Flavorings
- Foam regulators
- Incidental additives
- Leavening agents (Chapter 18)
- Nutrients
- Preservatives
- Texture modifiers: emulsifiers, gellers, thickeners, humectants

Anticaking Agents

Anticaking agents may be added in small amounts to finely powdered or crystalline food. Calcium silicate, calcium stearate, sodium aluminosilicate, and sodium ferrocyanide are some examples. These compounds act by either or both of two mechanisms. In the first, illustrated by sodium aluminosilicate, the anticaking agent acts as a dehydrating agent. It incorporates water molecules into its crystal structure, preventing the effects of water on the food product.

Figure 20.6 helps to understand this action. Silica has a complex, three-dimensional, repeating structure in which each silicon atom is surrounded by four oxygen atoms. Each oxygen atom acts as a bridge between two silicon atoms. Part (a) is a two-dimensional cross section of the structure and (b) is a portion of the full three-dimensional structure. The symbol $(SiO_2)_n$ indicates that silica contains two oxygen atoms for every silicon atom, and the structure contains a large number of repeating units, shown by n. In sodium aluminosilicate, some of the silicon atoms may be replaced by aluminum atoms. Each aluminum will carry a negative charge (anion) and require some cation (for instance, Na^+) to balance the charge. These charges give the crystal a great affinity for water, and the large holes in the crystal enable water molecules to enter and become trapped. Thus sodium aluminosilicate minimizes caking of a salt.

In a second mechanism of anticaking, sodium ferrocyanide acts on common table salt, NaCl, by altering the shape of the salt crystals. Normally, NaCl forms cubical crystals that can pack tightly together and are thus susceptible to caking. In the presence of small amounts of sodium ferrocyanide, star-shaped crystals form and do not cake so readily.

Clarifying Agents

The acceptability of certain liquid foods is decreased by turbidity. In brewing, the final product is chill-proofed by adding proteolytic enzymes that cleave the proteins in beer

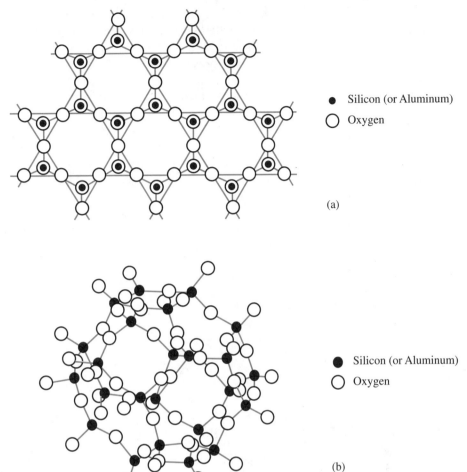

● Silicon (or Aluminum)
○ Oxygen

(a)

Figure 20.6
Silica, or aluminosilicate.
See text for discussion.
(From Cotton, F. A.;
Wilkinson, G. *Advanced
Inorganic Chemistry,* 5th
ed. Copyright 1988 by John
Wiley & Sons. Reprinted by
permission of John Wiley &
Sons.

● Silicon (or Aluminum)
○ Oxygen

(b)

(Chapter 17). Beer might become cloudy if these proteins were to come out of solution during chilling. Fruit juices, wine, and vinegar are other food products that may form sediments or suspensions if not clarified.

Coloring and Bleaching Agents

Coloring and bleaching agents ensure that a food product has whatever color is considered appetizing. Color can be controlled by proper packaging of uncured meat and by sodium nitrite in cured meats.

The orange pigment in carrots, β-carotene or provitamin A, is a natural substance often used as a color additive. For example, it gives the yellow color to margarine. Several other natural and synthetic pigments are used as food colorings.

Flavor Enhancers

Flavor enhancers have little or no taste of their own but accentuate the flavor of other foods. The best known is monosodium glutamate (MSG). It particularly enhances the flavor of meat-containing foods, which accounts for its use in frozen meat, fish, and chicken dinners; dry soup mixes, canned stews and sauces; and ham and chicken salad spreads.

The parent compound, glutamic acid, is one of the 20 common amino acids in proteins. It is theorized that MSG works either by increasing the sensitivity of the taste buds, or by stimulating saliva flow.

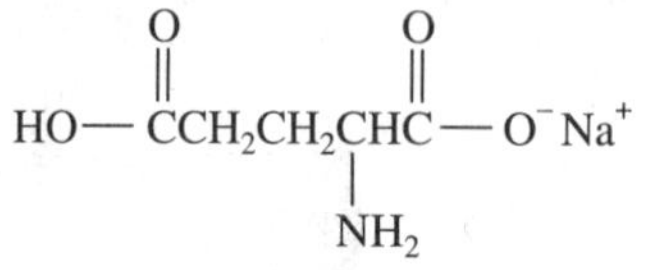

$$\text{HO}-\overset{\overset{\displaystyle O}{\|}}{\text{C}}\text{CH}_2\text{CH}_2\underset{\underset{\displaystyle \text{NH}_2}{|}}{\text{CH}}\overset{\overset{\displaystyle O}{\|}}{\text{C}}-\text{O}^-\text{Na}^+ \qquad \text{monosodium glutamate (MSG)}$$

Much controversy surrounds MSG. Some critics claim that the enhanced flavor masks a reduction in amount of more expensive nutritious ingredients. Concern is expressed over the use of MSG in baby foods, apparently to make them more appealing to mothers. Extensive studies have not shown MSG to be a hazardous substance, but its use in baby foods has been voluntarily discontinued by manufacturers.

In addition, MSG has been associated with symptoms called ''Chinese restaurant syndrome,'' so named because of the relatively heavy use of MSG in Chinese foods. Large quantities may cause a burning sensation in the neck, forearms, chest, and head.

Other flavor enhancers for meat and poultry are the sodium salts of two 5'-nucleotides, 5'-IMP and 5'-GMP, known as sodium inosinate and sodium guanylate. They have been marketed under such names as Mertaste, Ribotide, and Corral. They are more expensive than MSG and more effective (20 times as great).

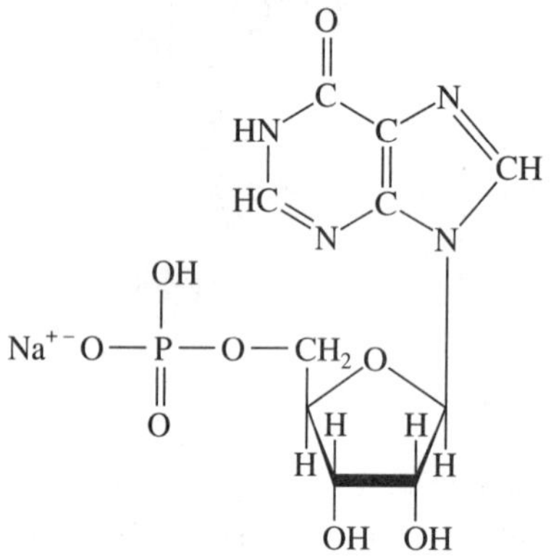

sodium inosinate 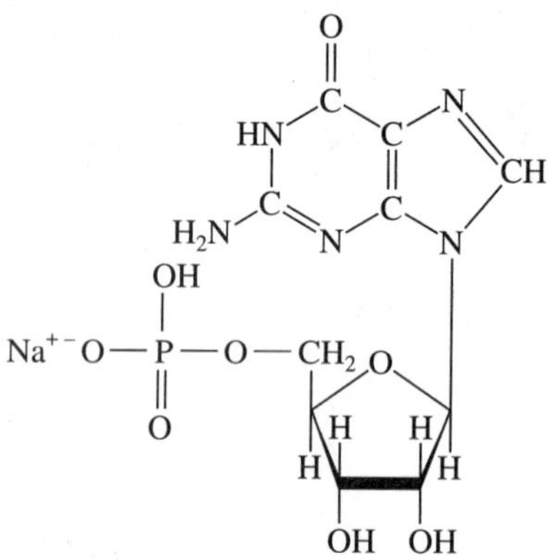   sodium guanylate

Maltol and ethyl maltol are flavor enhancers that are used in fruit-, vanilla-, and chocolate-flavored foods.

Sweeteners

Artificial sweeteners, often described as nonnutritive sweeteners, are the most studied, discussed, and treasured food additives. Overweight people, diabetics, and everyone else who wants to limit sugar intake are prospective consumers of these products.

Cyclamate

Calcium **cyclamate** was one of the most appealing of many artificial sweeteners marketed over the years. Its sweetness is about 30 times as great as sucrose, and it is not metabolized.

In 1962, Metrecal (with cyclamate) was the largest-selling food substitute for dieters. Cyclamate also was added to many other foods and beverages. In 1969, after more than 15 years of heavy use, the Food and Drug Administration removed cyclamate from the GRAS (generally recognized as safe) list and banned its use. Laboratory tests revealed an abnormally high frequency of bladder cancer in rats fed large amounts of the sweetener.

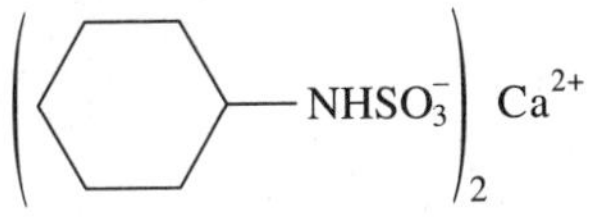

calcium cyclamate

Critics of the ban argued that the amount fed to the rats was so excessive that the test results were meaningless. Test animals are bred and kept under carefully controlled conditions (including diet) to prevent exposure to infectious diseases and other chemicals. Those favoring the ban argued that human exposure to diseases and chemicals outside the controlled laboratory environment might even enhance the dangers of cyclamate for humans.

It was also reported that injecting cyclamate or cyclohexylamine (CHA), a metabolic breakdown product of cyclamate, into fertilized chicken eggs caused deformities of embryos. It was also found that CHA could cause breakage of chromosomes in the cells of laboratory animals.

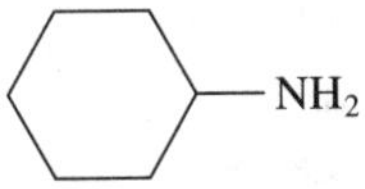

cyclohexylamine

Bladder cancer was the only disease cited in removing cyclamate from the GRAS list. Under the terms of the 1958 Delaney Amendment of the Food, Drug, and Cosmetic Act, anything found to cause cancer in humans or in animals should not be *added* to food. However, laboratory studies and their conclusions have been severely criticized and questioned ever since the ban became effective, because the amount of cyclamate used in the study was so large.

Saccharin

After cyclamate was banned, **saccharin** became the predominant artificial sweetener. Its sweetening effect was first reported in 1879, and it has been produced commercially in the United States since 1902.

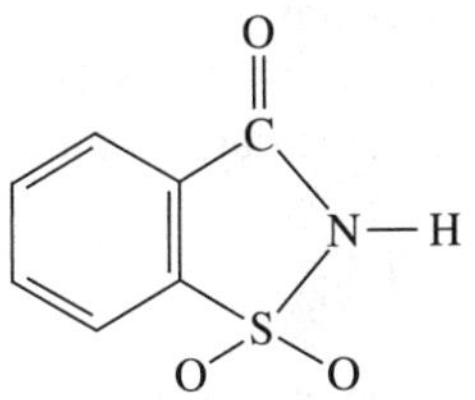

saccharin

Saccharin is about 300 times as sweet as sucrose. Before cyclamate was banned, artificially sweetened products often contained a ratio of 10 parts cyclamate to 1 part saccharin. This provided 50% of the sweetness from each ingredient *(5)*. The combination also masked the bitter aftertaste of saccharin.

Figure 20.7
Three well-known tabletop sweeteners containing artificial sweeteners. Sweet'N Low and Sugar Twin contain saccharin. Equal contains aspartame.

Saccharin is still used in foods and beverages, cosmetics, mouthwashes, toothpastes, drugs, and animal feed. As a particularly well-known tabletop sweetener in tablet, powder, and liquid forms, it is used in cooking and in coffee, tea, and cereal (Figure 20.7). Other uses include powdered juices and drinks, canned fruits, dessert toppings, cookies, gum, jams, candies, ice cream, and puddings.

In early 1977 the FDA announced its intention of banning saccharin as a food additive, because a Canadian study concluded that it caused bladder cancer in rats. This was not the first time that concern had been expressed over the product's safety. In 1972 the FDA removed it from the GRAS list. But instead of banning saccharin, the FDA issued an interim food additive regulation that froze uses at then-current levels and recommended certain limitations on its use. The interim regulation was issued pending completion of further studies, because no definitive studies linked saccharin and cancer at that time.

The Canadian study reported in early 1977 was severely criticized as invalid because the experimental design was like that used with cyclamate. In fact, the approach was a standard procedure in which laboratory animals (in this study, rats) are fed large doses of the suspected carcinogen. To be specific, 100 rats were fed saccharin at a level of 5% of their total diet for 2 years. Three of the first 100 rats developed bladder tumors. One hundred of these rats' offspring were fed the standard dose of saccharin for exposure both before and after birth. Of the 100 offspring studied, 14 developed bladder tumors. Thus, concern over the hazard to future generations is understandable.

On the other hand, critics believe the entire study is absurd. The 5% dietary level may not sound excessive until it is projected to a similar human intake of diet soda of 800 twelve-ounce bottles a day for a lifetime. Some critics even suggested that drowning would be far more likely than the development of cancer if such an experiment were conducted on humans. Other critics said saccharin-containing foods might be labeled, such as, *Warning: The Canadians have determined that saccharin is dangerous to your rat's health!*

Estimates showed that a person who used saccharin in coffee, drank diet soft drinks, and ate food sweetened with saccharin would ingest about 0.1% saccharin, that is, about 50 times less than the daily amount per body weight used in the Canadian study.

Aspartame

Marketed under the trade name NutraSweet, **aspartame** was first approved for use by the FDA in 1974. However, several challenges kept it off the market as a major competitor until it finally received approval for use in soft drinks in 1983. Like saccharin, aspartame is popular as a tabletop sweetener in products such as Equal.

The popularity of aspartame has led experts to be more particular about language. The term *artificial sweetener* is reserved for noncaloric or nonnutritive sweeteners—for exam-

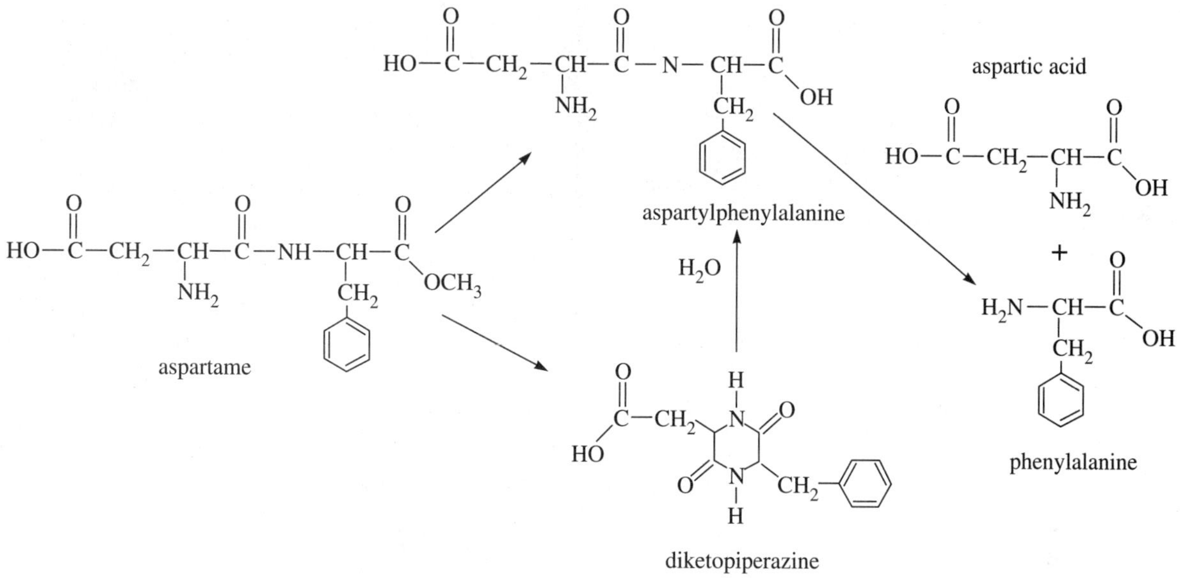

Figure 20.8   Aspartame and its breakdown products.

ple, cyclamate and saccharin. In contrast, aspartame is a *nutritive* sweetener with a caloric value approximately the same as sugars, at 4 Calories per gram. Aspartame and other nutritive sweetners are also called *alternative* sweetners. Aspartame's sweetness is about 180 times as great as that of sucrose, so aspartame can be present in amounts far lower than sucrose. This lower calorie content can be targeted to diabetics and others who must restrict their sugar intake.

Aspartame is an intriguing chemical compound sometimes symbolized APM. This emphasizes the three component parts of the molecule: aspartic acid, phenylalanine, and methyl alcohol. None has any sweetness as an individual compound or when combined physically. When combined chemically, as shown in Figure 20.8, the molecule exhibits a surprising sweetness that was discovered by accident in 1965 *(5, 6)*.

Aspartic acid and phenylalanine are amino acids present in all proteins. Aspartic acid is regarded as safe for all persons. Phenylalanine can be a problem for those with phenylketonuria (PKU), a deficiency of an enzyme required to metabolize phenylalanine. Therefore, all products containing aspartame contain warnings for these individuals, although it is generally believed that normal use does not supply enough phenylalanine to be a danger even for people with PKU.

Similarly, even the methyl alcohol released when aspartame is metabolized is not considered dangerous in the amounts involved in ordinary use. Dietary methyl alcohol also arises from some common fresh fruits and vegetables.

Not only is aspartame subject to breakdown by metabolism, it is also prone to hydrolysis (cleavage by water) to its component parts (Figure 20.8) when exposed to water. A cyclization reaction is possible also to form diketopiperazine, which cleaves to form the same three components of aspartame.

High temperatures encourage aspartame breakdown, as do high pH conditions. Thus aspartame is not effective as a sweetener of cooked foods, but is quite suitable in presweetened cereals; powdered soft drinks, coffee, and tea; chewing gum; gelatin; and pudding mixes. It has a lifetime of many months in bottled or canned soft drinks because these items are sufficiently acidic.

Thus far, the only significant drawback to aspartame is its cost, which is somewhat greater than the cost of sweetening with sucrose to the same level of sweetness. For this reason, saccharin is likely to remain a serious competitor, even though aspartame is generally regarded as more appealing. The superior heat stability of saccharin also gives it a distinct edge for use in foods that require heating.

Xylitol

Another alternative sweetener of commercial importance is the 4-carbon sugar **xylitol,** with sweetness about equal to that of sucrose. Its niche as a sweetener is in the prevention of dental caries, or tooth decay (Chapter 22), and as a suitable alternative for diabetics.

$$
\begin{array}{c}
CH_2OH \\
| \\
H-C-OH \\
| \\
HO-C-H \qquad \text{xylitol} \\
| \\
H-C-OH \\
| \\
CH_2OH
\end{array}
$$

Xylitol does not raise blood glucose levels. Of course, neither do the artificial sweeteners such as aspartame or saccharin. But xylitol provides bulk like other carbohydrates, so diabetics experience satisfaction from eating it.

The noncariogenic (antidental caries) effect results from the inability of bacteria in dental plaque to ferment xylitol. Fermentation would lead to the formation of acids that could attack tooth enamel. Xylitol is therefore used in sugarless chewing gum, candies, throat lozenges, cough syrups, toothpastes, and chewable tablets such as vitamins and other children's medications.

Other Additives

Foaming and **antifoaming agents** are used in certain food products. Dextrins and cellulose gums in canned whipped cream make it foam out of an aerosol can. Silicone antifoaming agents are added to liquids such as pineapple juice, which would otherwise tend to foam vigorously when shaken or poured.

Incidental additives that often find their way into food are pesticide residues and drugs such as antibiotics and growth hormones. The latter are used to improve the general health and development of farm animals. Another class of incidental food additives are substances that can leach out of packaging materials, such as antioxidants, driers, drying oils, plasticizers, and release agents.

Nutrients such as amino acids and vitamins are added to some foods. Amino acids are the building blocks for proteins (Section 8.6). **Preservatives** take different forms depending on the type of food. Table 20.2 provides a list of preservatives in the order in which they are covered in Part Four and some of the foods in which each is used.

Texture modifiers control the texture of some foods and improve shelf life by helping to prevent changes. Moisture is often a critical variable. Sometimes water can aid the breakdown of cell structure and cause foods to become soggy. Other times, the loss of moisture accompanies deterioration.

Table 20.2 Some important food preservatives

Preservative	Used In
Sulfur dioxide	Alcoholic beverages (including as antioxidants)
Sugar	Condensed milk, canned fruits, jams and jellies
Salt	Fermented foods (meats, cucumbers, sauerkraut)
Smoke	Meats
Acids	Fruits, cured meats, pickles
Antioxidants	Fats

Fat is a softener and lubricant in many baked goods and helps to retain freshness. Starches and gums are used as gellers and thickeners to increase viscosity of sauces, gravies, and puddings. Highly hydrophilic substances such as glycerol and sorbitol are often used as *humectants*—that is, they help to retain moisture.

$$\underset{\text{OH}}{\overset{\text{CH}_2}{|}} - \underset{\text{OH}}{\overset{\text{CH}}{|}} - \underset{\text{OH}}{\overset{\text{CH}_2}{|}} \qquad \text{glycerol}$$

$$\underset{\text{OH}}{\overset{\text{CH}_2}{|}} - \underset{\text{OH}}{\overset{\text{CH}}{|}} - \underset{\text{OH}}{\overset{\text{CH}}{|}} - \underset{\text{OH}}{\overset{\text{CH}}{|}} - \underset{\text{OH}}{\overset{\text{CH}}{|}} - \underset{\text{OH}}{\overset{\text{CH}_2}{|}} \qquad \text{sorbitol}$$

Section 9.2 contains a discussion of the properties of starch (amylose and amylopectin) as thickening agents.

Emulsifying agents stabilize water-oil emulsions such as mayonnaise and salad dressings (Section 21.4). The mode of action of emulsifiers will be considered further with soaps (Section 22.2).

SUMMARY

Food processing often converts fresh food into a stable form that can be stored safely for extended periods. Otherwise, microorganisms could cause spoilage of food and in some cases make it dangerous. *Clostridium botulinum* that causes botulism is most dangerous; infections from *Salmonella* and contamination by *Staphylococcus* and *Trichinella* also are serious. Ptomaine poisoning is a misnomer.

Cold storage, freezing, and dehydration are all convenient techniques for preventing or slowing the development of microorganisms. Heat-processing, or canning, kills microorganisms in food; the food container is then sealed to prevent further contamination. Spores make the botulism microorganism resistant to heat, although acidity increases its susceptibility to heat. Consequently,

high-acid foods often can be safely cold-packed. Large amounts of sugar present or added are an effective preservative.

Ionizing radiation can sometimes be used for food preservation, but undesirable side effects are possible. Prevention of sprouting is one beneficial effect of radiation.

In many cases, some degree of fermentation can preserve foods. Methods for curing meats (for example, corned beef), and for making pickles and sauerkraut parallel those used to ripen cheese, with salts to control microorganisms and acidity (added or produced) to prevent spoilage. Meat from animals that are well fed and rested before being slaughtered is most easily preserved. The various colors observed in meats can be attributed to myoglobin (purple), oxymyoglobin (red), metmyoglobin (brown), and nitrosomyoglobin (pink). In uncured meats, it is necessary to encourage oxygenation to avoid oxidation.

Although sometimes overused, additives greatly increase the safety, utility, shelf life, and quality of many foods. Controversy continues over the ban on cyclamates, but saccharin and aspartame have filled the void as alternative sweeteners. Nonnutritive sweeteners, called artificial sweeteners, are not metabolized and therefore have no caloric value. Aspartame is metabolized and has caloric value, but the quantity used is far less than the amount of sucrose required to give the same degree of sweetness. Xylitol also is metabolized but has special properties that make it useful in diabetic foods and for prevention of dental caries.

The long list of additives also includes flavor enhancers, anticaking agents, foam regulators, texture modifiers, and preservatives.

PROBLEMS

1. Give a definition or example of each of the following.

   a. Salometer
   b. Flavor enhancers
   c. Ptomaines
   d. MSG
   e. Botulism
   f. Vinegar
   g. Trichinosis
   h. Putrefaction
   i. NutraSweet
   j. PKU

2. Why is it necessary to use a pressure cooker when canning some foods?

3. What effect does acidity have on spore survival?

4. What food is known to cause a staphylococcal infection?

5. What is the value of quick freezing?

6. What is the strategy behind dehydration and freeze-drying?

7. How does a high sugar content act to preserve food?

8. Why does meat from a well-fed, rested animal keep much better than meat from a hungry, exercised animal?

9. How does a meat marinade work?

10. What is the cause of the bright red color of fresh meat? What is the cause of the pink color of cured meats, for example, ham or corned beef?

11. Why do some experts regard the use of $NaNO_2$ in cured meats as hazardous, and others consider it essential?

12. What is the purpose of $NaCl$ in preserving meat?

13. The overall production of vinegar requires first anaerobic and later aerobic conditions. Explain.

14. What is sauerkraut?

15. Name the preservative in each of the following: wines, jams and jellies, meats, pickles.

16. How is a pickle produced from a cucumber?

17. How does an anticaking agent work?

18. Describe the aspartame molecule.

19. Describe the advantages of the alternative sweeteners saccharin, aspartame, and xylitol compared with sucrose. Give an example of foods in which each is suitable or unsuitable as a substitute for sucrose, and explain. What are the advantages and disadvantages of each substance compared with the other two?

References

1. Lehninger, A. L. *Biochemistry;* Worth: New York, 1979; p. 4.

2. "Botulism Research." *Chemistry* **1964,** *37* (1), *27.*

3. Allen, J. C.; Hamilton, R. J. *Rancidity in Foods;* Elsevier: New York, 1989; pp. 228–230.

4. "Canning Tomatoes? Here's Something to Worry About." *Consumer Reports,* August **1974,** p. 569

5. Nabors, L. O.; Gelardi, R. C. *Alternative Sweeteners;* Marcel Dekker: New York, 1986, p. 77 and Chapter 3.

6. Grenby, T. H. *Progress in Sweeteners;* Elsevier: New York, 1989; Chapter 1.

Fats and Oils

21

S o far we have examined three main categories of biological molecules: proteins (Chapter 8), polynucleotides such as DNA (Chapter 8), and carbohydrates (Chapter 9). The structure and properties of fats and fat products will round out our discussion of foods.

Margarine and butter are among the most familiar food fat products. In the 1930s, sales of butter were about 90% of the table spread market, whereas today about 3 times more margarine than butter is consumed. Both products contain about 80% fat, but the sources of fat differ.

Salad oils, cooking oils, and shortenings are requirements for home and commercial food products, from potato chips to salads. Mayonnaises and salad dressings both have a viscous quality due to fat content. Unsaturation has important effects on the properties of various natural and processed fat products. The health aspects of saturated and unsaturated fats are discussed in Chapter 26.

21.1 **Chemical Structure**

GOAL: To learn the chemical structure of triglycerides and fatty acids.

Strains of bacteria that consume petroleum-based oils have been developed for use in cleaning up oil spills.

Consult Section 6.3 for a discussion of esters.

Petroleum-based oils are complex mixtures of linear and branched hydrocarbons (Chapter 6). As such, they can be burned to produce heat, but they have no fuel value for most living organisms. This characteristic is the primary distinction between these and food oils, since food oils, which are also predominantly hydrocarbon, contain ester groups that make them biodegradable; this means that they are susceptible to attack by living organisms that can thereby derive food value from them.

Both food oils and fats have the same general chemical structure. They differ only in melting point; an oil has a melting point below room temperature and is fluid, whereas a fat is solid under these conditions. Oils can be solidified if cooled enough, and solid fats (like butter) will change to oils if heated. Thus oils are sometimes described as liquid fats. The more general term *lipids* includes oils, fats, and fatlike substances such as steroids. All are predominantly hydrocarbons.

Unsaturation, including *cis-trans* (geometric) isomers, is discussed in Section 6.2.

A fat molecule is formed by the reaction of glycerin (glycerol) with three fatty acids. The product is commonly called a *triglyceride*. The symbol R describes the hydrocarbon portion of the fatty acid. The groups represented by R are linear chains of carbon atoms (plus hydrogens) of variable length. The length of these chains and the amount of unsaturation, if any, influence the physical properties of the fat, such as whether it will be a solid or a liquid.

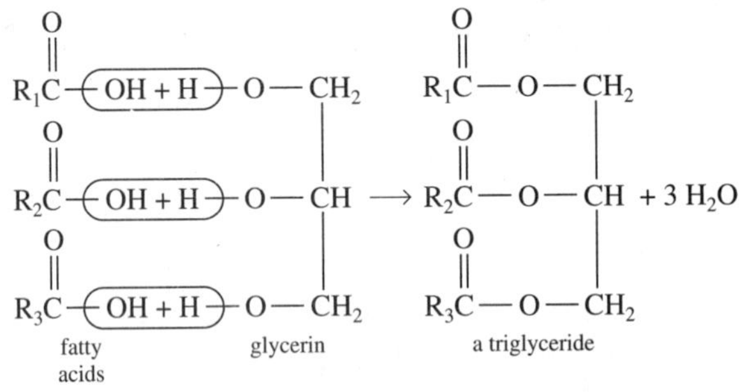

$$
\begin{array}{c}
R_1C\text{---}(OH + H)\text{---}O\text{---}CH_2 \quad R_1C\text{---}O\text{---}CH_2 \\
R_2C\text{---}(OH + H)\text{---}O\text{---}CH \rightarrow R_2C\text{---}O\text{---}CH + 3\,H_2O \\
R_3C\text{---}(OH + H)\text{---}O\text{---}CH_2 \quad R_3C\text{---}O\text{---}CH_2
\end{array}
$$

fatty acids   glycerin   a triglyceride

Table 21.1 shows that the fatty acids present in butterfat and coconut oil triglycerides tend to be relatively short-chain and saturated, whereas those predominant in soybeans, peanuts, and other plant seeds are long-chain acids with considerable unsaturation. The fatty acid composition of butterfat varies, ranging from C_4–C_{18}. Fatty acids almost always contain an even number of carbon atoms; that is, the hydrocarbon groups (R) contain an odd number of carbons. This is due to the series of reactions used for synthesis of fatty acids in which the long chains are built up two carbons at a time.

The wide variation of fatty acid content makes it impossible to draw a single molecular structure for a fat molecule from any particular source. One can draw a molecule such as trilaurin, which is the triglyceride made from glycerin and three lauric acid (C_{12}) residues, but even coconut oil, which has a particularly high percentage of lauric acid residues (about 50%), contains many triglyceride molecules in which one or more of the lauric acid groups

Table 21.1 Saturated fatty acids *(1)*

Common Name	Systematic Name	Notation	Typical Fat Source
Acetic	Ethanoic	2:0	. . . .
Butyric	Butanoic	4:0	Butterfat
Caproic	Hexanoic	6:0	Butterfat
Caprylic	Octanoic	8:0	Butterfat, coconut oil
Capric	Decanoic	10:0	Butterfat, coconut oil
Lauric	Dodecanoic	12:0	Coconut oil
Myristic	Tetradecanoic	14:0	Butterfat, coconut oil
Palmitic	Hexadecanoic	16:0	Most fats and oils
Stearic	Octadecanoic	18:0	Most fats and oils
Arachidic	Eicosanoic	20:0	Peanut oil
Behenic	Docosanoic	22:0	Rapeseed oil

Note: A notation such as 12:0 indicates a fatty acid with 12 carbons and no double bonds.

is replaced by another fatty acid. In fact, it is because fats, as well as petroleum oils, are mixtures of different compounds that they tend to form amorphous rather than crystalline solids when cooled below their melting points.

trilaurin

Bond-line formulas for organic compounds were introduced in Section 6.1.

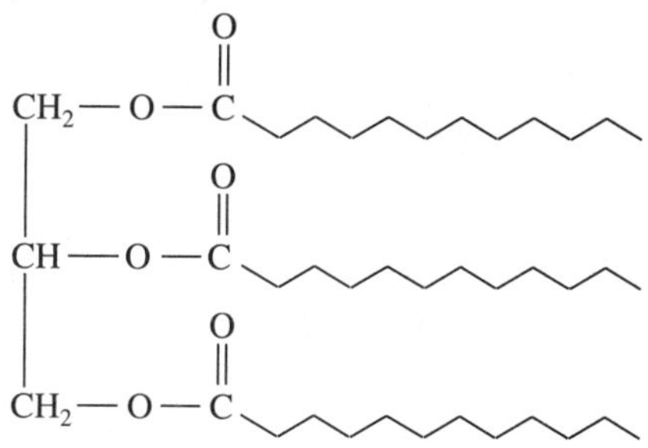

21.2   Unsaturation

GOAL: To appreciate the effects of unsaturation, the form it takes in natural fats, and the significance of iodine values.

The presence of carbon-carbon double bonds dramatically changes the physical properties of organic compounds such as fats. The popularly advertised *polyunsaturated* salad or cooking oils consist of triglycerides with a high content of unsaturated fatty acid residues. An important example is oleic acid (9-octadecenoic acid), which can be designated by the shorthand notation 18:1 to indicate it is an 18-carbon fatty acid containing one double bond. When it is important to specify the position of the double bond in oleic acid, the notation 18:1(9) may be used to indicate that the double bond follows carbon 9.

Stearic acid, 18:0, is the fully saturated 18-carbon fatty acid, whereas linolenic acid, 18:3 or 18:3(9,12,15), is at the other extreme—an 18-carbon triunsaturated fatty acid with double bonds following carbons 9, 12, and 15. The principal unsaturated fatty acids present in natural fats are listed in Table 21.2.

Table 21.2   Some unsaturated fatty acids in natural fats *(1,2)*

Common Name	Systematic Name	Notation	Typical Fat Source
Caproleic	9-Decenoic	10:1(9)	Butterfat
Lauroleic	9-Dodecenoic	12:1(9)	Butterfat, coconut oil
Myristoleic	9-Tetradecenoic	14:1(9)	Butterfat
Palmitoleic	9-Hexadecenoic	16:1(9)	Animal fats, seed oils
Petroselinic	6-Octadecenoic	18:1(6)	Parsley seed oil
Oleic	9-Octadecenoic	18:1(9)	Most fats and oils
Vaccenic	11-Octadecenoic	18:1(11)	Butterfat, beef fat
Linoleic	9,12-Octadecadienoic	18:2(9,12)	Most seed oils
Linolenic	9,12,15-Octadecatrienoic	18:3(9,12,15)	Soybean oil, linseed oil
Eleostearic	9,11,13-Octadecatrienoic	18:3(9,11,13)	Tung oil
Gadoleic	9-Eicosenoic	20:1(9)	Fish oils
Arachidonic	5,8,11,14-Eicosatetraenoic	20:4(5,8,11,14)	Fish oils
Erucic	13-Docosenoic	22:1(13)	Rapeseed oil

The *cis* form is the normal arrangement of all double bonds of fatty acids isolated from natural fats and oils. Two representations of oleic acid are shown.

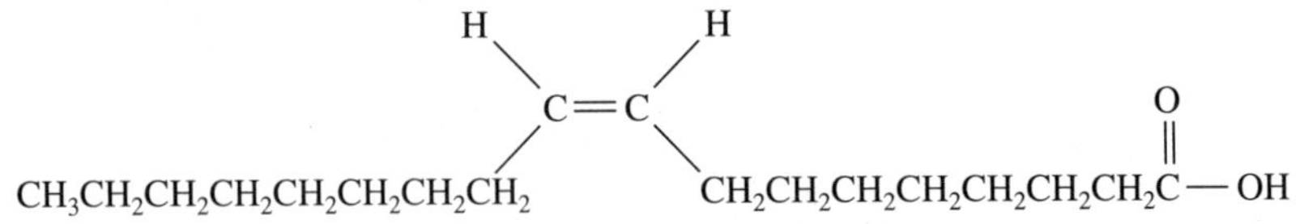

oleic acid, *cis* 18:1(9)

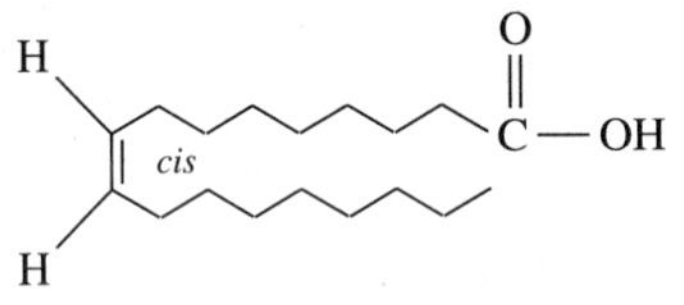

Since unsaturation is so important in determining the properties of the fat and since unsaturated fat is considered desirable in the diet, it is useful to have a quantitative measure of the amount of it. A convenient value for this purpose is the **iodine value,** which is based on the reaction between iodine chloride, ICl, and an unsaturated fatty acid.

In this reaction, ICl breaks one of the bonds of the carbon-carbon double bond and adds across the two carbon atoms. It is an instantaneous reaction that occurs quantitatively (100% efficiency). Generally, a measured excess of ICl is used. The leftover ICl is then measured to determine the amount that reacted with the fat. The iodine value is given in terms of the number of grams of iodine that react with 100 g of fat. Iodine, I_2, does not react with carbon-carbon double bonds, so the more reactive ICl or IBr is used. Then a calculation yields the amount of I_2 that would have reacted by the same pathway.

The higher the iodine value, the greater the amount of unsaturation. A so-called *polyunsaturated oil* is one with a relatively high iodine value. Table 21.3 contains a list of most of the important fats, their iodine values, and the fatty acids that make up 25% or more of the total. The iodine values of the 18:1, 18:2, and 18:3 fatty acids and their triglycerides are included for comparison.

Table 21.3 contains important facts about common fats and oils, including iodine values. Soybeans are the most important commercial source of oil. Soybean oil is one of several so-called polyunsaturated vegetable oils processed for direct use or converted to solid forms, such as margarine and shortening. The other principal oils converted into other forms are cottonseed, corn, palm, and safflower. Safflower oil is the most unsaturated of this group. As we will see in our discussion of margarines and shortenings, olive and peanut oils are not suitable for conversion but are used directly as salad and cooking oils. (Much of this chapter covers the uses of many of the fats and oils listed in Table 21.3.)

Linseed oil is the most highly unsaturated oil listed. It is not used as a food, either directly or in converted form, because it is very susceptible to oxidative spoilage (discussed in a later section). It is an important component of most oil-based paints.

The iodine values in Table 21.3 make it clear that vegetable oils are generally more unsaturated (have higher iodine values) than animal fats. Coconut oil and palm kernel oil are dramatic exceptions.

Table 21.3 Fats and oils (3)

Fat or Oil Source	Iodine Value	Important Fatty Acids[1] (>25%)	Order of Domestic Production[2]	Order of Domestic Use[2]
Soybean	120–141	(18:2) (18:1)	1	1
Cottonseed	97–112	(18:2) (16:0)	4	6
Lard (hogs)	58–68	(18:1) (16:0)	5	5
Lard (hogs fed peanuts)	>85			
Butterfat	25–42	(18:1) (16:0)		3
Tallow (beef)	35–48	(18:1) (16:0)	3[3]	4[3]
Tallow (lamb)	48–61	(18:1) (16:0)		
Corn	103–128	(18:2) (18:1)	2	2
Peanut	84–100	(18:1) (18:2)	7	9
Safflower	140–150	(18:2)	8	10
Coconut	7.5–10.5	(12:0) (14:0)		3
Palm	44–58	(16:0) (18:1)		8
Palm kernel	14–23	(12:0)		7
Cocoa butter	35–40	(18:1) (18:0)		
Sunflower	125–136	(18:2)	6	
Olive	80–88	(18:1)		
Chicken fat	64–76	(18:1) (16:0)		
Human fat	57–73	(18:1) (16:0)		
Linseed (flaxseed)[4]	175–202	(18:3)		

Fatty Acids	Iodine Value of Fatty Acid	Shorthand Notation	Iodine Value of Triglyceride
Oleic	90	(18:1)	86
Linoleic	181	(18:2)	173
Linolenic	274	(18:3)	262

[1]Listed in order of decreasing content.   [2]Order varies from year to year.   [3]Edible tallow only. Some tallow is used in making soap.   [4]A nonfood oil.

Diet is a factor in determining the characteristics of fat stored by an animal. Compare the iodine values for lard from hogs on a standard diet (mostly soybean meal, corn meal, and tallow) with lard from hogs fed peanuts. These animals store fat that becomes increasingly like the fat in peanuts, that is, more unsaturated.

In addition to edible tallow, inedible tallow is an important raw material in the United States fats and oils industry. It is used in animal feeds, soaps, and lubricants. Tallow mixed with paraffin (from petroleum) constitutes the main ingredients of candles. (Soaps and lubricants are discussed in Chapter 22.)

21.3 Isolation

GOAL: To see how fats and oils are obtained from natural sources.

Vegetable oils are isolated by pressing or extracting oilseeds. In some instances, pressing, which removes most of the oil, is followed by extraction of the residual cake to remove most of the remaining oil. Hydrocarbon solvents such as hexane are normally used for extracting vegetable oils; these oils are also predominantly hydrocarbon, so they readily dissolve in these solvents.

Isolation of animal fats is called **rendering.** It involves *grinding,* followed by *heating,* which coagulates the protein, and *centrifugation,* during which the lower-density fat rises and separates.

21.4   Processed Fat Products

GOALS: To appreciate the origin of and steps in making margarine.
To understand the major factors that influence the melting points of fats.
To consider the significance of plastic range.
To understand the composition of mayonnaise and salad dressing.

Margarine and Butter

Butter is made by churning cream to cause the fat globules to clump together. The remainder of butter is about 16% water plus a small percentage of milk solids (lactose, casein, minerals) and salt, which acts as both a flavoring and a preservative.

The fat content of margarines is regulated by law. In a conventional margarine, it must be a minimum of 80%. Like butter, margarine contains about 16% water plus small amounts of salt and nonfat milk solids. The law requires that margarine contain 15,000 international units of vitamin A per pound. Vitamin D is optional, but if present must have a minimum of 1500 international units per pound *(4).* Diet margarines, also called imitation margarines, may be marketed with as little as 40% fat.

Many reasons are cited for the increased popularity of margarine. Near the top of the list is cost, because it has always been much cheaper than butter. The implication of saturated fats as a factor in heart disease increased the popularity of the more unsaturated margarines. Improved techniques for processing vegetable oils into margarine are another reason, especially as they include some important chemical changes in the fatty acids in margarines, discussed in the following section.

Melting Points and Hardening

Both butter and margarine are described as *plastic fats* to signify that they retain their shape at room or refrigerator temperature but may be deformed when cut with a knife. The **plastic range** is an important property of any table spread; it is the temperature range over which the spread retains its shape but may be readily deformed. Many margarines are more convenient than butter because they can be spread easily even at refrigerator temperatures. Butter is not easily deformed at refrigerator temperatures. The upper limit of the plastic range for both margarine and butter must be just below body temperature so that they will ''melt in the mouth.''

Several chemical changes occur when processing a liquid oil into a solid margarine. This process, known as **hydrogenation,** is the addition of hydrogen, as H_2 gas, to double-bonded carbons. This is illustrated for ethylene, the simplest unsaturated compound:

ethylene   hydrogen   ethane
gas

Since the addition of hydrogen to a liquid oil increases the melting point and eventually causes the formation of a solid fat, the process is called **hardening.** The more saturated animal fats, such as lard and tallow, have high melting points and are solid at room temperature; this is in contrast to most vegetable oils, which are much more unsaturated and melt well below room temperature. Two commercially important vegetable oils that are exceptions are coconut oil (m. p. 24–26 °C) and palm kernel oil (m. p. 23–26 °C). They have unusually low iodine values (Table 21.3) compared with other vegetable oils. Their melting points are close to room temperature (about 22 °C).

Another factor that influences the melting points of fats is the length of the fatty acid chains. Short-chain fatty acids tend to lower the melting point, whereas long-chain fatty acids raise the melting point. This partly explains why butter and margarine have similar melting points even though butter is a more saturated fat. Butterfat has an iodine value in the range 25–42, whereas conventional margarines have iodine values of 78–90. One important difference between the two products is the relatively high percentage of short-chain fatty acids in butterfat (Table 21.1), which counteracts the greater saturation by lowering the melting point.

A second change that accompanies hydrogenation also makes margarine similar to butter. Partial hydrogenation of diunsaturated or triunsaturated fatty acids, in addition to saturating many of the double bonds, often causes remaining double bonds to change (isomerize) into the *trans* arrangement. Recall that all natural fatty acids have the *cis* arrangement about the double bonds, which causes them to have lower melting points. Consequently, any conversion of double bonds to the *trans* form will raise the melting point and contribute to hardening the oil into a margarine. Margarines are often labeled ''lightly hydrogenated,'' which means some increase of saturation, accompanied by the *cis* to *trans* isomerization.

An example would be the hydrogenation of soybean oil (iodine value 120–141) to bring it into the range of values normal for conventional margarines (78–90). At that point the oil becomes a plastic solid. In contrast, unhydrogenated peanut oil has about the same range of iodine values (84–100) but is liquid at room temperature. The chief difference between natural peanut oil and partially hydrogenated (hardened) soybean oil is that the latter contains some *trans* double bonds due to hydrogenation. Untreated peanut oil has only the natural *cis* double bonds. Since hardening the oil into a margarine would lower the iodine value and make it uncompetitive in markets in which there is a premium on polyunsaturated fats, peanut oil is not used to manufacture margarines.

To summarize the hardening process, the following all tend to give fats a higher melting point: (1) long-chain fatty acids, (2) less unsaturation, and (3) *trans* arrangements of carbon-carbon double bonds. In converting an oil to a margarine, a processor cannot alter item 1, except by choice of oil, but items 2 and 3 are altered by hydrogenation. The effects of 1, 2, and 3 also are illustrated by the melting points of the following fatty acids:

- Oleic acid, *cis* 18:1(9), has a melting point like an 8:0 fatty acid.
- Elaidic acid, *trans* 18:1(9) has a melting point like a 12:0 fatty acid.

In other words, the presence of one double bond lowers the melting point of a C_{18} fatty acid so that it melts at about the same temperature as the C_8 saturated fatty acid. But changing the C_{18} fatty acid to the *trans* isomer counteracts the effect of the unsaturation by raising the melting point *(4).*

Typical changes are illustrated in Figure 21.1 for the hydrogenation of soybean oil. In general, as the iodine value decreases due to addition of hydrogen, the melting point rises. As the iodine value decreases from 129 to 76, the predominant changes are a decrease in the 18:2 fatty acid (linolenic acid) and an increase in the *trans* 18:1 fatty acid (elaidic acid), with no significant increase in the amount of 18:0 fatty acid (stearic acid). Therefore,

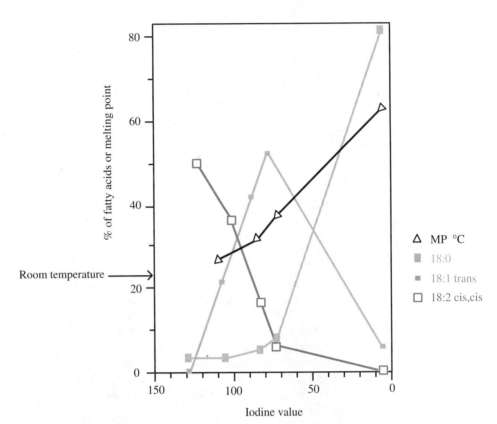

Figure 21.1
Changes during the
hydrogenation of
soybean oil *(4)*. See text
for discussion.

An iodine value of 86
corresponds to the amount
of unsaturation in an 18:1
fatty acid in a triglyceride
(Table 21.3).

we see a rather clean conversion of the 18:2 fatty acid to the *trans* 18:1 until most of the
18:2 is gone. We would predict the oil will begin to harden into a solid fat when the
melting point is at about room temperature (about 22 °C) and the iodine value is about
100.

In making margarines, vegetable oils are hydrogenated until the iodine value reaches
the range of 78–90. During hydrogenation, first the more unsaturated 18:2 fatty acids are
hydrogenated and isomerized until most of the double bonds present are those in the *cis*
and *trans* 18:1 fatty acids. At that point (iodine value about 80), the amount of 18:0 fatty
acid begins to rise at the expense of the 18:1 fatty acid, which undergoes hydrogenation.

The normal range of iodine values for margarines can also be achieved by blending
fats from different sources with different degrees of hydrogenation. This achieves the
necessary plastic range for the final product.

Salad Oils, Cooking Oils, and Shortening

Salad oils, cooking oils, and shortenings are fat products that are almost pure triglycerides.
Small amounts of additives improve appearance, ease of handling, and stability.

Shortening is a moderately hardened, pure fat. Like margarine, shortening with a
suitable plastic range can be made by hydrogenation or blending, or a combination of both.
Shortening is generally not refrigerated; when it is, it may become a rigid, nonplastic solid.
Since it is not used as a table spread but must be easily creamed when used for baking, it
can be soft at room temperature. Unlike margarine, complete loss of shape may not occur
until well above body temperature, since shortening is not consumed directly and therefore
need not melt in the mouth.

Mayonnaise and Salad Dressings

A popular brand of mayonnaise is pictured in Figure 21.2, with its counterpart, a spoonable salad dressing. Mayonnaise and salad dressing are important items in food oil consumption in the United States. Pourable dressings for salads bottled as French, Italian, Roquefort, and so on are covered by a federal standard of identity that requires that they contain at least 35% oil. They can have separate water and oil phases or be an emulsified viscous (semisolid) type.

Spoonable salad dressings and mayonnaise are in an emulsified and viscous form that prevents them from being poured but allows for convenient spreading. Either product is readily blended with other ingredients to form a wide variety of food toppings and sauces. Typical formulations of these two products are shown in Table 21.4. Mayonnaise must have a minimum oil content of 65%; salad dressings have at least 30%. Both generally contain more than the specified amounts, so the final products have the viscosity anticipated by the consumer. The lower oil content of salad dressings is counteracted by added starch, which acts as a thickener to provide the consistency of mayonnaise.

Egg yolk contains substances that act as emulsifiers to stabilize the suspension of oil and water, similar to the action of soaps (Chapter 22). The emulsion is formed by mixing

Figure 21.2
Typical examples of mayonnaise and spoonable salad dressing.

Table 21.4   Typical composition of mayonnaise and spoonable salad dressing *(5)*

Ingredients	% Weight in Mayonnaise	% Weight in Salad Dressings
Salad oil	77–82	35–50
Egg yolk	5.3–5.8	4.0–4.5
Vinegar	2.8–4.5	9.0–14.5
Salt	1.2–1.8	1.5–2.5
Sugar	1.0–2.5	9.0–12.5
Starch	None	5.0–6.5
Water to make 100%		

Source: Courtesy of AVI, Westport, CT.

Figure 21.3
Producing the mayonnaise emulsion.

with high-speed beaters (Figure 21.3). Salt concentrates in the water phase and acts as a preservative. Sugar, vinegar, and sometimes lemon juice are added to provide both flavor and preservative action. The acetic and citric acids of vinegar and lemon inhibit microbial growth. Several other spices and flavorings are also added. Salad dressings are generally much spicier and more acidic than mayonnaise. Products marketed as imitation and diet mayonnaise are milder flavored.

Cholesterol-free or eggless mayonnaise and spoonable salad dressings are relatively new products. They are formulated in the same way as the ordinary varieties are, except that the cholesterol-containing egg yolks are replaced by other emulsifying agents to stabilize the oil-water emulsion.

21.5 Applications of Fats and Oils

GOAL: To consider the major sources of triglycerides and some of the differences in fatty acid content that can determine the application of each fat.

So far, we have discussed the chemical structure of fats, and the ways of altering or using their physical properties to produce common products. We turn now to a fuller consideration of the fats listed in Table 21.3 in the foods already discussed and in certain other specialty items. The main vegetable oils are considered first, followed by animal fats.

Vegetable Oils

For almost 30 years, **soybean oil** has been the most common form of oil for direct use and for processing into margarine, shortening, mayonnaise, and salad dressings. An excellent salad oil, soybean oil is not as good for cooking unless it is partially hydrogenated. It has a relatively high content of the 18:2 fatty acid (51%). Fatty acids with two or more double bonds are susceptible to oxidative spoilage, called *reversion* or *rancidity* (Section 21.6).

At one time, more **cottonseed oil** was produced than any other vegetable oil. Its use is now limited, however, because the increased demand for synthetic fibers has decreased cotton production. Cottonseed oil has a small amount of 18:2 fatty acid, allowing it to be used in cooking. It is used heavily in frying potato chips. This illustrates a major problem associated with cooking fats: the oil itself becomes part of the finished food. Therefore, an oil used to cook a stored food must be quite resistant to spoilage.

Corn oil has popularity as a salad oil and for processing into margarine. Corn oil margarines are often highly touted in advertising campaigns, but the original oil differs very little from any other when hardened (hydrogenated) into a margarine. Like soybean oil, corn oil has a slight tendency to undergo reversion due to the high content (54%) of the 18:2 fatty acid. It is generally satisfactory for home cooking, however, since foods fried in the home usually are not stored for long periods of time.

Peanut oil has good qualities for use either in cooking or salads. It is sufficiently saturated to resist spoilage when used for cooking, and is sufficiently unsaturated to meet a desire for unsaturated fats in the diet. Its low iodine value cannot compete in a market that puts a premium on more unsaturated fats, so it is not processed into margarine or shortening.

The consumer is sometimes confronted with advertising that proclaims the special virtue of **safflower oil,** the most unsaturated of all the common edible oils. When hardened and blended into margarines, the products are characterized as having a ''strong oxidized flavor.'' This is apparently due to the unusually high content (75%) of 18:2 fatty acid. It also makes safflower oil less desirable for cooking *(6)*.

Until recently, **sunflower oil** was little used in the United States. However, elsewhere in the world, where other major oilseed crops will not grow, sunflowers are cultivated as a source of oil. Most sunflower oils have iodine values of 125–136.

Although **coconut oil** is not produced on the United States mainland, its use is heavy. It is highly saturated, with iodine values usually below 10, which makes it undesirable as a salad or cooking oil. However, the saturated fatty acids resist oxidation, so commercial food products that require a long shelf life are fried in coconut oil.

More than 70% of the fatty acids in coconut oil are 12:0, 14:0, and 16:0, with the first one predominating. Since all three melt over a range of only 19 degrees, coconut oil has a very narrow plastic range, which makes it particularly suitable for incorporation in coatings on ice cream bars. When cold, it is below the plastic range and therefore very hard, but it melts in the mouth quickly.

Somewhat the same behavior is observed for cocoa butter (from cocoa beans), which also has a narrow plastic range and melts just below body temperature. This property, plus its compatibility with chocolate, makes cocoa butter ideal for coatings on candy and other foods.

Coconut oil also has several important nonfood applications; soap production (Chapter 22) is a big one. It also is used in cosmetics, lipsticks, creams, lotions, hair oil, rubbing oil, and hydraulic brake fluids.

The oil palm tree is the source of both **palm oil** and **palm kernel oil** (Figure 21.4). The outer portion of the palm fruit is soft and contains the oil. The kernel is enclosed in a hard shell and is the source of the kernel oil. Palm kernel oil is very much like coconut oil and is used in certain types of candy coatings.

Palm oil has iodine values of 44–58, whereas the palm kernel oil has iodine values of 14–23. Palm oil is often blended with other oils for incorporation into margarines and shortenings. In Europe, where soybeans are not so plentiful, it is a major source of triglycerides, second only to fish oils *(7)*.

Olive oil is relatively expensive and is used directly, or blended with other oils, in salads and cooking. It is prized for its distinctive flavor.

Rapeseed oil was once a significant commercial oil in the United States but is no longer used due to a high erucic acid content, 22:1(13). Erucic acid has been associated

Figure 21.4   Important sources of triglycerides. In addition to fish, beef, peanuts, and corn, other items pictured are soybeans (left front); flaxseed (center front), which yields linseed oil; rapeseed (right front); palm kernels (second row right); and coconuts (center rear).

with heart problems in test animals. However, researchers in Canada developed new varieties of rapeseed with a low erucic acid content. The oil derived from one of these new varieties is **canola oil;** it is similar to soybean oil.

Animal Fats

At one time, **lard** was the cooking fat most used for baking and frying, both commercially and in the home. Its popularity has dropped in recent years because of the preference for unsaturated fats. Nevertheless, lard is still heavily used in commercial baking.

Beef **tallow** is a hard, plastic fat that is often blended with lard or vegetable oils to make shortening. It is also used in making soap.

Although herring and sardine oils are of little significance in the United States, they are used worldwide wherever fishing is a leading industry. **Fish oils** have an unusually high fatty acid content with four, five, or six double bonds; the oils are therefore susceptible to oxidative spoilage, which gives rise to the development of strong, fishy odors and flavors.

21.6 Additives in Fat Products

GOALS: To appreciate the use of additives in fat products, particularly antioxidants to
         prevent autoxidation.
         To realize the nature of microparticulated protein as a fat substitute.

Five categories of additives are common in marketed fats and fat products: antifoaming agents, colorings, flavorings, antioxidants, and emulsifiers. We have already discussed emulsifiers, used in stabilizing the water-oil emulsion in mayonnaise or salad dressings. They are considered further with soaps (Chapter 22).

Antifoaming Agents

Some cooking oils tend to foam when used for frying. Compounds such as methyl silicone are often added to counteract this action.

Colorings

We generally assume that shortenings have a clear, white color and margarines resemble butter. But the color in butter is due to β-carotene, or provitamin A, the same substance that causes the orange color in carrots. β-Carotene is a common pigment in most green vegetables (for example, spinach), although the green of chlorophyll masks the orange. The color appears in butter because β-Carotene is a fat-soluble hydrocarbon that is extracted from the cow's diet into milk, and then into the solid phase when butter is prepared. Often β-Carotene is added to margarine to achieve the color of butter.

Flavorings

In making margarines, manufacturers try to duplicate butter as much as possible. A compound known as biacetyl forms when butter is ripened. Since it is considered the main flavor ingredient, it is sometimes added to margarines.

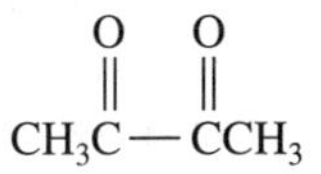

$$\underset{CH_3C-CCH_3}{\overset{\overset{\displaystyle O}{\|}\quad\overset{\displaystyle O}{\|}}{}}\qquad \text{biacetyl}$$

Antioxidants

The most important food additives are a group known as *antioxidants,* which are used to prevent the type of oxidative spoilage mentioned previously. The highly unsaturated fatty acids are most susceptible to development of oxidative rancidity. Carbon atoms adjacent to carbon-carbon double bonds are prone to oxidation, leading to formation of a *hydroperoxide,* which subsequently decomposes to foul-smelling products. The reaction is illustrated for a segment of a fatty acid chain.

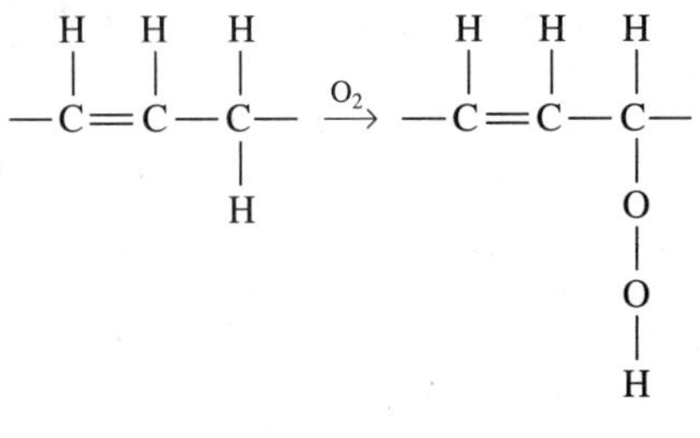

$$-\overset{\overset{\displaystyle H}{|}}{C}=\overset{\overset{\displaystyle H}{|}}{C}-\overset{\overset{\displaystyle H}{|}}{\underset{\underset{\displaystyle H}{|}}{C}}- \quad\overset{O_2}{\longrightarrow}\quad -\overset{\overset{\displaystyle H}{|}}{C}=\overset{\overset{\displaystyle H}{|}}{C}-\overset{\overset{\displaystyle H}{|}}{\underset{\underset{\underset{\underset{\displaystyle H}{|}}{O}}{|}}{\underset{\underset{\displaystyle O}{|}}{C}}}- \qquad \text{hydroperoxide}$$

This complex process is called *autoxidation.* An oxygen molecule is inserted into a carbon-hydrogen bond that is susceptible to attack by oxygen due to activation by the neighboring double bond. Hydroperoxides are prone to further chemical decomposition. The fatty acid chains cleave to form aldehydes, ketones, and lower-molecular-weight acids. These compounds tend to have strong odors, the smells of rancid fat.

Oxidation is less significant for monounsaturated fatty acids such as oleic acid, 18:1, than for those with multiple double bonds. Linoleic acid, 18:2, is a common component of most oils with high iodine values. Its autoxidation is shown in the following scheme.

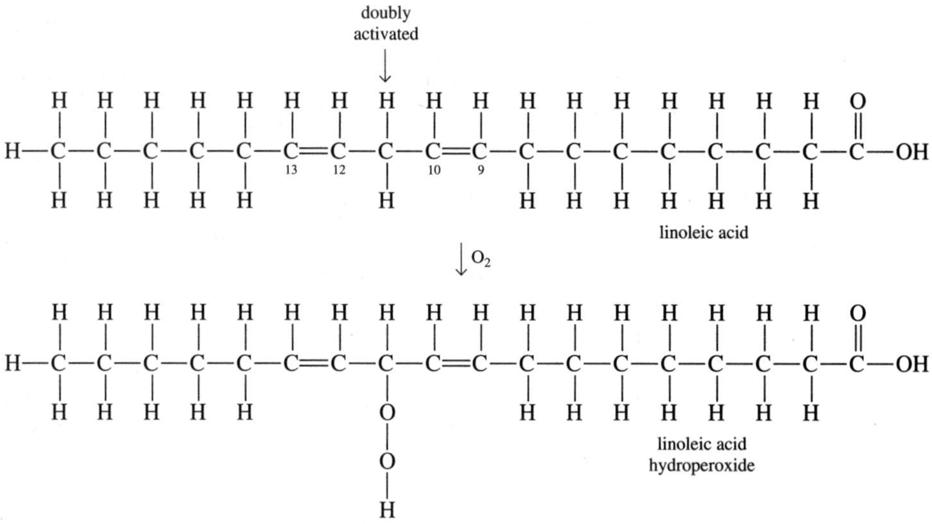

doubly
activated

linoleic acid

$\downarrow O_2$

linoleic acid
hydroperoxide

The fatty acid can be designated as 18:2(9,12) to signify the location of the two double bonds. Autoxidation occurs primarily at C_{11}, which is located between the double bonds. Both double bonds activate the C_{11} C-H bond, greatly encouraging the formation of a hydroperoxide and its subsequent decomposition to foul-smelling products.

As noted earlier, oils such as soybean and safflower are less suitable for cooking because of their high content of 18:2 fatty acid, 51% and 75%, respectively. The higher temperatures during cooking greatly increase the rate of autoxidation.

Hydrogenation can be used to decrease the linoleic acid content and thus improve the keeping quality of an oil high in that acid. It was shown previously (Figure 21.1) that hydrogenation conditions can be carefully selected to ensure that the 18:2 or 18:3 fatty acid will be affected the most. The resulting margarine or shortening has greatly decreased susceptibility to oxidative rancidity.

The ease of autoxidation of linoleic acid explains why simply blending saturated and unsaturated fats is not a satisfactory way to make margarine. Although a satisfactory plastic range is possible, the final unhydrogenated product tends to oxidize easily.

Given the prevalence of oils in the diet, their preservation is a concern. Cooking oils are commonly packaged in cans or brown bottles since light activates autoxidation. Furthermore, some bottled vegetable oils are capped under an atmosphere of nitrogen gas, so no oxygen comes in contact with them before they are opened.

Antioxidants are common additives that interfere with autoxidation. Two types of antioxidants are sometimes used together. The first is a sequestering agent, such as citric acid, that acts by tying up metal ions known to catalyze autoxidation. A second group of antioxidants directly interferes with autoxidation. In this group are butylated hydroxyanisole (BHA), butylated hydroxytoluene (BHT), and propyl gallate (PG).

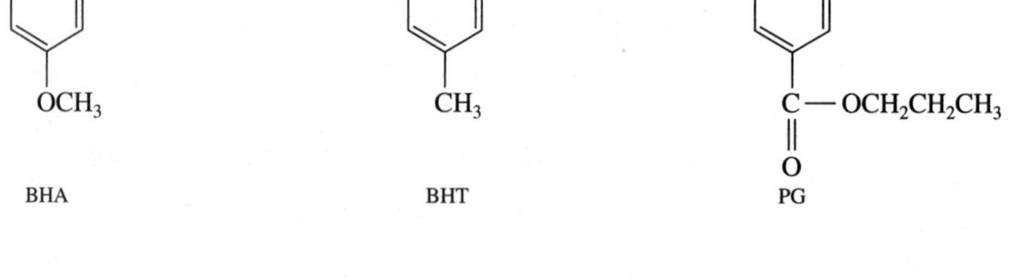

BHA

BHT

PG

Another antioxidant, vitamin E, also known as α-tocopherol, is one of several *tocopherols*. These widely distributed substances have been described as nature's antioxidants. They protect vegetable oils from harvest to processing, but the processing of oils into consumer products often destroys much of the tocopherols. Fats from animals and fish contain virtually no tocopherol, which explains their instability compared with vegetable oils *(7)*. Since the tocopherols are highly colored substances, they can only be added to processed fat products where a reddish brown tint is acceptable.

$$\text{HO} - \underset{\substack{CH_3 \\ | \\ CH_3}}{\underset{|}{\overset{CH_3}{\bigcirc}}} - (CH_2)_3 - \overset{CH_3}{\underset{|}{CH}} - (CH_2)_3 - \overset{CH_3}{\underset{|}{CH}} - (CH_2)_3 - \overset{CH_3}{\underset{|}{\underset{CH_3}{CH}}}$$

α-tocopherol

Vitamin C, ascorbic acid, is also an antioxidant, but it is water soluble and normally not present in oils.

Microparticulated Protein

In 1990 a product known as *microparticulated protein* was introduced as a fat substitute. It is used in foods such as frozen desserts, salad dressings, mayonnaise, sour cream, and certain cheeses (Figure 21.5). Marketed under the name Simplesse, the product is formed from a combination of egg white protein and milk protein (casein). It is processed into spherical particles with diameters of 0.1–3 microns (μ). At this size, the tongue cannot distinguish the particles individually, but rather perceives them as having the creamy taste and texture of fat. The round shape and uniform size allow the particles to roll easily over one another, creating the smoothness and richness normally associated with fat. Simplesse can provide 80% or greater reduction in fat content, depending on the food product, and a corresponding decrease in calorie content of more than 50%.

$1 \, \mu = 10^{-6} \, \text{m}$

Figure 21.5 A product in which 50% of the fat is replaced by Simplesse.

On the negative side, the microparticulated protein coagulates and loses its smooth consistency if heated at high temperatures. Therefore, it cannot be used in place of fat for cooking, frying, or baking. Furthermore, people who are allergic to milk protein or egg white may also have an allergic reaction to the microparticulated protein.

SUMMARY

Food fats and oils are triglycerides in which the acid portion consists of long, linear fatty acids. These fatty acids may be completely saturated or highly unsaturated. In general, animal fats are more saturated than vegetable oils. Their iodine values can be determined experimentally by reaction of iodine chloride (ICl).

Soybeans are the main source of food oil directly used in salads and cooking, in converted form as margarine and shortening, and as a principal ingredient in mayonnaise and salad dressings. Table 21.3 lists other sources of fats and oils.

Oils are generally isolated by solvent extraction or pressing or both. Animal fats are isolated by a process called rendering.

Margarine is widely used, produced by blending saturated and unsaturated fats, or more frequently by hydrogenation, to give a product with an acceptable plastic range. The melting point of a fat is increased by long-chain fatty acids, less unsaturation, and *trans* arrangement of the unsaturation. Natural fats contain only *cis* double bonds but hydrogenation can cause some *cis* to *trans* isomerization. Unsaturation also makes a fat more susceptible to autoxidation.

A wide variety of additives are used in fat products. The most usual and the most controversial are the synthetic antioxidants. Vitamin E is a natural, fat-soluble antioxidant often present in fats.

PROBLEMS

1. Give a definition or example of each of the following.

   a. Biodegradable
   b. Lipids
   c. Triglyceride
   d. Trilaurin
   e. Unsaturation
   f. Amorphous
   g. Iodine value
   h. Rendering
   i. Plastic range
   j. Hydrogenation
   k. Hardening
   l. Shortening
   m. Rancidity
   n. Canola oil
   o. Autoxidation
   p. Antioxidant

2. What is the primary difference between petroleum-based oils and food oils?

3. What is the main difference between food oils and fats?

4. Draw the structure of the triglyceride derived from lauric acid, 12:0, caproic acid, 6:0, and linoleic acid, 18:2(9,12).

5. What is the most important commercial source of food oil?

6. How do butter and margarine differ in composition?

7. What factors affect the melting point of a fat?

8. Most margarines have an iodine value of about 85, as does peanut oil. Explain why margarines are solid and peanut oil is liquid at room temperature.

9. What chemical changes take place when corn oil is hardened into a margarine? Why does it harden?

10. Butter and margarine have about the same melting temperature even though margarine has a much higher iodine value. How do you account for this?

11. Why is coconut oil especially useful for coatings on ice cream bars?

12. What are the sources of lard and tallow?

13. Why is cottonseed oil (iodine value 97–112) more suitable than safflower oil (140–150) for frying potato chips?

14. Write the chemical structure of a diunsaturated fatty acid hydroperoxide.

15. What is the main difference between a solid shortening and a margarine?

16. Compare the composition of the two products pictured in Figure 21.2.

17. Why can fish oils develop a bad odor?

References

1. Institute of Shortening and Edible Oils. *Food Fats and Oils;* Washington, DC, 1968.

2. McGilvery, R. W.; Goldstein, G. W. *Biochemistry: A Functional Approach;* W. B. Saunders: Philadelphia, 1983; pp. 204–205.

3. "1990/91 U. S. Soybean Outlook." *Oil Crops: Situation and Outlook Report* April 1990, p. 7.

4. Weiss, T. J. *Food Oils and Their Uses,* 2nd ed.; AVI: Westport, CT, 1983; pp. 4, 78, 188.

5. Weiss, T. J. *Food Oils and Their Uses;* AVI: Westport, CT, 1970; p. 25.

6. "Choosing a Margarine for Taste and Health." *Consumer Reports,* January 1975, pp. 42–47.

7. Allen, J. C.; Hamilton, R. J. *Rancidity in Foods,* 2nd ed.; Elsevier: New York, 1989; p. 90.

Consumer Chemistry: From Home Products to Health Issues

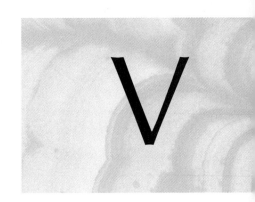

Dental Hygiene and Home Products

22

The importance of good dental hygiene is best appreciated if based on knowledge of the structure and properties of tooth enamel, including its vulnerability to acids formed in plaque. Fluoride is thought to have a protective effect on tooth enamel, but the benefits of water fluoridation are controversial.

Soaps are made from food oils and fats by treatment with a strong base. They act as emulsifying agents, called surfactants, to allow greasy dirt to be combined with water and washed away. Coconut oil is a particularly good raw material for making soap because of its unusual fatty acid content.

Synthetic detergents (syndets) are more suitable surfactants in hard water, but they are usually combined with builders to enhance their cleansing ability. Pollution due to phosphate builders was one of several problems associated with syndets. It led to development of alternative builders and surfactants.

Sodium hydroxide is the active ingredient in both oven cleaners and drain cleaners. In both applications, the strong base converts fat into soap, which is easily removed. Both products are potentially hazardous to handle.

CHAPTER PREVIEW
22.1 Dental Chemistry
22.2 Soaps
22.3 Detergents
22.4 Drain Cleaners and
     Oven Cleaners

22.1 Dental Chemistry

GOALS: To appreciate the crystal lattice structure of hydroxyapatite.
          To understand the processes of demineralization and remineralization and their relationship to dental caries.
          To consider the composition of dental plaque and its role in dental caries.
          To understand the importance of acidity as a cause of dental caries, and of fluoride as a means of preventing caries.
          To appreciate the controversy over water fluoridation.
          To consider the metals in dental amalgams.

The structure of tooth enamel, the chemical reactions of breakdown and reformation of the enamel, and the role of fluoride in prevention of decay are all part of dental chemistry. Chemistry is also involved when the dentist enters the picture.

Tooth Enamel

The thickness of tooth enamel from the surface of the tooth inward is about 2 mm. The enamel consists of hydroxyapatite, $Ca_5(PO_4)_3OH$. As with other salts, this does not suggest that enamel is made up of discrete molecules with this formula. Instead, a complex three-dimensional structure, a *lattice,* consists of a series of calcium ions, Ca^{2+}, phosphate ions, PO_4^{3-}, and hydroxide ions, OH^-, positioned around one another (Figure 22.1). The net charge on the lattice is zero, since there are 5 calcium ions for every 3 phosphate ions and every 1 hydroxide ion in the lattice.

In the environment of the mouth, hydroxyapatite can be dissolved or re-formed according to the reaction

$$Ca_5(PO_4)_3OH \underset{\text{remineralization}}{\overset{\text{demineralization}}{\rightleftharpoons}} 5Ca^{2+} + 3PO_4^{3-} + OH^-$$

The formation of hydroxyapatite is called *remineralization,* or simply mineralization. The breakdown of the enamel is called *demineralization.* The arrows in the equation confirm

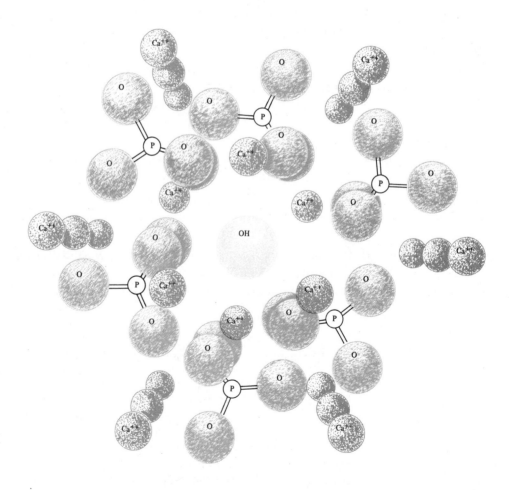

Figure 22.1 The lattice structure of hydroxyapatite. See text for discussion.

that remineralization is favored over demineralization; the enamel is therefore maintained. But acidic conditions can influence the equilibrium.

Localized demineralization of the teeth is known as *dental caries.* The most common reasons for tooth loss in individuals are caries and periodontitis (inflammation of the gums). These are the most widespread mouth diseases in humans; and despite our scientific accomplishments, their causes are not well understood. In both conditions a common factor is involved—*dental plaque,* a gelatinous mass of closely packed microorganisms attached to the tooth surface. Plaque is one example of microorganisms present in several parts of the body, often in a symbiotic (mutually beneficial) relationship.

The microorganisms present in dental plaque are part of the oral flora. Plaque also contains complex polysaccharides and traces of food, which nourish the microorganisms. Polysaccharides, which are produced by the bacteria, glue both right up against the tooth surface, where they can do maximum damage.

The *intestinal flora* synthesize vitamin K and one of the B vitamins. The excess can be absorbed and used by the body, while these microorganisms benefit from the body's supply of nutrients.

In the development of dental caries, plaque is important in demineralization. Most of the microorganisms are anaerobic rather than aerobic. They are favored by the stagnation and decreasing oxygen content that accompany plaque buildup and thickening. The gum line surfaces below the areas of contact between the teeth are the most stagnant and therefore most susceptible to caries. Also, anaerobic bacteria ferment carbohydrates to simple organic acids, particularly lactic acid. Aerobic bacteria metabolize carbohydrate further into carbon dioxide and water.

Starchy foods have very little tendency to promote dental caries, but sucrose can penetrate the plaque and undergo fermentation. Each time more sugar reaches the bacteria in the plaque, more acid is produced. Gradually the hydroxyapatite structure is attacked, since an acidic environment favors demineralization.

Under normal circumstances, the equilibrium between demineralization and remineralization favors the latter, but as acidity increases, demineralization is favored more and more. Look again at the demineralization reaction, and focus on the hydroxide ions that are produced. If the pH is low (acidic), the high concentration of hydrogen ions will remove (neutralize) the hydroxide ions and prevent remineralization. In addition, the phosphate ions, PO_4^{3-}, released during demineralization are converted to HPO_4^{2-} and $H_2PO_4^-$ as the pH is lowered, so PO_4^{3-} becomes less available for remineralization.

The pH of the mouth is about 6.8, varying with the foods recently eaten. But within plaque, at the surface of the tooth, the pH may be much lower. Then remineralization is hindered even more. Demineralization also supplies the microorganisms with calcium and phosphorus. Thus we see a never-ending cycle of more plaque causing more demineralization, which contributes to the development of more plaque.

Tooth enamel can remineralize using calcium and phosphate in saliva, but diet is a critical factor in determining whether the process can keep up. If a steady stream of sugar is supplied to the plaque, both during and between meals, demineralization is increasingly favored. Sticky carbohydrates like candies and confectionery goods may contribute to caries by adhering to areas between the teeth and along the gingival margins (between teeth and gums), and by packing into pits and fissures of the rear teeth. Caries are most common in these three areas.

In effect, cavities occur when saliva cannot supply calcium and phosphate fast enough to offset demineralization. Thus good dental hygiene is essential to minimize plaque buildup, lower acid production, and allow better access to calcium and phosphate in the saliva. At the other extreme, as more plaque builds up, saliva may not contain adequate nutrients for the organisms in the plaque; the bacteria may then attack gum tissue.

Some people develop very few caries. One or two per thousand remain free of caries indefinitely, even though they are exposed to cariogenic bacteria and high-sugar diets. These persons have been called caries-immune. The reason for this immunity has not been determined, but must include genetic familial influences. Studies show that such persons are about 40 times as numerous among relatives as in the general population. These same studies show that caries-free male adults outnumber females by 2 to 1.

Water Fluoridation

During the 1930s dental research suggested a relationship between fluoride, F^-, and decreased frequency of caries. This led to the introduction of fluoride into drinking water in Grand Rapids, Michigan, in 1945, followed by widespread fluoridation of public water supplies in many areas of the country. Approximately one-half of the United States population now drinks artificially fluoridated water.

Two factors seem to be involved in fluoride's effect. First, fluoride can inhibit certain enzymes that catalyze the fermentation of carbohydrates to lactic acid. Concentrations as low as 0.5 ppm (0.5 mg/kg of water) have a detectable inhibitory influence on the rate of acid production by the oral flora. Even when the fluoride concentration in saliva is less than 0.25 ppm, some may accumulate within the dental plaque to concentrations ranging from 6 ppm to nearly 180 ppm. This increases the inhibitory effect.

The second explanation for the anticariogenic properties is the finding that fluoride ions will substitute for hydroxide ions in hydroxyapatite:

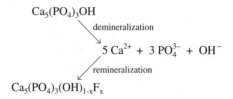

The formula $Ca_5(PO_4)_3(OH)_{1-x}F_x$ represents fluoridated hydroxyapatite, in which some of the hydroxide ions of the structure are replaced by fluoride ions. But the sum of the number of hydroxide ions $(1 - x)$ and the number of fluoride ions (x) still equals 1 for every 5 calcium and 3 phosphate ions in the enamel structure.

Whenever more Ca^{2+} and PO_4^{3-} become available through demineralization and from saliva, remineralization is increasingly favored. Available free F^- becomes incorporated into the hydroxyapatite structure. This shifts the equilibrium even farther in favor of remineralization. Fluoridated hydroxyapatite is less soluble in acid solution than hydroxyapatite, because fluoride ions help to lock the hydroxide ions more tightly in place. So, the chemistry explains both how fluoride is incorporated and why it helps prevent decay.

Other Sources of Fluoride

Although fluoridated water is the most common source, fluoride can be acquired through tablets and topical forms. The tablets are thought to be less effective than fluoride in water. A large portion of the higher dose in tablets is usually excreted in the urine. Regular ingestion of small quantities of fluoride in water leads to more complete absorption. Therefore, tablets provide full benefit only if taken every day. It is particularly difficult to ensure that youngsters will be conscientious about taking tablets, and they are thought to benefit the most from fluoride treatments.

Topical application of fluoride includes both the treatments given by dentists and the use of fluoride toothpastes (Figure 22.2). Toothpastes usually contain stannous fluoride,

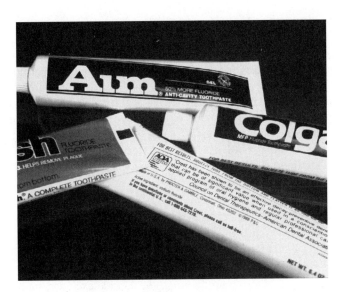

Figure 22.2
Fluoride toothpastes.

SnF_2, sodium fluoride, NaF, and monofluorophosphate (MFP), PO_3F^{2-}. The first two contain free fluoride ions, but MFP does not. Phosphate ions near the surface of the tooth, including those released by demineralizaiton, exist primarily as $H_2PO_4^-$, at the pH of the mouth. The similarity of PO_3F^{2-} and $H_2PO_4^-$ makes it possible for the MFP ion to replace some of the $H_2PO_4^-$. In addition, monofluorophosphate can penetrate and become bound in the plaque. There, at low pH due to organic acids, simple fluoride ions can be released by the reaction

$$PO_3F^{2-} + H_2O \underset{\longleftarrow}{\overset{\longrightarrow}{}} H_2PO_4^- + F^-$$

The Controversy Over Water Fluoridation

Although most health authorities in the United States maintain that fluoridation of water supplies is both beneficial and safe, scientific opinion is divided. Critics consider it a form of "compulsory mass medication."

There is limited disagreement about the protective effect of fluoride on tooth enamel. There is much more controversy over the possible *adverse* effects of water fluoridation. Various United States health groups have agreed since the 1950s that water fluoridated at about 1 ppm is a safe way to reduce dental caries. In other countries, health officials have either not allowed or have discontinued water fluoridation because of doubts about its safety. Even in the United States, support for fluoridation has diminished as the majority of referenda have rejected fluoridation. Many discussions of the subject have raised the following concerns (1–5):

1. There is one proven adverse effect of fluoride on teeth: Too much fluoride can cause a condition called *dental fluorosis*. In its mildest form this condition causes small, white, opaque areas on teeth. In its most severe form it causes a distinct brownish mottling. Fluoride reputedly makes teeth more resistant to caries. However, if dental fluorosis occurs, teeth often develop cavities anyway. Unfortunately, when dental fluorosis sets in, teeth are difficult to repair. Dental fluorosis can cause teeth to become very brittle, and brittle teeth don't hold fillings well. Dental fluorosis doesn't

result from artificial fluoridation alone because the levels are kept low enough to avoid this effect.

But the difference in concentration between the level believed necessary to achieve beneficial effects and the level that is expected to have harmful effects is not very great. Protection against dental caries does not increase above 1 ppm fluoride levels, but dental fluorosis then becomes a significant health problem. This sets the limits of effectiveness and safety for consumption of fluoridated water, but it does not account for large variations in fluoride consumption due to varying intake of water and of fluoride from all sources. Many bottled drinks and processed foods are prepared using fluoridated water. This fluoride plus that in dental hygiene products has led to an increase in dental fluorosis.

2. Over the last 40 years, corresponding to the period of use of fluoridated water, the rate of dental caries has decreased markedly, but many studies have shown that the improvement is often just as great in nonfluoridated areas as in fluoridated areas. The introduction of fluoride toothpastes, mouth rinses, and other products, as well as improved nutrition and oral hygiene, have all helped to improve dental health.

3. Most studies of the health effects of fluoride excluded subjects who might be particularly susceptible to adverse effects. People with kidney disease, who are likely to suffer from skeletal fluorosis, are among those excluded. Some studies suggest that fluoride accumulates in a person's bones, and may eventually reach toxic levels (at a water content of 0.97 ppm of fluoride). Unpleasant symptoms often disappear when fluoride consumption is discontinued.

4. The long-term effects of fluoride are unknown and are not easily studied.

5. Many researchers who have obtained results suggesting possible health risks due to fluoridation were discouraged by employers and journal editors from publishing their results.

6. The availability of fluoride in topical forms such as toothpaste and mouthwash, and in foods and beverages processed with fluoridated water, makes water fluoridation unnecessary. The use of topical fluoride preparations is optional, whereas fluoridated water supplies cannot be avoided.

Dental Amalgams

Cosmetic fillings have been found in Egyptian mummies, but most dental treatments at that time, and even into the 1700s, were extractions. The resultant space in the mouth was often filled with a false tooth fashioned from the tooth of an animal, or a piece of wood, ivory, or, in some cases, gold. Some restorative dentistry was performed in the Middle Ages by scooping out the decayed material and replacing it with some type of rosin or a metal oxide.

An Italian, Giovanni of Arcoli (1412–1484), is credited with being the first to use metals for filling teeth. He advised that the cavity be cleared of all decayed matter and then filled with gold leaf. For those who can afford it, this type of filling remains in use today.

In 1826, combinations of bismuth, lead, and tin were mixed with mercury to form the first dental amalgam. When metals are melted together, the resultant mixture is an *alloy.* An alloy containing mercury metal is an *amalgam,* and the process of combining the components of such an alloy is *amalgamation.* Mercury is a liquid at room temperature;

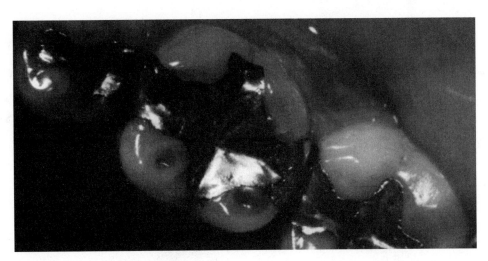

Figure 22.3
A dental amalgam.

Alloys are discussed in
Section 7.1.

when alloyed with other metals, the whole becomes a solid mass. The original amalgam of bismuth, lead, tin, and mercury had a melting range near the boiling point of water and had to be poured into the cavity at that temperature, which seems more painful than the toothache. Later, the formula was modified to melt at a temperature of about 66 °C so that it could be put into the cavity cold and then melted and adapted with a hot instrument.

In the late 1800s, G. V. Black (1836–1915) revolutionized dentistry. Although he had little formal education, he accumulated a vast store of knowledge of chemistry, dentistry, and metallurgy (the science and technology of metals). Black noticed that large amalgam fillings often became the source of other difficulties because they became chipped from biting and chewing. Therefore, he set out to develop a new filling material. He first determined the force required to chew various foods, the range of force as people clenched their teeth, and the amount of pressure fillings could withstand without cracking. He found that most jaws exerted 100–175 lb of pressure, some as much as 325 lb. Subsequent data indicate this to be approximately 30,000 psi (pounds per square inch).

Black developed an amalgam that would resist such a force, consisting of 65% silver, 27% tin, 6% copper, and 2% zinc. His basic formula is still used today (Figure 22.3). Most manufacturers increase the copper content to approximately 12%, with an accompanying reduction in tin content. When amalgamated with mercury, this newer alloy has superior physical and clinical properties.

The individual components of the amalgam add specific characteristics to the finished product. Silver provides tarnish resistance and strength, and generally slows the hardening of the amalgam, thus giving the dentist time to work the filling into the proper shape. Tin's high affinity for mercury is necessary for the early initiation of amalgamation. This decreases the setting time, so copper is used as a strengthening component, and in the 12% range it markedly improves strength at the margins of the restorations. Zinc aids in manufacturing the amalgam by reducing the tendency of the other metals to oxidize during melting.

Alloys for making amalgams are supplied in three forms—powder, tablets, and capsules containing the alloy and the mercury separated by some type of membrane. The dentist who chooses either the powder or the tablet adds the necessary mercury, combines them in a capsule, and places the capsule in a mechanical amalgamator to mix the mercury

Figure 22.4

A commercial amalgamator. An amalgamator is routinely used in a dentist's office to prepare an amalgam for filling a tooth. A premixed silver-tin-copper-zinc alloy is combined with mercury (liquid) in the capsule in the front right. When the capsule is activated, it is rapidly vibrated to ensure complete mixing of the amalgam components. An extra capsule, a funnel, and a plastic rod that is placed in the capsule to assist mixing appear in the top right. A 30-s timer and switch are on the left.

with the alloy (Figure 22.4). The amalgamator throws the mercury-alloy mixture back and forth with sufficient force to mix it properly. The amalgam is removed from the capsule, condensed (placed in the tooth), carved, and allowed to harden for 5–10 min. The amalgam continues to harden for months; however, 96–98% of the hardening occurs within 48 hours. After 24–48 hours, the alloy can be polished. Such restoration, if of superior quality, can be expected to serve for many years.

22.2 Soaps

GOALS: To learn the chemical structure and synthesis of soaps.
To understand the formation of soap micelles and their action as surfactants.
To understand the properties of soaps and the special virtues of coconut oil in making hand soaps.

Good hygiene is an important contributor to good health, and important to general good hygiene are soaps and detergents. Soaps are metal salts of fatty acids. The formula shows M^+ as the metal (usually sodium) and the anion derived from a fatty acid.

$$R - \overset{\overset{\displaystyle O}{\|}}{C} - O^- \ M^+$$

The first synthesis of soaps was carried out by the Romans about 2500 years ago. When they cooked animal fats with wood ashes, which contained potassium carbonate, K_2CO_3, they caused the set of reactions shown.

$$K_2^+CO_3^{2-} + H_2O \rightleftarrows K^+OH^- + K^+HCO_3^-$$

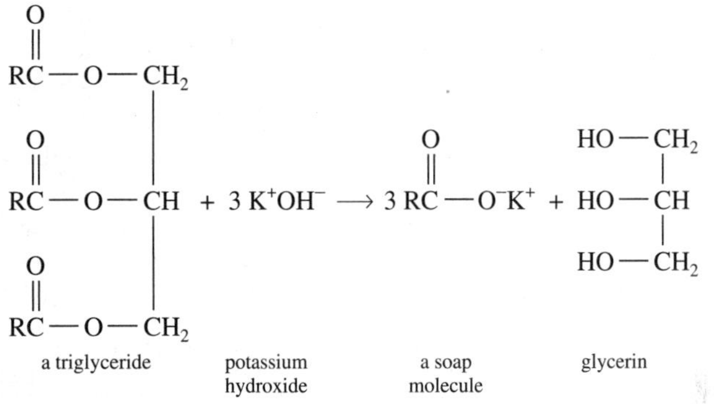

a triglyceride   potassium   a soap   glycerin
                    hydroxide   molecule

The first reaction forms potassium hydroxide, KOH, which causes (in a second reaction) the breakdown of triglycerides into the component parts, glycerin and fatty acids. The cleavage of triglyceride molecules by hydroxide ions is known as **saponification.** The fatty acid is neutralized by the strong base and ends up as a salt, a soap molecule.

In modern production, sodium hydroxide rather than KOH is normally used. Soap is isolated by adding NaCl, a procedure known as *salting out,* which lowers the solubility of the soap and causes it to solidify. Sodium soaps are hard solids that can be formed into familiar bars. Potassium soaps are soft and are usually used for making liquid soaps.

A typical soap molecule, the primary component of most hand soaps, is sodium myristate, the sodium salt of myristic acid, 14:0.

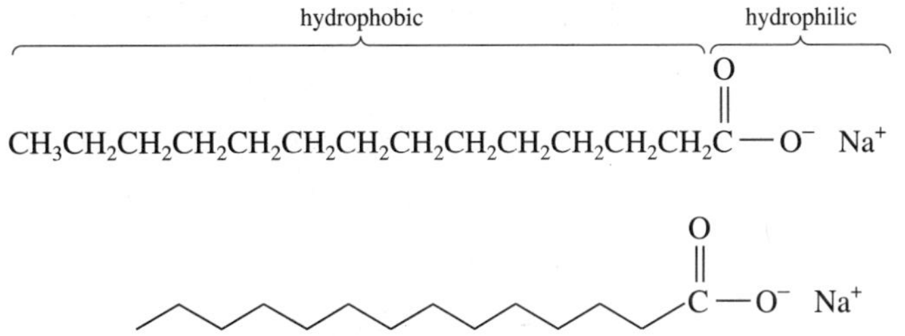

sodium myristate

Although it is correct to describe soap as a type of salt, the combination of the ionic group and the long hydrocarbon chain gives soap properties quite different from those of other salts. Salts in general are **hydrophilic,** or water-loving. They tend to dissolve in water readily, with cations and anions surrounded (solvated) by water molecules (Figure 7.4). Fats (and oils) are **hydrophobic,** or water-hating. When oil and water are mixed, they separate into two layers because of the repulsion between water and the hydrocarbon chains of the oil. Since a soap molecule combines both the hydrophilic nature of a salt and the hydrophobic nature of an oil, the result is a compound that can interact with both oil and water. More important, soap acts as an intermediary to allow oil and water to mix.

This is its most important function, since dirt invariably has some oil or grease incorporated into it. Oil causes the dirt to adhere to surfaces and prevents penetration by wash water, but soap breaks this barrier.

Surfactants

Soaps are described as surface-active agents, or **surfactants,** because they permit water to penetrate the surface of an oil droplet and divide it into many fine droplets. These then become suspended in water and are prevented from coalescing by interaction with soap in solution.

Milk is an analogous system in which the water-insoluble fat is kept in suspension by the action of milk protein (casein), which acts as a surfactant. Casein has both hydrophobic and hydrophilic groups. Usually casein is described as an emulsifier, and milk as an emulsion of oil in water. Similarly, soaps can be regarded as emulsifiers, although this word is generally not used in describing them.

The surfactant action of a soap can be explained on a molecular level. Soap differs from simple salts in another way: rather than separate molecules, they cluster together as **micelles** (Figure 22.5) in water. The hydrophobic ends of soap molecules clump together, avoiding interaction with water. The hydrophilic saltlike ends are on the surface exposed to the water.

In a cleansing action, each micelle traps a minute oil droplet. Rinsing washes away both soap and oil. Once the oil is removed, water can reach the surface that is to be cleaned, and very thorough cleansing is then possible.

It is necesary to strike a balance between the hydrophilic and hydrophobic properties in a soap molecule to produce the most effective surfactant. The best balance seems to come at the C-14 fatty acid, myristic acid. When the hydrocarbon chain is longer, the solubility in water decreases. Above C-18, the solubility is too low for the soap to dissolve, lather, and cleanse. When the chain is shorter than C-14, the ability to suspend oil droplets is decreased.

No natural oil source has myristic acid exclusively, but coconut oil has a complement of fatty acids very close to C-14 and is the most natural compromise. Many advertisements

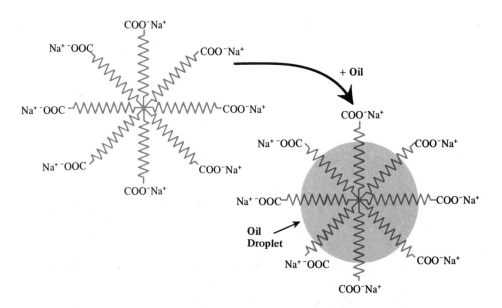

Figure 22.5
The surfactant action of a soap micelle.

for hand soaps call attention to coconut oil as an ingredient, and indeed, it is considered the most suitable oil for making soap. It has an unusually high content of lauric (12:0), myristic (14:0), and palmitic (16:0) acids. The high degree of saturation (low iodine value) is an extra bonus, because susceptibility to oxidation is extremely low.

Tallow from both beef and lamb frequently are used in making common soap, often in combination with coconut oil. Tallow contains more than 90% palmitic (16:0), stearic (18:0), and oleic (18:1) acids. In higher water temperatures used in commercial laundries, the reduced solubility of long-chain fatty acids is overcome due to increased solubility in hot water. This has made tallow a good source of soap for this application, even without blending with coconut oil.

Castile soap is made primarily from olive oil. Since olive oil has almost none of the 10:0 and 12:0 fatty acids, this product is reportedly less irritating to the skin than other soaps and is prized by some people for that reason.

Good lathering is an important characteristic in soaps. Here again the combination of hydrophilic and hydrophobic properties explains the action. Some hydrocarbon chains of the soap molecules align along the surface of the water, with the saltlike ends directed into solution exposed to the water. This weakens the attraction between the water molecules on the surface, promoting foaming and easy lathering. Soaps made from coconut oil have the best lathering characteristics.

The density of soap can be lowered by incorporating air to make the bars float. Creams, perfumes, deodorants, abrasives, antibacterial agents, and colorings are added to hand soaps.

Other Emulsifiers

Mayonnaise and salad dressings are described in Section 21.4.

Some commercially prepared foods contain emulsifying agents, too. Egg yolk is an important ingredient in mayonnaise and salad dressings, because it contains substances that act as emulsifiers and stabilize the oil-water emulsion in these products. The primary compounds involved are the lecithins and proteins known as lecithoproteins. Lecithins, cephalins, and certain other so-called phospholipids (also called phosphatides) are usually isolated from soybean oil, which contains about 2% phospholipids.

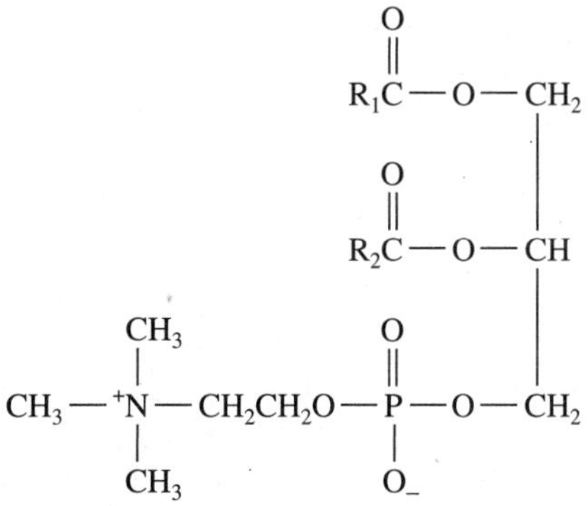

a lecithin 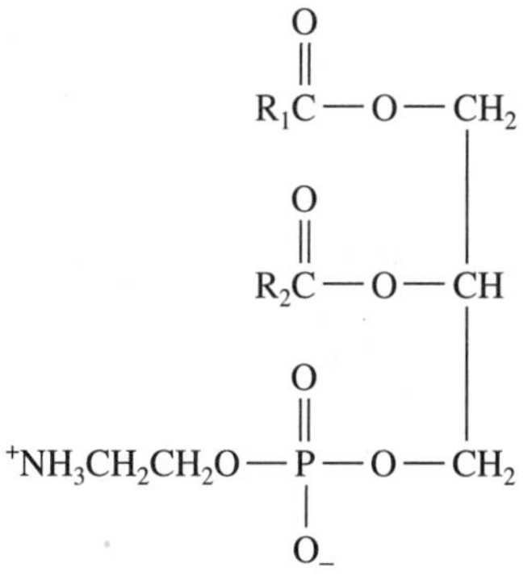   a cephalin

The charges on these substances give them hydrophilic character, while groups R_1 and R_2 are the hydrophobic portions of fatty acids.

Monoglycerides and diglycerides are also imporant emulsifiers that are sometimes added to shortenings. They help form a stable water-oil emulsion in cake batters, which allows for addition of more water to the batter; this permits incorporation of more sugar for a sweeter cake.

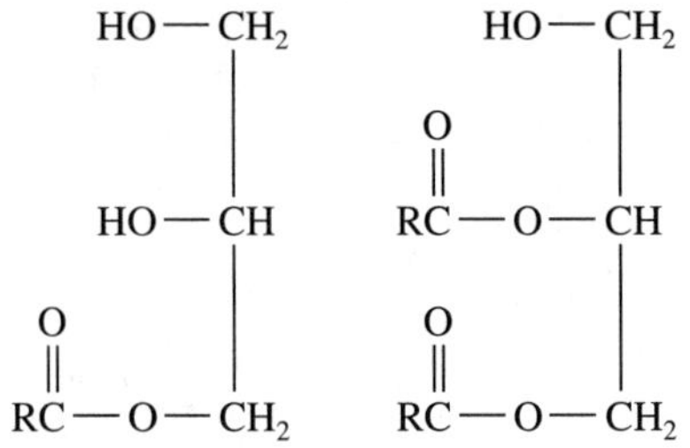

$$
\begin{array}{cc}
\text{HO}-\text{CH}_2 & \text{HO}-\text{CH}_2 \\
| & | \\
& \quad\ \ \text{O} \\
& \quad\ \ \| \\
\text{HO}-\text{CH} & \text{RC}-\text{O}-\text{CH} \\
| & | \\
\text{O} & \text{O} \\
\| & \| \\
\text{RC}-\text{O}-\text{CH}_2 & \text{RC}-\text{O}-\text{CH}_2 \\
\text{a monoglyceride} & \text{a diglyceride}
\end{array}
$$

Other Uses of Soap

Soaps of other metals such as aluminum, cadmium, cobalt, magnesium, nickel, zinc, and lead are used in a variety of applications. Their low water solubility permits appplications in which water might dissolve away a sodium soap. Copper soaps have been used as fungicides in paints in which their green color is acceptable. Zinc soaps are less active as fungicides and have very little color. Many soaps form gels with mineral oils and make excellent lubricating greases; a formulation containing 50% soap is typical.

22.3 Detergents

GOALS: To appreciate the problems, including environmental pollution, that have guided the development of detergents into their present form.
To know the structures of the surfactants that have been developed as alternatives to LAS.
To understand the strategies for keeping anionic surfactants separate from cationic fabric softeners.
To explain the action of optical brighteners.

Household detergents and related laundry products are continually changing as manufacturers strive to maximize their share of a huge consumer market. Several ingenious products have appeared over the last 30 years and some have disappeared because of environmental or other problems.

Water Hardness

Laundry detergents first became popular because soaps are unsuitable for laundering in hard water (Figure 22.6). Water hardness is attributable to the presence of metal ions, primarily calcium ions, Ca^{2+}, and magnesium ions, Mg^{2+}, which form insoluble precipitates with fatty acid anions. A ''bathtub ring'' is such a precipitate. Precipitates may be tolerable in the bathtub or wash basin, but are intolerable for washing clothes. They often end up as deposits on the clothes.

The solution for the precipitation problem came with the development of the alkylbenzene sulfonate (ABS) detergent. It is synthesized from benzene and propylene, which are available from petroleum. The reactions used for the synthesis are shown in Figure 22.7. Since four propylene units appear in the ABS, the product is commonly known as a *propylene tetramer* (PT) coupled to benzene, or as sulfonated PT-benzene.

Figure 22.6
Detergents.

Like soap, this synthetic detergent **(syndet)** is both hydrophilic and hydrophobic, so that it can stabilize water-oil emulsions. When sodium ions are replaced by calcium or magnesium, the detergent remains in solution. Unfortunately, merely remaining in solution does not ensure full surfactant action. The Mg^{2+} and Ca^{2+} can bind to the ABS anions and remain in solution but restrict the ability of the anion to suspend oil droplets.

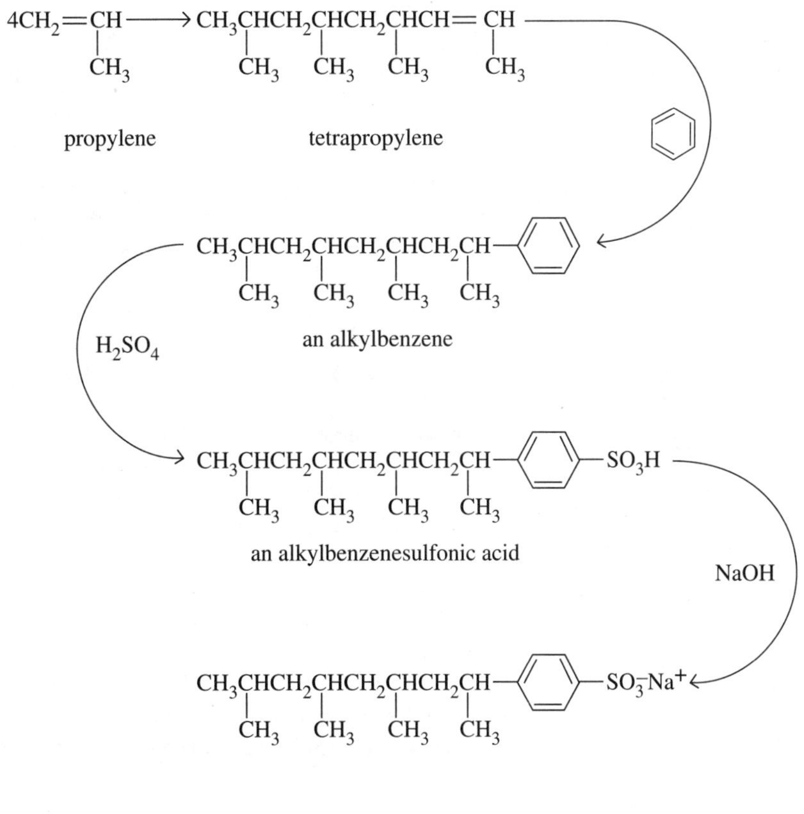

$$4CH_2{=}CH \longrightarrow CH_3CHCH_2CHCH_2CHCH{=}CH$$

propylene   tetrapropylene

$CH_3CHCH_2CHCH_2CHCH_2CH{-}\bigcirc$

an alkylbenzene

H_2SO_4

$CH_3CHCH_2CHCH_2CHCH_2CH{-}\bigcirc{-}SO_3H$

an alkylbenzenesulfonic acid

NaOH

$CH_3CHCH_2CHCH_2CHCH_2CH{-}\bigcirc{-}SO_3^-Na^+$

Figure 22.7
Sodium alkylbenzene sulfonate detergent (ABS), also known as sulfonated PT-benzene (PT = propylene tetramer).

In some areas, water is very hard, and the problem of detergents is intensified. Water softeners have been installed in some homes but are regarded as a luxury (Section 14.7).

Builders

Most homes do not have water softeners but do have some hardness in the water. In 1947, the syndet Tide combined the surfactant (15–20%) with a **builder** (variable 20–65%). Such a substance chemically softens water by binding to calcium and magnesium, so that the detergent remains free to function as a surfactant. From that time on, the popularity of syndets with builders increased dramatically. They surpassed soap in sales in 1953, and have since taken over more than 90% of the laundry product market.

The most common builders are sodium pyrophosphate and sodium tripolyphosphate, shown here binding Ca^{2+} and Mg^{2+}.

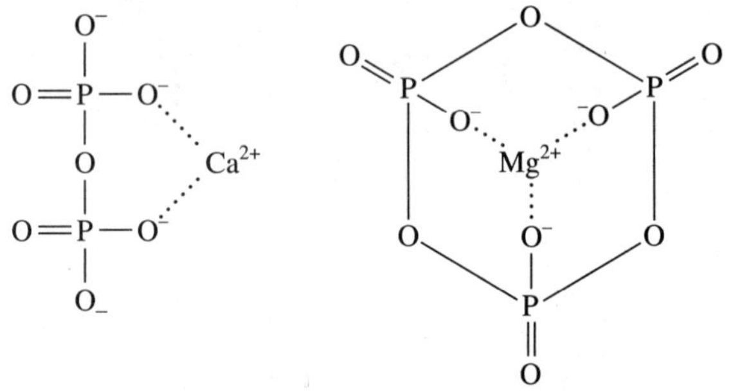

calcium
pyrophosphate

magnesium
tripolyphosphate

Together or separately these builders are called phosphates. Besides improving the washing properties of detergents, they have the virtue of being nontoxic. This is particularly important when the products are used in homes with small children.

The combination of ABS and phosphates was the chief detergent formulation throughout the 1950s. In the early 1960s, however, an environmental problem was traced to these products. Sudsy foam was building up in the water in many rivers and streams, and even flowing out of the faucets in many homes (Figure 22.8). It was found that the alkylbenzene sulfonates in detergents do not decompose fast enough during exposure to microorganisms in sewage treatment plants. In other words, they are not readily biodegradable, which means they remain in the water and may find their way into streams or back into the water supply of homes in some areas.

Consequently, the ABS-containing detergents disappeared from the market by 1965, and were replaced by detergents containing linear alkylbenzene sulfonate (LAS) surfactant. It is readily biodegradable and nonpolluting. The structure of LAS is shown.

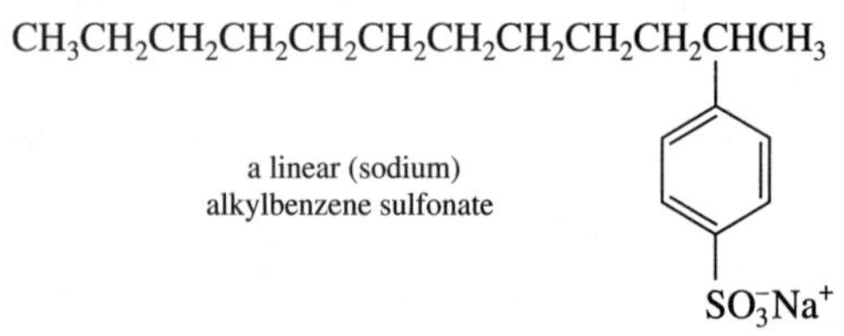

$$CH_3CH_2CH_2CH_2CH_2CH_2CH_2CH_2CH_2CH_2CHCH_3$$

a linear (sodium)
alkylbenzene sulfonate

$SO_3^-Na^+$

Like ABS, the LAS surfactant is synthesized from starting materials derived from petroleum. The important distinction between their structures is the branched versus the

Figure 22.8   The use of the branched-chain alkylbenzene sulfonate (ABS) detergents often caused foaming of fast-moving rivers and streams. This problem was solved by introducing biodegradable linear alkylbenzene sulfonate (LAS) detergents.

unbranched hydrocarbon chains. Like the fatty acid chains, the unbranched structure makes a hydrocarbon chain susceptible to attack by microorganisms in the environment and in sewage treatment plants. The branched chains of ABS detergents resist attack and are not readily metabolized by microorganisms. A biodegradable detergent is called soft, whereas a nonbiodegradable one is biochemically hard.

For a detergent, soft means biodegradable; hard means nonbiodegradable. For water, the terms hard and soft refer to mineral content.

In 1967 attention focused on another environmental problem attributed to the phosphate builders in detergents: the growth of aquatic plants, mostly algae, in lakes. Phosphorus, one of the principal nutrients for plant growth, frequently limits growth. If all required nutrients accumulate, a lake may be overrun by algal bloom, a condition called *eutrophication.* At first, this may be beneficial to a lake, since photosynthesis by the growing algae increases the supply of oxygen available to other aquatic plants and animals. But much of the oxygen produced by algae at the surface escapes into the atmosphere. As growth progresses, respiration by all living forms causes a net drain on the oxygen supply. When this **biochemical oxygen demand (BOD)** exceeds the supply, it causes the destruction of much aquatic life. The problem is accentuated as algal bloom blocks out sunlight required by other photosynthesizing aquatic plants. Even dead algae are troublesome, since decay bacteria may also consume oxygen. If the situation becomes extreme, a lake may deteriorate enough to be called dead.

Free-flowing streams and rivers, the site of the foam problem, capture atmospheric oxygen and do not suffer from oxygen deficiency.

Conceivably, sewage treatment plants could remove phosphates from waste water. But the level of purification required, known as tertiary sewage treatment, is still only experimental. Therefore, attention was focused on laundry detergents, since perhaps 40% or more of the phosphorus entering the aquatic environment came from these products at the peak of their use. A total ban on phosphates was undesirable because of their great convenience, efficiency, and safety. But the amount used, typically 20–65% sodium tripolyphosphate, or even higher for some presoaks, was decreased. Then in mid-1970, the big detergent manufacturers announced plans for replacing a large percentage of the phosphate with sodium nitriloacetate (NTA). It effectively binds hardness ions and functions as a builder.

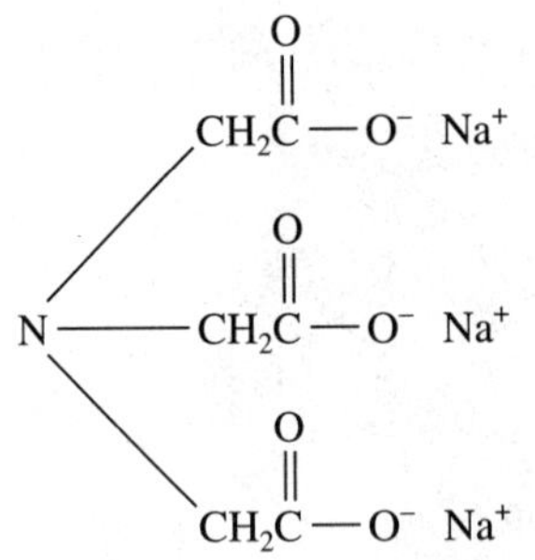

$$
N \begin{cases} CH_2C\overset{\displaystyle O}{\overset{\|}{}}-O^-\ Na^+ \\[1em] CH_2C\overset{\displaystyle O}{\overset{\|}{}}-O^-\ Na^+ \\[1em] CH_2C\overset{\displaystyle O}{\overset{\|}{}}-O^-\ Na^+ \end{cases}
$$

sodium nitriloacetate

By the end of 1970, NTA had already passed out of favor, because it also binds (sequesters) other metal ions such as cadmium and mercury. It was feared that these ions might be ingested and released at a site (for example, across the placental barrier) where the consequences could be very serious.

Thus, the combination of LAS surfactant and phosphate builders remained the dominant laundry detergent formulation well into the 1980s, using far less phosphates than were originally used. Environmental concerns slowly pressured detergent manufacturers, and by 1990 more than one-fourth of the states had passed legislation banning phosphate builders. Since then, phosphate-free products have become increasingly common.

At present, three builders dominate the market: sodium citrate, sodium carbonate, and zeolites. Sodium citrate is chemically similar to NTA with a structure well suited for binding Mg^{2+} and Ca^{2+}, but the binding is not as strong as with phosphates.

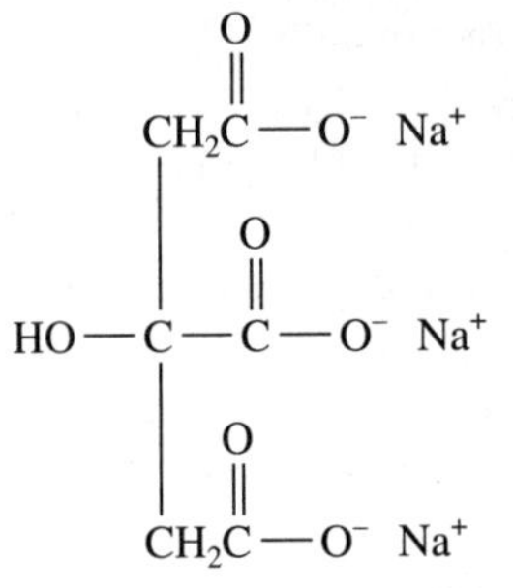

$$
HO-C \begin{cases} CH_2C\overset{\displaystyle O}{\overset{\|}{}}-O^-\ Na^+ \\[1em] C\overset{\displaystyle O}{\overset{\|}{}}-O^-\ Na^+ \\[1em] CH_2C\overset{\displaystyle O}{\overset{\|}{}}-O^-\ Na^+ \end{cases}
$$

sodium citrate

The high solubility of sodium citrate makes it particularly well suited for use in liquid laundry detergents, whereas sodium carbonate and the zeolites can be used only in powders.

Washing soda, or sodium carbonate, Na_2CO_3, is water soluble, but precipitates calcium and magnesium ions as calcium carbonate and magnesium carbonate.

$$Ca^{2+}\ +\ Na_2^+CO_3^{2-}\ \rightarrow\ CaCO_3\ +\ 2\ Na^+$$
$$Mg^{2+}\ +\ Na_2^+CO_3^{2-}\ \rightarrow\ MgCO_3\ +\ 2\ Na^+$$

Precipitates formed in a washing machine can deposit on clothes and even clog the machine. Nevertheless, one manufacturer of washing soda staged a "soap and soda" campaign as the answer to the environmental problem. One drawback to washing soda is the high alkalinity caused by the reaction with water to form NaOH. This makes its use hazardous around small children.

$$Na_2CO_3 + H_2O \rightleftharpoons NaOH + NaHCO_3$$

Zeolites, sodium aluminosilicates, are another interesting alternative to phosphates. Their use as anticaking agents was described in Section 20.4 and Figure 20.6. In detergents, zeolites function as *ion exchange* agents. Within the three-dimensional network of the zeolite, many Na^+ ions bind to the negatively charged aluminum atoms. The Ca^{2+} ions also are small enough to penetrate into the channels of the zeolite structure and, due to their greater positive charge, they replace the sodium ions and become trapped. Although Mg^{2+} ions are smaller than Ca^{2+} ions, they cannot move into the zeolite structure since they associate with a cluster of water molecules. The combination of Mg^{2+} ions plus water molecules is too large to enter the zeolite. Although there are no ecological or toxicological objections to zeolites, they are less effective than phosphate builders.

New Surfactants

Faced with less than ideal solutions to hard water and a desire to achieve new and improved characteristics, research has provided new surfactants, including anionics, nonionics, and cationics. These substances are now used in place of or in combination with the LAS ingredient.

Two alternatives to LAS are anionic surfactants—alcohol sulfates (AS) and ether sulfates (ES). Anionic detergents are particularly good for releasing soil from polyester fibers *(7)*. The AS surfactant can be made by reacting long-chain alcohols with sulfur trioxide, SO_3, and then with sodium hydroxide *(8,9)*. The alcohols are called fatty alcohols to note their similarity to fatty acids (Chapter 21). The conversion of a fatty alcohol to an alcohol sulfate is shown:

$$CH_3(CH_2)_nO-H \qquad \text{a fatty alcohol}$$
$$n = 11\text{--}17$$
$$\downarrow \begin{matrix} (1)\ SO_3 \\ (2)\ NaOH \end{matrix}$$
$$CH_3(CH_2)_nO-SO_3^-Na^+ \qquad \text{an alcohol sulfate (AS)}$$

Since AS surfactants do not irritate the eyes or skin, they have been in use in shampoos and hand dishwashing detergents for over 50 years. Although they are more expensive than the LAS surfactants, the AS function better in hard water. In many products, they are combined with an ether sulfate (ES) to achieve even better solubility in hard water.

The ES surfactant is synthesized from short-chain fatty alcohols, such as lauryl alcohol (C_{12}), by reaction with two to four units of ethylene oxide as shown. The lengthened chain is then reacted with SO_3 and NaOH to produce an anionic surfactant:

$$CH_3(CH_2)_{11}O-H \qquad \text{lauryl alcohol}$$

$$\begin{matrix} CH_2-CH_2 \\ \diagdown\ \diagup \\ O \end{matrix} \qquad \text{ethylene oxide}$$

$$CH_3(CH_2)_{11}O(CH_2CH_2O)_mCH_2CH_2O-H$$
$$m = 1\text{--}3$$
$$\downarrow \begin{matrix} (1)\ SO_3 \\ (2)\ NaOH \end{matrix}$$
$$CH_3(CH_2)_{11}O(CH_2CH_2O)_mCH_2CH_2O-SO_3^-Na^+$$
$$\text{an ether sulfate (ES)}$$

Because the ether group makes the chain more polar than the chain in the alcohol sulfate or the LAS, the ES surfactant is more soluble and therefore less susceptible to the effects of hard water. In states in which laws have banned the use of phosphates, combinations of AS and ES surfactants are generally used *(10)*.

A nonionic surfactant combined with the anionic ES surfactant is often the formulation for current liquid laundry detergents. The three most common nonionic surfactants are shown:

$$CH_3(CH_2)_{11}O(CH_2CH_2O)_mCH_2CH_2O-H$$

$$m = 1-3$$

a fatty alcohol ethoxylate

$$CH_3(CH_2)_nC(=O)-N \begin{array}{l} (CH_2CH_2O)_mCH_2CH_2O-H \\ (CH_2CH_2O)_mCH_2CH_2O-H \end{array}$$

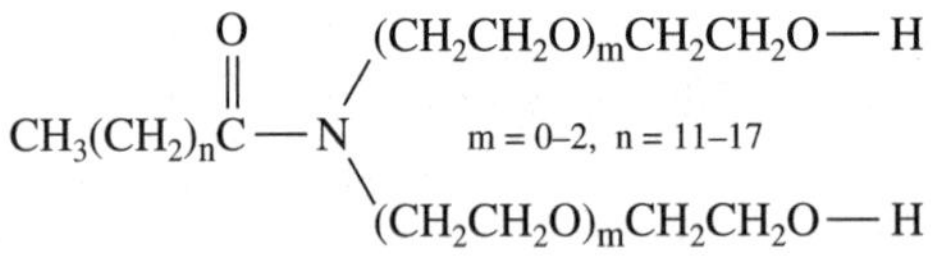

$$m = 0-2, \ n = 11-17$$

a fatty alkanol amide

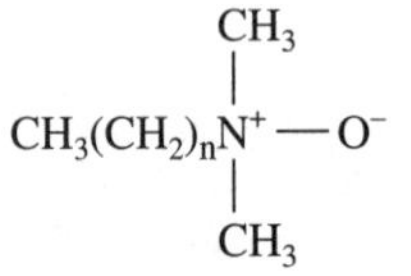

$$CH_3(CH_2)_nN^+(CH_3)_2-O^-$$

$$n = 12-18$$

an amine oxide

The fatty alcohol ethoxylate, which is not a good foam producer, is often used in liquid laundry detergents when heavy foaming is not required or acceptable. Because this type of surfactant is nonionic, it is unaffected by hardness ions, although the cleansing action is not as good as with the anionic surfactants.

The alkanol amide and amine oxide nonionic surfactants are relatively expensive but both are noted for heavy foaming. This makes them ideally suited for use in shampoos and in hand liquid dishwashing detergents.

The final class of surfactants is the cationics. An example is the *quaternary ammonium salt* shown:

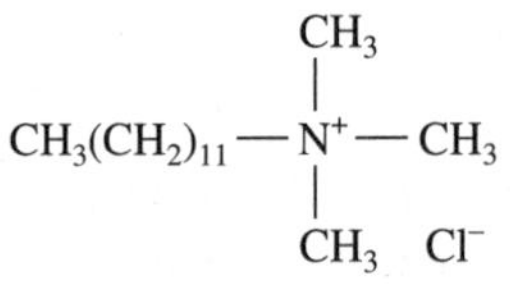

$$CH_3(CH_2)_{11}-N^+(CH_3)_3 \ Cl^-$$

a quaternary ammonium salt (quat)

Cationic surfactants are both expensive and inefficient, and thus are rarely used in detergents. Nevertheless, they do have antimicrobial properties and are included as antiseptics in cosmetics, as fungicides and germicides, and as fabric softeners.

Other Detergent Ingredients

Fabric Softeners

Cationic surfactants do not appear in detergents for cleaning purposes; they are widely used as fabric softeners and antistatic agents. Many washing machines are programmed to add a liquid fabric softener during the final rinse. Fabrics tend to carry negative charges that attract the quaternary ammonium ions. These cations remain bound to the fibers even after drying is complete, giving clothes a soft feeling and providing antistatic qualities.

Figure 22.9
Bounce fabric softener.

Figure 22.10
Bold laundry detergent.

The cationic softener should not be released into the wash water before all of the anionic surfactant has been removed in early rinse cycles. This ensures that the cationic and anionic materials do not combine and precipitate.

Many older washing machines do not have an automatic mechanism for introducing fabric softener in the final rinse. Therefore, some ingenious methods have been developed for introducing fabric softener. Products such as Bounce, Snuggle, and Downy are pieces of fabric in which quaternary ammonium salt has been absorbed and that are placed in the dryer with the wet clothes (Figure 22.9). As the clothes dry and the cooling effect of water evaporation ends, the dryer becomes hot enough to cause the fabric softener to sublime (vaporize) and deposit on the clothing.

Fab One-Shot is a small packet containing both an anionic surfactant and a cationic fabric softener. It consists of layers of ingredients that are kept separate from each other. In the washing machine, which reaches temperatures up to about 130 °F (55 °C), only the surfactant and other ingredients, such as whiteners, are released. The packet then is carried into the dryer, where temperatures approaching 200 °F (95 °C) cause the release of the fabric softener *(11)*.

Bold combines both anionic surfactant and cationic fabric softener in powder form (Figure 22.10). Here, the quaternary ammonium salt is combined with a solid fatty alcohol in the form of fine particles, size-selected so that they become trapped in the fibers of the clothes. The particles do not melt until about 150 °F (65 °C) and, therefore, remain intact in the washing machine. Once in the dryer, the higher temperatures cause them to melt and release the cationic fabric softener *(12)*.

Fillers

Powdered detergents must flow readily out of the box, be easy to measure, and cause little dust. Therefore, most of them contain large amounts (typically 60%) of inert filler, usually sodium sulfate, Na_2SO_4. Sodium sulfate is a white, crystalline material, well suited for mixing with various detergent ingredients. It is extremely water-soluble and is carried off in the early rinse cycle.

Optical Brighteners

Washed fibers tend to take on a slight yellow color. Substances alternately known as optical brighteners or fluorescent whitening agents are often present in detergents to counteract this problem. Several derivatives of a compound called stilbene have been produced for this purpose. One such compound is shown with the stilbene nucleus highlighted in color *(11)*.

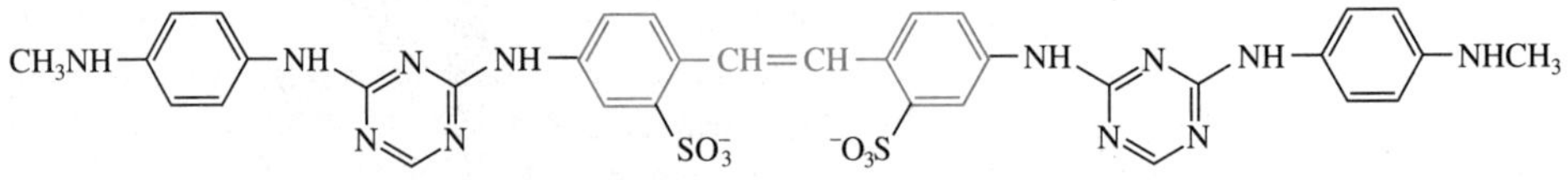

$$CH_3NH-\!\!\!\bigcirc\!\!\!-NH-\!\!\sqrt{N}-NH-\!\!\!\bigcirc\!\!\!-CH\!=\!CH-\!\!\!\bigcirc\!\!\!-NH-\!\!\sqrt{N}-NH-\!\!\!\bigcirc\!\!\!-NHCH_3$$

The strategy is that the optical brightener will fluoresce; this means that it will absorb invisible ultraviolet light and emit light in the blue region of the visible spectrum. This gives the perception of whiteness to the clothes, as the blue tinge complements the yellowishness on the fibers. Optical brighteners absorb into clothes and are not removed by rinsing.

Enzymes

Enzymes are sometimes included in detergents to aid in the removal of stains. The two most often used are *amylases,* which work on starch-based stains, and *proteases* or proteolytic enzymes, which break down protein-based soil.

22.4 Drain Cleaners and Oven Cleaners

GOAL: To understand the chemistry and hazards associated with drain cleaners and oven cleaners.

Two products related in both composition and mode of action are drain cleaners and oven cleaners. Clogged drains are generally attributable to fat. In studying soaps, we found that strong alkali will react with fat (triglycerides) to produce soap plus glycerin. Exactly this action explains the use of sodium hydroxide as both a solid and a concentrated water solution as drain cleaners. Some of the clogging fat may be converted to soap, which can help emulsify any remaining fat and be washed away.

Solid sodium hydroxide offers the additional advantage of liberating a large amount of heat when it comes in contact with water in the drain. The heat alone may be enough to break up the clog by melting some of the fat.

Oven cleaners also contain sodium hydroxide. They are usually dispensed in aerosol form with thickeners and propellants. Here again, a greasy residue can be converted to soap on the oven surface and easily washed away.

It cannot be overemphasized that any consumer product containing sodium hydroxide or related compounds is potentially very hazardous, particularly to the eyes. All too often, people become impatient and resort to mechanical means of unclogging a drain shortly after adding a drain cleaner. Sometimes, alkali is splashed into the eyes and onto the skin, with serious consequences.

SUMMARY

Tooth enamel is hydroxyapatite, $Ca_3(PO_4)_3OH$. It is susceptible to attack by acids formed both when hydroxide ions are released by demineralization and by the products of fermentation by microorganisms found in dental plaque.

Fluoride can be incorporated into tooth enamel by demineralization, followed by remineralization, to produce fluoridated hydroxyapatite; this form is much more resistant to acid, due to lower solubility and decreased tendency to release hydroxide ions. Fluoride can be applied topically by dentists and by the use of fluoride toothpastes. Stannous fluoride, SnF_2, sodium fluoride, NaF, and monofluorophosphate (MFP), PO_3F^{2-}, are the common sources of fluoride in toothpastes. The last of these provides a more prolonged release of fluoride ions, particularly when acidity is high. The addition of fluoride to public drinking water remains controversial (see Appendix A).

Alloys of silver, tin, copper, zinc, and mercury are amalgams. They are used by dentists for filling cavities.

Soaps are metal salts of fatty acids. As such, they are both hydrophobic and hydrophilic and act as surfactants (emulsifiers) by forming micelles to permit oil and water to mix, and allow efficient cleansing. Soaps are made by treating triglycerides with alkali (usually NaOH). A soap made from the 14-carbon fatty acid has the best properties. The hydrocarbon chain is short enough to allow good solubility, and long enough to give good surfactant properties. Of all the natural oils and fats, coconut oil is the most preferred for soaps, since it has an unusually high content of the 12-, 14- and 16-carbon fatty acids.

Many other compounds are important emulsifiers. Egg yolk contains lecithin and other emulsifiers that are responsible for the stability of various oil-water emulsions, such as mayonnaise.

Detergents are formulated with surfactants; they differ from plain soaps, which cause precipitation in hard water. The alkylbenezene sulfonate (ABS) was the first synthetic detergent (syndet). It did not precipitate in hard water, but neither did it clean very well in hard water until combined with one or more builders.

A builder is a substance that chemically softens water by binding to the hardness ions (calcium and magnesium). Builders that have been tried are sodium pyrophosphate and sodium tripolyphosphate, sodium nitriloacetate (NTA), washing soda (sodium carbonate), and zeolites (sodium aluminosilicates). The last acts by ion exchange, binding calcium ions but not magnesium ions.

Unfortunately, two environmental problems plagued the detergents formulated with ABS plus builder. The branched hydrocarbon portion of the ABS molecule is nonbiodegradable. This led to pollution of rivers and streams, which were sometimes found to foam vigorously. Replacement of ABS detergent with the linear alkyl sulfonates, LAS, solved the problem. Most successful builders used in syndets are phosphates, but the phosphates contribute to algal bloom and eutrophication of waterways. Attention has turned more toward alternative surfactants, including the anionics—alcohol sulfates and ether sulfates—as well as the nonionics and cationics.

The cationics are poor surfactants but are used as fabric softeners and antistatic agents. Some clever technologies use cationic fabric softeners to provide convenience while dealing with the need to keep them separated from the anionic surfactants. Other ingredients of commercial detergents are whiteners, fillers, and enzymes.

Drain cleaners and oven cleaners are useful but dangerous household chemicals containing sodium hydroxide.

PROBLEMS

1. Give a definition or example of each of the following.

   a. Demineralization
   b. Remineralization
   c. Hydroxyapatite
   d. Fluoridated hydroxyapatite
   e. Dental caries
   f. Dental plaque
   g. Amalgams
   h. Soap
   i. Hydrophobic
   j. Hydrophilic
   k. Surfactant
   l. Micelle
   m. Eutrophication
   n. Saponification

2. How does plaque affect teeth?

3. Why is sugar harmful to teeth?

4. Describe the mechanism for the incorporation of fluoride into tooth enamel.

5. How does fluoride affect tooth decay and why?

6. How safe is fluoride in drinking water? How important is concentration?

7. Write the reaction for the release of fluoride from MFP.

8. What are alternative methods of obtaining fluoride for the teeth? Compare these methods with fluoridated water.

9. How does a dentist prepare a dental amalgam?

10. How can soap be both hydrophilic and hydrophobic? Why is that important?

11. Why is coconut oil especially good for making soaps?

12. What is the advantage of castile soap?

13. How do detergents differ from soaps?

14. Why were ABS-containing detergents removed from the market?

15. Distinguish between hard water and a hard detergent.

16. Why are phosphates desirable in detergents? Why are they undesirable?

17. What are the major alternatives to phosphate builders? How does each work?

18. Write the structure of a typical ether sulfate. Why does it function better in hard water than an LAS surfactant does?

19. What is the strategy for avoiding contact between an anionic surfactant and a cationic fabric softener in the product Bold?

20. What is the purpose and mode of action of an optical brightener?

21. What is the active ingredient in drain cleaners and oven cleaners?

References

1. Hileman, B. "Fluoridation of Water." *Chemical and Engineering News*, August 1, 1988, pp. 26–42.

2. Waldbott, G. L.; Burgstahler, A. W.; McKinney, H. L. *Fluoridation: The Great Dilemma;* Coronado Press: Lawrence, KS, 1978.

3. Schultz, D. "Fluoride: Cavity-Fighter on Tap." *FDA Consumer* **1992**, *26* (1), 34.

4. "Public Health Service Report on Fluoride Benefits and Risks. (from the Centers for Disease Control)." *Journal of the American Medical Association* **1991**, *266* (8), 1061.

5. Shell, E. R. "An Endless Debate. (Fluoride in Water)." *The Atlantic* **1986**, *258*, 26.

6. Danielson, C.; Lyon, J. L.; Egger, M.; Goodenough, G. K. "Hip Fractures and Fluoridation in Utah's Elderly Population." *Journal of the American Medical Association* **1992**, *268* (6), 746.

7. Davidsohn, A. S.; Milwidsky, B. *Synthetic Detergents,* 7th ed.; John Wiley & Sons: New York, 1988; Chapter 2.

8. Myers, D. *Surfactant Science and Technology.* VCH: New York, 1988; Chapter 2.

9. Falbe, J. *Surfactants in Consumer Products;* Springer-Verlag: New York, 1987; Chapter 3.

10. Greek, B. F. "Detergent Components Become Increasingly Diverse, Complex." *Chemical and Engineering News*, January 25, 1988, pp. 21–53.

11. Stinson, S. C. "Consumer Preferences Spur Innovation in Detergents." *Chemical and Engineering News*, January 26, 1987, pp. 21–46.

12. Baskerville, R. J., Jr.; Schiro, F. G. *Detergent Compatible Fabric Softening and Antistatic Compositions.* U.S. Patent No. 3,936,537, February 3, 1976.

Immunochemistry

O ne of the biggest challenges of the past decade for medicine and science has been AIDS, the acquired immune deficiency syndrome. This is a disease caused by the human immunodeficiency virus (HIV), which is able to attack and inactivate the body's immune system. For most diseases caused by microorganisms, infection provokes the body's immune system to produce antibodies, whose chemistry is well known. Vaccines have been developed to fight off many viral and bacterial diseases such as measles, mumps, and diphtheria. Attempts to develop a vaccine against AIDS have been unsuccessful due to the viral attack on the immune system itself, as well as other factors that will be discussed in this chapter.

The chapter concludes with two special topics on immunity: the use of antibodies in diagnosis, and the response to allergens.

23.1 Microorganisms and Disease

GOALS: To distinguish viruses from other microorganisms and relate each to some common illnesses.
To appreciate why antibiotics are ineffective in combating viruses.

Microorganisms such as bacteria and fungi are capable of living by themselves as long as they have an adequate supply of nutrients. Viruses are an exception. They are purely parasitic particles that develop and reproduce only by invading cells of a host organism— microorganisms, plants, or animals. Viruses and other microorganisms may enter the body through the respiratory tract, digestive tract, genitourinary tract, or through the skin via open wounds or insect bites. If they enter the bloodstream, this paves the way for invasion of tissues throughout the body. The process by which an organism becomes established is called *infection*. Some prevalent infectious diseases are listed in Table 23.1.

Table 23.1 Some infectious diseases

Bacterial	Viral
Diphtheria	Measles (rubeola)
Meningitis	Hepatitis
Tetanus	German measles (rubella)
Whooping cough (pertussis)	Mumps
Tuberculosis	Poliomyelitis
Cholera	Rabies
Typhoid fever	Yellow fever
Plague	Influenza
Syphilis	Common cold
Gonorrhea	AIDS

Antibiotics are often effective against bacterial infections, but they are worthless against infections caused by viruses. The drugs act by selectively attacking bacterial cells because of the chemical differences between bacterial cells and host cells. A virus, however, relies on cellular metabolism in the host cells to reproduce itself. In general, drugs used against viruses attack both host *and* virus.

Many, but not all, infectious diseases confer resistance, or immunity, to repeated infection. For example, a person who has had measles, mumps, or polio develops lifelong immunity. Syphilis, gonorrhea, and most staphylococcal infections can be contracted repeatedly. The common cold may result from over a hundred different viruses. Immunity may result from infection by any of these, but there are plenty more to go around.

23.2 Immunochemistry

GOAL: To understand the chemical nature of the immune response.

The field of immunology studies the development of immunity, or the **immune response.** Often the field is divided into immunobiology, or cell-mediated response, and immunochemistry, or humoral (blood or tissue) response. Our attention focuses on the latter.

The humoral response is the ability to make **antibodies,** which are complex proteins that accumulate in the bloodstream in response to an infection. Substances that stimulate the production of antibodies are **antigens.**

The antibody proteins are called **γ globulins** or **immunoglobulins, Ig.** As seen in Figure 23.1, the immunoglobulin is a complex protein consisting of constant (black) regions and variable (color) regions. The constant regions are identical in all immunoglobulins, whereas the variable regions, which are unique for each antibody molecule, recognize different antigens.

The Ig proteins are able to neutralize an invading bacterium or virus particle and thus prevent repeated infection. At the initial infection, several days pass while synthesis of immunoglobulin reaches a level effective in combating disease. During that time, acute symptoms or complications may have serious effects. For example, whooping cough is accompanied by severe symptoms and is fatal to many infants under 1 year of age. Some symptoms result from the release of a toxin by disintegrating bacterial cells. Diseases such as mumps and German measles (rubella) are mild in infants; but mumps can be severe in adults, and rubella in a pregnant woman is a serious threat to the fetus.

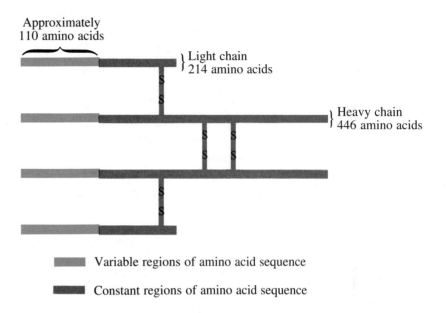

Figure 23.1
An antibody protein (immunoglobulin). The antibody molecule is a unit consisting of four protein chains (two heavy chains and two light chains) linked by disulfide bonds (—S—S—). All antibody molecules are the same in the portion shown in black, but each type is unique in the region of the molecule shown in color.

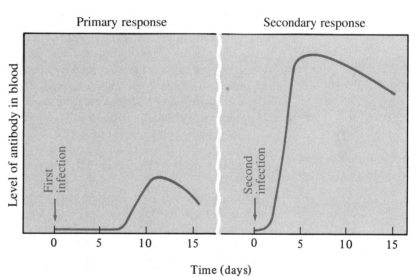

Figure 23.2
The immune response to an infection.

On a repeat infection, antibodies develop more rapidly. The cells responsible for producing immunoglobulins are known as memory cells, since they act against substances that previously caused an infection. A typical immune response curve (Figure 23.2) illustrates the rapid and higher level of antibody production at the time of a second infection.

23.3 Vaccines

GOAL: To appreciate the different types of immunity, both natural and those that are conferred by vaccines.

To provide protection against the diseases, attempts have been made to mimic the natural immune response. A vaccine causes the same immune response as the disease, but not the symptoms. It simulates an infection and stimulates the production of antibodies. A subsequent infection by the organism provokes the secondary immune response including the efficient synthesis of immunoglobulins, which neutralize the invading microorganism. Vaccines may take any of four forms:

1. Killed organisms (bacteria or viruses). Examples are vaccines against whooping cough (pertussis), typhoid, cholera, rabies, influenza, and polio (Salk vaccine).

2. Live attenuated organisms (usually viruses). Examples are vaccines against measles, German measles, mumps, polio (Sabin vaccine), and hepatitis.

3. Inactivated toxins, called toxoids (from bacteria). Examples are vaccines against diphtheria, tetanus, and *Haemophilus* b.

4. Recombinant DNA protein products.

Haemophilus influenza type b bacteria cause severe meningitis in children under 5 years of age. The vaccine, which is primarily a polysaccharide isolated from the surface of the bacteria, provokes an immune response.

Vaccines from Killed Organisms

A killed organism cannot reproduce in the host cell. However, it can evoke an immune response, resulting in antibody synthesis. Later exposure is greeted with a greatly enhanced level of antibodies—the memory response.

Live Attenuated Virus Vaccines

Some viruses are enough like the normal dangerous type to cause antibody production without causing a serious infection. These can be employed as *live attenuated virus vaccines.* When available, the vaccines are generally the most effective because they expose an individual to all components of a virus. Killed viruses, which cannot infect the cell, allow exposure only to surface components. The attenuated virus reproduces in the cell, even though its ability to cause severe symptoms has been removed.

The Salk vaccine, available since 1955, is a killed virus vaccine administered by injection. Since 1961, the oral Sabin vaccine, the classic example of a vaccine using live attenuated virus, has been administered because it provides more effective immunity (Figure 23.3).

The normal course of a polio infection begins in the digestive tract, invades the bloodstream, and attacks the spinal cord, resulting in muscle paralysis. Crippling results, and inability to use the chest muscles for breathing can be fatal. The attenuated virus can infect the digestive tract, where it multiplies and stimulates antibody production, but cannot invade the bloodstream. This one-stage infection causes mild symptoms, if any. The vaccine is a triple vaccine that provides immunity against all three polio virus strains.

In 1985 the Pan American Health Organization (PAHO) orchestrated a 4-year polio vaccination campaign in South and Central America to eradicate polio. Forty million children in 47 countries received either the Salk or Sabin vaccine. Only 11 cases of polio were reported for 1990, compared to 930 in 1986 *(1).*

Vaccines from Toxoids

A third type of vaccine is the *toxoid* or inactivated toxin. Large cultures of microorganisms are grown to isolate the toxic chemical substance, usually a protein, produced by the organism. The toxin is then inactivated, normally by treatment with some chemical. The resultant toxoid may be injected safely. Since the toxoid has much the same chemical structure as the toxin, it stimulates antibody production and provides protection in the event of a later infection by the microorganism when the toxin is released.

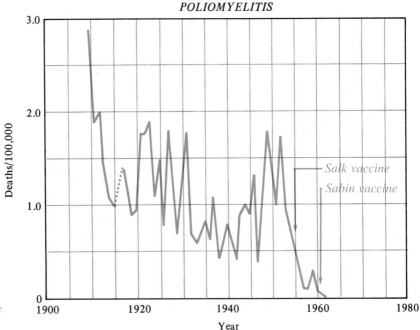

Figure 23.3
Poliomyelitis deaths per 100,000 population and effect of Salk and Sabin vaccines. (From Silverman, M.; Lee, P. R. *Pills, Profits, and Politics.* Copyright 1974 by the Regents of the University of California. Adapted by permission of the University of California Press.)

Table 23.2 (p. 492–493) is a typical vaccination schedule. Three doses of a combined diphtheria toxoid-tetanus toxoid-pertussis (killed organism) vaccine, designated DPT, is given to infants, as are vaccines against *Haemophilus* b and hepatitis B. Vaccination follows against the two kinds of measles, plus mumps, polio, smallpox, and other diseases for which we are currently at risk. Indications for boosters and adult doses are in the table.

There have been many expressions of concern in recent years over the high percentage of children who have not been vaccinated. A casual attitude toward vaccination is a logical consequence of the greatly reduced frequency of many diseases; the hazard seems small, even for an extremely contagious one such as measles, if most people are immunized against it. Therefore, it is commonly believed that not every individual should have to be vaccinated against measles; however, there are two flaws in this line of reasoning. For one thing, increased concern has been expressed by the United States Public Health Service (USPHS) and other sources that an increasing percentage of the population is not being adequately vaccinated. There is also the chance that not enough people may take the vaccines to retard the spread of some diseases. As many unimmunized persons reach older age without acquiring immunity, however, an epidemic becomes more possible, with symptoms that are more serious than they would have been in childhood.

Vaccines from Recombinant Techniques

Using techniques of recombinant DNA technology (Section 8.9), scientists construct live viruses that elicit an immune response (make antibodies) but are much less infectious or even noninfectious. In other words, the recombinant virus vaccines act as antigens that will stimulate production of antibodies, but lack the proteins necessary for infection. They can be produced in two ways. The dangerous virus can be genetically altered to eliminate

Table 23.2 Vaccines for major diseases

Disease	Age When Child Should Get the Shot	Dosage	Booster Needed?	For Adults?	Comments
Diphtheria	About 2 mo	3 shots, including tetanus and whooping cough, repeated at 4 and 6 mo	At 15 or 18 mo and again on entering kindergarten or first grade, and at ages 14 to 16	Adult toxoid is available for those facing unusual risk.	A booster may be recommended for child exposed to the disease.
Whooping cough (pertussis)	About 2 mo	As above	At 15 or 18 mo and again on entering kindergarten or first grade	Not indicated	Protection may be recommended for exposed adults with chronic illness.
Tetanus	About 2 mo	As above	As above, at ages 14 to 16, then every 10 yrs	Every 10 yrs; after an injury only if the wound is deep or dirty and last shot was over 5 yrs ago	Consult a doctor for tetanus-prone injuries (deep wounds contaminated by dirt).
Influenza	About 3 mo	Currently not recommended except for children with disabling or chronic disease; then 2 shots, spaced 2 mo apart, for those previously unimminized.	Once a yr	1 dose each fall for older people and those with chronic heart and lung disease	Immunization should be completed by mid-November.
German measles (rubella)	12 or 15 mo to puberty	1 shot, including measles and mumps	On entering kindergarten or first grade, or before middle school	Not pregnant women or those who might be pregnant within 2 mo of shot	Most adults are immune, but women of childbearing age can be tested for immunity.
Measles	12 or 15 mo	As above	On entering kindergarten or first grade, or before middle school	Rarely, since most adults already immune; a test can check this.	
Mumps	12 or 15 mo	As above	On entering kindergarten or first grade, or before middle school	Adults, particularly men who haven't had mumps, should receive vaccine, though half may have natural immunity.	To prevent spread, isolate patient until swelling is gone. Can cause sterility in adult men.
Haemophilus b	About 2 mo	Shots in 3 doses at 2, 4, and 6 mo	At 15 mo	Not recommended	Can be given simultaneously with DPT or polio vaccines.

Disease	Age When Child Should Get the Shot	Dosage	Booster Needed?	For Adults?	Comments
Hepatitis B	1–2 mo	Shots in 3 doses at 1–2 mo, 4 mo, 6–18 mo	None	Recommended in teenage years if not received earlier	
Polio	About 2 mo	Triple oral (Sabin) vaccine given in 2 doses at 2 and 4 mo	At 15 or 18 mo and again on entering kindergarten or first grade	Adults are reimmunized by handling newly immunized infants.	It is recommended that persons with impaired immune systems avoid newly immunized babies because infection is possible.

Based on data of the U.S. Center for Disease Control and the American Academy of Pediatrics (2).

DNA is the genetic material that codes for (contains the information to direct the synthesis of) cellular proteins. The piece of DNA that codes for a specific protein is called a gene. See Chapter 8 for details.

infectivity; or a new noninfectious virus, a *carrier virus,* can be used to carry genes (pieces of DNA) for the production of antigenic virus proteins. When the carrier virus infects the host, the antigens produced will elicit the production of antibodies.

Recombinant DNA technology, specifically gene cloning, provides a mechanism that avoids isolation of the toxic protein from the infectious organism. In this method, the gene that codes for the toxic protein is transferred out of the infectious organism and put into a second (vector) organism, usually bacterial cells. *Cloning* of a gene is completed when the host cells take up the transferred DNA and reproduce it. The bacterial cells can then produce large quantities of the toxin.

If the gene is first altered so the protein produced is a toxoid and not the toxin, chemical modification is avoided. This technology is being used to generate a new whooping cough vaccine since the current killed cell vaccine can cause neurological complications (3).

Another approach is being tried to generate a vaccine against malaria. Since the amino acid sequence of the proteins on the surface of malaria parasite is known, pieces of the proteins are being synthesized in the laboratory and tested as a vaccine to elicit an immune response (4).

Antitoxins: Passive Immunity

An animal (for example, a horse) can be dosed with an antigen to develop immunity. Then the immunoglobulins are isolated from the blood of the animal, purified, and injected into a human at risk of an infection. These antibodies are regarded as *antitoxins,* since they act against antigens, substances that cause toxic effects.

This procedure is said to confer **passive immunity** to the recipient. Compared with the natural immunity that results when an individual synthesizes his or her own antibodies after exposure to disease organisms or a vaccination, passive immunity is only temporary, typically 4–6 weeks. It provides no increased capacity for combating subsequent infections. In other words, there is no memory in the system, since individuals do not synthesize their own antibodies. This antitoxin-induced, passive immunity is not the best method for obtaining immunity, but at times no reliable and quick alternative is available. A good example is an unimmunized person with a deep tetanus-prone wound, in whom adminis-tration of tetanus antitoxin may be life-saving. Antitoxins also are employed for botulism and gas gangrene.

Rabies is another infection that can be treated with an antitoxin. Its unusually long development time (incubation period) of 1–3 months provides enough time for active immunization to be effective. The killed virus vaccine is normally given in daily injections for 2–3 weeks. Once symptoms have developed, there is no effective treatment, and death is inevitable. Usually, both antitoxin and vaccine are administered because of the potential severity of the disease *(5)*.

Breast-feeding newborns also conveys passive immunity. Many individuals are staunch supporters of breast-feeding, accurately claiming that the child acquires antibodies through the mother's milk, and thus has a greater resistance to any diseases for which the mother carries specific antibodies. Natural passive immunity is gained by passage of pre-formed antibodies through the placenta from mother to fetus. It is temporary, 4–6 months, but by then the child's immune system is functioning.

23.4 Influenza

GOAL: To understand the relationship between the structure of the influenza virus and its ability to bypass existing immunity.

Influenza viruses exist in different forms, usually symbolized as types A, B, or C and named for the location of first outbreak. Since the type A virus is most often associated with epidemics, such as the Asian flu and the Hong Kong flu, it is possible to produce a vaccine against it. Indeed, flu vaccines are available. Unfortunately, the type A virus is not a single virus but several, designated A_1, A_2, as well as by location. For example, it was the A_2 virus that caused the Asian flu beginning in 1957. More recent outbreaks resulted from influenza type A viruses. A-Taiwan (influenza A, first identified in Taiwan) and A-Shanghai (influenza A, first identified in Shanghai) were responsible for outbreaks in 1986 and 1987, respectively. A 1990 epidemic, primarily from A-Shanghai, resulted in over 8,000 deaths in America, primarily among the elderly. Influenza A-Taiwan and A-Beijing were responsible for an outbreak in 1991–1992.

Although viruses cannot be inhibited by antibiotics, such treatment is sometimes used to prevent complicating secondary bacterial infections such as pneumonia. The resistance of a virus to antibiotics is attributed to the parasitic nature of viruses. A virus is a relatively simple particle, consisting of strands of genetic material, either DNA or RNA, surrounded by a protein overcoat. Figure 23.4 represents the structure of an influenza virus particle, which resembles a ball studded with spikes. The core of the virus contains the genetic material and is surrounded by layers of protein and lipids. The spikes are composed of two kinds of glycoproteins (combination sugar and protein): hemagglutinin (H) and neur-aminidase (N).

Section 8.8 describes the difference between DNA and RNA.

When an influenza virus particle attacks a normal cell, the genetic material of the virus particle enters the host cell and takes control. The genetic material of the influenza virus is RNA. Once inside the host cell, RNA is used to make DNA. The genetic material directs the synthesis of proteins, including enzymes that control metabolism and other chemical processes (Section 8.8). In a viral infection, the cell machinery is turned over to the production of new virus particles that are then released for attack on other cells. The H glycoprotein allows the virus to enter host cells. The N glycoprotein permits virus particles to escape from host cells and infect other cells after the virus has been reproduced (replicated) within the host.

Immunity to the influenza virus presents a major challenge. Antibodies normally build up after an infection, or a vaccine can be administered. Either way, antibody proteins act by recognizing the surface proteins of a virus. Even in a killed virus vaccine, the surface proteins are not greatly altered, so that later exposure to the live virus evokes the memory response, causing rapid synthesis of immunoglobulins and neutralization of the virus.

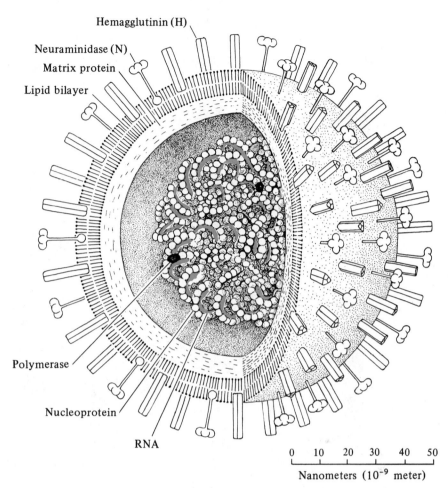

Hemagglutinin (H)

Neuraminidase (N)

Matrix protein

Lipid bilayer

Polymerase

Nucleoprotein

RNA

0   10   20   30   40   50

Figure 23.4

Nanometers (10^{-9} meter)

Cutaway diagram of an influenza virus particle reveals its external and internal construction. The hemagglutinin (H) and neuraminidase (N) spikes are embedded in a lipid, or fatty acid, bilayer that surrounds the core of the particle. Matrix proteins line the underside of the lipid bilayer and surround the core. Inside the core is a helical complex of molecules, consisting of RNA in association with enzymes and other proteins. The RNA contains the virus's genetic information, which is required for replication of the virus once it has entered (infected) a normal cell. (From Kaplan, M.; Webster, R. G. "The Epidemiology of Influenza." *Scientific American,* December 1977. Copyright by Scientific American.)

Unfortunately, the recognition step is the weak link in triggering the immune response. The surface glycoproteins (H and N) of influenza viruses undergo periodic mutations. Even minor mutations in either the protein or in the sugar portion can decrease recognition by the immune system. Therefore, the immune response is less effective and some degree of infection may occur. Minor mutations occur frequently, every year or two. Extensive mutation may totally obscure recognition of an invading virus, and thus all existing immunity may be bypassed. Therefore, vaccines developed against the influenza virus must be updated annually.

Major mutations typically occur every 10–15 years and are blamed for the serious flu epidemics in 1918–1919, 1933, 1947, 1957, and 1968. Figure 23.5 shows the migration

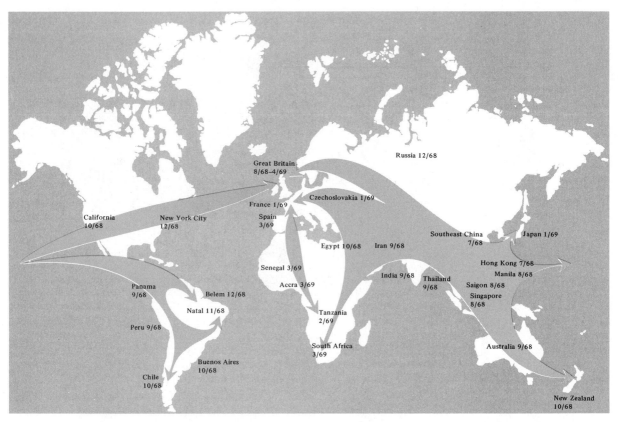

Figure 23.5 The 1968–1969 itinerary of Hong Kong flu, the last pandemic. Vaccines have prevented further pandemics. (Adapted with permission from E. D. Kilbourne, *Natural History* **1973**. Copyright 1973, by the American Museum of Natural History.)

of the Hong Kong flu around the world. This variation of the influenza virus is regarded as the last major global epidemic, or *pandemic*. No major epidemics have occurred recently due to the development of effective vaccines. Flu outbreaks occurred in 1986, 1987, 1990, and 1992, and primarily affected elderly persons, individuals with weak immune systems, and those who were not vaccinated.

One of the most devastating outbreaks of influenza A, known as the swine flu, occurred during 1918–1919. The disease struck in two waves. The first began in Spain in the spring of 1918 and spread to many parts of the world. It caused widespread illness but few deaths. In September 1918, the second wave began, presumably due to a slightly mutated form of the first virus. This illness was devastating, causing 20 million deaths throughout the world. Since vaccines were not available, people tried to protect themselves with face masks, as seen in Figure 23.6. Individuals affected by the first wave acquired immunity that protected them against the mutant virus. The mutant was similar enough to serve as an antigen and evoke a memory response. A January 1976 influenza outbreak was potentially dangerous. The causative organism was similar to the one that caused the 1918–1919 epidemic, so mass vaccination of the American population was carried out in the fall of 1976.

Flu vaccines usually consist of killed virus, but studies on recombinant live viruses, as well as recombinant toxoid vaccines, are under way.

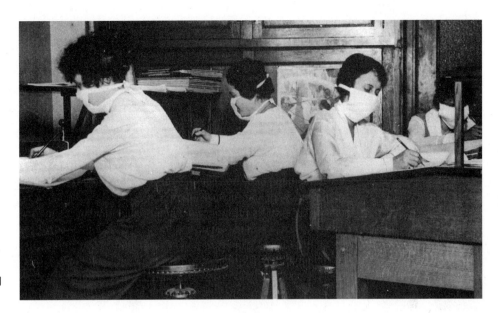

Figure 23.6
Face masks, not
vaccines, were the only
protection against the
deadly virus that caused
the 1918 flu pandemic.

23.5 AIDS

GOALS: To understand how HIV affects the immune system.
To appreciate the difficulties in developing a vaccine against HIV.
To understand the common strategy in using drugs against AIDS.

The first cases of acquired immune deficiency syndrome (AIDS) were reported in 1981. A few years later, researchers discovered that an RNA virus, the human immunodeficiency virus (HIV), causes AIDS. The World Health Organization estimates that 1.5 million adults and children worldwide have AIDS and 8–10 million adults are infected with HIV. Over 80,000 people in the United States have died of the disease since 1981.

The virus attacks and infects cells of the immune system, specifically white blood cells known as *helper T cells.* Their function is shown in Figure 23.7 and the effects of HIV infection in Figure 23.8. Helper T cells direct the memory cells (B cells) to produce antibodies. The infection causes a malfunctioning of the immune system. Sufferers with AIDS do not die from the HIV infection but from other infections that occur due to a suppressed immune system.

The HIV is a type of RNA virus called a retrovirus. Once inside the host cell, RNA is used to make DNA, which then becomes part of the host cell DNA. The incorporated viral DNA directs synthesis of new viral particles (Figure 23.8), but also has a second dangerous effect. Not only can viral particles infect helper T cells, but infected helper T cells also can attack uninfected cells. This process is a very unusual mechanism for continued infection.

The RNA of HIV is surrounded by a protein core, and this combination is surrounded by a coat consisting of proteins and glycoproteins. The antigenic glycoproteins in the protein coat recognize surface proteins on helper T cells, called *receptors.* The helper T cells thereby become targets for the virus, and AIDS results in a progressive decrease in helper T cells.

The HIV infection renders the immune system helpless and allows otherwise rare infections to invade and flourish. These infections, including *Pneumocystis carinii*

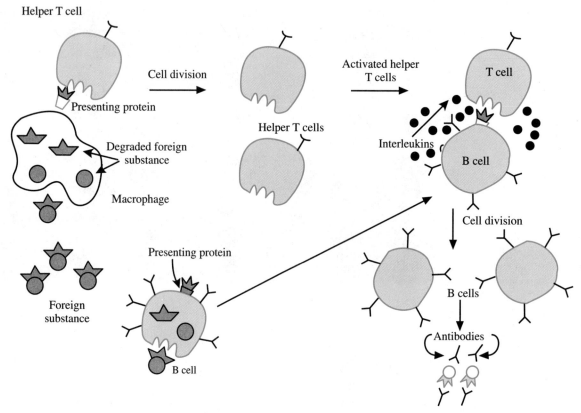

Figure 23.7 How helper T cells act. Macrophages and B cells engulf foreign substances such as the antigens on virus particles. Each type of cell then presents the degraded foreign substance as a receptor on the cell surface attached to a "presenting protein." Helper T cells bind to the antigens presented by the macrophages; this causes the helper T cells to multiply and secrete compounds called interleukins (represented as dots). The interleukins encourage B cells, the memory-activated antibody-producing cells, to multiply and release antibodies against the antigens. The antibodies bind to and neutralize the foreign substance, and mark it for destruction.

(pneumonia), cryptococcal meningitis, and toxoplasmosis (a brain infection), account for 90% of deaths due to AIDS. A rare skin cancer, Kaposi's sarcoma, also appears in many patients. One area of research focuses on developing agents effective against each of these infections.

Another area of research is attempting to develop anti-HIV drugs. The only drugs that have U.S. Food and Drug Administration (FDA) approval are zidovudine (originally known as azidothymidine, AZT), dideoxyinosine (DDI), and dideoxycytosine (DDC). The AZT works by mimicking deoxythymidine, a natural building block of DNA, required by HIV to make new viral particles. It blocks the synthesis of viral DNA and slows the progress of the disease. Both DDI and DDC work in a similar fashion. Currently, DDI is being administered in combination with AZT or given to individuals who are resistant to the effects of AZT.

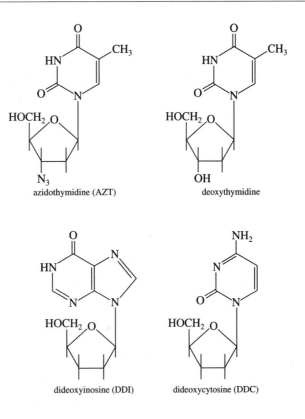

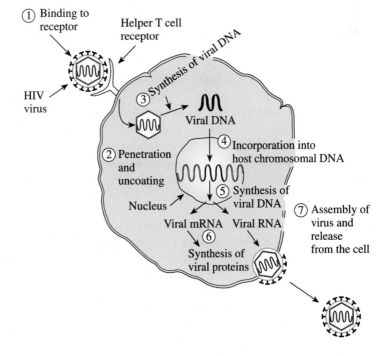

Figure 23.8
HIV infection of a helper T cell and virus replication. (1) HIV binds to receptor on helper T cell. (2) Core of virus enters cell. (3) Synthesis of DNA from viral RNA catalyzed by the enzyme reverse transcriptase. (4) Incorporation of viral DNA into host cell DNA. (5–7) Synthesis of viral RNA from DNA necessary to make new viral particles, which are assembled and exit to attack other cells.

No vaccine is available yet to protect against HIV infection, because several characteristics of the virus have made development of one very difficult. First, the virus attacks and kills cells that would be stimulated by the vaccine to produce antibodies. Second, the virus has a high mutation rate. It mutates frequently (within weeks), and minor changes in its DNA composition can affect the immune response. Third, infected helper T cells, which carry the HIV genetic material, can infect normal helper T cells, so a vaccine must also target infected cells.

Since AIDS is eventually always fatal, researchers are avoiding the most common type of vaccine, killed whole virus, for fear that some virus particles might escape being killed and end up in the vaccine. Live attenuated viruses are also being avoided for fear of reversion back to an active infectious virus, particularly since the mutation rate is so high. Vaccines currently being tested include those from recombinant DNA techniques that are being employed to produce the glycoproteins that make up the protein coat, which triggers an immune response. One approach uses the glycoproteins as the vaccine; another clones the glycoprotein genes into a carrier virus and uses the carrier virus as the vaccine. A third incorporates the use of virus-like particles (VLPs). They are made of protein (no DNA or RNA) and are approximately the same size as viral particles. The HIV surface glycoproteins can be inserted into a VLP.

Human trials are being conducted with each type of vaccine, but no product has been approved for large-scale use. Researchers have achieved some success with a vaccine to protect monkeys against simian immunodeficiency virus (SIV), a close relative to HIV that causes an AIDS-like disease in the animals. The vaccine is based on an envelope protein of SIV *(6)*.

The groups at highest risk for contracting AIDS are intravenous drug users who share needles and people who have sex with many partners. The disease has also been contracted through tainted blood in blood transfusions, but since 1985 a screening test has been available and now the blood supply should be safe.

23.6 Diagnostic Uses of Antibodies

GOAL: To consider the diagnostic applications of antibodies.

The **enzyme immunoassay (EIA)** test determines if a blood sample contains antibodies against the AIDS virus, or is HIV positive. The test exposes a blood sample to solid beads coated with HIV surface glycoproteins and other viral parts, the antigens (Figure 23.9A). Any HIV antibodies in the blood will bind to the solid beads. The beads are then washed and exposed to a second antibody, an antihuman immunoglobulin. This second antibody is developed in goats inoculated with human HIV antibody as an antigen. This second antibody recognizes and binds to the HIV antibody. Attached to it is an enzyme that causes a color to develop when treated with certain chemicals. Color indicates a positive test; HIV antibodies are present.

The screening test for AIDS is only one of the clinical uses for antibodies. A screening test for hepatitis B in blood also uses an EIA (Figure 23.9B). Blood is examined for the virus that causes the disease. Solid beads are coated with an antibody specific for the hepatitis B virus, a **monoclonal antibody.** Monoclonal antibodies are antibodies isolated from a single antibody-producing cell line. Therefore, they are structurally identical since they were produced by identical cells, and specifically recognize only one antigen.

If the blood sample contains hepatitis B virus, the antibody will recognize it and bind to form an antigen-antibody complex. The sample is then exposed to a second monoclonal antibody for hepatitis B that is attached to an enzyme. If the virus is present, the second

A. EIA - HIV Test

 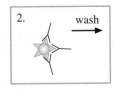 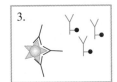 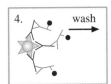 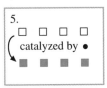

B. EIA - Hepatitis and HCG Test

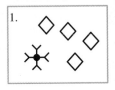 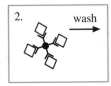 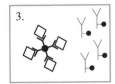 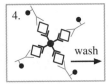 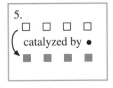

Figure 23.9 Enzyme immunoassays.
(A) For HIV: 1. A solid bead is coated with HIV protein coat and other HIV parts (⚝) to test a blood sample for the presence of HIV antibodies (ʎ). **2.** If the HIV antibodies are present, they will recognize and bind to the HIV parts on the bead. The beads are then washed to remove blood components that did not bind to them. **3.** The beads are exposed to a second antibody (Y), an antihuman immunoglobulin that has an enzyme attached to it (→•). **4.** If the HIV antibody is bound, the antihuman antibody will recognize and bind to it. The beads are washed again to remove any unbound antihuman immunoglobulin. **5.** Chemicals (ooooo) are added that undergo a color change in the presence of the enzyme.
(B) For hepatitis or HCG: 1. A solid bead is coated with hepatitis or human chorionic gonadotropin (HCG) monoclonal antibody (⚝). Blood or urine is tested for the hepatitis virus or HCG, respectively (◇), which is referred to as the antigen. **2.** If the antigen is present it will recognize and bind to the antibody. The beads are then washed to remove any blood or urine components that did not bind to them. **3.** The beads are exposed to a second monoclonal antibody (Y), to hepatitis or HCG, that has an enzyme (→•) attached to it. **4.** If the antigen is present, the second antibody will recognize it and bind to it. The beads are washed again to remove any unbound antibodies. **5.** Chemicals are added that cause a color change in the presence of the enzyme.

antibody binds to the virus particles in the complex. As in the AIDS test, the enzyme causes a color change.

A third example of the diagnostic uses of antibodies is the home pregnancy test, which can be bought in any drug store. It is also an EIA, but screens for *human chorionic gonadotropin (HCG)* in urine (Figure 23.9B). The HCG is a protein excreted in urine in the early stages of pregnancy (Section 26.7). The EIA is similar to the hepatitis screening test, except the monoclonal antibodies are specific for HCG. Such antibodies are used in many other diagnoses.

23.7 **Allergy**

GOAL: To understand the nature of allergy and the use of antihistamines.

Our coverage of immunochemistry so far has viewed antibodies as exerting positive effects. Unfortunately, there is another side to the story. Immunity can cause difficulties ranging from a nuisance like a runny nose for hay fever sufferers, to severe allergic reactions to drugs (for example, penicillin), called *anaphylactic reactions* or *anaphylaxis.* Anaphylaxis may even be fatal if not dealt with promptly.

In an allergic reaction, a person is initially exposed to an allergen (pollen, dust, ragweed, etc.). It acts as an antigen and triggers an immune response in which immunoglobulins (Ig) are produced. The Igs become anchored by receptor proteins to the surface of mast cells that line the nose, throat, and lungs. The mast cells are specialized to produce histamine.

When the person comes in contact with the allergen a second time, Ig recognizes and binds to the allergen, causing a series of events in which histamine is released. Histamine is a potentially potent toxin that can cause the contraction of certain types of muscle tissues, primarily in the lungs. In the extreme, suffocation may result. In less extreme cases, such as hay fever, an antigen may be inhaled in small quantities, causing the release of small amounts of histamine, leading to dilation of blood vessels, swelling, and increased secretion from the eyes and nose. Histamine release also is associated with allergic reactions to food, with bee stings, and with asthma attacks. Some very susceptible individuals prudently carry insect-sting kits containing an *antihistamine* at all times.

Antihistamines can block the effects of histamine on body tissues. The best-known antihistamine is chlorpheniramine (brand name Chlor-Trimeton). A common problem with the drugs is drowsiness, so nonsedating antihistamines were developed. One such drug, terfenadine (Seldane), has been commercially available since 1985.

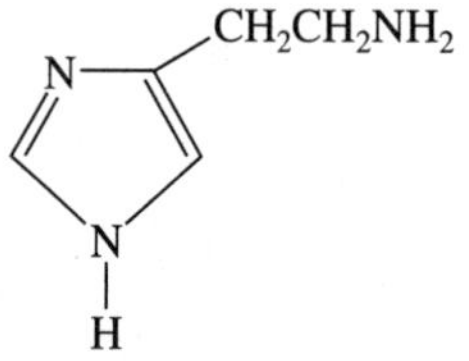

histamine

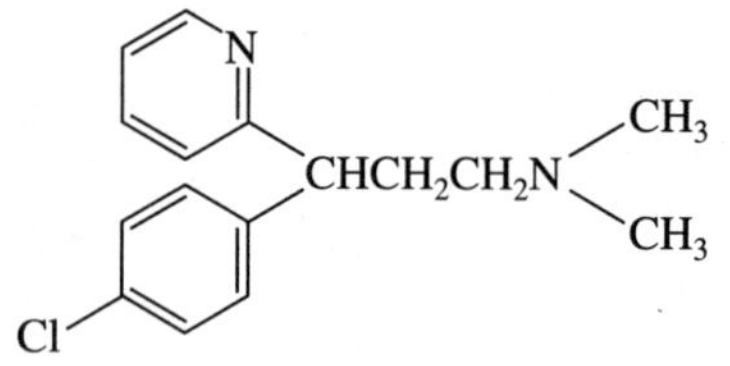

Chlor-Trimeton
generic name chlorpheniramine
an antihistamine

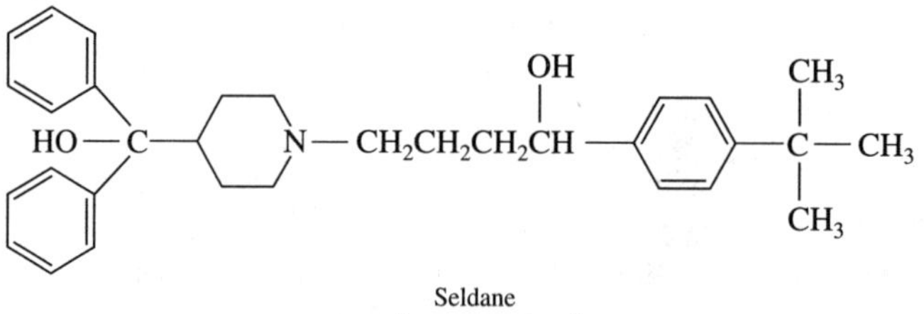

Seldane
generic name: terfenadine
an antihistamine

Why some people suffer from allergies and others do not depends on genetics. A child born to one allergic parent has a 25–40% chance of being allergic to something, and one with two allergic parents has a 30–60% chance.

Organ transplants highlight the normal role of immune systems, but here the effect is unfortunate. Since a transplanted organ is a foreign substance, it is usually rejected by an immune response. Suppressing the immunological system with drugs increases organ acceptance, but also increases the risk of infection.

SUMMARY

Diseases caused by microorganisms like bacteria and viruses also stimulate immunity toward later infection. Antibody proteins (immunoglobulins) neutralize infecting organisms. When an infection occurs for the first time (primary infection), the development of antibodies begins; if another infection (secondary) occurs, the efficient memory response can prevent it from taking hold. It is the variable region of each antibody that is responsible for recognizing and binding to the foreign antigen.

Three types of vaccines have been developed. Vaccines of dead organisms (for instance whooping cough vaccine), live attenuated viruses (for instance Sabin vaccine against polio), and inactivated toxins, or toxoids (for instance vaccines against diphtheria and tetanus), simulate the primary infection and stimulate the immune response. If a real infection should follow, it would be secondary, and the memory response would neutralize the invading organism. A fourth type, using recombinant DNA techniques, is in development. Despite recommended vaccinations, an alarming percentage of the population is not adequately immunized against some serious but preventable diseases.

Temporary passive immunity can be acquired by receiving antitoxins made from antibodies produced in another animal. Natural passive immunity passes antibodies from mother to fetus, and from mother to newborn in breast milk.

Since antibiotics and other drugs are ineffective in treating viral diseases such as the common cold and flu, it is important to establish immunity. Unfortunately, viruses occasionally undergo extensive mutation, especially of the exterior protein coat of the virus particle. When a person acquires immunity to a particular virus, his or her immune system recognizes the protein coat and a full-scale immune response occurs, neutralizing the virus. After a major mutation, the virus may not be recognized by the immune system, and any previously acquired immunity may be bypassed. Outbreaks of Asian flu and Hong Kong flu occurred. The Hong Kong flu was caused by an extensively mutated Asian flu virus, so that persons who had acquired an immunity to Asian flu were still susceptible to Hong Kong flu.

AIDS is a disease caused by a retrovirus called the human immunodeficiency virus or HIV. The virus attacks helper T cells of the immune system and renders them nonfunctional. AIDS is nearly always fatal, due not to the virus action but to secondary infections since the body's defense system is impaired. Problems arise when attempting to develop a vaccine against HIV. The virus attacks the cells that are crucial in antibody production, it mutates frequently, and HIV-infected helper T cells are themselves infectious.

Enzyme immunoassays are screening tests that use antibodies to diagnose conditions such as AIDS and hepatitis, and to detect proteins excreted early in pregnancy.

Hay fever, allergic reactions to insect bites or drugs (particularly penicillin), and rejection of transplanted organs are examples of an immune response with undesirable consequences.

PROBLEMS

1. Give a definition or an example of each of the following.

   a. Infection
   b. Immunology
   c. Antibodies
   d. Antigens
   e. Memory cells
   f. Toxoid
   g. Natural passive immunity
   h. Anaphylaxis
   i. AIDS
   j. Retrovirus
   k. EIA

2. How do microorganisms enter the body?

3. Why are antibiotics worthless against viral infections? Why are they sometimes given anyway?

4. What are some diseases that confer immunity? What diseases do not confer immunity?

5. How does a vaccine work?

6. What are the characteristics of three types of vaccines?

7. What is the difference between passive immunity and active immunity? Which is better and why?

8. Why do serious flu epidemics seem to occur every 10–15 years?

9. Compare and contrast the structures of the influenza and AIDS viruses.

10. Why have effective vaccines been developed against influenza, but none as yet against AIDS?

11. A brand name for AZT is Retrovir. Explain the significance of the name.

12. What is an EIA and what does it screen for?

13. What are some negative aspects of antibody production?

References

1. Gibbons, A. "Saying So Long to Polio." *Science* **1991,** *251,* 1020.

2. *Physicians' Desk Reference,* 44th ed. Medical Economics: Oradell, NJ, 1990; p. 880.

3. Pizza, M.; et al. "Mutants of Pertussis Toxin Suitable for Vaccine Development." *Science* **1989,** *246,* 497.

4. Marshall, E. "Malaria Vaccine on Trial at Last." *Science* **1992,** *255,* 1063.

5. Kaplin, M. M.; Koprowski, H. "Rabies." *Scientific American* **1980,** *241* (1), 120.

6. Hu, S. "Protection of Macaques Against SIV Infection by Subunit Vaccines of SIV Envelope Glycoprotein gp 160." *Science* **1992,** *255,* 456.

Chemotherapy for the Treatment of Disease

24

When good health deserts us, we turn to several types of drugs for curing disease. This chapter covers drugs such as antibiotics, antiviral agents, and anticancer agents. Drugs used by consumers for treating symptoms such as headache, fever, cold symptoms, and nervous tension are discussed in Chapter 25.

Throughout the chapter, the complete chemical structures will be given for each drug using abbreviations such as those described in Chapter 6. Although you should understand what the chemical formulas mean, it will not be necessary to learn these complex structures to understand the discussion of each.

24.1 Drug Names

GOAL: To understand the naming of drugs, and the distinction between generic and brand-name drugs.

Drug names are the cause of a long-term controversy among drug companies, physicians, and consumer organizations. The issue is largely economic and political, but the consumer who wants to evaluate it must understand the chemistry involved.

A drug can have three names. One is its *systematic name,* that is, the IUPAC name. It is seldom used outside the technical literature. Another is the *generic name,* which is unsystematic but still somewhat technical. Generic names are used frequently. Finally, the *brand name,* or *trademark name,* is the one adopted for use by each company that markets a drug. In other words, although a drug has only one generic name, it may have many brand names. An example is *amoxicillin,* one of the most frequently prescribed drugs in recent years; it is available under such brand names as Amoxil, Polymox, Trimox, and Wymox, as well as by its generic name. Amoxicillin is one member of the class of drugs known as *penicillins,* which will be discussed in Section 24.3.

Brand names are capitalized; generic names are not. The brand name frequently appears in the literature with the sign ® at the upper right to indicate that it is registered and that its use is restricted to its owner. Structural formulas are shown here with both names.

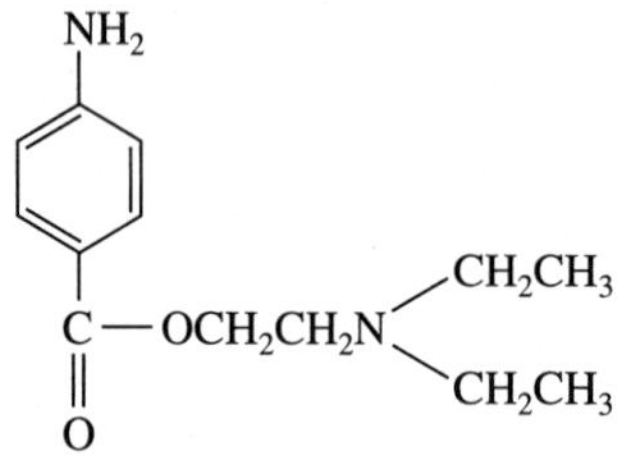

procaine
(Novocain)

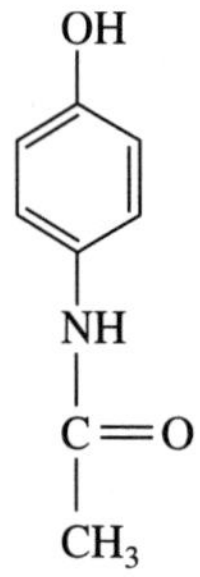

acetaminophen
(Tylenol, Tempra, Datril, Liquiprin, Trilium)

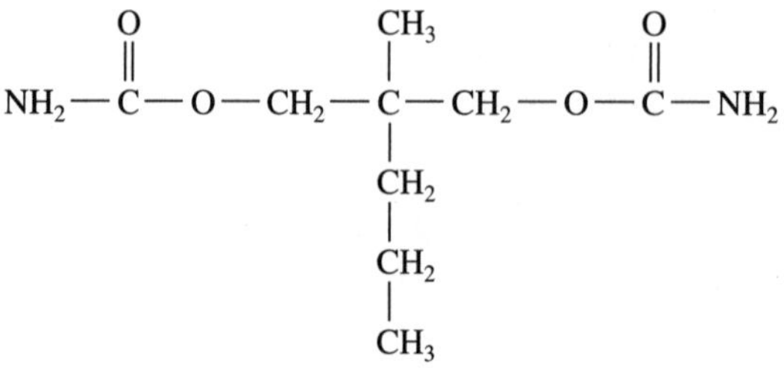

meprobamate
(Equanil, Miltown)

While developing a new drug, a company may apply for a patent that gives it exclusive rights for 17 years to produce and sell that drug. The patent holder also may sell rights to (that is, license) other companies to produce and market the drug. Even then, the first company maintains a great deal of control over the price of the drug. Each licensee must add the cost of the license to its costs of production and promotion in determining what it must charge to make a profit. In contrast, the company holding the patent not only avoids licensing expense but also derives income by granting licenses.

Patenting is meant to stimulate discoveries, in that the owner is guaranteed substantial rewards when a discovery is important. In the drug industry, a sizable portion of income is devoted to research. Even after a drug is successfully synthesized, much research must be done to determine its properties. Other problems and goals are involved in research, but none is more costly than proving the safety of the drug. Even after a patent has been granted, adequate testing is required and may consume several years of the patent period.

Many testing regulations have been put into effect since the early 1960s because of the thalidomide disaster. Thalidomide was used in Europe as a tranquilizer from 1957 to

1962. Large numbers of its consumers were pregnant women, and in 1962, serious birth defects were traced to it.

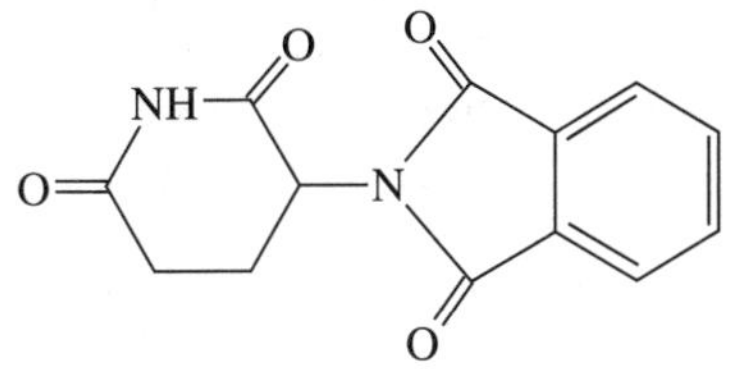

thalidomide

The pricing of drugs during the period in which a drug is protected by a patent raises little concern, but marketing practices after a patent expires are controversial. Some critics argue that the drug industry spends far more for promotion and advertising than it does for research, and that part of this promotion tries to convince physicians that expensive, brand-name drugs are superior to the cheaper, equivalent generics.

After a patent expires, the original brand name used by the patent holder is protected by the copyright law. But the drug itself may be marketed by any company using either its own brand name or the generic name. A drug sold under its generic name is invariably less expensive. Frequently, a generic drug is described as inferior to the more expensive, brand-name variety. How may a given drug differ from one manufacturer to another?

When a drug is formulated into a tablet, a capsule, or a liquid, certain inert ingredients are added. The active ingredient is always the same for all brands, but the inert ingredients may vary. For example, fillers, binders, and a coating contribute to a tablet's size, shape, flavor, and appearance. The inactive ingredients are *drug additives.*

Some additives could make a difference in how a drug performs (its pharmaceutical action), and the emphasis is centered on *bioavailability,* or **bioequivalence.** If one drug is better able than another to reach the target organ, it is superior. When a drug is administered orally, it follows the pathway pictured in the following scheme:

$$\text{stomach} \longrightarrow \text{intestine} \longrightarrow \text{blood} \longrightarrow \text{target organ}$$
$$\downarrow \qquad\qquad \downarrow$$
$$\text{excretion} \qquad \text{excretion}$$

A primary factor in determining the efficiency of action of an oral drug is the ease with which it enters the bloodstream. This is largely a matter of solubility of both the active ingredient and inert ingredients. First, particle size has a strong influence on solubility, since smaller particles have a greater surface area per given amount. The greater the surface area, the faster the drug will dissolve because of greater contact with the fluids in the digestive tract. Particle size is particularly important when a drug is given as a powder in a hard, gelatinous capsule (which itself may be the only inert ingredient). Second, inert ingredients can control the rate of absorption of a drug into the blood. According to the absorption scheme, if a drug is not readily absorbed, it will be excreted.

Thus, some chemically equivalent drugs may not be bioequivalent, but documented examples of differences between brand-name and generic-name drugs are rare *(1).* For one thing, particle size of a drug dispensed in capsules is simple to control. Second, the methods used to compress ingredients into a tablet, and the ingredients themselves, are standard. Each company strives to use the most economical formulation, and the least expensive method for one manufacturer is usually the least expensive for another. Third, a formulation can easily be copied. Chemical analysis of the ingredients makes it easy for one company to duplicate the formulation used by another. The fact that a company has the benefit of years of research on a given drug does not really give it much advantage over anyone else once the patent has expired. However, its brand-name product may be so well known as to have an edge in marketing, if not in quality.

Sometimes, one pharmaceutical company produces a drug in final form and sells it to several other companies, which then market it under their own brand names or under the generic name *(2).* Erythromycin, a frequently prescribed antibiotic for persons who cannot tolerate penicillin, can be bought under several names, at a wide variety of prices. Invariably, the product sold under the generic name is less expensive.

24.2   Drug Research and Development

GOALS: To appreciate the role of drug additives.
To understand the strategies of prodrugs and controlled-release formulations.

Drug Additives

Whereas food additives are controversial, drug additives are less so. Although inert, they may be critical in increasing the usefulness of a drug. Although many drugs are administered in liquid form, or as sprays, ointments, suppositories, and so on, the majority are capsules or tablets. Capsules are more expensive, so pharmaceutical manufacturers prefer to market their products in tablet form. On the other hand, capsules circumvent problems of taste, stability, and disintegration. A capsule is usually made of gelatin and has no flavor (Figure 24.1). It disintegrates rapidly, and the contents are quickly released. The contents may be quite stable, since they are not in contact with air. The drug can be in powdered form so that there is no delay in disintegration after it is consumed and therefore it can be readily dissolved and absorbed.

The chief additives in tablets are *fillers, binders, lubricants, disintegrators,* and *coatings.* Some drugs are so potent that only a small amount of the active ingredient is present in the tablet. In such cases, lactose, sucrose, or some other carbohydrate may be added to increase the size of the tablet to a reasonable volume. Tablets are made by mixing ingredients and compressing them in a mechanical device under high pressure (Figure 24.2). Starch paste provides adhesiveness and thus serves as a binder. Soaps such as calcium stearate and magnesium stearate are used as lubricants. They are present in small amounts and prevent the ingredients from sticking to the machinery used to compress the drug plus additives into a tablet.

$$(CH_3CH_2CH_2CH_2CH_2CH_2CH_2CH_2CH_2CH_2CH_2CH_2CH_2CH_2CH_2CH_2C\overset{\overset{\displaystyle C}{\|}}{}\!\!-O)_2Ca$$
calcium stearate

Dry starch is often added as a disintegrator. Starch expands when it becomes wet, and expansion of a tablet is the reverse of the process used to form (compress) the tablet in the first place. Therefore, starch speeds the breakdown of the tablet and aids in absorption.

Coatings have various uses. They solve problems of taste, appearance, and stability by keeping the contents from contact with air, and they can be designed to control the location and rate of release of the drug. For example, enteric-coated drugs withstand the acidic conditions in the stomach, but disintegrate rapidly and release the ingredients when they reach the alkaline conditions of the intestine. Thus, if a drug is sensitive to the stomach or if the stomach is sensitive to the drug, the coating avoids a problem. Once in the

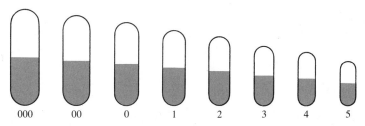

Figure 24.1
Various sizes and numbers of gelatin capsules. Reproduced showing actual sizes. (From Bergersen, B. *Pharmacology in Nursing,* 13th ed.; C. V. Mosby: St. Louis, 1976.)

Figure 24.2   Manufacturing medicine in tablet form.

intestine, a drug can be readily absorbed into the bloodstream and delivered to the intended site of action.

Goals of Drug Research

Scientists who develop a new drug must study many properties to evaluate its suitability and to find ways of increasing its efficiency. Some relevant properties are as follows:

1. Mode of action
2. Absorption and excretion characteristics

3. Side effects

4. Interaction with other drugs

5. Toxicity, including possible allergic reactions

6. Effective dosage

7. Patient appeal

An important but often poorly understood property of a drug is its mode of action. Once an investigator knows how a drug causes its desired effect (analgesic, antibiotic, etc.), ways of improving it or developing more effective agents often become apparent.

Painful injections are both undesirable and inconvenient; this has spurred a search for drugs suitable for oral administration. Items 2–6 also describe the action of a drug after administration. The active ingredient must have suitable solubility properties to permit absorption. Even if a drug has sufficient water solubility to be absorbed, it may not have adequate fat solubility to penetrate a target organ. In addition, the active drug may have adverse side effects, for example, stomach irritation. Or the drug itself may be adversely affected. In such cases, it may become necessary to abandon the drug, or to synthesize different forms of it in hopes of improving its properties while maintaining its effectiveness. An alternative strategy might be to synthesize a *prodrug,* or precursor form.

Prodrugs: Controlled Release

One of the challenging goals of drug research is the development of prodrugs. A prodrug, or drug precursor, is a less active or inactive form that is converted to the active form some time after administration. Generally, conversion is carried out by enzymes in the body; in some cases it occurs only after the prodrug has penetrated the target organ. A prodrug may overcome problems such as low solubility, instability, undesirable metabolism of the active drug before reaching the target, or factors relating to patient acceptance (taste, odor). A bitter medicine may be satisfactory for an adult, but oral medication for young children often must be chewable and therefore pleasant-tasting.

Aspirin is a type of prodrug. The compound known as salicylic acid is an effective analgesic (painkiller) and anti-inflammatory agent, but it is too corrosive for general use. The problem was largely overcome by converting salicylic acid to acetylsalicylic acid (aspirin). Once aspirin enters the bloodstream, the acetyl group is quickly removed and the salicylic acid provides the analgesic action *(3).*

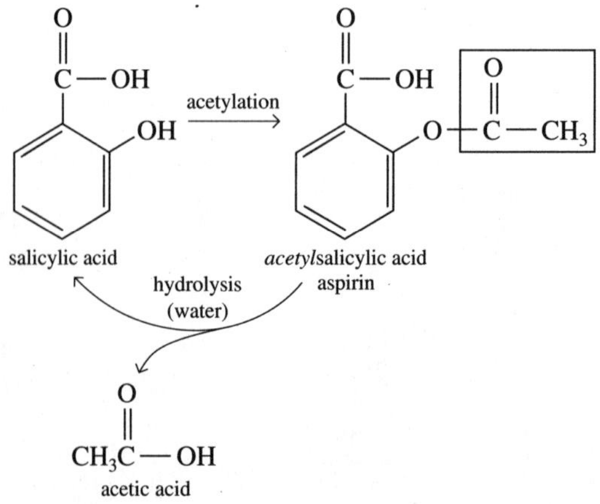

If a drug is absorbed too quickly, its effect may last only a short time, and its level (concentration) in the body may rise and fall quickly. This is generally described as a peak-and-valley effect. Such behavior in a painkiller would be undesirable. Prodrugs have been devised to achieve a delayed or sustained-release action. Layered tablets and coated slow-release beads and granules control the rate of release and therefore the rate of absorption. Thus the advantages of sustained-release formulations are reduced number and frequency of doses that must be administered; elimination of peak-and-valley effect; lesser amount necessary to achieve desired results; elimination of nighttime administration for which a patient may have to be awakened; and fewer side effects in the gastrointestinal tract.

In most instances of sustained or controlled release of a drug, polymers are involved. Also, tablets surrounded with a layer of a natural wax, such as beeswax, have had some success. When coated with layers of various thicknesses, some tablets break down rapidly and release their contents. Others disintegrate more slowly. This approach is known as the Spansule system. It has been applied widely since 1961, when Contac cold capsules were first marketed.

One of the vinyl polymers that is used for making soft contact lenses (Section 8.3) has been used for making a controlled-release system. The high affinity for water makes the poly(2-hydroxyethyl methacrylate) well suited for making soft contact lenses. It also can be used to deliver drugs at a slow rate; it is known as a *hydrogel* because of its ability to attain a high water content.

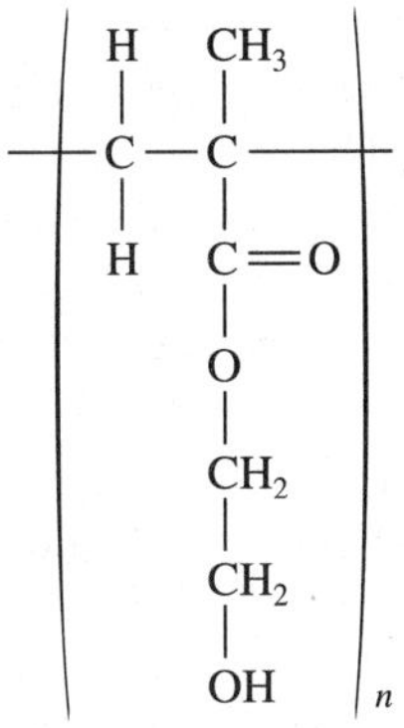

poly(2-hydroxyethyl methacrylate)
(soft contact lenses)

If a water-soluble drug is incorporated into a hydrogel, it will move slowly out of the polymer matrix as water moves in. If a drug tends to be released too quickly from the hydrogel, the polymer chains may be cross-linked to retard its movement. This type of polymer is nondegradable and passes through the gastrointestinal system unchanged, while the drug is absorbed.

In other systems, the controlled release relies on chemical breakdown of a polymer into monomer units. The drug molecules trapped inside are then released. In such a design, the breakdown products must be safe. One popular polymer is the polyester poly(lactic acid) as shown:

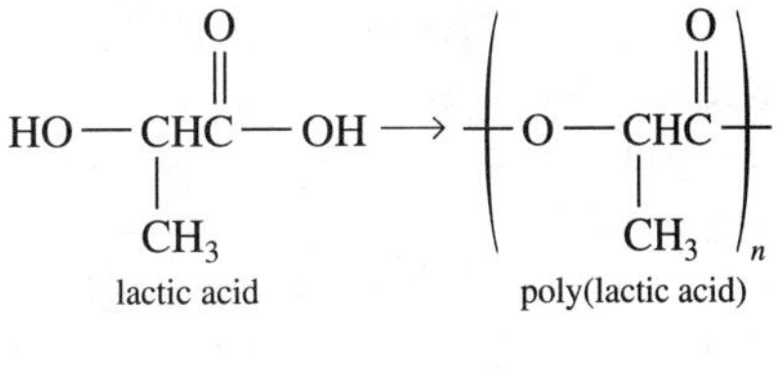

lactic acid   poly(lactic acid)

Poly(lactic acid) is one of several bioerodable polymers. It was first used for making degradable surgical sutures to replace catgut. Since lactic acid is readily metabolized and even shows up in the body during vigorous exercise, it is safe *(4)*.

24.3   Drugs for Curing and Controlling Disease

GOALS: To appreciate the history of antibiotics.
        To understand the general approaches to synthesizing antibiotics.
        To appreciate the chemistry of resistance.
        To understand the use of sulfa drugs.
        To review the history of the penicillins.

In studying some approaches to curing diseases, we emphasize the chemotherapy of infectious diseases caused by microorganisms and viruses. Cancer and heart disease are two prevalent conditions that do not fit this description. In Section 24.6 we consider cancer chemotherapy briefly. The treatment of heart disease is discussed in Chapter 26 in relationship to the cholesterol controversy.

Antibiotics

The development of antibiotics is one of the greatest human achievements, and it is relatively recent. In 1907 the German scientist Paul Ehrlich (1854–1915) discovered a substance called *salvarsan* in his search for a treatment for African sleeping sickness. Ehrlich coined the phrase ''magic bullet'' for salvarsan to denote its lethal effect on disease-causing organisms. This substance was also known as arsphenamine, because it contained arsenic, and salvarsan 606, because it was compound number 606 that Ehrlich tested *(5)*. Salvarsan also proved effective in treating syphilis. Ehrlich, who received the Nobel Prize in 1908, is generally recognized as the founder of chemotherapy.

Ehrlich's work was simple but ingenious. He followed up on the earlier (1884) work of the Danish physician H. C. Gram, who devised the Gram stain, a test commonly used to classify bacteria. After treatment with certain dyes and subsequent washing, *gram-positive* bacteria retain the dyes, whereas the dyes are easily removed from *gram-negative* bacteria. This suggests that certain chemical groups are present on the cell walls of gram-positive bacteria, whereas they are lacking in gram-negative bacteria. Ehrlich reasoned that it might be possible to incorporate a toxic element, such as arsenic, into a dye molecule that would then attach to bacterial cells and cause their death. Such was the case with his magic bullet.

Despite this early success, it was only after World War II that the first members of the penicillin class of antibiotics came into general use. In fact, one of the most widely used, *ampicillin,* was not available until 1961. The development of new and better antibiotics is a continuing goal of the pharmaceutical industry.

An antibiotic is defined as any substance produced by one microorganism that kills or inhibits the growth of other microorganisms. Those that kill bacteria are **bactericidal;** antibiotics that kill fungi are **fungicidal.** Those that merely inhibit microbial growth are said to be **bacteriostatic;** they usually act by inhibiting protein synthesis in the infecting microorganism. Although bactericidal activity would seem to be better than bacteriostatic, both give the body time to mobilize normal body defense mechanisms—to synthesize antibodies—while preventing the invading organism from overwhelming the body. Since Ehrlich's magic bullet was a synthetic substance, it is not strictly correct to classify it as an antibiotic.

Several new antibiotics have been obtained by genetic mutation. Just as yeast cells carry out a complex series of reactions in fermentation to produce ethyl alcohol, other microorganisms carry out reactions leading to formation of antibiotics. Ionizing radiation has been used to induce mutations that cause the microbes to synthesize other antibiotics or to synthesize the same ones more efficiently.

By changing the nutrient medium that sustains the life of microorganisms, other chemical pathways used for their metabolism can be selected. Certain components of the growth medium can be incorporated into an antibiotic molecule. It also has been possible to interfere with metabolic reactions so that a precursor to the normal antibiotic could be isolated. Such a precursor could then be converted to a different antibiotic. These are commonly designated *semisynthetic antibiotics,* to signify that a microorganism does part of the job of synthesis, but a chemist carries out the final step or steps.

There have been dramatic successes using mutants, alterations of growth media, and semisynthetic pathways in synthesizing new members of the penicillin family. These and other antibiotics have an impressive record in dealing with microbial infections despite their failure to combat viral infections. This has led many to describe them as miracle drugs.

Choosing an Antibiotic

Successful application of antibiotics requires the often critical decision about which to use. Sometimes, a sample of the invading microorganism can be obtained from an open wound, throat mucus, urine, blood, or other source. The microorganism can then be cultured and examined for sensitivity to antibiotics. **Culturing** involves growing the sample in a nutrient medium so it multiplies into a larger sample. Testing for antibiotic sensitivity is most easily done by mixing the cultured sample with a quantity of nutrients in melted agar (a complex polysaccharide). The agar is then poured into a culture plate and allowed to solidify to a gel. Drug companies market circular disks that are impregnated with known dosages of antibiotics. Placed on top of the agar gel, as shown in Figure 24.3, the antibiotics diffuse out of the disks. Each antibiotic that is effective against the microorganism inhibits growth in a ring around that disk. About 10 antibiotics are usually tested. The information about relative effectiveness is then transmitted to the physician, who chooses the antibiotic to prescribe for the patient.

Unfortunately, 72 hours or more can be required to grow a microorganism in sufficient quantity to test for antibiotic sensitivity. This period can sometimes be critical in the development of the bacteria in the body. Therefore, it usually becomes necessary for the physician to make an educated guess and administer an antibiotic before laboratory results are available. The choice is often a *broad-spectrum antibiotic,* that is, one that is effective against a wide variety of microorganisms. A *narrow-spectrum antibiotic* is preferable once the antibiotic sensitivity results are available, but the broad-spectrum drug is like an insurance policy with a wide range of coverage.

It is also possible to administer a combination of antibiotics, called *piggy-back antibiotics* when given in a single preparation, to provide defense against an even greater number of organisms. This seems like a logical approach when the susceptibility of the disease-causing organism is unknown. But it usually is considered a poor strategy for two reasons *(6).*

First, two antibiotics may be antagonistic toward one another. The prime example is a combination of bacteriostatic and bactericidal agents. Since the bactericidal antibiotic

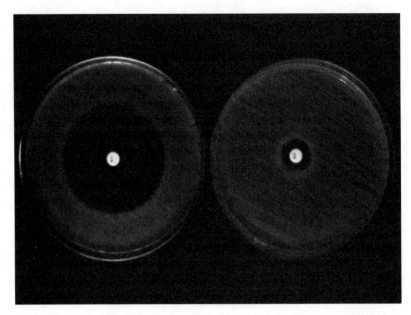

Figure 24.3

Testing for antibiotic sensitivity. Paper disks impregnated with different antibiotics are dropped onto the surface of plates inoculated with a microorganism to determine the sensitivity of the organism to different antibiotics. The antibiotic diffuses into the medium, inhibiting the sensitive organisms. Dark zones around the disks are regions in which microbial growth is prevented.

performs best when a microorganism is actively growing, a bacteriostatic drug may limit the growth enough to render the former ineffective.

Second, the development of resistant organisms is encouraged by the use of antibiotics. A resistant organism is commonly a mutant that can thrive and reproduce when competing organisms are killed by antibiotics. The mutant might be unable to compete with other microorganisms under normal circumstances, but when antibiotic is present, the microbe can develop on a large scale, sometimes with serious consequences.

A *resistant microorganism* has an enzyme that chemically changes the antibiotic into a harmless substance. For example, one that is resistant to penicillin has the enzyme *penicillinase*. Resistance is one reason for continuing research efforts to develop new antibiotics. As one source put it, ''Although the success of antibiotics may warrant their classification as miracle drugs, it is a miracle in constant need of renewal. Resistant variations of pathogenic, disease-causing bacteria, which can withstand effective antibiotic treatment, are continually arising, particularly in hospitals'' *(7).*

Broad-spectrum antibiotics and combinations also may have less serious but annoying consequences. Normally, a heavy concentration of beneficial bacteria live in the large intestine. These *intestinal flora* are influential in controlling the passage of wastes through the system and provide a source of vitamins. Because antibiotics tend to upset the flora, diarrhea (sometimes persistent) is a frequent side effect of antibiotic therapy. This argues strongly for the use of antibiotics only when needed, and explains why broad-spectrum antibiotics have the greatest effect on the intestinal flora.

Clearly, physicians are in a difficult situation. They should not administer antibiotics against a wide range of microbes, but they usually do not initially know what narrow-spectrum drug might be effective. If a disease is life-threatening or accompanied by severe symptoms, the choice must be one that has the greatest chance for success, despite potential problems. The result has been that many previously broad-spectrum antibiotics have become narrow-spectrum antibiotics because resistant microorganisms have been selected.

The Sulfa Drugs

The mid-1930s marked the real beginning of the wide-scale use of miracle drugs. In spite of the turn-of-the-century work of Ehrlich and observations on the activity of penicillin in 1928, it was the discovery of the properties of the sulfa drugs that stimulated well-organized attempts to develop penicillin and other antibiotics.

The story of the sulfa drugs began in 1931, with a patent on a new drug called Prontosil by a German dye manufacturer. Prontosil is a red substance that had been synthesized for use as a dye for wool but showed antibacterial action. In 1935, Gerhard Domagk published his results on Prontosil's effect in combating streptococcal infections in animals. Domagk was awarded the Nobel Prize in 1939 for his contributions to the development of the sulfa drugs. Soon after this report, it was found that Prontosil is converted to sulfanilamide (Figure 24.4) in the body and that sulfanilamide has the same antibacterial activity as Prontosil. Subsequently, a whole series of derivatives (modifications) of sulfanilamide, the sulfonamides or sulfa drugs, were synthesized.

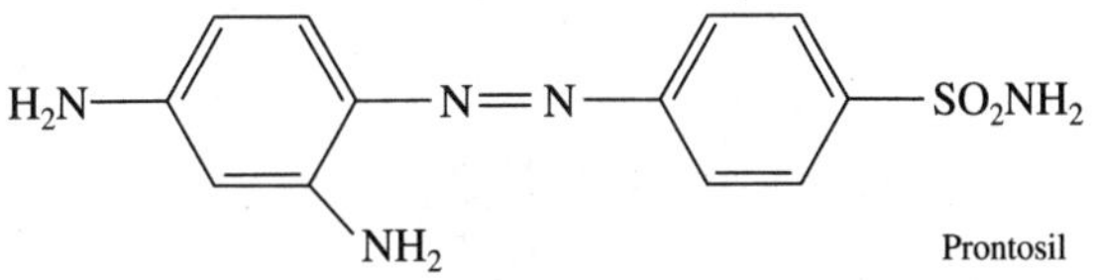

Prontosil

For animals, including humans, folic acid is a water-soluble B vitamin that we cannot synthesize, but must consume in the diet.

Strictly speaking, the drugs are not antibiotics since they are not produced by any living organisms. Their activity is bacteriostatic, and disease-causing microorganisms can become resistant, as they can to antibiotics. The drugs act as **antimetabolites**; that is, they interfere with metabolism in bacteria. Bacteria synthesize a substance called folic acid, which is essential for survival. An important precursor to folic acid is *para*-aminobenzoic acid (PABA), as shown in Figure 24.5. Sulfanilamide and other sulfa drugs can block the incorporation of PABA into folic acid. This process requires an enzyme, which is blocked because of the similarity between PABA and the sulfanilamide molecule.

Molecules interact with enzymes in a *lock-and-key* arrangement (Figure 24.6). If the key does not fit the lock, no reaction can take place. If the wrong key is inserted into the lock, it cannot open the lock but it can jam it, preventing other keys from entering the lock. In other words, the sulfa drugs are similar enough to fit in the lock and prevent the

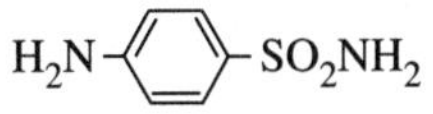

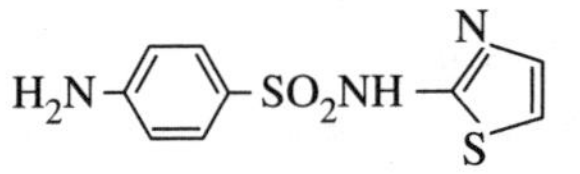

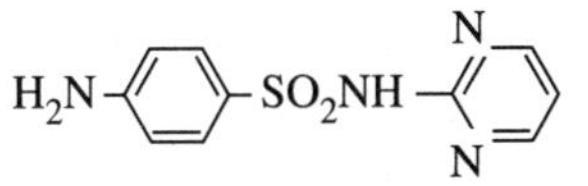

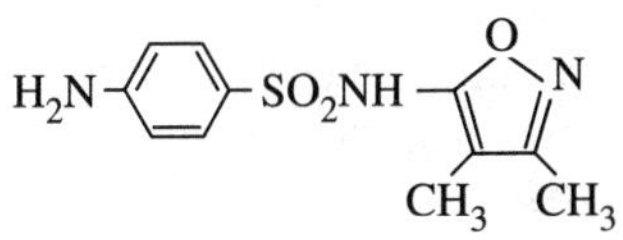

Figure 24.4
Sulfa drugs.

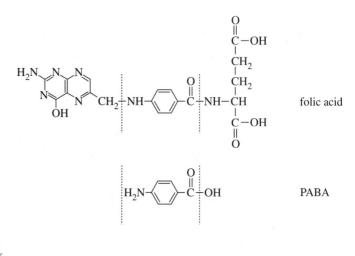

folic acid

PABA

sulfanilamide

Figure 24.5
Sulfa drugs can block the growth of bacteria by interfering with their ability to synthesize folic acid, which is essential for survival. The blockage results from the similarity of each of the sulfa drugs to PABA, a precursor to folic acid.

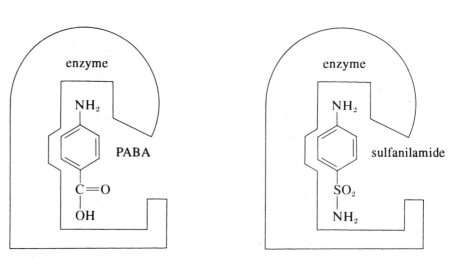

Figure 24.6

The lock-and-key model for the action of sulfa drugs. An enzyme acts on PABA during a sequence of reactions leading to folic acid, an essential compound for bacteria. The enzyme acts like a lock that can accept keys (PABA or sulfa drugs) that fit. When a sulfa drug enters, it blocks the enzyme from acting on PABA.

conversion of PABA to folic acid. Since folic acid is an essential compound, the invading bacteria cannot survive without it.

The sulfa drugs were once so widely used to combat a variety of infections caused by gram-positive bacteria that the period 1935–1948 is known as the sulfonamide era. Soldiers in World War II carried packets of them to sprinkle on open wounds to prevent infections. Unfortunately, the drugs cause many side effects in prolonged use, particularly kidney damage, so they have largely given way to the penicillins and other antibiotics. Since sulfa drugs are efficiently absorbed and later excreted in the urine, they still see some use in treating urinary tract infections *(8)*.

The Penicillins

Although almost everyone has heard of penicillin, few realize that the word applies to any of a number of different drugs, such as penicillin G, penicillin V, ampicillin, methicillin, oxacillin, cloxacillin, and nafcillin (Table 24.1). Differences in the structure of the R group (in the upper left of the formula in Table 24.1) distinguish the penicillins from each other.

This is not merely another example of generic and brand name proliferation. There actually are several similar but distinctly different penicillins. They can be distinguished from one another by the following properties: susceptibility to attack by stomach acid, spectrum of activity, and effect on resistant (penicillinase-forming) microorganisms. More important, these properties explain why continued effort is directed toward developing new antibiotics in general and new penicillins in particular.

The earliest penicillin to be used on a wide scale was penicillin G (generic). It is still an important antibiotic, but ranks low on all three properties, with a narrow spectrum of activity and susceptibility to resistance. In addition, it is attacked by acid in the stomach, which renders it inactive, thus limiting its usefulness as an oral antibiotic.

This effect of acid on penicillin G led scientists to synthesize acid-stable penicillins, which are true oral antibiotics. Penicillin V and ampicillin were the main products for many years, but amoxicillin is now regarded as the best. In fact, it has been the most prescribed of all drugs for all uses in recent years, principally due to its broad spectrum of action. Unlike some narrow-spectrum penicillins, which work only on gram-positive bacteria, amoxicillin also works on some gram-negative organisms.

The penicillins are a classic case of the development of resistance. Amoxicillin is among those that are deactivated by a penicillinase enzyme (Table 24.1 (pp. 518–519) and Figure 24.7 (p. 520)). Others, such as cloxacillin and nafcillin, are resistant to penicillinase, but as narrow-spectrum antibiotics, they are less often prescribed.

The penicillins and a group of antibiotics known as cephalosporins are classified as **β-lactam antibiotics,** where the β-lactam structure is the four-membered ring highlighted in color in Figure 24.7. The β-lactam ring is cleaved by penicillinase enzymes. In a few rare cases, another type of enzyme attacks the penicillin molecule in another way.

Changing the penicillin structure can sometimes prevent attack by a penicillinase enzyme. It is also possible to combine a penicillin with a penicillinase inhibitor in a single tablet. One such product is a combination of clavulanic acid and amoxicillin. Clavulanic acid inhibits penicillinase and thus allows amoxicillin to remain bactericidal even toward microbes that can destroy the antibiotic molecule. In this way, clavulanic acid serves as a *synergist* for amoxicillin *(7).*

Synergism is the improved action of two substances when used in combination.

amoxicillin

clavulanic acid

Table 24.1 Some penicillins (6, 8, 9)

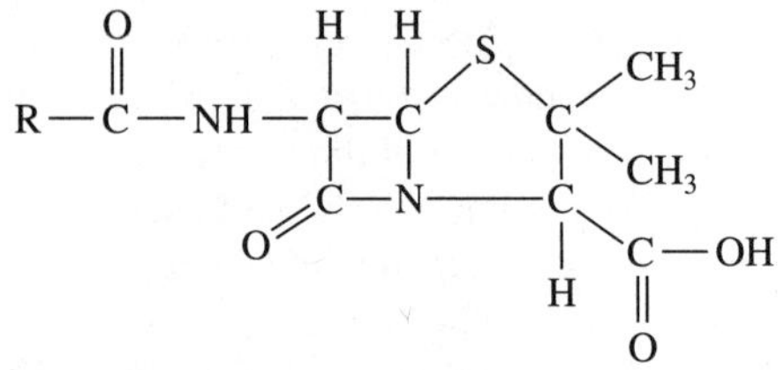

Name	Side Chain (R—)	Stability in Acid	Sensitivity to Penicillinase	Advantage
Penicillin G	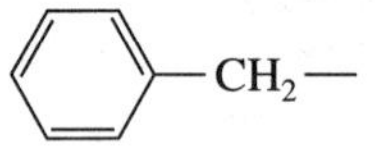 phenyl–CH₂–	Poor	Sensitive	Least expensive
Phenoxymethyl penicillin (penicillin V)	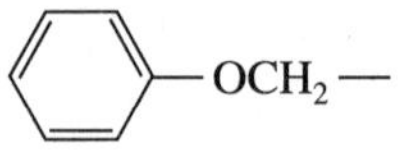 phenyl–OCH₂–	Good	Sensitive	Better oral absorption than penicillin G
Dicloxacillin	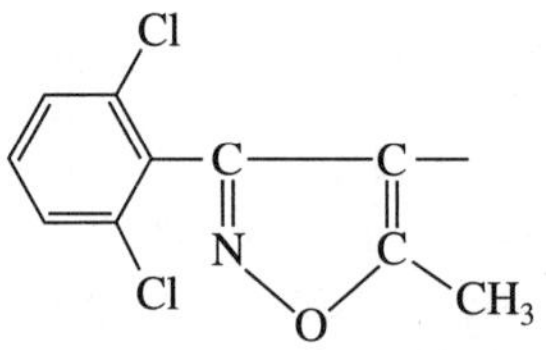 dichlorophenyl isoxazole	Good	Resistant	For infections due to penicillinase-producing organisms
Cloxacillin	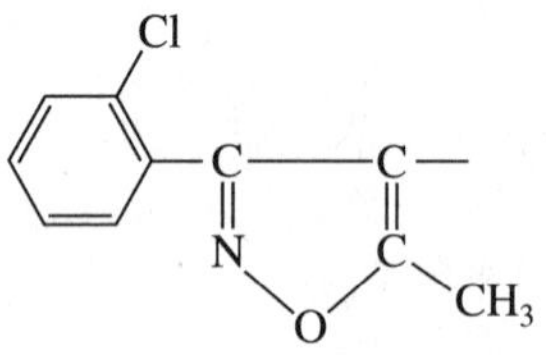 chlorophenyl isoxazole	Good	Resistant	For infections due to penicillinase-producing organisms
Nafcillin	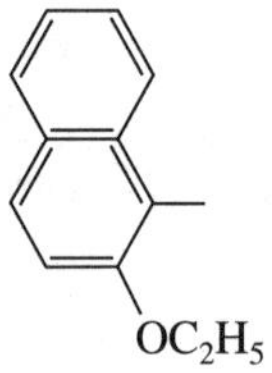 naphthyl–OC₂H₅	Poor	Resistant	No advantage over oxacillin or cloxacillins
Ampicillin	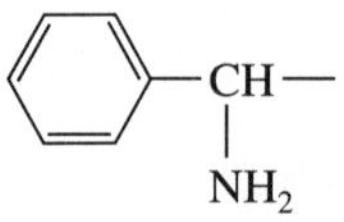 phenyl–CH(NH₂)–	Good	Sensitive	For infections due to gram-negative, nonresistant organisms
Carbenicillin	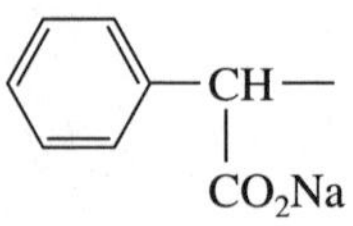 phenyl–CH(CO₂Na)–	Not given orally	Sensitive	For infections due to gram-negative, nonresistant organisms

Table 24.1   Some penicillins (*6, 8, 9*) (continued)

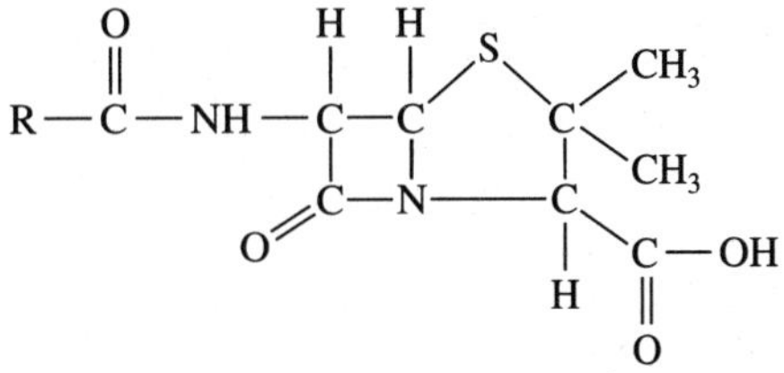

Name	Side Chain (R—)	Stability in Acid	Sensitivity to Penicillinase	Advantage
Amoxicillin	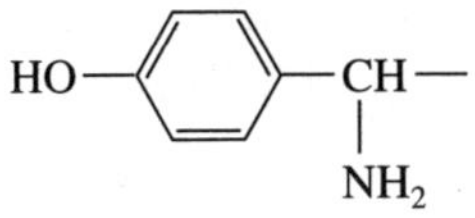	Excellent	Sensitive	Better absorption than ampicillin, broad spectrum
Ticarcillin	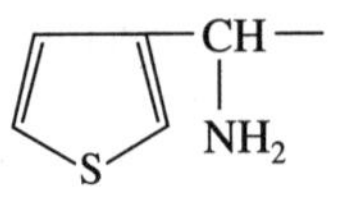	Not given orally	Sensitive	More active than carbenicillin, broad spectrum
Mezlocillin	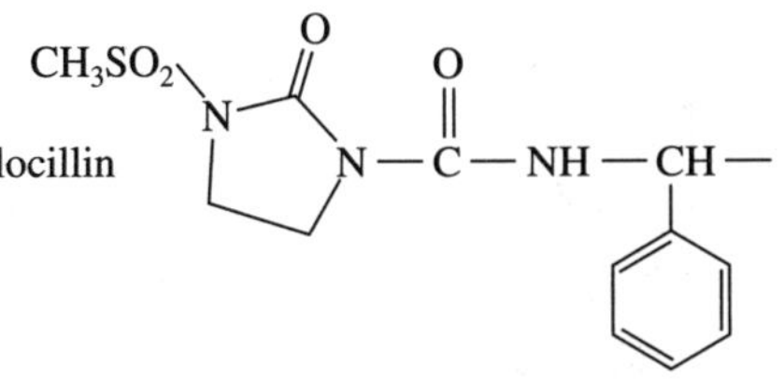	Not given orally	Sensitive	Exceptional activity and spectrum
Azlocillin	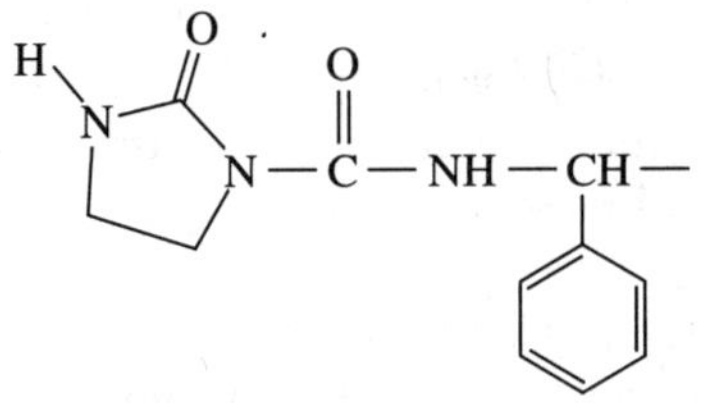	Not given orally	Sensitive	Exceptional activity and spectrum
Piperacillin	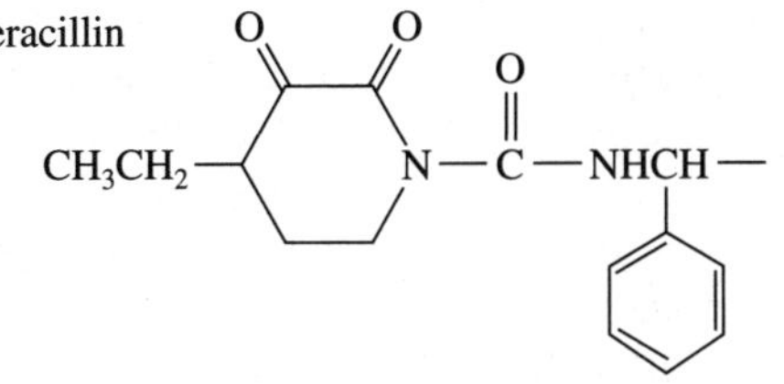	Not given orally	Sensitive	Exceptional activity and spectrum

Note: The newer drugs appear toward the bottom of the table.

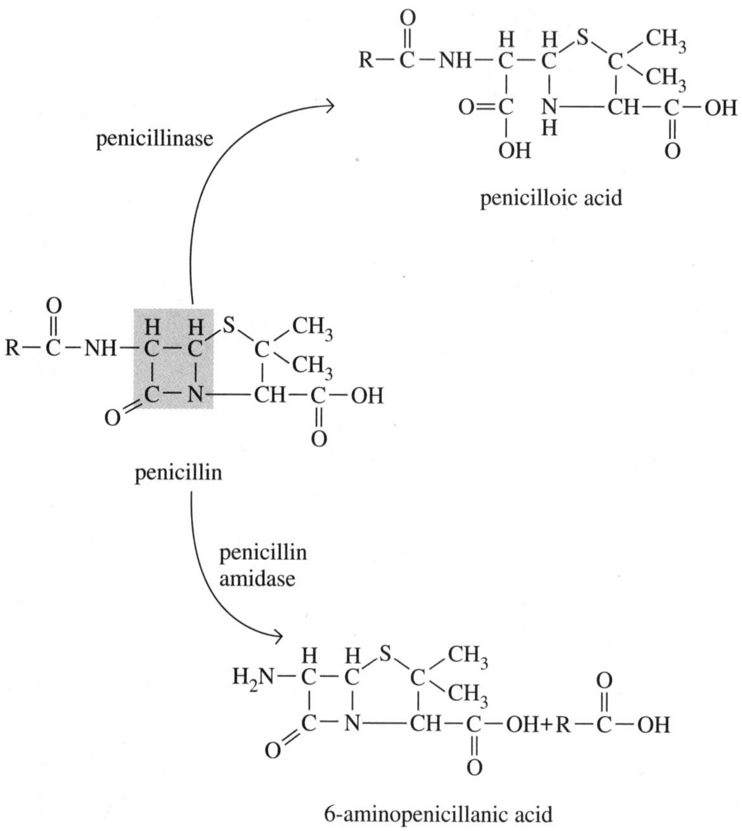

Figure 24.7
Resistance to the penicillins normally occurs due to production by bacteria of one of two enzymes: penicillinase and penicillin amidase. Penicillinase production is by far the more common form of resistance, attacking the β-lactam structure (in color) that is common to all penicillins. The action of amidase is shown.

The first observations of the antibiotic activity of penicillin are a classic example of an accidental discovery. Alexander Fleming was a bacteriologist at the University of London in 1928 who was experimenting with a *Staphylococcus* microorganism. His samples accidentally became contaminated with the mold *Penicillium notatum*. It not only grew on the nutrient provided for the bacteria, it inhibited the growth of the *Staphylococcus* organism near the mold. Fleming published his findings and named the unknown antibiotic penicillin because of its source. Unfortunately, he encountered technical problems in isolating penicillin.

Spurred by the early successes of the sulfa drugs, a group headed by Howard Florey and Ernst Chain picked up the work at Oxford University in 1939. Although their efforts were hindered by the war in Europe, the first clinical trial with a crude penicillin preparation was carried out in 1941. The treatment was a tremendous success, but the patient died. (The supply was inadequate to sustain the treatment.) Nevertheless, the potential for success was clear, and the work moved to the United States due to the war.

During the next few years, efforts were aimed at producing large quantities of penicillin. A new strain of *Penicillium* mold was obtained from the surface of a cantaloupe in a supermarket in Peoria, Illinois, in 1942. After this new strain was mutated with x-rays, some mutants were found to be unusually efficient in producing penicillin (actually penicillin G). This work contributed to the technique for producing the drug in large quantities. By 1946, conditions for its administration had been determined, and the supplies were adequate for civilian as well as military needs.

The highest degree of antibiotic activity was found in penicillin G, also known as benzyl penicillin. Its chemical structure was determined in later years. Penicillin G and other penicillins have since been synthesized by purely chemical methods, but the most economical approaches use microorganisms for all or part of the synthesis.

As an example, the synthesis of penicillin G is favored by adding phenylacetic acid to the mold's nutrient medium. This encourages incorporation of the benzyl group (Table 24.1) into the penicillin molecule. Since the mold carries out all the chemical reactions in the synthesis, penicillin G may be called a *biosynthetic* penicillin.

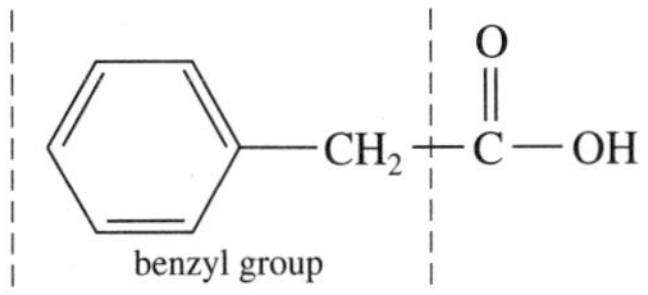

Even greater versatility is achieved by the synthesis of *semisynthetic* penicillins. A semisynthetic antibiotic is one synthesized partly by a microorganism and partly by chemists. The crucial compound was 6-aminopenicillanic acid, 6-APA, which is the penicillin molecule with no side chain group. By interrupting the normal fermentation, 6-APA can be isolated, after which a variety of side chain groups can be attached.

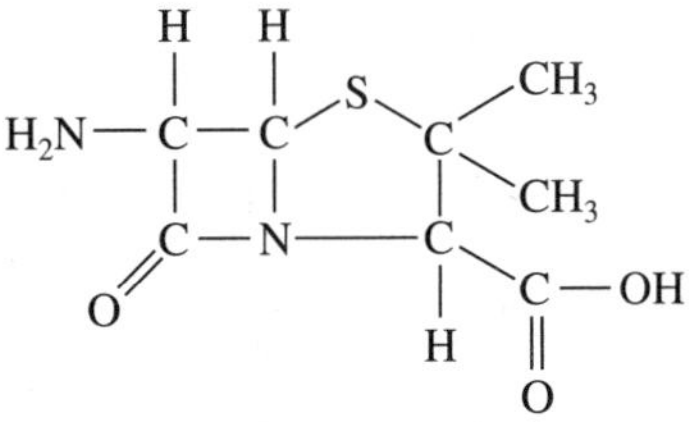

One of the most useful of the semisynthetic penicillins is phenoxymethyl penicillin, better known as penicillin V. It was the first true oral penicillin stable in the presence of stomach acid. Most of the others listed in Table 24.1 are semisynthetic also.

The drugs are unusually safe except to a small part of the population who are allergic to them and may suffer severe shock reactions (Section 23.7). Their mode of action is bactericidal. They are lethal to various microorganisms because they interfere with the formation of the cell walls in these bacteria, thus causing the contents of the cell to spill out. Animal cells do not have cell walls, only cell membranes, so the effect is restricted to bacterial cells. This explains the unusually low toxicity of the penicillins.

Other Antibiotics

As the potential of antibiotic chemotherapy became increasingly evident, efforts increased to isolate other antibiotic-producing microorganisms. Biosynthetic and purely synthetic compounds were produced and tested for antibiotic activity. The list of successes is too long to consider in any detail here, but we can look at some of the generalizations about antibiotics.

Although a penicillin is often the first choice for antibiotic therapy, problems of resistance, allergy, and spectrum of activity often require the use of other antibiotics. Streptomycin, erythromycin, chloramphenicol, the tetracyclines, and the cephalosporins often can be substituted effectively. The structures of some antibiotics are shown. It is not necessary for you to know these structures in any detail, yet it is interesting to note the great diversity of structure that can lead to antibiotic activity.

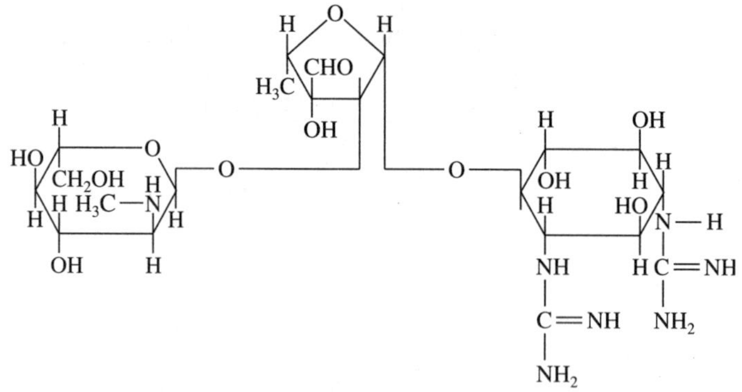

streptomycin

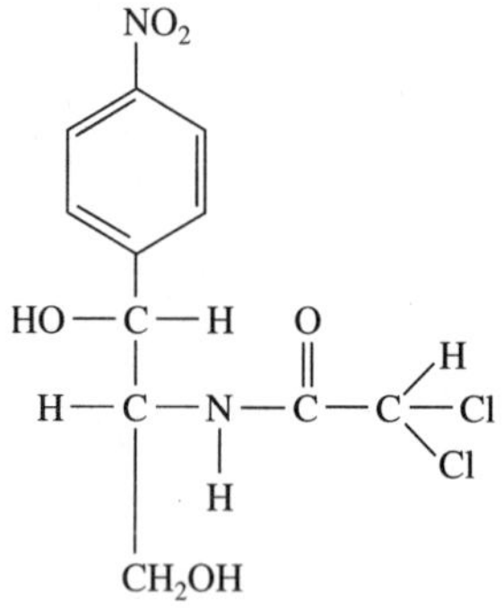

chloramphenicol

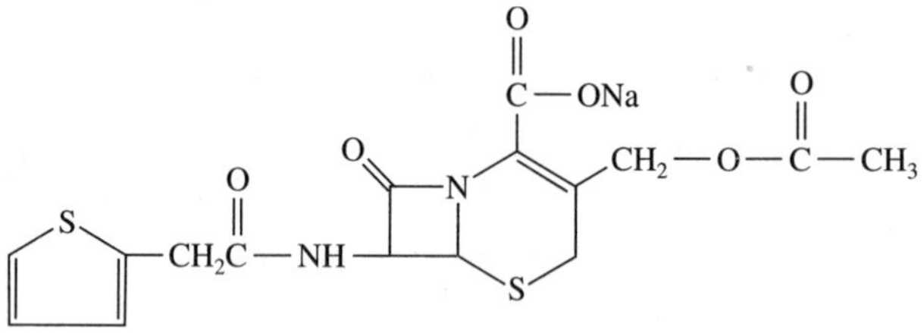

a cephalosporin
(cephalothin (sodium);
Keflin)

So far we have seen the use of antibiotics as antimetabolites (the sulfa drugs) and as inhibitors of cell wall synthesis (the penicillins). The cephalosporins, which now rival the penicillins in frequency of use, interfere with synthesis of bacterial cell walls and are therefore bactericidal. Like the penicillins, cephalosporins contain the β-lactam structure. The newest penicillins and cephalosporins have activities that extend to killing even gram-negative bacteria.

The third mode of action of antibiotics is inhibition of bacterial protein synthesis. As both microorganisms and animal cells grow and multiply, proteins are critical to both their structure and function. Differences in the chemical structures of the compounds they use for protein synthesis mean an antibiotic affects animal cells much less. However, side effects of inhibiting protein synthesis in human cells are not uncommon.

The tetracyclines are an important group of antibiotics that interfere with protein synthesis in microorganisms and thus are bacteriostatic. The first of the tetracyclines was chlortetracycline (Aureomycin), introduced in 1948. As the generic name suggests, the molecules have four rings. Tetracycline molecules differ in the variety of groups attached to the ring structure. Simple tetracycline (Achromycin) and oxytetracycline (Terramycin) are other important drugs in this class.

The tetracyclines are broad-spectrum agents. Some are semisynthetic modifications of the basic structure that is produced by several microorganisms. Many compounds have been found by large-scale screening of microbes obtained from soil samples. Widespread

The suffix *-mycin* indicates a substance that was isolated from microorganisms found in the soil.

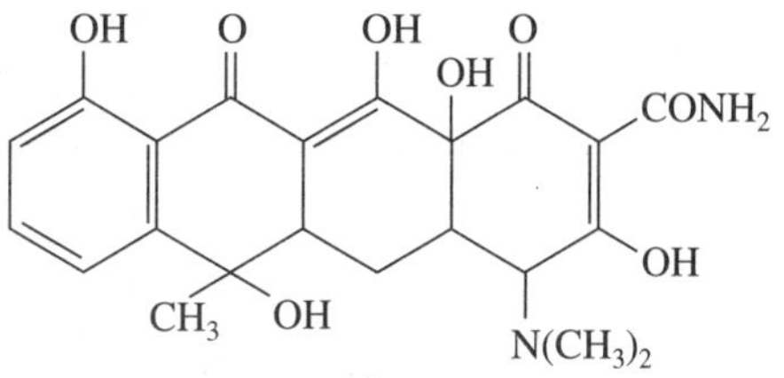

tetracycline

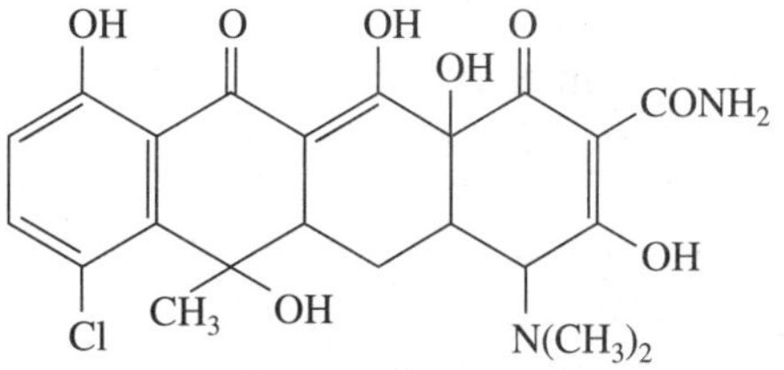

chlortetracycline

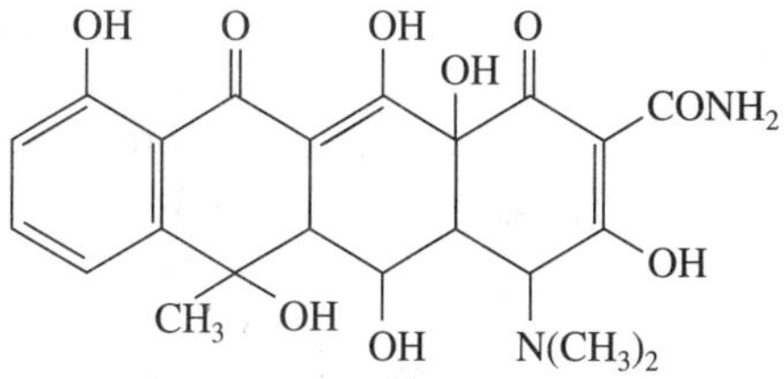

oxytetracycline

use of the tetracyclines has decreased their effectiveness, since resistant microorganisms have developed.

Other inhibitors of protein synthesis are streptomycin, chloramphenicol (Chloromycetin), and erythromycin. These antibiotics had their origin in microorganisms isolated from soil samples and have seen considerable use over many years. Each one has a broad spectrum of activity, although each is subject to development of resistant organisms. They are all bacteriostatic agents that act by inhibiting protein synthesis.

Streptomycin was first isolated in 1944. It was the first useful antibiotic against a broad spectrum of gram-negative bacteria. Erythromycin is often the antibiotic of choice for individuals with an allergy to penicillin.

24.4   Another Use of Antibiotics

GOAL: To appreciate the controversy over antibiotics as feed additives.

Antibiotics have been used as additives in animal feed for more than 30 years. During this time, many studies have confirmed their favorable effects, including faster growth with production of more meat, eggs, and milk for the same amount of feed.

Two theories suggest how antibiotics exert their effects on growth. First, the drugs may control the population of microorganisms in the intestines in a way that benefits the animals. Second, they may prevent disease so that the full metabolic capacity is geared to maximum growth rather than fighting disease. For example, the synthesis of antibody proteins (immunoglobulins) is an energy-consuming process. Energy (ATP) must be provided by metabolism of carbohydrates and other fuels. If an animal does not have to use its defenses against disease, it can be expected to grow faster.

Adding antibiotics to feed has the obvious advantage of increasing output of meat, eggs, and milk so that the costs to the consumer are potentially lower, but the practice has also been severely criticized for two reasons: resistance and allergy. Indiscriminate use of antibiotics encourages the development of microorganisms that are resistant to these antibiotics. Allergy is potentially an even more serious problem. Some individuals are allergic to antibiotics. For this reason, the drugs are not given to animals for several days before slaughter, providing time for their complete metabolism. This should prevent direct exposure of humans to antibiotics *(10)*.

24.5 Antiviral Chemotherapy

GOAL: To appreciate the limitations of antiviral chemotherapy.

Viruses are parasitic particles that can develop and reproduce only by invading the cells of organisms (Chapter 23). Because of this, most drugs with the potential for interfering with the development of viruses also tend to interfere with the functioning of normal cells. A virus particle invades a cell, multiplies, and then moves on to attack other cells, leaving behind increasing numbers of damaged and dead cells. Two serious diseases of viral origin are polio, which attacks the nervous system, and hepatitis, which damages liver cells.

For a drug to be effective in combating a virus attack, it must penetrate the cell and prevent multiplication of the virus particles without interfering with normal cellular chemistry. The latter requirement is the most challenging since the chemistry induced within the cell by the virus may not be much different from the normal cellular chemistry. Therefore, most drugs that are effective against viruses are also lethal to normal cells. As a result, most successes in combating viral diseases have been through vaccines that stimulate the cell's own defenses in combating the attack. Otherwise, most attempts at finding antiviral agents have been disappointing. Only a very few drugs are approved for use in the United States, each effective against only a very few viruses.

One of the successes is the compound *acyclovir*. The fully active form, acyclovir triphosphate, shown forming from simple acyclovir, inhibits the enzyme DNA polymerase.

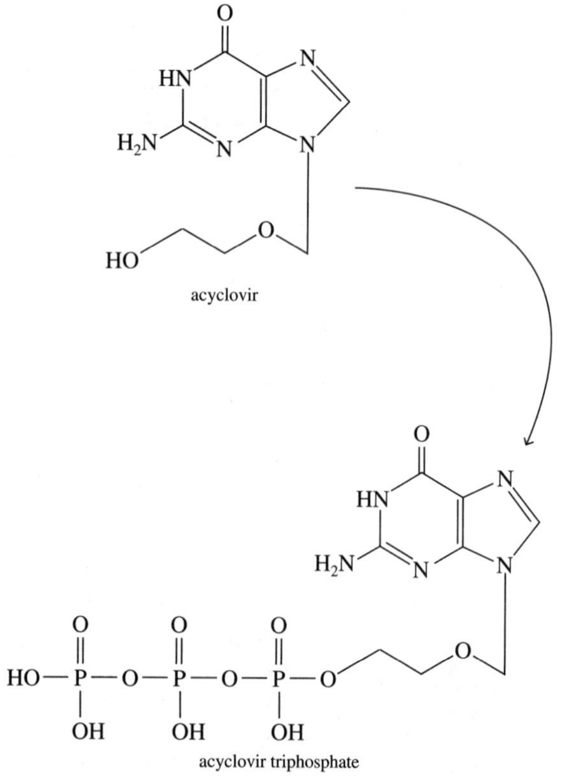

acyclovir

acyclovir triphosphate

This enzyme catalyzes the formation of DNA by the virus that is responsible for *Herpes simplex*. Acyclovir is only converted to its active form in virus-infected cells; the activation is limited in uninfected cells. Therefore, it is quite selective in interfering with the normal chemistry that takes place in cells that have been infected by the herpes virus. In effect, the virus contributes to its own self-destruction *(8, 11)*.

Despite what is known about the action of acyclovir, it has not been possible to produce any other antiviral agent that performs as effectively. In addition, acyclovir is ineffective against other viruses, and even some strains of herpes are resistant to it.

The compounds amantadine (Symmetrel) and rimantadine are effective against influenza A, which has caused several global epidemics such as Asian flu and Hong Kong flu. Unfortunately, they cause side effects, so they have not gained much popularity. The mode of action of the drugs is not understood.

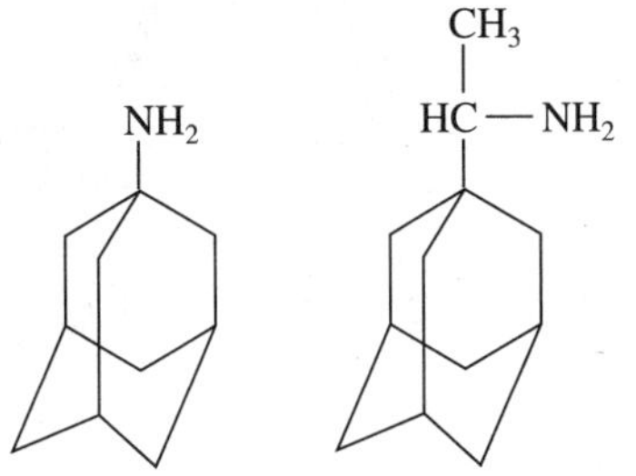

amantadine   rimantadine

AIDS chemotherapy is discussed in Section 23.5.

Zidovudine, originally known as azidothymidine (AZT), is approved for use in the treatment of AIDS. It is ineffective against most other viruses.

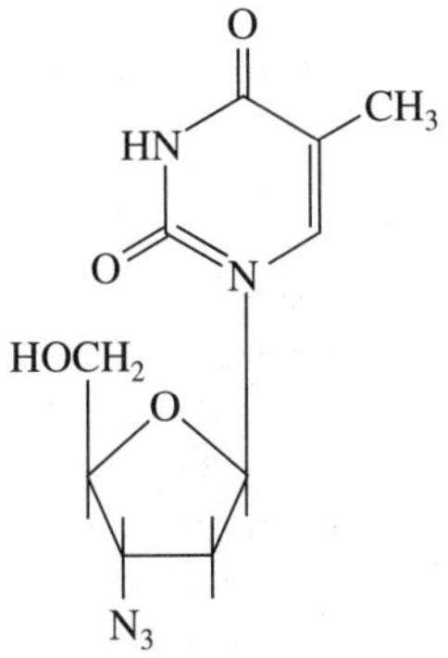

azidothymidine (AZT)

24.6 Anticancer Therapy

GOAL: To understand the complexities of cancer chemotherapy.

The major alternatives to radiation (Section 10.8) in the treatment of tumors are surgery and chemotherapy. Antibiotics and antiviral agents are forms of chemotherapy, but the general public normally associates the word with the treatment of cancer.

The choice of surgery, radiation, or chemotherapy is based on empirical observations. Solid tumors respond best to surgery, often followed by radiation therapy or chemotherapy to kill any cancer cells that are left behind. A *neoplasm* is defined as a new growth of cells having no physiologic function—that is, a tumor. One characteristic of neoplastic cells is proliferative (uncontrolled), sometimes rapid, growth. Drugs that combat tumors are known as *antineoplastic agents.* For them to be effective, they must show selective toxicity toward the faster-growing, less-controlled neoplastic cells. Antineoplastic drugs are sometimes the first form of attack on leukemias or on cells that have undergone *metastasis* (spread of cancer cells from the original location).

Like antiviral agents, the drugs have to attack abnormal cells while remaining nontoxic to normal cells. Because it is difficult to achieve selectivity of effects and avoid damage to normal cells, side effects are common. Nausea, vomiting, anorexia, and diarrhea follow disturbances of the stomach and intestinal lining. Bleeding and anemia follow bone marrow damage. Hair loss is common. All subside when the drugs are discontinued. Derivatives of tetrahydrocannibinol (THC), the active ingredient of marijuana, have been used successfully to deal with the nausea. Suppression of immunity is also a common side effect of cancer chemotherapy, and makes patients unusually susceptible to infections.

The similarity of cancer cells and human cells is best emphasized by the failure of the immune system to recognize cancer cells as foreign and to attack them. Surgery tends to make residual cells of solid tumors grow more actively; this makes them more susceptible to radiation or chemotherapy. Anticancer drugs, such as antibiotics, encourage development of resistant cells, making the disease harder to control as treatment continues.

SUMMARY

Drugs are usually identified by either the generic name or a trade name (brand name) chosen by the company holding the patent. A drug patent lasts for 17 years. During that time, a monopoly on the drug allows the company to make a reasonable profit and offset the costs of research, development, and testing. The drug industry is one of the most research-oriented industries. Heavy promotion of their products includes efforts to convince physicians and consumers that brand-name drugs are superior to less expensive generic brands. The evidence suggests that the brands are equivalent in most cases, however the variety of additives generally present in tablets and capsules provides possible differences. Sometimes a prodrug form is converted to the active form in the body. Controlled-release systems using wax-coated pellets (Contac), hydrogels, or nondegradable or degradable polyesters can increase the effectiveness of drug delivery.

The beginnings of antibiotic therapy date back to the early 1900s, but the biggest boost was the development of the sulfa drugs and the penicillins in the 1930s and 1940s. Some of the best-known antibiotics, with the suffix *-mycin,* were originally isolated from microorganisms in soil.

Most antibiotics are complex molecules that are difficult to synthesize. Even when they can be synthesized, microorganisms are normally used to produce them. These antibiotics are called *biosynthetics.* Semisynthetic antibiotics are produced by letting a microorganism do

part of the synthesis, after which the chemist completes the process.

The choice of an antibiotic by a physician is sometimes a critical decision, before laboratory analysis of the sensitivity of a disease-causing microorganism to various antibiotics is available. Culturing the microbe and determining its sensitivity routinely take a few days. Therefore, the physician will often choose an antibiotic with a broad spectrum of activity, or perhaps a combination of antibiotics. The risk is that favorable microorganisms also may be killed and dangerous resistant organisms may flourish.

Antibiotics against bacteria function as either bacteriostatic or bactericidal agents. Bacteriostatic agents inhibit a chemical process that is required for normal growth of microorganisms—for example, protein synthesis. Bactericidal agents kill susceptible microbes. Continuing research on antibiotics searches for an improved spectrum of activity and drugs against microbes that have become resistant.

Antibiotics also are used as supplements in animal feeds, to increase growth and production of meat, eggs, and milk. The reason for success is not known, and the practice has been criticized as contributing to the development of resistant microbes.

Chemotherapy against viruses and cancer has had very limited success owing to the nature of the diseases. Some cancers can sometimes be treated with drugs together with surgery or radiation therapy.

PROBLEMS

1. Give a definition or an example of each of the following.
   a. Bacteriostatic
   b. Bactericidal
   c. Generic name
   d. Brand name, or trade name
   e. Prodrug
   f. Antibiotic
   g. Broad-spectrum antibiotic
   h. Penicillinase
   i. Magic bullet
   j. Antimetabolite
   k. Miracle drugs
   l. Culturing
   m. Semisynthetic antibiotic
   n. Metastasis

2. How can two chemically equivalent drugs not be bioequivalent?

3. What are some of the principal drug additives?

4. What is the function of starch paste in the production of drug tablets?

5. Why are some drugs coated?

6. What is the peak-and-valley effect of drugs?

7. What are sulfa drugs? How do they work?

8. What does the suffix *-mycin* signify?

9. How can scientists cause changes in the synthesis of antibiotics?

10. Why is diarrhea a frequent side effect of antibiotic therapy?

11. Why is ampicillin marketed under the brand name Polycillin?

12. How is penicillin V more effective than penicillin G in treating disease?

13. Why do penicillins have unusually low toxicity?

14. What are the advantages and disadvantages of broad-spectrum antibiotics?

15. Why has it been difficult to produce a drug that is effective against viruses?

16. What does a penicillinase do to a penicillin?

17. What is the most popular penicillin and why?

18. How can clavulanic acid be used as a synergist for amoxicillin?

19. What is a hydrogel? How is it used for chemotherapy?

20. What is poly(lactic) acid? How is it used for chemotherapy?

References

1. Burack, R.; Fox, F. J. *The New Handbook of Prescription Drugs;* Ballantine Books: New York, 1975; p. 3.

2. "How to Pay Less for Prescription Drugs." *Consumer Reports,* January 1975, p. 48.

3. Shlafer, M.; Marieb, E. N. *The Nurse, Pharmacology, and Drug Therapy;* Addison-Wesley: Redwood City, CA; 1989; p. 1057.

4. Sanders, H. J. "Improved Drug Delivery." *Chemical and Engineering News,* April 1, 1985, p. 31.

5. Levinson, A. S. "The Structure of Salvarsan and the Arsenic-Arsenic Double Bond." *Journal of Chemical Education* **1977,** *54* (2), 98.

6. Lehne, R. A.; Crosby, L. J.; Hamilton, D. B.; Moore, L. A. *Pharmacology for Nursing Care;* W. B. Saunders: Philadelphia, 1990; p. 847.

7. Conn, P. M.; Gebhart, G. F. *Essentials of Pharmacology;* F. A. Davis: Philadelphia, 1989, pp. 375–400.

8. Pratt, W. B. *Fundamentals of Chemotherapy;* Oxford University: Toronto, 1973; pp. 100–101.

9. Buyske, D. A. "Drugs from Nature." *Chemtech,* June 1975, p. 361.

10. "Control of Drug Use in Food Animals Improves." *Chemical and Engineering News,* May 5, 1986, p. 21.

11. Robins, R. K. "Synthetic Antiviral Agents." *Chemical and Engineering News,* January 27, 1986, p. 28.

Chemotherapy for the Treatment of Symptoms and Drug Abuse

25

In daily life we often encounter pain—headaches, arthritis, upset stomach. In addition, psychic pain may accompany stress and unpleasant experiences. At times when pain develops we turn to over-the-counter or prescription drugs. If surgery is necessary, more severe pain can be blocked with anesthesia and preoperative and postoperative medications.

Chemotherapy is generally taken to mean the use of drugs to cure a disease, yet treating symptoms is often equally important. While our body defenses fight an infection, or when we recover from surgery, controlling symptoms may help healing. For viral diseases, treatment of symptoms is usually the only course of action until the body's immunity removes the invaders.

Blockage of physical or psychological pain can lead to drug abuse. This is an ever-increasing problem for people who have their own special symptoms or who cannot get by on life's ordinary pleasures. The slide from acceptable self-dosing to overuse of common remedies and perhaps to drug abuse is all too common. Worse yet, in 1985, "crack" cocaine emerged in the United States and quickly became the most dangerous drug up to that time. Shortly after, methamphetamine appeared in a form known as "ice." It is equally devastating.

The actions that several acceptable and abused drugs have on the body are the subjects of this chapter. Five categories of drugs are presented: analgesics for pain relief, anesthetics, antacids, tranquilizers, and antidepressants. Their effects on nerves, the digestive system, and other tissues are discussed. The variety and causes of drug abuse conclude the chapter.

25.1 Analgesics, the Pain Relievers

GOALS: To describe the relative advantages of the common NSAIDs.
To appreciate the use of various drugs in over-the-counter combinations.

Pain is often the first indication of malfunction or injury, and decreasing it may be the first item of concern. A vast array of analgesics (pain relievers) is available, both over the counter and by prescription.

Aspirin

Aspirin heads the list of over-the-counter drugs in popularity. It is available alone (plus binder and filler) and in combination with other analgesics or drugs. Sales of aspirin in all forms are estimated at 200 tablets of 5 grains (324 mg) each per person per year in the United States. Children's aspirin contains 1 1/4 grains (80 mg) of aspirin per tablet. About half the aspirin is consumed as combination pain relievers and cold remedies; they often contain a second analgesic, a decongestant, an antihistamine, and/or an antacid.

Besides having analgesic properties, aspirin is an effective antipyretic (fever-reducing) and anti-inflammatory agent. It also has an anticoagulant effect. One to three aspirin tablets per day reduce the chance of heart attack and stroke in high-risk patients by making formation of blood clots less likely (1).

Recall that acetylsalicylic acid is a prodrug (Section 24.2).

Aspirin is the generic name for the compound **acetylsalicylic acid.** It was first synthesized in 1853 but not marketed until 1899. Salicylic acid is an effective pain reliever, but it causes severe damage to the stomach, leading to significant bleeding. The addition of an *acetyl group,* by a reaction called *acetylation,* to the salicylic acid molecule greatly reduces the corrosive effect on the stomach.

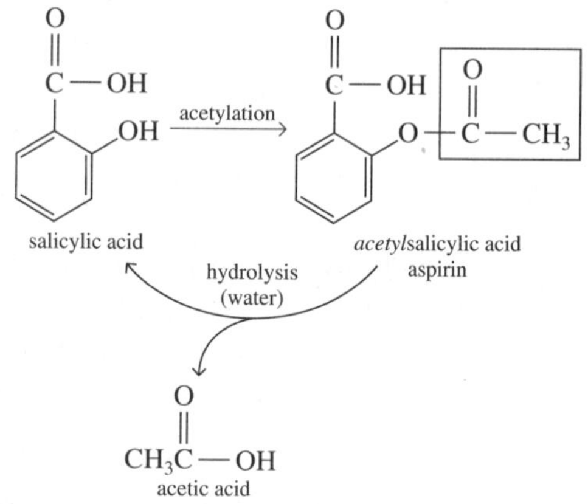

Aspirin can react with water to produce salicylic acid and acetic acid. This process is essentially the reverse of the reaction used to make aspirin from salicylic acid, as shown in the previous scheme. Acetic acid can be readily detected as the odor of vinegar (5% acetic acid) when you open an old bottle of aspirin that has had repeated exposure to humidity in the air. Although the smell may be strong, the amount of acetic acid present is normally small. In other words, the odor does not usually indicate significant decomposition of the aspirin.

Unfortunately, bleeding is not totally eliminated by acetylation. Most people do suffer enough damage to lose a small amount of blood (typically 0.5–2 mL) when they consume two regular aspirin. For a very small part of the population, the effect is more severe. For these individuals, **acetaminophen** (Datril, Tylenol, Tempra, Liquiprin, Panadol) is a satisfactory alternative, although more expensive.

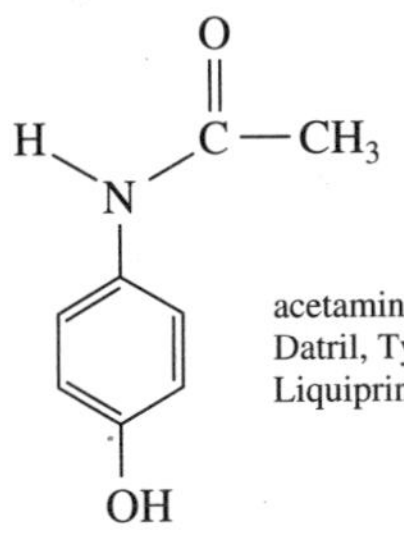

acetaminophen
Datril, Tylenol, Tempra,
Liquiprin, Trilium

Another risk of aspirin is Reye's syndrome, a rare, life-threatening illness of childhood. Epidemiological data suggest a link between the use of aspirin for treating chicken pox or influenza and Reye's syndrome. Since influenza usually begins with symptoms similar to those associated with the common cold, the usual warning is to avoid aspirin whenever children exhibit flu-like symptoms *(2,3)*.

Pharmacology textbooks classify aspirin as a nonsteroidal anti-inflammatory drug (NSAID) to distinguish it from cortisone and other steroids used to treat inflammation (see Chapter 26). Since injured tissues and rheumatic joints often have swelling, the anti-inflammatory property of aspirin is very important for relieving pain and treating arthritis.

Over-the-counter remedies for treating pain from arthritis, such as Arthritis Pain Formula and Arthritis Strength Bufferin, contain 7.5 grains (gr) of aspirin in each tablet, compared to the usual 5 gr. A simple calculation shows that three regular aspirin tablets provide the same dose as two arthritis-strength tablets. The latter usually contain an antacid to minimize stomach irritation (see discussion of combination remedies).

Alternatives

Most products marketed as aspirin-free contain acetaminophen as the analgesic. Acetaminophen acts against both pain and fever, but does not reduce inflammation. This drug is preferred for persons who are susceptible to gastric distress, heartburn, or nausea due to aspirin. It is also recommended as a substitute for aspirin for treating fever in children suspected of having chicken pox or influenza.

Another popular NSAID alternative to aspirin is **ibuprofen.** It may be obtained in prescription strength as Motrin or Rufen containing 300-800 mg per tablet, or in nonprescription tablets containing 200 mg in such products as Advil, Haltran, Midol 200, and Nuprin, as well as in generic form. Ibuprofen is not a good choice for individuals with high blood pressure (hypertension) or heart disease because it may cause sodium and water retention, whereas aspirin does not have these side effects. High blood pressure medication commonly includes a diuretic, often called a "water pill," to stimulate the loss of sodium and water. Ibuprofen would reverse the desired effect. A common use of ibuprofen is in the treatment of menstrual discomfort, for which it is reported to be particularly effective *(1)*.

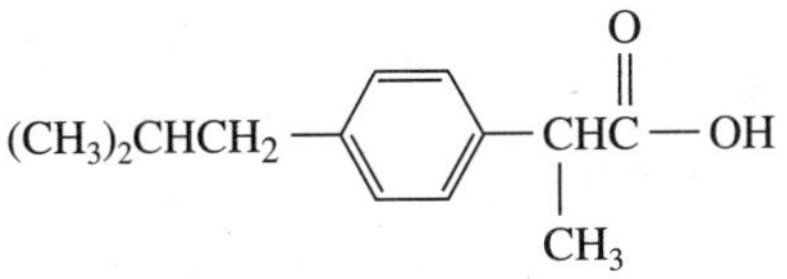

ibuprofen
Motrin, Rufen, Advil,
Haltran, Midol 200, Nuprin

A variation of the salicylic acid molecule is a compound with the generic name diflunisal (Dolobid). It has a more potent, longer-lasting analgesic effect than aspirin, allowing for use of lower doses to minimize side effects.

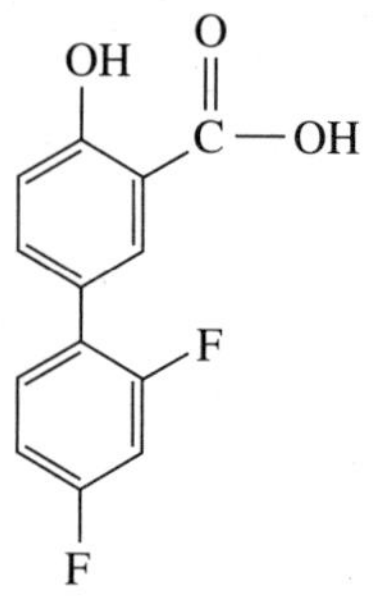

diflunisal

Combination Remedies

Many combination remedies are available. Most contain an analgesic—aspirin or acetaminophen, or both (Excedrin, for example). Remedies also may contain caffeine, an antacid, a decongestant, or an antihistamine. A list of combination drugs appears in Table 25.1.

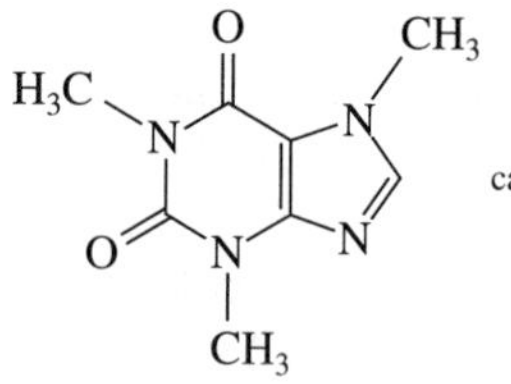

caffeine

Caffeine is discussed further in Section 25.6.

Caffeine is a vasoconstrictor; it causes blood vessels to contract. Since migraine headaches are attributed to dilation (expansion) of blood vessels in the head, caffeine has potential for treatment. However, the amount present in combination remedies is regarded as insignificant. Perhaps this is fortunate, since caffeine is a stimulant. Continual consumption might cause some sleepless nights when sleep might be the best medicine. It has been suggested that caffeine is present in some combination remedies to counteract drowsiness, a common side effect of many antihistamines (2).

For comparison, some cold remedies contain 15-30 mg of caffeine per tablet, whereas No Doz (100 mg), Quick Pep (150 mg), and Vivarin (200 mg) all contain much more. A cup of regular coffee contains about 100 mg of caffeine. Soft drinks normally contain 35-55 mg per 12 oz.

Inflammation of the nasal membranes leads to sneezing, runny nose, and congestion. Therefore, nasal decongestants and antihistamines are common components of most combination cold remedies. Decongestants such as phenylephrine and phenylpropanolamine provide relief from the stuffiness.

Section 23.7 considers the use of antihistamines in treating allergy.

The prefixes *rhin-* and *rhino-* refer to the nose. *Rhinitis* is inflammation of the nasal membranes. *Rhinorrhea* is the technical term for a runny nose.

Although they are frequently used, antihistamines are often regarded as ineffective in treating cold symptoms, since the release of histamine (Section 23.7) is associated with allergic reaction. Antihistamines do provide a slight reduction in rhinorrhea (runny nose) by decreasing secretions from nasal glands. Regardless of their effectiveness, antihistamines may cause drowsiness, so they can be potentially hazardous to a person who drives an automobile or operates heavy machinery after consumption.

Table 25.1 Some well-known simple and combination remedies

Product Name	Analgesic	Decongestant	Antihistamine	Antacid	Other
Alka-Seltzer (Original)	Aspirin			$NaHCO_3$	Citric acid[1]
Alka-Seltzer Plus	Aspirin	Phenylpropanolamine	Chlorpheniramine	$NaHCO_3$	Citric acid[1]
Alka-Seltzer Nighttime	Aspirin	Phenylpropanolamine	Diphenhydramine	$NaHCO_3$	Citric acid[1]
Actifed		Pseudoephedrine	Triprolidine		
Allerest		Phenylpropanolamine	Chlorpheniramine		
Anacin	Aspirin				Caffeine
Aspirin-Free Anacin	Acetaminophen				
Arthritis Pain Formula	Aspirin			$Al(OH)_3$ $Mg(OH)_2$	
Bufferin	Aspirin			$MgCO_3$ MgO $CaCO_3$	
CoAdvil	Ibuprofen	Pseudoephedrine			
Comtrex	Acetaminophen	Pseudoephedrine	Chlorpheniramine		Dextromethorphan[2]
Coricidin	Acetaminophen		Chlorpheniramine		
Coricidin D	Acetaminophen	Phenylpropanolamine	Chlorpheniramine		
CoTylenol	Acetaminophen	Phenylpropanolamine	Chlorpheniramine		Dextromethorphan[2]
Dristan	Acetaminophen	Phenylephrine	Chlorpheniramine		
Drixoral		Pseudoephedrine			
Excedrin	Acetaminophen + aspirin				Caffeine
Excedrin PM	Acetaminophen		Diphenhydramine		
NyQuil	Acetaminophen	Pseudoephedrine	Doxylamine		Dextromethorphan,[2] alcohol (25%)
Day Care	Acetaminophen	Pseudoephedrine, guaifenesin[3]			Dextromethorphan,[2] alcohol (10%)
Thera Flu	Acetaminophen	Pseudoephedrine	Chlorpheniramine		Dextromethorphan[2]
Tylenol Cold	Acetaminophen	Pseudoephedrine			
Vanquish	Acetaminophen + aspirin			$Al(OH_3)$ $Mg(OH)_2$	

[1]Causes an effervescent action.   [2]Cough suppressant.   [3]Expectorant.

Antacids are sometimes included in combination remedies or with aspirin alone. One example is buffered aspirin (Bufferin). It has been claimed that the antacid speeds absorption of the aspirin, but evidence for this is lacking. The antacid in the buffered aspirin tablet is said to react with stomach acid (HCl) thus causing more rapid disintegration of the tablet and release of the analgesic. However, a properly formulated aspirin tablet, usually containing acetylsalicylic acid plus starch binder and filler (Section 24.2), disintegrates rapidly on contact with water, so the presence of an antacid makes little difference.

Also, a buffer is a substance that resists a change of pH when acid or base is added. The alkaline substance in buffered aspirin does not meet this definition, since it may cause an increase in stomach pH.

It is also claimed that the antacid reduces stomach irritation. There is no clear evidence for or against such a claim since it is virtually impossible to measure this effect. Aspirin can irritate the stomach, but an antacid will not eliminate this simply by reacting with HCl.

One combination may help to reduce irritation caused by aspirin. The product, sold under the brand name Alka-Seltzer, contains the antacid sodium bicarbonate ($NaHCO_3$). The effervescent action is observed when Alka-Seltzer is dropped into water; the fizzing action helps to dissolve the aspirin. Since the analgesic is in solution when it is consumed, it passes through the stomach more quickly. This seems ideal, although regular use of antacids is not recommended (Section 25.3).

> Alka-Seltzer is available in several forms—with aspirin, with acetaminophen, or without an analgesic.

A number of drugs, including aspirin, are available in enteric-coated form. This prevents the tablet from breaking down in the acidic environment of the stomach, but allows for release of the aspirin once the tablet reaches the alkaline conditions in the small intestine.

Treatment of Severe Pain

When pain is severe, a physician normally enters the picture and prescribes from an array of potent prescription analgesics, most of which are addictive. Among them, propoxyphene is the prescription analgesic often used (4). It is available in a variety of forms under the brand names Darvon, Darvocet-N (in combination with acetaminophen), and Darvon Compound 65 (in combination with aspirin and acetaminophen), and under the generic name *propoxyphene*.*

Studies have shown that propoxyphene in either generic or brand-name form is not as effective as aspirin or acetaminophen despite its popularity; most studies show that it is no more effective than a placebo (5,6). But pain is a subjective condition, and it is often impossible to evaluate treatment. An expensive analgesic such as Darvon may seem to be a better choice than simple aspirin or acetaminophen, since drug companies routinely

---

*Like a great many drugs, propoxyphene is not marketed as the neutral compound. Compounds used as drugs can often be converted to salts by reaction with acids or bases. Propoxyphene is a typical example, as it can be converted to a hydrochloride salt by reaction with HCl. Like ammonia, NH_3, which is converted to ammonium chloride when it reacts with HCl, propoxyphene has a basic nitrogen atom that can accept a hydrogen ion.

$$NH_3 + HCl \rightarrow NH_4^+ + Cl^-$$

propoxyphene   propoxyphene hydrochloride

Although the propoxyphene reaction looks more complicated, it is a simple neutralization reaction in which an acid and a base react to produce a salt. After a substance is converted to a crystalline salt, its stability and ease of handling are often improved.

promote the idea that "you get what you pay for." Many individuals suffering from pain are willing and eager to believe it is true. However, the inexpensive aspirin and acetaminophen may provide sufficient analgesia. Although the potential for addiction does exist, Darvon has fewer risks than many other drugs. Therefore, it is not subject to narcotics regulations of the Controlled Substances Act.

The other main class of prescription painkillers is the opiates. These include natural substances that can be extracted from opium (morphine, codeine), a semisynthetic (heroin), and pure synthetics (Demerol, methadone). The synthetics are not derived in any way from opium, but they match the true opiates in action and potential for addiction (Section 25.6).

Morphine as a painkiller is about 50 times as potent as aspirin, but aspirin is not addictive, whereas morphine can be. Morphine was used widely in the Civil War as a painkiller, and more than 100,000 soldiers became addicts, a disorder known as "soldiers' disease." Recent evidence shows that morphine may not be addictive when it is used to treat severe pain *(7)*, so the problem may have had less to do with treating physical pain and more to do with the mental anguish of fighting the war.

Morphine must be injected because of erratic absorption from the intestinal tract and because an oral dose is rapidly metabolized in the liver *(1)*. Codeine is only about one-sixth as potent as morphine, but it is easier to administer because it can be taken orally.

Meperidine (Demerol) is the primary synthetic substance used to treat severe pain. It is between morphine and codeine in potency and is also addictive.

meperidine
(Demerol)

25.2 Drugs Used in Surgery

GOALS: To understand the approach to and difficulties associated with the treatment of severe pain.
To understand the application of local anesthetics.

A well-planned sequence of drugs normally accompanies major surgery, beginning the night before and extending throughout surgery in the operating room and in the immediate postoperative period. The night before surgery the patient is usually given a tranquilizer to encourage relaxation and a good night's sleep. During surgery, a combination of drugs is administered to achieve anesthesia, analgesia, and muscle relaxation. After surgery, morphine or Demerol may be given to relieve severe pain.

Of these drugs, the most interesting are the ones used during surgery. The old standby is ether, actually diethyl ether. It was first used in dentistry in 1847. It is an effective anesthetic and analgesic and also promotes good muscle relaxation. In addition, there is a wide gap between the amount that causes unconsciousness and the lethal dose, so it is extremely safe to administer. Like several of the common anesthetics, ether is easily vaporized and can be administered to a patient in a stream of oxygen. The stream usually contains 10-30% ether while inducing anesthesia and 5-15% to maintain the condition. Side effects such as nausea may accompany the use of ether. In addition, since it is extremely flammable and explosive, its use has been discontinued.

$$CH_3CH_2 \!-\! O \!-\! CH_2CH_3 \quad \text{diethyl ether}$$

Chloroform, CHCl$_3$, was first used for surgery in 1847. It is nonflammable, but it often causes liver damage. Also there is a narrow gap between the level required for anesthesia and the lethal dose, so it is rarely used in surgery.

At the present time, the primary drugs used during surgery are Sodium Pentothal, nitrous oxide (N$_2$O), enflurane, and isoflurane. Sodium Pentothal is a potent barbiturate-hypnotic that rapidly induces unconsciousness; however, it does not provide sufficient analgesia and muscle relaxation to be used alone. Enflurane and isoflurane are the two most popular anesthetics. They are etherlike molecules, highly substituted with halogen atoms (fluorine, chlorine, bromine), which makes them nonflammable. They are safe and potent, and can be used to maintain the anesthesia induced by Sodium Pentothal.

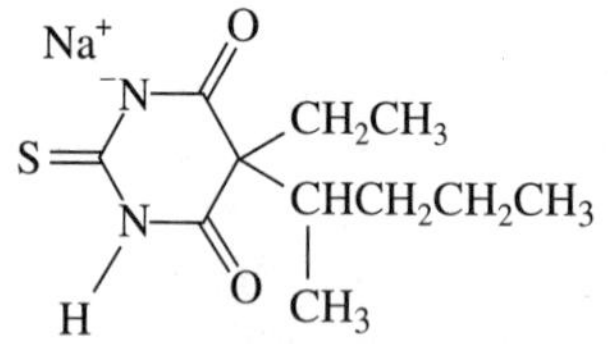

Sodium Pentothal 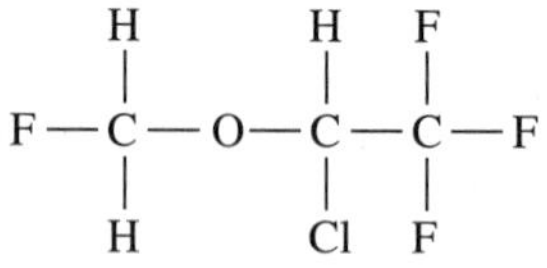   isoflurane 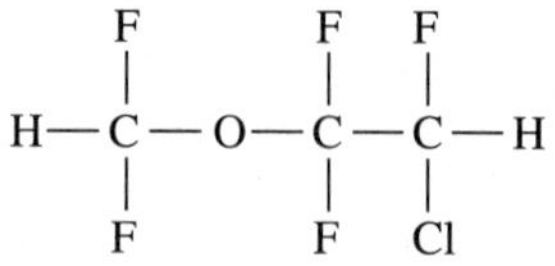   enflurane

Enflurane and isoflurane also can be used in place of Sodium Pentothal to induce anesthesia. Most people, however, prefer not to be put to sleep with a mask over the face. Once a patient is unconscious, enflurane or isoflurane can be administered through a mask in a stream of nitrous oxide and oxygen gases.

Nitrous oxide is also an excellent anesthetic and analgesic on its own. It is a colorless, sweet-tasting gas, sometimes known as laughing gas because it induces lightheadedness.

Local Anesthetics

For very minor surgery, it is often possible to use a local anesthetic, such as procaine (Novocain) or lidocaine (Xylocaine); of these, lidocaine is currently preferred. Either can be injected into the site to be treated or applied topically to the surface of the mucous membranes of the eye, nose, throat, or urethra.

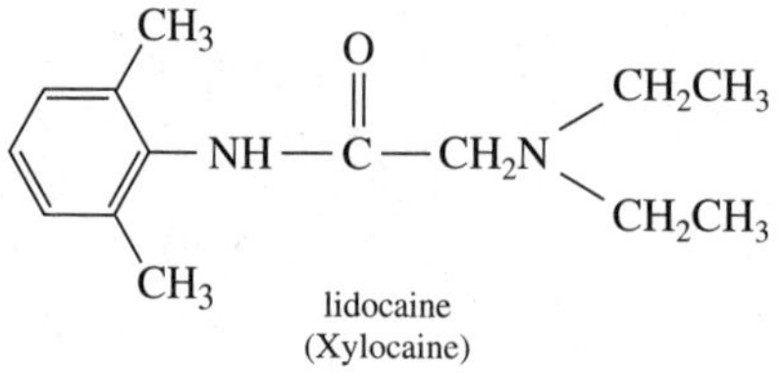

lidocaine
(Xylocaine)

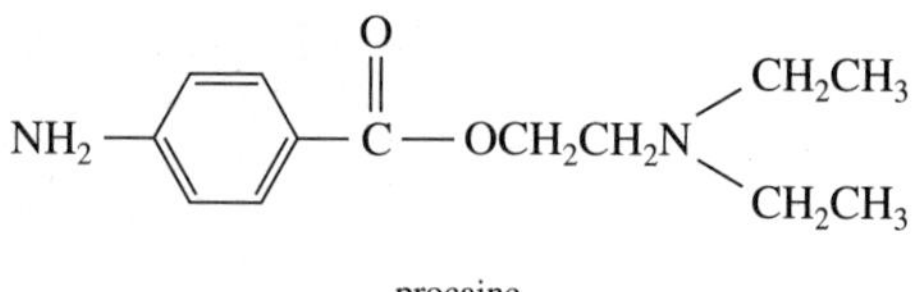

procaine
(Novocain)

Ethyl chloride is a colorless liquid that boils at 12 °C (54 °F), so it is a gas at room temperature, but it can be liquefied under pressure in a sealed container.

CH$_3$CH$_2$Cl ethyl chloride

When the liquid is sprayed on the surface of the skin, it quickly evaporates. When a liquid evaporates from a surface, the surface is cooled. This is a familiar phenomenon, noticed when one steps out of water after a shower or a swim. Some of the moisture evaporates, leading to a cold sensation. The heat lost during evaporation is called the **heat of vaporization.** Viewed another way, the heat of vaporization is the energy required to cause the change from liquid to gaseous state. Water has a very high heat of vaporization. That is one reason why it is an excellent body fluid, since evaporation of perspiration from the body surface provides an extremely efficient cooling mechanism.

The same strategy applies to using ethyl chloride as a local anesthetic. Due to its low boiling temperature, it evaporates rapidly and cools the surface so quickly that it freezes the tissues near the surface, making them insensitive to pain. It is often sprayed on a wound or other injured tissue to deaden pain, for example, in injuries in sporting events. It can also be used for minor surgery, such as draining carbuncles (boils).

25.3   Antacids

GOALS: To appreciate the variety of antacids that are available.
To learn the chemistry of effervescent antacids.

Next to pain, one of our most frequent symptoms is indigestion caused by overproduction of stomach acid, HCl. Many over-the-counter antacids, like those shown in Figure 25.1, are available to treat it, although the number of ingredients is small. The most common antacids contain sodium bicarbonate, $NaHCO_3$, calcium carbonate, $CaCO_3$, magnesium carbonate, $MgCO_3$, aluminum hydroxide, $Al(OH)_3$, magnesium hydroxide, $Mg(OH)_2$, and magnesium trisilicate, $Mg_2Si_3O_8$. Citrate and tartrate are less common. The antacid action of each is shown in the accompanying reactions. The ingredients of several commercial products are given in Table 25.2.

$$HCO_3^- + HCl \longrightarrow H_2CO_3 + Cl^-$$

bicarbonate   unstable
as $NaHCO_3$
or $KHCO_3$ $\downarrow$

$$H_2O + CO_2$$

$$CO_3^{2-} + 2\,HCl \longrightarrow H_2CO_3 + 2\,Cl^-$$

carbonate   unstable
as $CaCO_3$
or $MgCO_3$
or $NaAl(OH)_2CO_3$

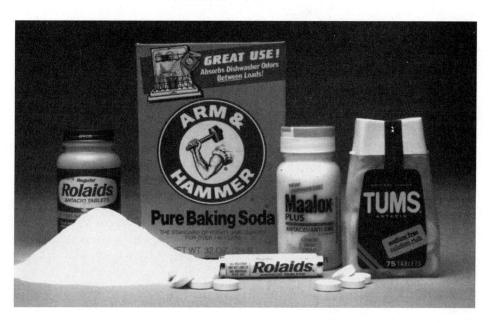

Figure 25.1
Some popular antacids.

$$OH^- + HCl \longrightarrow H_2O + Cl^-$$
hydroxide
as $Mg(OH)_2$
or $Al(OH)_3$
or $NaAl(OH)_2CO_3$

$$Si_3O_8^{4-} + 4\,HCl \longrightarrow 3\,SiO_2 + 2\,H_2O + 4\,Cl^-$$
trisilicate
as $Mg_2Si_3O_8$

$$C_6H_5O_7^{3-} + 3\,HCl \longrightarrow H_3C_6H_5O_7 + 3\,Cl^-$$
citrate
as $Na_3C_6H_5O_7$

$$C_4H_4O_6^{2-} + 2\,HCl \longrightarrow H_2C_4H_4O_6 + 2\,Cl^-$$
tartrate
as $Na_2C_4H_4O_6$

Antacids differ in advantages and disadvantages. Many persons prefer effervescent products. High doses of magnesium salts may have a laxative effect. In fact, $Mg(OH)_2$, milk of magnesia, is sold for that purpose. Aluminum salts may be constipating, again depending on the dose. Several antacids are formulated with combinations of magnesium and aluminum salts to prevent such side effects. Some contain a pain reliever. Finally, some antacids contain sodium and should be avoided by individuals on low-sodium diets. Several well-known products have been reformulated in recent years by substituting calcium carbonate for a sodium salt. Advertisements often emphasize that they are ''low in sodium'' or ''calcium rich.''

Table 25.2 Popular commercial antacid products

Product Name	Antacid Ingredients	Other Ingredients
Alka-Seltzer Original	$NaHCO_3$ + citric acid	Aspirin
Alka-Seltzer Advanced Formula	$KHCO_3$ + $NaHCO_3$ + $CaCO_3$ + citric acid	Acetaminophen
Bromo-Seltzer	$NaHCO_3$ + citric acid	Acetaminophen
Brioschi	$NaHCO_3$ + tartaric acid	
Milk of Magnesia	$Mg(OH)_2$	
Tums	$CaCO_3$	
Alka 2	$CaCO_3$	
Titralac	$CaCO_3$	
Maalox	$Mg(OH)_2$ + $Al(OH)_3$	
Mylanta	$Mg(OH)_2$ + $Al(OH)_3$	Silicone defoaming agent
Gaviscon	$Mg(OH)_2$ + $Al(OH)_3$ + $MgCO_3$	
Di-Gel	$Mg(OH)_2$ + $Al(OH)_3$	Silicone defoaming agent
Gelusil	$Mg(OH)_2$ + $Al(OH)_3$	Silicone defoaming agent
Rolaids (Regular)	$NaAl(OH)_2CO_3$	
Rolaids (Calcium Rich)	$CaCO_3$	

Effervescent Antacids

Antacids that contain sodium bicarbonate and citric acid (or tartaric acid) have a bubbling or effervescent action when dissolved in water. This is because carbon dioxide gas evolves, as shown.

$$
3\ Na^{+}HCO_{3}^{-} + \underset{\substack{\text{citric}\\\text{acid}}}{\begin{array}{c}\quad\\ CH_2\overset{\displaystyle O}{\overset{\|}{C}}\!-\!OH \\ | \\ HO\!-\!C\!-\!\overset{\displaystyle O}{\overset{\|}{C}}\!-\!OH \\ | \\ CH_2\overset{\displaystyle O}{\overset{\|}{C}}\!-\!OH \end{array}} \longrightarrow 3\ \underset{\text{unstable}}{H_2CO_3} + \underset{\substack{\text{sodium}\\\text{citrate}}}{\begin{array}{c}\quad\\ CH_2\overset{\displaystyle O}{\overset{\|}{C}}\!-\!O^{-}Na^{+} \\ | \\ HO\!-\!C\!-\!\overset{\displaystyle O}{\overset{\|}{C}}\!-\!O^{-}Na^{+} \\ | \\ CH_2\overset{\displaystyle O}{\overset{\|}{C}}\!-\!O^{-}Na^{+} \end{array}}
$$

sodium bicarbonate

$$3\ H_2O + 3\ CO_2 \ \text{(gas)}$$

The reaction as shown does not occur until the ingredients dissolve in water. But even at normal humidity, there is plenty of moisture in the air to enter the tablet and cause the antacid and the acid to combine and react. For that reason, Alka-Seltzer comes sealed in a foil package that prevents contact with moisture in the air before the container is opened. After standing in contact with air, an Alka-Seltzer will appear flat when added to water, much like a carbonated soda that has been left open and has lost its carbonation. One explanation for how Alka-Seltzer may provide relief from the discomfort of overeating is that the release of CO_2 by action of acid on $NaHCO_3$ induces belching, which aids in expelling swallowed air *(8)*.

See Common Acids and Bases (Section 7.5) for further discussion of the reactions of bicarbonate and carbonate salts.

The sodium citrate produced in the reaction of sodium bicarbonate with citric acid may also accept hydrogen ions, H^{+}, and revert to citric acid. Therefore, even the sodium citrate will act as an antacid. In fact, all the citric acid in an Alka-Seltzer tablet is neutralized to sodium (or potassium) citrate. More than 85% of the bicarbonate is consumed by the effervescent action, so relatively little is left to act as an antacid.

Sodium tartrate is formed from the reaction of tartaric acid and sodium bicarbonate when the antacid Brioschi is dissolved in water.

$$
\underset{\text{tartaric acid}}{\begin{array}{c} \overset{\displaystyle O}{\overset{\|}{C}}\!-\!OH \\ | \\ H\!-\!C\!-\!OH \\ | \\ H\!-\!C\!-\!OH \\ | \\ \underset{\displaystyle O}{\underset{\|}{C}}\!-\!OH \end{array}} + 2\ Na^{+}HCO_{3}^{-} \longrightarrow 2\ \underset{\text{unstable}}{H_2CO_3} + \underset{\text{sodium tartrate}}{\begin{array}{c} \overset{\displaystyle O}{\overset{\|}{C}}\!-\!O^{-}Na^{+} \\ | \\ H\!-\!C\!-\!OH \\ | \\ H\!-\!C\!-\!OH \\ | \\ \underset{\displaystyle O}{\underset{\|}{C}}\!-\!O^{-}Na^{+} \end{array}}
$$

sodium bicarbonate

$$2\ H_2O + 3\ CO_2 \ \text{(gas)}$$

The sodium tartrate acts in the same way as sodium citrate in accepting hydrogen ions from stomach acid to form the weaker tartaric acid.

25.4 Tranquilizers

GOALS: To distinguish between major and minor tranquilizers.
To understand the use of some antihistamines as sleeping aids.

Diazepam (Valium) and chlordiazepoxide (Librium) are among the most prescribed drugs. They are tranquilizers, or more specifically, minor tranquilizers, a description meant to distinguish them from drugs that are used to treat severe mental disorders. Tranquilizers sedate without inducing sleep, and are used to relieve anxiety, excitement, and restlessness. Miltown and Equanil are brand names for another important minor tranquilizer, which is also available under the generic name *meprobamate*. Meprobamate was first used as a muscle relaxant.

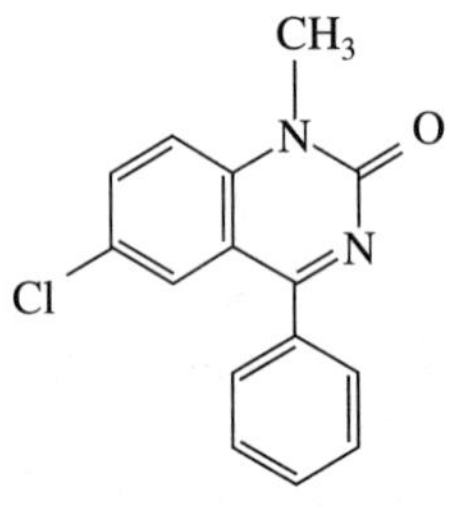

diazepam
Valium

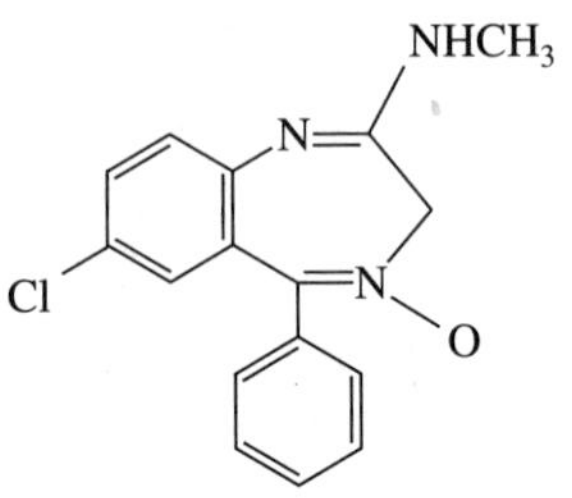

chlordiazepoxide
Librium

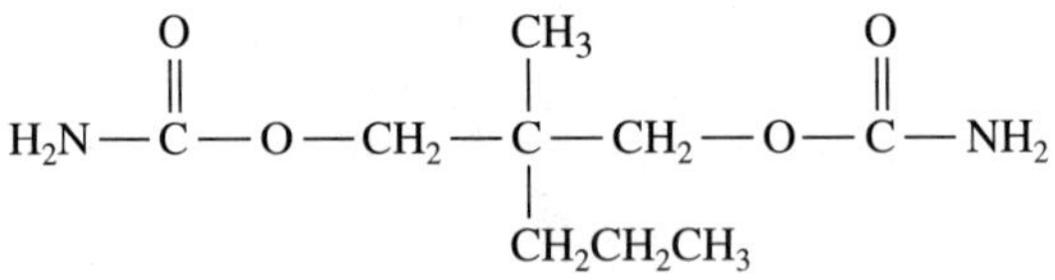

meprobamate
Miltown, Equanil

Chlorpromazine, sold under the brand name Thorazine, was the first and best-known of the major tranquilizers. Many more are now available. Since chlorpromazine was first tried on psychotic patients in 1952, more than 50 million people throughout the world have received it.

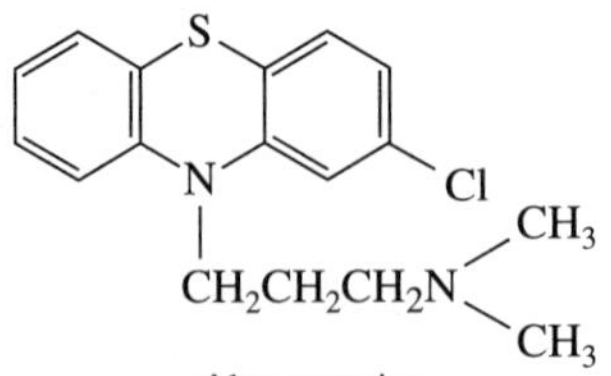

chlorpromazine
(Thorazine)

All tranquilizers are available only by prescription. Over-the-counter remedies that claim to be tranquilizers are generally mild sedatives or antihistamines, which may cause drowsiness as a side effect. Driving should be avoided when taking any of these drugs.

The antihistamine most often used is diphenhydramine, which appears in such well-known products as Sleep-Eze, Nytol, Sominex, Compoz, Alka-Seltzer Nighttime, and Excedrin PM. Doxylamine is another compound that causes drowsiness; it is used in Unisom.

Two relatively new antihistamines for which drowsiness is not a side effect are terfenadine (Seldane) and astemizole (Hismanal). Neither compound can cross the blood-brain barrier and affect the central nervous system.

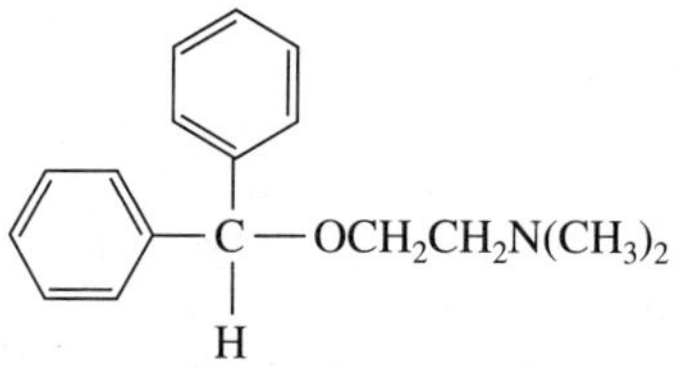

diphenhydramine

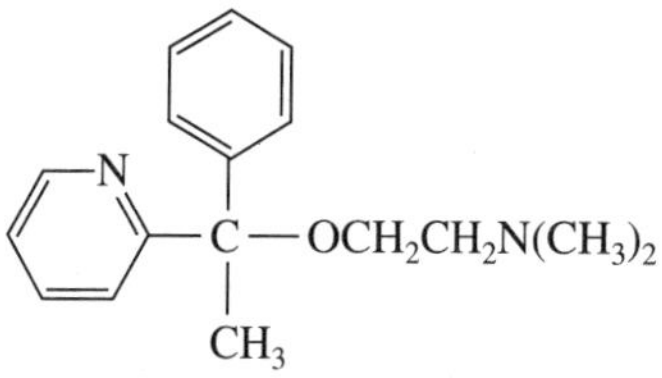

doxylamine

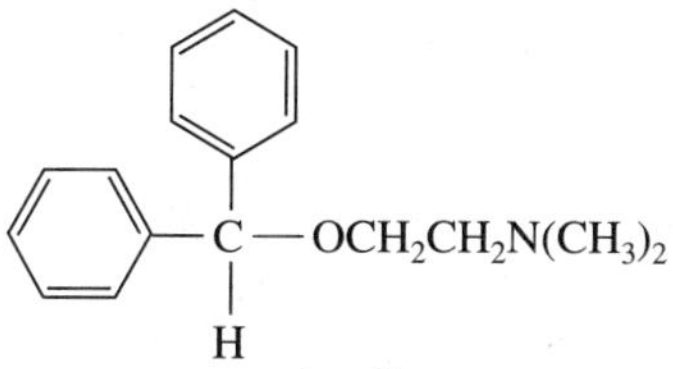

terphenadine
(Seldane)

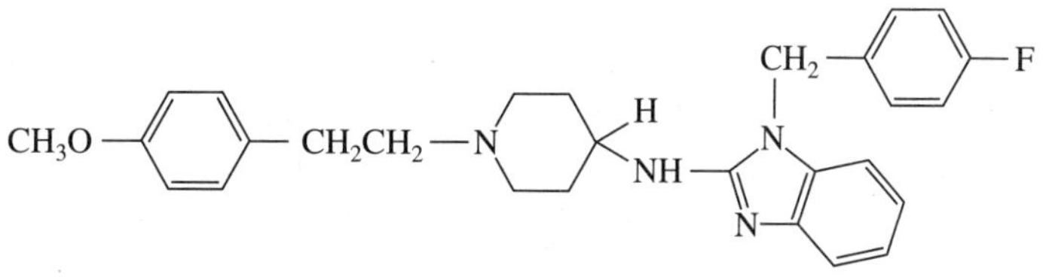

astemizole
(Hismanal)

25.5 Antidepressants

GOAL: To appreciate the existence of antidepressants, including Prozac.

Although many individuals seek the peace and tranquility that they think may be obtained by taking some sort of pill, some require the opposite. For this reason the tricyclic antidepressants Elavil and Tofranil are often prescribed by physicians.

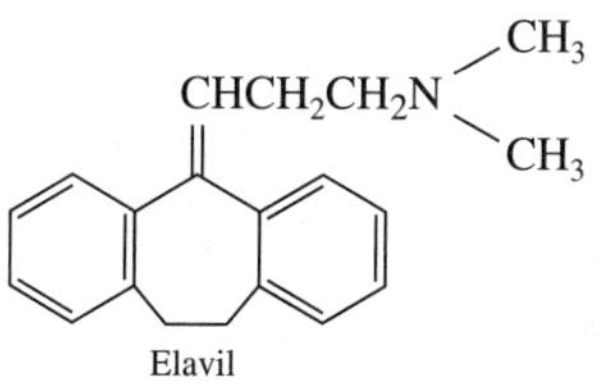

Elavil

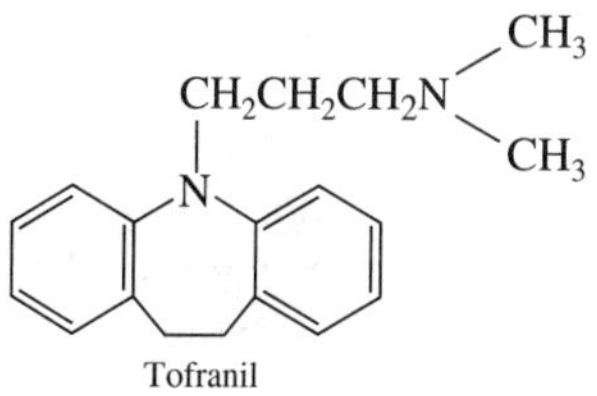

Tofranil

The drug Prozac (fluoxetine) has had a stormy history as an antidepressant. Introduced in 1987, it was proclaimed a ''wonder drug'' for treating depression by late 1989. Within a year, it was the subject of millions of dollars of lawsuits by persons claiming to have become suicidal while taking it. Given a suicide rate of 30,000 Americans a year before Prozac came along, it is not clear whether the drug encourages suicidal tendencies or any increase in rate. Strangely, much of its popularity is due not to its great effectiveness, but to the absence of side effects common to most antidepressants. The full truth about Prozac may never be known, but its future role likely will be decided by litigation and headlines.

$$F_3C \!-\!\!\!\left\langle \bigcirc \right\rangle\!\!\!- O - CHCH_2CH_2\overset{+}{N}H_2CH_3 \qquad \text{fluoxetine hydrochloride} \\ \qquad\qquad\qquad (\text{Prozac})$$

Cl⁻

The amphetamines (see following section) also stimulate the central nervous system and were once thought to be safe and nonhabit-forming. This has proved untrue.

25.6 Drug Abuse

GOALS: To understand drug addiction.
To appreciate the structure of neurons and the role of neurotransmitters.
To understand how some drugs may interfere with neuron activity.
To recognize the problems associated with the major abused drugs—alcohol, the opiates, cocaine, amphetamines, barbiturates, hallucinogens, marijuana, caffeine, and nicotine.

To understand drug abuse, we have to appreciate the meaning of drug addiction, which is normally described in terms of two phenomena: *dependence* and *tolerance. Dependence* signifies a condition in which a user experiences withdrawal symptoms if the use is discontinued. *Tolerance* not only describes the body's increased ability to handle a drug without ill effects, but it also signifies the condition in which the user requires increasing amounts of the drug to experience the same effects. Sometimes tolerance results from increased levels of liver enzymes that metabolize the drug, so a given dose is cleared out of the body more quickly. In other cases, the tolerance is attributed to a chemical alteration of the brain response to the drug.

Dependence may be physical or psychological or both. If a physical dependence develops, withdrawal may lead to severe symptoms and may even be fatal. Withdrawal should always be carried out with medical supervision. Psychological dependence, sometimes called *habituation,* is a more vague condition but a very real one. An individual who suffers from it simply finds life unbearable unless he or she is experiencing the effects of a drug. Many individuals who are critical of addicts seem unable to accept the realities of psychological dependence.

As knowledge of the effects of drugs has progressed, it has become increasingly clear that the distinction between psychological and physical dependence is not clear-cut. An addict may not exhibit dramatic withdrawal symptoms when stopping a drug. But the chemical effects that a drug has on the central nervous system (CNS) may lead to some profound and long-lasting consequences for the user's physical condition.

Drug Effects on Neurotransmitters

To appreciate the possible chemical effects of drugs on the CNS, it is necessary to know how nerve impulses are transmitted. The brain, with a mass of about 1.4 kg (3 lbs), contains some 100 billion nerve cells, or *neurons*. Drugs stimulate or depress neurons by altering processes by which these cells communicate with one another.

Within individual neurons, electrical impulses are transmitted through distances ranging from very short to as long as 3 feet. Nerve cells are not in physical contact with one another. The effects of the electrical activity in one neuron are felt by other nerve cells, not by transmission of an electrical impulse between cells, but by transmission of chemicals from one cell to as many as 10,000 other neurons. The chemicals involved in transmitting nerve impulses are **neurotransmitters.**

Two well-known neurotransmitters are *dopamine* and *norepinephrine* (sometimes called noradrenaline). Their origin and interrelationship are shown in Figure 25.2. Both compounds are derived from the amino acid *tyrosine,* which is a component of most proteins. Note that they are part of a scheme of reactions leading to dopamine and then to norepinephrine.

The transmission process is symbolized in Figure 25.3. A neuron consists of three parts:

1. A *cell body* that processes and receives electrical signals

2. An *axon* that carries electrical impulses for delivery to other neurons and releases neurotransmitters from its vesicles

3. *Dendrites*—branches that receive incoming signals from other neurons by interacting with neurotransmitters

The ends of the axons come within 200 nm (10^{-9} m) of the dendrites of other neurons. This region is called the *synapse.* The synapse consists of the terminus (end) of the axon, the neighboring dendrite, and the space between the axon and the dendrite, called the *synaptic gap* or *junction.* The neurotransmitter chemicals are released from the axon of the first neuron, move within a millisecond across the synaptic gap, and interact with the surface of the dendrite *(9).*

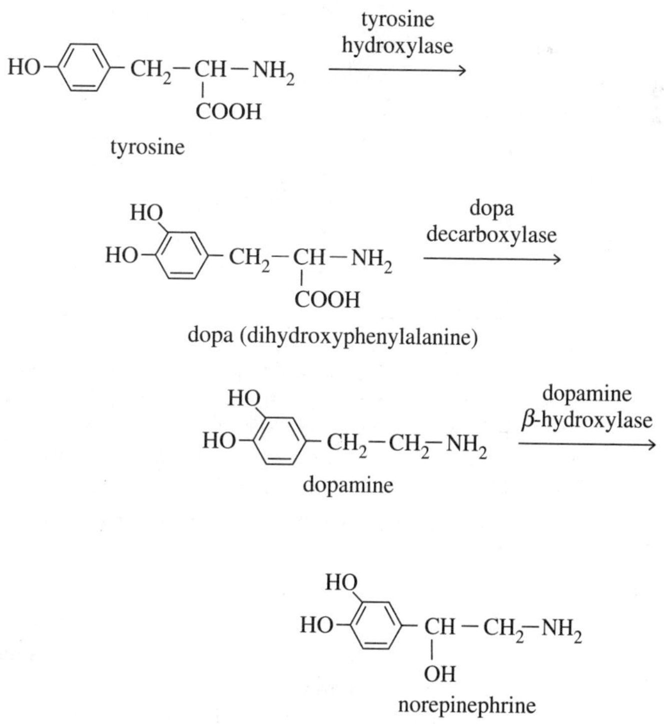

Figure 25.2 Synthesis of the neurotransmitter norepinephrine from the amino acid tyrosine. The intermediate dopamine also is an important neurotransmitter.

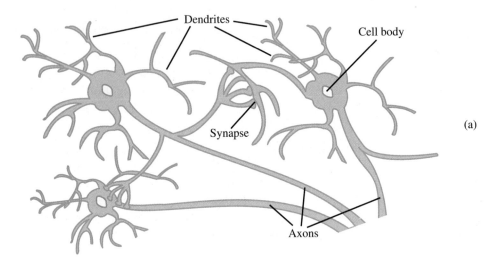

(a)

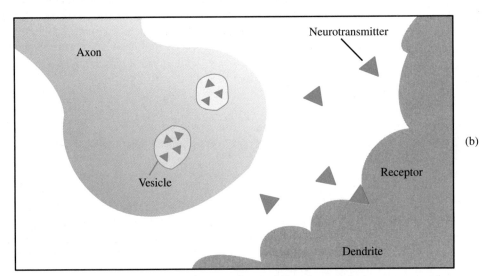

(b)

Figure 25.3

Neurons interacting at a synapse. When an electrical impulse travels down an axon from the nerve cell body, a neurotransmitter is released from vesicles in the axon and then from the axon into the synaptic cleft between the two neurons. The neurotransmitter molecules (pictured as triangles and squares) move across the synaptic gap to interact with receptor sites on the dendrites of the neighboring neuron. Thus the nerve impulse is transmitted between neurons. The neurotransmitter molecules are then released to move back to the axon for reuptake and incorporation into the protective vesicles. Some neurotransmitter molecules are degraded.

On the surface of the dendrites are *receptors,* regions of chemical structure that recognize a particular neurotransmitter. For example, receptors that recognize dopamine are *dopamine receptors.*

Prior to release from the axon, the neurotransmitter molecules are encapsulated in *synaptic vesicles.* Thus they are protected from destruction by enzymes and remain available for release in response to an electrical impulse from the cell body.

Once stimulation causes release of the neurotransmitter molecules, they move to receptors on the dendrites of neighboring neurons. Then the chemical is released back into the synaptic gap. What happens next depends on the particular neurotransmitter. Some neurotransmitter molecules are attacked by enzymes so that their effects dissipate. Dopamine and norepinephrine are mostly reabsorbed into the axon and reinserted into vesicles; the absorption is termed *reuptake*. Under normal circumstances, very little enzymatic destruction of the neurotransmitters occurs. But if reuptake is blocked, an abnormal loss of dopamine may result.

Drugs that affect the process of nerve transmission are classified as *psychoactive (10)*. The list of psychoactive drugs is enormous; it includes tranquilizers (for example, Valium and chlorpromazine), antidepressants, amphetamines, cocaine, opiates (morphine, codeine, and heroin), and many others. Several of these substances have already been mentioned. Others are discussed in the remainder of this chapter.

To illustrate how a drug may alter the normal transmission of impulses between neurons, consider cocaine, perhaps the most serious drug problem in the United States. Cocaine is classified as a central nervous system stimulant because it blocks the reuptake of dopamine into the axon of the neuron that is sending a signal. As a result of this blockage, dopamine molecules remain in the synaptic gap longer and in greater numbers than normal. This results in continuing stimulation of the neurons. Thus the neurons are overstimulated, a condition that the drug user describes as a ''high'' or a ''rush.'' If the dose of cocaine reaching the brain cells is very large, the stimulation also is great, giving the user great pleasure.

Since dopamine is blocked from reuptake, many of its molecules are destroyed and the supply is diminished. Thus, euphoria is replaced by severe depression as normal levels of dopamine become unavailable. The only recourse for the user who experiences the pain of these chemical changes is likely to be more cocaine *(11)*. Further details of cocaine use are presented in a later section.

Thus dependence on cocaine and some other drugs may be classified as psychological because of the absence of observable physical symptoms. But the physical consequences of the chemical alterations may be severe, and may cause the user to return to a drug for relief with far more urgency than the psychological need to return to a state of euphoria. It may be the only way the addict knows to relieve the physical pain that has resulted from chemical abuse of the nerve cells. That is why drug addiction is so often difficult to treat. When dealers find drugs, such as crack (cocaine) and ice (methamphetamine), that have such profound effects on the brain chemistry, they are assured of steady and profitable business.

Table 25.3 (pp. 546–549) is a detailed compilation of the properties of the commonly abused drugs.

Alcohol

Some of the most serious drug problems result from the abuse of substances that are depressants of the central nervous system. These include the analgesic narcotics (heroin, morphine, etc.), barbiturates, tranquilizers, and alcohol. Of these, alcohol is the most frequent cause for concern. It is the drug most often used legally by adults and illegally by young people. Alcoholism is the greatest medical problem resulting from the use of drugs, with estimates of the number of alcoholics ranging as high as 10 million for the United States alone. It has been estimated that there are about 40 times as many alcoholics in the United States as there are heroin addicts.

Alcohol addiction is not a condition that arises from short-term use of alcohol, whereas addiction to heroin and other drugs may develop quickly. For this reason, alcoholism is

Table 25.3   Reference guide to some drugs that are subject to abuse

	Examples	Slang Names	Pharmacological Classification	Medical Uses
Narcotics (analgesics)	Opium derivatives			None
	Heroin	H, horse, smack	Central nervous system depressants	To relieve pain
	Morphine	M, Mary, schoolboy		To relieve pain and coughing
	Codeine			To relieve pain
	Synthetic			To relieve pain
	Methadone			
	Demerol			
	Cocaine hydrochloride	Snow, coke	Central nervous system stimulant	Local anesthetic
	Cocaine	Crack, rock, freebase		
Hallucinogens (psychedelics)	Synthetic		Central nervous system stimulants and/or depressants	None[1]
	LSD	Acid, cubes, trips		
	Psilocybin			
	Mescaline			
	PCP	Angel dust		
	Derivatives of the hemp plant *Cannabis sativa*			None
	Marijuana	Pot, grass, Mary Jane		
	Hashish	Hash		
Stimulants (pep pills)	Amphetamines	Ups, co-pilots, pep pills	Central nervous system stimulants	To relieve mild depression and fatigue
	Benzedrine	Bennies		To reduce appetite
	Dexedrine	Dexies		To treat narcolepsy (a disease characterized by an overwhelming desire to sleep)
	Methedrine	Speed, ice, crank, crystal		
Depressants (sedatives and hypnotics)	Barbiturates	Downs, barbs, goofballs	Central nervous system depressants	To treat insomnia, anxiety, nervous tension, and epilepsy; also mental disorders
	Amytal	Blue heavens		
	Seconal	Red devils, red birds		
	Nembutal	Yellow jackets, nemmies		
	Tuinal	Rainbows		
	Methaqualone (Quaalude)	Quads, ludes sopor, Colt 45		

[1]PCP has been used as a veterinary anesthetic.

Table 25.3 Reference guide to some drugs that are subject to abuse (*Continued*)

How Taken When Abused	Physical Symptoms and Behavior Patterns	Chief Dangers
By injection or sniffed By injection or orally Orally (usually as a cough syrup) By injection or orally By injection or orally By injection or inhaled	"High" feeling (euphoria) followed by depression; impaired coordination; pinpoint pupils; watery eyes, running nose, chills, sweating, loss of appetite, weight; drowsiness, sleepiness, stupor. The need for narcotics has driven many users to crime. Excitation, dilated pupils, restlessness, tremors (especially of the hands), and hallucinations.	A very strong physical and psychological dependence may develop. General physical deterioration, possible social deterioration. Painful withdrawal symptoms. Infections, abscesses, tetanus, hepatitis from nonsterile injections. Death from overdose due to respiratory depression. The stimulation is more pronounced than with amphetamines (stimulants). Very strong psychological dependence may develop. Mental confusion and dizziness, depression, convulsions, death from overdose.
Orally or by injection Smoking or orally (hashish 5 times as potent as marijuana)	Effects vary greatly with dose and individual. May cause restlessness, exhilaration or depression, dilated pupils; illusions, delusions, hallucinations; distortion or intensification of sensory perceptions; nausea, vomiting, decreased ability to discriminate between fact and fantasy; unpredictable behavior, psychotic reactions, acute anxiety. Effects vary with the method of ingestion (whether smoked or eaten). Can produce reddening of eyes, dry mouth-throat, coughing spells, talkativeness, laughter. "High" feeling, possible hallucinations, altered perception of time and space; visual distortions. Exaggerated sensory perceptions can precipitate psychotic acts.	Permanent personality changes may occur. Bizarre mental effects. Unpredictable behavior. Exhibit dangerous acts of invulnerability. Suicidal or homicidal tendencies, flashbacks or same type of reactions may occur months later, after drug discontinued. Can lead to aggressive and antisocial behavior. Moderate psychological dependence liability. Distortion in sense perceptions may lead to accidents. Can lead to more serious drug abuse through contact with pushers of other drugs as well as contact with persons using more dangerous drugs.
Orally or by injection	Diluted pupils, loss of appetite, excitation, talkativeness, jumpiness, and irritability. Dry nose, lips, and mouth; bad breath; extreme fatigue, sleeplessness. With large doses intravenously, delusions, hostility, dangerously aggressive behavior, hallucinations, induced psychosis with panic.	Can develop high blood pressure or heart attacks. Potential for brain damage, malnutrition, exhaustion, pneumonia. Very strong psychological dependence develops quickly. Engenders reckless behavior. Can cause coma and death—"Speed Kills."
Orally	Constricted pupils, drunk appearance, slurred speech, incoherency, depression, drowsiness, dullness. Overdose produces unconsciousness, coma, pinpoint pupils, respiratory paralysis, and death.	High psychological dependence liability and physical dependence development with continued use. Hazards from faulty judgment and coordination. Painful withdrawal symptoms. Some indication of kidney damage, possible brain and liver damage. Death from overdose (alone or from combination with alcohol and barbiturates).

Table 25.3   Reference guide to some drugs that are subject to abuse (*Continued*)

	Examples	*Slang Names*	*Pharmacological Classification*	*Medical Uses*
Tranquilizers (minor)	Miltown Equanil Valium Librium	Downs	Central nervous system depressants	To counteract tension or anxiety and to overcome insomnia, depressing the central nervous system less than barbiturates do. Used also as muscle relaxants
Organic solvents (deliriants)	Toluene Airplane glue Plastic cement Acetone Nail polish remover Carbon tetrachloride Dry cleaner fluid Gasoline Paint thinners Lighter fuels	None	Central nervous system depressants	None (too toxic for use)
Alcohol (ethyl)	Wine Beer Whiskies	Booze, sauce	Central nervous system depressants	To sedate, promote sleep, dilate blood vessels, and be a food source for energy
Nicotine (tobacco)	Cigarettes Cigars Pipe tobacco Chewing tobacco Snuff	Cancer sticks, coffin nails, weeds	Central nervous system stimulants and/or depressants	None

Source: *Drug Reference Chart*, Council on Drug Abuse, Toronto, Canada. Reprinted with permission.

Table 25.3   Reference guide to some drugs that are subject to abuse (*Continued*)

How Taken When Abused	*Physical Symptoms and Behavior Patterns*	*Chief Dangers*
Orally	Similar to hypnotics in biological activity. In some people produce unusual feeling of cheerfulness and well-being. Sweating, skin rash, depression, mental sluggishness, urinary retention, anger, anxiety, tension, agitation, excitability, slurred speech.	As with hypnotics, but less so. Less likely to produce psychological and physical dependence. Have an augmented effect with alcohol, barbiturates and opiates (narcotics). Vision disturbances, dizziness, drowsiness. Withdrawal can produce agitation, nausea, depression, and sometimes convulsions.
Inhalation (sniffing)	Depending on exposure, the reactions may last from about 5 minutes to half an hour. Effects include enlarged pupils, confusion, slurred speech, dizziness, and a "high" feeling. Distortion of sights and sounds and hallucinations are also reported. Excessive secretions (running nose, watering eyes) and poor muscular control may be noted. Drunk appearance. Irritability, drowsiness, unconsciousness.	Moderate psychological dependence liability. Hazards from impaired judgment. Can induce aggressive behavior, antisocial acts. Possible permanent damage to the brain, liver, and kidneys. Accidental death from overdose (choking or suffocation).
Orally	The effects of a given amount of alcohol depend on the rate of drinking, the emotional state of the drinker, the size of the drinker, etc. Heartburn, gastritis, nausea, vomiting, increased urinary flow, malnutrition, various mood states, various diseases, anger, anxiety, tension, fear, belligerence.	Potential for physical and psychological dependence. Hazards from faulty judgment and coordination, emotional liability, and increased aggressiveness; accidental death from overdose alone or in combination with other depressants (e.g., barbiturates). Social and personal deterioration. Antisocial acts. Irreversible damage to brain, liver, and kidneys.
Chewed, snuffed, or smoked	Tars and smoke irritate the tissues, increasing saliva and bronchiolar secretions. Increased blood pressure and heart rate, and enlarged pupils. With increasing doses causes tremors, vomiting, accelerated respiration, slowing of water excretion by kidneys, paralysis of respiration, and convulsions.	High psychological dependence. With long-term use cancer of lungs, larynx, and mouth; irritative respiratory syndrome, chronic bronchitis, and pulmonary emphysema, damage to heart, blood vessels; impaired vision. An overdose of nicotine can cause convulsions, respiratory failure, and death.

most often regarded as a disease of middle age. Other forms of drug addiction are more prominent among young people, although teenage alcoholism has become a widespread problem in recent years. Withdrawal from alcohol can be quite severe for an alcoholic, more serious than withdrawal from heroin and several other drugs. The addiction persists, and abstinence is the only remedy, although it is difficult to maintain.

Crime is a frequent consequence of drug abuse. Addicts often turn to crime to support their habit. Alcoholics and those who abuse alcohol to a lesser degree may commit even more serious crimes, both intentional and accidental (for example, while driving), when they are under the influence.

Throughout the previous paragraphs, the word *alcohol* was used several times. To a chemist, the word is vague, since an alcohol can be any compound with the $-OH$ functional group. Organic acids also contain the $-OH$ group, but they are readily distinguished from alcohols due to the presence of the carbonyl functional group.

$$\underset{\underset{\displaystyle C}{\|}}{\overset{\displaystyle O}{}}$$

Consult Alcohols, Acids, and Esters in Section 6.3 for more information.

The specific compound referred to simply as alcohol is ethyl alcohol. It has a very different structure and properties from those of acetic acid, even though both substances have two carbon atoms and the $-OH$ functional group.

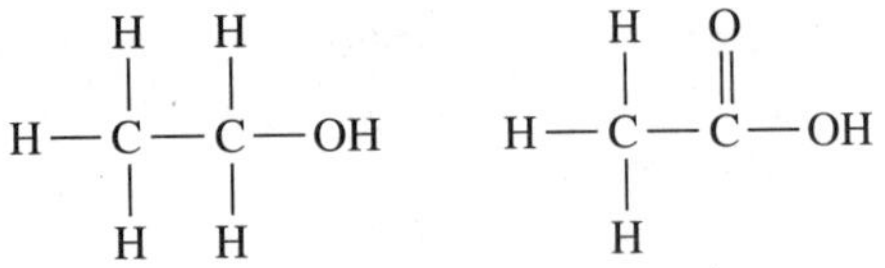

ethyl alcohol

acetic acid

The production of alcohol by fermentation was considered in our study of alcoholic beverages (Chapter 17). An industrial process accounts for more than 75% of all of the ethyl alcohol produced. It is outlined in the following reaction starting with ethylene.

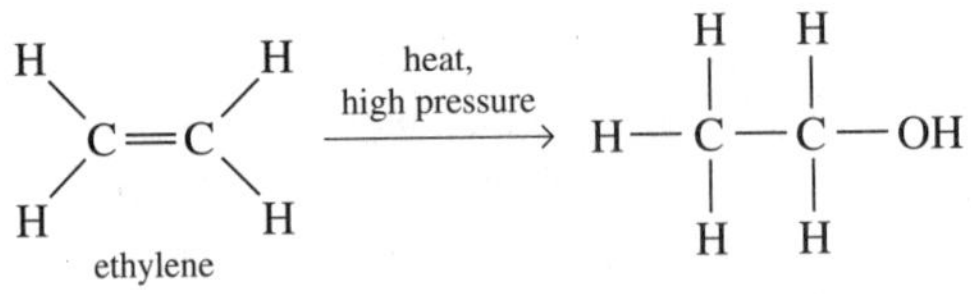

ethylene → heat, high pressure

In industry ethanol is widely used as a solvent for lacquers, varnishes, perfumes, and flavorings, and as a medium for chemical reactions.

Metabolism

Since alcohol does not have to be digested in the stomach or intestine, it can be absorbed into the bloodstream from either location. Champagne, sparkling wines, and whiskey mixed with carbonated beverages have a faster effect, since carbon dioxide relaxes the pyloric valve, which serves as a gate between the stomach and the small intestine. Once alcohol has been ingested, it is oxidized (mostly in the liver) according to the accompanying sequence.

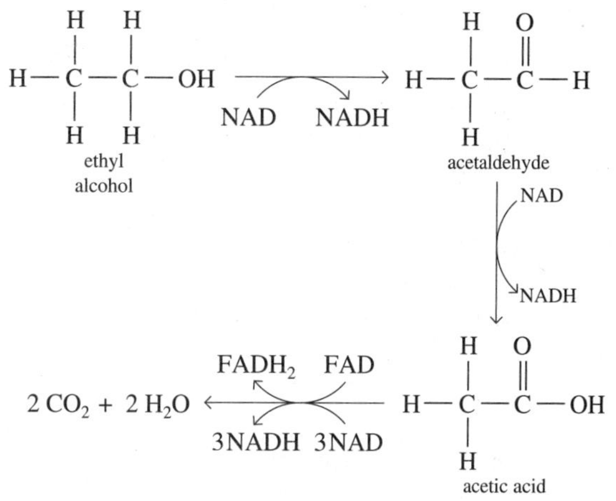

A heavy drinker builds up tolerance to increasing amounts of ethyl alcohol as increasing levels of liver enzymes are produced. Liver damage is a frequent result of long-term alcohol use.

Ethyl alcohol has a relatively high caloric value at 7.1 calories per gram. A large amount of useful energy (ATP) is derived from the metabolism of alcohol. The bulk of the ATP is produced from the regeneration of active niacin (NAD) and active riboflavin (FAD) as described in Chapter 17.

The steps for formation of ATP are **oxidative phosphorylation.** They begin with the NADH and $FADH_2$ produced in the previous sequence of reactions.

The final step in this scheme is the series of reactions known as the citric acid cycle (Krebs cycle) discussed in Section 17.1 and shown in detail in Figure 17.4.

$$NADH + O_2 \longrightarrow NAD + H_2O$$
$$3\,ADP \quad 3\,ATP$$

$$FADH_2 + O_2 \longrightarrow FAD + H_2O$$
$$2\,ADP \quad 2\,ATP$$

Before alcohol is metabolized, it exerts its effects on the body. Ethyl alcohol actually depresses the nervous system, although it often appears to act as a stimulant by depressing certain inhibitory effects of the brain. The full range of effects of alcohol are described in Table 25.4.

Treatment

Since both physical and psychological dependence on alcohol may develop, withdrawal should be carried out carefully and under medical supervision. Withdrawal symptoms from alcoholism are generally much more severe than those resulting from withdrawal from narcotics such as heroin, and may be fatal. Chlordiazepoxide (Librium) is sometimes used to treat the symptoms. Once an individual has been safely withdrawn, a drug with the

Table 25.4 Symptoms of alcohol consumption

Blood Alcohol (%)	Effects
Up to 0.03	Sedation and tranquility
0.03–0.05	Changes in reflex response, reaction time, and some skills (e.g., diminished automobile driving and athletic activity)
0.05–0.15	Loss of coordination
0.15–0.20	Obvious intoxication, release of inhibition leading to talkativeness and boisterous behavior
0.20–0.30	Blurred vision, dilation of pupils, slurred speech, staggering gait
0.30–0.40	Depression, stupor, unconsciousness
>0.50	Death

generic name disulfiram and the brand name Antabuse can be taken to help stay ''on the wagon.'' The drug prevents (inhibits) the step in the metabolism of alcohol in which acetaldehyde is converted to acetic acid. When this happens, acetaldehyde builds up in the bloodstream and causes extreme discomfort. Since it only builds up as a result of alcohol ingestion, a person taking disulfiram is discouraged from further intake of alcohol.

disulfiram
(Antabuse)

Other Alcohols

Some other common alcohols are listed in Table 25.5. Methyl alcohol (wood alcohol) can be obtained by destructive distillation of wood. It is quite toxic, and even in doses far below lethal amounts it causes blindness.

See Section 13.3 for more information on sources and uses of methanol.

Isopropyl alcohol has a high bactericidal activity and finds considerable use as rubbing alcohol. It also is toxic. Ethylene glycol, used as antifreeze, is about half as toxic as ethyl alcohol. Glycerin is less toxic than ethyl alcohol.

The Opiates

The drugs most often associated with abuse and addiction are the opiates—heroin, morphine, and codeine—so called, because morphine and, to a lesser extent, codeine are derived from opium. Opium is the residue resulting from evaporation of the juice of poppies grown in the Orient and the Middle East. It contains about 10% morphine and about 0.5% codeine.

The chemical structures of the opiates are similar. Morphine is the simplest; codeine has a methyl group on one of the alcohol groups; heroin is diacetylmorphine (Figure 25.4, p. 554).

The three substances are classified as **narcotics** to signify that they cause sedation (narcosis) and relieve pain (analgesia). The term narcotic may also be used to describe a drug that has the potential for causing dependence.

In addition to narcosis and analgesia, opiates cause an elevation of mood (euphoria), and a general feeling of peace and tranquility. Constipation is a common side effect, often

Table 25.5 Important alcohols

Formula	Common Name	IUPAC Name
H | H—C—OH | H	Methyl alcohol (wood alcohol)	Methanol
H H | | H—C—C—OH | | H H	Ethyl alcohol	Ethanol
H H H | | | H—C—C—C—OH | | | H H H	*n*-Propyl alcohol	1-Propanol
H H H | | | H—C—C—C—H | | | H O H | H	Isopropyl alcohol (rubbing alcohol)	2-Propanol
H | H—C—OH | H—C—OH | H	Ethylene glycol (antifreeze)	1,2-Ethanediol
H | H—C—OH | H—C—OH | H—C—OH | H	Glycerin (glycerol)	1,2,3-Propanetriol

used to advantage in cases of severe diarrhea. Then opium is administered as a camphorated solution containing alcohol, known as paregoric. Camphor gives the mixture a bitter taste to discourage abuse of the drug.

Opium has a long history of use as a painkiller. It was legally and conveniently available during most of the nineteenth century through physicians, drug stores, grocery stores, and general stores, and could be purchased through the mail in many patent medicines. Morphine (from Morpheus, the Greek god of dreams) was widely used as a painkiller during the Civil War and was produced in several of the Confederate states.

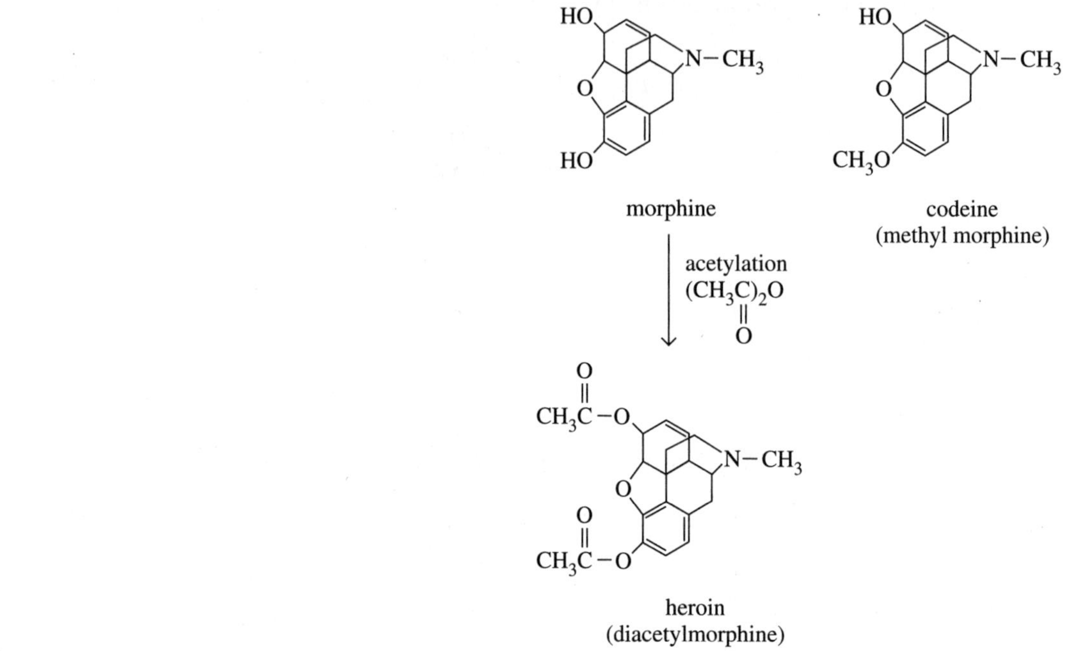

Figure 25.4   The opiates. Morphine and codeine are natural substances, whereas heroin is a synthetic opiate made from morphine.

Codeine is the least potent of the three opiates. It is frequently used as a cough suppressant. Although it is naturally present in opium, it is usually synthesized from the more abundant morphine.

Heroin has the same physical and psychological effects as morphine and codeine but is far more potent and much more addictive. The unusually high addiction potential explains why it has been rejected for use by the medical profession and at the same time finds such great popularity among drug pushers, who want to keep their customers eager to buy again and again.

Heroin is a synthetic opiate produced by acetylation of morphine. Addicts most often take heroin because that is what smugglers smuggle. Since it is so potent, it is far more economical to transport and deliver than the less potent morphine and other drugs. The potency of a sample varies since the narcotic is commonly "cut" (diluted) with lactose or some other substance several times before it reaches the user.

High prices force users to administer the drug in the manner that is most effective and consequently least costly. Usually, this means intravenous injection, a practice called main-lining. Infectious hepatitis, due to injections given under unsanitary conditions, is common among drug users and is a frequent cause of death. The AIDS epidemic has lessened somewhat the number of heroin users, as some individuals want to avoid injecting with unsanitary needles. Unfortunately, cocaine use has more than made up the difference.

Treatment

Treating opiate abuse can be said to have two separate areas of emphasis—treatment of overdose episodes and treatment of addiction. Both are significant. The first requires short-term emergency treatment. The second deals with drug addiction in general and heroin addiction in particular.

Death from a heroin overdose is rare. In most cases, several hours pass between administration of heroin and death, and during this time the effects of the drug can be countered by using a *narcotic antagonist.* The best-known narcotic antagonists are naltrexone and naloxone. Both are thought to work by attaching to sites on the neurons known as morphine receptors. Since morphine antagonists have a greater affinity for the receptors than narcotics do, the effects of heroin on the nervous system are blocked.

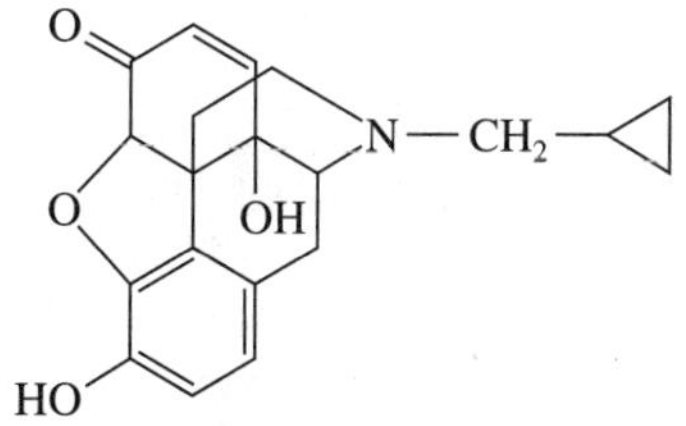

naltrexone 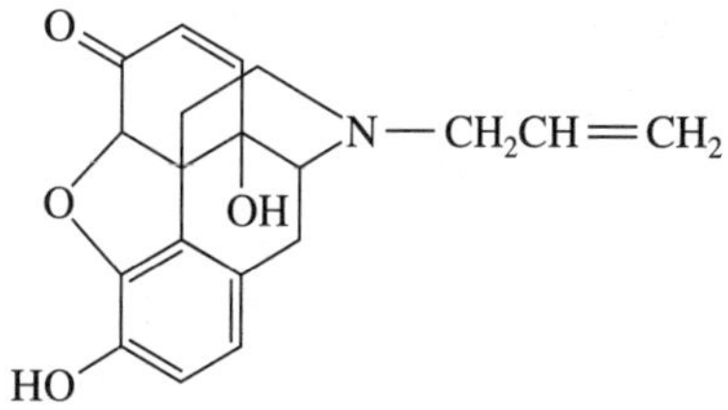   naloxone

Heroin overdose is a frequent cause of death in addicts, but good evidence suggests that it is not an overdose at all. This unusual type of death has been called *syndrome X.* Its cause is unknown, although speculation suggests that it may be caused by quinine, an adulterant often added to the drug by dealers. Deaths also may be the result of a combined action of alcohol and heroin *(12).*

Long-term therapy for heroin addicts focuses on the realities of addiction and its consequences. The reality of addiction is, ''once an addict, always an addict.'' It is because of this fact that *methadone maintenance* programs have come into prominence. Methadone is the generic name for a drug with the brand name Dolophine. It resembles the opiates in depressing the central nervous system, and it is a very potent analgesic. But, it does not cause drowsiness or euphoria (unless mainlined) as heroin and morphine do. Therefore, an addict who switches to methadone is able to hold a productive job and return to a more normal way of life. Another advantage of methadone is that it can be taken orally, so the dangers of infection due to injection are avoided.

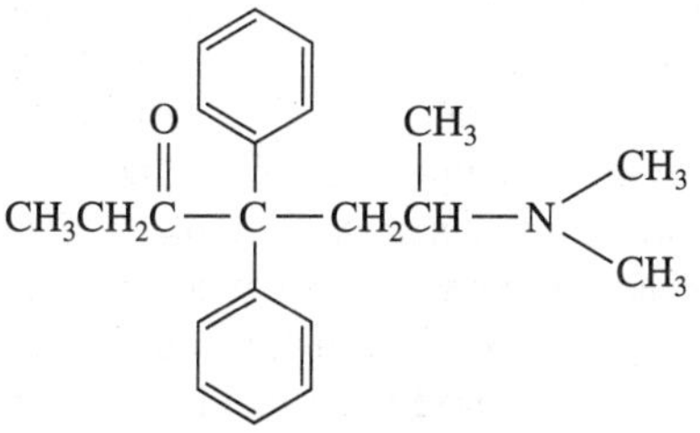

methadone
(Dolophine)

Methadone is an addictive drug. This may seem undesirable except that it increases the likelihood that the user will keep coming back for more, rather than reverting to heroin. That is, dependence on heroin is transferred to methadone. The addict does not experience heroin withdrawal symptoms as long as he or she receives a regular dose of methadone. At health clinics for addicts, methadone is normally provided in orange juice or other beverages to ensure that it will not be mainlined. Its half-life is about 15 hours in the body, which explains why a daily dose is sufficient *(9, 13).*

Methadone, like most drugs, is not without some unpleasant side effects. But withdrawal symptoms when it is discontinued are far less severe than those with heroin. Unfortunately, the success rate of ending the use of heroin is not very high.

Some risks are associated with methadone maintenance. Normal doses are often lethal for nonaddicts. Since much of the heroin sold on the streets is so weak that individuals who use it regularly may not actually be physically addicted, regardless of what they think, overdosage can be a real danger. In addition, there have been many reports of fatal poisonings of children who accidentally took a parent's methadone that had been premixed with fruit juice.

Cocaine

Coke and snow are two slang names for another frequently abused drug. Until the mid-1980s, cocaine was available exclusively as the fluffy white powder, cocaine hydrochloride. It is obtained from the leaves of the coca plant, *Erythroxylon coca,* which comes from Peru, Bolivia, and Colombia. Cocaine was a major drug of abuse in the late nineteenth century. Even Sigmund Freud was a staunch supporter of its use in treating morphine addicts, for which he was eventually severely criticized.

cocaine   + HCl →   cocaine hydrochloride

Coca-Cola was originally developed in 1886. At that time, it was made from a syrup consisting of coca and caffeine, but in 1903, the cocaine was removed. By the 1970s, cocaine hydrochloride had become a symbol of affluence due to its high cost. Sniffing the powder (known as snorting) using soda straws or even rolled-up dollar bills was regarded as a glamorous way to get high.

In late 1985 the nature of cocaine use in the United States took a dramatic turn with the introduction of freebase. Commonly known as crack or rock, this substance is just plain cocaine, rather than cocaine hydrochloride. Cocaine is a basic molecule because it can accept a hydrogen ion (H^+) onto the nitrogen atom. When HCl is the source of the hydrogen ions, the product is cocaine hydrochloride.

Before crack was readily available, it was common to treat cocaine hydrochloride with baking soda, $NaHCO_3$, to form cocaine. This was done by dissolving cocaine hydrochloride in water and adding the baking soda. The baking soda would fizz due to the formation and breakdown of carbonic acid H_2CO_3 (Section 25.3), leaving behind NaCl in solution and cocaine, which precipitates as a colorless, opaque, solid mass. The solid mass could then be broken into small pieces, called rocks, and smoked.

Although it is common to say that crack is smoked to distinguish it from being snorted or injected, it is not ignited. Instead, the rocks are placed in a glass pipe, called a camoke (Figure 25.5), and heated from the outside with a butane lighter or torch. This vaporizes the cocaine, which is then inhaled. In contrast, the hydrochloride salt form will not readily vaporize. Since cocaine that is inhaled is in its basic form, the common jargon for it is freebase.

Crack bought on the street comes in small pieces, usually in a small vial containing three to five rocks. Per gram of cocaine, it is quite expensive when bought this way, although each vial may sell for only a few dollars. This makes it accessible even to young

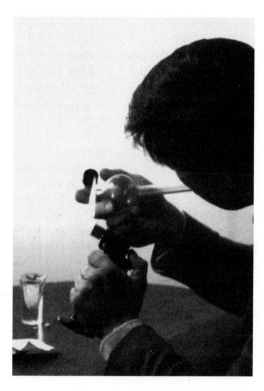

Figure 25.5
A camoke is commonly used for inhaling crack cocaine. The volatile freebase cocaine is vaporized by heating the glass bulb with a butane lighter.

children who could not afford to buy cocaine hydrochloride. In fact, a trafficker trying to build up business might even offer first-time customers free samples, hoping to get them addicted.

When cocaine hydrochloride is snorted, it passes into the bloodstream by absorption into blood vessels in the nasal passages. As noted earlier, it is a local anesthetic because it shrinks blood vessels. Therefore, it not only numbs the nasal passages, but it also restricts the blood flow through the vessels into which it will be absorbed. As a result, absorption is slow and inefficient, although the effects may be clearly felt by the user.

In contrast, when freebase cocaine is inhaled, it enters the lungs where the enormous surface area ensures very rapid and efficient absorption into the blood. It reaches the brain in less than 10 seconds and causes an intense state of euphoria, or a rush *(14)*.

Now that we know how cocaine is produced and consumed, let us consider the chemistry that takes place when it reaches the neurons in the brain cells. The intense rush felt by the user is attributed to the inhibition of the reuptake of dopamine from the synaptic gap (see earlier section ''Drug Effects on Neurotransmitters''). Under normal circumstances there are continual release and reuptake of dopamine by the axon of each neuron such that nerve impulses occur and then fade. When the neurotransmitter chemical is not removed, that is, when reuptake is blocked, the result is a powerful stimulation of the central nervous system.

As the minutes pass, two things happen, both of which set the user up for intense feelings of depression within about 20 minutes. Since the dopamine molecules are not reabsorbed in the normal way, many are lost due to attack by enzymes in the synapse. Also, the diester structure of the cocaine molecule causes it to be very susceptible to metabolism, with a blood plasma half-life of about 12 minutes. Therefore, within about 20 minutes the supplies of both dopamine and cocaine drop sharply. The euphoria is replaced by intense dysphoria (poor feeling) and a craving for more cocaine.

Consult Section 6.3 for background information about esters.

This is the first phase of withdrawal, which may be so unpleasant as to make many users almost instantly addicted to crack. Since the inhaled cocaine reaches the brain in such a large amount and in such a short time, both the euphoria and the dysphoria are magnified, together with the potential for addiction. This information is commonly regarded as a marketing strategy by drug traffickers. The huge doses of cocaine associated with the use of crack result in a more severe depletion of dopamine. The user finds it increasingly difficult to achieve the same high that was first experienced. This frustration leads to ever increasing use.

Even after an addict has stopped using cocaine, the condition of *anhedonia* occurs in which the person is unable to experience even normal pleasures in life. This form of depression leads many back to cocaine. Or they may seek medical help, which sometimes involves the use of drugs to help elevate the mood by altering the action of neurotransmitters. Neurons do not seem to return to the normal state that preceded cocaine use because of the severe effects of the drug. Although an addict may not show visible signs of withdrawal, the unseen physical effects can be devastating.

Amphetamines

Among the other drugs that have been subject to abuse are the amphetamines and the barbiturates, which are often called uppers and downers, respectively. Amphetamine users often have to resort to barbiturates to be able to sleep.

The chemical structure of amphetamine is shown. The two structures are **stereoisomers** of the same compound, where the prefix *stereo-* signifies that they differ in the three-dimensional arrangement of the groups.

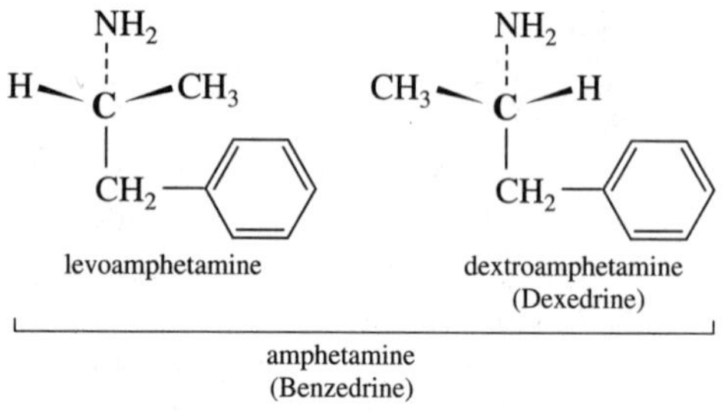

levoamphetamine

dextroamphetamine
(Dexedrine)

amphetamine
(Benzedrine)

The mixture of isomers has been marketed under the name Benzedrine. Dextroamphetamine (Dexedrine) is the name given to the stereoisomer of amphetamine shown on the right. Dexedrine alone is more potent than Benzedrine.

Amphetamines are stimulants of the central nervous system, a fact that is consistent with a chemical structure similar to epinephrine, better known as adrenaline. Amphetamines induce the same effect as adrenaline but have a more sustained action. They increase the heart rate and blood pressure, and increase wakefulness, postpone fatigue, and increase drive and energy. They produce a temporary elevation of mood, which is usually followed by increased fatigue, irritability, and depression.

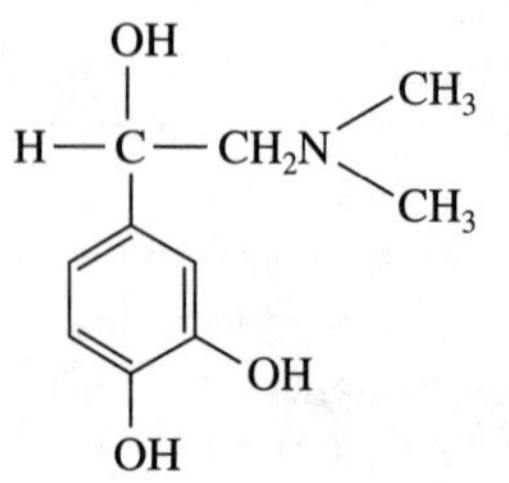

epinephrine
(adrenaline)

The amphetamines were once thought to be safe and nonhabit-forming. They are sometimes known as pep pills. Students who are studying for exams and truck drivers who have to stay awake are often cited as users. Truck drivers' slang names for the drugs include coast-to-coasts and copilots. Amphetamines have been used extensively as diet pills. They cause anorexia (loss of appetite) and at the same time elevate the mood; these tend to encourage physical effort, and consequently help in losing weight. Benzedrine was once used in inhalers for relief of nasal congestion. Now dangers of these compounds are clearer.

Although all of the amphetamines have similar modes of action in the body, the most potent one is methamphetamine (Methedrine), or speed. The ''speed freak'' is an individual who injects it intravenously for several days at a stretch (called a run) and then sleeps for an extended period, often with the help of barbiturates or heroin. Amphetamines can be taken orally with less dramatic effects.

Like simple amphetamine, methamphetamine exists in two isomeric forms that differ in the three-dimensional arrangement of the groups as shown. Of the two stereoisomers, the one on the right is several times more potent in its pharmacological action.

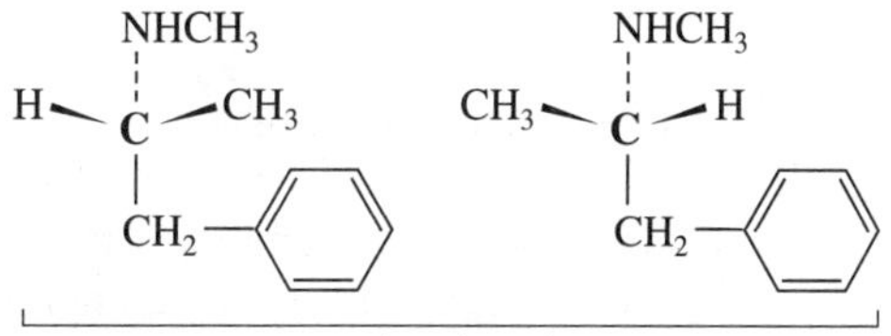

methamphetamine
(speed)

Like most amines, methamphetamine is susceptible to air oxidation. For this reason, it is usually handled in the form of the hydrochloride salt, shown below.

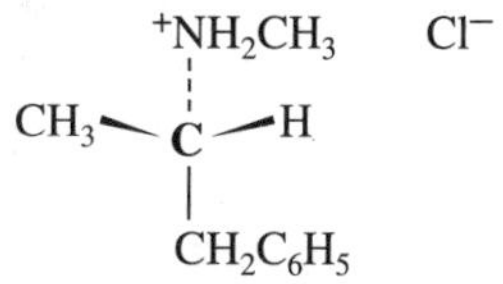

methamphetamine
hydrochloride

Methamphetamine hydrochloride is like cocaine hydrochloride, whereas plain methamphetamine is comparable to freebase cocaine.

In the late 1980s a form of methamphetamine known as ''ice'' appeared on the drug scene. It is methamphetamine hydrochloride, consisting of only the more potent stereoisomer. It has been available for many years as an equal mixture of the two stereoisomers called a *racemate*. Ice appears as transparent, sheetlike crystals. Its common street names are ''speed,'' ''crank,'' and ''crystal.'' Unlike cocaine hydrochloride, which must be converted to freebase to be vaporized and inhaled, methamphetamine hydrochloride is somewhat volatile. Therefore, it can be vaporized in a glass pipe by heating it with a butane lighter, even without converting to freebase *(14–16)*.

Unlike cocaine, which is a natural product, amphetamine in all its forms is a synthetic produced by a series of chemical reactions. Therefore, changes in synthetic methods or starting materials can lead to a variable product. In the past, only the mixture of two isomers of methamphetamine was commonly available because of the limitations of starting materials required to synthesize the molecule. But South Korea and Taiwan have abundant supplies of the plant chemicals ephedrine and norpseudoephedrine, both of which can be used to synthesize methamphetamine and yield only the more potent isomer. The U.S.

Drug Enforcement Administration has issued rules aimed at regulating the sale and import of these two chemicals, among others *(17)*.

$$NHCH_3$$
$$CH_3 \diagdown C \diagup H$$
$$CHC_6H_5$$
$$OH$$
ephidrine

$$NHCH_3$$
$$CH_3 \diagdown C \diagup H$$
$$CH_2C_6H_5$$
methamphetamine

$$NH_2$$
$$CH_3 \diagdown C \diagup H$$
$$CHC_6H_5$$
$$OH$$
norpseudoephidrine

Like cocaine, methamphetamine has profound chemical effects on the neurons of brain cells. Its three specific effects are that it stimulates the release of the neurotransmitters norepinephrine and dopamine, it inhibits storage of neurotransmitters in the protective storage vesicles, and it directly stimulates the norepinephrine receptors, which are meant to detect impulses from neighboring neurons. Thus the receptor neurons remain in an excited state. Now we can understand, on a molecular level, the increased alertness, elevation of mood, mild euphoria, increased physical performance, decreased fatigue, and improved concentration associated with these drugs.

Cocaine is rapidly metabolized in the body, but most of a methamphetamine dose is eliminated from the body unchanged. Since only a portion is metabolized, the half-life is about 12 hours *(15)*. Therefore, its effects are felt for much longer than with crack.

An acute tolerance develops, probably due to the effect that methamphetamine has on the release of neurotransmitters. Doses that would be lethal to an occasional user may have no effect once tolerance develops. The rate of supply of neurotransmitter molecules is limited and gradually decreases. Subsequent doses release smaller amounts of the neurotransmitters. In addition, when use is discontinued, a lengthy period of severe depression is likely to result from the depletion of neurotransmitters.

Barbiturates

Alcohol and barbiturates have similar effects on the body. One can get drunk on barbiturates and can become addicted to them, both physically and psychologically. Withdrawal symptoms are among the most severe experienced by addicts, usually more severe than those with heroin. Convulsions, delirium tremens, and even death typically result from abrupt withdrawal. Only decreasing doses of barbiturates themselves are considered suitable for use in aiding withdrawal, since many anticonvulsant drugs are ineffective in treating these symptoms.

The barbiturates are a family of chemically related compounds. Each has a structure that is a slight alteration, or *derivative,* of barbituric acid. Barbituric acid was first synthesized in 1864 from urea and malonic acid.

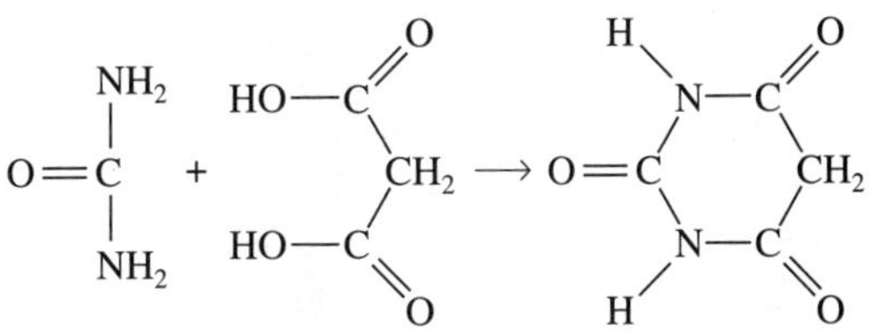

Barbital was the first barbiturate used in medicine. It was introduced under the name Veronal in 1903. Phenobarbital, sold also as Luminal, was the second. The structures of these and several of the other common barbiturates, including the popular anesthetic Pentothal and its sodium salt, are shown in Figure 25.6.

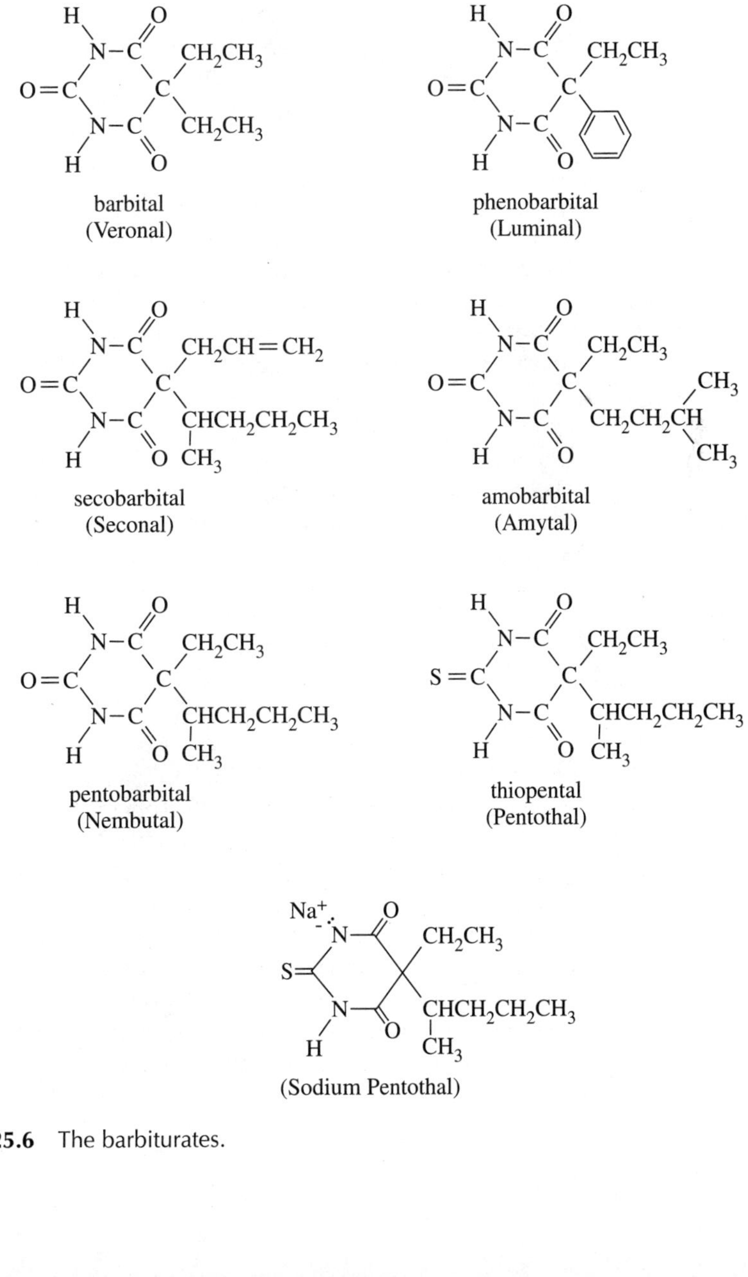

barbital
(Veronal)

phenobarbital
(Luminal)

secobarbital
(Seconal)

amobarbital
(Amytal)

pentobarbital
(Nembutal)

thiopental
(Pentothal)

(Sodium Pentothal)

Figure 25.6 The barbiturates.

Like the opiates, barbiturates are CNS depressants (downers). The symptoms of barbiturate intoxication are slowed reflexes, impaired coordination, slurred speech, and drowsiness. Medically, the drugs are used in several ways: as sleeping pills (sedatives), anticonvulsants, anesthetics (Sodium Pentothal) and preanesthetics, and in psychiatry.

Besides differences in chemical structure, barbiturates are identified by their speed and duration of action. They are classified as long-acting (Veronal and phenobarbital), intermediate-acting (Amytal), short-acting (Nembutal and Seconal), or ultra-short-acting (Sodium Pentothal).

On the drug scene, nemmies (Nembutal), seccies (Seconal), and tuies (Amytal plus Seconal, called Tuinal) are some slang names. So are yellow jackets (Nembutal), red devils (Seconal), and ''blue heavens'' (Amytal), describing by the colors of the capsules.

The effects of relatively short-acting barbiturates begin within a few minutes and last for only a few hours. For medical use, the ultra-short action of Sodium Pentothal makes it a frequent choice for inducing anesthesia in surgery. Seconal and Nembutal are prescribed as sleeping pills since they are relatively short-acting. They also see frequent use as preanesthetic sedatives, given to patients 30–60 minutes before surgery. Long-acting barbiturates typically take up to an hour to exhibit any effects, but the effects may last for 6 hours or more. Phenobarbital (Luminal) is often used in treating epilepsy and other brain disorders.

Hallucinogens

Hallucinogens' primary effects are changes in perception, especially visual, with vivid hallucinations that are often brightly colored. They may cause stimulation, depression, or neither. No physical dependence develops, but bizarre mental effects and permanent personality changes sometimes occur, often long after use of the drug.

Lysergic Acid Diethylamide

The best known and most popular of the hallucinogens is LSD. It affects those neurons that use the neurotransmitter serotonin. Convulsions, psychosis, uncontrollable rage, coma, and death can result.

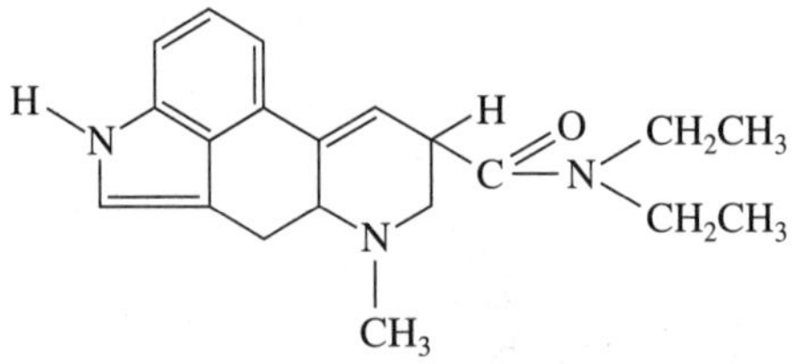

lysergic acid
diethylamide
LSD

Mescaline and Psilocybin

Mescaline is a component of the peyote cactus that grows in Mexico and southwestern United States. It has been used in religious ceremonies by Native Americans for many centuries. It is classified as a norepinephrine psychedelic.

Psilocybin is another hallucinogenic drug, derived from the Mexican mushroom *Psilocybe mexicana.* It affects the neurotransmitter serotonin.

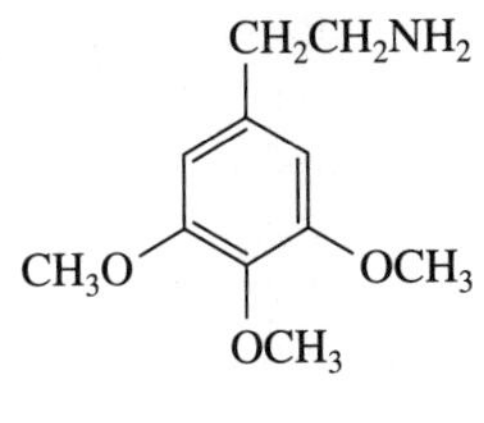

mescaline

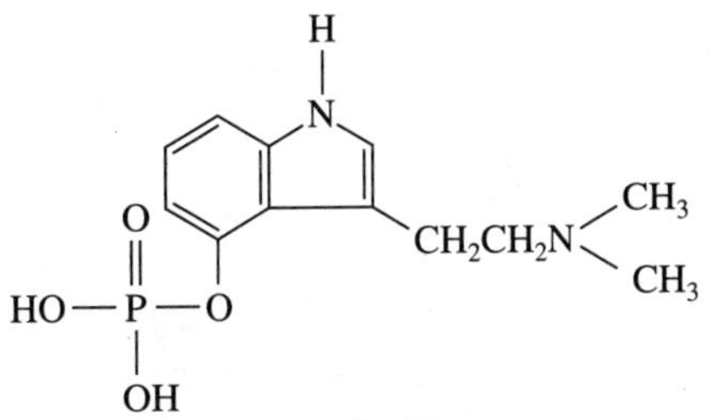

psilocybin

Marijuana

One of the most controversial of the abused drugs is marijuana, also spelled marihuana. Questions of safety, potential for addiction, and possible legalization are the key issues. Marijuana is composed of the dried leaves, flowering tops, stems, and seeds of the Indian hemp plant *Cannabis sativa.* The stems yield tough fibers that are used for making rope. Marijuana is often identified by the name *cannabis,* although grass, pot, and weed are common slang. The active component of cannabis is tetrahydrocannibinol, or THC. The structure of THC does not resemble any known neurotransmitter, and its action remains unknown. The potency of marijuana is related to its THC content.

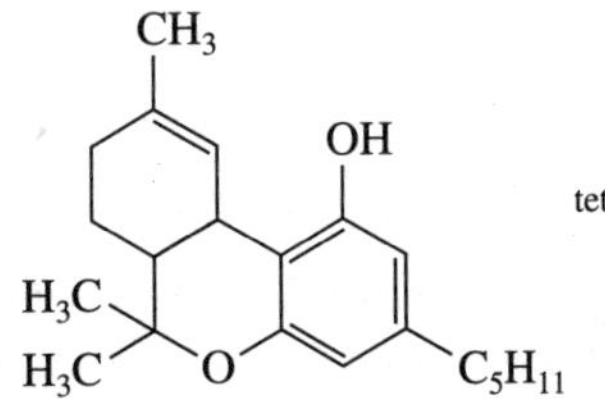

tetrahydrocannabinol
(THC)

Hashish is the dried resin obtained from the flowering parts of the plant. Its THC content was once much higher than that of marijuana, ranging to 20%, compared to only about 1% in marijuana. More recently, marijuana tends to contain 2–3% THC, whereas most hashish has about 4–5%. Thus, the distinction between the two is not very great.

Powdered marijuana or hashish can be consumed orally, but the side effects are felt more rapidly when either is smoked in pipes or cigarettes. The effects typically last for 2–3 hours. Cannabis may act as either a stimulant or a depressant and is sometimes classified as a hallucinogenic. The effects of cannabis intoxication are feelings of well-being, excitement, disturbance of associations, alterations in appreciation of time and space, emotional upheaval, illusions, and hallucinations. Well-being sometimes alternates with depression.

Some experts place marijuana in a class by itself because of the absence of cross-tolerance with all other categories of drugs. In other words, normally a person who develops tolerance to any one stimulant or depressant acquires tolerance to other drugs of the same type. Marijuana does not lead to any cross-tolerance, even to drugs normally classified as hallucinogens *(9).*

Users of marijuana may show negative tolerance. Breakdown products remain in the body for many days, and THC persists in the bloodstream for more than 3 days after a dose. A regular user may get high on a smaller dose than the occasional or novice user.

Marijuana is not regarded as addictive; that is, tolerance and withdrawal symptoms do not appear, but it can produce psychological dependence that results in restlessness, anxiety, irritability, insomnia, or other symptoms. The more direct problems are often overlooked, such as driving performance, work performance, and relations with other people, which often are adversely affected. Marijuana cigarettes are known to contain about 50% more tar than commercial cigarettes, and tar causes skin tumors in rats. Heavy use seems to have adverse effects on the bronchial tract and lungs.

One well-known positive application of THC is in the treatment of nausea caused by cancer chemotherapy.

Phencyclidine

Another chemical on the drug scene is phencyclidine (generic name), usually identified as PCP or angel dust. It is a veterinary anesthetic that has been used illicitly as a hallucinogen. It is a white, water-soluble powder that can be smoked, injected, or taken orally. Sometimes it is dusted on parsley or marijuana and rolled into cigarettes.

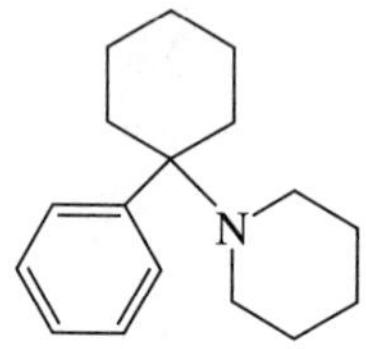

Because PCP is inexpensive, it is often deceptively sold as THC, the active ingredient of marijuana. It has also been used by dealers to cut more expensive drugs such as mescaline, cocaine, and THC.

It is among the most dangerous drugs to hit the streets since LSD became available in the late 1960s, being linked to hundreds of murders, suicides, and accidental deaths. The effects of the drug are very unpredictable, but irrational or violent action is commonplace, particularly for frequent users. A high resembles drunkenness and can lead to euphoria, depression, or hallucinations. Large doses can cause convulsions, psychosis, uncontrollable rage, coma, and death, resembling the effects associated with LSD.

Caffeine and Nicotine

Surely, you might say, the section on drug abuse is hardly the place to discuss caffeine and nicotine. The next thing you know, we will be criticizing Mom's apple pie. Yet, caffeine and nicotine are two of the most abused drugs, and each is an interesting case.

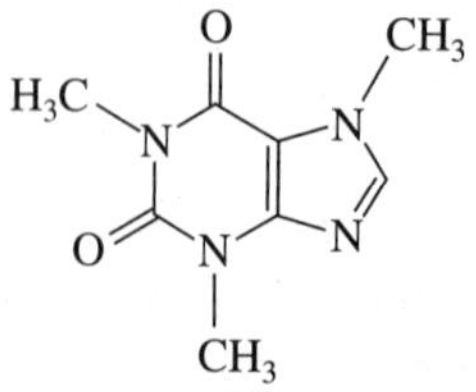

caffeine 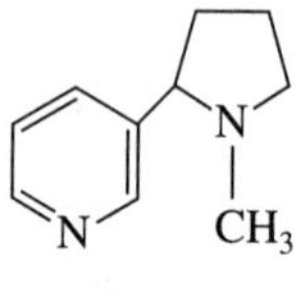   nicotine

Caffeine causes stimulation of the central nervous system (that should sound familiar). It is commonly consumed in coffee, tea, cocoa, and cola drinks. It produces a more rapid and clear flow of thought and decreases drowsiness and fatigue. After taking caffeine, one

Figure 25.7

In the interest of cancer research, the rabbit at the Tbilisi Institute of Oncology in Russia learned to smoke. All the rabbits in the experiment, consuming up to nine cigarettes a day, had chronic pneumonia and emphysema by the end of the fourth year. This did not affect their enthusiasm for smoking, however.

The presence of caffeine in combination remedies and alone in tablet form is discussed in Section 25.1.

can carry out a greater sustained intellectual effort, although physical coordination and timing may be adversely affected. The effective dose of caffeine is about 200 mg. There is no evidence of physical dependence, although some psychological dependence is indicated, for example, in individuals who "cannot face the day without their morning cup of coffee."

The problem is much less amusing with nicotine, since the consequence of smoking may be much more severe. Some people smoke for enjoyment or to alleviate stress, and others are addicted to nicotine. This latter circumstance cannot be underestimated, since some experts believe that cigarette dependence develops more easily than it does with alcohol or barbiturates. The dependence is psychological only. Many individuals smoke strictly for the dose of nicotine. Studies have shown that an addict will take fewer puffs on a high-nicotine than on a low-nicotine cigarette. Withdrawal reportedly is characterized by craving, anxiety, irritability, hunger, restlessness, decreased concentration, drowsiness, and sleep disturbance.

Worse, cigarette smoking is the greatest cause of fatal cancer among all environmental cancer-causing factors. It causes 15–20% of all cancer deaths—almost 70,000 a year from lung cancer alone. Per capita consumption of cigarettes varies but now shows a sharp increase, especially among young people. Fifty million Americans are smokers, of which 17 million try to stop each year. Only 1.3 million are successful. (See Figure 25.7.)

The products resulting from the incomplete combustion of tobacco and tobacco paper are the potential carcinogens, not the nicotine. An example is 3,4-benzpyrene, which forms from cellulose by burning cigarette paper. In other words, nicotine attracts the smoker and may lead to addiction, but other components of the smoke are harmful to tissues. It has also been reported that although nicotine itself is not a carcinogen, it can act synergistically with other carcinogens present in cigarette smoke and increase the frequency of tumors in mice.

3,4-benzpyrene

Overall, nicotine is generally rated second only to alcohol as a serious drug problem. As a way of dealing with it, some experts have drawn a parallel to methadone, suggesting that high-nicotine cigarettes would cut down consumption and minimize risks. In other words, the strategy would be to counteract addiction by supplying a drug, in this case the drug that causes the addiction, or a substitute such as methadone.

Nicotine gum, with 2 mg per piece (recently approved by the FDA at 4 mg per piece), has been available as replacement therapy in the United States since the mid-1980s. Many gum users did not like the flavor, and many failed to use it properly, resulting in a poor success rate. The success of the gum depends on absorption of nicotine into blood vessels in the mouth. This can occur only under alkaline conditions, and requires slow chewing to prevent swallowing much of the nicotine. If swallowed, nicotine is metabolized due to absorption from the gastrointestinal tract and inactivation in the liver. If the gum is chewed while eating or drinking coffee, fruit juices, or other acidic substances, nicotine cannot be absorbed into the bloodstream from the mouth.

In 1992 a new product, the nicotine transdermal delivery system (nicotine patch), was introduced to help treat tobacco dependence *(18)*. The patch reduces nicotine withdrawal symptoms, such as negative moods and craving for cigarettes, but increased hunger and weight gain are frequent consequences when smoking is stopped. The patch seems to have a much higher success rate than nicotine gum in curbing smoking. Since it provides a continuous supply of nicotine without any conscious effort by the user, it is superior in satisfying the needs of the user.

SUMMARY

The number one over-the-counter drug is aspirin. It is available alone or in combination with other analgesics or other ingredients such as antacids, decongestants, antihistamines, and caffeine. Acetaminophen is the generic name for an aspirin substitute and is available under a variety of brand names (Datril, Tylenol, etc.). Like aspirin, it is an analgesic and antipyretic, but does not have the antiinflammatory action of aspirin.

Aspirin and acetaminophen are classified as nonsteroidal antiinflammatory drugs (NSAIDs). Ibuprofen is another NSAID that is available both over the counter and in various prescription drugs.

Propoxyphene (Darvon) is an important prescription painkiller. The opiates (usually codeine and morphine) and meperidine (Demerol) are more potent painkillers, but all have addiction potential. A sequence of drugs is used in surgery to induce anesthesia (unconsciousness), analgesia, and muscle relaxation. Enflurane and isoflurane are etherlike substances that are administered in a stream of oxygen and nitrous oxide (N_2O) after anesthesia is induced by Sodium Pentothal. Novocaine and Xylocaine are often used as local anesthetics for minor surgery. Ethyl chloride can be used locally to freeze small areas of surface tissue and make them more insensitive to pain.

Neurotransmitters such as dopamine and norepinephrine carry nerve impulses among neurons. Many drugs affect nerve transmission, including all steps, beginning with the electrical impulse from the cell body of the neuron to the axon, and ending with the reuptake of neurotransmitter molecules into protective vesicles. In between, neurotransmitters are released into the synaptic gap, where they travel to neighboring neurons and interact with receptor sites. The neurotransmitter molecules then may return to the original neuron and are reabsorbed or may be enzymatically destroyed. Drugs may stimulate or inhibit the release of neurotransmitters, block reuptake or insertion into vesicles, or directly interact with receptors. Both physical and psychological effects may result and may be indistinguishable.

The opiates—heroin, morphine, and codeine—are classified as narcotics since they cause narcosis and analgesia. Morphine and codeine have some legitimate medical uses. Heroin is a major abused drug. Methadone maintenance programs are an important part of heroin addiction treatment.

Although alcohol remains the most abused drug, dangerous agents such as crack (cocaine) and ice (methamphetamine) have profound effects on neurons. They are usually inhaled, thus allowing rapid absorption of large amounts. Cocaine has a short half-life and must be consumed over and over to avoid the dysphoria that always follows the euphoria. Methamphetamine has a much longer half-life. The drug called "ice" is a single stereoisomer of methamphetamine with much greater potency than other amphetamines or even the isomer mixture.

Many different barbiturates are available for both legitimate use and abuse. Barbiturate withdrawal symptoms are among the most severe.

The hallucinogens, including LSD, mescaline, psilocybin, and phencyclidine, cause various forms of stimulation and depression. Each acts on a particular neurotransmitter. Marijuana is sometimes classified as a hallucinogen, although its mode of action is not understood.

PROBLEMS

1. Give a definition or an example of each of the following.
   a. Chemotherapy
   b. Analgesic
   c. Antipyretic
   d. NSAID
   e. Vasoconstrictor
   f. Dependence
   g. Tolerance
   h. Neuron
   i. Neurotransmitter
   j. Axon
   k. Dendrites
   l. Synapse
   m. Dopamine receptors

2. What are the three common over-the-counter analgesics? How do their actions differ?

3. What is the side effect of antihistamines that makes them dangerous?

4. What class of drugs is usually used to treat severe pain?

5. How does morphine compare with aspirin as a painkiller?

6. What are the drugs used in surgery, beginning the night before and through the postoperative period?

7. Why is chloroform a poor anesthetic?

8. How are isoflurane and enflurane administered?

9. How does ethyl chloride function as a local anesthetic?

10. What is the cause of the effervescent action of some popular antacids?

11. Why are hydrochloride salts often the preferred form of some drugs?

12. Describe the action of Antabuse on an alcoholic.

13. Why do sparkling wines or whiskey mixed with carbonated beverages have a faster effect than the same amount of alcohol alone?

14. What is an addictive narcotic?

15. What are the advantages and disadvantages of methadone maintenance programs?

16. How do cocaine and methamphetamine compare in their effects on neurotransmitters?

17. Describe the action of baking soda on cocaine hydrochloride.

18. Why is inhaling cocaine more dangerous than snorting?

19. Describe the differences in the duration of effects of crack and ice.

20. What are some side effects of marijuana?

21. Why is marijuana regarded as a drug in a class by itself?

22. What is the greatest cause of fatal cancer?

23. It has been advertised that "Anacin contains the pain reliever most prescribed by doctors." Explain.

24. It has been advertised that Bufferin (1) contains an added protection ingredient not present in Bayer aspirin or Anacin, and (2) provides an extra relief action not available with Tylenol. Are these claims justified?

25. The nicotine in nicotine gum can only be absorbed from within the mouth if alkaline conditions are present. Write a reaction to show how nicotine changes in the presence of acetic acid.

References

1. Shlafer, M.; and Marieb, E. N. *The Nurse, Pharmacology, and Drug Therapy;* Addison-Wesley: Redwood City, CA, 1989, pp. 333, 1058.

2. Lehne, R. A.; Crosby, L. J.; Hamilton, D. B.; and Moore, L. A. *Pharmacology for Nursing Care;* W. B. Saunders: Philadelphia, 1990; pp. 704, 746.

3. Kolata, G. "Study of Reye's-Aspirin Link Causes Concern." *Science* **1985,** *227,* 31.

4. Silverman, H. M. and Simon, G. I. *The Pill Book;* Bantam Books: New York, 1990, p. 946.

5. Burack, F.; and Fox, F. J. *The New Handbook of Prescription Drugs;* Ballantine Books: New York, 1975; p. 178.

6. Smith, R. J. "Federal Government Faces Painful Decision on Darvon." *Science* **1979,** *203,* 857.

7. Melzack, R. "The Tragedy of Needless Pain." *Scientific American* **1990,** *262,* 27.

8. Hem, S. L. "Physicochemical Properties of the Antacids." *Journal of Chemical Education* **1975,** *52* (6), 383.

9. Julien, R. M. *A Primer of Drug Action, Fifth Edition;* W. H. Freeman: New York, 1988, pp. 141–144, Chapter 5, Chapter 10, Appendix I.

10. Stinson, S. S. "Psychoactive Drugs." *Chemical and Engineering News,* October 15, 1990; p. 33.

11. Zurer, P. S. "Scientists Struggling to Understand and Treat Cocaine Dependency." *Chemical and Engineering News,* November 21, 1988, p. 7.

12. Brecher, E. M.; and the Editors of Consumer Reports. *Licit and Illicit Drugs;* Little, Brown: Boston, 1972.

13. Inturrisi, C. E.; and Verebely, K. "Disposition of Methadone in Man After Single Oral Dose." *Clinical Pharmacology and Therapeutics* **1972,** *13,* 923.

14. Allen, D. *The Cocaine Crisis;* Plenum Press: New York, 1987; pp. 25–36.

15. Cho, A. K. "Ice: A Dosage Form of an Old Drug." *Science* **1990,** *249,* 631.

16. Cotton, P. "Medium Isn't Accurate 'Ice Age' Message." *Journal of the American Medical Association* **1990,** *263,* 2717.

17. Hanson, D. "Rules Aim at Halting Diversion of Chemicals to Make Illicit Drugs." *Chemical and Engineering News,* February 27, 1989, p. 17.

18. Fiore, M. C.; Jorenby, D. E.; Baker, T. B.; and Kenford, S. L. "Tobacco Dependence and the Nicotine Patch: Clinical Guidelines for Effective Use." *Journal of the American Medical Association* **1992,** *268,* 2687.

Steroids and Contraception

26

The steroid family of compounds includes cholesterol, sex hormones, bile acids, and antiinflammatory agents. They are amazingly diverse in chemical structure and biological function. In this chapter, we will consider some of the important natural and synthetic steroids.

Understanding the chemistry of steroids has been one of the challenges of modern chemistry. Since the discovery of cholesterol, research and development in this field have led to significant changes in medical technology. It was the success in synthesizing cortisone that brought the steroids to public awareness; later, everyone became aware of the link between cholesterol and heart disease.

Great public attention was focused in the early 1960s when steroid hormones were developed for use as oral contraceptives, commonly known as the Pill. During the 1970s and 1980s these same chemicals received much adverse publicity when controversial reports linked them to certain cancers and heart disease. The use and abuse of anabolic steroids by athletes in the late 1970s and 1980s brought further negative public awareness. Most notably, the 1988 Olympic 100-meter champion, Ben Johnson of Canada, was disqualified for taking a muscle-building steroid. Then, in 1991, Norplant, the first major contraceptive to be approved by the Food and Drug Administration (FDA) in 25 years, brought renewed hope for a safe, reliable, effortless method of contraception that would be effective for 5 years.

26.1 Steroids

GOAL: To recognize the steroid nucleus and the way that groups may be attached.

To understand the mode of action of steroid hormones and oral contraceptives, we must first look at the chemistry of steroids. The steroids, or sterols as they are sometimes called, are part of the class of compounds called *lipids*. Lipids also include fatty acids, fats and

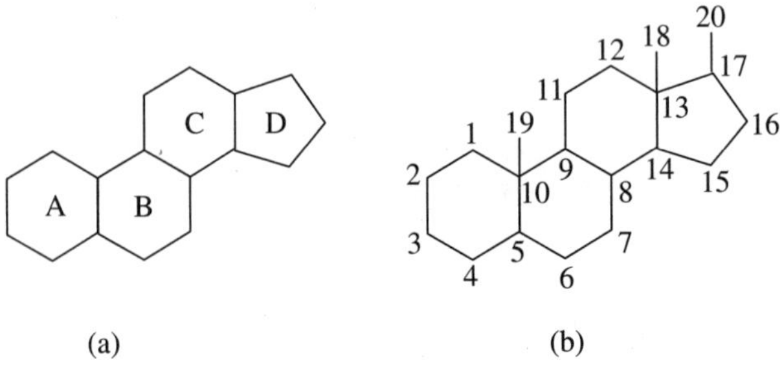

Figure 26.1
The steroid nucleus.   (a)   (b)

oils, fat-soluble vitamins, and waxes. A lipid is predominantly a hydrocarbon. It has physical properties typical of fats and oils (for instance, high solubility in hydrocarbon solvents and low solubility in water).

Food fats and oils are discussed in Chapter 21.

Lipids are abundant in nature, since they are present in all plant and animal tissues, and in the cells of microorganisms. Steroids exist in the free state and in combination with other chemicals, such as fatty acids or carbohydrates. In the human body, they are principally present in nerve tissue (including the brain), cell membranes, reproductive organs and glands, bile, and blood.

Steroidal compounds exist in plants from which they can be extracted for human use. For example, digitoxin and digoxin are two important drugs. Collectively called digitalis, they consist of two components, a sugar and a steroid, or a cardenolide. Digitalis has been used with striking success since the late 1700s in treating heart disorders.

Digitoxin comes from the dried leaves of the plant *Digitalis purpurea,* purple foxglove. The name is derived from the flowers that have the shape of a finger. Digoxin is extracted from the leaves of *Digitalis lanata,* white foxglove. Lanoxin, the brand name of digoxin, has become the number 1 drug for treating heart disease in conjunction with newer medications.

All steroids, both natural and synthetic, have a common structure, the steroid nucleus, pictured in Figure 26.1. The basic building block is the fused four-ring system shown in (a). It is joined in a way that makes it relatively rigid compared with other compounds. By convention, the rings are lettered A, B, C, and D in the order shown.

The carbon atoms of the steroid nucleus are numbered from 1 to 17, as shown in Figure 26.1(b). The numbers 18 and 19 represent the carbon atoms of methyl groups usually attached at positions 13 and 10, respectively. Each methyl group is represented by a solid line (using the type of abbreviation introduced in Chapter 6). Figure 26.1 shows a methyl group attached to carbon 17, but usually more complex groups are present at this position.

Because of the way in which the rings are joined, the steroid nucleus is nearly flat, all carbons lying in the same plane. All of the atoms attached from below this ring system are said to be in the alpha (α) configuration, and all those attached from above in the beta (β) configuration. Atoms in the α position are designated by a broken line to signify that the bond is projected below the plane of the ring system, and atoms in the β position are designated by a solid line to indicate that the bond projects above the plane of the ring system. For example, the synthetic steroid mestranol, a frequent component of oral contraceptives, has an $-OH$ at C-17 above the ring and the ethinyl group ($-C\equiv C-H$) below the ring.

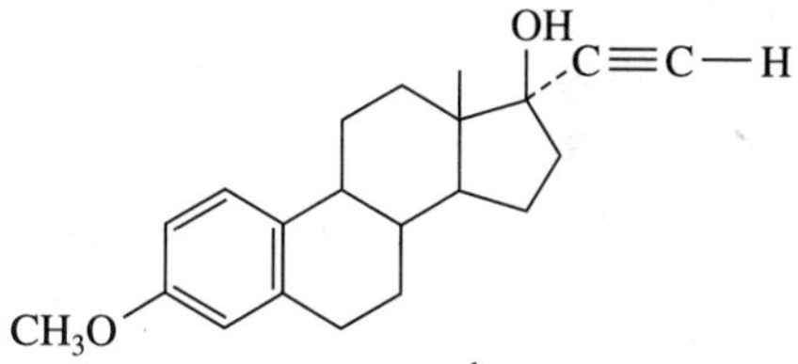

mestranol

The biochemical activity of any steroid depends on *where* (that is, at which carbon) and *how* (above or below) each group is attached to the steroid nucleus. The methyl groups located at positions 10 and 13 are known as *angular methyl groups;* they lie above the plane of the molecule. Together with other groups or double bonds that are present, they determine the properties of different steroids.

26.2   Cholesterol in the Body

GOALS: To describe the facts concerning the relationship between cholesterol and
        atherosclerosis, including the significance of HDLs and LDLs.
        To describe the action of anticholesterol drugs.

Cholesterol is a white, crystalline solid first isolated from gallstones, from which it got its name (Greek, *chole,* "bile"; *steros,* "solid"). It is identified by the β-hydroxyl group at C-3, a hydrocarbon chain of eight carbons attached to C-17, and a double bond in ring B between C-5 and C-6.

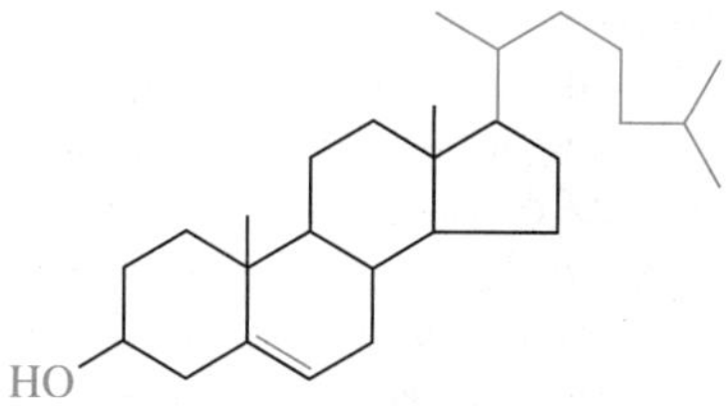

cholesterol

Cholesterol is notorious as contributing to **atherosclerosis,** better known as hardening of the arteries, and to heart disease. These diseases are characterized by the buildup of deposits of cholesterol and other lipids on the walls of the arteries. As they thicken and harden, blood flow becomes partially blocked. Blood pressure rises, and muscles, especially the heart, receive too little oxygen for normal activity.

The chain that links saturated fats to cholesterol, cholesterol to atherosclerosis, and atherosclerosis to high blood pressure and heart attacks leads to the current preference for low amounts of saturated fats and cholesterol in the diet. Yet, the body uses cholesterol as the precursor for synthesizing many essential steroids, for example, the adrenal and sex hormones. Humans (and other animals) can synthesize cholesterol, and experiments using radioactive tracers show most of the cholesterol in human tissues is synthesized rather than absorbed from the diet.

Thus cholesterol is essential to life. It is not classified as a vitamin since the body can synthesize it. The amount synthesized is determined to some extent by the amount consumed in the diet. High dietary cholesterol acts as a *feedback inhibitor.* Feedback inhibition is a common phenomenon in which the product of a series of reactions prevents its own synthesis by interfering with some step earlier in the pathway.

High-Density Lipoproteins and Low-Density Lipoproteins

About 1975, studies on atherosclerosis began to focus attention on how cholesterol exists in the bloodstream. Because it is a solid substance and insoluble in water, blood (which is mostly water) does not dissolve it. Cholesterol interacts with other chemicals, namely, *lipoproteins,* which are soluble in blood. These substances are HDLs, *high-density lipoproteins,* LDLs, *low-density lipoproteins,* and VLDLs, *very low-density lipoproteins.* Most blood cholesterol is associated with either HDL or LDL.

Both HDL and LDL consist of proteins, triglycerides, and cholesterol in various amounts. The HDLs have the highest density because they typically contain about 50% protein; LDLs usually contain only about 20% protein.

The HDLs have been characterized as ''good'' cholesterol and very beneficial to health. The LDLs, the major carriers of cholesterol, have been classified as ''bad'' and detrimental to health. The matter is much more complex, but it has become popular to assume that HDLs tend to mobilize cholesterol so that it is carried from the tissues to the liver for removal from the body. In contrast, LDLs tend to carry cholesterol to the tissues and to encourage the buildup of plaques (atherosclerotic deposits) that narrow blood vessels *(1).* When dietary habits are changed to lower blood cholesterol levels, the goal is to decrease LDL cholesterol the most.

Much attention has been focused on the apparent protective effect of HDLs. High HDL levels may counteract high blood cholesterol levels and minimize the occurrence of atherosclerosis. In fact, some experts believe that the risk of heart attack can be predicted by comparing only the total blood cholesterol and HDL cholesterol levels; when the ratio of total blood cholesterol to HDL-cholesterol is high, the risk is also high.

Researchers have tried to find out what factors increase or decrease HDL levels. The balance of saturated and unsaturated fats seems not to be a factor. Low-fat, low-cholesterol diets result in a small reduction of LDL cholesterol levels. But the same diet lowers HDL cholesterol levels by a similar amount, and therefore cancels out much of the benefit. In fact, such a diet may even undercut the benefits of an increase of HDL cholesterol levels achieved by drug therapy *(2).* The major factors that are known to increase HDL levels are being female, exercise, and moderate amounts of alcohol. Smoking, obesity, and lack of exercise are generally associated with low levels.

It is well known that women have a much lower risk of heart attack than men, particularly before menopause, possibly due to hormonal factors. The female hormone estrogen is believed to increase HDL levels. During natural menopause, estrogen levels drop. A recent 5-year study indicates this is responsible for an increase in LDL cholesterol and a decrease in HDL cholesterol. Therefore, as a woman's blood cholesterol level worsens, the risk of heart disease rises *(3).* Studies show that the male hormone testosterone and the anabolic steroid stanozolol reduce HDL levels *(4),* which may explain why men have a higher risk of heart attack at all ages.

Increased risk of heart attack is a side effect of oral contraceptives. Even though these drugs contain an estrogen, the other component, a progestin, depresses HDL levels. A complete discussion of oral contraceptives appears in Section 26.8.

The influence of exercise on HDL levels is somewhat uncertain. Runners consistently show elevated levels, but they generally smoke less, drink more alcohol, and are leaner. All three factors are generally linked to elevated HDL levels.

A little information may allow a simple conclusion, whereas more information may confound the matter. Consider the effect of alcohol. Moderate alcohol consumption is thought to provide protection against heart disease by HDL elevation. Nevertheless, even moderate consumption has so many negative health effects that the likelihood of death from all causes actually increases.

Intense research on HDLs since 1975 has led to some rather astonishing drugs that can alter levels of specific blood lipids, including cholesterol. In cases where a diet low in saturated fat and cholesterol does not sufficiently lower blood cholesterol, several drugs can help control the level. Their long-term safety will not be known for some time, however. Physicians can employ four strategies for reducing total and LDL cholesterol in the blood:

1. Inhibit biosynthesis of cholesterol in the liver.
   Available product: lovastatin (Mevacor)
   Lovastatin is the newest and most widely used drug for lowering blood cholesterol; it is a prodrug converted in the body to an active drug that inhibits an essential enzyme in the biosynthesis of cholesterol in the liver. Once the mechanism for making cholesterol is shut down, the liver absorbs more LDL cholesterol from the blood to meet the demand.

2. Prevent cholesterol absorption from the intestine into the bloodstream.
   Available products: cholestyramine resin (Questran) and colestipol hydrochloride (Colestid)
   These two drugs are commonly known as bile acid sequesterants since they tie up, and thus prevent the action of, bile acids (see Section 26.3), which emulsify lipids in the intestine so they can be absorbed into the bloodstream.

3. Inhibit LDL production.
   Available product: nicotinic acid (Nicolar)
   This product decreases the production of VLDL, a precursor of LDL.

4. Increase LDL metabolism.
   Available products: probucol (Lorelco) and gemfibrozil (Lopid)
   Both increase LDL metabolism. Probucol also has a slight inhibitory effect on cholesterol biosynthesis and absorption of dietary cholesterol. Gemfibrozil causes an increase in HDL cholesterol.

26.3   The Bile Acids and Bile Salts

GOAL: To study the structure and action of the bile acids and bile salts.

The *bile acids* are shown in Figure 26.2. They are emulsifying agents needed for absorption of cholesterol, other lipids, and fat-soluble vitamins so they can pass from the intestine into the bloodstream. Such steroids are derived from cholesterol by removal of the three terminal carbon atoms from the C-17 side chain and saturation (hydrogenation) of the double bond in ring B. Thus the bile acids are identified by the presence of 24 carbon atoms and the oxidation (insertion of oxygen) at the 3 and sometimes the 12 and 7 positions.

Cholic acid and chenodeoxycholic acid are the primary bile acids; they are synthesized in the liver. The others shown in Figure 26.2 are synthesized by bacteria from the primary bile acids in the intestine. They are combined with amino acids, such as glycine or taurine, to form glycocholate and taurocholate, respectively. These compounds are known as *bile salts* since the acid groups, $-COOH$ and $-SO_3H$, are readily ionized to $-COO^-$ and

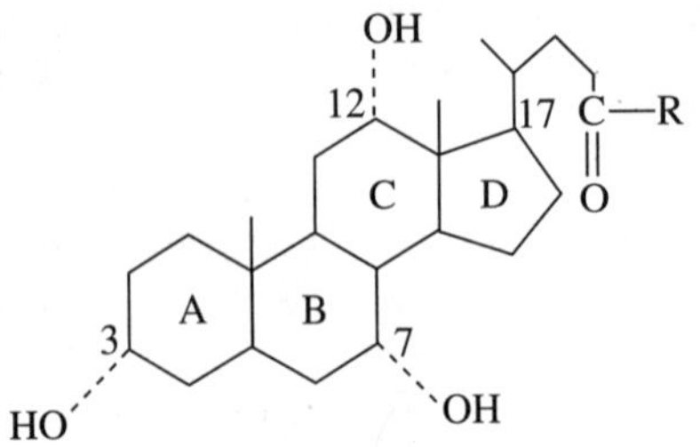

cholic acid (R = OH)
glycocholate (R = NHCH$_2$COOH)
taurocholate (R = NHCH$_2$CH$_2$SO$_3$H)

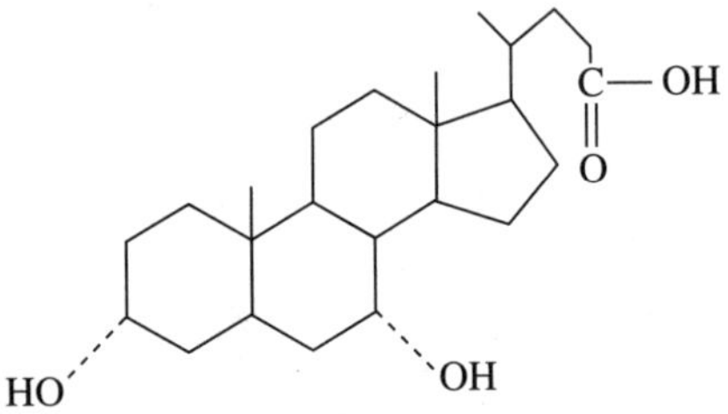

chenodeoxycholic acid

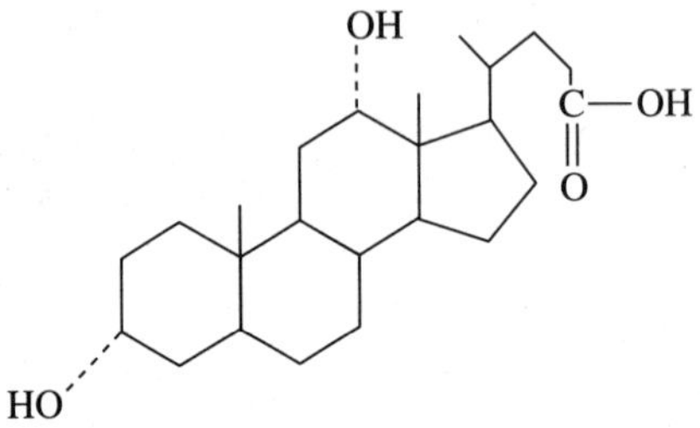

deoxycholic acid

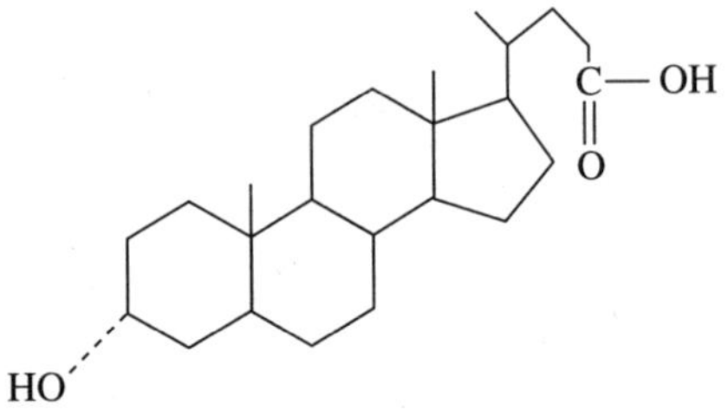

lithocholic acid

Figure 26.2   The bile acids.

$-SO_3^-$ in the alkaline conditions of the intestinal tract. Bile salts are stored in the gall-bladder and are released into the intestine to aid in the digestion and absorption of dietary lipids, fats, and cholesterol by promoting the formation of water-soluble micelles (Figure 22.2). Thus, a breakdown product of cholesterol helps the body to digest cholesterol.

Bile salts act as emulsifiers for dietary lipids. The steroid nucleus itself is very hydrophobic. When very hydrophilic groups are introduced at several positions of the molecule, the structure created is like that of soaps, detergents, and other emulsifying agents; this structure helps hydrophobic (nonpolar) lipids enter the bloodstream (mostly water) for delivery to the tissues.

Bile is a liquid secreted by the liver, stored in the gallbladder, and discharged into the intestine. The liver produces 500-700 mL of bile per day. Gallbladder bile is composed primarily of bile salts, free cholesterol, bile acids, and biliverdin, a green pigment. Although most of the bile salts and cholesterol are recycled and reabsorbed, some are excreted in the feces. Thus bile is the major route for the body to eliminate cholesterol.

Several diseases can disrupt the normal detergent-like function of the bile salts in digestion. A deficiency of bile salts delivered to the intestine leads to impaired micelle formation and inadequate absorption of fat. In gallbladder obstruction by stones, and after gallbladder removal, fat in the diet must be reduced. Surgical removal of a section of intestine also can prevent reabsorption of bile salts for reuse, leading to a deficiency. Significant changes in the acidity of bile and a change in the bile salt/cholesterol ratio in bile solution may result in precipitation of cholesterol as gallstones.

26.4   The Action of Hormones

GOAL: To explain the general mode of action of hormones and the role of tropic hormones.

The word hormone (Greek *horman,* ''to excite'') has been mentioned many times, but we have not considered exactly what a hormone is. All hormones function in much the same way. They are produced by *endocrine* glands and released into the bloodstream to circulate and exert their effects on target tissues elsewhere in the body. A hormone affects a specific tissue by regulating the rates of various functions, such as the activity of enzymes or the transport of ions.

Most hormones are steroids, proteins, small peptides, or amines. Steroid hormones are synthesized in the adrenal cortex, gonads (testes and ovaries), and placenta. Protein and peptide hormones, such as insulin, are produced by the hypothalamus, pituitary, placenta, parathyroid, and pancreas. The thyroid gland is the exclusive source of thyroid hormone, and the adrenal medulla synthesizes the catecholamines, such as epinephrine (adrenaline). The effects of steroid hormones are summarized in Table 26.1.

In several cases, a *tropic* hormone is released from one endocrine gland and stimulates or suppresses the production of a hormone by another gland. For example, the anterior pituitary gland (located below the brain) produces protein hormones *gonadotropic hormones* or simply *gonadotropins.* The word **gonadotropic** indicates the ability to act on the sex glands. Thus, the protein hormones released by the anterior pituitary stimulate the production of steroid hormones by the sex glands—the ovaries and testes. Conversely, the sex hormones suppress (by feedback inhibition) further production of gonadotropins when an adequate supply of steroid hormones is present in the blood. The protein hormones of the anterior pituitary gland and their biological effects are listed in Table 26.2.

Before we study oral contraceptives, let us consider some other steroids. We do so by considering these substances in the order in which they are synthesized from cholesterol in the body.

Table 26.1   Steroid hormones and their primary physiological effects

Hormone	Target Organ/Tissue	Effect
Adrenal corticosteroids		
Glucocorticoids (cortisone)	Many	Regulate carbohydrate and protein metabolism; anti-inflammatory and immunosuppresive; promote water excretion; increase blood pressure and cardiac output.
Mineralocorticoids (aldosterone)	Kidneys; parotid gland; gastrointestinal tract	Regulate sodium and potassium blood levels; elevate blood pressure.
Sex hormones		
Androgens (testosterone)	Male gonads, brain	Cause spermatogenesis; maintain male sex characteristics.
Estrogen (estradiol)	Female gonads, uterus, brain	Regulates menstrual cycle changes; maintains female sex characteristics.
Progestogens (progesterone)	Uterus, brain	Regulate menstrual cycle changes; maintain pregnancy.

Table 26.2 Hormones of the anterior pituitary and their physiological effects

Hormone	Target Organ/Tissue	Effect
Adrenocorticotropin (corticotropin; ACTH)	Adrenal cortex	Forms and/or secretes adrenal cortical steroids.
Luteinizing hormone (LH)	Ovaries	Luteinizes site of ovulation; secretes progesterone.
	Testes	Develops interstitial tissue, which secretes androgens such as testosterone.
Follicle-stimulating hormone (FSH)	Ovaries	Develops follicles; with LH, secretes estrogen, and ovulation.
	Testes	Develops seminiferous tubules; spermatogenesis.
Thyrotropin (TSH)	Thyroid	Synthesizes and secretes thyroid hormones.
Somatropin (growth hormone)	Most tissues	Bone and muscle growth; anabolic effect on calcium, phosphate, and nitrogen metabolism; metabolizes carbohydrate and lipid.
Prolactin	Breasts	Proliferation of ducts, initiates milk secretion.

26.5 Steroid Hormones

GOALS: To understand the source and importance of the adrenal corticosteroids.
To appreciate the sources and the roles of the three sex hormones.
To know the facts about anabolic steroids.

The Adrenal Corticosteroids

The adrenal glands (located above the kidneys) are composed of two functionally distinct structures, the medulla and the adrenal cortex. The medulla (inner part), which is part of the sympathetic nervous system, secretes two catecholamines with hormonal activity—epinephrine (adrenaline) and norepinephrine. The cortex (outer part) is part of the hypothalamic-pituitary-adrenal endocrine system, and secretes the group of hormones known collectively as *adrenal corticosteroids.* The two major classes, *glucocorticoids* and *mineralocorticoids,* are both synthesized from cholesterol.

The mineralocorticoids, particularly aldosterone (Figure 26.3), regulate the body's balance of sodium and potassium ions. The glucocorticoids, better known as cortisone and hydrocortisone (or cortisol), influence a variety of essential body functions. For example, they regulate the metabolism of carbohydrates and proteins during periods of stress. They also are antiinflammatory and immunosuppressive.

Cortisone and hydrocortisone (cortisol) are shown in Figure 26.3. These natural glucocorticoids and their synthetic analogs are the main drugs used to treat inflammatory and allergic conditions. They interfere with all stages of the inflammatory response by constricting small blood vessels, which localizes the response and also reduces redness. They also suppress cell chemicals, such as histamine and serotonin, that would be released during the inflammatory process.

The release of histamine in allergic reactions and the use of antihistamines are described in Section 23.7.

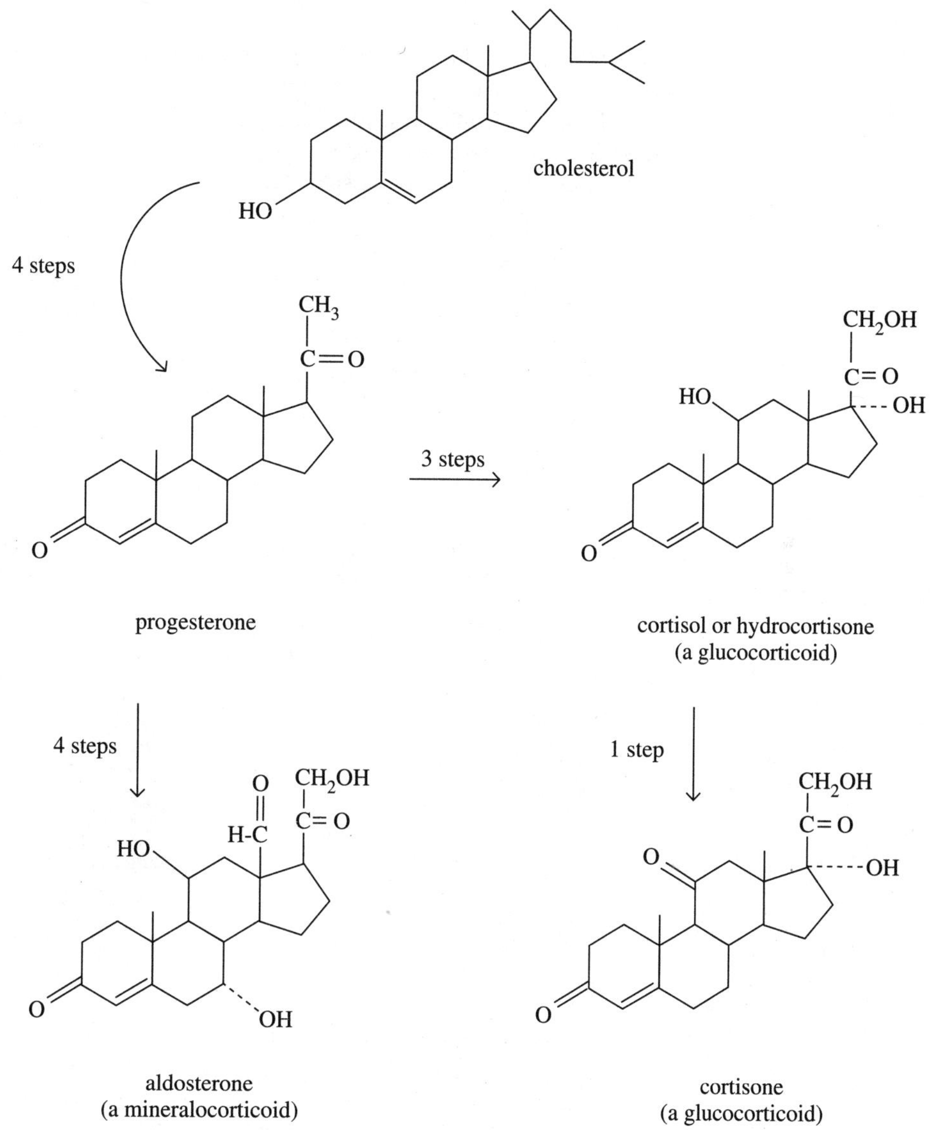

Figure 26.3 Adrenal corticosteroids synthesized from cholesterol by way of progesterone.

In 1938 E. C. Kendall at the Mayo Clinic first isolated cortisone, for which he received the Nobel Prize in 1950. Hailed as a ''miracle drug'' in the early 1950s, cortisone produces dramatic relief in patients with rheumatoid arthritis. Its antiinflammatory action, which relieves arthritic symptoms, is just one of its many effects. It was quickly discovered that cortisone was as dangerous as it was useful, however, due to side effects. These side effects parallel symptoms of excessive secretion of the hormone: fluid retention, unwanted hair, a moon-face appearance, and mental depression. In its antiinflammatory action, cortisone also suppresses the body's ability to fight infection and respond to stress.

Extensive research has been conducted to moderate chemically the adverse effects of cortisone. The pharmaceutical industry has successfully synthesized cortisone derivatives that are much more potent and safer. In 1980, over-the-counter preparations of hydrocortisone acetate were marketed for the first time. Cream, lotion, and ointment forms should be used only on the skin. They provide temporary relief from irritations and itching due to eczema or dermatitis, allergic rashes such as poison ivy and poison oak, and rashes due to soaps, detergents, cosmetics, and jewelry. Current research into the chemistry of the corticosteroid molecule aspires to make it specific for a certain tissue, thus eliminating unwanted effects on other tissues *(5)*.

The Sex Hormones

The sex hormones also are steroids, related to cholesterol as shown in Figure 26.4. Most of the arrows signifying a conversion from one to the next actually represent several steps. Each steroid represents a whole class of related compounds.

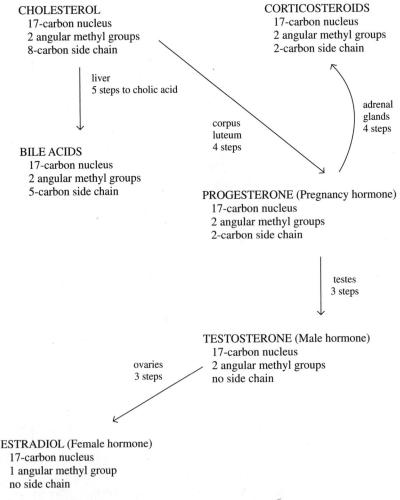

Figure 26.4   The metabolism of cholesterol to form the bile acids, corticosteriods (e.g., cortisone), and sex hormones. See Figure 26.5 for structures. Arrows represent pathways.

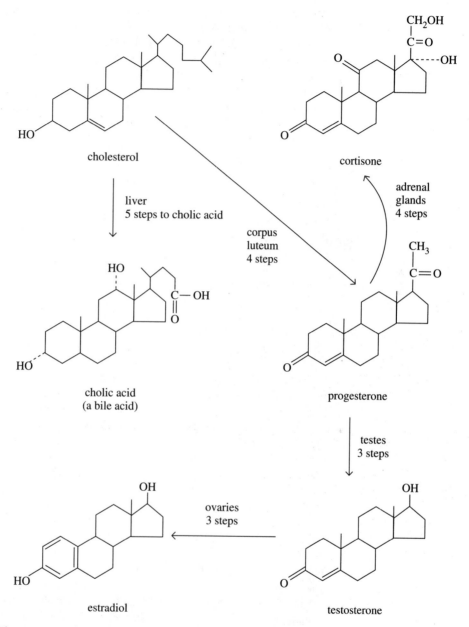

Figure 26.5 The metabolism of cholesterol to form the bile acids, corticosteriods (e.g., cortisone), and sex hormones.

The same information is expanded in Figure 26.5, including the structure of each steroid. The number of carbon atoms in the four rings of the nucleus remains the same throughout the pathway of reactions. The side chain at C-17 and the angular methyl groups are the reaction sites in this scheme.

Figure 26.4 also names the organ or gland in which each steroid is an end product. Although progesterone is a precursor to cortisone (from the adrenal glands), as well as

Ovulation is the release of an egg (ovum) from the ovary.

testosterone (a male hormone) and estradiol (a female hormone), the pathway ends with a different steroid in different sites. The full reaction series to estradiol happens in the ovaries before ovulation; the corpus luteum (formed in the ovary after ovulation) converts cholesterol as far as progesterone. In men, the testes carry the synthetic sequence from cholesterol to testosterone, but not to estradiol.

Each of the three sex hormones is actually just one member of the three classes of hormones—*androgens* (male hormones), *progestogens* (pregnancy hormones), and *estrogens* (female hormones). Each member of each class differs from the others in the kinds and positions of groups present.

The Androgens: Anabolic Steroids

The male hormones, androgens, are produced from cholesterol in the testes. The most active is *testosterone,* high levels of which support sperm production from the time of puberty on. During fetal development, its presence causes differentiation of the male sex organs.

The androgens promote secondary sex characteristics associated with masculinizing (androgenic) qualities such as muscle strength, deepening voice, and facial and body hair. They also have anabolic (tissue-building) qualities such as increased skeletal muscle mass, increased hemoglobin concentration and red blood cell mass, increased calcium deposition in the bones, and increased retention of total body nitrogen. Androgens are also normally secreted by the adrenals in men, and in very small amounts by the ovaries and adrenals in women.

Androgenic-anabolic compounds can be prescribed as drugs. When taken in proper dosages, they have beneficial therapeutic uses. First developed in the 1930s, they were used to build tissue while an individual was recovering from an injury, surgery, or even starvation. Today the drugs have therapeutic uses such as increased protein tissue building after surgery, an injury, severe burns, or a bone fracture; treatment of muscular dystrophy, diabetes, osteoporosis, starvation, and malnutrition; and replacement therapy for testosterone deficiency. They are most widely used in treating pituitary dwarfism, invasive breast cancer, and some kinds of anemia *(6)*.

The anabolic steroids have been the center of much controversy recently due to their heavy misuse by professional and Olympic athletes. In the 1950s, weight lifters and bodybuilders in the Soviet Union first started taking these powerful agents as ''performance aids'' to build up muscle size and strength. This practice spread to Eastern Europe and finally to the West as the results convinced athletes of their value in winning. Today, anabolic steroids are the most commonly abused drugs in sports. Many athletes think they are just ''training drugs'' or ''nutritional supplements'' that will provide them with a winning edge in sports where explosive, short-term muscular strength is important *(7)*. Examples of some of the steroids in use are provided in Figure 26.6. Chemically, they are synthetic derivatives of testosterone that have been modified for oral ingestion or intramuscular injection. Combinations of the oral and injectable doses are often taken, a practice known as ''stacking.''

Many athletes start using anabolic steroids under a physician's care, usually for therapeutic purposes. Once therapy is stopped, they may continue using them without supervision, frequently at much higher dosages. A 1990 report in the *Journal of the American Medical Association* indicated that physicians still prescribe 20% of all steroids to athletes, even though the drugs are harmful to the health of the users and their use is illegal in certain competitive sporting events *(8)*. When used in dosages that are sometimes 10–100 times larger than therapeutic amounts, the desired muscle building and strengthening do

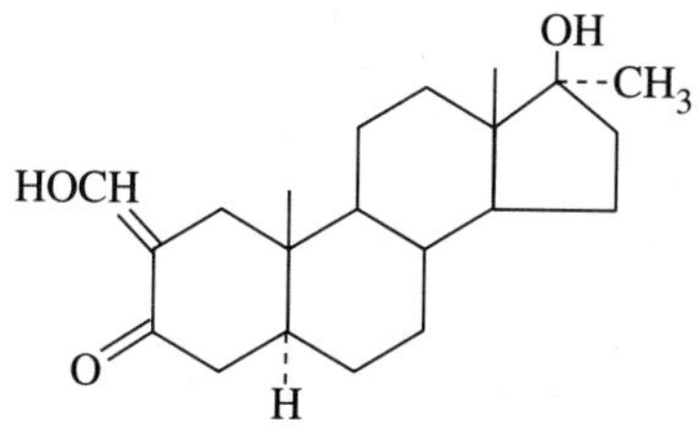

oxandrolone
(Anavar)
17*β*-Hydroxy-17-methyl-
2-oxa-5*a*-androstan-3-one

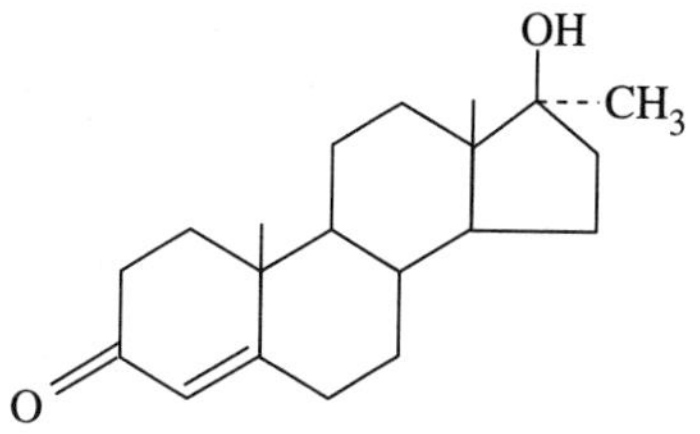

methandrostenolone
(Dianabol)
17ß-Hydroxy-17-methyl-
androsta-1,4-dien-3-one

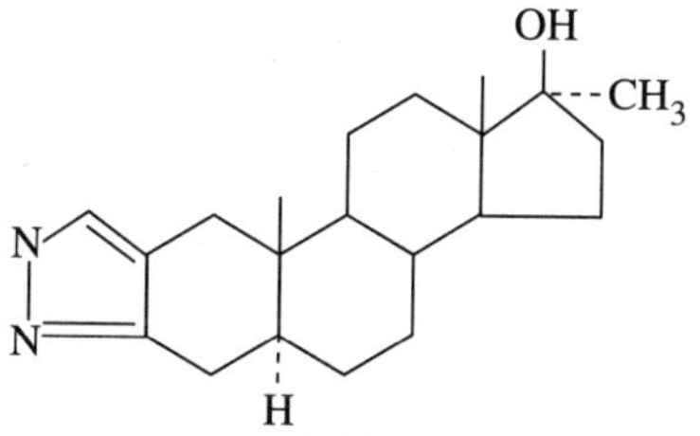

stanzolol
(Winstrol)
17-Methyl-5*a*-androstano-
[3,2-c]-pyrazol-17ß-ol

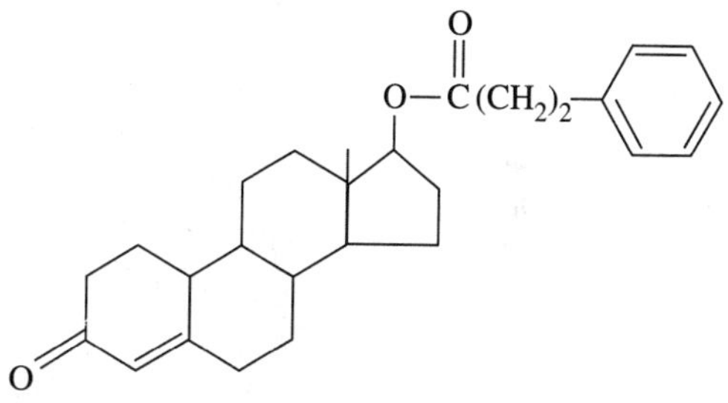

nandrolone phenylpropionate
(Durabolin)
17ß-phenylpropionate estr-4-en-3-one

Figure 26.6 Some anabolic steroids commonly abused by athletes for muscle building and strengthening.

occur *(9)*. However, many harmful side effects have resulted from this overuse. Men risk suppression of natural hormone production, causing breasts to develop and testes to shrink. The risk of heart disease, kidney and liver damage, and liver cancer increases. Women risk irreversible masculinizing traits such as facial hair, a deep voice, and a male physique. The long-term risks in adolescents are still unknown. Steroid abusers also may become very aggressive or even violent; this behavioral reaction is known as a ''roid'' rage. Severe acne also is common, and can often be a sign of continuing anabolic steroid abuse.

Support for research on steroid effects lags, at a time when some say we have a ''steroid epidemic'' among athletes, including adolescents and college students. In 1990 the Department of Health and Human Services estimated that from 5–10% of boys and 0.5–2.5% of girls in high school have some involvement with these drugs *(10)*. Also, a 1990 study among college athletes revealed much higher use than previously thought, with projected rates of use of 14.7% for men and 5.9% for women.

Professional and Olympic-caliber athletes have been cited for steroid use. German women's swim-team coaches admitted that the agents helped achieve domination of the sport for some 20 years *(11)*. Ben Johnson of Canada had his 1988 gold medal-winning performance disqualified when traces of the drug were found in his urine. As anabolic steroid abuse continues to receive headlines, the public will increasingly ask for a solution to this contemporary problem.

The Progestogens

Of the several progestogens, the most active is progesterone. It is synthesized in the ovaries (by the corpus luteum), by the placenta during pregnancy, and by the adrenals. It is often called the pregnancy hormone. It is an intermediate in testosterone and estradiol synthesis, as shown in Figure 26.4.

Plasma progesterone levels increase 100-fold as the corpus luteum matures after ovulation has occurred (see Figure 26.7). Under the stimulation of progesterone (and estradiol), blood vessels develop in the lining of the uterus, which thickens in preparation for possible implantation of a fertilized egg. The increased blood supply brings nutrients for the developing embryo if a fertilized egg has become implanted. Progesterone also prevents subsequent ovulation, which could result in a second fertilization. Thus it is a built-in protective agent for preventing a second pregnancy. Progesterone-like compounds are major components of oral contraceptives for this reason.

The Estrogens

Estrogens are the female hormones that stimulate the growth and maturation of female reproductive organs such as the breasts and the lining of the uterus. In the latter function, they collaborate with progesterone. Estradiol is the principal estrogen, although estrone also is quite active.

All estrogens are C-18 steroids that have undergone desaturation of ring A and the loss of the C-19 methyl group of the androgen precursor (see Figure 26.5). It is important to note that cholesterol is converted to testosterone and then to estradiol in the ovaries. After menopause, or when there is a defect in this conversion, as estrogen levels decrease, a buildup of androgens results in hirsutism (increased facial hair) and increased storage of fat on the abdomen and in the skeletal muscles. The effects of an estrogen deficiency can also be caused by abuse of anabolic steroids, as discussed in the previous section.

26.6 Biology of Ovarian and Uterine Cycles

GOAL: To be aware of the changes that take place in the ovaries and the uterus during the monthly female cycle, ending in either menstruation or pregnancy.

Figures 26.7 and 26.8 show the reproductive system in detail, including all steps leading up to and following ovulation. The first event is the development of a mature ovum in a cavity of the ovary, the *Graafian follicle*. The ovum and follicle mature in 12-14 days. Then the follicle ruptures and a mature ovum is released from the ovary at *ovulation*. After ovulation the follicle becomes an endocrine gland, the *corpus luteum,* which secretes progesterone. The entire process to this point is shown in Figure 26.7.

Figure 26.8 shows these events in more detail. The cycle occurs in alternate ovaries approximately every 28 days, terminating in menstruation unless fertilization takes place. Usually only one ovum is released in any cycle. After ovulation, the mature, unfertilized ovum passes into the fallopian tube and moves toward the uterus.

Fertilization is possible only in the fallopian tube and only for a short period after the ovum is released. Sperm released into the female reproductive tract must make their way up the fallopian tube to fertilize the ovum.

An unfertilized ovum travels down the fallopian tube into the uterus, where it cannot be fertilized. The sperm that enter the system survive only a short time, since the supply of nutrients is not sufficient to sustain them. If it is fertilized, the ovum moves down the fallopian tube into the uterus. It then becomes implanted in the lining of the uterus, which has become saturated with blood vessels carrying nutrients that sustain the embryo.

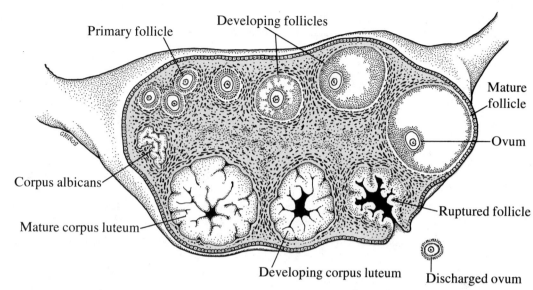

Figure 26.7   Growth and rupture of an ovarian (Graafian) follicle and formation of the corpus luteum. (Adapted from Philip Rhodes. *Birth Control.* London: Oxford University Press, 1971.)

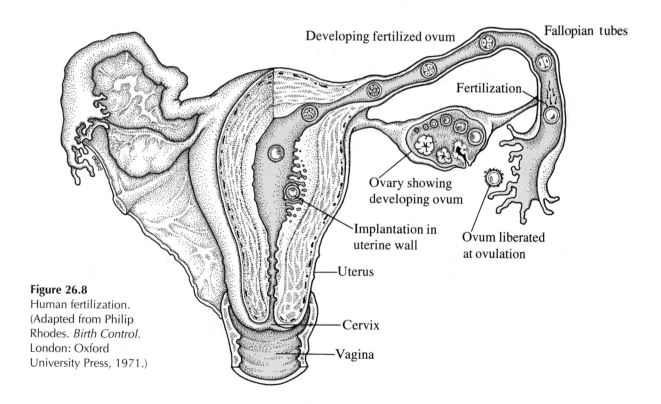

Figure 26.8
Human fertilization.
(Adapted from Philip
Rhodes. *Birth Control.*
London: Oxford
University Press, 1971.)

The monthly development of the spongy inner layer of the uterus (endometrium) begins shortly after the follicle begins to develop in the ovary; that is, it occurs before, and does not depend on, successful fertilization. If a fertilized ovum becomes implanted, the uterine lining remains intact, and the embryo develops into a fetus. If fertilization does not happen, the ovum does not become implanted in the uterus, and the uterine lining is sloughed off, and menstrual flow begins.

26.7 Chemistry of Ovarian and Uterine Cycles

GOAL: To appreciate the role of the tropic and steroid hormones in controlling physical changes that take place during the female cycle.

Hormones are active in the events of the ovarian and uterine cycles. The entire process is triggered by the release of a single hormone from the hypothalamus region of the brain. The hypothalamic hormone is the protein *gonadotropin-releasing hormone,* or GnRH. Released into the bloodstream, GnRH travels to the pituitary gland and stimulates the release of pituitary hormones into the bloodstream. The two principal hormones released by the pituitary are *follicle-stimulating hormone* (FSH) and *luteinizing hormone* (LH). Both are proteins. They are gonadotropic hormones, and stimulate the sex glands (gonads).

As the name suggests, FSH travels to the ovaries and stimulates the development of an ovum within a Graafian follicle. This is designated as day 1 (same as day 29) in Figure 26.9. As it develops, the follicle produces increasing amounts of estrogen. A peak concentration is reached on about day 12, while the ovum matures to the point of ovulation. The second gonadotropin, LH, is released about this time; it causes the ripened follicle to release the ovum; that is, ovulation occurs. Under the stimulation of LH, the empty follicle is transformed into the corpus luteum, a temporary endocrine gland that produces progesterone. As the level of sex hormones (estrogen and progesterone) builds up, the hypothalamus and pituitary recognize the change and stop the release of FSH and LH, preventing development of another ovum. This is an example of *negative feedback,* in which the end products, the sex hormones, suppress the repetition of the events that caused them to form.

When fertilization and implantation do not occur, the corpus luteum degenerates and the supply of progesterone is cut off. The drop in circulating progesterone signals the hypothalamus to release GnRH and begin a new cycle. Since the continued growth of the lining of the uterus also depends on progesterone and estrogen, the lining deteriorates and is lost in menstruation.

If fertilization and implantation occur, the corpus luteum persists for several months. By then the placenta produces progesterone at levels high enough to sustain the inhibition of the hypothalamus-pituitary chain, to maintain the uterine lining, and to prevent ovulation.

Progesterone also causes the glands of the cervix (the lining of the entrance to the uterus) to produce a sticky secretion that makes it difficult for sperm to enter the cervical opening. This phenomenon can also contribute to the effectiveness of oral contraceptives, as we will see.

The continued output of progesterone is stimulated by one more hormone. *Human chorionic gonadotropin* (HCG) is produced by the placenta of the developing fetus. Its presence in the urine of a pregnant woman forms the basis for several pregnancy tests, including the numerous home testing kits. By the sixth week of a pregnancy, the placenta is producing enough HCG to be detected in a sample of the first urine a woman passes in the morning.

The pituitary is commonly called the master gland because the release of pituitary hormones initiates the complex series of chemical events; yet once the cycle is set in motion, even it is under the control of chemicals produced in other glands.

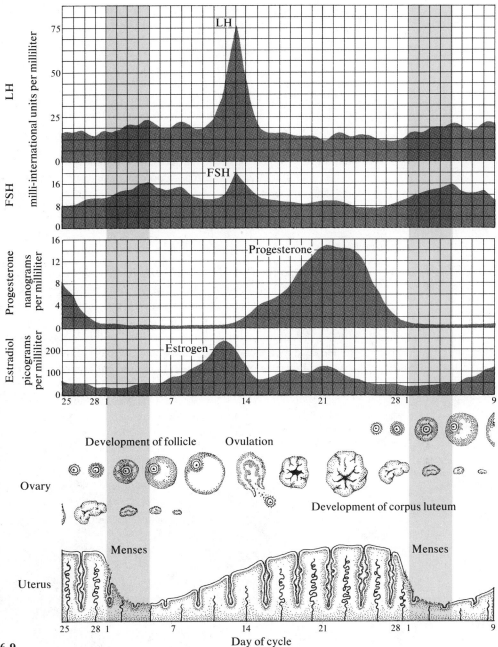

Figure 26.9

Ovarian and uterine cycles, showing variation of hormone levels and events on same time scale. At about day 26, the progesterone level is dropping sharply, and a new follicle is beginning to develop in one of the ovaries. Day 1 is the beginning of menstruation. Levels of the gonadotropins (FSH and LH) peak at day 13 and cause ovulation. As the levels of these hormones subside, progesterone levels elevate and remain high until about day 25 when they drop, unless fertilization has occurred. See text for further discussion. (Adapted from S. J. Segal, "The Physiology of Human Reproduction," *Scientific American*, September 1974. Copyright © 1974 by Scientific American, Inc.)

26.8 Strategy of the Pill

GOALS: To understand the various forms of oral contraceptives.
To understand the two contraceptive actions of these agents.
To see the nature of the synthetic steroids used in oral contraceptives.

Fertility control and contraceptive technology have improved with better understanding and knowledge of the female cycle. Although oral contraceptives (or birth control pills) are commonly known as the Pill, actually three types of pills have been developed and used, all chemically similar to natural steroids. From its introduction in the early 1960s, the Pill became the most popular method of contraception in the United States. Since the 1970s, continuing research and development have resulted in modifications of the dosages and delivery regimens.

The Combination Birth Control Pill

The term *progestin* is used for steroids that are synthetic modifications of the progesterone molecule and that have the same hormonal properties. The term *estrogen* refers to both natural and synthetic estrogens.

The first type (first-generation) of oral contraceptive is known as the conventional or combination pill. It contains two synthetic steroids—an estrogen and a progestin.

The first combination pills contained a constant amount of estrogen and progestin, and one was taken every day beginning on day 5 (day 1 being the first day of menstrual flow) and continuing for 21 days. At the end of the 21 days (on day 25), the routine was stopped and menstruation could begin. After 7 days of taking no pills, the same steps were followed in each subsequent month for as long as contraception was desired.

Some preparations on the market contained seven more inactive pills, or *placebos,* taken beyond day 25 (Figure 26.10). One placebo tablet was taken every day, making it less likely that a woman would lose track of the day on which the next cycle began and an active pill should be taken. In some products, the seven additional pills contained an iron supplement to counteract the loss of iron that occurs during menstruation.

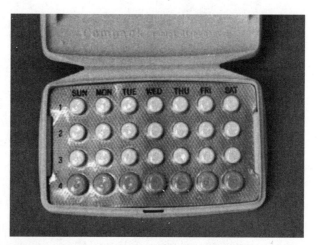

Figure 26.10

Oral contraceptives. Demulen-28 is one of the original combination (monophasic) oral contraceptives, which consist of 21 pills with constant levels of progestin and estrogen, plus 7 pills that are placebos to help the user keep an accurate count of the days of the cycle. Similar products contain an iron supplement in the seven placebos to counteract the loss of iron in menstruation.

Figure 26.11
Ortho-Novum 7/7/7 28-day regimen is a triphasic oral contraceptive, consisting of seven pills containing 0.5 mg of estrogen, seven containing 0.75 mg of estrogen, and seven containing 1 mg of estrogen. The estrogen in this product is ethinyl estradiol. All 21 pills also contain a constant amount of the progestin norethindrone. The final seven pills contain only inert ingredients.

Over the years, dosages in the combination pills were modified from the same-dose-every-day *monophasic* schedule for 21 days to separate dosing routines. Two new schedules are the single-dose *biphasic* and *triphasic* pills, in which the strength of each steroid is varied during the different phases of the 21-day cycle. The monophasic preparation remains the most widely used.

The biphasic regimen uses pills of one strength for 10 days (the first phase), a second strength for the next 11 days (the second phase), and seven inactive pills to complete the 28-day cycle. Each phase uses a different-colored pill. The triphasic regimen prescribes three different strengths, with pills of four different colors (Figure 26.11). Depending on the brand, the first-phase pills are taken for 6 or 7 days, the second-phase pills for 5 to 9 days, and the third-phase pills for 5 to 10 days. Seven inactive pills may follow for a 28-day cycle.

The strategy of the biphasic and triphasic schedules is to mimic more closely the hormone levels in the body during the normal monthly cycle. For example, the development of the lining of the uterus and menstruation occur while under the influence of the synthetic steroids, but ovulation and fertilization are prevented. The constant high level of synthetic steroids in the blood causes a negative feedback inhibition of the release of FSH and LH. Thus a pseudopregnancy condition is created.

The glands of the cervix produce the sticky secretion that develops in response to the natural progesterone. This condition retards sperm entry into the uterus. *The sperm are unable to reach an ovum, but an ovum is not there anyway.* The areas of influence within the body of the combination pill and the other two types of contraceptives to be discussed are compared in Figure 26.12.

Several undesirable side effects led to further modifications, including the development of the sequential pill.

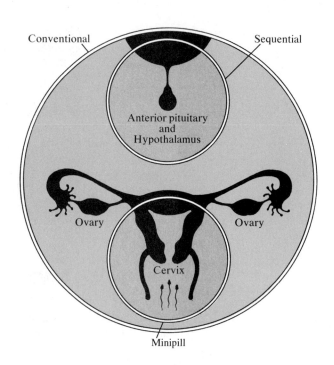

Figure 26.12
The modes of action of oral contraceptives. Circles indicate areas affected by different tablets. (1) Large circle encloses the brain-pituitary network, the cervix, and the uterine lining, all affected by the combination pill. (2) Upper circle includes only the brain-pituitary network, which is primarily affected by the sequential pills to prevent ovulation. (3) Lower circle includes only the cervix and uterus, the primary areas of influence of the minipill. This approach permits normal ovulation but causes development of a barrier to sperm. (Courtesy of Syntex Laboratories, Inc.)

The Sequential Birth Control Pill

The second generation of oral contraceptives was introduced in 1965. Often referred to as the sequential pill, the approach was developed to mimic even more closely the natural levels of ovarian hormones. The routine began with pills containing only estrogen on day 5; after 16 days a combination of estrogen and progestin was taken for another 5 days. This elevated the level of progestin at about the normal time in the cycle, but the continued high levels of estrogen suppressed the release of FSH and LH (Figure 26.9). Thus, no ovulation occurred, as is the case during pregnancy.

As with the conventional pill, the buildup of the lining of the uterus and menstruation occurred normally during each cycle. The absence of the progestin in the sequential pill in the early stages of the cycle caused the fluid in the cervix to remain clear. Sperm can penetrate normally, but ovulation is inhibited. *The sperm now would be able to reach an ovum, but none is there* (Figure 26.12).

Sequential pills were removed from the United States market because their estrogen-related side effects were serious and the pills were not as effective as the combination pill.

The Minipill Birth Control Pill

The third generation of oral contraceptives in the early 1970s was the minipill. Developed to minimize the side effects associated with high-dose estrogen-containing combinations, the only chemical agent in these preparations is a small dose of synthetic progestogen. This pill must be taken every day of the month. Because of the low dose of progestin and the absence of estrogen, the release of FSH and LH is not blocked and ovulation occurs normally when taking these pills. But progestin stimulates the cervix to secrete sticky mucus that blocks passage of the sperm. In addition, it alters the uterine lining so a fertilized egg cannot implant. Thus, *the sperm are unable to reach an ovum that is available for fertilization, and implantation of a fertilized egg is prevented* even if sperm does reach the ovum (Figure 26.12).

By the end of the 1970s the minipill had been removed from the United States market because of a high frequency of liver disorders and irregular bleeding patterns. Its contraceptive effectiveness was lower than that of the combination pills, especially if one day was missed. Later studies showed that it caused significant adverse changes in lipid and carbohydrate metabolism. The minipill was reintroduced for estrogen-sensitive women in the 1980s, but the number of women using it is very small *(12)*.

Synthetic Hormones Used in Oral Contraceptives

Now we consider the changes made at the molecular level to produce effective synthetic steroids. The problems are the same as for most drugs: side effects, cost, proper dosage, and effectiveness when taken orally.

The possible side effects have received much notoriety. Undesirable side effects are often considered an acceptable risk when the drug cures a disease or counteracts some severe symptoms, but when they affect a healthy person, they are less likely to be regarded as acceptable. Like many other drugs, a synthetic steroid is often more potent than a natural one, and if the potency is directed to the target tissue, a lower dosage may result in fewer side effects. Also, the level of side effects judged as acceptable can vary greatly.

Two other concerns are cost and dosage. The potency of a drug determines how much is necessary, and is a factor in determining cost and whether a product will be profitable to market. Finally, there was the need to develop an oral contraceptive, since repeated injections are undesirable.

Eight steroids have come into prominence in oral contraceptives—two estrogens and six progestins. The structures of the synthetic steroids are shown in Figure 26.13. Both synthetic estrogens differ little from natural estradiol; the only significant change is the *ethinyl group* ($-C\equiv C-H$) at C-17. The ethinyl group, which is located α, as signified by the dotted line, is a key feature of all synthetic sex hormones. It enhances two properties of steroids—potency and stability in the gastrointestinal tract. Stability is important so the product can be taken orally. Natural hormones are largely inactivated during digestion.

The differences between the natural hormone progesterone and progestins are more dramatic. The ethinyl group is present for the reasons previously cited. In addition, there is no methyl group (C-19) at the number 10 position of the steroid nucleus, as indicated by the prefix *nor-*. See, for example, norethindrone, also known as 17-α-ethinyl-19-nor-testosterone. Note also, as reflected in the name, that norethindrone and all other progestins resemble testosterone in structure. The carbon-containing chain is missing from the 17-β position and the hydroxy group is present as in testosterone.

testosterone

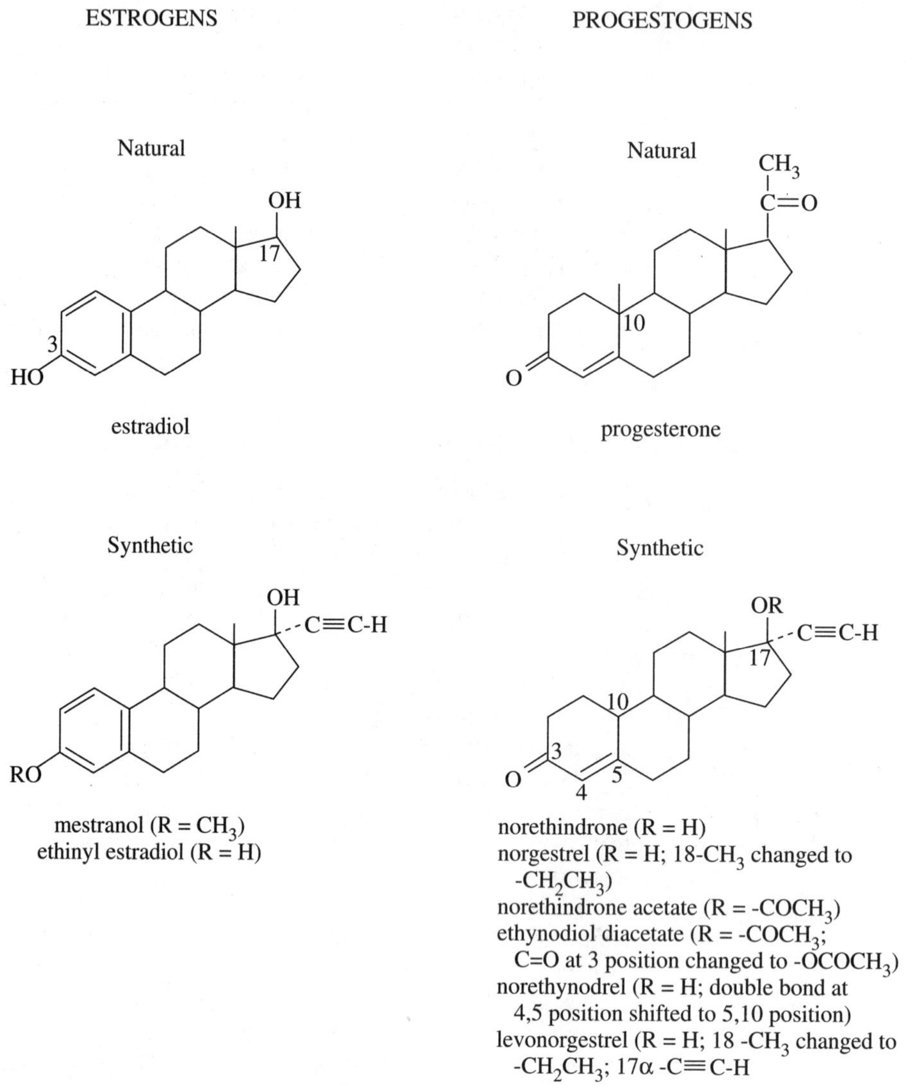
ESTROGENS
PROGESTOGENS
Natural
OH
17
3
HO
estradiol
Natural
CH3
C=O
10
O
progesterone
Synthetic
OH
C≡C-H
RO
mestranol (R = CH3)
ethinyl estradiol (R = H)
Synthetic
OR
C≡C-H
17
10
3
5
4
O
norethindrone (R = H)
norgestrel (R = H; 18-CH3 changed to -CH2CH3)
norethindrone acetate (R = -COCH3)
ethynodiol diacetate (R = -COCH3; C=O at 3 position changed to -OCOCH3)
norethynodrel (R = H; double bond at 4,5 position shifted to 5,10 position)
levonorgestrel (R = H; 18 -CH3 changed to -CH2CH3; 17α -C≡C-H

Figure 26.13 Natural and synthetic female sex hormones.

Side Effects

Estrogen and progestogen are highly potent hormones that are meant to suppress almost completely the natural synthesis of gonadal steroids as well as the pituitary gonadotropins. Consequently, oral contraceptives with fixed amounts of synthetic hormones, which do not mimic the natural balance of hormone levels, are almost certain to cause undesirable side effects. A woman and her physician can choose among different products, testing different combinations to fit her body chemistry—either different steroids or different amounts (Table 26.3). Even a proper balance does not ensure that all side effects will be eliminated, since synthetic steroids have chemical structures that differ slightly from the natural hormones and may therefore affect certain body processes differently. Some of the side effects are symptoms that are often experienced during pregnancy: bloating, nausea,

Table 26.3   Oral contraceptives that have been available in the United States

Trade Names	Manufacturer	Tablet Composition			
		Progestin	Dose (mg)	Estrogen	Dose (mg)
Combination Pills (monophasic dose)					
Enovid-E	Searle	Norethynodrel	2.5	Mestranol	0.1
Demulen ⅓₅, ¹⁄₅₀ᵃ	Searle	Ethynodiol diacetate	1.0	Ethinyl estradiol	0.035–0.05
Ovral, Lo/Ovral	Wyeth	Norgestrel	0.3–0.5	Ethinyl estradiol	0.03–0.05
Ovcon-35, -50	Mead-Johnson	Norethindrone	0.4–1.0	Ethinyl estradiol	0.035–0.05
Brevicon	Syntex	Norethindrone	0.5	Ethinyl estradiol	0.035
Norinyl 1 + 50	Syntex	Norethindrone	1.0	Mestranol	0.05
Ortho-Novum ¹⁄₅₀	Ortho	Norethindrone	1.0	Mestranol	0.05
Norlestrin ¹⁄₅₀	Parke-Davis	Norethindrone acetate	1.0	Ethinyl estradiol	0.05
Norlestrin Fe ᵇ1	Parke-Davis	Norethindrone acetate	1.0	Ethinyl estradiol	0.05
Loestrin ¹⁄₂₀	Parke-Davis	Norethindrone acetate	1.0–1.5	Ethinyl estradiol	0.02–0.03
Levlen-21, -28	Berlex	Levonorgestrel	0.15	Ethinyl estradiol	0.03
Nordette-21, -28	Wyeth	Levonorgestrel	0.15	—	—
		Ethynodiol diacetate	0.03		
Combination Pills (biphasic dose)					
Ortho-Norum 10/11/-21, -28	Ortho	Norethindrone	0.5(10W) 1.0(11P)	Ethinyl estradiol	0.035 in all tablets
Combination Pills (triphasic dose)					
Ortho-Novum 7/7/7/-21, -28	Ortho	Norethindrone	0.5(7W) 0.75(7LP) 1.0(7P)	Ethinyl estradiol	0.035 in all tablets
Tri-Norinyl-21	Syntex	Levonorgestrel	0.5(7B) 1.0(9G) 0.5(5B)	Ethinyl estradiol	0.035 in all tablets
Triphasil-21, -28	Wyeth	Levonorgestrel	0.05(6B) 0.075(5W) 0.125(10Y)	Ethinyl estradiol	0.03(6B) 0.04(5W) 0.03(10Y)
Minipill					
Micronor	Ortho	Norethindrone	0.35	None	—
Nor QD	Syntex	Norethindrone	0.35	None	—
Ovrette	Wyeth	Norgestrel	0.075	None	—

ᵃSymbolism such as 1 + 50 or ¹⁄₅₀ indicates that the product contains 1 mg progestogen and 50 μg (0.050 mg) estrogen.
ᵇContains iron.

and mood swings. This is not surprising, since some formulations create a pseudopregnancy condition by artificially elevating steroid hormone levels.

When the Pill was introduced in the United States in 1960, widespread controversy arose over adverse reactions. The original higher-dose pills showed a statistical probability of an increased risk of blood clots, high blood pressure (hypertension), depression and psychiatric disturbances, gallbladder disease, benign liver tumors, urinary tract infections, and others. Also a study in New York from 1968 to 1973 showed an increased risk of birth defects from breakthrough pregnancy while using oral contraceptives or from supportive hormone therapy during pregnancy. The hormonal effects on fetuses were somewhat sex-specific, with males affected most frequently.

Most studies before 1975 showed the increased risk of adverse effects to be associated with the estrogen component. In response, manufacturers developed the sequential pill and the minipill. Failures of these products finally resulted in the low-dose and varied-dose combination pills and minipill now on the market. Risks are listed in package inserts and most literature, together with warnings that continued use of the Pill during pregnancy could increase the risk of birth defects. Other potential side effects and complications are also listed.

Initially, apprehension was great that oral contraceptives might increase the risk of breast, ovarian, or uterine cancer. As evidence has been gathered on the newer, low-dose agents, it appears that they have protective properties against endometrial and ovarian cancers, ovarian and breast cysts, uterine fibroids, and ectopic pregnancies. However, the question of their link to breast cancer remains unanswered *(13, 14)*.

The only research evidence demonstrating an increased risk of malignancy appears to be liver cancer, reported in 1986. Larger multiyear studies are still in progress and the results will not be known for some time. There is also a higher risk of benign liver tumors from use of these drugs.

Research has also confirmed the increased risk of blood clots, increasing blood-clotting factors, and increasing blood pressure with the use of oral contraceptives, especially for anyone who smokes or has a history of any of these medical problems. In contrast, some unexpected benefits have been revealed—for example, protection from pelvic inflammatory disease (PID), regulation of menstrual flow and amount, and elevation of HDL cholesterol (with some estrogen preparations).

The Pill is one of the most studied drugs in history. After 3 decades of research and experience in this country and in Europe, experts continue to caution that long-term evaluation of the newer, low-dose products is still necessary for conclusive answers regarding their benefits and risks. From the evidence available so far, the absolute risk of developing problems is small compared with the benefits of avoiding unwanted pregnancy, except for susceptible individuals. Nevertheless, it is advisable to be aware of other forms of contraception.

26.9 Alternative Methods of Contraception

GOAL: To consider the status of subdermal implants, postcoital agents ("morning-after pills"), injectable contraceptives, IUDs, spermicidal agents, and sterilization methods.

The alternatives available to American women are far fewer than those available to women in Europe and much of the rest of the world. No new methods or chemical ingredients have been introduced in the United States since 1960, for many reasons: bad publicity on side effects in the 1970s; increased product liability exposure in a changing legal environment; women's health network concerns about side effects; and stricter federal animal toxicity tests. In addition, by 1989 the pharmaceutical industry had completely withdrawn from contraception research and development. Today, in fact, most of the estrogenic and progestational substances in the products are made in Western Europe and only marketed through United States companies *(15)*.

The alternative contraceptive methods available to women include several safe and effective methods. Among these are a subdermal (under the skin) implant, postcoital agents or morning-after pills, injectable contraceptives, intrauterine devices, physical barriers, and surgical sterilization.

Subdermal Implants

After successful use in 14 countries including Great Britain, Finland, Sweden, and Thailand, the FDA approved in 1990 the first major new contraceptive method to be introduced in this country in 30 years. Norplant is actually an old contraceptive in a new package. The active ingredient is the progestin levonorgestrel.

Norplant consists of six rod-shaped Silastic (a rubber-like material commonly used in surgical implants) capsules, each the size of a matchstick (Figure 26.14). Using a local anesthetic, they are implanted under the skin in the upper arm where they continuously release minute amounts of progestin into the bloodstream. The implants are effective for as long as 5 years before they must be removed and replaced with new ones. They can also be removed at any time before that, with full return of fertility.

The levonorgestrel released from Norplant capsules works much like minipills by making the cervical mucus sticky and altering the uterine lining so a fertilized egg cannot implant. It has also been reported that ovulation is prevented in almost 60% of users'

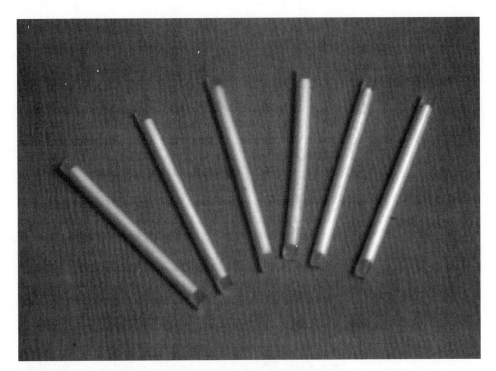

Figure 26.14   Norplant capsules are implanted under the skin in the upper arm. They release a progestin to provide contraception for up to 5 years *(16)*.

menstrual cycles. The failure rate of Norplant has been reported to be one-tenth to one-twentieth that of combination pills, making it the most effective contraceptive method yet. Its major advantage is its effortless protection from unwanted pregnancy for 5 years *(17)*.

The long-term safety of Norplant has yet to be studied, but it appears that irregular menstrual bleeding and spotting are its only side effects. Therefore, it is suggested for use by women over age 35 and those unable to use the combination pills.

Postcoital Methods

The search for suitable postcoital agents led to the prostaglandins. These are long-chain fatty acids produced in several tissues. They induce uterine contractions that stimulate menstruation. At the end of pregnancy, by the same action, two prostaglandins, PGF_2 and PGE_2, have been used to induce labor. The prostaglandins are administered by injection or as suppositories. Their high cost makes them impractical at this time.

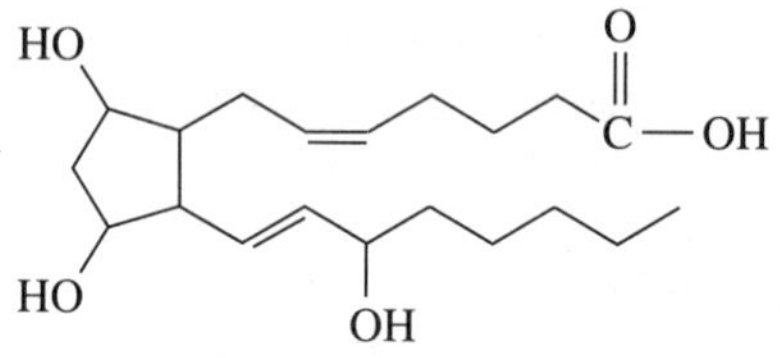

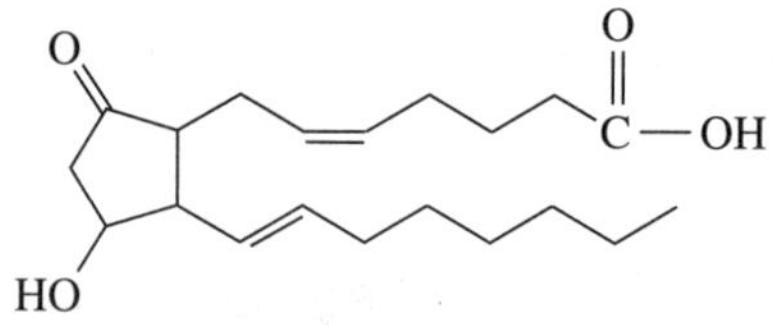

Other contraceptives that use chemical substances are available. Postcoital agents have been developed, but harmful side effects have been associated with their continued use. Oral administration of large amounts of estrogens, particularly diethylstilbestrol (DES), for 5 days beginning within 72 hours after insemination, has been used for many years.

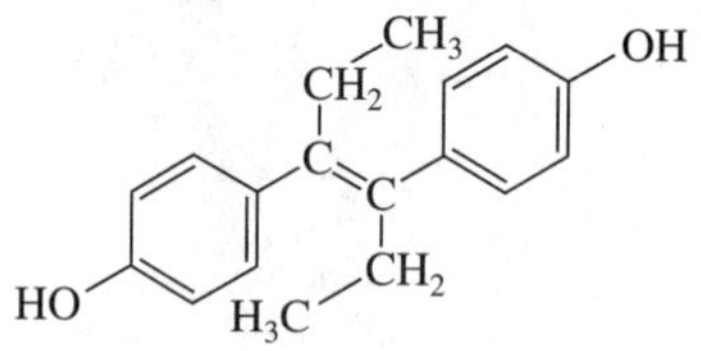

Originally, DES was prescribed in low doses to improve mood and prevent miscarriages. Studies strongly implicated it as a cause of vaginal or cervical cancer in daughters of mothers who received it during pregnancy. As a result, the FDA approved DES only for emergencies such as rape or incest *(18)*.

Among the less harmful agents is the oral contraceptive Ovral, which contains ethinyl estradiol and norgestrel. When taken as a double dose it has the same effect as a morning-after pill.

An innovative but controversial new postcoital method approved for use in France in 1988 is RU-486, or *mifepristone*. Developed and clinically tested in 1985, the highly successful steroid competes with natural progesterone in the uterus. It is classified as an *antiprogesterone,* or progesterone blocker. When RU-486 is taken in pill form with a small dose of a prostaglandin after fertilization, it causes breakdown of the endometrium (Figure 26.8), increased contractions of the uterine muscles, and finally, expulsion of the embryo and endometrium. Because of its effectiveness and possible use up to 5 weeks after fertilization, it is often called an *abortifacient (19).*

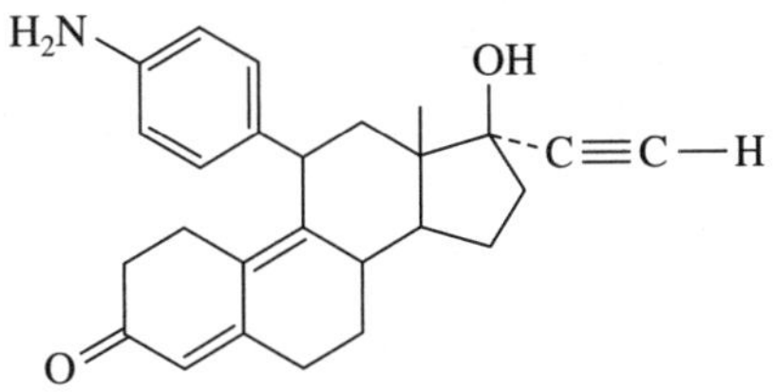

mifepristone
RU-486

Antiabortion groups have blocked introduction of RU-486 in many countries around the world. Because of this opposition, the FDA is not likely to approve it for quite some time, even though in 1990 the American Medical Association endorsed its testing and possible use *(20–22).*

Injectable Contraceptives

A second agent widely used in more than 80 countries around the world is Depo-Provera. It is given by intramuscular injections and it is effective for 90 days. Medroxyprogesterone acetate is its synthetic progestin. Since it resembles progesterone, it produces very few undesirable side effects. Depo-Provera's contraceptive action is to block ovulation and alter the cervical mucus and uterine lining. After 1 year's use most women experience no menstruation for as long as use is continued. Menstruation ends much earlier for many women.

Numerous studies worldwide demonstrated Depo-Provera to be an extremely safe, potent, and effective contraceptive since it was introduced in the late 1960s. However, powerful lobbying groups in the United States stopped its approval by the FDA. The major concern is its long-acting effect. Because normal menstruation does not occur, infertility sometimes lasts up to 18 months after stopping the injections; for this reason, Depo-Provera is recommended only for women who do not want to become pregnant soon after stopping its use *(23).*

Intrauterine Devices

The intrauterine device (IUD), originally developed and used in the 1920s and 1930s, received revived interest in the 1960s. Most IUDs have been made of polyethylene that is coated with barium so they can be seen on X-ray films. Nylon tails extend from the uterus into the vagina for easy removal.

The most important advantage of the IUD over other methods is that after insertion by a physician, the user is generally unaware of it and need not adhere to a pill-taking schedule. The device can be left in place for some time. Once it is removed, fertility returns within a few days.

The IUD does not stop ovulation. But it does create an inflammatory response in the endometrium that prevents implantation of a fertilized egg. The addition of copper and progestins further enhances contraceptive effectiveness by altering cervical mucus and the uterine lining through continuous release of small amounts of these substances.

A rise in adverse side effects was caused by some of the earlier IUDs, particularly the Dalkon shield. All but one, Progestasert, were withdrawn from the United States market by 1986 because of increasing product liability lawsuits against the manufacturers. The most common adverse effects were expulsion, perforation of the uterus, irregular bleeding, uterine cramps, tubal infection, and PID. Improvements in the shapes, sizes, and materials led to a reduction in most of these reactions. The copper in some IUDs helped reduce tubal infections, but not PID. It is hoped that this alternative will become more acceptable to users who cannot tolerate, or choose to avoid, chemical methods of birth control.

Barrier and Sterilization Methods

Spermicidal (sperm-killing) agents, in the form of creams, foams, jellies, suppositories, and sponges, prove to be less effective than other methods because high motivation is required for routine use. Mechanical barrier methods, such as the condom and diaphragm, do not require a chemical agent since the mechanism of action is simply to block the passage of sperm. However, spermicidal jelly in conjunction with a diaphragm adds protection and lubrication. A new female condom, which fills the vagina and covers the cervix, has been developed primarily to protect against sexually transmitted diseases, namely, HIV, which causes AIDS *(24)*.

The most drastic methods of contraception are the sterilization techniques—tubal ligation (cutting the fallopian tubes) in women and vasectomy (cutting the *vas deferens*) in men. Sterilization is the number one contraceptive method chosen by couples over age 30. Both methods were once essentially irreversible, but modern microsurgery and laser techniques have greatly improved their reversibility.

26.10   Future Methods of Contraception

GOAL: To appreciate the potential use of immunization and tropic hormone analogs for contraception.

When use of oral contraceptives and intrauterine devices became widespread in the 1960s, it was hoped that new, more effective chemical contraceptive compounds would soon be marketed. The much-publicized side effects led to stringent screening of new agents, with consequent failure to produce new ones that meet rigid safety standards. Still, research continues.

Immunization against the pregnancy hormone HCG (Section 26.7) has been studied for more than 20 years. Results of clinical trials have been encouraging, but there is no evidence that a pregnancy hormone vaccine will appear soon. Natural production of the hormone as a result of pregnancy would evoke a memory response resulting in antibody production and destruction of the hormone. The goal is to create an immunization that would only last for 1-3 years. After that time, normal fertility would return, or a booster shot could be given to extend the contraceptive effects.

An antisperm vaccine for either men or women is also under study. Several proteins from sperm have been tested for antibody response in animals, but antibody production has been unreliable.

Another technique has undergone extensive trials for both men and women in Sweden and Canada. A synthetic analog of the stimulatory protein luteinizing hormone-releasing hormone (LHRH), also known as GnRH (Section 26.7), was given to women in high daily doses by injection or as a nasal spray *(14, 23)*. Since the LHRH analog is a protein, it is destroyed in the digestive tract if taken orally. Ovulation was inhibited during all but 2 of the 89 total treatment months. The two failures resulted from technical problems with the nasal spray. The synthetic hormone produced no significant side effects, and ovulation resumed rapidly after treatment was stopped.

The LHRH analog administered in daily injections or nasal spray also has shown promise as a male contraceptive. The drug suppresses sperm production to levels of infertility, and reduces testosterone production. Sperm density returns to normal levels after treatment is discontinued. Side effects such as loss of sex drive, hot flashes, and temporary impotence due to testosterone suppression continue to trouble researchers. Other analogs are under study, as are more convenient methods of administering the drug, such as a pill, suppository, or subdermal implant.

The hormone analog has a direct inhibitory effect on the pituitary and the release of gonadotropins in both women and men. Scientists are optimistic that it may eventually replace oral contraceptives as well as become the first male contraceptive *(25)*.

26.11   Chemical Methods of Male Contraception

GOAL: To be aware of the proposed methods for interfering with production, development, and transport of sperm, as well as the potential risks of certain methods.

Male contraceptives under development are even now mostly hypothetical. Three strategies have been proposed: interference with sperm production (spermatogenesis), inhibition of sperm development after spermatogenesis, and blockage of the transport of sperm to the woman. Existing methods involve the last of these strategies: physical means such as the condom and coitus interruptus, and surgical sterilization (vasectomy). Progress is very slow in male contraceptive technology because less is known about the male reproductive system compared with the female system. Even spermicidal jellies are questioned for the possibility that they might be responsible for birth defects if a surviving sperm fertilized an ovum.

The remaining method, inhibition of sperm production, is the subject of the greatest amount of research. An effective male oral contraceptive was developed in China and has been available there since 1972. The chemical agent *gossypol* is extracted from cottonseed oil. The nonsteroidal substance works by inhibiting an enzyme that is critical in the metabolism of sperm and sperm-producing cells of the testes. The enzyme is lactate dehydrogenase X, present only in sperm and testis cells. It does not affect the production of testosterone and, therefore, does not lessen a man's sex drive.

gossypol

Large-scale clinical studies of gossypol have shown that the drug is 99% effective in producing reversible infertility. Pills are taken for about 2 months until sperm counts are down. Thereafter, a maintenance dose is taken every 1-2 weeks for as long as contraception is desired. Fertility usually returns about 3 months after discontinuing the pills, although some research has shown that as many as 20% of men using gossypol experience permanent sterility. It is unlikely that gossypol will be available to men in the United States for some time because of concern over damage to sperm-producing cells, which may lead to sterility *(26)*.

Finally, a research study funded by the United Nation's World Health Organization reported that testosterone injections are 99% effective in male contraception. There are few side effects, making this a very promising development *(25)*.

SUMMARY

The 17-carbon steroid nucleus is the primary characteristic of all steroids. Two methyl groups, called angular methyl groups and located at C-10 and C-13, plus functional groups determine the biochemical properties of each steroid. An important steroid is cholesterol, which has a β-hydroxy group at C-3, a double bond in ring B, and an 8-carbon side chain at C-17.

Cholesterol is the precursor for many other important steroids. The bile acids, produced in the liver, assist in digestion of lipids by detergent action. Cortisone and related steroids are produced in the adrenal glands, as are glucocorticoids that affect carbohydrate metabolism, and mineralocorticoids for water and mineral balance. The sex hormones are also synthesized from cholesterol in the sex glands (ovaries and testes), the gonads. There are three types of sex hormones—progestogens (pregnancy hormones), androgens (male hormones), and estrogens (female hormones).

Cholesterol is a solid substance that would be insoluble in water and blood if it did not interact with high-density lipoproteins (HDLs) and low-density lipoproteins (LDLs) in the blood. The HDLs have a role in preventing, and LDLs in contributing to, heart disease. Diet affects primarily the LDL levels, whereas smoking, exercise, hormones, obesity, and alcohol consumption influence HDL levels.

A hormone is a substance (usually steroid or protein) that is released into the bloodstream from one site in the body and exerts its effects elsewhere. Release of steroid adrenal and sex hormones is controlled by protein hormones released by the hypothalamic region of the brain and by the pituitary gland. The pituitary hormones (FSH and LH) are called gonadotropic because they stimulate the gonads (ovaries and testes) to become active and release the sex hormones.

Testosterone (an androgen) is necessary for the development of the male reproductive system. Synthetic androgens and anabolic steroids have health benefits when prescribed in therapeutic dosages. Misuse of anabolic steroids by athletes has resulted in serious health consequences.

Progesterone (a progestogen) and estradiol (an estrogen) are necessary to maintain a woman's menstrual cycle. They also prepare the lining of the uterus for implantation of a fertilized egg (ovum). Progesterone also causes the glands of the cervix to produce a sticky secretion that blocks sperm and thus reduces the probability of fertilization. Since high progesterone levels also prevent ovulation, progesterone-like compounds are an important ingredient in all oral contraceptives.

The interplay between the gonadotropic hormones and the sex hormones is the basis for the action of most oral contraceptives on the market. Gonadotropic hormones are released during the early days of the female cycle during and after menstruation, and peak sharply at about midcycle. The sharp rise triggers ovulation, after which the levels of sex hormones rise. Elevation of the sex hormones suppresses further release of the gonadotropic hormones by feedback inhibition. The cycle ends in either menstruation or pregnancy. As menstruation occurs, the sex hormones drop to low levels and the gonadotropins rise in preparation for another cycle. In pregnancy, the level of sex hormones, particularly progesterone, remains high and suppresses the gonadotropins.

Since the early 1960s, several types of oral contraceptives have been developed. Combinations of synthetic estrogen and progestogen act by suppressing ovulation by feedback inhibition of the gonadotropic hormones, and by increasing cervical secretions. Improved combination pills use monophasic, biphasic, and triphasic dosage schedules. They mimic the natural cycle of hormone release, similar to sequential pills, which have been discontinued. The minipill acts only on the secretions of the cervix and uterine lining to prevent pregnancy. All of these agents have certain side effects.

Other methods of contraception are available, but the choice in America is limited compared with the rest of the world. The Norplant system of subdermal implants that last up to 5 years is the only new contraceptive method approved in the United States since the Pill in the 1960s. Contraception by injection and a morning-after pill, Depo-Provera and RU-486, respectively, may be approved in the future. Several alternatives are under investigation, including a vaccination for both women and men and a safe male birth control pill. A male pill using gossypol or testosterone may be adopted in the future.

PROBLEMS

1. Give a definition or an example of each of the following.
   a. Steroid nucleus
   b. Angular methyl group
   c. Bile acid
   d. Hormone
   e. Progestogens
   f. Androgens
   g. Estrogens
   h. Gonadotropins
   i. Feedback inhibition
   j. "Roid" rage
   k. Stacking
   l. Graafian follicle
   m. FSH and LH
   n. Anabolic steroid
   o. GnRH
   p. HCG
   q. Combination pill
   r. Norplant
   s. RU-486
   t. IUD

2. Draw the steroid nucleus and label the rings and carbon atoms.

3. How does one specify the location and configuration of functional groups attached to the steroid nucleus?

4. Since the basic chemical structure is the same for all steroids, what causes the various classes to function in different ways?

5. What substance is essential for the synthesis of many steroids required by the body?

6. Explain the differences between the two types of lipoproteins that transport most of the cholesterol that is carried in the blood.

7. Which steroid is the common precursor to those produced by the sex glands and adrenal glands?

8. Describe the effects that abuse of anabolic steroids have on the body. What are the benefits of therapeutic dosages of anabolic drugs?

9. Describe the source and function of the hormones in the menstrual cycle.

10. Draw the chemical structures for estradiol and testosterone. What is the main structural difference between these steriod hormones?

11. Draw the chemical structures for a synthetic estrogen and a progestin. What modifications of these molecules cause the enhanced chemical activity that makes it possible to take them orally?

12. Explain the strategy of each of the following including the intended advantages of each:
    a. Three dosing schedules of combination pills
    b. Minipill
    c. Norplant system

13. What is gossypol and how does it work as a male contraceptive?

14. What is the strategy of using LH-RH?

References

1. Roth, E. M.; Streicher, S. L. *Good Cholesterol, Bad Cholesterol;* Prima: Rocklin, CA, 1989; p. 339.

2. Hunninghake, D. B.; et al. "The Efficacy of Intensive Dietary Therapy Alone or Combined with Lovastatin in Outpatients with Hypercholesterolemia." *New England Journal of Medicine* **1993,** *328,* 1213–1219.

3. Matthews, K. A.; Meilahn, E.; Kuller, L.; Kelsey, S. F.; Caggiula, A. W.; Wing, R. R. "Menopause and Risk Factors for Coronary Heart Disease." *New England Journal of Medicine* **1989,** *321,* 641.

4. Thompson, P. D.; Cullinane, E. M.; Sady, S. P.; Chenevert, P. N.; Saritelli, A. L.; Sady, M. A.; Herbert, P. N. "Contrasting Effects of Testosterone and Stanozolol on Serum Lipoprotein Levels." *Journal of the American Medical Association* **1989,** *261,* 1165.

5. Brody, J. F. "Hailed and Feared Cortisone Now Safer and More Varied." *New York Times,* July 13, 1981.

6. Taylor, W. N. *Anabolic Steroids and the Athlete;* McFarland: New York, 1982.

7. Mohun, J. *Understanding Drugs: Drugs, Steroids, and Sports;* Watts: New York, 1988.

8. Breo, D. L. "Of MDs and Muscles—Lessons from Two Retired Steroid Doctors." *Journal of the American Medical Association* **1990,** *263,* 1697.

9. Bower, B. "Pumped Up and Strung Out: Steroid Addiction May Haunt the Quest for Bigger Muscles." *Science News,* July 13, 1991, p. 30.

10. Cowart, V. S. "Blunting 'Steroid Epidemic' Requires Alternatives, Innovative Education." *Journal of the American Medical Association* **1990,** *264,* 1641.

11. Janofsky, M. "Coaches Concede that Steroids Fueled East Germany's Success in Swimming." *New York Times,* December 3, 1991, (Section B) p. 15.

12. Nourse, A. E. *Birth Control;* Watts: New York, 1988.

13. Hays, C. L. "Confusion Renewed Over Pill and Risks." *New York Times,* January 18, 1989, (Section C) p. 1.

14. Kase, L. M. "The Pill Is Safe." *Self,* April **1992,** 134.

15. Djerassi, C. "The Bitter Pill." *Science* **1989,** *245,* 356.

16. Purvis, A. "A Pill that Gets Under the Skin." *Time,* December 24, 1990, p. 66.

17. Holzman, D. "Birth Control on the Five-Year Plan." *Good Housekeeping,* March 1992, p. 92.

18. "DES (Diethylstilbestrol)." *Good Housekeeping,* January 1975, p. 122.

19. Ullman, A.; Teutsch, G.; Philibert, D. "RU 486." *Scientific American,* June **1990,** 42.

20. Baum, R. M. "RU-486: Abortion Controversy in U. S. Clouds Future of Promising Drug." *Chemical and Engineering News,* March 11, 1991, p. 7.

21. Rosenfield, A. "Mifepristone (RU-486) in the United States." *New England Journal of Medicine* **1993,** *328,* 1560–1561.

22. "Here Comes RU-486." *Time,* May 3, 1993, p. 25.

23. Shapiro, H. I. *The New Birth Control Book;* Prentice-Hall: New York, 1988; pp. 256, 276.

24. Ilgen, L. "In Control." *Weight Watchers Magazine,* January 1989, p. 14.

25. "One Step Closer to a Pill for Men." *New Scientist,* June 5, 1986, p. 28.

26. Konig, U. "The Male Pill Is on Its Way." *World Press Review,* October 1991, p. 48.

Photo Credits

Chapter 1 p. 20, Courtesy of Thexton; **Chapter 2** p. 26 (top and bottom), The Granger Collection; p. 31, The Granger Collection; **Chapter 4** p. 65, Courtesy of U.S. Steel; p. 67, Ken O'Donoghue; p. 71, Courtesy of Ziebart; p. 72, Courtesy of William R. Stine; p. 78, Tom Hollyman/Photo Researchers, Inc.; **Chapter 5** p. 84, J. Kane; **Chapter 6** p. 99 (top and bottom), J. Kane; p. 100, J. Kane; p. 106 (top), Courtesy of J. Bohning, (bottom), Paul Silverman/Fundamental Photographs; p. 107, Courtesy of Mobil Oil Corporation; p. 111 (top and bottom), J. Kane; p. 116 (top and bottom), J. Kane; **Chapter 7** p. 124, J. Kane; p. 126, Richard Megna/Fundamental Photographs; p. 131, J. Kane; p. 136, E.R. Degginger; p. 137, Charlie Winters/Timeframe Photography; p. 140, Courtesy of Fisher Scientific; p. 145, Bruce Roberts/Photo Researchers; **Chapter 8** p. 152, Tony Stone Images; p. 154, Focus on Sports; p. 158 (top), Courtesy of Goodyear, (bottom figure, top left, right, and left), Courtesy of General Tire; p. 161, Courtesy of The Plastics Society; **Chapter 10** p. 194, Courtesy of The Nucleus; p. 195, Courtesy of Dosimeter Corporation; p. 196, The Granger Collection; p. 202 (bottom), Courtesy of EPA; p. 210, Photo Researchers; p. 212, Courtesy of Charles Nortmann; p. 213, Courtesy of National Institute of Health; p. 214, Courtesy of New England Nuclear; **Chapter 11** p. 227, Courtesy of Joint European Torus (JET) Joint Undertaking; p. 228, Courtesy of Pennsylvania Power and Light Company; p. 230 (inset and main photo), Courtesy Westinghouse Corporation; p. 233, Courtesy of General Electric Company; p. 234, Courtesy of General Electric Company; **Chapter 13** p. 278, Reprinted, with permission, from the Annual Book of ASTM Standards, © American Society for Testing and Materials, 1916 Race Street Philadelphia, PA 19103.; p. 288, Courtesy of Union Carbide Corporation; p. 291, Courtesy of Pacific Gas & Electric Company; p. 294, Courtesy of Siemens; p. 298, Courtesy of Consumer Power Company; p. 302 (left), Peter A. Simon/Photo-

take, (right), Courtesy of General Motors Inc.; **Chapter 14** p. 312, Field Museum of Natural History, GN 40263 & GN 83213.3; p. 327, Betz/Visuals Unlimited; **Chapter 15** p. 344, Grant Heilman; p. 346, Courtesy of Rodale Press; **Chapter 16** p. 359, Edward S. Ross; p. 365, Edward S. Ross; p. 366, Edward S. Ross; p. 367 (top, (a), (b), and (c)), Thomas Eisner, (bottom), Thomas Eisner & Daniel Aneshansky; p. 368, Edward S. Ross; p. 371, Department of Agriculture; p. 373, Department of Agriculture; p. 374, Edward S. Ross; **Chapter 17** p. 382, Courtesy of Anheuser-Busch Corporation; p. 383, Courtesy of The Boston Beer Company; p. 394 Courtesy of Scott Laboratories; p. 397, Richard Gross/The Stock Market; **Chapter 18** p. 404 (a), (b), (c), Courtesy of US Department of Agriculture; p. 408, J. Kane; p. 411 (a), (b), (c), Courtesy of Campbell Taggart, Inc.; **Chapter 19** p. 417, Courtesy of Sterling Winthrop; p. 418, Courtesy of Gaulin Corporation; p. 421, Courtesy of Switzerland Cheese Association; p. 422 (1–6), Courtesy of Pauly Cheese Company; **Chapter 20** p. 426, Courtesy of Presto Industries; p. 430, Courtesy of Brookhaven National Laboratories; p. 432, Courtesy of Adolph's Ltd.; p. 433, Courtesy of Percy A. Brown & Company; p. 435, J. Kane; p. 440, J. Kane; **Chapter 21** p. 453, Courtesy of Krafts Foods; p. 454, Courtesy Gaulin Corporation; p. 456, Courtesy of Ashland Chemical Company; p. 459, Courtesy of Kaukauna Cheese; **Chapter 22** p. 469, Fundamental; p. 471, Courtesy of Johnson & Johnson Dental Products Company; p. 472, Courtesy of Caulk Division/Dentsply; p. 477, J. Kane; p. 479, Courtesy of U.S. Department of Agriculture; p. 483 (top and bottom), J. Kane; **Chapter 23** p. 497, The Bettmann Archive; **Chapter 24** p. 509, Courtesy of Merck Company; p. 514, John Durham/Photo Researchers; **Chapter 25** p. 537, Robert Mathena/Fundamental Photographs; p. 557, Tony Stone Images; p. 565, Sovfoto/Eastfoto; **Chapter 26** p. 586, J. Kane; p. 587, Courtesy of Ortho/Novum; p. 593, Custom Medical Stock

Glossary

acid substance that turns litmus red; acid solutions in water have a pH less than 7; a proton donor (Brönsted-Lowry definition)

active niacin (NAD) a form of the B vitamin (niacin), which is often needed for the oxidation of organic compounds, e.g., in the fermentation of glucose

addition polymer a high-molecular-weight compound formed when unsaturated monomers combine by simple addition without the involvement of any functional groups other than the unsaturation

aerobic functioning in the presence of oxygen

agricultural lime *see* limestone

albumin water-soluble protein of blood plasma, or serum, muscle, the whites of eggs, milk, and other animal substances, and of many plant tissues and fluids

alkali metals elements in Group IA in the periodic table

alkaline having a pH greater than 7; basic

alkaline earth metals elements found in Group IIA of the periodic table

allelochemicals nonnutrient substances from one organism that affect the behavior and welfare of organisms of other species; also called allelochemics

allomone a chemical substance released by a plant or animal that provides an advantage to the producer, e.g., a defensive secretion

alloy a solution of solids

amalgam an alloy containing mercury

amino acids the repeating unit in proteins; this unit is characterized by having an amino group ($-NH_2$) and an acid group ($-COOH$)

amorphous having no definite shape

anaerobic functioning in the absence of oxygen

analgesia relief from pain

androgens male sex hormones

anion negatively charged particle

antibiotic drug that interferes with the growth of microorganisms

antibody protein molecule that provides immunity by inhibiting the growth of microorganisms

antigen substance that stimulates the release of antibodies

antihistamines chemicals that block the effects of histamine on body tissues

antimetabolite a chemical substance that interferes with normal metabolism

antimone a chemical substance released by a plant or animal that is harmful to both the producer and the recipient

antineutrino uncharged particle with zero mass that is released during β^- decay

antioxidant a chemical substance that prevents oxidation

antipyretic drug that reduces fever

antitoxin a chemical substance that acts against poisons (toxins)

aqueous relating to water

atherosclerosis hardening of the arteries mainly from fat deposition

atom the smallest unit of an element that possesses all the properties of that element

atomic number the number of protons in the nucleus

atomic weight the average relative mass of an element compared with a standard, i.e., carbon, with an atomic mass of 12 atomic mass units (amu); usually known as the average atomic mass

autoxidation oxidation by atmospheric oxygen

average atomic mass the average relative mass of an element compared with a standard, i.e., carbon, with an atomic mass of 12 atomic mass units (amu); sometimes called the atomic weight

β-galactosidase *see* lactase

bactericidal able to kill bacteria

bacteriostatic able to inhibit the growth of bacteria

base substance that turns litmus blue; basic solutions in water have a pH greater than 7; a proton acceptor (Brönsted-Lowry definition)

basic having a pH greater than 7; alkaline

Becquerel (Bq)   SI (System International) unit that describes the amount of radioactive material in a sample

bifunctional   containing two functional groups

biochemical oxygen demand (BOD)   amount of oxygen consumed by a sample during a five-day period at 20 °C

biodegradable   able to be decomposed by organisms in the biosphere

bioequivalent   having the same biological effect

biomass   living matter

biosphere   the part of the world in which life can exist

bitumen   a gooey organic material obtained from tar sands by heating

boiling point   the temperature at which a liquid is transformed into a vapor

bottled gas   a mixture of propane and butane that is removed from natural gas

brachytherapy   the technique of implanting radioactive material at a disease site

breeder reactor   a type of nuclear reactor that produces fissionable fuel

brine   water solution containing a very high concentration (>100,000 ppm) of dissolved minerals

British thermal unit   the amount of energy required to raise one gram of water by 1 °C

broad-spectrum antibiotic   drug capable of interfering with the growth of a wide variety of microorganisms

builder   substance added to detergents to tie up "hardness" ions

calorie   the amount of energy required to raise one gram of water by 1 °C

Calorie   1000 calories; also called kilocalorie

Cannabis sativa   the Indian hemp plant, the source of marijuana

caramelization   the heating of sugar to produce a brown amorphous substance

carbohydrates   any of various neutral compounds of carbon, hydrogen, and oxygen (like sugars, starch, cellulose); most are formed by green plants and constitute a primary class of animal food

carbonation   the addition of carbon dioxide, usually under pressure

carbonization   the process of producing char by removing volatile matter from coal

carat   the unit of mass used for precious stones, equal to 200 mg

caries   progressive destruction of bones or teeth

casein   a group of proteins found in milk

catalyst   substance that increases the speed of a reaction and is not consumed by the reaction

cation   positively charged ion

cellulose   a polysaccharide consisting of glucose units joined by beta linkages

char   the residue that remains following removal of volatile matter from coal

chemotherapy   treatment of disease with drugs (chemicals)

chlorination   treatment with chlorine

cis   a form of isomerism of unsaturated organic compounds in which two designated atoms or groups attached to the two carbon atoms of the double bond are on the same side of the molecule

cloning   the transfer and reproduction of DNA (deoxyribonucleic acid) in host cells

coenzyme   substance that acts as a catalyst in cooperation with an enzyme

cogeneration   a system for energy conversion in which a primary energy resource is consumed for production of both heat and electric power

compost   new topsoil produced by the decay of organic matter

compound   substance formed by combining elements

condensation polymer   a high-molecular-weight compound formed by reactions involving functional groups found in the monomers

congener   substance produced as a side product of fermentation

corpus luteum   an endocrine gland formed in the ovaries following the release of an ovum; it releases the hormone progesterone, which aids in the development of a fertilized ovum

cosmic radiation   radiation that penetrates the earth's atmosphere from outer space

covalent bond   a nonionic chemical bond formed by shared electrons

critical   in referring to a nuclear reactor, a condition in which an average of one product neutron becomes a bombarding neutron

critical mass   the weight of the isotope required for the reactor to achieve the supercritical condition in which more than one product neutron becomes a bombarding neutron

cross section   probability of capture of a particle by a nucleus

cryogenic   very low-temperature

culturing   growing an organism in a nutrient medium

Curie (Ci)   a unit that describes the amount of radioactive material that is found in a sample

demineralization   the breakdown of minerals, e.g., hydroxyapatite (tooth enamel)

density   the weight of a given volume of a substance

deoxyribonucleic acid (DNA)   a polymeric substance that is located in cell nuclei and contains the information to direct the functioning of the cell

destructive distillation   exposure to high temperatures without burning so that some components of a sample are vaporized

deuterium   the isotope of hydrogen that contains one neutron in the nucleus

dialysis   the separation of substances in solution by means of their unequal diffusion through semipermeable membranes

dilute   containing less solute than that required to reach saturation

disaccharide   a carbohydrate consisting of two monosaccharide units bonded covalently

disparlure   the sex attractant of the gypsy moth

divalent   having the capacity to combine with two other atoms

doubling time   referring to energy, the time during which consumption of an energy resource equals all previous use of the resource

effervescent   releasing bubbles of gas

electrolysis   the use of electric current to drive a chemical reaction

electron   elementary particle with a negative charge numerically equal to that of a proton

electronegativity   attraction of an atom for electrons in a molecule

element   substance that contains only one kind of atom

emulsifying agent   substance that stabilizes a suspension of two immiscible liquids, e.g., oil and water

emulsion   suspension of fine particles of one liquid in another liquid

endocrinology   study of glandular secretions

endothermic   energy-releasing

enteric drug   substance treated to pass through the stomach unchanged for ultimate disintegration in the intestines

entomology   study of insects

enzyme   protein that catalyzes chemical reactions in living organisms

equilibrium   a reversible reaction in a state where the rates of forward and reverse reactions are equal

estrogens   female sex hormones

eutrophication   the change of a body of water into a condition stimulating the growth of algae to the detriment of animal life

exothermic   energy-absorbing

exponential notation   shorthand notation used in writing very small and very large numbers, e.g., 1×10^6 = one million

fast reactor   a reactor that uses unmoderated bombarding particles

fermentation   breakdown of carbohydrates by microorganisms, such as the conversion of sugar to alcohol and carbon dioxide by the action of enzymes from yeast

fertile isotope   an isotope that can be used to breed a fissionable isotope

fission   the splitting of an atom as a result of bombardment of the nucleus by neutrons

fixed carbon   residue remaining following removal of volatile matter from coal

formula mass   the mass in atomic mass units of individual units, called formula units, of a substance that does not exist as discrete molecules, such as NaCl

fortification   addition of alcohol

functional group   an atom or a group of atoms other than carbon or hydrogen usually responsible for the chemical and physical properties of the compound

fusion   combination of two small nuclei to produce one larger nucleus

galvanization   coating the surface of iron with another metal to inhibit corrosion

gene   segment of a DNA molecule that is responsible for directing the synthesis of one protein

generic name   a technical, unsystematic name used to describe drugs and other products

genetic code   a three-letter code that is formulated from the chemical building blocks of DNA; it directs the placement of the 20 common amino acids in their proper sequence in proteins

genetics   branch of biology that deals with the heredity and variation of organisms

geometric isomers   molecules that differ in the three-dimensional arrangement of atoms, e.g., *cis* and *trans* isomers

glycolysis   metabolic breakdown of glucose

gonads   sex glands

gram-negative   refers to a kind of bacterium to which certain dyes will not adhere

gram-positive   refers to a kind of bacterium to which

certain dyes will adhere

Gray (Gy)   SI unit of absorbed dose

ground state   the lowest possible energy state of an atom or molecule

group   vertical row of elements in the periodic table

habituation   psychological dependence on a drug

half-life   length of time needed for half a sample of a radioactive isotope to decay

hallucinogen   substance that produces perceptions of things that do not exist

hard water   water that contains ions, e.g., Ca^{2+}, Mg^{2+}, that form insoluble compounds with soap

hard wheat   wheat that has a high content of the proteins that go into the making of high-gluten doughs for baking

hardening   the formation of a solid fat due to the addition of hydrogen

heat of vaporization   amount of energy (in calories) required to change 1 g of liquid to vapor at its boiling point

heavy water ^2H_2O or D_2O; water in which the element hydrogen is present as the isotope deuterium with one neutron in the nucleus

herbicides   weed killers

high-Btu gas   gas whose combustion releases 1000 Btu per standard cubic foot of gas

homogenization   process for reducing fat particles of an emulsion to uniform size

hormone   substance secreted by endocrine glands, carried by the blood, and used for regulating other tissues in the body

horsepower   unit of power equal to 746 watts

humectants   substances that help to retain moisture

hybrid   an offspring of two animals or plants of different species

hydration   the interaction between solute particles and water molecules

hydrocarbons   organic compounds containing only hydrogen and carbon

hydrogenation   addition of hydrogen

hydrolysis   the cleavage of a compound by water

hydrometer   a device used to measure density of liquids

hydronium ion H_3O^+

hydrophilic   having an attraction for water

hydrophobic   repelled by water

hydroxyapatite   tooth enamel; a crystalline substance containing ions of calcium, hydroxide, and phosphate; symbolized $Ca_5(PO_4)_3OH$

hypothalamus   portion of the brain that controls the function of many organs in the body and is involved in many sensations, e.g., thirst, hunger, blood pressure, body temperature

immune response   release of antibodies in response to invasion by an antigen

immunity   ability to resist development of pathogenic organism due to the acquired ability to produce antibodies that act against such an organism

immunochemistry   branch of chemistry dealing with immunity

immunology   branch of science dealing with immunity

inactive niacin (NADH)   a form of the B vitamin niacin, which is sometimes used directly in metabolism but is usually converted to active niacin (NAD)

inert gas   *see* noble gases

influenza   a virus disease; flu

invert sugar   an equal mixture of glucose and fructose, usually obtained by cleavage of sucrose

in vitro   in a test tube or other artificial environment outside the body

in vivo   in the living organism

iodine value   the number of grams of iodine, I_2, consumed by 100 g of a fat; usually determined by absorption of ICl

ion   an atom existing as a charged particle

ionizing radiation   any particulate or electromagnetic radiation that will cause a target substance to become charged

iron pyrites FeS_2; also called pyritic sulfur

isomer   one of two or more compounds having the same formula but a different arrangement of the atoms within the molecule

isotopes   atoms of the same element differing only in number of neutrons in the nucleus; nuclides

IUPAC   acronym for the International Union of Pure and Applied Chemistry, an organization that formulates rules for naming chemical compounds

joule   a unit of energy equal to 9.5×10^{-4} Btu or 0.24 calorie

kairomone   chemical substance released by a plant or animal that provides an advantage to the recipient, e.g., repellents form toxic substances

karat   the gold content of an alloy expressed in 24ths or parts per 24; for example, 12-karat gold contains 50% (m/m) gold

kerogen a waxy organic material obtained from oil shale by heating

lactose intolerance the inability to metabolize lactose due to the absence of lactase

lactase the enzyme required to metabolize lactose; also called β-galactosidase

lactic acid fermentation metabolism of glucose under anaerobic conditions to produce lactic acid

leavening the rising of baked goods from the release of CO_2, NH_3, or other gases trapped in the product before and during baking

legumes plants inhabited by microorganisms that fix nitrogen

light water H_2O; water containing only the isotope of hydrogen with no neutrons in the nucleus

lime calcium oxide, CaO; also known as quicklime

limestone calcium carbonate, $CaCO_3$; also known as "agricultural lime"

limit dextrins starch fragments produced when the amylases have acted to the limit of their ability

limiting reagent the reactant that is present in a molar amount that is not sufficient to cause complete reaction of all other reactants

low-Btu gas gas whose combustion releases 80–170 Btu per standard cubic foot of gas

lye sodium hydroxide, NaOH

mass the amount of material contained in a substance

mass number the sum of the number of protons and neutrons in the nucleus

medium-Btu gas gas whose combustion releases about 320 Btu per standard cubic foot of gas

messenger RNA a form of ribonucleic acid that carries a coded message to direct protein synthesis

metabolism the chemical changes in living cells by which energy is provided for vital processes and activities

metastasis secondary growth of a malignant tumor in another part of the body

micelle a cluster of molecules in solution, oriented to permit maximum interaction of water with the hydrophilic portion of the molecules and minimum interaction with the hydrophobic portion

microbiology study of microorganisms

micron 1×10^{-6} meter

mineralization formation of inorganic ions from organic compounds

miscible mutually soluble

moderator substance used to slow neutrons so they can come under the influence of nuclear forces and cause reactions

molar mass the mass in grams of one mole of a substance

molarity the concentration of a solution defined as the number of moles of solute dissolved in one liter of the solution

mole the quantity of a substance that contains the same number of particles or formula units as there are atoms of ^{12}C in exactly 12 grams of a pure sample of ^{12}C

molecular biology the field of science that is an outgrowth of genetics and biochemistry

molecular mass the sum of the masses of the atoms in a molecule expressed in atomic mass units

molecular weight *see* molar mass

molecule smallest unit of a compound that possesses all the properties of that compound

monoclonal antibodies antibodies isolated from a single antibody-producing cell line

monomers chemical building blocks of polymers

monosaccharide a sugar that cannot be cleaved into simpler sugars

monovalent having the capacity to combine with one other atom

must juice from grapes before fermentation

mutation change in the amino acid sequence in a protein causing the appearance of a hybrid form of the organism

narcotic substance that causes sedation (narcosis) and relieves pain

narrow spectrum antibiotic a drug capable of interfering with the growth of only a few microorganisms

negatron a negative electron

neoplasm abnormal growth or tumor

neuron a nerve cell

neurotransmitters chemicals involved in transmitting nerve impulses

neutralization reaction of an acid and a base to produce a salt

neutrino uncharged particle with zero mass that is released during β^+ decay

neutron uncharged particle found in the nucleus with a mass nearly equal to that of a proton

nitrification the conversion of ammonia, NH_3, to nitrate, NO_3^-

noble gases any of the inert or nonreactive gases: helium, neon, argon, krypton, xenon, and radon, which appear in group VIIIA of the periodic table

nuclear fission the splitting of an atomic nucleus accompanied by the release of large amounts of energy

nuclear fusion the union of atomic nuclei to form heavier nuclei accompanied by the release of large quantities of energy when certain light elements unite

nuclides *see* isotopes

octet rule any element is most stable when it has a set of eight valence electrons

orbital a region in space where there is a high probability of finding an electron

osmosis the flow of solvent molecules through a semipermeable membrane to bring two solutions on either side of the membrane to equal concentration

ovulation release of an egg from the ovaries

oxidation combination of a substance with oxygen; loss of electrons; dehydrogenation

oxidative phosphorylation combination of the processes by which active niacin is regenerated and ATP is formed; phosphate + NADH + O_2 + ADP → NAD + H_2O + ATP

oxygenation treatment with oxygen

pandemic global epidemic

parts per million (ppm) number of parts of a component of a sample per million parts of the total sample

passive immunity type of temporary immunity acquired by transferring preformed antibodies to a person to provide protection against a particular infection

pathogenic causing disease

penicillinase an enzyme that breaks down penicillin molecules

period a horizontal row of elements in the periodic table

pH a numerical scale used to describe the level of acidity or basicity of a water solution

pharmacology study of the action of drugs on the body

pheromone chemical compound used for communication among animals, e.g., a sex attractant

photosynthesis process by which chlorophyll-containing plants use energy from the sun to convert carbon dioxide and water into oxygen and sugar

plaque collection of microorganisms on teeth

plasticizer substance that can be added to a rigid polymeric material in order to impart flexibility

plastic range temperature range in which a fat product retains its shape but is readily deformed

polymer substance of high molecular weight, synthesized from repeating units called monomers

polypeptide a polymer with amino acids as monomer units

polyunsaturated fat triglycerides containing a high level of unsaturated fatty acids

positron a positive electron

potash potassium carbonate, K_2CO_3; also the K_2O content of a fertilizer

power the rate of delivery of energy

precursor compound that serves as a source of another compound, e.g., glucose as a precursor to alcohol via fermentation

prodrug inactive form of the drug that is converted to the active drug after it is consumed

proof the concentration of alcohol double the percent by volume

protein a polymer with amino acids as monomer units

proteolytic protein-cleaving

protium an isotope of hydrogen with no neutrons in the nucleus

proton a positively charged particle in an atomic nucleus

psychedelic conveying intensely pleasureful sensual perception

pyritic sulfur FeS_2; also called iron pyrites

quad metric prefix equal to 10^{15}; often used as a unit of energy equal to 10^{15} Btu

rad unit of radiation absorbed dose

radical form of a compound resulting from the removal of a hydrogen atom

radionuclide radioactive isotope

radiopharmaceutical radioactive drug

reduction gain of electrons; hydrogenation

rem unit of absorbed dose expressed in terms of the biological effect of radiation, which varies from one type of tissue to another and from one type of radiation to another

remineralization formation of tooth enamel (hydroxyapatite)

rendering isolation of a fat from an animal source

resin the polymeric component of a plastic

resistance ability to avoid alteration by an enzyme that causes a chemical change in that substance

respiration the use of oxygen by living organisms to produce energy

reverse transcription the synthesis of DNA using

information coded in RNA

reversion spoilage of a fat by oxidation

roentgen unit describing the quantity of radioactivity emitted by a sample (only applicable to X− or gamma radiation)

ruminant a cud-chewing mammal that has several stomach compartments, one of which is called the rumen

salometer unit used to describe salt concentration; 1% NaCl = 4 ° salometer

salt compound containing a cation and an anion

saturated containing only carbon-carbon single bonds; solution that contains all the solute it can dissolve at a given temperature

sequestering agent substance that binds metal ions

smelting process for extracting metals from their ores that includes melting

soft radiation radiation of very low energy

soft wheat wheat with a low content of the proteins that go into the making of high-gluten doughs for baking

solute substance dissolved in a solvent

solution homogeneous mixture of two or more substances

solvation the interaction between solute particles and solvent molecules

solvent substance that dissolves another substance

specific energy the amount of energy that can be stored per kilogram of battery weight

specific gravity the density of a substance compared with another used as a standard

specific heat the amount of energy (in calories) needed to raise the temperature of one gram of a substance 1 °C

spectator ions ions present as counterions, but which do not undergo a change during a reaction

spermicide substance capable of killing sperm

starch a polysaccharide consisting of repeating units of glucose

steady state the condition under which the rate of formation and the rate of decay of a substance are equal

stratosphere layer of atmosphere ranging from 10 to 30 miles into space

surfactant substance that is both hydrophobic and hydrophilic and can bring about mixing of hydrophobic and hydrophilic substances

synapse a region of close approach of two neurons where signals are transmitted by neurotransmitters

syndet synthetic detergent

synergist substance that works in cooperation with another

synomone a chemical substance released by a plant or animal that provides an advantage to the producer and the recipient

teletherapy exposure to an external source of radiation

teratogen substance that causes birth defects

tetravalent having the ability to combine with four other atoms

thermal neutrons slow neutrons

thermodynamics the study of heat, work, and energy, and the changes they produce

thermoplastic resin a polymeric material that can be heated to melting and restored to its original chemical form by cooling

thermosetting resin a polymeric material that cannot be heated to melting without destruction of its chemical structure; contains cross-links between polymer chains

tolerance ability to tolerate increasing amounts of a drug without ill effects; condition of requiring increasing amounts of a drug to experience the same effects

toxin a poison

toxoid an inactivated toxin

tracers substances that can be used to monitor the movements of chemical compounds

trans a form of isomer of unsaturated organic compounds in which two designated atoms of groups attached to two carbon atoms of a double bond lie on opposite sides of the molecule

transition elements metallic elements in the center portion of the periodic table in the B groups

transcription synthesis of messenger RNA using information coded in DNA

translation the synthesis of protein using information coded in messenger RNA

transmutation conversion of one element to another either naturally or artificially

transuranium elements elements following uranium in the periodic table

triglyceride a triester resulting from the combination of glycerin and three fatty acid molecules

tritium an isotope of hydrogen with two neutrons in the nucleus

trivalent having the capacity to combine with three other atoms

troposphere layer of atmosphere ranging from

ground level to 10 miles into space

unsaturation   double or triple bonds; solution that contains less solute than it can dissolve at a given temperature

valence   combining capacity of an atom

valence electrons   electrons found in the outermost shell of an atom

volatile   having a low boiling point

volatile matter   components of a sample that can be removed by heating

vulcanization   the treatment of rubber with sulfur to introduce cross-links between polymer chains

watt   unit of power equal to 3.412 Btu per hour

weight   the force with which gravity attracts a body to the earth

whey   the liquid remaining after separation of the curd from milk

wood alcohol   methanol or methyl alcohol

Working Level (WL)   unit that describes the level of radon gas in an air sample

wort   the product of the mashing step in brewing

Answers to Selected Even-Numbered Problems

Chapter 1

2. a. 227 g
   b. 2.6 qt
   c. 161 km
   d. 16.4 ft
   e. 1.06 qt
   f. 2.2 lb
   g. 39.4 in.
4. a. 2×10^{-2}
   b. 8.3×10^{-4}
   c. 9.07×10^2
   d. 0.0096
   e. 18,500
6. a. 15 °C
   b. 14 °F
   c. 298 K; 323 K; 258 K

Chapter 2

2. a. atoms of the same element differing in the number of neutrons in the nucleus
   b. the isotope of hydrogen with one neutron in the nucleus
   c. the isotope of hydrogen with two neutrons in the nucleus
   d. the average of the masses of the isotopes of any element, taking account of the natural abundance of each isotope
   e. same as average atomic mass
   f. compounds containing nitrogen and oxygen that vary in the ratio of the two elements
   g. a low-energy form of an atom or other particle
   h. a high-energy form of an atom or other particle
   i. an electron found in the highest energy level (valence level) of an atom
4. a. The law of conservation of mass confirms the existence of atoms, which are not created or destroyed during chemical reactions.
   b. The law of definite proportions confirms the existence of atoms by describing the limitations on how atoms combine to form molecules.
   c. The law of multiple proportions, like the law of defi-nite proportions, confirms the existence of atoms by describing the limitations on how atoms combine to form molecules.
6. atomic number; average atomic mass
   a. 6; 12.01
   b. 13; 26.98
   c. 56; 137.3
   d. 92; 238.0
8. The average atomic mass of an element reflects the natural abundance of the isotopes of an element.
10. Each model places electrons in levels of increasing energy. The Bohr model places electrons in circular orbits around the nucleus. In the wave mechanical model, the electrons are found in regions of space called orbitals that are calculated as having the highest probability (lowest energy) of containing electrons. The shapes of the orbitals vary.
12. a. regions in space that contain electrons
    b. s, p, d, f
    c. when there is only one orbital in a sublevel
    d. when there is more than one orbital in a sublevel
14. a. 7; nitrogen
    b. 12; sodium
    c. 10; fluorine
    d. 20; potassium
    e. 18; chlorine
16. a. 2 e⁻ in first level; 4 e⁻ in second level
    b. 2 e⁻ in first level; 7 e⁻ in second level
    c. 2 e⁻ in first level; 8 e⁻ in second level; 1 e⁻ in third level
    d. 2 e⁻ in first level; 8 e⁻ in second level; 3 e⁻ in third level
    e. 2 e⁻ in first level; 8 e⁻ in second level; 8 e⁻ in third level; 2 e⁻ in fourth level
18. a. $4p^3$
    b. $4s^2$
    c. $3d^5$
    d. $5p^5$
    e. $5f^3$
    f. $4p^6$
20. a. $1s^2 2s^2$
    b. $1s^2 2s^2 2p^6 3s^2 3p^6 4s^2 3d^{10} 4p^5$
    c. $1s^2 2s^2 2p^6 3s^2$

d. $1s^22s^22p^63s^23p^64s^23d^5$
e. $1s^22s^22p^63s^23p^64s^23d^{10}4p^5$
f. $1s^22s^22p^63s^23p^64s^23d^{10}4p^65s^24d^{10}5p^66s^24f^{14}5d^{10}$
g. $1s^22s^22p^63s^23p^64s^1$
h. $1s^22s^22p^63s^23p^64s^23d^{10}4p^65s^24d^{10}5p^66s^24f^{14}5d^{10}6p^2$

Chapter 3

2. a. the combining capacity of an atom or ion
b. an electron in the outermost (highest) energy level
c. a horizontal row of the periodic table
d. a vertical row of the periodic table
e. An atom is most stable when its valence level contains eight electrons.
f. an element in group VIIIA, with a filled valence level
g. a group IA element, such as Na
h. a group IIA element, such as Mg
i. a group VIIA element, such as Cl
j. an element lying between groups IIA and IIIA, such as Fe
k. H_2, N_2, Cl_2
l. an ion formed by combining two or more elements, such as SO_4^{2-}
m. Consult the periodic table.
n. Consult the periodic table.
o. Consult the periodic table.
p. the relative attraction of an atom for electrons in a molecule
q. three covalent bonds between two atoms, as in $N\equiv N$
4. Other elements achieve noble gas structures by gaining or losing valence electrons, including by covalent bond formation.
6. All have seven valence electrons and tend to form monovalent ions (e.g., Br^-) or exist in monovalent combinations with other elements (e.g., HBr).
8. Group number equals valence for the A group elements, except for a few elements that have different ionic forms, such as lead and tin.
10. by loss of electron(s) to form a cation, such as Mg^{2+}, or gain of electron(s) to form an anion, such as Cl^-; by formation of a covalent bond, such as in HCl
12. a. ICl
b. BaO
c. NH_4Br
d. H_2S
e. $CuNO_2$
f. $AgNO_3$
g. $KHCO_3$
h. $Al_2(SO_4)_3$
i. SnO_2
j. KH_2PO_4
k. $Mg(CN)_2$
l. $CaCl_2$

m. $SrCO_3$
n. $NaNO_2$
o. $NaHSO_4$
p. SO_3
q. $SiCl_4$
r. $Al(OH)_3$
s. $Fe_2(SO_4)_3$
t. $(NH_4)_2SO_3$

Chapter 4

2. a. CH_4 (g) + 2 O_2 (g) → CO_2 (g) + 2 H_2 (g)
b. 2 H_2 (g) + O_2 (g) → 2 H_2O (l)
c. NaOH (aq) + HCl (aq) → NaCl (aq) + H_2O (l)
d. 2 $KClO_3$ (s) → 2 KCl (s) + 3 O_2 (g)
e. Ag^+ (aq) + Cl^- (aq) → AgCl (s)
f. Ca (s) + 2 H_2O (l) → H_2 (g) + $Ca(OH)_2$ (s)
g. 3 H_2 (g) + N_2 (g) → 2 NH_3 (g)
h. 2 SO_2 (g) + O_2 (g) → 2 SO_3 (g)
i. 2 Mg (s) + O_2 (g) → 2 MgO (s)
j. $Al(OH)_3$ (aq) + 3 H_2SO_4 (aq) → Al_2SO_4 (aq) + 6 H_2O (l)
k. Fe_2O_3 (l) + 3 CO (g) → 2 Fe (l) + 3 CO_2 (g)
l. Cu (s) + 2 $AgNO_3$ (aq) → Ag (s) + $Cu(NO_3)_2$ (aq)
4. Neither oxidation nor reduction can occur without the other.
6. light weight and resistance to corrosion (oxidation)
8. As zinc cations form at the anode, sulfate ions are needed to maintain a balance of charges. Likewise, copper cations are lost at the cathode so that an excess of anions develops in the cathode compartment.
10. Cations are required in solution to support the formation of anions (e.g., OH^-) at the cathode. Similarly, anions must be present to support the formation of cations (e.g., H^+).
12. A tin coating protects iron from corrosion since tin is more susceptible to oxidation.
14. Oxidation at an aluminum surface forms a coating of aluminum oxide that adheres strongly and protects against further oxidation.
16. Anions encourage formation of Fe^{2+}. Cations encourage formation of hydroxide ions.

Chapter 5

2. a. 34 amu; 34 g
b. 111 amu; 111 g
c. 46 amu; 46 g
d. 32 amu; 32 g
e. 342 amu; 342 g
f. 188 amu; 188 g
4. a. 3.6 g
b. 191 g

c. 1104 g

d. 5.9 g

6. a. $2\ AgNO_3 + MgCl_2 \rightarrow 2\ AgCl + Mg(NO_3)_2$

   b. 10 mol

   c. 3.5 mol

   d. 86.8 g

   e. 144 g

8. a. 5.0 mol

   b. 5.4 g

10. a. $2\ SO_2 + O_2 \rightarrow 2\ SO_3$

    b. 1 mol

    c. 2 mol

    d. SO_2

    e. O_2; 1.5 mol; 48 g

    f. 1 mol SO_2; 0.5 mol O_2

    g. 64 g SO_2; 16 g O_2

    h. additional 192 g; 256 g (4 mol) total

12. a. NaCl

    b. 0.16 mol NaCl; 0.08 mol $Pb(NO_3)_2$

    c. 9.28 g NaCl; 26.48 g $Pb(NO_3)_2$

    d. 22.24 g

14. a. 0.02 mol

    b. 2.2 g

16. a. 6.9 g

    b. 5.4 g

18. 72 g

20. a. H_2

    b. 7.0 g (3.5 mol)

    c. 9.0 g (0.5 mol)

22. 142 g

24. a. 0.15 mol

    b. 0.30 mol

    c. 0.2 mol

    d. HCl; 0.1 mol

    e. 0.15 mol Zn; 0.30 mol HCl

    f. 0.15 mol

    f. 2.7 g

Chapter 6

2. a. hexane

   b. 2-methylhexane

   c. 4-chloro-2-methylhexane

   d. 2,4-dimethylhexane

   e. 4-chloro-2,2-dimethylpentane

   f. 2,2,4-trichloropentane

   g. *cis*-2-pentene

   h. *trans*-2-pentene

   i. l-butene

4. a. 2-methylheptane

   b. 4-ethyl-2-methyloctane

   c. *trans*-2-pentene

   d. 2,3-dimethylhexane

e. *trans*-3-octene

f. *cis*-5,6-dimethyl-2-heptene

g. *cis*-4-octene

h. *trans*-4,4-dimethyl-2-hexene

6. $CH_3(CH_2)_{10}CH_3$, $CH_3(CH_2)_{12}CH_3$, $CH_3(CH_2)_{14}CH_3$, $CH_3(CH_2)_{16}CH_3$, along with the odd-numbered, unbranched hydrocarbons containing 9–19 carbon atoms and many branched hydrocarbons with boiling points ranging 175–325 °C (Table 6.3)

8. The tetrahedral angle (109.5 °) provides maximum separation of the four atoms attached to carbon.

10. The highly branched structure leads to a slower, smoother burning.

12. Benzene is a triene; cyclohexane is saturated.

14. —OH versus —OR

16. Natural gas is almost exclusively methane with a small amount of ethane; bottled gas contains mostly propane and butane.

Chapter 7

2. Due to the large difference in electronegativities of hydrogen and oxygen, the bonds are very polar. The dipoles in separate molecules attract one another, causing the molecules to cluster together and raise the boiling point. Furthermore, many polar compounds are soluble in water due to hydrogen bonding.

4. The dipoles (described for problem 2) attract sodium and chloride ions; Na^+ to the oxygen atoms and Cl^- to the hydrogen atom.

6. Oil is nonpolar; water is polar.

8. The solubility of a gas in a liquid decreases sharply when the solvent freezes, and could cause a buildup in pressure.

10. 1% (m/m) = 10^4 ppm; therefore, 1 ppm = 10^{-4}% (m/m) = 10^3 ppb

12. 0.200 g

14. They are identical; both consist of a solution of ammonia gas in water containing a small amount of ammonium and hydroxide ions.

16. Ammonium hydroxide releases ammonia gas (NH_3); hydrochloric acid releases hydrogen chloride (HCl). The gases react to form solid NH_4Cl.

18. red, lemon juice; blue, milk of magnesia; red, vinegar; blue, ammonia water; blue, $NaHCO_3$; blue, Na_2CO_3; red, H_2CO_3

20. Concentration describes the amount of solute present; strength describes the ability of the solute to separate into ions.

22. Low pH is not accurate for such a product.

24. 10 g

26. 1.675 g NaCl + water to a total volume of 250 mL

28. 1492 mL

30. 990 g

32. $4.4 \times 10^{-8}\%$ (m/m)
34. 1.0 g

Chapter 8

2. Addition polymers form by reaction at double bonds. Condensation polymers form by reaction of functional groups, and a small molecule, e.g., water, is usually released during polymerization.
4. $(CH_2CHCN)_n$ for polyacrylonitrile
6. foams in many applications
8. Twenty different monomers (amino acids) must be assembled in exactly the right sequence.
10. The polypeptide may be defective or even nonfunctional.
12. a set of three-letter signals (found in mRNA) that direct the placement of the 20 amino acids during polypeptide synthesis
14. Transcription is the synthesis of mRNA using the sequence of nucleotides in DNA to direct the synthesis. In reverse transcription, RNA directs the synthesis of DNA. Translation is the synthesis of polypeptides using the sequence of nucleotides in mRNA, i.e., the genetic code, to direct the synthesis.

Chapter 9

2. a. lactose
  b. sucrose
  c. maltose
  d. amylose or cellulose
  e. amylopectin or glycogen
4. Cotton is almost pure cellulose and the beta linkages make it indigestible to humans. Ruminants have microbes with enzymes necessary to cleave β linkages.
6. Read some labels.

Chapter 10

2. a. 22 p, 35 n, 27 e; copper cation
  b. 79 p, 119 n, 79 e; gold atom
  c. 92 p, 143 n, 92 e; uranium atom
  d. 53 p, 78 n, 54 e; iodide ion (anion)
  e. 50 p, 63 n, 50 e; tin atom
  f. 80 p, 123 n, 78 e; mercury(II) cation
  g. 33 p, 41 n, 33 e; arsenic atom
  h. 31 p, 37 n, 28 e; gallium cation
  i. 38 p, 52 n, 36 e; strontium cation
  j. 9 p, 10 n, 10 e; fluoride ion (anion)
4. a. $^{241}_{94}Pu \rightarrow {}^{241}_{95}Am + {}^{0}_{-1}\beta$
  b. $^{15}_{8}O \rightarrow {}^{15}_{7}N + {}^{0}_{+1}\beta$
  c. $^{232}_{90}Th \rightarrow {}^{228}_{88}Ra + {}^{4}_{2}He$
  d. $^{59}_{28}Ni + {}^{0}_{-1}e \rightarrow {}^{59}_{27}Co$
  e. $^{74}_{35}Br + {}^{0}_{-1}e \rightarrow {}^{74}_{34}Se$

  f. $^{59}_{26}Fe \rightarrow {}^{59}_{27}Fe + {}^{0}_{-1}\beta$
  g. $^{205}_{82}Pb \rightarrow {}^{205}_{81}Tl + {}^{0}_{+1}\beta$
6. Soft radiation (most α and β) has too much bulk and too little energy to penetrate a radiation counter; liquid scintillation counting avoids this problem.
8. 7
10. The event triggered worldwide investigation of radon.
12. Consult Figure 10.10.
  a. $^{238}_{92}U \rightarrow {}^{234}_{90}Th + {}^{4}_{2}He$
  b. $^{226}_{88}Ra \rightarrow {}^{222}_{86}Rn + {}^{4}_{2}He$
  c. $^{222}_{86}Rn \rightarrow {}^{218}_{84}Po + {}^{4}_{2}He$
  d. $^{210}_{84}Po \rightarrow {}^{206}_{82}Pb + {}^{4}_{2}He$
14. Isotopes of polonium (^{218}Po, ^{214}Po, ^{210}Po), bismuth (^{214}Bi, ^{210}Bi), and lead (^{214}Pb, ^{210}Pb) are the radon daughters. Consult Figure 10.10.
16. an incremental risk of 1% to 3% lung cancer deaths with long-term exposure
18. Radon enters basements from surrounding soil; it is more dense than air; basement ventilation is usually poor. In colder weather, closed-up houses trap radon gases, and higher inside temperatures create a chimney effect.
20. MACS allows a determination of the age of samples up to 100,000 years old, whereas traditional carbon-14 dating has a limit of usually less than 50,000 years.
22. Only gamma rays are normally able to penetrate body tissues from an external source. If an alpha-emitting isotope can be deposited at a site of disease, the effects of the radiation will be most effective due to the very short-range penetration.

Chapter 11

2. a. $^{2}_{1}H + {}^{3}_{1}H \rightarrow {}^{4}_{2}\alpha + {}^{1}_{0}n$
  b. $^{6}_{3}Li + {}^{1}_{0}n \rightarrow {}^{4}_{2}\alpha + {}^{3}_{1}H$
  c. $^{235}_{92}U + {}^{1}_{0}n \rightarrow {}^{94}_{37}Rb + {}^{140}_{55}Cs + 2\,{}^{1}_{0}n$
  d. $^{27}_{13}Al + {}^{2}_{1}H \rightarrow {}^{25}_{12}Mg + {}^{4}_{2}\alpha$
  e. $^{235}_{92}U + {}^{1}_{0}n \rightarrow {}^{138}_{56}Ba + {}^{95}_{36}Kr + 3\,{}^{1}_{0}n$
  f. $^{238}_{92}U + {}^{1}_{0}n \rightarrow {}^{239}_{92}U$
  g. $^{99m}_{43}Tc \rightarrow {}^{99}_{43}Tc + \gamma$
  h. $^{99}_{42}Mo \rightarrow {}^{99m}_{43}Tc + {}^{0}_{-1}\beta$
  i. $^{98}_{42}Mo + {}^{1}_{0}n \rightarrow {}^{99}_{42}Mo + \gamma$
  j. $^{108}_{49}In + {}^{0}_{-1}e \rightarrow {}^{108}_{48}Cd + \gamma$
  k. $^{197}_{79}Au + {}^{1}_{0}n \rightarrow {}^{198}_{79}Au + \gamma$
  l. $^{27}_{13}Al + {}^{1}_{0}n \rightarrow {}^{25}_{12}Mg + {}^{1}_{1}H$
  m. $^{99}_{43}Tc + {}^{1}_{0}n \rightarrow {}^{100}_{43}Tc$
    $^{100}_{43}Tc \rightarrow {}^{100}_{44}Ru + {}^{0}_{-1}\beta$
4. Due to the lack of charge, neutrons enter nuclei easily and tend to pass through unaffected. Slower neutrons are usually more readily captured into a nucleus. Because charged particles have great difficulty penetrating a positively charged nucleus or a cloud of electrons, they require acceleration.
6. Heavy water captures few neutrons, allowing for less

enriched fuel, but it is far more expensive.

8. a. $^{235}_{92}U + ^{1}_{0}n \rightarrow ^{131}_{53}I + ^{103}_{39}Y + 2\,^{1}_{0}n$

 b. $^{235}_{92}U + ^{1}_{0}n \rightarrow ^{138}_{56}Ba + ^{95}_{36}Kr + 3\,^{1}_{0}n$

10. Radioactive decay of fission products accounts for about 7% of the energy released in a reactor during and after fission. This decay continues even after a reactor has been shut down.

12. To increase the rate of fission, neutron-absorbing control rods would be removed from the reactor.

14. The α particles released during fusion carry about 20% of the energy of fusion and remain in the reactor as a source of energy for additional fusion reactions.

16. At the time of the accident, the fuel was near the end of its useful lifetime; therefore the level of radioactive fission products was at a maximum. The core was uncovered when the stuck-open PORV and indications of low pressure led the operators to decrease the supply of cooling water. Steam bubbles that formed caused the core to be uncovered.

18. A negative void coefficient presents the danger of a runaway reactor since a loss of coolant allows fission to continue and even accelerate.

22. The common feature of both is the absence of a critical mass prior to detonation. A nonnuclear explosive forms a critical mass in both designs.

24. for a sample with a radius = 5 cm

$$\frac{V}{SA} = \frac{523 \text{ cm}^3}{314 \text{ cm}^2} = 1.67 \text{ cm}$$

for a sample with a radius = 25 cm

$$\frac{V}{SA} = \frac{65,416 \text{ cm}^3}{7,850 \text{ cm}^2} = 8.33 \text{ cm}$$

Chapter 12

2. *primary:* fossil fuels (coal, gas, petroleum, shale oil, tar sands), nuclear fission and fusion, solar energy, geothermal energy, wind, falling water; *secondary:* electricity and hydrogen

4. The kilowatt (kW) is a unit of power; the kilowatt-hour (kW-h) is a unit of energy. It takes 3 kW-h$_t$ to generate 1 kW-h$_e$ due to low energy conversion efficiency.

6. a. 1 Q (10^{15} Btu)

 b. 1 kW-h$_e$ (3 kW-h$_t$)

 c. They are equal.

 d. 2.52×10^{19} cal (100 Q)

 e. 1000 kW-h$_e$ (3.4×10^6 Btu)

 f. 10^6 kJ (2.4×10^2 cal)

8. from 1950 (31 Q), about 19 years (1969 @ 62 Q); from the 1979 peak, current use is back to about the same levels after a decline in the early 1980s

10. (20.49 Q/29.70 Q) $\times$ 100 = 69% lost (See also Figure 13.9.)

12. See Figure 12.7.

14. For older plants (operating before 1979), no more than 4 lb of SO_2 may be released per 1 million Btu consumed. For newer plants, 1.2 lb of SO_2 may be released per 1 million Btu consumed, plus reduction of SO_2 emissions by 70% or more.

16. Only high-Btu gas can be economically transferred through pipelines over long distances.

18. as water (H_2O) in the form of steam

20. $SO_2 + H_2O + CaCO_3 \rightarrow CaSO_3 + H_2CO_3 \rightarrow H_2O + CO_2$

Chapter 13

2. *liquids:* synthetic liquids from coal, methanol, ethanol, shale oil, oil from tar sands; *gases:* low-, medium-, and high-Btu gases from coal, hydrogen

4. small, branched, and aromatic hydrocarbons, oxygenates (particularly MTBE and ETBE)

6. to inhibit corrosion and stabilize the blend against separation caused by water

8. easier starting, discourages ingestion, less hazardous fire due to color and smoke

10. natural gas, electrolysis of water

12. compact storage of hydrogen

14. on the Rance River in France

16. *dry cell:* common battery used in flashlights and small appliances (not to be confused with the alkaline battery), electrons released by oxidation of zinc container; *lead storage battery:* electrons released by oxidation of lead

18. Secondary cells (e.g., the lead storage battery) can be recharged; primary cells (e.g., the dry-cell battery) cannot be recharged.

20. The zinc can is oxidized as the battery discharges.

22. At 35 W-h/kg, the lead storage battery would weigh more than 11 times (400 ÷ 35) as much as gasoline of equivalent energy content.

Chapter 14

2. From burning of fossil fuels, particularly coal-containing sulfur; the oxides of sulfur ultimately become sulfuric acid which is irritating to the throat and lungs; $CaCO_3 + H_2SO_4 \rightarrow CaSO_4 + H_2CO_3$; then $H_2CO_3 \rightarrow CO_2 + H_2O$

4. aerosol propellants, blowing agents, fire extinguishing materials, solvents

6. once started, two reactions form a chain: $Cl\bullet + O_3 \rightarrow ClO\bullet + O_2$; $ClO\bullet + O \rightarrow Cl\bullet + O_2$

8. $Cl\bullet + O_3 \rightarrow ClO\bullet + O_2$; the presence of $ClO\bullet$ indicates that destruction of ozone has occurred.

10. because of dissolved gases and minerals, some of which are actually desirable

12. It is caused by water-containing cations, such as calcium and magnesium, which can be precipitated by heating as $CaCO_3$ and $MgCO_3$. The cations form soap scums and deposit in boilers and kettles.
14. Two ways are oxygenation and chlorination; chlorination may form dangerous chlorinated organic compounds. There are no drawbacks to oxygenation.
16. Oxygen is required for the formation of ''yellow boy.''

Chapter 15

2. $CO_2 + H_2O + light \rightarrow O_2 + carbohydrate$
4. the size of the soil particles: sand > silt > clay
6. plant and animal residues plus microorganisms, which provide the fuel and the mechanism for formation of inorganic nutrients (by mineralization) needed for plant growth
8. *advantages:* inexpensive, most concentrated form of N, quickly converted to NO_3^-, which is required by plants; *disadvantages:* toxic, difficult to apply to soil, adds no organic matter
10. A herbicide known as 2,4,5-T was used as a defoliant in Vietnam as a component of Agent Orange and as a common weed killer in the U.S. It was banned due to concerns over possible contamination of it by TCDD, which can form during production of 2,4,5-T if conditions are not properly controlled. TCDD is a potentially dangerous substance.
12. Changes b, c, and d occur, whereas the carbon content decreases. Since N, P, and K are the three macronutrients that must be supplied by the soil, concentration of these three during composting produces a highly effective fertilizing material.

Chapter 16

2. See Table 16.2.
4. by addition of a synergist that blocks the enzyme that is responsible for the resistance
6. moderation, saturation, and multiple attack
8. Dichlorvos is somewhat volatile; this makes it suitable for use as an insecticidal organophosphate fumigant.
10. Smoke reduces the honeybee's response to alarm pheromone.
12. The boll weevil does tremendous damage to cotton crops.
14. monarch butterfly
16. Confusion uses the sex attractant pheromone which is distributed over a selected area to interfere with normal mating by means of confusing males. Masking is the use of a chemical that blocks the attractiveness of the pheromone.
18. Ionizing radiation or chemosterilization may produce insects that cannot mate productively. It is done to males

since they often mate many times, whereas females usually mate only once.

Chapter 17

2. barley, because of the ease of malting
4. See Figure 17.3; each reaction is catalyzed by an enzyme provided by yeast.
6. See Tables 17.1 and 17.2.
8. amylopectin from starches that are not broken down by amylases
10. Heavy brewing uses higher-than-normal concentrations of ingredients; it encourages formation of yeast fermentation products (also called congeners) that contribute to the aroma and flavor of beer so that a light beer will taste more like a regular beer.
12. any fruits, but usually grapes
14. Campden tablets are solid sodium metabisulfite ($Na_2S_2O_5$) used as a source of SO_2 for home wine making.
16. In a ''good year'' wines mature fully. In a ''bad year'' the grapes may have too little sugar and too much acid. Fortification (addition of sugar) and malo-lactic fermentation (for reduction of acidity) counteract these defects.
18. The carbonation lasts far longer in sparkling wines since some of the carbon dioxide is chemically bound and is thus released more slowly.
20. barley malt dried over peat fires

Chapter 18

2. wheat, due to the high gluten content of wheat flour
4. Gluten traps gases inside dough and causes leavening.
6. yeast (CO_2), baking powder (baking soda plus acid), steam, air
8. Ammonium bicarbonate releases CO_2, H_2O, and NH_3 when heated. The ammonia cannot be tolerated in the final product, so this leavening agent can be used only when the product is baked to dryness.
10. Sugar provides food for yeast.
12. to provide richness (shortness) in both flavor and texture

Chapter 19

2. 87% water, 5% carbohydrate (lactose), 4% fat, 3% protein, 1% minerals
4. lactose: β linkage; maltose: α linkage
6. The lactose is metabolized to lactic acid.
8. *cottage cheese:* from skim milk (low fat content), although some butterfat is added in making creamed cottage cheese, unripened; *cheddar cheese:* from whole milk, cured
10. usually correct, but some cheeses are not cured, e.g., cottage cheese

Chapter 20

2. The higher pressure makes it possible to reach the higher temperatures sometimes required to kill spoilage microorganisms or their spores.
4. Creamy foods such as cream-filled pastries are most susceptible since they are not heated during preparation.
6. to remove water required by spoilage microorganisms
8. A well-fed, rested animal will have meat with a higher glycogen content at the time of slaughter when anaerobic fermentation produces lactic acid, which helps preserve the meat.
10. oxymyoglobin (MbO_2); nitrosomyoglobin (MbNO)
12. Spoilage microorganisms do not tolerate salt. Those that produce lactic acid tolerate salt.
14. fermented cabbage
16. by fermentation in the presence of increasing concentrations of salt
18. aspartic acid, phenylalanine, and methyl alcohol linked covalently

Chapter 21

2. Food oils contain functional groups that make them biodegradable.
6. Butter has a lower iodine value, shorter fatty acid chains, and only *cis* double bonds.
8. During production of margarine, hydrogenation causes some *cis* to *trans* isomerization, which increases the melting point and encourages solidification. Peanut oil has only *cis* double bonds.
10. The greater unsaturation in the margarine is offset by longer-chain fatty acids and some *trans* double bonds in margarine.
12. lard from hogs; tallow from cattle and lambs
14. See Antioxidants, Section 21.6.
16. Differences include those that affect flavor, such as vinegar and sugar (both higher in salad dressing). The major differences are the amount of oil (much less in salad dressing) and the amount of starch (none in mayonnaise). The oil provides viscosity, which is maintained by the starch when the oil content is lowered.

Chapter 22

2. Dental plaque is a collection of microorganisms that can carry out lactic acid fermentation. Lactic acid, as well as other acids, contribute to demineralization of tooth enamel.
4. remineralization (following demineralization) in which some fluoride ions replace hydroxide ions
6. It is regarded as safe at low levels, but effective levels (about 1 ppm) are not far below levels at which dental

fluorosis becomes significant. Other side effects of fluoride also may be severe for hypersensitive individuals.
8. Alternative methods include topical fluoride by dentists and fluoride in toothpastes, mouthwashes, salt, chewing gum, milk and breakfast cereals. A low, frequent dosage in fluoridated water is regarded as superior to other sources of fluoride.
10. Soap molecules have long hydrophobic hydrocarbon chains and very hydrophilic salt groups, both of which are necessary to bring about a mixing of oil and water, the surfactant action.
12. Olive oil (used to make castile soap) has very little 10:0 and 12:0 fatty acids, which are more irritating to the skin when present in soaps.
14. They are not biodegradable.
16. They act as water softeners by binding hardness ions, and they are very safe. They contribute to the growth of algae in waterways.
18. The ES surfactant is more soluble (than the LAS) and thus has less tendency to precipitate in hard water.
20. to absorb light and reemit it as blue light that cancels out any yellow color that clothes develop from repeated washing

Chapter 23

2. through the respiratory tract, the digestive tract, genito-urinary tract, and the skin
4. Measles, mumps, influenza, and polio confer immunity. Syphilis, gonorrhea, and staphylococcus do not.
6. Killed organisms retain similarities to the normal organisms to elicit an immune response. Live attenuated viruses cause a mild version of the normal infection and thus confer immunity. A toxoid evokes immunity toward the toxin released by an organism.
8. due to periodic major mutations which enable the virus to bypass existing immunity
10. HIV attacks antibody-producing cells, mutates frequently, and HIV-infected T cells are infectious.
12. EIAs are diagnostic tests which take advantage of the specificity of antibodies to test for viruses (such as HIV and hepatitis) and/or proteins (such as HCG in pregnancy).

Chapter 24

2. They may not dissolve or be absorbed at the same rate or to the same degree.
4. disintegrator
6. When it is first administered, the concentration of drug in the bloodstream peaks, after which it eventually drops to a low level (valley) until it is taken again.
8. that the antibiotic was isolated from a soil microorganism

10. Antibiotics may alter the population of intestinal bacteria, called the intestinal flora.

12. Penicillin G is susceptible to destruction by stomach acid. Penicillin V is resistant to attack and makes a better oral penicillin.

14. Before the antibiotic sensitivity of disease-causing bacteria is known, a broad-spectrum antibiotic increases the probability of effectiveness in combating infection. Broad-spectrum antibiotics encourage development of resistant microorganisms.

16. The enzyme causes cleavage of the (β-lactam) ring of a penicillin.

18. Clavulanic acid has a synergistic effect because it blocks the attack of penicillinase on amoxicillin.

20. Poly(lactic acid) is a bioerodable polymer that can be used to deliver a drug by slow release which occurs as the polymer degrades.

Chapter 25

2. aspirin (also antipyretic, anti-inflammatory, and anticoagulant), acetaminophen (also antipyretic), and ibuprofen

4. opiates

6. tranquilizer, anesthetic-analgesic, analgesic

8. during surgery through a mask in a stream of oxygen and nitrous oxide

10. the release of CO_2 due to the action of an acid on $NaHCO_3$ when the drug is dissolved in water

12. It blocks the complete metabolism of ethyl alcohol and causes a buildup of acetaldehyde.

14. causes sedation and analgesia, and may cause dependence

16. Cocaine inhibits the reuptake of dopamine. This leads to a powerful stimulation of receptors (euphoria). Later, excessive loss of dopamine causes dysphoria. Amphetamines stimulate release of norepinephrine and dopamine and inhibit storage of neurotransmitters in vesicles so that receptors remain excited.

18. The amount and rate of absorption of cocaine is far greater when it is inhaled.

20. Some side effects are restlessness, anxiety, irritability, insomnia, apathy, sluggishness, reduced physical performance (e.g., poor driving); high tar content adversely affects lungs.

22. cigarette smoking

24. Bufferin contains a mixture of antacids; this is supposed to decrease stomach irritation from aspirin by buffering, but an antacid is not a buffer. The extra relief action is the anti-inflammatory action of the aspirin; this is not available from the Tylenol (acetaminophen).

Chapter 26

2. See Figure 26.1.

4. different functional groups plus the presence or absence of angular methyl groups

6. HDLs (with a higher protein content) reduce the risk of atherosclerosis. LDLs (with a lower protein content) increase the risk of plaque formation. A high HDL/LDL ratio is desirable.

8. *Harmful effects:* High dosages cause severe acne, oily skin, degeneration of testicles, coronary artery disease, liver tumors and disease, stunted growth in youth, water retention, aggressive behavior (called "roid rage") and irreversible masculinization (voice deepening, pattern baldness, hirsutism) of women.
Useful effects: Steroids help build tissue following injury, surgery, burns, or starvation, and effectively treat testosterone deficiency.

10. The aromatic A ring and the absence of the angular methyl (C-19) between rings A and B are the major differences that affect the hormonal responses to the two steroids.

12. a. Ovulation and fertilization are prevented by the effect of steroid hormones on FSH and LH.
Monophasic: 21 pills containing constant amounts of progestin and estrogen, followed by 7 pills containing either nothing or iron; lower level of steroids than older pills to minimize side effects
Biphasic: estrogen constant for 21 days; progestin of one strength for 10 days, followed by different strength of progestin for 11 days; an additional 7 pills similar to the monophasics to mimic more closely the natural levels of hormones in the menstrual cycle
Triphasic: variable with brand; approximately 7 days of progestin of one strength, followed by two additional phases (about 7 days each) of two other levels of progestin; estrogen content constant throughout; an additional 7 pills like the monophasics to mimic more closely the natural levels of hormones in the menstrual cycle

b. Minipill is used by estrogen-sensitive women in constant progestin dosage every day of the year. Progestin stimulates sticky cervical secretions and alters uterine lining, which blocks sperm and inhibits implantation, respectively.

c. Norplant delivers a constant low dosage of progestin, like the minipill. It eliminates the need to take daily medication.

14. A synthetic analog of LH-RH, by injection or nasal spray, suppresses ovulation or sperm production.

Index

Boldfaced terms are defined in the glossary.

Abortifacient, 595
ABS. *See* Alkylbenzenesulfonates
Accelerator mass spectroscopy, 206
Accelerator transmutation of waste (ATW), 241
Accelerators, cyclic or linear, 222
Acetaldehyde, 323, 397, 431, 551–552
Acetaminophen, 506, 531, 533, 534
Acetate, 55
Acetic acid, 141, 365, 397, 431, 454, 510, 530, 550, 551
Acetic anhydride, 179
Acetobacter, 431
Acetone, 548–549
Acetylation, 510, 530, 554
Acetylene, 96, 109
Acetylsalicylic acid. *See* Aspirin
Achromycin, 522–523
Acidity. *See also* pH
    of food, 429, 431–435, 441, 443
    in soil, 341–343
    of wines, 393–395
Acid mine drainage, 258
Acid rain, 309–310, 313, 315, 324
Acids, 113, 115–117, 135–145, 393–395, 407–408, 429, 431–435, 441–443, 465, 467–469
    in baking powder, 407–408
    in dental caries, 465, 467–469
Acquired immune deficiency syndrome. *See* AIDS
Acrilan, 149. *See also* Polyacrylonitrile
Acrylic acid, 153
Acrylic polymers, 153
Acrylonitrile, 153
Actinide elements, 48, 240–241
Action level, 203
Active niacin. *See* NAD
Active riboflavin. *See* FAD
Acyclovir, 524–525
Addiction. *See* Drug addiction
Addition polymers, 149–155
Additives. *See* Food additives
Adenosine diphosphate. *See* ADP
Adenosine monophosphate. *See* AMP
Adenosine triphosphate. *See* ATP

Adhesives, 179
Adjuncts, 382
ADP, 385, 387–388, 551
Adrenal cortex, 575
Adrenal corticosteroids, 575–579
Adrenaline, 558
Advil. *See* Ibuprofen
Aerobic fermentation. *See* Fermentation
Aerosols, 314, 321
Africanized honeybee, 370
Agent Orange, 351
Aggregation pheromone, 372
Agrichemicals, 337–355
Agricultural lime, 342
AIDS, 487, 497–501, 525, 554
Air, 267, 309–320
    expansion in baking, 407–410
    pollution, 309–320
Alarm pheromone, 369–370
Albumin, 207
Alcohol, 85, 123–124, 279, 281–282, 381–400, 407, 410, 431, 545, 548–552, 560, 572
    beverages, 381–400, 423, 548–549
    energy content, 282
    fermentation, 384–386, 388–391, 398
    in light beer, 392–393
    production, 550
Alcohols, 113, 115–117
Alcohol sulfates. *See* Surfactants
Aldrin, 360
Ale, 382
Alfalfa, 346
Alfalfa looper, 374
Alkali metals, 47, 50
Alkaline. *See* pH
Alkaline cell battery, 299
Alkaline earth metals, 47, 51
Alkalinity in soils, 341–342
Alkanes, 109
Alka-Seltzer, 143–144, 533–534, 538–540
Alkenes, 109–111
Alkylbenzenesulfonates (ABS), 476–479
Alkyl group, 113
Allelochemicals, 363–368

Allergy, 501–502, 523, 576, 578
Allomone, 364
Alloy, 120, 132, 470. *See also* Amalgams
Alpha decay. *See* Radioactivity
Alpha particle, 222, 227
    penetration, 222
Alternative sweeteners, 441
Alum, 341–342
Aluminosilicates, 283–285
Aluminum, 54, 72, 251
Aluminum chloride, 45
Aluminum hydroxide, 46, 533, 537–538
Aluminum oxide, 72
Aluminum sulfate, 46, 342
Amalgamators, 471–472
Amalgams. *See* Dental amalgams
Amantidine, 525
Americium, 223
Amide, 114, 159, 160
Amine, 114, 159, 160, 162
Amino acids, 163, 339
6-Aminopenicillanic acid, 521
Ammonia, 52, 57, 68, 119–120, 137, 262, 298, 311, 339, 343–344, 352, 408
    water, 137
Ammonium bicarbonate, 408
Ammonium bromide, 119, 123, 126
Ammonium chloride, 137
Ammonium hydroxide, 137
Ammonium ion, 52, 58, 298–299, 339, 343
Ammonium nitrate, 58, 345
Ammonium phosphate, 46, 348
Ammonium sulphate, 59, 344, 345
Amoxicillin, 505–506, 517, 519
AMP, 165–167
Amphetamines, 542, 545–547, 558–560
Ampicillin, 512, 517–518
Amylase, 381–382, 391, 484
    alpha, 381, 384
    beta, 381, 384
Amyloglucosidase, 392
Amylopectin. *See* Starch
Amylose. *See* Starch
Amytal, 546–547, 562

Common Energy Units and Relationships

$$
\begin{aligned}
1 \text{ quadrillion Btu (Q)} &= 8.79 \times 10^{11} \text{ kilowatt-hours thermal (kW-h}_t) \\
&= 2.93 \times 10^{11} \text{ kilowatt-hours electric (kW-h}_e) \\
&= 2.52 \times 10^{14} \text{ kilocalories (kcal)} \\
&= 1.00 \times 10^{15} \text{ British thermal units (Btu)} \\
&= 1.06 \times 10^{15} \text{ kilojoules (kJ)} \\
&= 2.52 \times 10^{17} \text{ calories (cal)} \\
&= 1.06 \times 10^{18} \text{ joules (J)}
\end{aligned}
$$

Table 1.1 Some English-metric conversions

Length

1 mile (mi)	= 1.61 kilometers (km)
1 yard (yd)	= 0.914 meter (m)
1 inch (in.)	= 2.54 centimeters (cm)

Mass

1 pound (lb)	= 454 grams (g)
1 pound (lb)	= 0.454 kilogram (kg)
1 ounce (oz)	= 28.4 grams (g)

Volume

1 gallon (gal)	= 3.78 liters (L)
1 quart (qt)	= 0.946 liter (L)
1 pint (pt)	= 0.473 liter (L)
1 fluid ounce (fl oz)	= 29.6 milliliters (mL)

Table 1.2 Prefixes used in the metric system

Multiple	Prefix	Abbreviation
10^9	Giga-	G
10^6	Mega-	M
10^3	Kilo-	k
10^2	Hecto-	h
10^1	Deka-	da
10^{-1}	Deci-	d
10^{-2}	Centi-	c
10^{-3}	Milli-	m
10^{-6}	Micro-	μ
10^{-9}	Nano-	n